GLOBAL PERSPECTIVES ON THE BIOLOGY AND LIFE HISTORY OF THE WHITE SHARK

Global Perspectives on the Biology and Life History of the White Shark

Edited by MICHAEL L. DOMEIER

CRC Press
Taylor & Francis Group
Boca Raton London New York

CRC Press is an imprint of the
Taylor & Francis Group, an **informa** business

CRC Press
Taylor & Francis Group
6000 Broken Sound Parkway NW, Suite 300
Boca Raton, FL 33487-2742

Version Date: 20111107

International Standard Book Number-13: 978-1-4398-4841-8 (eBook - PDF)

Visit the Taylor & Francis Web site at
http://www.taylorandfrancis.com

and the CRC Press Web site at
http://www.crcpress.com

To Tom Pfleger and Paxson Offield, who enabled and assisted me both on and off the water, and to my wife, Amy, who has helped me through the reality that working with sharks is not just about sharks

Contents

Preface xiii
Acknowledgments xv
Editor xvii
Contributors xix

Section I
Biology, Behavior, and Physiology

Chapter 1
Captive Feeding and Growth of Young-of-the-Year White Sharks, *Carcharodon carcharias*, at the Monterey Bay Aquarium 3

Juan M. Ezcurra, Christopher G. Lowe, Henry F. Mollet, Lara A. Ferry, and John B. O'Sullivan

Chapter 2
Oxygen Consumption Rate of Young-of-the-Year White Sharks, *Carcharodon carcharias*, during Transport to the Monterey Bay Aquarium 17

Juan M. Ezcurra, Christopher G. Lowe, Henry F. Mollet, Lara A. Ferry, and John B. O'Sullivan

Chapter 3
Size-Based Analysis of Diet and Trophic Position of the White Shark, *Carcharodon carcharias*, in South African Waters 27

Nigel E. Hussey, Heather M. McCann, Geremy Cliff, Sheldon F. J. Dudley, Sabine P. Wintner, and Aaron T. Fisk

Chapter 4
White Sharks and Cephalopod Prey: Indicators of Habitat Use? 51

Malcolm J. Smale and Geremy Cliff

Chapter 5
Heavy Metals, Trace Elements, and Organochlorine Contaminants in Muscle and Liver Tissue of Juvenile White Sharks, *Carcharodon carcharias*, from the Southern California Bight 59

Christopher G. Mull, Mary E. Blasius, John B. O'Sullivan, and Christopher G. Lowe

Chapter 6
Boat-Strike Wound Healing in *Carcharodon carcharias* 77

Alison Towner, Malcolm J. Smale, and Oliver Jewell

Chapter 7
A Summary of Observations on the Maximum Size Attained by the White Shark, *Carcharodon carcharias* 85

José I. Castro

Chapter 8
Investigatory Behavior toward Surface Objects and Nonconsumptive Strikes on Seabirds by White Sharks, *Carcharodon carcharias*, at Seal Island, South Africa (1997–2010) 91

Neil Hammerschlag, R. Aidan Martin, Chris Fallows, Ralph S. Collier, and Rob Lawrence

Chapter 9
Comparisons between White Shark–Pinniped Interactions at Seal Island (South Africa) with Other Sites in California 105

Chris Fallows, R. Aidan Martin, and Neil Hammerschlag

Section II
Migratory Patterns and Habitat Use

Chapter 10
Fine-Scale Habitat Use by White Sharks at Guadalupe Island, Mexico 121

Michael L. Domeier, Nicole Nasby-Lucas, and Chi H. Lam

Chapter 11
Sex-Specific Migration Patterns and Sexual Segregation of Adult White Sharks, *Carcharodon carcharias*, in the Northeastern Pacific 133

Michael L. Domeier and Nicole Nasby-Lucas

Chapter 12
The Northeastern Pacific White Shark Shared Offshore Foraging Area (SOFA): A First Examination and Description from Ship Observations and Remote Sensing 147

Michael L. Domeier, Nicole Nasby-Lucas, and Daniel M. Palacios

Chapter 13
Connectivity among White Shark Coastal Aggregation Areas in the Northeastern Pacific 159

Salvador J. Jorgensen, Taylor K. Chapple, Scot Anderson, Mauricio Hoyos, Carol Reeb, and Barbara A. Block

Chapter 14
Historic Fishery Interactions with White Sharks in the Southern California Bight 169

Christopher G. Lowe, Mary E. Blasius, Erica T. Jarvis, Tom J. Mason, Gwen D. Goodmanlowe, and John B. O'Sullivan

Chapter 15
Incidental Catch and Ecological Observations of Juvenile White Sharks, *Carcharodon carcharias*, in Western Baja California, Mexico: Conservation Implications 187

Omar Santana-Morales, Oscar Sosa-Nishizaki, Miguel A. Escobedo-Olvera, Erick C. Oñate-González, John B. O'Sullivan, and Daniel Cartamil

Chapter 16
A New Life-History Hypothesis for White Sharks, *Carcharodon carcharias*, in the Northeastern Pacific 199

Michael L. Domeier

Chapter 17
Habitat Use and Spatial Dynamics of Juvenile White Sharks, *Carcharodon carcharias*, in Eastern Australia 225

Barry D. Bruce and Russell W. Bradford

Chapter 18
Identifying Juvenile White Shark Behavior from Electronic Tag Data 255

Russell W. Bradford, Alistair J. Hobday, and Barry D. Bruce

Chapter 19
Beach Areas Used by Juvenile White Shark, *Carcharodon carcharias*, in Eastern Australia 271

Jonathan M. Werry, Barry D. Bruce, Wayne Sumpton, Dennis Reid, and David G. Mayer

Chapter 20
Seasonal Sexual and Size Segregation of White Sharks, *Carcharodon carcharias,* at the Neptune Islands, South Australia 287

Rachel L. Robbins and David J. Booth

Chapter 21
Regional Population Connectivity, Oceanic Habitat, and Return Migration Revealed by Satellite Tagging of White Sharks, *Carcharodon carcharias*, at New Zealand Aggregation Sites 301

Clinton A. J. Duffy, Malcolm P. Francis, Michael J. Manning, and Ramon Bonfil

Chapter 22
The Third Dimension: Vertical Habitat Use by White Sharks, *Carcharodon carcharias*, in New Zealand and in Oceanic and Tropical Waters of the Southwest Pacific Ocean 319

Malcolm P. Francis, Clinton A. J. Duffy, Ramon Bonfil, and Michael J. Manning

Chapter 23
New Caledonia (South Pacific) as a Potential Tropical Wintering Ground for the White Shark, *Carcharodon carcharias* 343

Eric Clua and Bernard Séret

Section III
Population Monitoring, Policy, and Review

Chapter 24
Application of Molecular Genetics for Conservation of the White Shark, *Carcharodon carcharias*, L. 1758 357

Chrysoula Gubili, Clinton A. J. Duffy, Geremy Cliff, Sabine P. Wintner, Mahmood Shivji, Demian Chapman, Barry D. Bruce, Andrew P. Martin, David W. Sims, Catherine S. Jones, and Leslie R. Noble

Chapter 25
Use of Photo Identification to Describe a White Shark Aggregation at Guadalupe Island, Mexico 381

Nicole Nasby-Lucas and Michael L. Domeier

Chapter 26
Problems with Photo Identification as a Method of Estimating Abundance of White Sharks, *Carcharodon carcharias*: An Example from Guadalupe Island, Mexico 393

Oscar Sosa-Nishizaki, Enrique Morales-Bojórquez, Nicole Nasby-Lucas, Erick C. Oñate-González, and Michael L. Domeier

Chapter 27
Implications of Increasing Pinniped Populations on the Diet and Abundance of White Sharks off the Coast of Massachusetts 405

Gregory B. Skomal, John Chisholm, and Steven J. Correia

Chapter 28
Back to the Wild: Release of Juvenile White Sharks from the Monterey Bay Aquarium........ 419

Kevin C. Weng, John B. O'Sullivan, Christopher G. Lowe, Chuck E. Winkler, Mary E. Blasius, Kerri A. Loke-Smith, Timothy J. Sippel, Juan M. Ezcurra, Salvador J. Jorgensen, and Michael J. Murray

Chapter 29
Shark Spotters: A Pioneering Shark Safety Program in Cape Town, South Africa........ 447

Alison Kock, Sarah Titley, Wally Petersen, Monwabisi Sikweyiya, Sakhile Tsotsobe, Darryl Colenbrander, Howard Gold, and Gregg Oelofse

Chapter 30
The Use of a Nonlethal Technique to Assess the Reproductive Biology of the White Shark, *Carcharodon carcharias* 467

James A. Sulikowski, Laura J. Williams, and Michael L. Domeier

Chapter 31
Responding to the Risk of White Shark Attack: Updated Statistics, Prevention, Control Methods, and Recommendations 477

Tobey H. Curtis, Barry D. Bruce, Geremy Cliff, Sheldon F. J. Dudley, A. Peter Klimley, Alison Kock, Robert N. Lea, Christopher G. Lowe, John E. McCosker, Gregory B. Skomal, Jonathan M. Werry, and John G. West

Chapter 32
A Review of Research on the White Shark, *Carcharodon carcharias*, in Southern Africa......... 511

Sheldon F. J. Dudley

Preface

The history of White Shark research is rather recent and punctuated by the publication of two important volumes, *The Biology of the White Shark* (Sibley et al., 1985) and *Great White Sharks: The Biology of Carcharodon carcharias* (Klimley and Ainley, 1996). Both of these books were inspired by special symposia, brief gatherings of all the best White Shark researchers to share their experiences and research findings. Although many scientific meetings result in books, such books are usually only mildly interesting because researchers tend to save their best works for top-rated journals. However, White Shark science is unique in that the first two books included all the best research and became the foundation of knowledge, White Shark Bibles if you will, for all that followed. *Global Perspectives on the Biology and Life History of the White Shark* follows in the footsteps of its predecessors. It was inspired by a 2010 International White Shark Symposium held in Honolulu, Hawaii, and it encompasses the most important contemporary research findings in a single peer-reviewed volume. This is not a proceedings, because researchers who could not attend the meeting were invited to contribute their valuable research results.

With this book comes a historic change in the context of White Shark research. The relatively low abundance of this species, combined with its exceptional size and predatory nature, equated to a fish that was very hard to study. The introduction of White Shark research papers always included an obligatory statement that this species is rarely the focus of comprehensive studies and little is known of its basic biology and life history. In just over a decade, the tools available for White Shark research have evolved from a few cameras and perhaps a shark cage to a quiver of sophisticated electronic tags that can provide intimate behavioral records at a sampling interval of just a few seconds. With the publication of this book, White Sharks have earned the title of the most studied shark in the ocean. Only a few pelagic bony fishes, such as tunas and billfish, have carried more of these electronic data recorders through our oceans. These tags have led to the startling discovery that White Sharks are more of a pelagic fish than a coastal fish, capable of crossing ocean basins and spending more than a year in the deep ocean habitat. Although this book marks this change in our basic understanding of their life history, we still have no idea how White Sharks can survive in the open ocean for such lengthy periods—and why. Perhaps the next scientific White Shark book will again rewrite what we think we know.

Acknowledgments

The International White Shark Symposium that convened in Honolulu, Hawaii (February 7–10, 2010), provided the momentum and interest for this book. The symposium would not have been possible without the help of my fellow organizing committee members Alison Kock, John B. O'Sullivan, and Kevin C. Weng or the funding provided by the Marine Conservation Science Institute, Save our Seas Foundation, Change Happens Foundation, Guy Harvey Ocean Foundation, Pelagic Fisheries Research Program, and the Monterey Bay Aquarium. Finally, I must recognize the hard work and assistance with this book from my longtime associate Nicole Nasby-Lucas.

Editor

Michael L. Domeier, PhD, is the founding president of the Marine Conservation Science Institute, a Southern California nonprofit organization dedicated to bridging the gap between research and conservation. Dr. Domeier earned his BS in marine biology from the Florida Institute of Technology in 1987, followed by a PhD in marine biology and fisheries from the University of Miami's Rosenstiel School of Marine and Atmospheric Science in 1992. He has made important contributions to a wide variety of specialized fields, including coral reef fish ecology, pelagic fish ecology, and the advancement of electronic tagging technology and methods. He was the organizing chair of the Fourth International Billfish Symposium held in 2004, and he also chaired the 2009 International White Shark Symposium. He is recognized as a leader in White Shark research based upon his pioneering work at Guadalupe Island, Mexico.

Contributors

Scot Anderson
Inverness, California

Mary E. Blasius
Institute for Integrated Research in Materials, Environments, and Society
California State University, Long Beach
Long Beach, California

Barbara A. Block
Department of Biology
Stanford University
Pacific Grove, California

Ramon Bonfil
The Bronx High School of Science
Brooklyn, New York

David J. Booth
Department of Environmental Sciences
University of Technology
Broadway, Australia

Russell W. Bradford
CSIRO Marine and Atmospheric Research
Hobart, Australia

Barry D. Bruce
CSIRO Marine and Atmospheric Research
Hobart, Australia

Daniel Cartamil
Marine Biology Research Division
Scripps Institution of Oceanography
University of California, San Diego
San Diego, California

José I. Castro
National Oceanic and Atmospheric Administration at Mote Marine Laboratory
Sarasota, Florida

Demian Chapman
Guy Harvey Research Institute
and
Save Our Seas Foundation Shark Centre
Nova Southeastern University
Dania Beach, Florida

Taylor K. Chapple
Department of Wildlife, Fish, and Conservation Biology
University of California, Davis
Davis, California

John Chisholm
Massachusetts Division of Marine Fisheries
Shark Research Project
New Bedford, Massachusetts

Geremy Cliff
KwaZulu-Natal Sharks Board
Umhlanga Rocks, South Africa
and
Biomedical Resource Unit
University of KwaZulu-Natal
Durban, South Africa

Eric Clua
Secretariat of the Pacific Community
Noumea, France

Darryl Colenbrander
Environmental Resources Department
City of Cape Town, South Africa
Cape Town, South Africa

Ralph S. Collier
Shark Research Committee
Van Nuys, California

Steven J. Correia
Massachusetts Division of Marine Fisheries
Shark Research Project
New Bedford, Massachusetts

Tobey H. Curtis
National Marine Fisheries Service
National Oceanic and Atmospheric Administration
Gloucester, Massachusetts

Michael L. Domeier
Marine Conservation Science Institute
Fallbrook, California

Sheldon F. J. Dudley
KwaZulu-Natal Sharks Board
Umhlanga Rocks, South Africa
and
Biomedical Resource Unit
University of KwaZulu-Natal
Durban, South Africa

Clinton A. J. Duffy
Department of Conservation
Auckland, New Zealand
and
Marine Science
University of Auckland
Auckland, New Zealand

Miguel A. Escobedo-Olvera
Department of Biological Oceanography
Centro de Investigación Científica y de Educación Superior de Ensenada
Baja California, Mexico

Juan M. Ezcurra
Monterey Bay Aquarium
Monterey, California

Chris Fallows
Apex Expeditions
Simonstown, South Africa

Lara A. Ferry
Division of Mathematics and Natural Sciences
Arizona State University
Phoenix, Arizona

Aaron T. Fisk
Great Lakes Institute for Environmental Research
University of Windsor
Windsor, Ontario, Canada

Malcolm P. Francis
National Institute of Water and Atmospheric Research
Greta Point, New Zealand

Howard Gold
Environmental Resources Department
City of Cape Town, South Africa
Cape Town, South Africa

Gwen D. Goodmanlowe
Department of Biological Sciences
California State University, Long Beach
Long Beach, California

Chrysoula Gubili
Institute of Biological and Environmental Sciences
Aberdeen University
Aberdeen, United Kingdom
and
Marine Biological Association of the United Kingdom
Plymouth, United Kingdom

Neil Hammerschlag
Rosenstiel School of Marine and Atmospheric Science
and
Leonard and Jayne Abess Center for Ecosystem Science and Policy
and
R. J. Dunlap Marine Conservation Program
University of Miami
Miami, Florida

Alistair J. Hobday
CSIRO Marine and Atmospheric Research
Hobart, Australia

Mauricio Hoyos
Centro Interdisciplinario de Ciencias Marinas
Baja California Sur, Mexico

Nigel E. Hussey
College of Natural Sciences
Bangor University
Menai Bridge, United Kingdom
and
Great Lakes Institute for Environmental Research
University of Windsor
Windsor, Ontario, Canada

Erica T. Jarvis
Department of Biological Sciences
California State University, Long Beach
Long Beach, California

Oliver Jewell
Oceans Research
Mossel Bay, South Africa

Catherine S. Jones
Institute of Biological and Environmental Sciences
Aberdeen University
Aberdeen, United Kingdom

Salvador J. Jorgensen
Department of Biology
Stanford University
Pacific Grove, California
and
Monterey Bay Aquarium
Monterey, California

A. Peter Klimley
Department of Wildlife, Fish, and Conservation Biology
University of California, Davis
Davis, California

Alison Kock
Department of Zoology
University of Cape Town
Rondebosch, South Africa
and
Shark Spotters
Soetwater Education Centre
Kommetjie, South Africa
and
Save Our Seas Foundation Shark Centre
Kalk Bay, South Africa

Chi H. Lam
Marine Environmental Biology
University of Southern California
Los Angeles, California

Rob Lawrence
African Shark Eco-Charters
Simonstown, South Africa

Robert N. Lea
California Academy of Sciences
San Francisco, California

Kerri A. Loke-Smith
California State University, Long Beach
Long Beach, California

Christopher G. Lowe
California State University, Long Beach
Long Beach, California

Michael J. Manning
National Institute of Water and Atmospheric Research
Wellington, New Zealand

Andrew P. Martin
Department of Ecology and Evolutionary Biology
University of Colorado
Boulder, Colorado

R. Aidan Martin
Zoology Department
University of British Columbia
Vancouver, British Columbia, Canada

Tom J. Mason
Department of Biological Sciences
California State University, Long Beach
Long Beach, California

David G. Mayer
Department of Employment, Economic Development, and Innovation
EcoSciences Precinct
Agri-Science Queensland
Brisbane, Australia

Heather M. McCann
Great Lakes Institute for Environmental Research
University of Windsor
Windsor, Ontario, Canada

John E. McCosker
California Academy of Sciences
San Francisco, California

Henry F. Mollet
Monterey Bay Aquarium
Monterey, California
and
Moss Landing Marine Laboratories
Moss Landing, California

Enrique Morales-Bojórquez
Centro de Investigaciones Biológicas del Noroeste
Baja California Sur, Mexico

Christopher G. Mull
Department of Biology
Simon Fraser University
Burnaby, Canada
and
Department of Biological Sciences and Institute for Integrated Research in Materials, Environments, and Society
California State University, Long Beach
Long Beach, California

Michael J. Murray
Monterey Bay Aquarium
Monterey, California

Nicole Nasby-Lucas
Marine Conservation Science Institute
Fallbrook, California

Leslie R. Noble
Institute of Biological and Environmental Sciences
Aberdeen University
Aberdeen, United Kingdom

Gregg Oelofse
Environmental Resources Department
City of Cape Town, South Africa
Cape Town, South Africa

Erick C. Oñate-González
Centro de Investigación Científica y de Educación Superior de Ensenada
Baja California, Mexico

John B. O'Sullivan
Monterey Bay Aquarium
Monterey, California

Daniel M. Palacios
Joint Institute for Marine and Atmospheric Research
University of Hawaii at Manoa
Honolulu, Hawaii
and
Environmental Research Division
National Oceanic and Atmospheric Administration/National Marine Fisheries Service/Southwest Fisheries Science Center
Pacific Grove, California

Wally Petersen
Shark Spotters
and
Kommetjie Environmental Awareness Group
Soetwater Education Centre
Kommetjie, South Africa

Carol Reeb
Department of Biology
Stanford University
Pacific Grove, California

Dennis Reid
New South Wales Industry and Investment
Cronulla Fisheries Centre
Cronulla, Australia

Rachel L. Robbins
Department of Environmental Sciences
University of Technology and Fox Shark Research Foundation
Aberfoyle Park, South Africa

Omar Santana-Morales
Department of Biological Oceanography
Centro de Investigación Científica y de Educación Superior de Ensenada
Baja California, Mexico

Bernard Séret
Research Institute for Development
Department of Systematics and Evolution
National Museum of Natural History
Paris, France

Mahmood Shivji
Guy Harvey Research Institute
and
Save Our Seas Foundation Shark Centre
Nova Southeastern University
Dania Beach, Florida

Monwabisi Sikweyiya
Shark Spotters
Soetwater Education Centre
Kommetjie, South Africa

David W. Sims
Marine Biological Association of the United Kingdom
and
Marine Biology and Ecology Research Group
School of Marine Sciences and Engineering
University of Plymouth
Plymouth, United Kingdom

Timothy J. Sippel
Pelagic Fisheries Research Program
University of Hawaii at Manoa
Honolulu, Hawaii

Gregory B. Skomal
Massachusetts Division of Marine Fisheries
Shark Research Project
New Bedford, Massachusetts

Malcolm J. Smale
Port Elizabeth Museum at Bayworld
Humewood, South Africa
and
Zoology Department
Nelson Mandela Metropolitan University
Port Elizabeth, South Africa

Oscar Sosa-Nishizaki
Centro de Investigación Científica y de Educación Superior de Ensenada
Baja California, Mexico

James A. Sulikowski
Marine Science Center
University of New England
Biddeford, Maine

Wayne Sumpton
Department of Employment, Economic Development, and Innovation
EcoSciences Precinct
Agri-Science Queensland
Brisbane, Australia

Sarah Titley
Shark Spotters
and
Kommetjie Environmental Awareness Group
Soetwater Education Centre
Kommetjie, South Africa

Alison Towner
Dyer Island Conservation Trust
and
Animal Demography Unit Department of Zoology
University of Cape Town
Rondebosch, South Africa

Sakhile Tsotsobe
Sport, Recreation and Amenities
City of Cape Town, South Africa
Cape Town, South Africa

Kevin C. Weng
Pelagic Fisheries Research Program
University of Hawaii at Manoa
Honolulu, Hawaii
and
Hopkins Marine Station
Stanford University
Pacific Grove, California

Jonathan M. Werry
Australian Rivers Institute—Coasts and Estuaries
Griffith School of Environment
Griffith University
Southport, Australia
and
Ocean and Coast Research
Gold Coast, Australia

John G. West
Taronga Zoo
Taronga Conservation Society Australia
Mosman, Australia

Laura J. Williams
Marine Science Center
University of New England
Biddeford, Maine

Chuck E. Winkler
Southern California Marine Institute
Terminal Island, California

Sabine P. Wintner
KwaZulu-Natal Sharks Board
Umhlanga Rocks, South Africa
and
Biomedical Resource Unit
University of KwaZulu-Natal
Durban, South Africa

SECTION I

Biology, Behavior, and Physiology

CHAPTER 1

Captive Feeding and Growth of Young-of-the-Year White Sharks, *Carcharodon carcharias*, at the Monterey Bay Aquarium

Juan M. Ezcurra*
Monterey Bay Aquarium

Christopher G. Lowe
California State University, Long Beach

Henry F. Mollet
Monterey Bay Aquarium and Moss Landing Marine Laboratories

Lara A. Ferry
Arizona State University

John B. O'Sullivan
Monterey Bay Aquarium

ABSTRACT

The Monterey Bay Aquarium developed a program with the support of colleagues from Stanford University, California State University Long Beach, and the Southern California Marine Institute to display young-of-the-year (YOY) White Sharks (*Carcharodon carcharias*), culminating in the display of five White Sharks in the 3.8-million-L Outer Bay exhibit between 2004 and 2009. The Outer Bay exhibit displays a variety of pelagic fishes, including Yellowfin Tuna (*Thunnus albacares*), Dolphinfish (*Coryphaena hippurus*), and Scalloped Hammerhead Sharks (*Sphyrna lewini*) and is maintained at 20°C. Four of the White Sharks fed consistently while on display (70–198 d), eating mostly King Salmon (*Onchorhynchus tshawytscha*), Pacific Mackerel (*Scomber japonica*), and Sablefish (*Anoplopoma fimbria*) at a mean (±SE) daily ration of 747 ± 46 g or 1.62 ± 0.15% body mass d^{-1} (% BM d^{-1}). One shark did not feed regularly and was released after 11 d. Daily ration peaked between 3.1 and 3.5% BM d^{-1}, which is among the highest reported for any shark species. The captive White Sharks grew in mass at a rate of 71.6 ± 8.2 kg yrs.$^{-1}$, yielding a mean gross conversion efficiency of 27.1 ± 3.8%. They grew at a mean rate of 64.9 ± 8.5 cm yrs.$^{-1}$, approximately twice the growth rate estimated from a von Bertalanffy growth function for White Sharks (Cailliet et al., 1985). A simplified bioenergetics model was used to determine parameter estimates for consumption, growth, and metabolism. This model assumed that 27% of energy intake was lost to waste, and it suggested that a mean 26.8 ± 2.9% of energy intake was invested into somatic growth, and 46.2 ± 2.9% of energy was consumed by metabolism. YOY White Sharks showed high growth capacity at optimal conditions in captivity; however, the energetic demands of White Sharks in the wild remain unknown.

INTRODUCTION

Although the White Shark, *Carcharodon carcharias*, is the focus of much interest from both the research community and the public, the display and study of a living specimen have been difficult to achieve. Many studies have focused on predatory behavior (e.g., Anderson et al., 1996; Long et al., 1996; Klimley et al., 1996, 2001), reproductive biology (Pratt, 1996; Uchida et al., 1996; Francis, 1996; Saidi et al., 2005; Chapter 30, this book), age and growth (Cailliet et al., 1985), and more recently on migration patterns of adult White Sharks (e.g., Boustany et al., 2002; Bonfil et al., 2005; Bruce et al., 2006; Domeier and Nasby-Lucas, 2008; Chapters 11, 13, 16, and 21, this book). Very little is known about neonates and juveniles of this species. However, satellite archival tag technology has recently been used to study the swimming behavior and thermal niche of young-of-the-year (YOY) and juvenile White Sharks, which were bycatch of commercial fisheries in the Southern California Bight (Dewar et al., 2004; Weng et al., 2007; Chapters 14 and 16, this book). This information on juvenile White Sharks has enabled the Monterey Bay Aquarium (MBA) to conduct a program to continue the study of juvenile White Sharks in the wild (Chapters 14 and 15, this book) and to place living specimens on display to the public.

Historically, the long-term display of a living White Shark has been attempted by many public aquariums with little success because of the difficulty in acquiring healthy specimens and the challenges in transport (Hewitt, 1984; http://homepage.mac.com/mollet/Cc/Cc_captive.html); however, since 2004 the MBA has displayed five juvenile White Sharks. The White Shark is a top-level predator that has cosmopolitan distribution in temperate and tropical seas (Compagno, 1984). These sharks can attain large size (Mollet et al., 1996) and are active swimmers that undergo large-scale geographic migrations (Boustany et al., 2002; Bonfil et al., 2005; Bruce et al., 2006; Domeier and Nasby-Lucas, 2008; Chapters 11, 13, 16, and 21, this book). These factors make acquisition and display of this species difficult. With this in mind, the program undertaken by the MBA to display a White Shark was a multiyear, incremental approach to study the behavior and movements of YOY White Sharks in the Southern California Bight and to display a specimen in the 3.8-million-L Outer Bay exhibit (OBE) (Figure 1.1). We report the captive feeding and growth and the energy budgets of YOY White Sharks displayed at the Monterey Bay Aquarium.

MATERIALS AND METHODS

The YOY White Sharks were captured in the Southern California Bight between August 2004 and August 2009 with the intent to place them on public display at the MBA. The YOY White Sharks, ranging in size from 137 to 164 cm total length (TL) and 25.2 to 47.0 kg body mass, were either bycatch of the commercial gillnet fishery or targeted catch by the MBA collecting staff. The

Figure 1.1 A YOY White Shark on public display in the 3.8-million-L Outer Bay exhibit at the Monterey Bay Aquarium. (Courtesy of Randy Wilder, Monterey Bay Aquarium.)

sharks were transported in water tanks on the fishing vessels to a 40-m diameter by 11-m deep ocean pen anchored off Malibu, California to allow them to recover from capture stress and begin feeding (Figure 1.2). Upon introduction to the pen, some sharks were tagged with a pop-off archival satellite tag (PSAT) to record their swimming behavior with respect to depth and water temperature. During the period that the sharks were in the pen (10–25 d), MBA husbandry department staff made observations on the condition of the sharks and offered fresh fish—Pacific Mackerel (*Scomber japonicus*), White Croaker (*Genyonemus lineatus*), Bonito (*Sarda chiliensis*), and King Salmon (*Oncorhynchus tshawytscha*)—to stimulate feeding.

The YOY White Sharks were transported to the MBA after they were determined to be healthy and feeding regularly in the pen. Food was generally withheld from the sharks 24 h prior to

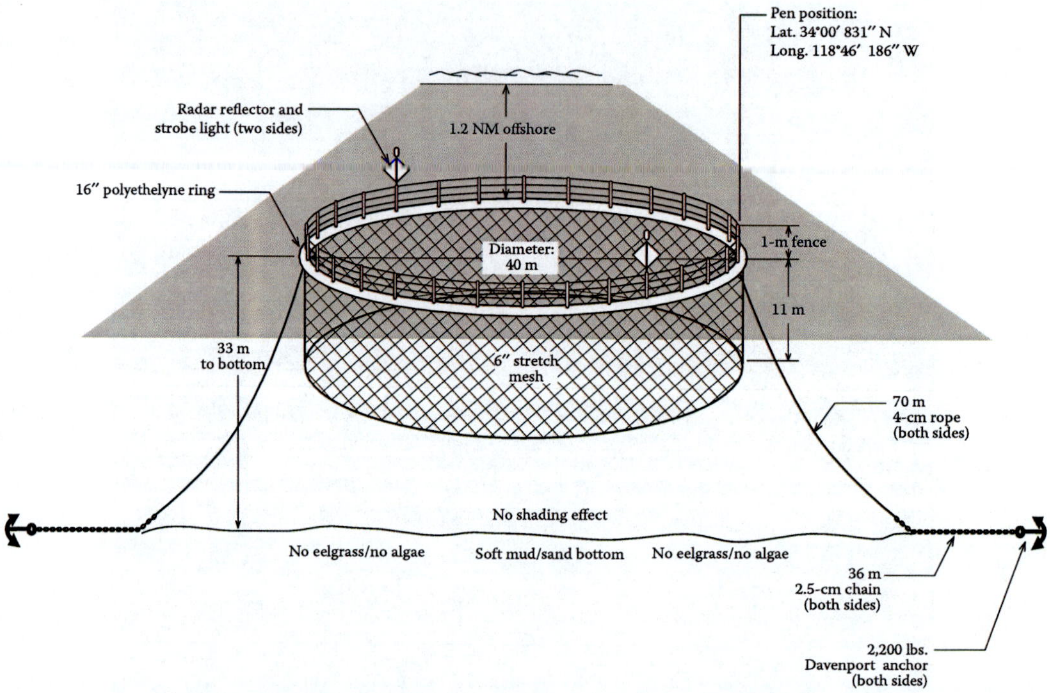

Figure 1.2 The ocean net pen (40 m diameter × 11 m depth) used to hold YOY White Sharks after capture in the Southern California Bight. The sharks were allowed to recover from capture stress and begin feeding prior to transport to the Monterey Bay Aquarium for public display. (Courtesy of John O'Sullivan, Monterey Bay Aquarium.)

transport, except in 2006 when the shark was fed within approximately 6 h prior to transport. The sharks were netted out of the ocean pen and placed unrestrained in a 250-L vinyl shark box with oxygenated seawater (~125% saturation) at 16°C, with a recirculating submersible pump (4164 lph; Rule Industries, MA) to ventilate the sharks during the 30–90-min. transport via boat to the shore. The sharks were then transferred to an 11,356-L pelagic fish transport tank mounted on the trailer of a commercial tractor for the approximately 6 h duration transport to the MBA (Chapter 2, this book). Upon arrival at the MBA, except for shark #06-01, the sharks were weighed and measured (measurement was taken over the curve of the body for most animals but straight length for sharks #07-01, 08-01, and 9-01) and placed on public display in the 3.8-million-L Outer Bay exhibit. Shark #06-01 was measured over the body curve in the field on August 17, 2006.

While on display, the YOY White Sharks were offered food daily. Food items were individually weighed, tethered with cotton string to attach them to the feeding pole, and fed to the shark. This method reduced the potential of the shark biting the feeding stick or of other tank inhabitants taking food items offered to the White Sharks. If a food item was shredded and dropped, then an estimate of the weight of the food ingested was made. The mean daily ration for each week was calculated as wet weight (g) and as % BM d^{-1} for each 7-d period that the sharks were on display. Food items fed to the White Shark were also sent for caloric analysis at NP Analytical Laboratories (St. Louis, MO). The energy equivalent of total food consumption was determined by the feeding rate (% BM d^{-1}), energy content of food type, and total duration in captivity. In addition, two dead YOY White Sharks (whole fish) that were bycatch of the commercial fishery were sent to NP Analytical Laboratories for caloric analysis to determine the energy content of White Shark tissue.

Conversion from total length over the curve of the body to straight length for the sharks was done via linear regression analysis. Straight total length and mass for the YOY White Sharks were log-transformed and plotted, allowing a linear regression analysis to determine the intercept and slope, which were used to calculate mass with the back-transformed allometric equation Mass = $a\mathrm{TL}^b$, where log (a) is the intercept and b is the slope. Feeding data were recorded and calculated as the mean daily feeding ration for each week, both as wet weight (g) and % BM d^{-1}. Feeding and growth data were analyzed using Microsoft Office Excel 2003 and graphed using Sigma Plot 9.0.

Simplified energy budgets were created by using food consumption data (C) and calculated gain in mass (G) by using assumptions regarding energy loss to waste (W) and by solving for metabolic costs (M) and were described by the formula $C = G + M + W$. The total number of calories from food ingested by each shark while in captivity was used to calculate consumption (C). The energy invested into somatic growth (G) was calculated as the gain in mass converted to energetic equivalent by multiplying the mass by the caloric content of YOY White Shark tissue. Energy loss to waste such as feces and urine (W) was estimated to be 27% of consumption (Wetherbee and Gruber, 1993; Wetherbee and Cortes, 2004). Metabolic costs were solved as consumption minus energy loss and energy invested into growth ($M = C - (G + W)$).

RESULTS

Feeding Ration

After transport to the MBA, most YOY White Sharks fed within 7 d and continued feeding regularly; however, individual behavior influenced feeding in the OBE, which contained many other larger fishes. Shark #06-01 did not feed for 7 d, presumably because of being fed at the ocean pen until the day of transport. Feeding for this shark was initiated after a live California Skate (*Raja inornata*) offered as food was wounded by one of the Galapagos Sharks (*Carcharhinus galapagensis*) in the OBE. Feeding was also difficult to initiate with shark #09-01 because it swam at the bottom of the OBE for the first month in captivity. Fresh, dead Pacific Mackerel was used to initiate feeding; however, at first this shark was intimidated by the Galapagos and Scalloped Hammerhead Sharks in the exhibit. This was most likely the cause for the slower feeding rate initially for this shark. Larger Bluefin Tuna (*Thunnus orientalis*), Yellowfin Tuna (*T. albacares*), and Common Dorado (*C. hippurus*), ranging in body mass from approximately 40 to 140 kg, also posed significant challenges because of their aggressive competition for food. Feeding the relatively small YOY White Sharks in this setting necessitated surface feedings to reduce the potential of collisions with other fishes and to accurately record feeding amounts. However, within 1 month, the YOY White Sharks became the most aggressive animals in the OBE and at times would charge the other fishes if they came close to the feeding station. Sharks #04-01 and 09-01 began to chase and attack other sharks prior to release; however, feedings on other fishes were not observed by staff or recorded by cameras monitoring the OBE. The White Sharks did not feed on tunas that were fatally wounded by collisions with the exhibit window and were consumed by the other sharks. Unrecorded feedings would not have been significant enough to change the feeding rates reported in this study.

Four of the five YOY White Sharks fed within 1 d to 1 week of introduction into the OBE and fed consistently while on display, ranging from 70 to 198 d. Three sharks (#04-01, #06-01, and #07-01) showed a strong feeding preference for King Salmon (*O. tshawytscha*), which comprised 80.9–96.6% of the diet as wet weight (Table 1.1). One shark (#09-01) ate primarily Pacific Mackerel (*Scomber japonicus*; 99.1% of the diet as wet weight). Shark #04-01 had the widest diet range in addition to salmon, consisting of 13.6% whole Pacific Mackerel, 4.1% whole Albacore (*Thunnus alalunga*), and 1.4% whole Bonito (*Sarda chiliensis*). Shark #06-01 also fed on Sablefish (*A. fimbria*), comprising 14.9% of the diet as wet weight, along with a variety of other food items comprising less than 2% of the total diet. Sharks #07-01 and #09-01 had the narrowest diet range, with the

Table 1.1 Diet of Four Captive YOY White Sharks while on Display in the Outer Bay Exhibit at the Monterey Bay Aquarium from September 14, 2004 until November 4, 2009

Food Type	Shark #04-01	Shark #06-01	Shark #07-01	Shark #09-01
King Salmon (*Onchorynchus tshawytscha*)	80.9%	83.1%	96.6%	0.9%
Pacific Mackerel (*Scomber japonicus*)	13.6%	0.3%	0.9%	99.1%
Sablefish (*Anoplopoma fimbria*)	0%	14.9%	2.5%	0%
Albacore (*Thunnus alalunga*)	4.1%	0.8%	0%	0%
Pacific Bonito (*Sarda chiliensis*)	1.4%	0.2%	0%	0%
Mahi-Mahi (*Coryphaena hippurus*)	0%	0.1%	0%	0%
California Skate (*Raja inornata*)	0%	0.6%	0%	0%

Food types are indicated as percentages of the diet on a wet-weight basis.

preferred food item comprising 96.6–99.1% of the diet as wet weight (King Salmon and Pacific Mackerel, respectively).

The YOY White Shark displayed in 2008 fed only once during the 11 d it was on display and was released because of concern regarding its health. This shark was feeding in the ocean pen, and upon arrival at the MBA, it navigated the OBE very well; however, this shark only fed once on approximately 400 g of salmon. While it was on display, the shark lost 2.6 kg of mass from its original body mass of 25.2 kg, or 10.3% of its body mass. Body mass loss because of starvation can be used to estimate a maintenance ration of 0.94% BM d^{-1} for YOY White Sharks, which is equivalent to 405 kcal d^{-1}. This shark was transported back to the Southern California Bight, where it was tagged with a PSAT and released. Data from the PSAT showed that this shark survived for at least 30 d after release, at which time the PSAT separated from the shark and began reporting data of the shark's movements.

Mean daily ration for each week as wet weight (g) showed an oscillating pattern with feeding peaks and troughs that generally diminished over the time the four sharks were on display (Figure 1.3). Mean daily ration for each week ranged from a low of 0 g for the first week of shark #06-01 to a high of 1718.6 g, also for shark #06-01 (Figure 1.3). The grand mean daily ration for all four sharks was 747 ± 46 g d^{-1}. A period of approximately 3–8 weeks corresponded to a single cycle of feeding peak and decline, with larger differences in feeding peaks and troughs tending toward longer cycles as seen with shark #06-01. Observations showed that total evacuation time after the first feeding (on salmon fillet) for shark #04-01 was approximately 36 h. After 15–23 weeks in captivity, there was a large decrease in the oscillations in feeding ration for sharks #04-01, 06-01, and 07-01. Shark #09-01 was released after approximately 10 weeks because of predatory behavior in the exhibit and did not follow the general feeding pattern of the other sharks.

The mean daily ration for each week as % BM d^{-1} for three of the four White Sharks (#04-01, #06-01, and #07-01) showed a significant decreasing trend over the period that they were on display (Figure 1.4). Mean daily ration for each week for the three sharks ranged between a low of 0.2% BM d^{-1} once the sharks began to feed to a high of 3.5% BM d^{-1}. The decreasing trend in mean daily ration as % BM d^{-1} was similar for all three sharks, and the slopes for the linear regressions were statistically significant ($p < 0.0013$). Mean daily rations during the time on display for sharks #04-01, #06-01, and #07-01 were 1.68 ± 0.11, 1.32 ± 0.10, and 1.46 ± 0.10% BM d^{-1}, respectively.

White shark #09-01 had a different swimming and feeding pattern, spending most of the time near the bottom of the exhibit, which made feeding attempts much more difficult because of

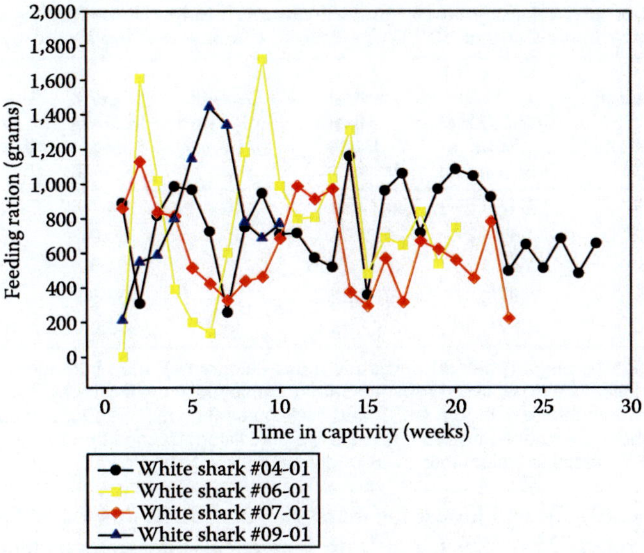

Figure 1.3 Feeding ration as wet weight of food (grams) for four captive YOY White Sharks displayed at the MBA from September 14, 2004 until November 4, 2009.

competition from the other fishes. The feeding pattern for shark #09-01 showed a slowly increasing mean daily ration for each week until reaching a peak (3.5% BM d^{-1}) at week 6 and then decreasing until release at week 10 on display. Mean daily ration for shark #09-01 was somewhat higher than for the three other sharks at $2.03 \pm 0.16\%$ BM d^{-1}. The grand mean daily ration for each week for all four sharks was $1.62 \pm 0.15\%$ BM d^{-1}.

The energy equivalent of food consumption showed a general pattern of highest total consumption and mean daily consumption for the entire time on display (kcal d^{-1}) for YOY White Sharks

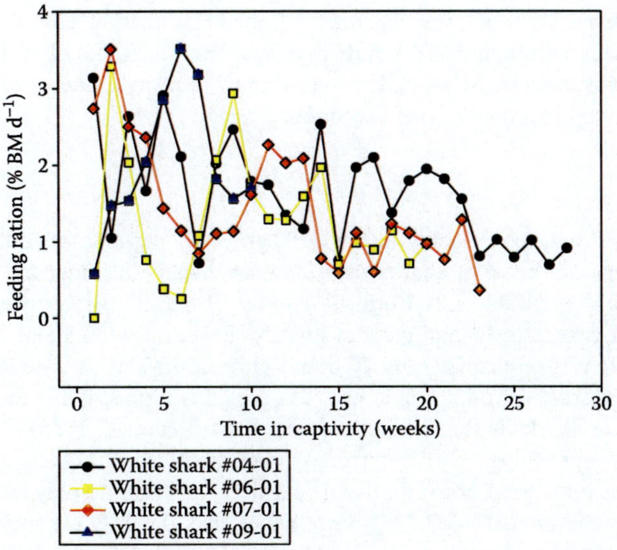

Figure 1.4 Feeding ration as % BM d^{-1} for four captive YOY White Sharks displayed at the MBA from September 14, 2004 until November 4, 2009.

Table 1.2 Total Length and Body Mass upon Introduction to the Outer Bay Exhibit and upon Release of Five Captive YOY White Sharks Displayed at the Monterey Bay Aquarium

White Shark Number	Duration in Captivity (d)	Initial/Final Total Length (m)	Initial/ Final Body Mass (kg)	Growth in Length (cm yrs.$^{-1}$)	Growth in Mass (kg yrs.$^{-1}$)	K_1 Wet Weight (%)	K_1 Energetic Equivalent (%)
#04-01	198	1.41/1.84	28.0/73.4	80.8	83.7	30.9	29.7
#06-01	138	1.64/1.87	47.0/77.6	56.3	80.9	28.2	26.1
#07-01	161	1.43/1.76	30.6 /63.6	76.2	74.8	33.3	31.4
#08-01	11	1.37/1.37	25.2/22.6	0	−86.3	—	—
#09-01	70	1.57/1.66	36.2/45.4	45.2	48.0	16.0	24.3

The table notes the total length (meters) and body mass (kilograms) upon introduction to the Outer Bay exhibit and upon release of five captive YOY White Sharks displayed at the Monterey Bay Aquarium. Duration in captivity (d), growth in total length (cm yrs.$^{-1}$) and body mass (kg yrs.$^{-1}$), and gross conversion efficiency (K_1) on a wet-weight and energetic-equivalent basis are also listed. Dashes for gross conversion efficiency of White Shark #08-01 indicate weight loss (−86.3 kg yrs.$^{-1}$) rather than growth.

#04-01, #06-01, and #07-01, and lowest for shark #09-01. Sharks #04-01, #06-01, and #07-01 fed at a relatively high rate (1.32–1.68% BM d^{-1}) on energetically dense foods (energy content: Black Cod > King Salmon > Albacore > Pacific Mackerel) (Table 1.1). Mean daily food consumption for the entire time in captivity was greatest for the sharks #04-01 and #06-01 and decreased for sharks #07-01 and 09-01; mean daily food consumption values were 1329.9 ± 72.5, 1470.2 ± 106.6, 1119.5 ± 71.4, and 940.3 ± 85.2 kcal d^{-1}, respectively. Shark #09-01 fed at a higher rate (2.03% BM d^{-1}) but on energetically less dense Pacific Mackerel because of a feeding preference, and this shark was on display (70 d) less than half the time as the other sharks (Table 1.2).

Predatory behavior was observed in two of the five YOY White Sharks while on display in the OBE. During the final 5 weeks on display, shark #04-01 fatally attacked two Soupfin Sharks (*Galeorhinus galeus*), consuming the caudal fin and caudal peduncle of one of the Soupfin Sharks. However, this feeding was not included in this study because the estimated weight of this meal was less than 1 kg and amounted to less than 1% of the diet on a wet-weight basis. This shark was also observed chasing Scalloped Hammerhead Sharks and Galapagos Sharks prior to being released. White shark #09-01 was also observed chasing Scalloped Hammerhead Sharks and attacked a male Galapagos Shark, which prompted MBA staff to release this shark. No other predatory behavior of White Sharks was observed by MBA staff or verified by security cameras mounted in the exhibit that recorded constantly to a digital video recorder.

Captive Growth

The YOY White Sharks on display in the OBE grew in length at twice the calculated growth for YOY White Sharks in the wild. Shark #04-01 was on display the longest (198 d) and grew at the fastest rate (80.8 cm yrs.$^{-1}$; Table 1.2). Both sharks #06-01 and 07-01 were held for a shorter durations (138 and 161 d, respectively) and grew at slightly slower rates (56.3 and 77.1 cm yrs.$^{-1}$, respectively). Shark #09-01 was on display only 70 d and grew at the slowest rate (45.2 cm yrs.$^{-1}$). Mean growth for all four sharks was 64.9 ± 8.5 cm yrs.$^{-1}$, which is almost twice the calculated first-year growth rate for White Sharks in the wild (35 cm yrs.$^{-1}$; Cailliet et al., 1985).

Growth in body mass followed the expected general pattern of highest growth for the sharks that had the highest total food consumption (kcal d^{-1}) while on display. Shark #04-01 had the highest annual growth in mass (83.7 kg yrs.$^{-1}$) and had the highest total food consumption (261,979 kcal), followed by shark #06-01 (80.9 kg yrs.$^{-1}$ and 201,410 kcal). Shark #07-01 grew at a slightly slower rate (74.8 kg yrs.$^{-1}$ and 180,232 kcal), and shark #09-01 grew most slowly and consumed the lowest amount of calories while on display (48.0 kg yrs.$^{-1}$ and 64,882 kcal). Mean

Table 1.3 Simplified Energy Budgets of Four Captive YOY White Sharks at the Monterey Bay Aquarium

White Shark Number	Consumption (C) (kcal)	Growth (G) (kcal)	Metabolism (*M*) (kcal)	Energy Budget $C = G + M + W$
#04-01	261,980	77,861	113,384	100 = 29.7 + 43.3 + 27
#06-01	201,411	52,479	94,551	100 = 26.1 + 46.9 + 27
#07-01	180,232	56,595	74,974	100 = 31.4 + 41.6 + 27
#09-01	64,883	15,778	31,586	100 = 24.3 + 48.7 + 27
				Mean: 100 = 27.9 + 45.1 + 27

The table shows the simplified energy budgets of four captive YOY White Sharks at the Monterey Bay Aquarium as described by the equation $C = G + M + W$, where C = consumption, G = energy investment into somatic growth, M = metabolic costs, and W = energy loss to waste, which is set at 27% of C. C is based on the energetic equivalent (kcal) of total feedings for each shark, G is based on growth in mass for each shark converted to energetic equivalent (kcal) by using the conversion factor of 1,715 kcal kg^{-1} as determined from caloric analysis of two YOY White Sharks. M was determined by solving the simplified energy budget equation.

growth in mass for all four sharks was 71.6 ± 8.2 kg yrs.$^{-1}$ Gross conversion efficiency (K_1) on a wet-weight basis was lowest for shark #09-01 (16.0%) and much higher for sharks #04-01, #06-01, and #07-01 (30.9, 28.2, and 33.3%, respectively; Table 1.3). Mean K_1 on a wet-weight basis for all four sharks was 27.1 ± 3.8%. K_1 on an energy-equivalent basis was slightly lower than K_1 on a wet-weight basis but showed high agreement to the general pattern of K_1 for all four sharks. The largest difference between the two K_1 values (8%) was for shark #09-01. This shark had a higher K_1 (24.3%) on an energy-equivalent basis because of the lower caloric value of its primary food item, Pacific Mackerel.

Mass in relation to TL was very similar for sharks #04-01, #06-01, and #07-01 and was described by the allometric equation $M = a\text{TL}^b$. Sharks #04-01, #06-01, and #07-01 had very similar a and b values (a = 8.34, 7.37, and 8.42, respectively; and b = 3.55, 3.74, and 3.47, respectively). However, growth was much lower for shark #09-01, and the a and b values for the allometric equation describing total length and mass were somewhat different (a = 5.25 and b = 4.26).

Energy Budgets

Simplified energy budgets for the YOY White Sharks showed a general pattern of the greatest energy expenditure from metabolic costs (M) and large energy investment into somatic growth (G) while the sharks were on display (Table 1.3). Energy loss to waste (W) was held constant (27% of C), and the amount of energy invested into metabolic costs as a proportion of C ranged from 41.6 to 48.7% of C. The mean consumption rate for all four White Sharks was 1215.0 ± 116.5 kcal d^{-1}. The energetic content of the two YOY White Sharks was determined to be 1.715 kcal g^{-1} of tissue. This value was multiplied by the gain in body mass for each shark while they were on display to determine the energy investment into G, which ranged between 24.3 and 29.7% of C. A mean energy budget for all four sharks while in captivity was described by the following equation:

$$C = G + M + W{:}100 = 27.9\% \ (\pm 1.6) + 45.1\% \ (\pm 1.6) + 27\% \quad (1.1)$$

DISCUSSION

The peak in mean daily ration recorded for the White Sharks in this study were similar to those reported for active pelagic sharks (Salini et al., 1999; Bush and Holland, 2002) but were slightly lower than the daily ration calculated for the Shortfin Mako Shark (*Isurus oxyrinchus*; Wood et al., 2009). Daily rations reported for juveniles of three species of carcharhinid sharks (2.9–3.44% BM d^{-1} for *Carcharhinus dussumieri*, *C. tilstoni*, and *Negaprion acutidens*; Salini et al., 1999) were similar to the initial peaks in mean daily ration of YOY White Sharks in our study (3.1–3.5% BM d^{-1}). However, the feeding experiments done by Salini et al. (1999) were of very short duration (10 d) and therefore do not give any indication of the range in daily ration over time in captivity for these species. The highest daily ration calculated for an ectothermic, obligate ram-ventilating shark is for the Scalloped Hammerhead Shark (3.54% BM d^{-1}; Bush and Holland, 2002), which is also similar to the peak in mean daily ration in our study. The peak in mean daily ration of the White Sharks in our study is lower than the daily ration calculated for the Shortfin Mako Shark (4.6% BM d^{-1}; Wood et al., 2009); however, the daily ration in that study was estimated by calculating the energetic needs of the Shortfin Mako Shark and then calculating the amount of food needed to satisfy those energetic needs. The dominant food item in the diet of Shortfin Mako Sharks in the northeast Atlantic ocean as determined by gut content analysis is Bluefish (*Pomatomus saltatrix*), which is 63% lower in energy content (4,800 kJ kg^{-1}) compared with the preferred food for White Sharks in our study, King Salmon (7,536 kJ kg^{-1}). If the main prey item of Shortfin Mako Sharks had been more energetically dense, then the estimate of daily ration for the Shortfin Mako Shark would have been lower. Indeed, shark #09-01 fed on the least energetically dense food item (99% Pacific Mackerel by weight, 4,312 kJ kg^{-1}), and of all the White Sharks in our study, it fed at the highest daily ration for the entire period it was on display (mean = 2.03% BM d^{-1}).

Captive growth of the YOY White Sharks in our study support the assertion by Van Dykhuizen and Mollet (1992) that captive growth may be higher than growth in the wild by factors of two to three. We report a mean growth rate of captive YOY White Sharks (64.9 ± 8.5 cm $yrs.^{-1}$) that is twice the first-year growth calculated from a von Bertalanffy growth function for this species (35 cm $yrs.^{-1}$; Cailliet et al., 1985). In addition, a YOY White Shark tagged and released as part of MBA's field research program in Southern California was recaptured 405 d later, yielding a growth rate of 33.4 cm $yrs.^{-1}$ (C. G. Lowe, personal observation). This observed growth rate is in close agreement with the growth rate for a YOY White Shark calculated using the von Bertalanffy growth function from Cailliet et al. (1985), and it further supports that our captive growth rate is approximately twice that of growth in the wild. We also report a mean growth rate in mass (71.6 ± 8.2 kg $yrs.^{-1}$) that is up to three times greater than the calculated rate (23 kg $yrs.^{-1}$) for first-year growth using the von Bertalanffy growth function (Cailliet et al., 1985) and allometric equations for TL and mass for White Sharks from Mollet and Cailliet (1996). Growth of sharks in captivity can be variable and depends on a variety of factors, with feeding ration thought to be a main determinant (Taylor and Wisner, 1989). In addition, lamnid sharks in the wild have high growth rates in the first year of life (39 cm $yrs.^{-1}$ for Shortfin Mako Sharks in New Zealand waters; Bishop et al., 2006), and a diet of energy-rich salmonids has been correlated with faster growth in lamnid sharks (Goldman and Musick, 2003). Many studies of growth in captivity report faster growth for species such as the Sandbar Shark (*Carcharhinus plumbeus*), Bull Shark (*Carcharhinus leucas*), Lemon Shark (*Negaprion brevirostris*), and Scalloped Hammerhead Shark (reviewed in Mohan et al., 2004); however, studies on the Sevengill Shark (*Notorynchus cepedianus*; Van Dykhuizen and Mollet, 1992) and the Sand Tiger Shark (*Carcharias taurus*; Govender et al., 1991) have reported growth in captivity that may be similar to growth in the wild. For this reason, caution should be used when citing growth rates in captivity. However, these captive growth studies can add to our understanding of the biology of elasmobranchs that are difficult to acquire or need large enclosures for a captive environment, such as the White Shark (Mollet et al., 2002; Cailliet and Goldman, 2004; Mohan et al., 2004).

Mean gross conversion efficiency (K_1) on a wet-weight basis for captive YOY White Sharks (27.1% ± 3.8) was similar to reported values for sharks in the first year of life. Van Dykhuizen and Mollet (1992) reported K_1 values of 25–40% for captive Sevengill Sharks (*Notorynchus cepedianus*) in the first year of life. K_1 values of 10–25% have also been reported for other elasmobranchs and teleosts (Wetherbee and Cortes, 2004). In addition, we report K_1 as energetic equivalents, which show very good agreement with K_1 on a wet-weight basis for three of the four captive White Sharks (#04-01, #06-01, and #07-01). The largest difference in K_1 values (wet weight versus energetic equivalents) was for shark #09-01, which fed almost entirely on Pacific Mackerel, which was lower in caloric value to the main food item of the other sharks, King Salmon. K_1 values also decrease with increasing age and are affected by the level of consumption (Wetherbee and Cortes, 2004). Endothermy in lamnid sharks has been theorized to speed up digestive processes, shortening the time for gastric evacuation and enhancing food intake (Cortes and Gruber, 1990: Carlson et al., 2004). Total evacuation time (36 h) observed for White Sharks in this study is faster than the values reported for gastric evacuation for ectothermic sharks (Wetherbee and Cortes, 2004), and peak daily feeding ration and K_1 are among some of the highest reported for any shark species to date.

The simplified energy budgets constructed for captive YOY White Sharks at the MBA reflect a high investment into somatic growth (mean ± SE = 27.9% ± 1.6). The energetic content of YOY White Shark tissue as determined by caloric analysis yielded a value (7.18 kJ g^{-1}) that is slightly higher than those reported to date for the Lemon Shark (5.41 kJ g^{-1}; Cortes and Gruber, 1990), the Scalloped Hammerhead Shark (6.07 kJ g^{-1}; Lowe, 2002), or the Shortfin Mako Shark (5.56 kJ g^{-1}; Wood et al., 2009). These differences in energy content of shark tissue may be the result of interspecific differences in the relative abundance of tissues with different caloric content (i.e., liver mass). Another possibility is that the differences are due to the methods used to determine energy content, which is why we chose to send whole specimens to be homogenized and sampled. Investment into somatic growth as high as 24.3–31.4% of consumption are most likely the result of higher growth rates caused by the high growth capacity of YOY sharks compared with mature animals, such as the Bull Shark studied in captivity (G = 7%; Schmid and Murru, 1994). Higher growth rates and potentially lower metabolic costs because of captive conditions are also presumably the cause for greater investment into G.

The largest and most variable component of an energy budget is the energy invested into metabolic costs (Lowe, 2002), which ranged for White Sharks from 41.6 to 48.7% of C (mean ± SE = 45.1% ± 1.6). The shark that did not feed (#08-01) incurred a body mass loss that was 0.94% BM d^{-1}; this is equivalent to 405 kcal d^{-1}, or 33% of the mean consumption for all sharks (1215 kcal d^{-1}). This estimate of metabolic costs is similar to our estimate from the energy budget, especially when taking into account the added energy consumption because of specific dynamic action for the feeding sharks. Our estimate of metabolic costs for White Sharks in this study is also quite similar to the metabolic costs (44.9% of C) reported from an energy budget for captive Pelagic Stingrays (*Dasyatis violacea*; Ezcurra, 2001). However, caution should be used when comparing the metabolic costs determined from an energetics model derived for captive elasmobranchs with those reported from the more common indirect calorimetry method because of the many differences between the two methods. Schmid and Murru (1994) also reported lower estimates of M for the Bull Shark from a bioenergetics approach compared with reported metabolic rates for carcharhinid sharks derived from the method of indirect calorimetry.

Our estimates of metabolic costs in a captive setting may be useful as a starting point toward better understanding of the energetic demands of YOY White Sharks in the wild. Migratory patterns of YOY White Sharks in the wild include large-scale horizontal (thousands of kilometers) and vertical movements (hundreds of meters) (Dewar et al., 2004; Weng et al., 2007), which would be much more energetically costly than the activity levels of captive White Sharks. A bioenergetics study of wild White Sharks similar to the one reported for another lamnid shark, the Shortfin Mako (Wood et al., 2009), may now be possible. Energy investment into growth may be estimated with existing information on growth in the wild (Cailliet et al., 1985), and determining the metabolic costs for YOY White Sharks in the wild is now possible because of technological advances in the field of physiological telemetry (Lowe and Goldman, 2001). This step is crucial to a better understanding of daily ration and energetic demands of this top-level predator.

ACKNOWLEDGMENTS

Many thanks go to a very talented and diverse group of people who have made this study possible, including the crew of the F/V *Barbara H*, Dave Powell, Dr. Chuck Winkler, Randy Hamilton, Chuck Farwell, Jon Hoech, Joe Welsh, Dr. Mike Murray, Bobby Bettancourt, Kevin Lewand, and the staff of the applied research and husbandry departments at the MBA. J. M. E. is especially grateful to Roxane Buck-Ezcurra for her constant support and encouragement during field work and for her review of this manuscript.

REFERENCES

Anderson, S. D., R. P. Henderson, P. Pyle, D. G. Ainley. 1996. White shark reactions to unbaited decoys. In: A. P. Klimley, D. G. Ainley (eds). *Great white sharks: The biology of Carcharodon carcharias* (pp. 223–228). Orlando, FL: Academic Press.

Bishop, S. D. H., M. P. Francis, C. Duffy, J. C. Montgomery. 2006. Age, growth, longevity and natural mortality of the shortfin mako shark (*Isurus oxyrinchus*) in New Zealand waters. *Aus J Mar Freshwater Res* 57: 143–154.

Bonfil, R, M. Meyer, M. C. Scholl, R. Johnson, S. O'Brien, H. Oosthuizen, S. Swanson, D. Kotze, M. Paterson. 2005. Transoceanic migration, spatial dynamics, and population linkages of white sharks. *Science* 310: 100–103.

Boustany, A. M., S. F. Davis, P. Pyle, S. D. Anderson, B. J. Le Boeuf, B. A. Block. 2002. Expanded niche for white sharks. *Nature* 415: 35–36.

Bruce, B. D., J. D. Stevens, H. Malcolm. 2006. Movements and swimming behaviour of white sharks (*Carcharodon carcharias*) in Australian waters. *Mar Biol* 150: 161–172.

Bush, A., K. Holland. 2002. Food limitation in a nursery area: Estimates of daily ration in juvenile scalloped hammerheads, *Sphyrna lewini* (Griffith and Smith, 1834) in Kaneohe Bay, Oahu. *J Exp Mar Biol Ecol* 278: 157–178.

Cailliet, G. M., L. J. Natanson, B. A. Welden, D. A. Ebert. 1985. Preliminary studies on the age and growth of the white shark, *Carcharodon carcharias*, using vertebral bands. *Memoirs Southern California Academy of Sciences* 9: 49–60.

Cailliet, G. M., K. J. Goldman. 2004. Age determination and validation in chondrichthyan fishes. In: J. C. Carrier, J. A. Musick, M. R. Heithaus (eds). *Biology of sharks and their relatives* (pp. 339–347). Boca Raton, FL: CRC Press, Inc.

Carlson, J. K., J. K. Goldman, C. G. Lowe. 2004. Metabolism, energetic demand, and endothermy. In: J. C. Carrier, J. A. Musick, M. R. Heithaus (eds). *Biology of sharks and their relatives* (pp. 203–224). Boca Raton, FL: CRC Press, Inc.

Compagno, L. J. V. 1984. *Vol. 4: Sharks of the world: An annotated and illustrated catalogue of shark species known to date. Part 1: Hexanchiformes to Lamniformes*. Rome, Italy: United Nations Development Programme.

Cortes, E., and S. H. Gruber. 1990. Diet, feeding habits and estimates of daily ration of young lemon sharks, *Negaprion brevirostris* (Poey). *Copeia* 1990: 204–218.

Dewar, H., M. Domeier, N. Nasby-Lucas. 2004. Insights into young of the year white sharks (*Carcharodon carcharias*) behavior in the Southern California Bight. *Environ Biol Fishes* 70: 133–143.

Domeier, M. L., N. Nasby-Lucas. 2008. Migration patterns of White Sharks *Carcharodon carcharias* tagged at Guadalupe Island, Mexico, and identification of an eastern Pacific shared offshore foraging area. *Mar Ecol Prog Ser* 370: 221–237.

Ezcurra, J. M. 2001. The mass-specific routine metabolic rate of captive pelagic stingrays, *Dasyatis violacea*, with comments on energetics. MS thesis, Moss Landing Marine Laboratory, California State University, Stanislaus.

Francis, M. P. 1996. Observations on a pregnant white shark with a review of reproductive biology. In: A. P. Klimley, D. G. Ainley (eds). *Great white sharks: The biology of Carcharodon carcharias* (pp. 157–172). Orlando, FL: Academic Press.

Goldman, K. J., J. A. Musick. 2003. Growth and maturity of salmon sharks (*Lamna ditropsis*) in the eastern and western North Pacific, and comments on back-calculation methods. *Fish Bull* 104: 278–292.

Govender, A., N. Kistnasami, R. P. Van Der Elst. 1991. Growth of spotted ragged-tooth sharks *Carcharias Taurus* (Rafinesque) in captivity. *S African J Mar Sciences* 11: 15–19.

Hewitt, J. C. 1984. The great white shark in captivity: A history and prognosis. *AAZPA Annual Proceedings* 1984: 317–324.

Klimley, A. P., B. J. Le Beof, K. M. Cantara, J. E. Richert, S. F. Davis, S. Van Sommeran, J. T. Kelly. 2001. The hunting strategy of white sharks (*Carcharodon carcharias*) near a seal colony. *Mar Biol* 138: 617–636.

Klimley, A. P., P. Pyle, S. D. Anderson. 1996. The behavior of white sharks and their pinniped prey during predatory attacks. In: A. P. Klimley, D. G. Ainley (eds). *Great white sharks: The biology of Carcharodon carcharias* (pp. 175–191). Orlando, FL: Academic Press.

Long, D. J., K. D. Hanni, P. Pyle, J. Roletto, R. E. Jones, R. Bandar. 1996. White shark predation on four pinniped species in central California waters: Geographic and temporal patterns inferred from wounded carcasses. In: A. P. Klimley, D. G. Ainley (eds). *Great white sharks: The biology of Carcharodon carcharias* (pp. 263–274). Orlando, FL: Academic Press.

Lowe, C. G. 2002. Bioenergetics of free-ranging juvenile scalloped hammerhead sharks (*Sphyrna lewini*) in Kaneohe Bay, Oahu, HI. *J Exp Mar Biol Ecol* 278: 141–156.

Lowe, C. G., K. J. Goldman. 2001. Thermal and bioenergetics of elasmobranchs: Bridging the gap. *Environ Biol Fishes* 60: 251–266.

Mohan, P. J., S. T. Clark, T. H. Schmid. 2004. Age and growth of captive sharks In: M. Smith, D. Warmolts, D. Thoney, R. Hueter (eds). *The elasmobranch husbandry manual captive care of sharks, rays and their relatives* (pp. 201–226). Columbus, OH: Ohio Biological Survey.

Mollet, H. F., G. M. Cailliet. 1996. Using allometry to predict body mass from linear measurements of the white shark. In: A. P. Klimley, D. G. Ainley (eds). *Great white sharks: The biology of Carcharodon carcharias* (pp. 81–89). Orlando, FL: Academic Press.

Mollet, H. F., G. M. Cailliet, A. P. Klimley, D. A. Ebert, A. D. Testi, L. J. V. Compagno. 1996. A review of length validation methods and protocols to measure large white sharks. In: A. P. Klimley, D. G. Ainley (eds). *Great white sharks: The biology of Carcharodon carcharias* (pp. 91–108). Orlando, FL: Academic Press.

Mollet, H. F., J. M. Ezcurra, J. B. O'Sullivan. 2002. Captive biology of the pelagic stingray, *Dasyastis violacea* (Bonaparte, 1830). *Mar Freshwater Res* 53: 531–541.

Pratt, H. L., Jr. 1996. Reproduction in the male white shark. In: A. P. Klimley, D. G. Ainley (eds). *Great white sharks: The biology of Carcharodon carcharias* (pp. 131–138). Orlando, FL: Academic Press.

Saidi, B., M. N. Bradai, A. Boijain, O. Guelorget, C. Capape. 2005. Capture of a pregnant female white shark, *Carcharodon carcharias* (Lamnidae) in the Gulf of Gabes (southern Tunisia, central Mediterranean) with comments on oophagy in sharks. *Cybium* 29: 303–307.

Salini, J. P., M. Tonks, S. J. Blaber, J. Ross. 1999. Feeding of captive, tropical carcharhinid sharks from the Embley River estuary, Northern Australia. *Mar Ecol Prog Ser* 184: 309–314.

Schmid, T. H., F. L. Murru. 1994. Bioenergetics of the bull shark, *Carcharhinus leucas*, maintained in captivity. *Zoo Biol* 13: 177–185.

Taylor, L., M. Wisner. 1989. Growth of captive blacktip reef sharks (*Carcharhinus melanopterus*). *Bull Inst Oceanogr (Monaco)* 5: 211–217.

Uchida, S., M. Toda, K. Teshima, K. Yano. 1996. Pregnant white sharks and full-term embryos from Japan. In: A. P. Klimley, D. G. Ainley (eds). *Great white sharks: The biology of Carcharodon carcharias* (pp. 139–155). Orlando, FL: Academic Press.

Van Dykhuizen, G., H. F. Mollet. 1992. Growth, age estimation and feeding of captive sevengill sharks, *Notorynchus cepedianus*, at the Monterey Bay Aquarium. *Aus J Mar Freshwater Res* 43: 297–318.

Weng, K. C., J. B. O'Sullivan, C. G. Lowe, C. E. Winkler, H. Dewar, B. A. Block. 2007. Movements, behavior and habitat preferences of juvenile white sharks *Carcharodon carcharias* in the eastern Pacific. *Mar Ecol Prog Ser* 338: 211–224.

Wetherbee, B. M., E. Cortes. 2004. Food consumption and feeding habits. In: J. C. Carrier, J. A. Musick, M. R. Heithaus (eds). *Biology of sharks and their relatives* (pp. 225–246). Boca Raton, FL: CRC Press, Inc.

Wetherbee, B. M., S. H. Gruber. 1993. Absorption efficiency of the lemon shark *Negaprion brevirostris* at varying rates of energy intake. *Copeia* 1993: 416–425.

Wood, A. D., B. M. Wetherbee, F. Juanes, N. E. Kohler, C. Wilga. 2009. Recalculated diet and daily ration of the shortfin mako (*Isurus oxyrinchus*), with a focus on quantifying predation on bluefish (*Pomatomus saltatrix*) in the northwest Atlantic Ocean. *Fish Bull* 107: 76–88.

CHAPTER 2

Oxygen Consumption Rate of Young-of-the-Year White Sharks, *Carcharodon carcharias*, during Transport to the Monterey Bay Aquarium

Juan M. Ezcurra*
Monterey Bay Aquarium

Christopher G. Lowe
California State University, Long Beach

Henry F. Mollet
Monterey Bay Aquarium and Moss Landing Marine Laboratories

Lara A. Ferry
Arizona State University

John B. O'Sullivan
Monterey Bay Aquarium

ABSTRACT

In 2002, the Monterey Bay Aquarium (MBA) started a project to display young-of-the-year (YOY) White Sharks (*Carcharodon carcharias*) in the 3.8-million-L Outer Bay exhibit and to conduct field research on the movement patterns of this highly active predator in the Southern California Bight. A pelagic fish transport tank used to bring YOY White Sharks to the MBA between 2004 and 2009 allowed the first direct measurements of oxygen consumption rates (MO_2) of free-swimming White Sharks during transport, ranging in duration between 5 and 7.5 h. We obtained MO_2 data during five transports of four individual White Sharks [137–157 cm of total length (TL) and 22.6–36.2 kg of body mass] at water temperatures ranging from 15.2 to 17.9°C. The mass-specific MO_2 calculated from 16 measurements for the four sharks was 246 ± 13 mg O_2 kg^{-1} h^{-1} (mean ± SE), which is among the highest reported for any shark species. Total MO_2 values of White Sharks from our study were pooled with the routine metabolic rate of the Shortfin Mako Shark to estimate the mass-scaling relationship for lamnid sharks, which is described by the allometric equation $MR = 458.5 \times M^{0.79}$.

INTRODUCTION

The White Shark is a large, highly active predator that is known to undergo long-distance migrations that can span thousands of kilometers (Boustany et al., 2002; Bonfil et al., 2005; Bruce et al., 2006; Weng et al., 2007a; Domeier and Nasby-Lucas, 2008; Jorgensen et al., 2010; Chapters 11, 12, 13, 16, and 21, this book). It is a member of the family Lamnidae and is a regional endotherm using vascular, countercurrent heat exchangers (*retia mirabilia*) to maintain elevated tissue temperatures (Carey and Teal, 1969; Carey et al., 1981, 1982, 1985; Block and Carey, 1985; Goldman et al., 1996; Goldman, 1997; Bernal et al., 2001b, 2005). Elevated tissue temperatures and increased activity levels in this species have lead researchers to theorize that White Sharks have high metabolic demands (Lowe and Goldman, 2001; Carlson et al., 2004). Because of the difficulty of obtaining live specimens and their high mobility and large size, field or laboratory studies on the metabolic rate of the White Shark have not been conducted. However, the capture of small White Sharks has enabled measurements comparable to those obtained for another lamnid, the Shortfin Mako (*Isurus oxyrhinchus*) (Graham et al., 1990; Bernal et al., 2001b; Sepulveda et al., 2007).

In 2002, the MBA initiated a field project to study young-of-the-year (YOY) White Sharks in the Southern California Bight and took steps toward placing a live specimen on public display. Studies using pop-up satellite archival tags of YOY White Sharks that were incidentally captured by commercial fishers have allowed researchers to gain a better understanding of the migratory patterns and thermal niche of these juvenile sharks (Dewar et al., 2004; Weng et al., 2007b). Since 2004, access to these YOY White Sharks has allowed the MBA to place four sharks on public display in the 3.8-million-L Outer Bay exhibit, which houses a community of pelagic fishes, including the Yellowfin (*Thunnus albacares*) and Bluefin (*Thunnus orientalis*) tunas. The opportunity to handle live YOY White Sharks allowed staff at the MBA to record oxygen consumption rates during transport of this species, which has not been previously maintained successfully in aquaria (Chapter 1, this book). This chapter presents an estimate of the mass-specific metabolic rate as determined from oxygen consumption rates (MO_2) of four YOY sharks. The objective was to determine the allometric equation that describes the increase in metabolic rate with increasing body mass of the White Shark and the closely related Shortfin Mako Shark (*I. oxyrinchus*) to allow comparison between these lamnid sharks and other pelagic species.

MATERIALS AND METHODS

Oxygen consumption rate data (MO_2) were obtained for four YOY White Sharks during transport from a 40-m-diameter ocean pen anchored in the Southern California Bight to the MBA between September 14, 2004 and August 26, 2009. These sharks ranged in size from 137 to 157 cm of TL and 22.6 to 36.2 kg of body mass (Table 2.1) and were held from 10 to 25 d in the ocean pen to allow their recovery from capture stress and begin feeding prior to transport. Sharks scheduled for transport were not fed for at least 24 h prior to their being netted from the ocean pen, placed unrestrained in a 250-L vinyl shark box containing oxygenated seawater (~125% saturation) at 16°C, and equipped with a recirculating submersible pump (4164 lph; Rule Industries, Gloucester, MA), which provided ventilation during the 30–90-min. transport via boat to the shore (15–30 km). Each shark was then transferred to an 11,356-L pelagic fish transport tank (Figure 2.1) mounted on the trailer of a commercial tractor for the drive to the MBA (524 km). Upon placement in the transport tank, the sharks were able to swim unimpeded, and at times they would rest on the bottom of the tank for periods of 30–90 sec during the approximately 6-h trip to Monterey. One individual, shark

Table 2.1 **White-Shark Total Length (cm), Mass (kg), Mean Temperature (°C), and Oxygen Consumption Rate (MO_2) as Calculated for Each Individual Run during the Transports of Four Sharks to the Monterey Bay Aquarium and One Shark Transported from the Aquarium to Santa Barbara, CA to Be Released (#08-01 on September 7, 2008)**

White Shark Number	Transport Destination Date: Run	Shark Total Length (cm)/ Mass (kg)	MO_2 (mg O_2 kg^{-1} h^{-1})	Mean Temperature (°C)
		To MBA		
04-01	September 14, 2004: #1	141/28.0	348.0	15.5 ± 0.02
04-01	September 14, 2004: #2	141/28.0	226.3	16.5 ± 0.01
07-01	August 28, 2007: #1	143/30.6	267.2	17.1 ± 0.01
07-01	August 28, 2007: #2	143/30.6	202.6	18.3 ± 0.01
		To MBA		
08-01	August 27, 2008: #1	137/25.2	370.4	16.6 ± 0.01
08-01	August 27, 2008: #2	137/25.2	256.9	17.1 ± 0.01
08-01	August 27, 2008: #3	137/25.2	267.7	17.5 ± 0.01
08-01	August 27, 2008: #4	137/25.2	281.2	17.8 ± 0.01
		To Release Site		
08-01	September 7, 2008: #1	137/22.6	189.9	16.8 ± 0.01
08-01	September 7, 2008: #2	137/22.6	223.1	17.2 ± 0.01
08-01	September 7, 2008: #3	137/22.6	235.2	17.4 ± 0.01
08-01	September 7, 2008: #4	137/22.6	189.9	17.5 ± 0.01
		To MBA		
09-01	August 26, 2009: #1	157/36.2	247.0	16.8 ± 0.02
09-01	August 26, 2009: #2	157/36.2	226.2	17.0 ± 0.01
09-01	August 26, 2009: #3	157/36.2	198.0	17.5 ± 0.02
09-01	August 26, 2009: #4	157/36.2	215.0	17.6 ± 0.01

Figure 2.1 Young-of-the-year White Shark swimming in the pelagic fish transport tank (volume, 11,356 L) during transport to the Monterey Bay Aquarium for public display. (Courtesy of Randy Wilder, Monterey Bay Aquarium.)

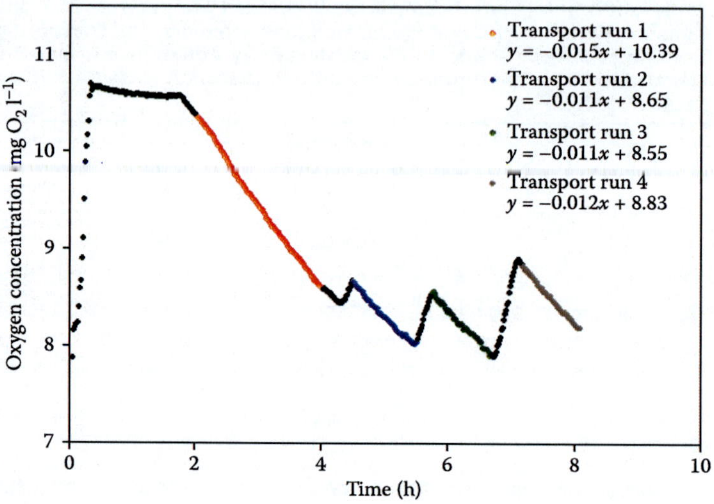

Figure 2.2 The seawater dissolved oxygen concentration (mg O_2 L^{-1}) profile in the 11,356-L transport tank during the transport of a 25.2-kg female White Shark (#08-01) to the Monterey Bay Aquarium from Southern California on August 27, 2008. Oxygen consumption rates (MO_2) for the White Shark were determined by using linear regression analysis of oxygen concentration (mg O_2 L^{-1}) versus time (minutes) for each of the four periods (depicted in color) when oxygen was not injected into the transport tank.

#08-01, was also transported back to Southern California for release on September 7, 2008 because of inconsistent feeding while on display.

During transport, oxygen levels in the transport tank were maintained between 8 and 11 mg L^{-1} by delivering pure oxygen from an oxygen cylinder and regulator through flexible airline tubing to Venturi injectors in the filtration piping (Figure 2.2). Water flow in the chamber was driven through a filter loop by a ¾ HP pool pump (Hayward Industries, Inc., Elizabeth, NJ) operated during transport. The transport tank could not be shut with an airtight seal because of the presence of baffles, which were used to reduce motion at the surface of the water. Water temperature, pH, and oxygen concentration data were logged with a YSI Model 556 Multi Probe system that sampled water off the main filtration loop. Transport water pH declined (maximum decrease was 0.4 units pH) because of the production of CO_2 by the shark during transport; however, the pH remained above 7.4 because of the large water volume in the transport tank.

Total MO_2 values for each shark were calculated from the rate of decline in oxygen concentration of the transport water during periods when oxygen was not added to the system. Linear regression was used to calculate the rates of total oxygen decline, which were corrected by subtracting estimates of background oxygen depletion obtained prior to each transport. Background oxygen depletion could not be measured at the end of the transports because the water was drained from the tank to remove the shark. Total MO_2 was calculated as Total MO_2 = −slope ((mg O_2 L^{-1}) $min.^{-1}$) × water volume (11,356 L) × 60 min. h^{-1}. Each shark was measured and weighed using an Intercomp model CS 1500 hanging scale prior to being placed on display, and shark #08-01 was reweighed prior to transport for release. At the time of release, all of the sharks were tagged with pop-up archival satellite tags according to the methods used by Weng et al. (2007b). The satellite tagging results demonstrated that the sharks survived at least 30–180 d following release.

RESULTS

Water temperature and pH did not change significantly during the transports. Surface water temperature at the ocean pen was approximately 20°C (±1°C), and upon capture all sharks were transferred to the 250-L shark box, which contained water at a temperature of 16°C. Water temperature in the pelagic fish transport tank ranged from 15.2 to 17.9°C (mean ± SE = 17.1 ± 0.3°C) during transports. Slight temperature increases (1.03 ± 0.12°C) occurred during individual transports (Table 2.1). Linear regressions of time and oxygen concentration, made during specific periods when oxygen was not being injected into the transport tank, showed a significant negative relationship ($p < 0.0001$; Figure 2.2).

In most cases, the trend for the mass-specific MO_2 values was to decline with time. Background oxygen decline versus time without the shark in the transport tank yielded values of 1.5–69.3 mg O_2 h^{-1}, which were subtracted from each result to yield adjusted total MO_2 values for each run. The mass-specific MO_2 values for the four sharks ranged between 189.9 and 370.4 mg O_2 kg^{-1} h^{-1} (Table 2.1). The mean mass-specific MO_2 values for each shark were very similar, ranging from 221.5 ± 10.3 to 287.2 ± 60.8 mg O_2 kg^{-1} h^{-1}, and the mean for all four sharks was 246.5 ± 13.1 mg O_2 kg^{-1} h^{-1}.

The routine metabolic rate (RMR) for the Shortfin Mako Shark (Sepulveda et al., 2007) and the total MO_2 of the White Shark were combined and plotted as a function of body mass to calculate the allometric equation ($MR = a \times M^b$), which describes the increase in metabolic rate with increasing body mass and the mass-scaling coefficient (b) for lamnid sharks. For comparison of lamnid sharks with other elasmobranchs and endothermic teleosts, the metabolic rates for other species of elasmobranchs were temperature adjusted to 17°C using a Q_{10} of 2.3 (Carlson et al., 2004) and then analyzed in the same manner described above.

DISCUSSION

Because measurement of metabolic rate was secondary to the husbandry and display of White Sharks, the metabolic data reported in this study were obtained during the transport of specimens from the field to the MBA. This approach, while limited, has added to our understanding of the metabolic rate for lamnid sharks. The ratio of White Shark mass to water volume (2.0–3.2 × 10^{-3} kg L^{-1}) of our transport tank is almost identical to that of the Shortfin Mako Sharks studied in a swim tunnel respirometer (1.5–3.2 × 10^{-3} kg L^{-1}) used by Graham et al. (1990) and Sepulveda et al. (2007). However, the inability of our system to be closed with an airtight seal was a drawback to our method, as necessitated by the transports. It is possible that the unsealed surface of the tank allowed oxygen to be lost to the surrounding environment; therefore, to account for this loss, our background oxygen decline was measured at the highest oxygen differential between the water (~120% saturation) and atmosphere prior to the sharks being placed into the tank. This oxygen loss from the system, in addition to variation in background respiration, may have contributed to some of the variability in oxygen consumption rates of the sharks. Alternatively, it is possible that the background oxygen decline measured is an underestimate of background respiration because it was measured at the beginning of the transports. The good fit between time and oxygen decline during the respiration runs to a linear model (r^2 values greater than 0.96) suggests that no oxygen was entering the water system via leaks in the plumbing during the respirometry runs. Swimming of the White Sharks was erratic at times and less efficient than straight-line swimming because of the amount of turning necessitated by the confines of the transport tank (Weihs, 1981). At times, the sharks would stall and rest on the bottom of the tank for brief periods; however, rough estimates of swimming speed from video of sharks #08-01 and #09-01 taken immediately after introduction to the transport tank ranged from approximately 58 to 81 cm s^{-1} or 0.4 to 0.5 TL s^{-1}. These estimates of approximate swimming speed fall within the range of swimming speeds for juvenile Shortfin Mako Sharks studied by Graham et al. (1990) and Sepulveda et al. (2007). Although swimming speed could not be controlled, our study provides the only oxygen consumption rates for YOY White Sharks obtained by direct measurement to date.

Despite the limitations of this study, the oxygen consumption rate reported for YOY White Sharks appears to fit well with the RMR reported for the Shortfin Mako Shark, which are among the highest reported for any pelagic shark (Graham et al., 1990; Sepulveda et al., 2007). The mean mass-specific MO_2 from this study (246 ± 13 mg O_2 kg^{-1} h^{-1}) is lower than the values reported by Graham et al. (1990) and Sepulveda et al. (2007) for the Shortfin Mako at similar experimental temperatures (369 ± 11 and 344 ± 22 mg O_2 kg^{-1} h^{-1}), as would be expected for White Sharks that were four to five times greater in body mass. To account for these differences in body mass, metabolic rate data (mg O_2 h^{-1}) for the Shortfin Mako (Sepulveda et al., 2007) and White Sharks (this study) were pooled to estimate the scaling relationship for lamnid sharks described by the allometric equation $MR = 458.5 \times M^{0.79}$ (Figure 2.3). The mass-scaling coefficient, b, for lamnids is very similar to the range of b values (mean = 0.8) reported for other elasmobranch and teleost species to date (Parsons, 1990; Ezcurra, 2001; Korsmeyer and Dewar, 2001; Dowd et al., 2006). The similarity in the mass-scaling coefficient from our study to those reported for other species suggests that oxygen consumption rates in these two closely related species are very similar. An elevated metabolic rate for both of these endothermic, lamnid sharks would be expected based on their high activity level and capacity for high-performance swimming (Bernal et al., 2001a; Donley et al., 2004). In addition, endothermy increases the efficiency of the aerobic, red muscles used in continuous swimming (Bernal et al., 2005) and is theorized to provide a selective advantage for these species during "bounce diving" forays into cool waters during vertical oscillatory swimming patterns associated with prey search (Dewar, 2004; Sepulveda et al., 2004; Weng et al., 2007b).

Lamnid sharks have metabolic rates that are much higher than those of ectothermic, pelagic sharks and more similar to those of endothermic tunas (Figure 2.3). The experimental temperature

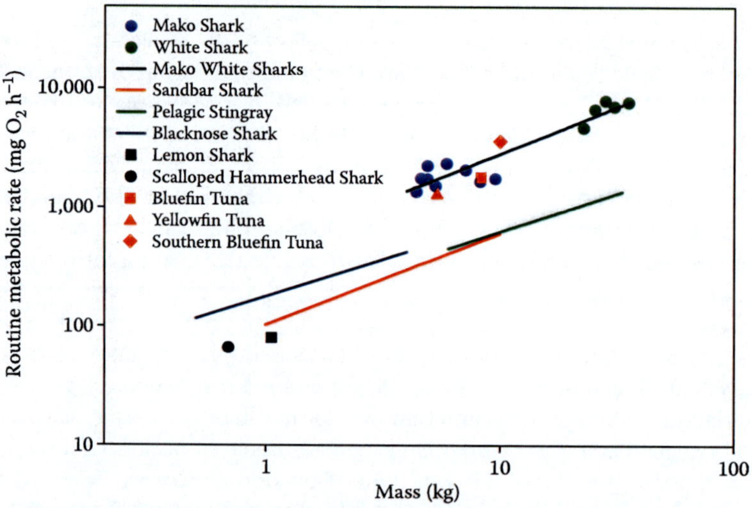

Figure 2.3 RMR (mg O_2 h^{-1}) of lamnid sharks (Shortfin Mako Shark, *I. oxyrinchus*, Sepulveda et al., 2007; White Shark, *C. carcharias*, present study) in relation to mass (kilograms) compared with other active pelagic sharks and tunas. The lines are for RMR calculated over a range of mass with the allometric equation RMR = aM^b. The line describing the RMR in relation to mass for lamnid sharks is 458.5 $M^{0.79}$, from log (a) = 2.66 ± 0.08 (SE) and slope (b) = 0.79 ± 0.08 (SE). Lemon Shark (*Negaprion brevirostris*, Bushnell et al., 1989), Scalloped Hammerhead Shark (*Sphyrna lewini*, Lowe, 2001), Pelagic Stingray (*Dasyatis violacea*, Ezcurra, 2001, RMR = 140.1 $M^{0.65}$), Blacknose Shark (*Carcharhinus acronotus*, Carlson et al., 1999, RMR = 446.8 $M^{0.62}$), and Sandbar Shark (*Carcharhinus plumbeus*, Dowd et al., 2006, RMR = 102.7 $M^{0.77}$) were temperature-adjusted to 17°C by using a Q_{10} of 2.3 (Carlson et al., 2004). Metabolic rates for Yellowfin and Bluefin Tuna (*T. albacares* and *T. orientalis*, Blank et al., 2007 at 20°C) and Southern Bluefin Tuna (*Thunnus maccoyii*, Fitzgibbon et al., 2006 at 19°C) were for tunas swimming at swimming speeds of 0.65–1.0 body lengths s^{-1}.

in our study (17°C) was generally lower than the temperatures in other metabolic studies, which necessitated temperature adjustment of RMR reported for other sharks (Q_{10} of 2.3; Carlson et al., 2004). In addition, the body mass of the White Sharks (22.6–36.2 kg) was 2–30 times greater than that of sharks in other metabolic studies (Bushnell et al., 1989; Carlson et al., 1999; Lowe, 2001; Dowd et al., 2006). However, when differences in temperature and body mass are taken into account, the RMR for lamnids is approximately five times greater than the RMR for species of ectothermic, obligatory ram-ventilating sharks (Figure 2.3). Although they are more divergent taxonomically, the metabolic rate for lamnid sharks is closer to that of Bluefin and Yellowfin Tuna at minimal swimming speeds (0.65–1.0 body lengths s^{-1}) at similar experimental temperatures (Fitzgibbon et al., 2006; Blank et al., 2007; Figure 2.3). The convergent evolution of high-performance swimming and endothermy in the lamnid sharks and tunas has resulted in specialized morphology and physiology (streamlined body shape; internalized aerobic, red muscle capable of retaining metabolic heat; elevated enzyme activities associated with aerobic and anaerobic metabolism in white muscle; large gill surface area; low blood-water barrier thickness; and a circulatory system with a high-oxygen delivery capacity to the tissues) in these distant groups (Dickson et al., 1993; Bernal et al., 2001a; Korsmeyer and Dewar, 2001; Donley et al., 2004; Sepulveda et al., 2007; Wegner et al., 2010). Until now, comparisons of metabolic rate and swimming kinematics between these two groups have been limited, because studies focusing on lamnids have been restricted to the Shortfin Mako Shark (Graham et al., 1990; Bernal et al., 2001b; Donley et al., 2004; Sepulveda et al., 2007). By using methods developed to transport highly active, ram-ventilating tunas (Farwell, 2001), we have been able to report the first direct measurement of oxygen consumption rates of YOY White Sharks. Continued implementation of these methods could lead to further studies of White Shark metabolic rates and swimming kinematics under laboratory conditions to advance our understanding of energetics in lamnid sharks.

ACKNOWLEDGMENTS

Many thanks go to a very talented and diverse group of people who have made this study possible, including the crew of the *F/V Barbara H*, Randy Hamilton, Dr. Mike Murray, Chuck Winkler, Chuck Farwell, Joe Welsh, Bobby Bettancourt, and the staff of the husbandry and applied research departments at the MBA. Thanks also to Ellen Freund, Dan Cartamil, Nick Wegner, Jeffrey Graham, and an anonymous reviewer for their helpful comments and suggestions for this manuscript. J. M. E. especially thanks his family for their understanding and support during his absences while working in the field.

REFERENCES

Bernal, D., K. A. Dickson, R. E. Shadwick, J. B. Graham. 2001a. Analysis of the evolutionary convergence for high performance swimming in lamnid sharks and tunas. *Comp Biochem Physiol* 129A: 695–726.

Bernal, D., J. M. Donley, R. E. Shadwick, D. A. Syme. 2005. Mammal-like muscles power swimming in a cold-water shark. *Nature* 437: 1349–1357.

Bernal, D., C. A. Sepulveda, J. B. Graham. 2001b. Water-tunnel studies of heat balance in swimming mako sharks. *J Exp Biol* 204: 4043–4054.

Blank, J. M., C. J. Farwell, J. M. Morrissette, R. J. Schallert, B. A. Block. 2007. Influence of swimming speed on metabolic rates of juvenile Pacific bluefin tuna and yellowfin tuna. *Physiol Biochem Zool* 80: 167–177.

Block, B. A., F. G. Carey. 1985. Warm brain and eye temperatures in sharks. *Comp Biochem Physiol* 156B: 229–236.

Bonfil, R., M. Meyer, M. C. Scholl, R. Johnson, S. O'Brien, H. Oosthuizen, S. Swanson, D. Kotze, M. Paterson. 2005. Transoceanic migration, spatial dynamics, and population linkages of white sharks. *Science* 310: 100–103.

Boustany, A. M., S. F. Davis, P. Pyle, S. D. Anderson, B. J. Le Boeuf, B. A. Block. 2002. Satellite tagging: Expanded niche for white sharks. *Nature* 415: 35–36.

Bruce, B. D., J. D. Stevens, H. Malcolm. 2006. Movements and swimming behaviour of white sharks (*Carcharodon carcharias*) in Australian waters. *Mar Biol* 150: 161–172.

Bushnell, P. G., P. L. Lutz, S. H. Gruber. 1989. The metabolic rate of an active, tropical elasmobranch, the lemon shark (*Negaprion brevirostris*). *J Exp Biol* 48: 279–283.

Carey, F. G., J. G. Casey, H. L. Pratt, D. Urquhart, J. E. McCosker. 1985. Temperature, heat production and heat exchange in lamnid sharks. *Mem S Calif Acad Sci* 9: 92–108.

Carey, F. G., J. W. Kanwisher, O. Brazier, G. Gabrielson, J. G. Casey, H. L. Pratt, Jr. 1982. Temperature and activities of a white shark, *Carcharodon carcharias*. *Copeia* 1982: 254–260.

Carey, F. G., J. M. Teal. 1969. Mako and porbeagle: Warm bodied sharks. *Comp Biochem Physiol* 28A: 199–204.

Carey, F. G., J. M. Teal, J. W. Kanwisher. 1981. The visceral temperatures of mackerel sharks (lamnidae). *Physiol Zool* 54: 334–344.

Carlson, J. K., J. K. Goldman, C. G. Lowe. 2004. Metabolism, energetic demand, and endothermy. In: J. C. Carrier, J. A. Musick, M. R. Heithaus (eds.), *Biology of sharks and their relatives* (pp. 203–224). Boca Raton, FL: CRC Press Inc.

Carlson, J. K., C. L. Palmer, G. H. Parsons. 1999. Oxygen consumption rate and swimming efficiency of the blacknose shark, *Carcharhinus acronotus*. *Copeia* 1999: 34–39.

Dewar, H., M. Domeier, N. Nasby-Lucas. 2004. Insights into young of the year white sharks (*Carcharodon carcharias*) behavior in the Southern California Bight. *Environ Biol Fishes* 70: 133–143.

Dickson, K. A., M. O. Gregorio, S. J. Gruber, K. L. Loefler, M. Tran, C., Terrrel. 1993. Biochemical indices of aerobic and anaerobic capacity in muscle tissues of California elasmobranch fishes differing in typical activity level. *Mar Biol* 117: 185–193.

Domeier, M. L., N. Nasby-Lucas. 2008. Migration patterns of white sharks *Carcharodon carcharias* tagged at Guadalupe Island, Mexico, and identification of an eastern Pacific shared foraging area. *Mar Ecol Prog Series* 370: 221–237.

Donley, J. M., C. A. Sepulveda, P. Konstantinidis, S. Gemballa, R. E. Shadwick. 2004. Convergent evolution in mechanical design of lamnid sharks and tunas. *Nature* 429: 61–65.

Dowd, W. W, R. W. Brill, P. G. Bushnell, J. A. Musick. 2006. Standard and routine metabolic rates of juvenile sandbar sharks (*Carcharhinus plumbeus*), including the effects of body mass and acute temperature change. *Fish Bull* 1104: 323–331.

Ezcurra, J. M. 2001. The mass-specific routine metabolic rate of captive pelagic stingrays, *Dasyatis violacea*, with comments on energetics. MS thesis, Moss Landing Marine Laboratory, California State University, Stanislaus.

Farwell, C. J. 2001. Tunas in captivity. In: B. A. Block, E. D. Stevens (eds.), *Tunas: Physiology, Ecology and Evolution* (pp. 391–412). San Diego, CA: Academic Press.

Fitzgibbon, Q. P., R. S. Seymour, D. Ellis, J. Buchanan. 2006. The energetic consequence of specific dynamic action in southern bluefin tuna *Thunnus maccoyii*. *J Exp Biol* 210: 290–298.

Goldman, K. J. 1997. Regulation of body temperatures in the white shark, *Carcharodon carcharias*. *J Comp Physiol Biochem* 167: 423–429.

Goldman, K. J., S. D. Anderson, J. E. McCosker, A. P. Klimley. 1996. Temperature, swimming, depth, and movements of a white shark at the South Farallon Islands, California. In: A. P. Klimley, D. G. Ainley (eds.), *Great white sharks: The biology of Carcharodon carcharias* (pp. 111–120). San Diego, CA: Academic Press.

Graham, J. B., H. Dewar, N. C. Lai, W. R. Lowell, S. M. Arce. 1990. Aspects of shark swimming performance determined using a large water tunnel. *J Exp Biol* 151: 175–192.

Jorgensen, S. J., C. A. Reeb, T. K. Chapple, S. Anderson, C. Perle, S. R. Van Sommeran, C. Fritz-Cope, A. C. Brown, A. P. Klimley, B. A. Block. 2010. Philopatry and migration of Pacific white sharks. *Proc R Soc B Biol Sci* 277: 679–688.

Korsmeyer, K. E., H. Dewar, 2001. Tuna metabolism and energetics. In: B. A. Block, E. D. Stevens (eds.), *Tunas: Physiology, ecology and evolution* (pp. 35–78). San Diego, CA: Academic Press.

Lowe, C. G. 2001. Metabolic rates of juvenile scalloped hammerhead sharks (*Sphyrna lewini*). *Mar Biol* 139: 447–453.

Lowe, C. G., K. J. Goldman. 2001. Physiological telemetry of elasmobranchs: Bridging the gap. *Environ Biol Fishes* 60: 251–256.

Parsons, G. R. 1990. Metabolism and swimming efficiency of the bonnethead shark, *Sphyrna tiburo*. *Mar Biol* 104: 363–367.

Sepulveda, C. A., J. B. Graham, D. Bernal. 2007. Aerobic metabolic rates of swimming juvenile mako sharks, *Isurus oxyrinchus*. *Mar Biol* 152: 1087–1094.

Sepulveda, C. A., S. Kohin, C. Chan, R. Vetter, J. B. Graham. 2004. Movement patterns, depth preferences, and stomach temperatures of free-swimming juvenile mako sharks, *Isurus oxyrinchus*, in the Southern California Bight. *Mar Biol* 145: 191–199.

Wegner, N. C., C. A. Sepulveda, K. R. Olson, K. A. Hyndman, J. B. Graham. 2010. Functional morphology of the gills of the shortfin mako, *Isurus oxyrinchus*, a Lamnid shark. *J Morphol* 27: 937–948.

Weihs, D. 1981. Effects of swimming path curvature on the energetics of fish motion. *Fish Bull* 79: 171–176.

Weng, K. C., A. M. Boustany, P. Pyle, S. D. Anderson, A. Brown, B. A. Block. 2007a. Migration and habitat of White Sharks (*Carcharodon carcharias*) in the eastern Pacific Ocean. *Mar Biol* 152: 877–894.

Weng, K. C., J. B. O'Sullivan, C. G. Lowe, C. E. Winkler, H. Dewar, B. A. Block. 2007b Movements, behavior and habitat preferences of juvenile white sharks *Carcharodon carcharias* in the eastern Pacific. *Mar Ecol Prog Ser* 338: 211–224.

CHAPTER 3

Size-Based Analysis of Diet and Trophic Position of the White Shark, *Carcharodon carcharias*, in South African Waters

Nigel E. Hussey*
Bangor University and Great Lakes Institute for Environmental Research

Heather M. McCann
Great Lakes Institute for Environmental Research

Geremy Cliff
KwaZulu-Natal Sharks Board and University of KwaZulu-Natal

Sheldon F. J. Dudley
KwaZulu-Natal Sharks Board and University of KwaZulu-Natal

Sabine P. Wintner
KwaZulu-Natal Sharks Board and University of KwaZulu-Natal

Aaron T. Fisk
Great Lakes Institute for Environmental Research

ABSTRACT

Understanding the diet and trophic ecology of apex predators is critical for effective ecosystem management, but analysis is generally restricted by their low abundance, threatened status, and, hence, available data. To address this knowledge gap for the White Shark (*Carcharodon carcharias*), we examined stomach content data ($n = 225$) and undertook preliminary stable isotope analysis ($\delta^{15}N$ and $\delta^{13}C$) of muscle tissue ($n = 21$) of sharks sampled from KwaZulu-Natal, South

Africa. Summarized stomach content analysis (SCA) data indentified a size-based transition in diet, an increase in mammalian prey, and a corresponding decrease in teleost and elasmobranch prey with increasing size of shark. Multivariate analysis of SCA data by size class was significant, but a high degree of overlap of functional prey groups across size classes was observed. The smallest sharks containing whale and seal prey were 194 and 209 cm precaudal length (PCL), respectively. Trophic position calculated from SCA data (TP_{SCA}) ranged from 4.2 to 5.0 with sharks feeding across 0.8 trophic levels. Although TP_{SCA} data were highly variable, there was a significant linear relationship with increasing size of shark. Similarly, there was a significant increase in $\delta^{15}N$ values of muscle tissue with size of shark and feeding across 1.4 trophic levels was predicted. Sharks >200 cm PCL had highly variable $\delta^{15}N$ muscle tissue values (minimum, 14.7; maximum, 16.8), possibly indicating multiple feeding strategies. For $\delta^{13}C$, the two smallest animals (124 and 134 cm PCL) had lower $\delta^{13}C$ values relative to all other sharks except the largest individual. Because these sharks are newborn animals, this suggests interference of the maternal signature and may indicate that large individuals, including pregnant females, switch forage base and/or remain offshore for extended periods of time.

INTRODUCTION

With the advent of industrial fishing and the initial absence of multispecies management directives, stocks of marine predators have undergone drastic declines (Myers and Worm, 2003, 2005; Hutchings and Baum, 2005). For terrestrial systems, "top-down" effects of predator removal have been historically documented (Elton, 1927; Leopold et al., 1947; Paine, 1980), but our understanding of equivalent processes in marine systems has been limited (Baum and Worm, 2009). Recent work, however, has started to recognize the need to define the role of predators to better understand the processes maintaining ecosystem function, stability, and resilience and to determine the consequences of continued overexploitation (Stevens et al., 2000; Myers et al., 2007; Heithaus et al., 2008; Baum and Worm, 2009).

Quantifying diet provides an important tool to examine the role of an organism within the system it inhabits and the influence it exerts on potential prey populations (Cortés and Gruber, 1990; Wetherbee et al., 1990). Diet data are also required to estimate trophic position and to generate size-based trophic profiles (Pauly et al., 1998; Cortés, 1999). These baseline data are required to develop predictive frameworks for monitoring prey or predator removal and/or population fluctuations (Pace et al., 1999; Myers et al., 2007; Heithaus et al., 2008). Traditionally, stomach content analysis (SCA) has been used to examine diet/trophic interactions through providing insights into the type of prey consumed and the diversity of prey consumed and enabling the calculation of standardized trophic position (Cortés, 1997 and 1999). Importantly, SCA can be used to determine how these parameters vary with sex and size (Lowe et al., 1996). More recently, chemical tracers, for example the stable isotopes of nitrogen and carbon (^{15}N and ^{13}C), have been employed as a complementary tool to SCA (Vander Zanden et al., 1997; Fisk et al., 2002). Stable isotopes of a predator's muscle tissue reflect that of the prey consumed and foraging location and are generally long-term integrated measures (MacNeil et al., 2005; Hussey et al., 2011). Nitrogen isotopes are viewed as the most reliable measure of trophic position (Post, 2002), whereas carbon isotopes typically reflect variation in baseline producers and therefore the foraging habitat of the predator (DeNiro and Epstein, 1978).

The White Shark (*Carcharodon carcharias*) is one of the largest of all extant marine predators, attaining a maximum recorded total length of 6 m (see Chapter 7, this book). It is distributed throughout temperate waters but is reported from both tropical and temperate regions (Compagno, 2002). It is principally an epipelagic predator inhabiting coastal waters but is also known to reside in oceanic waters (Weng et al., 2007a; Nasby-Lucas, 2009; Jorgensen et al., 2010; Chapters 10,

11, 16, and 21, this book) and undertake large-scale migrations (Bonfil et al., 2005, 2010; Chapter 21, this book). With concerns over localized depletions of this species, the White Shark was afforded protection in South Africa, Australia, and the United States. It is listed as vulnerable by the International Union for Conservation of Nature and Natural Resources and included on Appendix II of the Convention on International Trade in Endangered Species, to restrict international trade in body parts.

Because large predatory sharks can influence prey population dynamics and their range encompasses different management subareas and fishing pressures, regional White Shark populations can potentially exert significant control across multiple components of the marine ecosystem (Hussey et al., 2011). To date, most research focused on the diet of White Sharks has examined either limited stomach content data (Bass et al., 1975; Klimley, 1985) or the localized feeding patterns and predatory behavior at seal colonies (Klimley et al., 1992, 1996, 2001; Martin et al., 2005; Laroche et al., 2008; Chapter 9, this book). Our knowledge of general diet and size-based diet/trophic position has remained limited with the exception of stomach content data presented by Tricas and McCosker (1984) and Cliff et al. (1989) and more recently nitrogen stable-isotope profiles reported by Estrada et al. (2006). This is not surprising considering the logistical difficulties of studying free-ranging marine predators and the low natural abundance, threatened status, and, hence, available data for White Sharks (Tricas and McCosker, 1984; Malcolm et al., 2001). The objectives of this study were as follows:

1. Examine size-based diet shifts and calculate ontogenetic trophic profiles using a comprehensive archived stomach content data set.
2. Compare ontogenetic trophic profiles generated from stomach content data with those from nitrogen stable-isotope data ($\delta^{15}N$) of white muscle tissue.
3. Investigate size-based switches in foraging location using carbon stable isotope data ($\delta^{13}C$).

This study was focused on the White Shark population off the coast of Southern Africa.

MATERIALS AND METHODS

Stomach Content Sampling

Samples (225 sharks with reliable stomach content data) were obtained from White Sharks incidentally caught in beach protection nets set at popular bathing beaches along the coast of KwaZulu-Natal (KZN), South Africa between 1978 and 2009 (Figure 3.1). The nets were set approximately 300–500 m parallel to the beach in water 10–14 m deep (Dudley et al., 2005). For specific details regarding net-servicing operations, refer to Cliff et al. (1988).

PCL was measured as the straight-line distance from the tip of the snout to the precaudal notch as defined by Dudley et al. (2005). Maturity stage was determined by the state of the reproductive organs according to Cliff et al. (1989). Stomach contents were sorted; prey was identified to the lowest possible taxonomic level, counted, and weighed to the nearest 0.1 g. Suspected scavenging of prey items was documented.

Diet Composition

Cumulative prey curves were used to assess the sample size sufficiency for accurately describing total diet and diet by size class. The order of stomachs sampled was randomized 999 times, and the mean cumulative exponential of new prey items was plotted as a function of sample size. Diet is considered to be adequately described when the curve approaches an asymptote (Ferry and Cailliet, 1996). To standardize our results with previous elasmobranch-diet studies, diet composition was

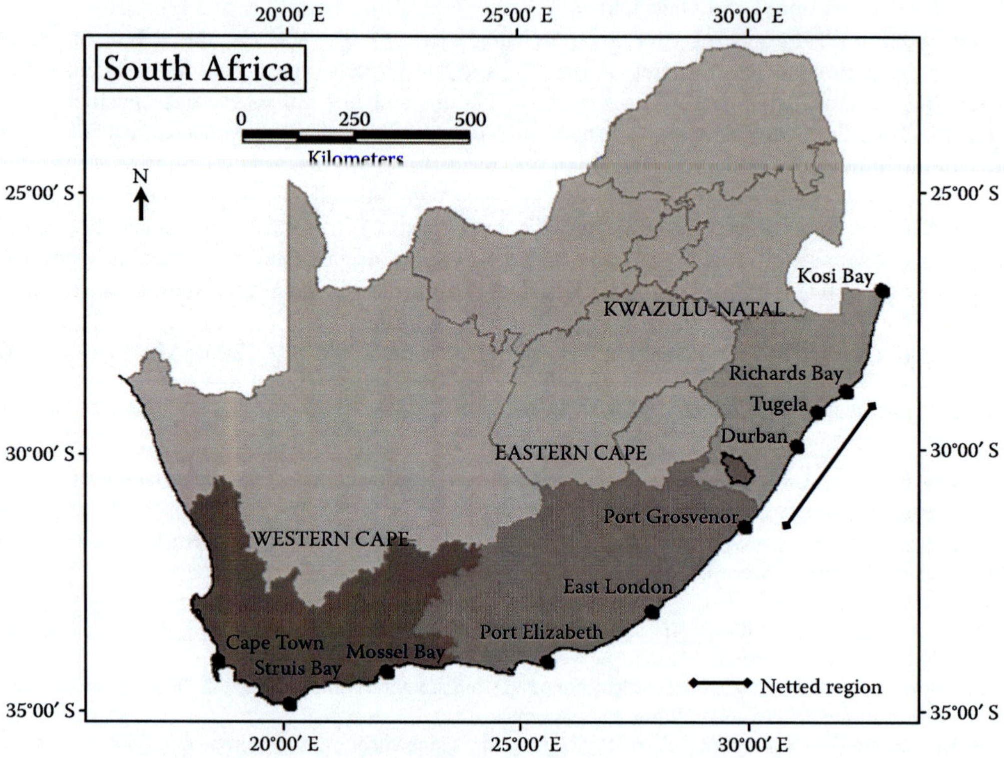

Figure 3.1 Map of South Africa showing the KwaZulu-Natal coastline where White Sharks were incidentally sampled from KwaZulu-Natal Sharks Board beach protection nets.

calculated as percentage number (%N), percentage mass (%M), percentage occurrence (%O), and percentage index of relative importance (%IRI) (Hyslop, 1980; Cortés, 1997).

For statistical analysis, prey were grouped according to functional prey categories defined by family and habitat (Table 3.1). All analysis was conducted with data for both sexes pooled, because a one-way analysis of similarities (ANOSIM) found no significant difference in overall diet composition between sexes ($R = 0.023$, $p = 0.122$).

Size-Based Diet Shift

To examine whether White Sharks undergo a size-based diet shift, we evaluated differences in dietary composition of sharks in four successive size classes; <185 cm, 185–234.9 cm, 235–284.9 cm, and >285 cm. Sizes were defined to provide maximum resolution to examine size-based diet shifts, while maintaining sample numbers per size class. Division by maturity was deemed unsuitable because of limited catches of mature sharks. Each dietary index, %N, %M, %F, and %IRI size class data were subjected to nonmetric multidimensional scaling based on a Bray-Curtis similarity coefficient and one-way ANOSIM (White et al., 2004; Huveneers et al., 2007). Accepting that individual stomachs of large sharks typically contain only one or a few of the many indentified prey items, dietary data for groups of three to five individual sharks were pooled per size class prior to the above analyses, herein referred to as dietary samples (Platell and Potter, 2001; White et al., 2004). Similarity percentage analysis was used to explore the dietary categories that contributed most to dissimilarity in diet between size classes (Clarke and Warwick, 2001).

Table 3.1 **Percentage Number (%N), Percentage Mass (%M), Percentage Frequency (%F), and Percentage Index of Relative Importance (%IRI) of Functional Prey Groups, Identified to Species Level Where Possible, in the Diet of Successive Size Classes of White Shark Sampled from South African Waters**

	1 (<185 cm PCL)				2 (185–234.9 cm PCL)				3 (235–284.9 cm PCL)				4 (≥285 cm PCL)			
Functional Prey Groups	%N	%M	%F	%IRI	%N	%M	%F	%IRI	%N	%M	%F	%IRI	%N	%M	%F	%IRI
Elasmobranch	38.33	71.81	44.44	67.72	7.23	46.31	37.21	34.59	3.36	44.88	36.17	29.92	28.57	24.18	33.33	17.84
Squaliformes																
Dogfish, *Squalus* sp.	13.33	2.48	2.78	1.98	0	0	0	0	0	0	0	0	0	0	0	0
Orectolobiformes																
Whale Shark, *Rhincodon typus*	0	0	0	0	0	0	0	0	0	0	0	0	4.76	0.1	8.33	1.16
Carcharhiniformes																
Requiem Sharks, unidentified Carcharhinidae	0	0	0	0	0.11	0.94	0.78	0.08	0.35	1.01	4.26	0.39	0	0	0	0
Dusky, *Carcharhinus obscurus**	3.33	13.16	5.56	4.14	2.63	18.14	10.85	20.89	1.06	10.57	10.64	8.27	0	0	0	0
Sandbar, *Carcharhinus plumbeus*	0	0	0	0	0.11	1.94	0.78	0.15	0	0	0	0	0	0	0	0
Milk, *Rhizoprionodon acutus*	0	0	0	0	0.33	2.52	2.33	0.61	0	0	0	0	0	0	0	0
Copper, *Carcharhinus brachyurus**	0	0	0	0	0	0	0	0	0.18	13.92	2.13	2.01	0	0	0	0
Hammerhead Sharks, unidentified sphyrnid spp.	0	0	0	0	0.11	1.14	0.78	0.09	0.18	6.28	2.13	0.92	4.76	9.48	8.33	3.4
Lamniformes																
Thresher, unidentified Alopiidae	0	0	0	0	0.11	1.63	0.78	0.13	0	0	0	0	0	0	0	0
Spotted Ragged Tooth, *Carcharias taurus*	0	0	0	0	0	0	0	0	0.35	6.16	4.26	1.85	0	0	0	0
Unidentified shark																
Unidentified shark	3.33	3.56	5.56	1.73	0.22	2.33	1.55	0.37	0.35	3.26	4.26	1.03	4.76	10.38	8.33	3.61
Unidentified small shark	11.67	34.47	19.44	40.51	2.3	2.14	15.5	6.38	0.18	0.02	2.13	0.03	4.76	1.96	8.33	1.61
Unidentified large shark*	1.67	14.83	2.78	2.07	0.55	11.22	3.88	4.23	0.18	3.14	2.13	0.47	4.76	2.24	8.33	1.67
Benthic rajiformes																
Guitarfishes, unidentified Rhinobatidae	0	0	0	0	0.11	0.06	0.78	0.01	0.18	0.49	2.13	0.10	0	0	0	0

continued

Table 3.1 (Continued) Percentage Number (%N), Percentage Mass (%M), Percentage Frequency (%F), and Percentage Index of Relative Importance (%IRI) of Functional Prey Groups, Identified to Species Level Where Possible, in the Diet of Successive Size Classes of White Shark Sampled from South African Waters

	1 (<185 cm PCL)				2 (185–234.9 cm PCL)				3 (235–284.9 cm PCL)				4 (≥285 cm PCL)			
	%N	%M	%F	%IRI	%N	%M	%F	%IRI	%N	%M	%F	%IRI	%N	%M	%F	%IRI
Giant Guitarfish, *Rhynchobatus djiddensis*	0	0	0	0	0.11	0.78	0.78	0.06	0	0	0	0	0	0	0	0
Stingray, unidentified Dasyatidae	1.67	0.16	2.78	0.23	0	0	0	0	0.18	<0.01	2.13	0.03	0	0	0	0
Benthopelagic rajiformes																
Spotted Eagle Ray, *Aetobatus narinari*	0	0	0	0	0.11	3.23	0.78	0.24	0	0	0	0	0	0	0	0
Bull Ray, *Pteromylaeus bovinus*	0	0	0	0	0.11	0.01	0.78	0.01	0	0	0	0	0	0	0	0
Manta Ray, *Manta birostris*	1.67	2.5	2.78	0.52	0	0	0	0	0	0	0	0	0	0	0	0
Unidentified ray																
Unidentified batoid	0	0	0	0	0.11	0.16	0.78	0.02	0	0	0	0	0	0	0	0
Unidentified elasmobranch																
Unidentified elasmobranch	1.67	0.65	2.78	0.29	0.22	0.07	1.55	0.04	0.18	<0.01	2.13	0.03	4.76	0.02	8.33	1.14
Mammal	1.67	6.49	2.78	0.31	4.82	24.54	33.33	16.99	3.36	29.14	40.43	22.66	47.62	70.13	66.67	79.67
Pinnipedia																
Unidentified seal	0	0	0	0	0.99	5.4	6.98	4.13	2.12	14.86	25.53	28.97	19.05	21.13	33.33	38.38
Odontoceti																
Unidentified dolphin*	0	0	0	0	2.74	11	19.38	24.67	0.88	7.64	8.51	4.85	4.76	0.01	8.33	1.14
Bottlenose Dolphin, *Tursiops aduncus**	0	0	0	0	0.11	0.53	0.78	0.05	0.18	0.10	2.13	0.04	4.76	12.83	8.33	4.2
Common Dolphin, *Delphinus delphis**	1.67	6.49	2.78	1.02	0.11	5.45	0.78	0.4	0	0	0	0	4.76	1.39	8.33	1.47
Risso's Dolphin, *Grampus griseus*	0	0	0	0	0	0	0	0	0.18	3.72	2.13	0.55	0	0	0	0
Mysticeti																
Unidentified whale	0	0	0	0	0.33	0.15	2.33	0.1	0.18	2.83	2.13	0.43	14.29	34.83	25	35.19
Unidentified mammal																
Cetacean	0	0	0	0	0.11	1.56	0.78	0.12	0	0	0	0	0	0	0	0
Mammal	0	0	0	0	0.33	0.44	2.33	0.17	0	0	0	0	0	0	0	0

Unidentified land animal	0	0	0	0	0.11	0.01	0.78	0.01	0	0	0	0	0	0	0	0
Teleost	25	21.25	30.56	19.55	82.04	29	24.03	46.33	90.64	25.98	23.40	46.79	19.05	5.55	8.33	2.08
Reef																
Herrings, Clupeidae	0	0	0	0	0	0	0	0	0.18	0.03	2.13	0.03	0	0	0	0
Halfmoon Rockcod, *Epinephelus rivulatus*	0	0	0	0	0.11	0.25	0.78	0.03	0	0	0	0	0	0	0	0
Spotted Grunter, *Pomadasys commersonnii*	0	0	0	0	1.75	0.08	0.78	0.13	0	0	0	0	19.05	5.55	8.33	5.87
Santer, *Cheimerius nufar*	0	0	0	0	0.11	0.16	0.78	0.02	0	0	0	0	0	0	0	0
Englishman, *Chrysoblephus anglicus*	0	0	0	0	0.11	0.03	0.78	0.01	0	0	0	0	0	0	0	0
Eels, Anguilliformes	0	0	0	0	0.11	0.04	0.78	0.01	0	0	0	0	0	0	0	0
Pelagic																
South African Pilchard, *Sardinops ocellatus*	0	0	0	0	76.89	23.65	3.1	28.89	88.87	19.21	6.38	46.10	0	0	0	0
Tunas, Scombridae	0	0	0	0	0.22	0.45	1.55	0.1	0	0	0	0	0	0	0	0
Chub Mackerel, *Scomber japonicus*	5	1.04	2.78	0.76	0	0	0	0	0	0	0	0	0	0	0	0
Benthic																
Seacatfishes, Ariidae	0	0	0	0	0.11	0.03	0.78	0.01	0	0	0	0	0	0	0	0
Sea breams, Sparidae	3.33	1.05	5.56	1.1	0.11	0.11	0.78	0.02	0	0	0	0	0	0	0	0
White Musselcracker (brusher), *Sparodon durbanensis*	0	0	0	0	0.33	1.35	1.55	0.24	0	0	0	0	0	0	0	0
Slinger, *Chrysoblephus puniceus*	0	0	0	0	0	0	0	0	0.18	0.04	2.13	0.03	0	0	0	0
Bronze Bream, *Pachymetopon grande*	0	0	0	0	0.11	0.48	0.78	0.04	0	0	0	0	0	0	0	0
Red Tjor-Tjor, *Pagellus natalensis*	0	0	0	0	0.11	0.01	0.78	0.01	0	0	0	0	0	0	0	0
Karenteen, *Sarpa salpa*	0	0	0	0	0.11	0.02	0.78	0.01	0	0	0	0	0	0	0	0
Kob, *Argyrosomus hololepidotus*	0	0	0	0	0.11	0.21	0.78	0.02	0.18	5.76	2.13	0.84	0	0	0	0
Geelbek, *Atractoscion aequidens*	1.67	3.73	2.78	0.68	0.22	0.3	0.78	0.04	0	0	0	0	0	0	0	0

continued

Table 3.1 (Continued) Percentage Number (%N), Percentage Mass (%M), Percentage Frequency (%F), and Percentage Index of Relative Importance (%IRI) of Functional Prey Groups, Identified to Species Level Where Possible, in the Diet of Successive Size Classes of White Shark Sampled from South African Waters

	1 (<185 cm PCL)				2 (185–234.9 cm PCL)				3 (235–284.9 cm PCL)				4 (≥285 cm PCL)			
	%N	%M	%F	%IRI	%N	%M	%F	%IRI	%N	%M	%F	%IRI	%N	%M	%F	%IRI
Unidentified teleost	15	15.43	22.22	30.53	1.53	1.83	10.08	3.14	1.24	0.94	14.89	2.16	0	0	0	0
Cephalopod	28.33	0.43	30.56	12.16	5.81	0.15	20.16	2.08	2.47	<0.01	14.89	0.63	4.76	0.08	8.33	0.41
Octopodiformes																
Unidentified octopus	0	0	0	0	0.33	<0.01	2.33	0.07	0.35	0	4.26	0.10	0	0	0	0
Decapodiformes, teuthida																
Unidentified squid	13.33	0	11.11	6.69	3.4	<0.01	8.53	2.69	0.71	<0.01	6.38	0.30	0	0	0	0
Loligo squids, Loligo spp.	5	0.11	5.56	1.28	0.55	<0.01	2.33	0.12	0.53	<0.01	2.13	0.08	0	0	0	0
Chiroteuthid squids, Chiroteuthidae	0	0	0	0	0	0	0	0	0.18	<0.01	2.13	0.03	0	0	0	0
Decapodiformes, sepiida																
Unidentified cuttlefish	8.33	0.31	13.89	5.42	1.53	0.14	9.3	1.45	0	0	0	0	4.76	0.08	8.33	1.16
Unidentified cephalopods	1.67	<0.01	2.78	0.21	0	0	0	0	0.71	<0.01	2.13	0.10	0	0	0	0
Mollusk	6.67	0.02	2.78	0.26	0.11	<0.01	0.78	0	0	0	0	0	0	0	0	0
Unidentified bivalves	6.67	0.02	2.78	0.84	0	0	0	0	0	0	0	0	0	0	0	0
Unidentified invertebrate	0	0	0	0	0.11	<0.01	0.78	0.01	0	0	0	0	0	0	0	0

* Evidence for prey items scavenged by White Sharks.

Stable Isotope Sampling and Analysis

Muscle tissue was sampled from sharks caught in beach protection nets between 2005 and 2008 (n = 21). All sharks were removed from nets, transported to the KwaZulu-Natal Sharks Board main laboratory, and stored frozen until public dissection. White muscle tissue (~5 g) was excised anterior to the first dorsal fin adjacent to the vertebral column and stored frozen. Prior to stable isotope analysis, the muscle tissue was freeze dried and homogenized using a SPEX CertiPrep 8000-D ball milling unit (SPEX CertiPrep, Metuchen, NJ). Accepting the known effect of lipid content on $\delta^{13}C$ values (DeNiro and Epstein, 1977), all muscle tissue samples were lipid extracted following the technique detailed by Hussey et al. (2010a). Between 400 and 600 μg of both nonlipid-extracted (BULK) and lipid-extracted (LE) tissue per individual shark were weighed into tin capsules, and stable carbon and nitrogen isotope ratios were provided from a continuous flow isotope ratio mass spectrometer (Finnigan MAT Delta^{plus}; Thermo Finnigan, San Jose, CA).

Stable isotope abundances are expressed in δ values as the deviation from standards in parts per thousand (‰) using the following equation:

$$\delta X = [(R_{sample}/R_{standard}) - 1] \times 1000 \qquad (3.1)$$

where X is ^{15}N or ^{13}C, and R is the ratio $^{15}N/^{14}N$ or $^{13}C/^{12}C$. The standard reference material was atmospheric nitrogen for N_2 and Pee Dee Belemnite carbonate for CO_2. The analytical precision (standard deviation) based on two standards (n = 59 for each standard): NIST 8414 and internal lab fish muscle were 0.11‰ and 0.19‰, respectively, for $\delta^{15}N$ and were 0.05‰ and 0.06‰, respectively, for $\delta^{13}C$. Paired t tests were used to examine directional changes in $\delta^{15}N$, %N, $\delta^{13}C$, %C, and C:N ratio between BULK and LE muscle tissue.

Size-Based Trophic Profile

Trophic position of White Sharks using stomach content data (TP_{SCA}) was calculated following Cortés (1999):

$$TP_{SCA} = 1 + \left(\sum_{i=1}^{7} p_i \times TP_i\right) \qquad (3.2)$$

where TP_{SCA} is diet-calculated trophic position per dietary sample, p_i is the proportion of each prey category in the total diet (expressed as %M), and TP_i is the trophic position for each functional prey category. The trophic positions of functional prey categories were defined as elasmobranch (3.65), teleost (3.24), cephalopod (3.2), crustacean (2.52), mollusk (2.1), bird (3.87), and mammal (4.02) based on Cortés (1999). The miscellaneous functional prey group was excluded from all trophic position calculations.

Both TP_{SCA} for all dietary samples within the four defined size classes and mean TP_{SCA} (± S.D.) per size class were calculated. Accepting that dietary samples consisted of consecutive increasing size animals, the mean PCL for each dietary sample was also calculated and plotted against the TP_{SCA} estimate.

To examine $\delta^{15}N$ ontogenetic trophic shifts and to facilitate a comparison with TP_{SCA}, $\delta^{15}N$ data were plotted versus PCL for each shark, and mean (±S.D.) data were presented per size class. A trophic level was defined as 2.29‰ $\delta^{15}N$ according to Hussey et al. (2010a, 2010b) based on a semicontrolled experiment on large sharks.

Size-based $\delta^{13}C$ profiles were examined to explore possible switches in foraging location/base with increasing size of shark. This was undertaken based on reported gradients in $\delta^{13}C$ values in the marine environment off southern Africa (Hill et al., 2006; Hill and McQuaid, 2008):

1. An increase in $\delta^{13}C$ values moving from KZN to the Western Cape
2. Depleted $\delta^{13}C$ values in offshore waters compared with coastal waters

RESULTS

Stomach Content Data

Diet Composition

The cumulative prey curve for all sharks combined indicated that the data were describing a general asymptotic relationship but stabilization was not achieved (Figure 3.2a). Cumulative prey curves for data divided by size class did not approach an asymptote. This indicated that more individuals would be required to provide a more accurate representation of diet by size class, particularly for size class 4 animals (>285 cm PCL; Figure 3.2b–e).

A diverse range of prey items were consumed within the main functional prey groups: elasmobranch ($n = 14$), mammal ($n = 5$), teleost ($n = 18$), and cephalopod ($n = 4$) (Table 3.1 and Figure 3.3). Identified prey items were of varying size, ranging from large unidentified whale (mysticeti) and Whale Shark (*Rhincodon typus*) to small loligo squids, South African Pilchards (*Sardinops ocellatus*), and unidentified bivalves (Table 3.1).

In terms of %M, %F, and %IRI, elasmobranchs were the principal diet component of the smallest White Sharks followed by teleost prey. Elasmobranch prey was dominated by the Dusky Shark (*Carcharhinus obscurus*), whereas teleost prey consisted of Chub Mackerel (*Scomber japonicus*) and sea breams (Sparidae) (Table 3.1). When considering %N, %F, and %IRI, cephalopods were also important to the diet of small sharks, specifically unidentified squids (Figure 3.4a and Table 3.1). With increasing size class of White Shark, there was a marked increase in mammal prey for all calculated diet indices and a corresponding decrease in teleost, elasmobranch, and cephalopod prey (Figure 3.4; cephalopods for %N, %F, and %IRI only). Of the elasmobranch prey, Dusky Sharks remained the most important prey item of size class 2 and 3 sharks, but overall elasmobranch prey diversity increased (Table 3.1). Size class 2 White Sharks (185–234.9 cm) consumed the most diverse range of reef, pelagic, and benthic teleosts (Table 3.1). Within the diet of the large White Sharks (>285 cm), seal was the most numerous mammal prey item (%N), whereas whale contributed the most by mass (%M) (Figure 3.4, a and b).

Size-Based Diet Shift

For all of the diet indices examined, White Sharks underwent a significant size-based shift in diet (Figure 3.5). Multidimensional scaling ordination of the data showed a level of diet separation between size class 1 and 4 sharks, but a high degree of overlap between all size class sharks was evident as indicated by the low global *R* statistic values (Figure 3.5). Pairwise comparisons found that the diet of size class 1 sharks was significantly different from all other size classes for all calculated indices (Figure 3.5). Similarity percentage analysis found that mammal prey was the principal dietary component driving this diet separation.

Shark and cephalopod prey principally occurred in the diet of sharks between 140 and 270 cm PCL (Figure 3.6). All of the main functional prey groups occurred in the diet of sharks measuring ~200 and 250 cm PCL. Of the marine-mammal prey, dolphin was first consumed by a shark of 177 cm

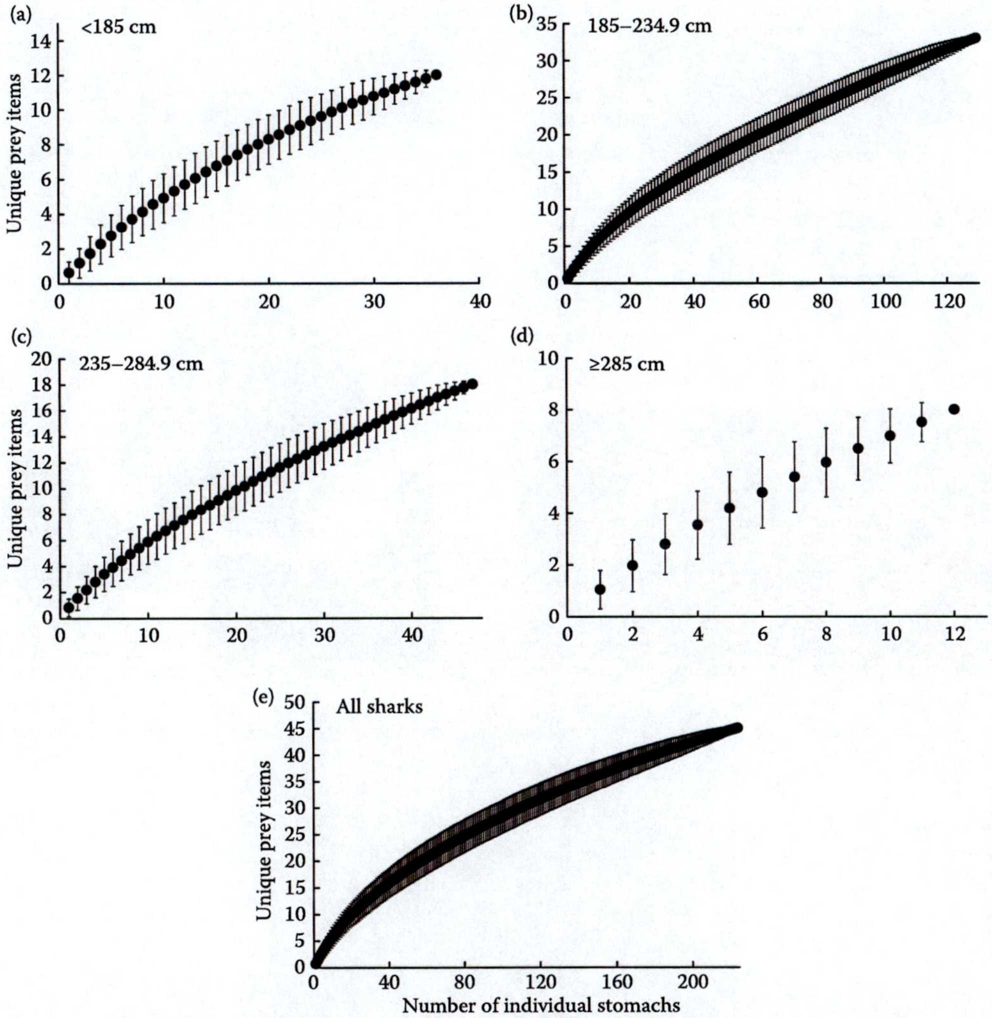

Figure 3.2 Randomized cumulative prey curve for White Sharks sampled from beach protection nets. (a) Size class 1 (<185 cm PCL). (b) Size class 2 (185–234.9 cm PCL). (c) Size class 3 (235–284.9 cm PCL). (d) Size class 4 (≥285 cm PCL). (e) All White Sharks combined.

PCL, seal by a shark of 194 cm PCL, and whale by a shark measuring 208 cm PCL (Figure 3.6). Seal and whale were the dominant prey, occurring in the diet of sharks >300 cm PCL (Figure 3.6).

Ontogenetic Trophic Profile

Stomach Content Data (TP_{SCA})

Stomach content-calculated trophic position (TP_{SCA}) varied from 4.2 to 5.0 (Figure 3.7a). Although TP_{SCA} estimates were highly variable across the size range of animals examined (Figure 3.7a), there was a significant increase in TP_{SCA} with PCL (linear regression: $r^2 = 11.7\%$; $F_{1,75} = 11.05$, $p = 0.001$). Overall TP_{SCA} predicted White Sharks of all sizes feeding across 0.8 trophic levels. Mean size class TP_{SCA} values predicted feeding across 0.3 trophic levels (Figure 3.7a).

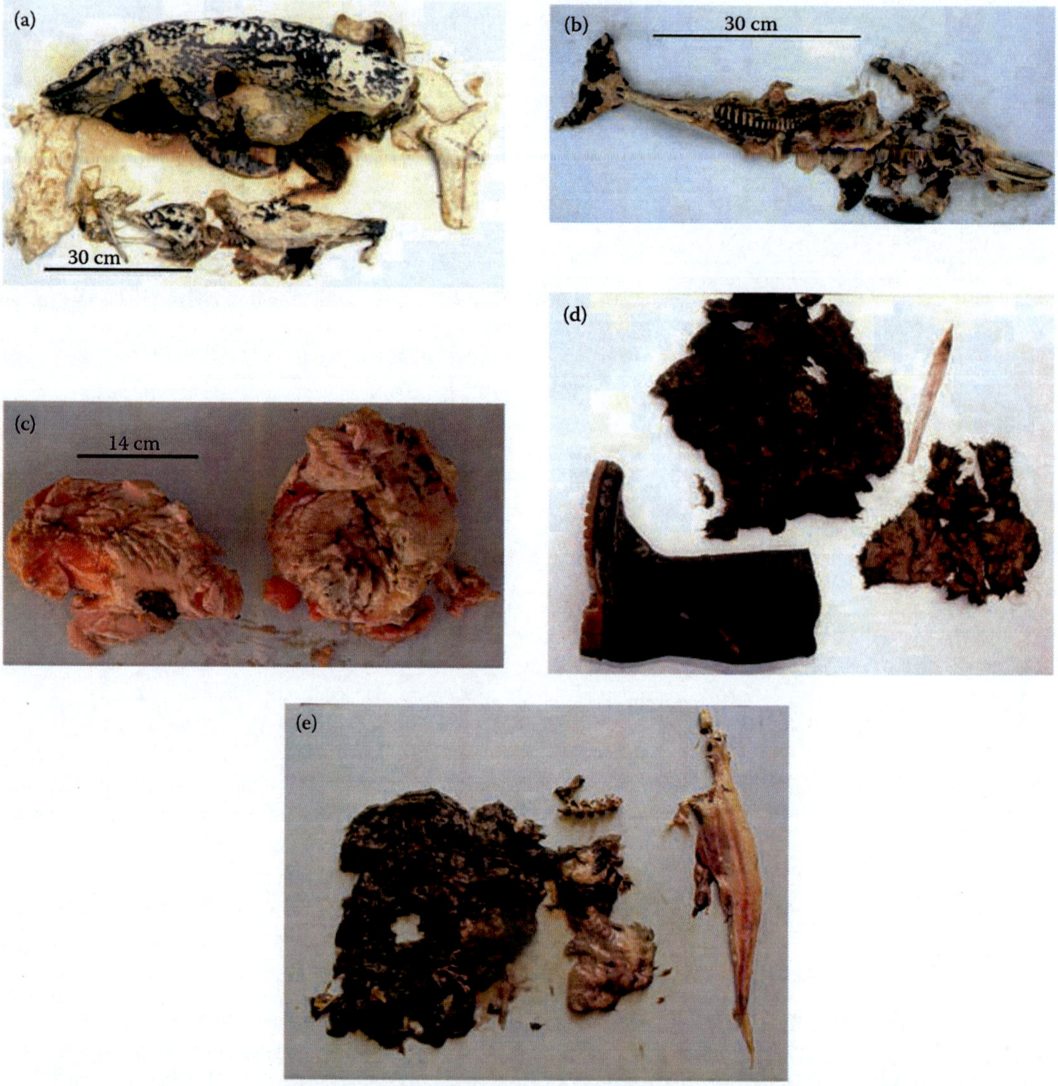

Figure 3.3 Stomach contents retrieved from White Sharks incidentally caught in beach protection nets in KwaZulu-Natal, South Africa. (a) Unidentified whale blubber and unidentified dolphin (302-cm PCL female). (b) Unidentified dolphin (177-cm PCL female). (c) Unidentified whale blubber (325-cm PCL female). (d) Unidentified seal and rubber boot (235-cm PCL male). (e) Unidentified seal and milk shark (*Rhizoprionodon acutus*) (231-cm PCL female). PCL, precaudal length of the dissected shark.

Stable Isotopes ($\delta^{15}N$ and $\delta^{13}C$): Lipid Extraction Effects

Lipid extraction of White Shark muscle tissue resulted in a significant increase in $\delta^{13}C$, %C, and C:N ratio; a significant decrease in %N; and no change in $\delta^{15}N$ values (Figure 3.8). Because lipid extraction resulted in a significant increase in $\delta^{13}C$, LE data were used in all subsequent analysis.

Stable Isotope Data (TP_{SIA})

The values of $\delta^{15}N$ increased from 13.4‰ to ~15.4‰ between animals of 124–200 cm PCL (Figure 3.7b). For animals >200 cm PCL, $\delta^{15}N$ values were variable, ranging between 14.9‰ and

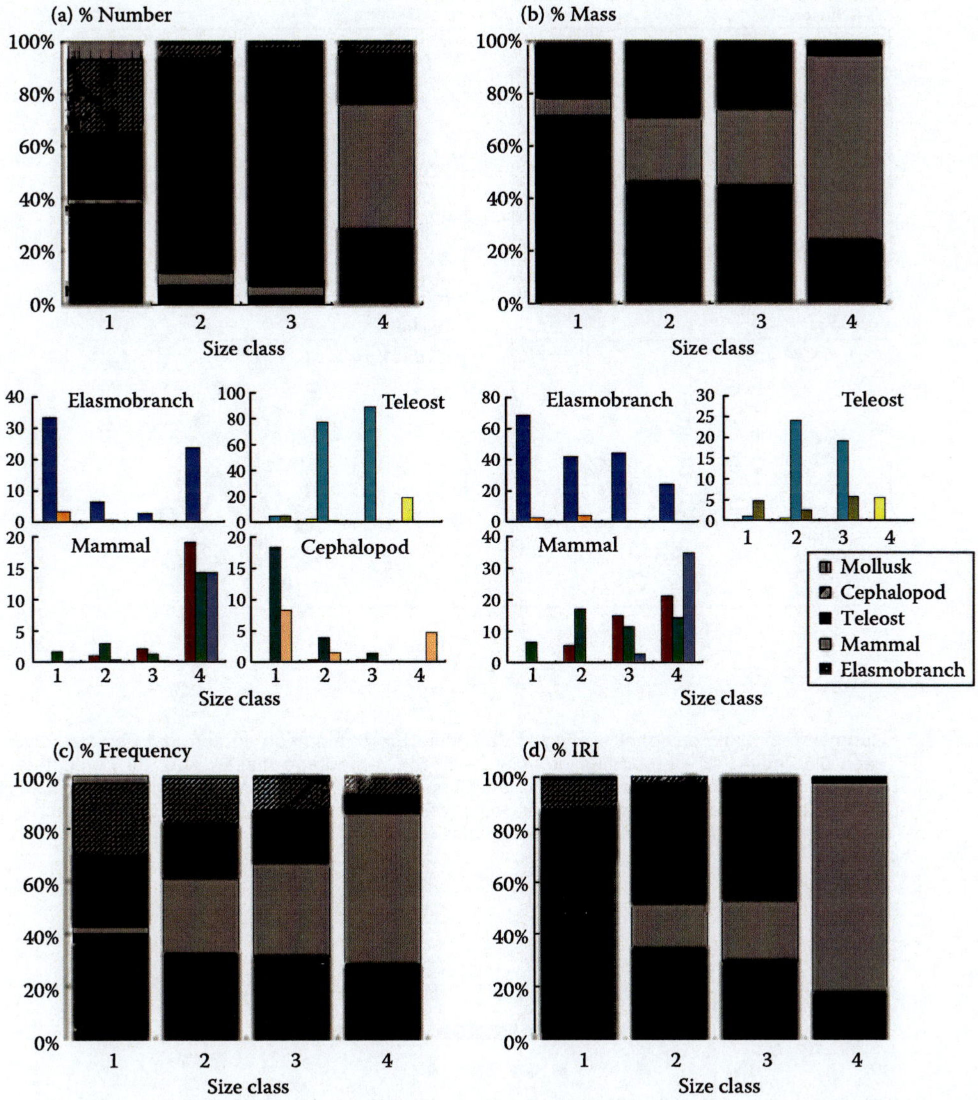

Figure 3.4 Graphic summary of White Shark stomach content data. Percentage number (%N) (a) and percentage mass (%M) (b) contribution to total diet of each size class of shark, including detailed breakdown of main functional prey groups: elasmobranch (shark ■, ray ■), teleost (reef ■, pelagic ■, benthic ■), mammal (dolphin ■, seal ■, whale ■), and cephalopod (octopus ■, squid ■, cuttlefish ■). (c and d) Percentage frequency (%F) (c) and percentage index of relative importance (%IRI) (d).

16.6‰ (Figure 3.7b). Overall there was a significant linear increase in $\delta^{15}N$ with PCL ($r^2 = 24.6\%$; $F_{1,19} = 7.51$, $p = 0.013$). Nitrogen stable isotopes predicted that White Sharks of all sizes were feeding across 1.4 trophic levels. Mean size class $\delta^{15}N$ values predicted feeding across 0.7 trophic levels.

Foraging Location ($\delta^{13}C$)

The $\delta^{13}C$ values of the two smallest White Sharks, a female of 124.0 cm PCL (–16.8‰) and a male of 134.0 cm PCL (–16.0‰), were depleted and similar to the value for the largest animal,

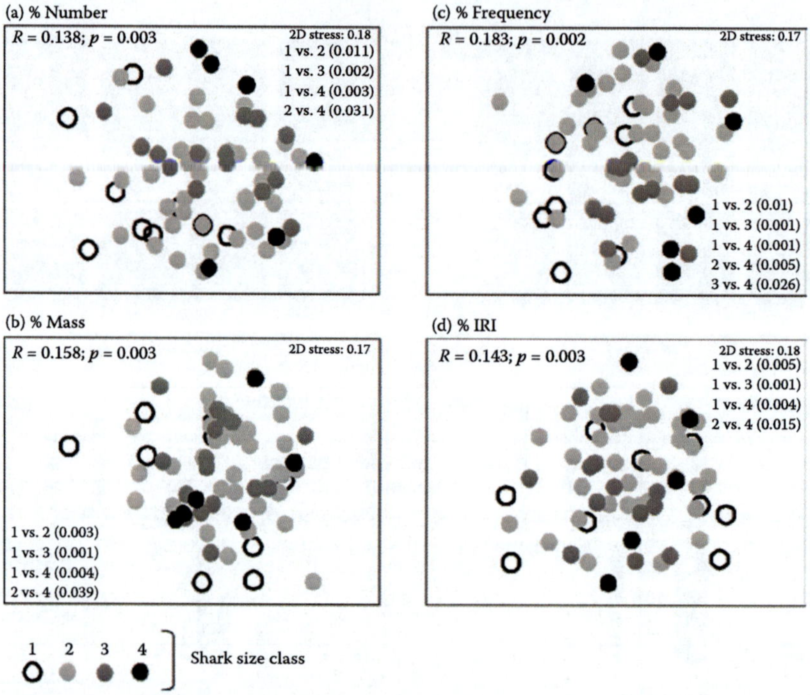

Figure 3.5 Nonmetric multidimensional scaling of the White Shark stomach content data by size class for each diet index. (a) Percentage number (%N). (b) Percentage mass (%M). (c) Percentage frequency (%F). (d) Percentage index of relative importance (%IRI). *R*, ANOSIM global *R* statistic and associated *p* value. Significant pairwise tests (with *p* value in brackets) are detailed in each figure. The level of statistical significance was set at $\alpha = 0.05$.

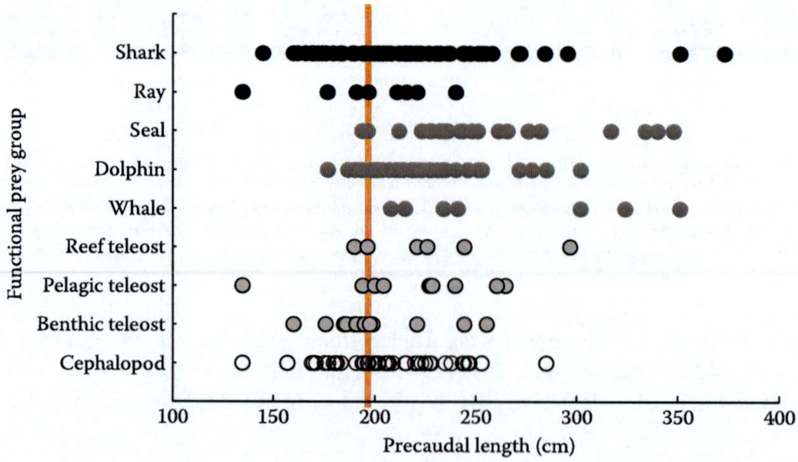

Figure 3.6 The relationship between White Shark length and occurrence of the functional prey groups. The vertical orange line indicates the smallest Australian/American White Shark with pinniped stomach content remains according to Tricas and McCosker (1984) [240 cm TL = ~198 cm PCL according to Cliff et al. (1989)].

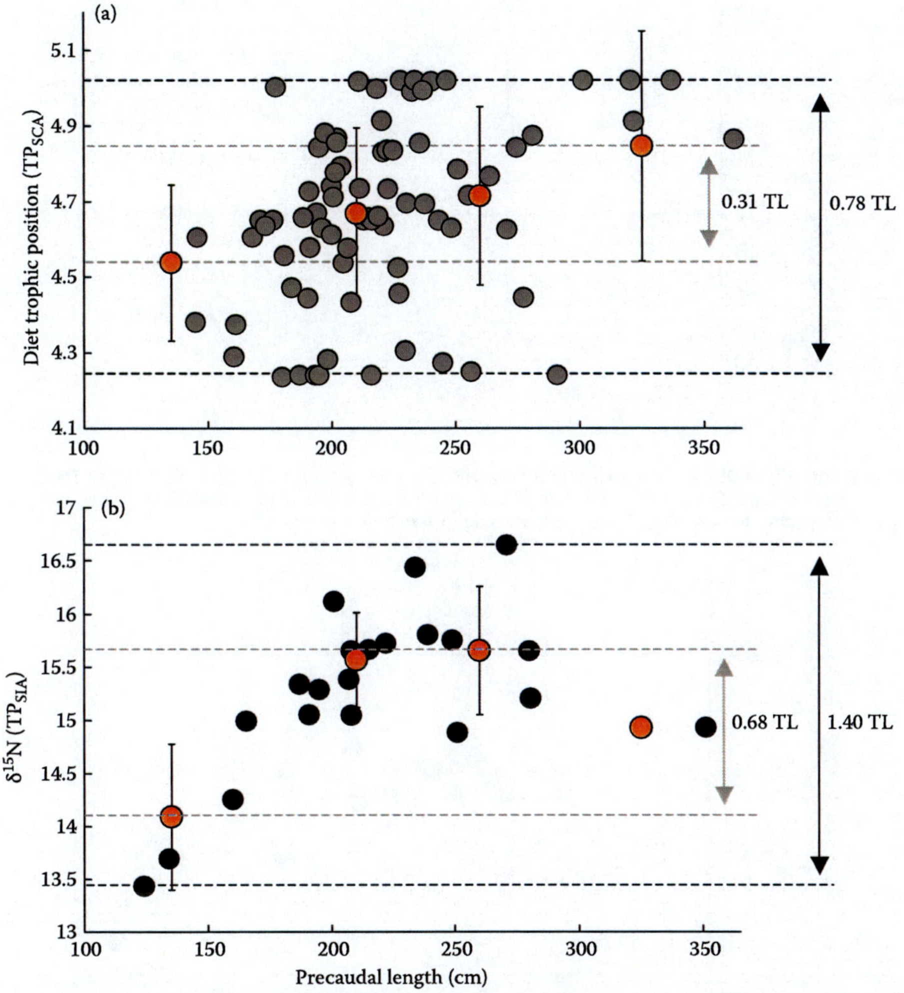

Figure 3.7 The relationship between diet calculated trophic position (TP_{SCA}) (a) and $\delta^{15}N$ (TP_{SIA}) trophic position (b) with increasing size of White Shark. The gray dots in (a) are the calculated TP_{SCA} for each dietary sample, and the black dots in (b) are $\delta^{15}N$ values (TP_{SIA}) for individual White Sharks. Mean (±SD) TP_{SCA} and $\delta^{15}N$ (TP_{SIA}) per size class are shown (orange dots). The arrows indicate the trophic range of the White Shark for individual animals and for mean size-class data.

which measured 351.2 cm PCL (male shark: –16.0‰) (Figure 3.9). Overall, the data followed a parabolic shape, with a rapid increase in $\delta^{13}C$ with size, followed by a plateau, with evidence for a decline in the largest animal (Figure 3.9). Sharks measuring between 160.0 and 280.0 cm PCL had a mean (±SD) $\delta^{13}C$ value of –14.66‰ ± 0.43 with a minimum and maximum of –15.28‰ and –14.02‰, respectively.

DISCUSSION

By analyzing the most comprehensive archival stomach content data available combined with preliminary stable isotope data, it was possible to define size-based diet shifts and to generate ontogenetic trophic profiles of South African White Sharks. These data complement previous

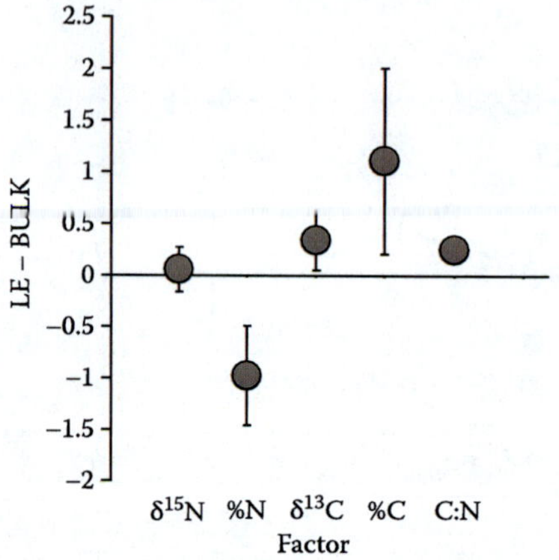

	t	p
$\delta^{15}N$	1.81	$p = 0.085$
%N	9.17	$p < 0.0001$
$\delta^{13}C$	4.65	$p < 0.0001$
%C	4.84	$p < 0.0001$
C:N	11.01	$p < 0.0001$

Figure 3.8 The effect of the lipid extraction process on $\delta^{15}N$, total percentage of nitrogen (%N), $\delta^{13}C$, total percentage of carbon (%C), and C:N ratio of muscle tissue of White Sharks. LE, values for lipid-extracted tissue; BULK, values for nonlipid-extracted tissue.

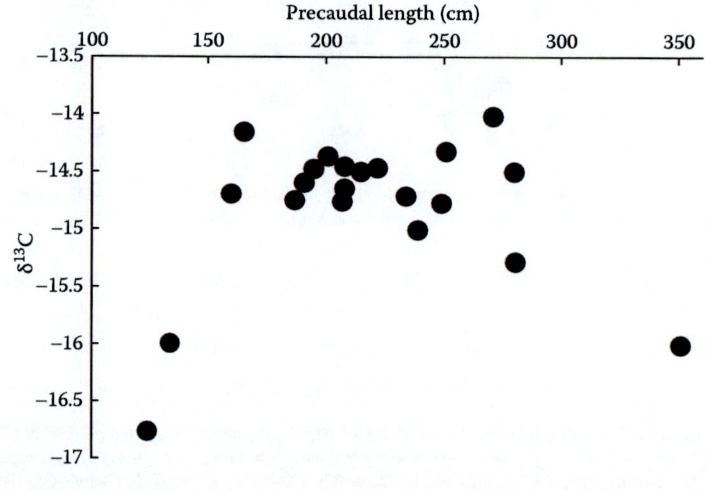

Figure 3.9 The relationship between $\delta^{13}C$ and increasing size of White Shark as an indicator of size-based shifts in foraging location.

stomach content data from South Africa (Bass et al., 1975; Cliff et al., 1989, 1996; the data presented in this study amalgamate those in the latter two studies) and other geographic localities (Tricas and McCosker, 1984; Bruce, 1992; Fergusson et al., 2000; Malcolm et al., 2001) and stable isotope data for the eastern North Atlantic (Estrada et al., 2006) and Pacific Oceans (Kerr et al., 2006).

Diet data derived from stomach contents for White Sharks have typically been restricted by sample numbers (Bass et al., 1975; Tricas and McCosker, 1984; Bruce, 1992) and the occurrence of empty or everted stomachs (Bass et al., 1975; Cliff et al., 1989). Cumulative prey curves indicated that the overall data set provided a reasonable indication of the total diet of the White Shark but that diet by size class will require further investigation. As with many large predatory sharks, the White Shark exhibited generalist feeding behavior, incorporating a large number of prey items from several functional prey categories. This type of feeding is reported for other closely related lamnids

(Stillwell and Kohler, 1982; Joyce et al., 2002) and for many large carcharhinids (Wetherbee et al., 1990; Simpfendorfer et al., 2001). Within the functional prey categories, prey diversity in terms of habitat of occurrence was also high; for example, the teleost group included reef, pelagic, and demersal species. This suggests that White Sharks forage in a number of habitats and/or opportunistically forage if prey is abundant or easily available. Both Tricas and McCosker (1984) and Klimley (1985) reported near-shore pelagic and demersal teleost species in the stomachs of White Sharks. Furthermore, diving behavior documented from pop-up archival satellite tags indicates that White Sharks adopt various foraging modes, including feeding in the surface mixed layer, on the seabed, and at depth (Dewar et al., 2004; Weng et al., 2007a, 2007b; Nasby-Lucas, 2009; Chapters 10 and 22, this book). The generalist feeding strategy of White Sharks is likely related to the availability of preferred prey, seasonal prey abundance, and/or the energetic requirements of an individual related to life stage.

Typical of large predatory fish (Scharf et al., 2000), the White Shark demonstrated asymmetric feeding behavior, whereby larger prey were consumed with increasing predator size, but small prey items were retained in the diet. Squid, loligo squid, chiroteuthid squid, and cuttlefish, for example, were found in the stomachs of sharks >244 cm PCL, as well as those of smaller animals. The fact that the diet of small White Sharks is a subset of the diet of larger individuals was a principal factor contributing to the high degree of overlap of functional prey categories across size classes. The occurrence of small prey in the diet of large sharks, however, may indicate that abundant small fish such as the South African Pilchard, *Sardinops sagax*, may be energetically viable prey when compared with the energetic costs of chase down, manipulation, handling, and the probability of prey capture associated with marine-mammal prey. The stomachs of three White Sharks, measuring, 227.8, 228.5, and 239.8 cm PCL contained 300, 300, and 477 Pilchards each. These feeding events are linked to the annual sardine run off the South African coast (Cliff et al., 1996; Dudley and Cliff, 2010).

A clear size-based shift in diet was evident, even when considering the high degree of overlap in functional prey categories across size classes. The diet of the smallest White Sharks was focused on elasmobranches and teleosts, and the first evidence for foraging on seals was an individual measuring 194 cm PCL. This is in agreement with Tricas and McCosker (1984) for the size of White Sharks first feeding on seal off Australia and the United States. Tricas and McCosker (1984) postulated that an ontogenetic development in dentition at approximately 3 m TL may account for the shift in predatory behavior. Whale and seal were the dominant prey occurring in the diet of White Sharks >3 m in this study, in agreement with Tricas and McCosker (1984) and Casey and Pratt (1985). Further comparative study into the dentition of White Sharks off South Africa relative to California and other geographically separated populations and how this relates to diet is required. The occurrence of seals in the diet of White Sharks >200 cm PCL is in agreement with observations on White Shark–seal predation events in the western cape (Martin et al., 2005; Chapter 9, this book) and supports the hypothesis of White Sharks showing preference for marine mammals with increasing size (Tricas and McCosker, 1984; Casey and Pratt, 1985; Klimley, 1985). It is important to note, however, that White Sharks feeding on whale prey are typically scavenging events (McCosker, 1985; Long and Jones, 1996; Dudley et al., 2000) and are therefore opportunistic in nature in contrast to directed predation on pinnipeds (Martin et al., 2005; Chapter 9, this book). Stomach content data therefore indicate that White Sharks are opportunist, generalist predators similar to many large sharks (Ebert, 1994) but show size-based preference for large-bodied, energetically valuable prey. The abundance of seal prey (i.e., seal colonies) influences the movement and residency pattern of White Sharks >200 cm PCL, but when considering the diverse habitats occupied by prey identified from stomach contents, it does not constrain movement (Bruce et al., 2006; Chapters 10, 11, 16, and 21, this book).

As would be expected, there was an increase in diet-calculated trophic position (TP_{SCA}) and $\delta^{15}N$ (TP_{SIA}) with increasing size of shark, correlated with the size-based shift to mammal prey. TP_{SCA} calculated from dietary samples was highly variable across the size range of sharks sampled, similar to the findings of Cortés (1999). This was likely a result of the high degree of overlap of functional prey groups across size classes. For $\delta^{15}N$ and accepting the small sample size, the initial increase in TP_{SIA} was fairly rapid in contrast to TP_{SCA}. Midsize range sharks, however, had variable $\delta^{15}N$ values. Because $\delta^{15}N$ values of muscle tissue represent long-term integrated dietary signatures (MacNeil et al., 2005), this could correspond to variable feeding strategies, whereby some sharks selectively feed on marine mammals, whereas other individuals roam between shelf and oceanic waters. Variation in habitat therefore drives differences in foraging behavior. This is supported by variable residency rates for individual White Sharks at seal colonies (Bonfil et al., 2005). Alternatively, the observed variation in $\delta^{15}N$ values could be an indication of individual foraging success. Certain White Sharks may be more skilled and able to feed on seals, whereas other less successful individuals incorporate other prey within their diet to meet energetic requirements. Intrapopulation variation in foraging success is well documented across a broad range of species. We cannot rule out, however, that elevated $\delta^{15}N$ signatures or the observed $\delta^{15}N$ variation of the midsize sharks may be a result of a cessation of feeding. It is well documented that the tissues of animals that cease feeding for extended periods of time increase in ^{15}N (Hobson et al., 1993). Previously, Carey et al. (1982) estimated that a White Shark could survive for 1.5 months on 30 kg of whale blubber. When considering the documented transoceanic migration speed of a White Shark (Bonfil et al., 2005), it is also possible that feeding is limited during these long-distance movements. Consequently, the observed $\delta^{15}N$ variation may be a result of interval-based feeding patterns or sharks that have returned from a migration and/or are in poor condition. Recent observations of the feeding behavior of young-of-year White Sharks in captivity, however, indicate that daily feeding ration and metabolic activity are higher than previously thought for this species (Chapter 1, this book). Although the captive shark data may not be truly representative of wild animals (Chapter 1, this book), these data may indicate that White Sharks cannot go very long periods without feeding, suggesting that the variation in the $\delta^{15}N$ values of midsize White Sharks are instead a result of differing intrapopulation feeding strategies.

Habitat partitioning between small and large animals may also contribute to the observed size-based shifts in diet and associated trophic position. Weng et al. (2007b) reported that small White Sharks off the coast of California occupied coastal nursery regions that were partially separated from the habitat occupied by larger individuals and were not characterized by large seal populations. Based on catches of small individuals (Bass et al., 1975; Cliff et al., 1989) and a scavenging event (Dicken, 2008), the suggested nursery region for White Sharks off southern Africa is in eastern cape waters. Larger White Sharks are known to transit through this area (Bonfil et al., 2005) but are more commonly sighted in the vicinity of highly productive seal colonies in the western cape (Martin et al., 2005). Size-based habitat partitioning and associated diet separation are widely reported in sharks (Simpfendorfer et al., 2005). Furthermore, the eastern cape and KwaZulu-Natal provide highly productive nursery habitat for numerous carcharhinid species, including the Dusky Shark (Hussey et al., 2009). This may explain the high incidence of elasmobranch prey, particularly the Dusky Shark, in the diet of White Sharks. The $\delta^{13}C$ data, however, are contradictory to this proposed size-based habitat model because the two smallest animals had the most depleted $\delta^{13}C$ values, indicative of offshore foraging (Schell et al., 1998). The smallest sharks, however, were estimated to be only a few months old, and their isotope values were most probably confounded by the maternal signature (Olin et al., 2011).

White sharks >4 m PCL are not commonly reported from the coastal waters of the cape region or from captures in beach protection nets in KZN (Cliff et al., 1989, 1996). In the eastern Pacific Ocean, large White Sharks are known to be seasonal residents during the Northern Elephant Seal (*Mirounga angustirostris*) pupping season (Weng et al., 2007b; Jorgensen et al., 2010; Chapters 10 and 16, this book) but then move offshore (Weng et al., 2007b; Domeier and

Nasby-Lucas, 2008; Chapter 11, this book), with large females remaining resident in offshore waters for extended periods of time (Jorgensen et al., 2010; Chapter 11 and 16, this book). The largest animal included in this diet analysis was 373 cm PCL, and therefore we have limited data to comment on the prey of large mature White Sharks. The data for the largest size class would suggest that whale and seal are primary dietary components, but stomach content data may be inaccurate if these sharks inhabit offshore waters for extended periods of time, because oceanic prey may be digested and therefore not detected when the animal returns inshore. The fact that the $\delta^{13}C$ values of the largest shark and the smallest shark were depleted relative to all the mid-size sharks may indicate that large animals in South African waters, including pregnant females, are foraging in offshore waters and remain offshore. Malcolm et al. (2001) found that White Sharks >5 m TL contained only teleosts and elasmobranchs, and Smale and Cliff (Chapter 4, this book), using cephalopod beaks from this stomach content data, suggested that oceanic cephapods may be an important diet component of the largest White Sharks. Considering the abundance of large cephalopods in oceanic waters and their importance in the diet of carnivorous whales (Fiscus et al., 1989), cephalopods in conjunction with pelagic teleosts may be an underestimated component in the diet of large White Sharks.

In conclusion, the White Shark off Southern Africa is an opportunist, generalist predator that shows size-based preference for marine-mammal prey, similar to observations on the diet and feeding behavior of this species in other geographic localities. Both stomach content and stable isotope data ($\delta^{15}N$), however, showed that White Sharks >200 cm PCL may not all feed exclusively on marine mammals, indicating (1) individual specialization within a generalist population and (2) interindividual foraging success. Further work, including coupling genetic and stable isotope analyses, is required to examine these points. The White Shark is an apex predator that exerts influence across multiple components of marine ecosystems. Considering the inability of such large predatory sharks to sustain exploitation even at moderate levels (Garcia et al., 2008), understanding and documenting the diet and trophic ecology of this species is a requirement for ecosystem- and fisheries-management and conservation initiatives.

REFERENCES

Bass, A.J., J.D. D'Aubrey, and N. Kistnasamy. 1975. Sharks of the east coast of southern Africa. V. The families Odontaspididae, Scapanorhynchidae, Isuridae, Cetorhinidae, Alopiidae, Orectolobidae and Rhiniodontidae. *Oceanographic Research Institute Investigational Reports* 39: 22–26.

Baum, J.K., and B. Worm. 2009. Cascading top-down effects of changing oceanic predator abundances. *Journal of Animal Ecology* 78: 699–714.

Bonfil, R., M. Meyer, M.C. Scholl, R. Johnson, S. O'Brien, H. Oosthuizen, S. Swanson, D. Kotze, and M. Paterson. 2005. Transoceanic migration, spatial dynamics and population linkages of white sharks. *Science* 310: 100–103.

Bonfil, R., M.P. Francis, C. Duffy, M.J. Manning, and S. O'Brien. 2010. Large-scale tropical movements and diving behaviour of white sharks *Carcharodon carcharias* tagged off New Zealand. *Aquatic Biology* 8: 115–123.

Bruce, B.D. 1992. Preliminary observations on the biology of the white shark, *Carcharodon carcharias*, in South Australian waters. *Australian Journal of Marine and Freshwater Research* 43: 1–11.

Bruce, B.D., J.D. Stevens, and H. Malcolm. 2006. Movements and swimming behaviour of white sharks (*Carcharodon carcharias*) in Australian waters. *Marine Biology* 150: 161–171.

Carey, F.G., G. Gabrielson, J.W. Kanwisher, and O. Brazier. 1982. The white shark, *Carcharodon carcharias*, is warm-bodied. *Copeia* 1982: 254–260.

Casey, J.G., and H.L. Pratt, Jr. 1985. Distribution of the white shark, *Carcharodon carcharias*, in the western North Atlantic. *Memoirs of the Southern California Academy of Sciences* 9: 2–14.

Clarke, K.R., and R.M. Warwick. 2001. *Changes in Marine Communities: An Approach to Statistical Analysis and Interpretation*. Plymouth Marine Laboratory, Plymouth, UK.

Cliff, G., S.F.J. Dudley, and B. Davis. 1988. Sharks caught in the protective gill nets off Natal, South Africa. 1. The sandbar shark *Carcharhinus plumbeus* (Nardo). *South African Journal of Marine Science* 7: 255–265.

Cliff, G., S.F.J. Dudley, and B. Davis. 1989. Sharks caught in the protective gill nets off Natal, South Africa. 2. The great white shark *Carcharodon carcharias* (Linnaeus). *South African Journal of Marine Science* 8: 131–144.

Cliff, G., S.F.J. Dudley, and M.R. Fry. 1996. Catches of white sharks in KwaZulu-Natal, South Africa and environmental influences. In: *Great White Sharks: The Biology of Carcharodon carcharias* (A. P. Klimley and D. G. Ainley, eds.). Academic Press, San Diego, CA, pp. 351–362.

Compagno, L.J.V. 2002. Sharks of the world. *FAO Fishery Synopsis*, Vol. 2. Rome, Italy: Food and Agriculture Organization of the United Nations.

Cortés, E. 1997. A critical review of methods of studying fish feeding based on analysis of stomach contents: Application to elasmobranch fishes. *Canadian Journal of Fisheries and Aquatic Sciences* 54: 726–738.

Cortés, E. 1999. Standardised diet compositions and trophic levels of sharks. *ICES Journal of Marine Science* 56: 707–717.

Cortés, E., and S.H. Gruber. 1990. Diet, feeding habits and estimates of daily ration of young lemon sharks, *Negaprion brevirostris* (Poey). *Copeia* 1990: 204–218.

DeNiro, M.J., and S. Epstein. 1977. Mechanism of carbon isotope fractionation associated with lipid synthesis. *Science* 197: 261–263.

DeNiro, M.J., and S. Epstein. 1978. Influence of diet on the distribution of carbon isotopes in animals. *Geochimica et Cosmochimica Acta* 42: 495–506.

Dewar, H., M. Domeier, and N. Nasby-Lucas. 2004. Insights into young of the year white sharks (*Carcharodon carcharias*) behaviour in the Southern California Bight. *Environmental Biology of Fishes* 70: 133–143.

Dicken, M.L. (2008) First observations of young of the year and juvenile great white sharks (*Carcharodon carcharias*) scavenging from a whale carcass. *Marine and Freshwater Research* 59: 596–602.

Domeier, M.L., and N. Nasby-Lucas. 2008. Migration patterns of white sharks *Carcharodon carcharias* tagged at Guadalupe Island, Mexico, and identification of an eastern Pacific shared offshore foraging area. *Marine Ecology Progress Series* 370: 221–237.

Dudley, S.F.J., and G. Cliff. 2010. Influence of the annual sardine run on catches of large sharks in the protective gillnets off KwaZulu-Natal, South Africa, and the occurrence of sardine in shark diet. *African Journal of Marine Science* 32: 383–397.

Dudley, S.F.J., M.D. Anderson-Reade, G.S. Thompson, and P.B. McMullen. 2000. Concurrent scavenging off a whale carcass by great white sharks, *Carcharodon carcharias* and tiger sharks, *Galeocerdo cuvier*. *Fishery Bulletin* 98: 646–649.

Dudley, S.F.J., G. Cliff, M.P. Zungu, and M.J. Smale. 2005. Sharks caught in protective gill nets off KwaZulu-Natal, South Africa. 10. The dusky shark, *Carcharhinus obscurus* (Lesueur 1818). *African Journal of Marine Science* 27: 107–127.

Ebert, D.A. 1994. The diet of the sixgill shark *Hexanchus griseus* off Southern Africa. *South African Journal of Marine Science* 14: 213–218.

Elton, C.S. 1927. *Animal Ecology*. Sidgwick and Jackson, London, UK.

Estrada, J.A., A.N. Rice, L.J. Natanson, and G.B. Skomal. 2006. Use of isotopic analysis of vertebrae in reconstructing ontogenetic feeding ecology in white sharks. *Ecology* 87: 829–834.

Fergusson, I.K., L.J.V. Compagno, and M.A. Marks. 2000 Predation by white sharks *Carcharodon carcharias* (Chondrichthyes: Lamnidae) upon chelonians, with new records from the Mediterranean Sea and a first record of the ocean sunfish *Mola mola* (Osteichthyes: Molidae) as stomach contents. *Environmental Biology of Fishes* 58: 447–453.

Ferry, L.A., and G.M. Cailliet. 1996. Sample size and data analysis: Are we characterizing and comparing diet properly? In: *Feeding Ecology and Nutrition in Fish: Proceedings of the Symposium on the Feeding Ecology and Nutrition in Fish* (D. MacKinlay and K. Shearer, eds.). American Fisheries Society. International Congress on the Biology of Fishes, San Francisco, CA, pp. 71–80.

Fiscus, C.H., D.W. Rice, and A.A. Wolman. 1989. Cephalopods from the stomachs of sperm whales taken off California. *NOAA Technical Report NMFS* 83: 1–12.

Fisk, A.T., S.A. Tittlemier, J.L. Pranschke, and R.J. Norstrom. 2002. Using anthropogenic contaminants and stable isotopes to assess the feeding ecology of Greenland sharks. *Ecology* 83 : 2162–2172.

Garcia, V. B., L.O. Lucifora, and R.A. Myers. 2008. The importance of habitat and life-history to extinction risk in sharks, skates, rays and chimaeras. *Proceedings of the Royal Society, Part B, Biological Sciences* 275: 83–89.

Heithaus, M.R., A. Frid, A.J. Wirsing, and B. Worm. 2008. Predicting ecological consequences of marine top predator declines. *Trends in Ecology and Evolution* 4: 202–210.

Hill, J.M., and C.D. McQuaid. 2008. $\delta^{13}C$ and $\delta^{15}N$ biogeographic trends in rocky intertidal communities along the coast of South Africa: Evidence of strong environmental signatures. *Estuarine and Coastal Shelf Science* 80: 261–268.

Hill, J.M., C.D. McQuaid, and S. Kaehler. 2006. Biogeographic and nearshore-offshore trends in isotope ratios of intertidal mussels and their food sources around the coast of Southern Africa. *Marine Ecology Progress Series* 318: 63–73.

Hobson, K.A., R.T. Alisauskas, and R.G. Clarke. 1993. Stable nitrogen isotope enrichment in avian-tissues due to fasting and nutritional stress: Implications for isotopic analyses of diet. *Condor* 95: 388–394.

Hussey, N.E., I.D. McCarthy, S.F.J. Dudley, and B. Mann. 2009. Movement and growth of dusky sharks (*Carcharhinus obscurus*): A long-term tag and release study in South African waters. *Marine and Freshwater Research* 60: 571–583.

Hussey, N.E., J. Brush, I.D. McCarthy, and A.T. Fisk. 2010a. $\delta^{15}N$ and $\delta^{13}C$ diet discrimination factors for large sharks under semi-controlled conditions. *Comparative Biochemistry and Physiology Part A* 155: 445–453.

Hussey, N.E., M.A. MacNeil, and A.T. Fisk. 2010b. The requirement for accurate diet-tissue discrimination factors for interpreting stable isotopes in sharks. *Hydrobiologia* 654: 1–5.

Hussey, N.E., S.F.J. Dudley, I.D. McCarthy, G. Cliff and A.T. Fisk. 2011. Stable isotope profiles of large marine predators: Viable indicators of trophic position, diet and movement in sharks? *Canadian Journal of Fisheries and Aquatic Sciences*. In press.

Hutchings, J.A., and J.J. Baum. 2005. Measuring marine fish biodiversity: Temporal changes in abundance, life history and demography. *Philosophical Transactions of the Royal Society, Series B, Biological Sciences* 360: 315–338.

Huveneers, C., N.M. Otway, S.E. Gibbs, and R.G. Harcourt. 2007. Quantitative diet assessment of wobbegong sharks (genus *Orectolobus*) in New South Wales, Australia. *ICES Journal of Marine Science* 64: 1272–1281.

Hyslop, E.J. 1980. Stomach content analysis: A review of methods and their application. *Journal of Fish Biology* 17: 411–429.

Jorgensen, S.J., C.A. Reeb, T.K. Chapple, S. Anderson, C. Perle, S.R. Van Sommeran, C. Fritz-Cope, A.C. Brown, A.P. Klimley, and B.A. Block. 2010. Philopatry and migration of Pacific white sharks. *Proceedings of the Royal Society Part B* 277: 679–688.

Joyce, W.N., S.E. Campana, L.J. Natanson, N.E. Kohler, H.L. Pratt, Jr., and C.F. Jensen. 2002. Analysis of stomach contents of the porbeagle shark (*Lamna nasus* Bonnaterre) in the northwest Atlantic. *ICES Journal of Marine Science* 59: 1263–1269.

Kerr, L.A., A.H., Andrews, G.M. Cailliet, T.A. Brown, and K.H. Cole. 2006. Investigations of Δ15N, $\delta^{13}C$ and $\delta^{15}N$ in vertebrae of white shark (*Carcharodon carcharias*) from the eastern North Pacific Ocean. *Environmental Biology of Fishes* 25: 337–353.

Klimley, A. P. 1985. The areal distribution and autoecology of the white shark, *Carcharodon carcharias*, off the west coast of North America. *Memoirs of the Southern California Academy of Sciences* 9: 15–40.

Klimley, A.P., S.D., Anderson, P. Pyle, and R.P. Henderson. 1992. Spatiotemporal patterns of white shark (*Carcharodon carcharias*) predation at the South Farallon Islands, California. *Copeia* 1992: 680–690.

Klimley, A.P., P. Pyle, and S.D., Anderson. 1996. The behaviour of white sharks and their pinniped prey during predatory attacks. In: *Great White Sharks: The Biology of Carcharodon carcharias* (A. P. Klimley and D. G. Ainley, eds.). Academic Press, San Diego, CA, pp. 175–191.

Klimley, A.P., B.J. Le Boeuf, K.M. Cantara, J.E. Richert, S.F. Davis, S. Van Sommeran, and J.T. Kelly. 2001. The hunting strategy of white sharks (*Carcharodon carcharias*) near a seal colony. *Marine Biology* 138: 617–636.

Laroche, R.K., A.A. Kock, L.M. Dill, and W.H. Oosthuizen. 2008. Running the gauntlet: A predator-prey game between two age classes of seals. *Animal Behaviour* 76: 1901–1917.

Leopold, A., L.K. Sowls, L.K., D.L. Spencer. 1947. A survey of over-populated deer ranges in the United States. *Journal of Wildlife Management* 11: 162–183.

Long, D. J., and R.E. Jones. 1996. White shark predation and scavenging on cetaceans in the Eastern North Pacific Ocean. In: *Great White Sharks: The Biology of Carcharodon carcharias* (A. P. Klimley and D. G. Ainley, eds.). Academic Press, San Diego, CA, pp. 293–307.

Lowe, C.G., B.M. Wetherbee, G.L. Crow, and A.L. Tester. 1996. Ontogenetic dietary shifts and feeding behaviour of the tiger shark, *Galeocerdo cuvier*, in Hawaiian waters. *Environmental Biology of Fishes* 47: 203–211.

MacNeil, M.A., G.B. Skomal, and A.T. Fisk. 2005. Stable isotopes from multiple tissues reveal diet switching in sharks. *Marine Ecology Progress Series* 302: 199–206.

Malcolm, H., B.D. Bruce, and J.D. Stevens. 2001. *A Review of the Biology and Status of White Sharks in Australian Waters*. CSIRO Marine Research, Hobart, Tasmania, Australia.

Martin, R.A., N. Hammerschlag, R.S. Collier, and C. Fallows. 2005. Predatory behaviour of white sharks (*Carcharodon carcharias*) at Seal Island, South Africa. *Journal of the Marine Biological Association of the United Kingdom* 85: 1121–1135.

McCosker, J.E. 1985. White shark attack behaviour: Observations of and speculations about predator and prey strategies. *Memoirs of the Southern California Academy of Science* 9: 123–135.

Myers, R.A., and B. Worm. 2003. Rapid worldwide depletion of predatory fish communities. *Nature* 423: 280–283.

Myers, R. A., and B. Worm. 2005. Extinction, survival or recovery of large predatory fishes. *Philosophical Transactions of the Royal Society B* 360: 13–20.

Myers, R.A., J.K. Baum, T.D. Shepherd, S.P. Powers, and C.H. Peterson. 2007. Cascading effects of the loss of apex predatory sharks from a coastal ocean. *Science* 315: 846–850.

Nasby-Lucas, N., H. Dewar, C.H. Lam, K.J. Goldman, and M. Domeier. 2009. White shark offshore habitat: A behavioural and environmental characterisation of the Eastern Pacific shared offshore foraging area. *PLoS One* 4: e8163

Olin, J.A., N.E. Hussey, M. Fritts, M.R. Heupel, C.A. Simpfendorfer, G.R. Poulakis, and A.T. Fisk. 2011. Maternal meddling in neonatal sharks: Implications for interpreting stable isotopes in young sharks. *Rapid Communications in Mass Spectrometry* doi:10.1002/rcm.4946.

Pace, M.L., J.G. Cole, S.R. Carpenter, and J.F. Kitchell. 1999. Trophic cascades revealed in diverse ecosystems. *Trends in Ecology and Evolution* 14: 483–488.

Paine, R.T. 1980. Food webs: Linkage, interaction strength and community infrastructure. *Journal of Animal Ecology* 49: 667–685.

Pauly, D., A. Trites, E. Capuli, and V. Christensen. 1998. Diet composition and trophic levels of marine mammals. *ICES Journal of Marine Science* 55: 467–481.

Platell, M. E., and I.C. Potter. 2001. Partitioning of food resources amongst 18 abundant benthic carnivorous fish species in marine waters on the lower west coast of Australia. *Journal of Experimental Marine Biology and Ecology* 261: 31–54.

Post, D.M. 2002. Using stable isotopes to estimate trophic position: Models, methods, and assumptions. *Ecology* 83: 703–718.

Scharf, F.S., F. Juanes, and R.A. Rountree. 2000. Predator size-prey size relationships of marine fish predators: Interspecific variation and effects of ontogeny and body size on trophic-niche breadth. *Marine Ecology Progress Series* 208: 229–248.

Schell, D.M., B.A. Barnett, and K.A. Vinette. 1998. Carbon and nitrogen isotope ratios in zooplankton of the Bering, Chukchi and Beaufort seas. *Marine Ecology Progress Series* 162: 11–23.

Simpfendorfer, C.A., A.B. Goodreid, and R.B. McAuley. 2001. Sex, size and geographical variation in the diet of the tiger shark, *Galeocerdo cuvier*, from Western Australian waters. *Environmental Biology of Fishes* 61: 37–46.

Simpfendorfer, C.A., G.G. Freitas, T.R. Wiley, and M.R. Heupel. 2005. Distribution and habitat partitioning of immature bull sharks (*Carcharhinus leucas*) in a southwest Florida estuary. *Estuaries and Coasts* 28: 78–85.

Stevens, J.D., R. Bonfil, N.K. Dulvy, and P. Walker. 2000. The effects of fishing on sharks, rays and chimaeras (chondrichthyans) and the implications for marine ecosystems. *ICES Journal of Marine Science* 57: 476–494.

Stillwell, C.E., and N.E. Kohler. 1982. Food, feeding habits, and estimates of daily ration of the shortfin mako (*Isurus oxyrinchus*) in the northwest Atlantic. *Canadian Journal of Fisheries and Aquatic Sciences* 39: 407–414.

Tricas, T.C. and J.E. McCosker. 1984. Predatory behaviour of the white shark (*Carcharodon carcharias*), with notes on its biology. *Proceedings of the California Academy of Sciences* 43: 221–238.

Vander Zanden, M.J., G. Cabana, and J.B. Rasmussen. 1997. Comparing the trophic position of freshwater fish calculated using stable nitrogen isotope ratios $\delta^{15}N$ and literature dietary data. *Canadian Journal of Fisheries and Aquatic Sciences* 54: 1142–1158.

Weng, K.C., A.M. Boustany, P. Pyle, S.D., Anderson, A. Brown, and B. Block. 2007a. Migration and habitat of white sharks (*Carcharodon carcharias*) in the Eastern Pacific Ocean. *Marine Biology* 152: 877–894.

Weng, K.C., J.B. O'Sullivan, C.G. Lowe, C.E. Winkler, H. Dewar, and B.A. Block. 2007b. Movements, behaviour and habitat preference of juvenile white sharks *Carcharodon carcharias* in the eastern Pacific. *Marine Ecology Progress Series* 338: 211–224.

Wetherbee, B.M., S.H. Gruber and E. Cortés. 1990. Diet, feeding habits, digestion and consumption in sharks, with special reference to the lemon shark, *Negaprion brevirostris*. In: *Elasmobranchs as Living Resources: Advances in the Biology, Ecology, Systematics, and the Status of the Fisheries* (H.L. Pratt, Jr., S.H. Gruber, and T. Taniuchi, eds.). *NOAA Technical Report. NMFS* 90: 29–47.

White, W.T., M.E. Platell, and I.C. Potter. 2004. Comparisons between the diets of four abundant species of elasmobranchs in a subtropical embayment: Implications for resource partitioning. *Marine Biology* 144: 439–448.

CHAPTER 4

White Sharks and Cephalopod Prey

Indicators of Habitat Use?

Malcolm J. Smale*
Port Elizabeth Museum at Bayworld and Nelson Mandela Metropolitan University

Geremy Cliff
KwaZulu-Natal Shark Board and University of KwaZulu-Natal

ABSTRACT

The stomach contents of White Sharks (*Carcharodon carcharias*) caught in large mesh gill-nets off the KwaZulu-Natal coast of South Africa were examined for prey remains during routine biological necropsies. Of 220 White Sharks with prey remains in their stomachs sampled between 1983 and 2008, cephalopod beaks, which are highly digestion resistant, were found in 24. This sample was split according to predator size to investigate whether these prey remains could be used to test the hypothesis that smaller individuals [<2.5 m total length (TL)] preyed on coastal prey to a greater extent than larger individuals (>2.5 m TL), which may be a proxy for habitat use. These preliminary results supported the hypothesis in that the cephalopods from small sharks were composed almost exclusively of coastal and benthic species typical of the continental shelf, such as octopods and sepiids, and coastal pelagic prey, such as loliginid squids. Cephalopod prey of large sharks were dominated by mesopelagic and oceanic prey (e.g. *Ancistrocheirus*, *Octopoteuthis*, and Ommastrephidae), with few coastal taxa. This suggests that larger sharks are using deeper offshore waters to a greater extent than smaller individuals, supporting the findings from tagging research on these apex predators. This study has shown the potential of stomach content analyses to provide an alternative approach to studying habitat use by large predators, although there are some potential pitfalls of such investigations.

INTRODUCTION

White sharks are apex predators of the marine food webs where they occur. They have been the focus of several studies of shark and marine mammal interactions, particularly pinnipeds (Ainley et al., 1985; McCosker, 1985; Klimley et al., 1996; Laroche et al., 2008; Martin et al., 2009), and scavenging on cetaceans (Long and Jones, 1996; Dudley et al., 2000; Dicken, 2008). Dietary change and consequent trophic level increase have been demonstrated with stable isotope analyses (Estrada et al., 2006; Chapter 3, this book), supporting previous observations documenting a concomitant change in the dental morphology at 3–4 m of length (Tricas and McCosker, 1984; McCosker, 1985). Few stomach content analyses have been undertaken to date, possibly because White Sharks often avoid or escape fishing gear, are rarely caught, and, even when landed, are infrequently subjected to careful necropsies. Cliff et al. (1989) showed that although marine mammals are important in the diet of these sharks off KwaZulu-Natal, especially in sharks of >3 m TL, elasmobranchs and teleosts are also important prey, particularly in sharks <3 m TL, whereas cephalopods are minor components of their diets.

Large subadult and adult White Sharks are known to travel widely, as is evident from isolated offshore island records from the Pacific and Indian Oceans (Taylor, 1985; Cliff et al., 2000; Zuffa et al., 2002; Chapter 23, this book) and the use of photographically identified individuals (Domeier and Nasby-Lucas, 2007); satellite technology confirmed that they cross ocean basins (Boustany et al., 2002; Bonfil et al., 2005, 2010; Chapter 1, this book). Small juvenile White Sharks are known to appear in coastal waters (Chapters 14, 15, 16, and 19, this book), but their use of offshore habitats is poorly known. Recent work (Dewar et al., 2004; Bruce et al., 2006; Chapters 14, 15, 16, and 17, this book) has shown that although small White Sharks are coastal, they may move more widely and penetrate deeper waters than previously thought.

Cephalopods are infrequent prey of White Sharks caught inshore (Cliff et al., 1989; Chapter 3, this book), and consequently they have not received much attention. Because most feeding studies have been on individuals captured from shallow coastal waters, they cannot adequately reflect the entire prey spectrum. Cephalopod beaks, being digestion resistant (Clarke, 1986), have the potential to broaden prey studies to include a wider habitat range of White Sharks, as has successfully been done for other shark species (Clarke and Stevens, 1974; Smale and Cliff, 1998; Markaida and Sosa-Nishizaki, 2010).

In this study we examine the cephalopod prey of White Sharks to investigate predation events that may have happened away from the site of capture. We use these results to test the hypothesis that the stomach contents of large White Sharks would include offshore prey from feeding events either while migrating or during offshore forays. We predicted that smaller individuals would have preyed on cephalopods typical of shelf waters and that such differences could be used to better understand differences in habitat use between smaller and larger individuals.

MATERIALS AND METHODS

Large-mesh gillnets are deployed at popular swimming beaches along the KwaZulu-Natal coast to provide protection against shark attack; details of this operation, run by the KwaZulu-Natal Sharks Board (KZNSB, formerly the Natal Sharks Board) are provided by Cliff and Dudley (1992). Sharks caught in these nets are released when found alive; when dead, they are brought to a laboratory for routine necropsy, which includes a careful examination of the stomach contents and preservation of digestion-resistant items, such as teleost otoliths and cephalopod beaks (Cliff et al., 1989). Of 220 White Sharks with prey remains in their stomachs sampled between 1983 and 2008, cephalopod beaks, which are highly digestion resistant, were found in 24.

To date, cephalopod beaks have not been included in published feeding analyses (Cliff et al., 1989, 1996). In this study, we used the Port Elizabeth Museum cephalopod beak collection and the literature (Clarke, 1986; Smale et al., 1993) to identify cephalopod beaks preserved from stomachs of White Sharks since 1983. Lower beaks were used throughout, although in some cases upper beaks could be identified reliably. They were identified to the lowest taxon possible, and lower beaks were measured. Body lengths [dorsal mantle lengths (DML)] and original body mass were estimated using either lower-beak crest length (Octopoda and Sepiidae) or lower-beak rostral length (other taxa) by using regressions available in the literature or calculated from Port Elizabeth Museum reference material. A linear regression was fitted to the data relating dorsal mantle length and shark length, and cephalopod body weight and shark length, respectively, to investigate the relationship between predator size and prey size using the statistical package Prism 4.

To test the hypothesis that smaller and larger sharks use the environment differently, we arbitrarily split the sample by size to sharks smaller than, or larger than, 2.5 m TL. This size was used because it approximates the midpoint of the size range of material available.

RESULTS

Cephalopod beaks were found in 13 small (four male and nine female) and 11 large (seven male and four female) White Shark stomachs. Small White Sharks preyed mainly on octopods, *Sepia* spp. and *Loligo* spp., and consequently coastal and benthic groups dominated the cephalopoda recorded. They ranged in size from 81 to 68 mm DML and 15 to 683 g. Only one small White Shark had remains of *Ancistrocheirus leseuri* (122 mm DML and 115 g), a mesopelagic species (Figure 4.1a). Large sharks had taken *Octopus*, *Sepia*, and *Loligo*, typical of coastal areas (eight prey items), and they ranged in size from 70 to 344 mm DML and 85 to 592 g. *Ancistrocheirus*, *Octopoteuthis*, *Lycoteuthis*, *Ornithoteuthis*, and *Chiroteuthis* and *Argonauta* (a pelagic octopod) were more numerous (16 items; Figure 4.1b). These oceanic and mesopelagic species were estimated to range in size from 99 to 220 mm DML and 44 to 611 g. A comparison of the prey habitats of cephalopod prey clearly indicates the dominance of coastal and benthic cephalopods in White Sharks <2.5 m, whereas mesopelagic and oceanic taxa dominated the cephalopod prey of those >2.5 m (Figure 4.2).

There was no significant relationship between either cephalopod length or mass when regressed against shark length (t test on slopes, $p > 0.05$). The absence of a trend may be explained by the restricted size range of the cephalopod prey (few get very large), and soft-bodied cephalopods are easily consumed by powerful predators such as White Sharks.

DISCUSSION

Understanding of the behavior and movement patterns of White Sharks has advanced significantly in the last decade. Boustany et al. (2002) showed that six adult sharks of 3.7–5 m length spent time in coastal waters to 75 m deep and that four of these sharks moved into oceanic waters and exhibited bimodal swimming depths of 0–5 and 300–500 m and little time at intermediate depths. Boustany et al. were uncertain whether the oceanic phase of at least 5 months represented a feeding or breeding migration. Similarly, Bonfil et al. (2005) demonstrated that not only can White Sharks of ~3.8 m TL undertake extensive return migrations across ocean basins, they may also dive as deep as 980 m. Recently, philopatry to coastal hotspots has been demonstrated in the eastern Pacific, with evidence of extensive offshore movements and associated diving to 100–200 m over ocean depths of 400 m, which may be related to feeding or mating behavior (Domeier and Nasby-Lucas, 2008; Jorgensen et al., 2010; Chapters 11 and 16, this book).

Bonfil et al. (2005) recorded a shark of ~2–2.3 m TL moving into oceanic waters 242 km southeast of Port Elizabeth, South Africa. Although juvenile White Sharks are known to inhabit shallow inshore areas, some studies (Dicken, 2008; Chapters 17 and 19, this book) revealed that individuals of 1.9–2.1 m TL can make offshore and open-ocean forays, diving to 984 m, far deeper than the 240–400 m recorded previously by Dewar et al. (2004) and Weng et al. (2007), and that they make frequent slope and off-shelf excursions. Therefore, although juvenile White

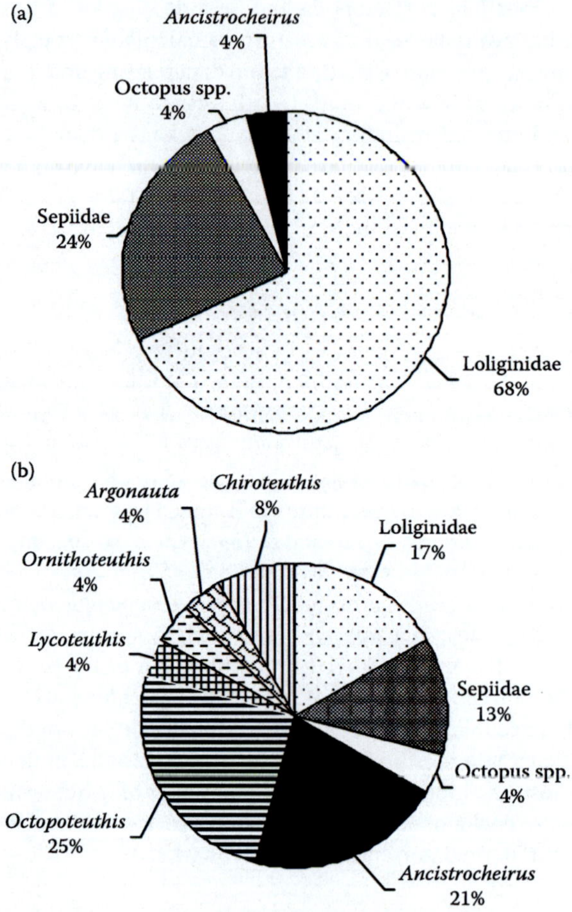

Figure 4.1 (a) Cephalopod-prey composition by number of 13 *Carcharodon carcharias* <2.5 m TL. (b) Cephalopod-prey composition by number of 11 *C. carcharias* >2.5 m TL.

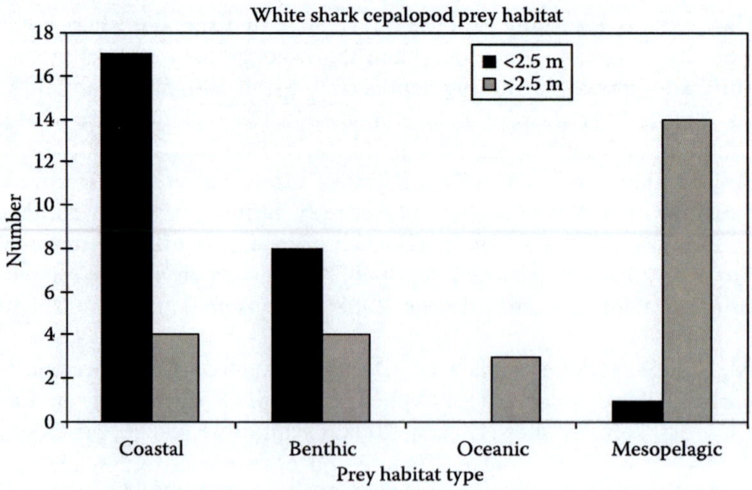

Figure 4.2 Cephalopod prey by habitat of *C. carcharias* <2.5 and >2.5 m TL.

Sharks frequent coastal areas, recent studies have discovered that they are not confined to shallow waters.

An earlier feeding study of *C. carcharias* of 1.6–4.2 m TL captured in large-mesh gillnets off KwaZulu-Natal, found elasmobranchs, teleosts, and marine mammals as prey in the stomach contents of 58 White Sharks with prey (Cliff et al., 1989). Cephalopods were recorded in 4% of stomachs, and beaks alone were found in a further four stomachs. All of the identified fish and squid prey were coastal species. The cephalopod beaks found in White Shark stomachs in this study could have been secondary prey. That is prey of lower-level predators consumed by the White Sharks, such as Common Dolphins (*Delphinus delphis*) and Dusky Sharks (*Carcharinus obscurus*), which range widely and could hunt beyond the neritic zone and consume oceanic and mesopelagic prey. Although this cannot be entirely ruled out, it is not necessarily the reason for squid beaks being found in White Shark stomachs because White Sharks are seen at squid-spawning aggregations (Smale, unpublished observations). Indeed, Domeier et al. (Chapter 12, this book) found three species of spawning squid within the Shared Offshore Foraging Area, a pelagic region in the northeastern Pacific frequented by adult White Sharks. Furthermore, the remains of fresh cephalopods flesh in some White Shark stomachs (Cliff et al., 1989) suggest that direct consumption is more likely. Finally, beaks are only thought to remain in shark stomachs for a matter of days (Dunning et al., 1993), limiting the time in which accumulation could occur.

Cephalopods taken by small White Sharks in this study were mainly neritic, namely Sepiidae, Octopoda, and Loliginidae (Roeleveld, 1998; Reed et al., 2005). Loliginid squids could not be identified to species, partly because the systematics of this family needs revision (Roeleveld, 1998). Nevertheless, Loliginidae are important prey of a suite of predators off the east coast, and White Sharks are occasionally seen at spawning aggregations of *Loligo vulgaris reynaudii*, but less commonly than a number of other chondrichthyan predators (Smale et al., 2001). Despite neritic cephalopods dominating the prey of small White Sharks, *Ancistrocheirus* was found in one female of 2.3 m TL, indicating that they occasionally hunt in deeper waters. This is supported by the finding of Bonfil et al. (2005) describing a 2–2.3-m TL shark moving into oceanic waters 242 km southeast of Port Elizabeth. It is also supported by recent findings of White Sharks as small as young-of-the-year ranging out to waters 240 m deep (Dewar et al., 2004), and sharks of 1.9–2.1 m TL making offshore forays and open-ocean excursions (Chapter 17, this book).

Most of the cephalopods taken by large White Sharks came from deep waters, including *Octopoteuthis*, a gelatinous squid found in mesopelagic-bathypelagic and benthic-bathyal zones in tropical and temperate waters (Nesis, 1987). Lycoteuthidae are small to medium-sized muscular squids with numerous light organs on their heads, arms, and bodies and are found in mesopleagic and benthic-bathyal zones in tropical and subtropical waters. *Ancistrocheirus lesueuri* is found in epi-meso-pelagic and benthic-bathyal waters of temperate and tropical waters (Nesis, 1987). The dominance of mesopelagic cephalopods over neritic species in large sharks suggests that they regularly hunt offshore.

The pattern of larger sharks taking more offshore prey species than smaller conspecifics is consistent with a previous study of cephalopods preyed on by Tiger and Hammerhead Sharks off KwaZulu-Natal (Smale and Cliff, 1998), and an earlier study of Blue Sharks off Looe, UK (Clarke and Stevens, 1974). More recently, Markaida and Sosa-Nishizaki (2010) found that Blue Shark prey varies between areas and that mesopelagic cephalopods were dominant prey in offshore waters, although their findings on the relationship of Blue Shark size and prey were inconclusive. Clearly, the diving depths recorded during tracking studies show that mesopelagic prey are vulnerable to predation by White Sharks and may explain some of the tracks recorded by White Sharks in offshore areas (Chapters 11 and 12, this book).

In conclusion, we believe that this preliminary study has shown that stomach content analyses can provide indirect evidence of habitat use. The results obtained here supported the hypothesis that smaller sharks had consumed more neritic prey and larger sharks took more oceanic and mesopelagic cephalopods. When additional material is obtained from areas beyond the neritic zone, greater insights into the life histories and behavior of these apex predators offshore will be obtained. Complementing this is the potential for more sophisticated ultrasonic and satellite tags to monitor parameters such as tail beat, gape opening, and stomach temperatures to directly monitor more details of their life history and ecology.

ACKNOWLEDGMENTS

Thanks to the operations staff of the KZNSB for providing the sharks and to the laboratory staff of the KZNSB for collecting and preserving the beaks. Constructive comments of two anonymous reviewers are gratefully acknowledged.

REFERENCES

Ainley, D.G., R.P. Henderson, H.R. Huber, R.J. Boekelheide, S.G. Allen, and T.L. McElroy. 1985. Dynamics of white shark/pinniped interactions in the Gulf of the Farallones. *Mem. South. Calif. Acad. Sci.* 9: 109–122.

Bonfil, R., M.P. Francis, C. Duffy, M.J. Manning, and S. O'Brien. 2010. Large-scale tropical movements and diving behavior of white sharks *Carcharodon carcharias* tagged off New Zealand. *Aquat.* Biol. 8: 115–123.

Bonfil, R., M. Meÿer, M.C. Scholl, R. Johnson, S. O'Brien, H. Oosthuizen, S. Swanson, D. Kotze, and M. Paterson. 2005. Transoceanic migration, spatial dynamics and population linkages of white sharks. *Science* 310: 100–103.

Boustany, A.M., S.F. Davis, P. Pyle, S.D. Anderson, B.J. Le Boeuf, and B.A. Block. 2002. Expanded niche for white sharks. *Nature* 415: 35–36.

Bruce, B.D., Stevens, J.D., and H. Malcolm. 2006. Movements and swimming behaviour of white sharks (*Carcharodon carcharias*) in Australian waters. *Mar. Biol.* 150: 161–172.

Clarke, M.R. (ed.). 1986. *A handbook for the identification of cephalopod beaks*. Oxford, UK: Clarendon Press.

Cliff, G., and S.F.J. Dudley. 1992. Protection against shark attack in South Africa, 1952 to 1990. *Aust. J. Mar. Freshwater Res.* 43: 263–272.

Cliff, G., S.F.J. Dudley, and B. Davis. 1989. Sharks caught in the protective gill nets off Natal, South Africa. 2. The great white shark *Carcharodon carcharias* (Linnaeus). *S. Afr. J. Mar. Sci.* 8: 131–144.

Cliff, G., S.F.J. Dudley, and M.R. Jury. 1996. Catches of white sharks in KwaZulu-Natal, South Africa and environmental influences. In *Great white sharks: The biology of Carcharodon carcharias* (pp. 351–362), A.P. Klimley and D.G. Ainley (eds.). San Diego, CA: Academic Press.

Cliff, G., L.J.V. Compagno, M.J. Smale, R.P. van der Elst, and S.P. Wintner. 2000. First records of white sharks, *Carcharodon carcharias*, from Mauritius, Zanzibar, Madagascar and Kenya. *S. Afr. J. Sci.* 96: 365–367.

Dewar, H., M. Domeier, and N. Nasby-Lucas. 2004. Insights into young of the year white shark, *Carcharodon carcharias*, behavior in the Southern California Bight. *Environ. Biol. Fishes* 70: 133–143.

Dicken, M.L. 2008. First observations of young of the year and juvenile white sharks (*Carcharodon carcharias*) scavenging from a whale carcass. *Mar. Freshwater Res.* 59: 596–602.

Domeier, M.L., and N. Nasby-Lucas. 2007. Annual re-sightings of photographically identified white sharks (*Carcharodon carcharias*) at an eastern Pacific aggregation site (Guadalupe Island, Mexico). *Mar. Biol.* 150: 977–984.

Domeier, M., and N. Nasby-Lucas. 2008. Migration patterns of white sharks *Carcharodon carcharias* tagged at Guadalupe Island, Mexico, and identification of an eastern Pacific shared offshore foraging area. *Mar. Ecol. Prog. Ser.* 370: 221–237.

Dudley, S.F.J., M.D. Anderson-Reade, G.S. Thompson, and P.B. McMullen. 2000. Concurrent scavenging off a whale carcass by great white sharks *Carcharodon carcharias* and tiger sharks *Galeocerdo cuvier*. *Fish. Bull.* 98: 646–649.

Dunning, M.C., M.R. Clarke, and C.C. Lu. 1993. Cephalopods in the diet of oceanic sharks off Eastern Australia. In *Recent Advances in Fisheries Biology* (pp. 119–131), T. Okutani, R.K. O'Dor, and T. Kubodera (eds.). Tokyo: Tokai University Press.

Estrada, J.A., A.N. Rice, L.J. Natanson, and G.B. Skomal. 2006. Use of isotopic analysis of vertebrae in reconstructing ontogenetic feeding ecology in white sharks. *Ecology* 87: 829–834.

Jorgensen, S.J., C.A. Reeb, T.K Chapple, S. Anderson, C. Perle, S.R. Van Sommeran, C. Fritz-Cope, A.C. Brown, A.P. Klimley, and B.A. Block. 2010. Philopatry and migration of Pacific white sharks. *Proc. Roy. Soc. B*. 277: 1–10.

Klimley, A.P., P. Pyle, and S.D. Anderson. 1996. The behavior of white sharks and their pinniped prey during predatory attacks. In *Great white sharks: The biology of Carcharodon carcharias* (pp. 175–191), A.P. Klimley and D.G. Ainley (eds.). San Diego, CA: Academic Press.

Laroche, K.R., A.A. Kock, L.M. Dill, and W.H. Oosthuizen. 2008. Running the gauntlet: A predator-prey game between sharks and two age classes of seals. *Anim. Behav.* 76: 1901–1917.

Long, D.J., and R.E. Jones. 1996. White shark predation and scavenging on cetaceans in the eastern north Pacific Ocean. In: *Great white sharks: The biology of Carcharodon carcharias* (pp. 293–307), A.P. Klimley and D.G. Ainley (eds.). San Diego, CA: Academic Press.

Markaida, U., and O. Sosa-Nishizaki. 2010. Food and feeding habits of the blue shark *Prionace glauca* caught off Ensenada, Baja California, Mexico, with a review on its feeding. *J. Mar. Biol. Ass. U.K.* 90: 977–994.

Martin, R.A., D.K. Rossmo, and N. Hammerschlag. 2009. Hunting patterns and geographic profiling of white shark predation. *J. Zool.* 279: 111–118.

McCosker, J.E. 1985. White shark attack behavior: Observations of and speculations about predator and prey strategies. *Mem. South. Calif. Acad. Sci.* 9: 123–135.

Nesis, K.N. 1987. *Cephalopods of the world*. Neptune City, NJ: TFH Publications.

Reed, A, P. Jereb, and C.F.E. Roper. 2005. Cuttlefishes. In *Cephalopods of the world: Chambered nautiluses and sepioids*, P. Jereb and C.F.E. Roper (eds.). *FAO Species Catalogue for Fisheries Purposes*, No. 4. Vol. 1. Rome: FAO.

Roeleveld, M.A.C. 1998. The status and importance of cephalopod systematics in southern Africa. *S. Afr. J. Mar. Sci.* 20: 1–16.

Smale, M.J., and G. Cliff. 1998. Cephalopods in the diets of four shark species (*Galeocerdo cuvier*, *Sphyrna lewini*, *S. zygaena*, and *S. mokarran*) from KwaZulu-Natal, South Africa. *S. Afr. J. Mar. Sci.* 20: 241–253.

Smale, M.J., M.R. Clarke, N.T. Klages, and M.A.C. Roeleveld. 1993. Octopod beak identification: Resolution at a regional level (Cephalopoda, Octopoda: southern Africa). *S. Afr. J. Mar. Sci.* 13: 269–293.

Smale, M.J., W.H.H. Sauer, and M.J. Roberts. 2001. Behavioural interactions of predators and spawning chokka squid off South Africa: Towards quantification. *Mar. Biol.* 139: 1095–1105.

Taylor, L. 1985. White sharks in Hawaii: Historical and contemporary records. *Mem. South. Calif. Acad. Sci.* 9: 41–48.

Tricas, T.C., and J.E. McCosker. 1984. Predatory behavior of the white shark (*Carcharodon carcharias*), with notes on its biology. *Proc. Calif. Acad. Sci.* 43: 221–238.

Weng, K. C., J.B. O'Sullivan., C.G. Lowe, C.E. Winkler, H. Dewar, and B.A. Block. 2007. Movements, behavior and habitat preferences of juvenile white sharks *Carcharodon carcharias* in the eastern Pacific. *Mar. Ecol. Prog. Ser.* 338: 211–224.

Zuffa, M., D. Van Grevelynghe, A. De Maddalena, and T. Storai. 2002. Records of the white shark, *Carcharodon carcharias* (Linnaeus, 1758) from the western Indian Ocean. *S. Afr. J. Sci.* 98: 347–349.

CHAPTER 5

Heavy Metals, Trace Elements, and Organochlorine Contaminants in Muscle and Liver Tissue of Juvenile White Sharks, *Carcharodon carcharias*, from the Southern California Bight

Christopher G. Mull*
Simon Fraser University and California State University, Long Beach

Mary E. Blasius
California State University, Long Beach

John B. O'Sullivan
Monterey Bay Aquarium

Christopher G. Lowe
California State University, Long Beach

ABSTRACT

The Southern California Bight (SCB) is thought to be a nursery ground for the White Shark (*Carcharodon carcharias*) in the eastern north Pacific. Because of their proximity to highly urbanized coastline in the SCB, juvenile White Sharks are potentially exposed to relatively high levels of trace elements, heavy metals, and organochlorines. To assess potential exposure levels, concentrations of 17 elements, 6 isomers of dichlorodiphenyltrichloroethane (DDT), 54 congeners of polychlorinated biphenyls (PCBs), and 24 chlorinated pesticides were measured in the muscle and liver tissue of juvenile White Sharks using inductively coupled plasma mass spectrometry (ICP-MS) and gas chromatography-mass spectrometry (GC-MS). Average total mercury concentrations were significantly higher in muscle (3.01 ± 0.61 μg/g wet weight) than liver (0.206 ± 0.07 μg/g wet weight) ($p < 0.01$). Mean total mercury level in muscle was six times higher than the established wildlife screening value of concern (0.5 μg/g). Lipophilic DDT, PCBs, and chlordanes were the predominant organic contaminants found in liver tissue. Levels of DDT (72.37 ± 72.86 μg/g lipid) and PCBs (17.53 ± 12.75 μg/g lipid) in the liver were the highest observed in any elasmobranch reported to date and were significantly higher than those observed in red muscle and white muscle. These levels are known to have deleterious effects in marine vertebrates, suggesting that there is the potential for physiological impairment caused by contaminant exposure.

INTRODUCTION

It is well known that long-lived, apex marine predators such as tunas, billfishes, and sharks accumulate high levels of bioavailable contaminants via the food web (Gelsleichter and Walker, 2010). Because of consumer health-risk concerns, public attention has been focused on commercially important species and the levels of contaminants in muscle; however, there is limited information on tissue-contaminant loads of nontarget species, especially for elasmobranchs. Long life spans, typically high trophic position and large lipid-rich livers make sharks particularly susceptible to contaminant uptake and accumulation. High tissue levels of contaminants have been shown to have deleterious effects on aquatic vertebrates from impairment of reproduction and growth (Cross, 1988; Hose et al., 1989) to possible immune suppression (Gelsleichter et al., 2006); however, little is known of baseline contaminant loads in elasmobranchs. With global populations of elasmobranchs in decline, understanding the extent of contaminant exposure and potential physiological effects of exposure is critical to future management and conservation.

White Sharks (*Carcharodon carcharias*) are globally distributed apex predators with a population residing in the northeastern Pacific (Jorgensen et al., 2010). The nursery grounds for this population are thought to include coastal and shelf waters ranging from Baja California, Mexico to Point Conception, CA (Klimley, 1985; Weng et al., 2007; Chapters 14, 15, and 16, this book). The nursery area in the SCB is a highly urbanized region with a legacy of contamination (Schiff et al., 2000), increasing the likelihood for sharks and their prey to be exposed to anthropogenic contaminants and xenobiotics. From the 1950s to the 1970s, large amounts of organic contaminants were introduced to the marine environment through waste-water treatment system outfalls off Los Angeles from residential and industrial sources (Lee et al., 2002). Currently, an estimated 10 metric tons of PCBs and 110 metric tons of DDT remain in the sediments off the Palos Verdes Shelf (Lee et al., 2002; Murray et al., 2002) with evidence of these contaminants now spread throughout the entire SCB (Schiff et al., 2000; Connolly and Glaser, 2002; Jarvis et al., 2007; Blasius and Goodmanlowe, 2008). The vast majority of trace-elemental contamination (i.e. arsenic, cadmium, and lead) comes from urban runoff, although a significant amount of mercury is introduced through atmospheric deposition from local power plants, oil refineries, and motor vehicles (Schiff et al., 2000).

Because of low rates of metabolic turnover and a high affinity for lipids, anthropogenic compounds such as methyl mercury and organochlorines (OC) tend to accumulate in liver tissue and can increase throughout the lifetime of an animal. Higher-trophic-level organisms with large amounts of lipid stores, such as sharks, are particularly susceptible to the bioaccumulation of lipophilic anthropogenic contaminants. Previous studies of other elasmobranchs have demonstrated accumulation of mercury (Hg), selenium (Se), and cadmium (Cd) in muscle and liver tissues with increasing body size (Storelli et al., 2003; Branco et al., 2007; McMeans et al., 2007; Endo et al., 2008). Numerous studies have also quantified OC contaminants in the lipid-rich livers of elasmobranchs (Serrano et al., 2000; Storelli and Marcotrigiano, 2001; Fisk et al., 2002; Storelli et al., 2003; Gelsleichter et al., 2005, 2006, 2008; Schlenk et al., 2005; Cornish et al., 2007; Strid et al., 2007; Gelsleichter and Walker, 2010).

Although many studies have focused on the levels of contaminants in tissues of adults, few have focused on tissue loads in juveniles or across age classes. Despite typically feeding at lower trophic levels than adults, juveniles often reside in nursery habitats, commonly coastal bays or estuaries, and are more likely to be exposed to higher concentrations of anthropogenic pollutants because of their proximity to coastal urban development (Gelsleichter and Walker, 2010). Because juvenile sharks reside and feed within these nursery habitats, they are more likely to be exposed to contaminants resulting in potential physiological impairment. Although previous research has demonstrated that juvenile teleosts may be more susceptible to impacts of contaminants because of their rapid growth compared with adults (Soffientino et al., 2010), this has yet to be quantified in elasmobranchs. The goals of our study were to (1) characterize contaminant loads in young White Sharks from the SCB and (2) determine whether bioaccumulation is occurring in young White Sharks by examining the relationships between tissue contaminant concentrations and age/size. As upper trophic-level predators, even during early life stages, White Sharks in the SCB are vulnerable to environmental and dietary contaminant exposure. This study is the first to examine levels of trace elements, heavy metals, and OCs in the muscle and livers of juvenile White Sharks in the SCB using ICP-MS and GC-MS. Baseline contaminant levels of White Sharks sampled from the SCB are compared with sharks sampled from other locations to determine the relative degree of exposure and trophic transfer.

MATERIALS AND METHODS

Sample Collection

Juvenile White Sharks were collected as incidental bycatch in Southern California White Seabass (*Atractoscion nobilis*) and California Halibut (*Paralichthys californicus*) gillnet fisheries in cooperation with the Monterey Bay Aquarium juvenile White Shark research program. All of the sharks captured alive were released and tagged if in good condition. Animals that were dead upon capture were brought back to California State University Long Beach Shark Lab for dissection and tissue collection. Information about sex, total length (TL), weight, and date of capture was recorded. Red- and white-muscle samples were collected from the axial musculature, and liver samples were collected from the middle of the left lobe. Tissue samples were collected from 19 sharks collected within the SCB (TL ranging from 173 to 283 cm) and one shark from central Oregon (TL = 330 cm). The estimated ages of individuals sampled were calculated using a von Bertalanffy growth curve for White Sharks from the northeastern Pacific population (Cailliet et al., 1985).

Elemental Analysis

The samples were prepared for analysis by ICP-MS in accordance with Environmental Protection Agency method 3052. Wet tissue samples (~500 mg) were digested in 9 ml of concentrated optima grade nitric acid (Fisher Scientific, Pittsburgh, PA) in a MARS 5 microwave reaction system (CEM Corporation, Matthews, NC) and preserved in 4% nitric acid until analysis.

The samples were then analyzed using an ELAN 6100 DRC ICP-MS (PerkinElmer, Waltham, MA). For quantification, the samples were diluted to instrument calibration range, and instrument responses (counts per second; CPS) were compared with standard curves from 0 to 500 ppb using multielemental instrument-calibration solution 2 (Spex Certiprep, Metuchen, NJ). Mercury standard curves were created using a certified mercury standard (Spex Certiprep, Metuchen, NJ) and ranged from 0 to 5 ppb. To control for sample carryover during analysis, the washout time was increased to 2 minutes, and acid blanks were run after each sample. All standards, blanks, and samples were spiked with 25 ppb of indium for use as an internal standard to account for instrument drift. All of the samples were standardized to their indium responses prior to determination of concentration.

There is an isobaric interference for arsenic (As) caused by ArCl ions created with ICP-MS. This interference was corrected for using the algorithm:

$$As^{75} = As^{75} - (Se^{77} * 3.132) + (Se^{82} * 2.736) - (Kr^{83} * 2.761) \tag{5.1}$$

Prior to analysis, the PE 6100 was tuned using a 10-ppb dilution of tuning solution 1 (Spex Certiprep, Metuchen, NJ). The instrument was tuned to achieve maximum sensitivity to Mg^{25}, In^{115}, and U^{238}, while minimizing the creation of oxides (Ce:CeO) and doubly charged species ($Ba:Ba^{++}$). For each set of 20 samples, a complete set of quality assurance/quality control was done, including duplicates, reagent blanks, blank spikes, matrix spikes, and certified reference materials. The certified reference materials DORM-2 and DOLT-3 (National Research Council Canada, Ottawa, Canada) were used to assess extraction and detection efficiency of muscle and liver, respectively. The minimum detection limits for the assay were determined using the 95% confidence interval of the reagent blanks.

Organic Analysis

To obtain samples suitable for contaminant analysis, the microwave extraction procedure of U.S. Environmental Protection Agency method 8260C was followed (U.S. Environmental Protection Agency, 2006). Approximately 1.0 g of liver or 10.0 g of muscle (red and white) was placed in a double-walled Teflon vessel, PCB and pesticide surrogate recovery standards (tetrachloro-m-xylene, PCB30, PCB112, and PCB198; AccuStandard, New Haven, CT) were added to the sample to monitor the efficiency of the extraction process (Muir and Morita, 2003), and finally 25 mL of a 3:1 dichloromethane (DCM):acetone was added. The extraction was performed in the microwave at 100°C with a heating time of 15 minutes and repeated two additional times to ensure 99.9% extraction efficiency (Sun et al., 2004). Lipid content was determined gravimetrically from split aliquots of the extracts after evaporating the DCM (Bligh and Dyer, 1959). Sample cleanup of the extracts was accomplished by elution through a chromatographic column packed with 3% water-deactivated silica gel capped with basic alumina. The target compounds were eluted from the column with 15 mL of hexane and 30 mL of 30% DCM/hexane.

The samples were concentrated by rotovap and injected using an autosampler (7683B series; Agilent Technologies, Santa Clara, CA) onto an Agilent gas chromatograph (6890N series) equipped with a mass selective detector (GC-MS; Agilent 5973 inert series). The GC column employed was a ZB-5 (Phenomenex, Torrance, CA) fused silica capillary (0.25-mm inner diameter × 60 m) with 0.25-μm film thickness. The temperature profile of the GC oven was programmed from 45 to 125°C at 20°C/min. and then to 295°C at 2.5°C/min. and held for 10 minutes. Injector and transfer line temperatures were set at 285 and 300°C, respectively. The source and quadrupole temperatures were set at 230 and 150°C, respectively. Helium was used as the carrier gas at a flow velocity of 40 cm/sec. The MS was used in the electron ionization mode and scanned from 45 to 500 atomic mass units at a rate of 1.66 scans/sec. Quantification of each target analyte was based on the largest single ion with confirmation from at least two additional ions (Cochran and Frame, 1999).

The individual chlorinated pesticides and fifty-four PCB congeners were determined against corresponding components in standards obtained from AccuStandard (New Haven, CT). Tetrabromobiphenyl was used as an internal standard and was added to the samples prior to GC-MS analysis of the extracts (Muir and Morita, 2003). Concentrations of individually resolved peaks of PCBs isomers and congeners were summed to obtain total PCB concentrations (Schulz et al., 1989). Similarly, total DDT was calculated from the sum of 2,4′-DDE, 4,4′-DDE, 2,4′-DDD, 4,4′-DDD, 2,4′-DDT, and 4,4′-DDT; total chlordanes refers to sum of heptachlor, heptachlor epozide, chlordane-α, chlordane-γ, *cis*-nonachlor, and *trans*-nonachlor; and hexachlorocyclohexane.

The GC-MS was connected to a Hewlett Packard (Intel Pentium 4) PC equipped with the software program Agilent ChemStation (Rev. D.01.02.16; June 15, 2004), which was used to integrate the various analyte peaks of interest. Continuous calibration verification standards, internal standards/surrogates, method blanks, and sample duplicates were used for quality assurance/quality control purposes. Detection limits were defined as three times the standard deviation of the background noise. Detection limits for all of the OCs analyzed in this study were in the range of 1.0 ng/g on a wet-weight basis.

Data Analysis

No corrections were made to sample values based on recovery factors because they were all in the acceptable range (70–130%). The values for duplicate samples analyzed were averaged for data summaries and statistical analyses. The values below detection limits were treated as zero values. All of the elemental analyte values for samples were expressed on a wet-weight basis (μg of analyte per g wet weight tissue), whereas organics were expressed on a lipid-weight basis (μg analyte per g of lipid). All of the analyses were carried out using Sigmaplot 11.0 (Systat Software Inc., San Jose, CA).

RESULTS

Elemental Analysis

The results for all elements analyzed according to tissue type can be seen in Table 5.1. There was no distinctive pattern of elemental distribution between red and white muscle and liver, for either essential or nonessential elements. For the majority of elements, there were no significant differences between red and white muscle, and the values were pooled for subsequent analyses, with the exception of Fe ($p < 0.0001$) and Cu ($p < 0.0001$) (Table 5.1). As ($p < 0.0001$), Zn ($p < 0.01$), and Cd ($p < 0.001$) were all significantly higher in liver than both muscle tissue types. Hg ($p < 0.0001$) and Se ($p < 0.0001$) were both significantly higher in both red and white muscle tissue than liver.

No elemental levels found in red or white muscle correlated with total length. Only concentrations of As ($r = 0.4767, p = 0.038$) and Cd ($r = 0.5452, p = 0.0158$) found in liver tissue were positively correlated with TL. Hg was not significantly correlated with total length in red muscle ($r = 0.3171$, $p = 0.186$), white muscle ($r = 0.2611, p = 0.347$), or liver ($r = 0.3968, p = 0.0925$) (Figure 5.1).

There was a nonlinear relationship between Hg concentration and Hg:Se molar ratio, with a ratio approaching 1:1 with increasing concentrations of Hg (Figure 5.2). Average ± SD Se:Hg ratios for red muscle (0.967 ± 0.72) and white muscle (1.27 ± 1.01) were significantly lower than liver ratios (24.25 ± 30.3) ($p < 0.0001$).

Organics

PCBs, DDT, and chlordane-related compounds (CHLs) were the major OCs found in the young-of-the-year (YOY) and juvenile White Sharks of the SCB relative to other contaminant groups such as the chlorinated pesticides, hexachlorobenzene, and mirex. Hexachlorohexanes, aldrin, chlorthal-dimethyl, dieldrin, perthane, endosulfan II, endrin aldehyde, endosulfan sulfate, and endrin ketone concentrations were below the detection limits in all samples. Endrin, endosulfan I, and methoxychlor were also below detection limits. DDTs were the predominant OC contaminant in YOY and juvenile White Shark liver, with a mean level of 72.37 ± 72.86 μg/g on a lipid weight (lw) basis, followed by PCBs (17.53 ± 12.75 μg/g lw) and CHLs (1.42 ± 0.88 μg/g lw) (see Table 5.3).

Mean concentrations of both PCBs and DDT in the three different tissues (liver, white muscle, and red muscle) of YOY and juvenile White Sharks were significantly different from each other ($F_{2,42} = 3.506$; $p = 0.039$ for PCBs; $F_{2,42} = 11.780$; $p < 0.001$ for DDT; Figure 5.3). The only significant difference in PCBs was between the liver tissue and red muscle, where liver had significantly higher concentrations of PCBs than red muscle (Tukey's comparisons; $p = 0.031$ for liver versus red muscle; $p = 0.390$ for liver versus white muscle; $p = 0.371$ for white muscle versus red muscle;

Table 5.1 Tissue Concentrations (μg/g ww tissue) for All Elements Measured in the Present Study

Tissue	Cr	Co	As	Sr	Ag	Cd	Sn	Ba	Hg	Pb
Nonessential										
Red muscle	1.17 (0.32–2.03)	0.007 (N.D. to 0.09)	1.63 (0.42–3.14)	0.66 (0.14–1.16)	0.204 (N.D. to 27.3)	0.005 (N.D. to 0.44)	0.058 (N.D. to 0.24)	0.562 (0.05–1.39)	3.3 (0.84–8.64)	0.045 (N.D. to 0.272)
White muscle	1.09 (0.72–2.01)	0.037 (N.D. to 0.65)	2.58 (0.69–6.61)	0.56 (0.13–3.28)	0.406 (0.009–2.73)	0.033 (N.D. to 0.63)	0.101 (N.D. to 0.82)	1.44 (0.04–16.4)	2.78 (0.41–10.3)	0.077 (N. D. to 0.93)
Liver	2.56 (0.33–3.37)	0.002 (N.D. to 0.02)	6.51 (0.68–10.9)	0.39 (0.02–0.98)	0.02 (N.D. to 0.23)	0.047 (N. D. to 0.24)	0.067 (N.D. to 0.23)	0.962 (0.02–5.24)	0.206 (0.016–0.708)	0.046 (N.D. to 0.2)
	Ca	**Mn**	**Fe**	**Ni**	**Cu**	**Zn**	**Se**			
Essential										
Red muscle	46.6 (6.43–83)	0.49 (0.15–1.48)	46.5 (11.6–82.5)	0.17 (N.D. to 1.2)	2.56 (0.78–4.6)	9.42 (1.66–17)	0.7 (0.35–1.05)			
White muscle	41.2 (8.85–137.3)	0.189 (0.03–1.2)	12.8 (3.07–34.2)	0.18 (N.D. to 1.1)	0.52 (0.18–2.9)	3.19 (1.91–7.8)	0.64 (0.34–1.19)			
Liver	27.52 (N.D. to 91.4)	0.38 (0.02–1.2)	77.2 (6.9–156.8)	0.097 (0.008–0. 27)	3.33 (0.87–5.23)	9.22 (1.15–13.8)	1.5 (0.26–4.7)			

Elements have been separated into nonessential and essential elements. Numbers represent the mean μg/g ww tissue; the numbers in parentheses represent the data range. N.D., sample concentrations below detection limits.

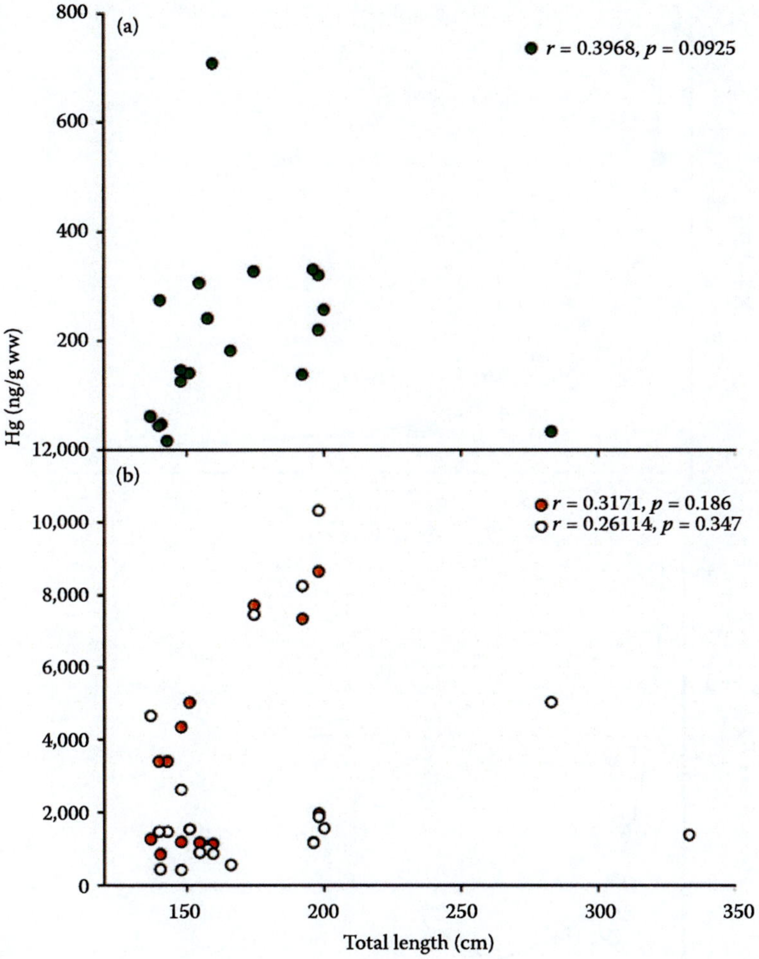

Figure 5.1 Total Hg concentrations in liver (•) (a) and red (•) and white muscle (o) (b) versus total length (cm TL).

Figure 5.3). Liver had significantly higher concentrations of DDT than both white and red muscle, but there were no significant differences between white and red muscle (Tukey's comparisons; $p < 0.001$ for liver versus red muscle; $p = 0.029$ for liver versus white muscle; $p = 0.075$ for white muscle versus red muscle; Figure 5.3).

There was no significant correlation between any of the OCs and total length in any tissue type (Figure 5.4). There was a trend toward increasing PCBs and DDT concentrations with age, with the year 1 and year 2 age classes both having higher levels than the YOYs (Figure 5.5b), although there was no significant difference among age groups.

DISCUSSION

This study is the first concurrent assessment of elemental- and organic-contaminant loads in the muscle and liver of White Sharks, and one of only a few studies in elasmobranchs that have included samples from YOY and juvenile sharks. YOY and juvenile White Sharks from the SCB exhibited

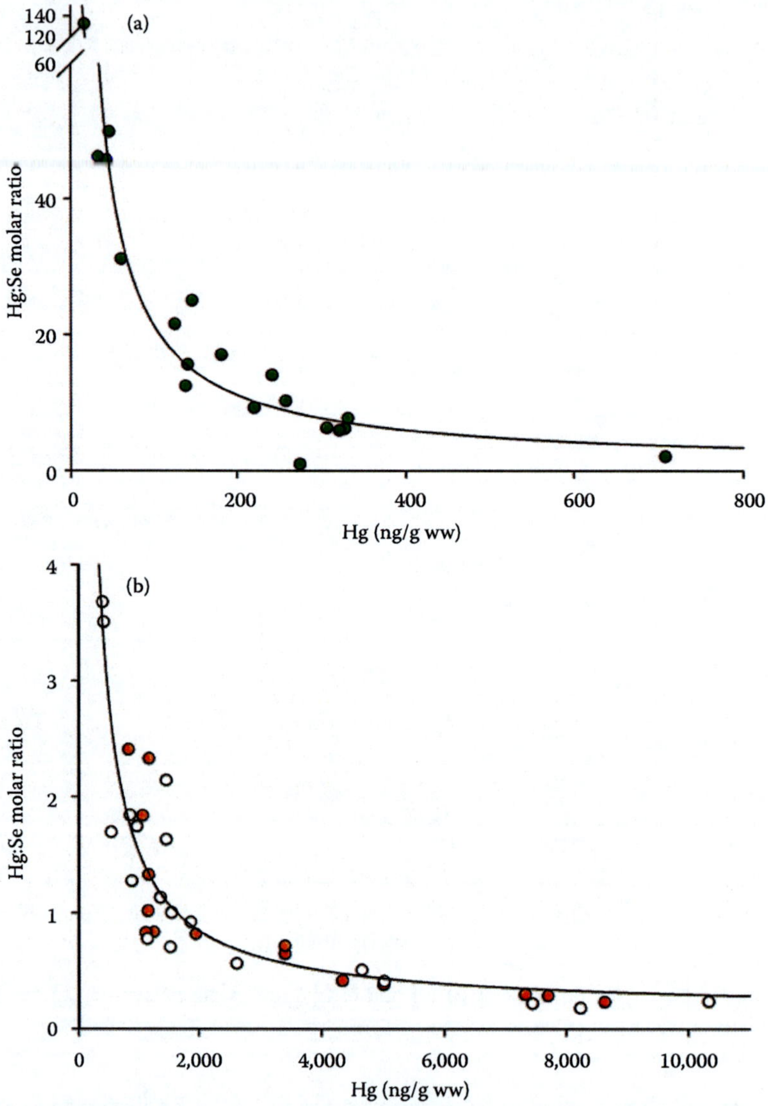

Figure 5.2 Hg:Se ratios in liver (•) (a) and red (•) and white muscle (○) (b). The solid line represents nonlinear best fits for the data.

the highest liver levels of DDT and PCBs, and the second highest level of mercury reported for any elasmobranch to date (Tables 5.2 and 5.3). These relatively high contaminant loads are especially surprising because of the young age of the sharks, whereas in most species for which comparisons are possible, data only exist for adults.

Interspecific Comparison

There was no clear pattern of elemental distribution between muscle and liver tissue, regardless of whether they were essential or nonessential elements. Of note were the relatively high concentrations of Hg in the muscle of the young sharks (Table 5.2). Young White Sharks from the SCB exhibited the highest levels of muscle Hg [3.01 μg/g wet weight (ww)] observed in any

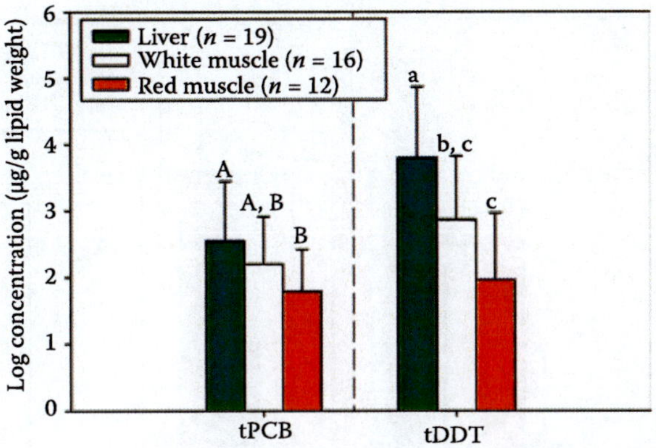

Figure 5.3 Log-transformed mean concentrations of total PCBs and total DDT for liver (green), white muscle (white), and red muscle (red). The error bars represent standard deviation, and the letters at the top of the bars denote significant grouping (α = 0.05).

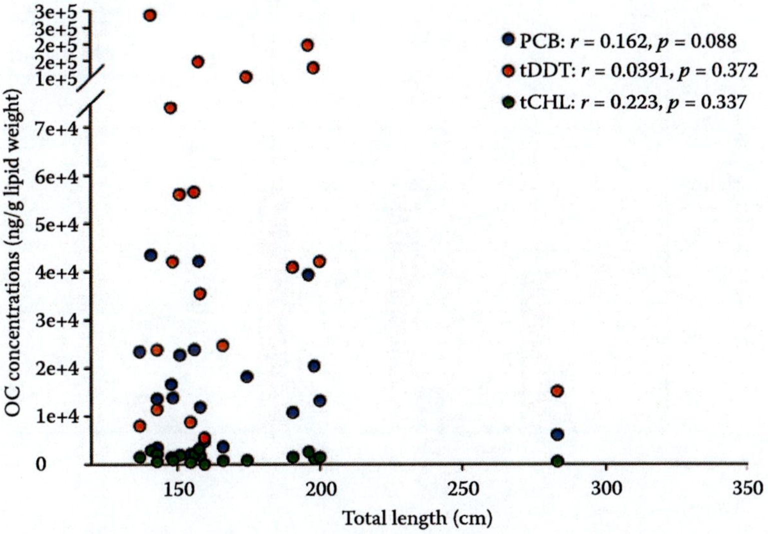

Figure 5.4 Concentrations of PCBs (•), total DDT (•), and total CHLs (•) versus total length (cm TL).

shark species reported to date outside of the Ionian Sea, a global hot spot for Hg (Storelli et al., 2003). Storelli et al. (2003) found mean levels of 12.15 μg/g ww Hg in muscle of adult Smooth Hammerhead Sharks (*Sphyrna zygaena*), whereas muscle Hg levels were below 2 μg/g ww for all of other species examined (Table 5.2). Endo et al. (2008) reported a muscle Hg concentration of 3.65 μg/g ww from a single Bull Shark (*Carcharhinus leucas*) specimen examined (Table 5.2). Liver Hg levels from young White Sharks in the SCB (0.206 μg/g ww) were lower than all other species examined with the exception of *Somniosus pacificus* (Table 5.2). Mean muscle Cd concentrations (0.02 μg/g ww) were lower in young White Sharks than adult Smooth Hammerhead Sharks, although several individuals exhibited elevated tissue levels (~0.6 μg/g ww). White Shark liver Cd concentrations (0.047 μg/g ww) were lower than all other species examined (Table 5.2). White Shark muscle Se concentrations (0.67 μg/g ww) were higher than that of Blue Sharks (*Prionace glauca*) but much lower than Smooth Hammerhead Sharks, whereas liver Se concentrations in

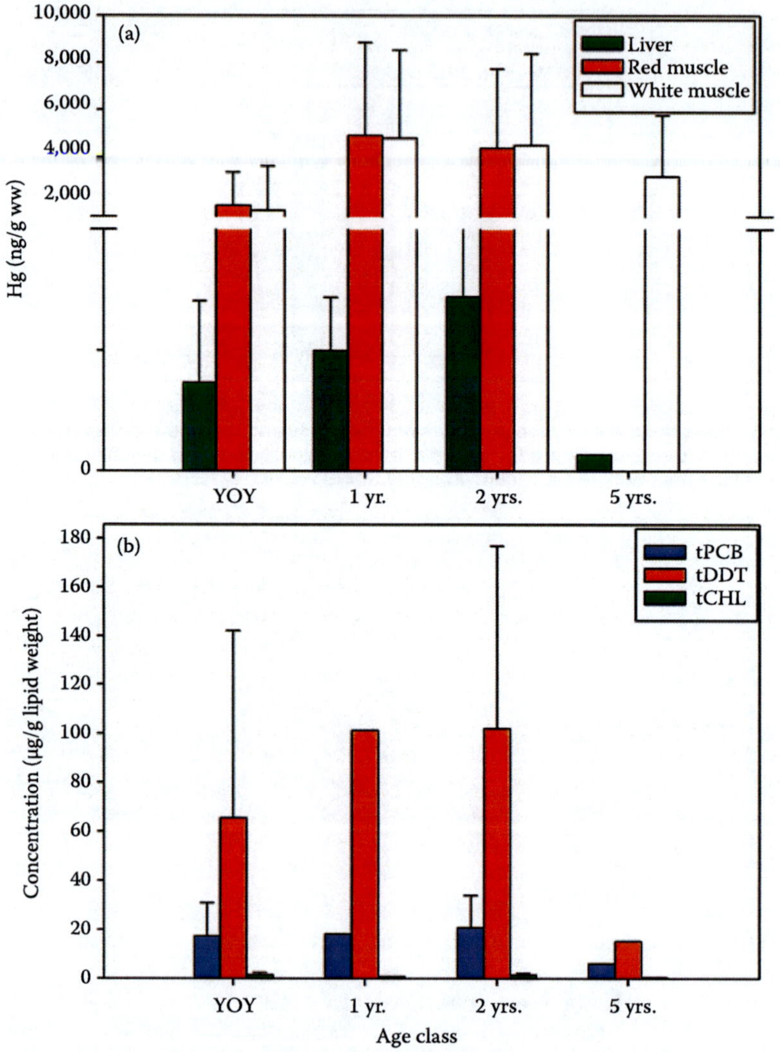

Figure 5.5 (a) Mean concentrations of total Hg by age for liver (green), red muscle (red), and white muscle (white). (b) Mean concentrations of total PCBs (blue), total DDT (red), and total CHL (green) between age classes. Ages were estimated using length at age information from Cailliet et al. (1985). Error bars represent standard deviation.

young SCB White Sharks were higher than all species examined with the exception of Smooth Hammerhead Sharks (Table 5.2).

Most studies regarding tissue loads of elemental contaminants have focused on liver (Table 5.2). Because of physiological mechanisms of detoxification, liver concentrations of nonessential elements, specifically Hg, tend to be more elevated relative to muscle tissue, although a potential ontogenetic pattern may exist. Recent evidence has suggested that methyl mercury, the more bioavailable form of Hg, may be elevated in muscle tissues because of a high affinity for sulfhydryl groups in thiol-containing amino acids, abundant in muscle proteins (Gelsleichter and Walker, 2010), although this is not clearly established across all elasmobranch species. Contrary to many studies, we observed higher concentrations of total Hg in muscle than liver in young SCB White Sharks (Table 5.1). A similar pattern was observed in Tiger Sharks (*Galeocerdo cuvier*) from Japan (Endo

Table 5.2 Tissue Concentrations (μg/g ww tissue) for Arsenic (As), Selenium (Se), Cadmium (Cd), and Mercury (Hg) from the Present Study and Previously Published Studies

					Se		Cd		Hg		
Species	**Location**	**Sample Year**	***n***	**Age Class**	**Muscle**	**Liver**	**Muscle**	**Liver**	**Muscle**	**Liver**	**Source**
Carcharodon carcharias	SCB	2005–2009	19	YOY to juvenile 137–330	0.67 (0.34–1.2)	1.5 (0.26–4.7)	0.02 (N.D. to 0.63)	0.047 (N.D. to 0.24)	3.01 (0.41–10.3)	0.206 (0.016–0.71)	Current study
Galeocerdo cuvier	Japan	2007	42	Juvenile to Adult 119–320	N.M.	N.M.	N.D.	0.14 (0.02–1.45)	0.772 (0.38–1.34)	1.16 (0.1–20.1)	Endo et al. (2008)
Carcharhinus albimarginatus	Japan	2007	8	Adult 95–129	N.M.	N.M.	N.D.	0.36 (0.02–1.4)	1.79 (1.31–2.35)	0.7 (0.26–1.41)	Endo et al., (2008)
Carcharodon leucas	Japan	2007	1	Adult 299	N.M.	N.M.	N.D.	2.97	3.65	28.1	Endo et al. (2008)
Carcharhinus plumbeus	Japan	2007	1	Adult 129	N.M.	N.M.	N.D.	0.73	1.29	3.62	Endo et al. (2008)
Prionace glauca	western Africa	2004–2005	64	Juvenile to adult 95–265	0.084–0.3	0.47–3.0	N.M.	N.M.	0.22–2.5	0.032–2.2	Branco et al. (2007)
Sphyrna zygaena	Ionian Sea	2001	4	Adult 277–303	3.24 (2.86–3.58)	8.09 (6.54–9.45)	0.03 (0.02–0.03)	19.81 (18.4–21)	12.15 (8.55–21.07)	35.89 (32.31–39.48)	Storelli et al. (2003)
Somniosus pacificus	Prince William Sound and Cumberland Sound	1999–2000	14	Adult 139–244	N.M.	0.542	N.M.	2.64	N.M.	0.122	McMeans et al. (2007)
Somniosus microcephalus	Prince William Sound and Cumberland Sound	1999–2000	24	Adult 234–322	N.M.	0.516	N.M.	3.91	N.M.	0.492	McMeans et al. (2007)

The numbers at the top represent mean values in μg/g ww tissue, whereas the values in parentheses represent the concentration ranges. N.M. denotes elements that were not measured, and N.D. denotes elements that were below the detection limit.

et al., 2008) with young sharks having higher concentrations of total Hg in the muscle than the liver, whereas an inverse relationship was seen in adult sharks. Young Tiger Sharks (119–229 cm TL) displayed liver concentrations of total Hg similar to those of our study (~200 ng/g ww in both species), although muscle concentrations of total Hg were significantly higher in young White Sharks in our study compared with young Tiger Sharks (3.01 μg/g ww versus 0.65 μg/g ww). Adult Tiger Sharks from Japan had muscle-tissue Hg concentrations of 1.01 μg/g ww but much higher concentrations in liver tissue (3.89 μg/g ww) (Endo et al., 2008) (Table 5.2). This observed pattern of total Hg distribution in young sharks may be the result of differential growth rates between juvenile and adult stages (Endo et al., 2008), with more energy being utilized for muscle development during juvenile growth. Some species of sharks possess enlarged livers at the time of birth, which they rely on for growth until individuals become proficient at feeding (Bush and Holland, 2002; Lowe, 2002; Hussey et al., 2009), and increased metabolism of liver tissue for muscle growth could contribute to the observed pattern of total Hg distribution in young White Sharks and Tiger Sharks.

Young White Sharks from the SCB exhibited high loads of DDT and PCB compared with other populations of White Sharks and other species. Average liver concentrations of DDT in YOY White Sharks from our study were 72.37 μg/g lw, 50 times higher than those observed from juvenile White Sharks from South Africa's east coast (1.50 μg/g lw; Schlenk et al., 2005). Furthermore, DDT and PCB liver concentrations in YOY White Sharks (DDT: 72.37 μg/g lw, PCB: 17.53 μg/g lw) were much greater when compared with other larger, older shark species, such as the Greenland Shark [*S. microcephalus*, DDT: 7.16 μg/g lw, PCBs: 3.44 μg/g lw (Fisk et al., 2002); PCBs: 4.40 (Strid et al., 2007)] and the Smooth Hammerhead [PCB: 4.26 μg/g lw (Storelli et al., 2003)] (Table 5.3). These different contaminant loads are likely attributed to exposure to local contaminants in each location, the degree of movement for each species, and bioaccumulation through diet. YOY White Sharks from the SCB exhibited the highest liver levels of DDT and PCBs reported for any elasmobranch to date, even when compared with much older individuals (Table 5.3).

Age/Length Trends

The majority of data used for comparative analysis of tissue elemental concentrations was only available from adult sharks of other species (Table 5.2), making the observed high levels of trace elements in young White Sharks more surprising. These types of contaminants are typically found in higher concentrations in older individuals because of bioaccumulation and are characterized by positive correlations between TL and tissue concentrations of contaminants. This pattern has not been observed in White Sharks acquired from the SCB (Figures 5.1 and 5.5a). Liver total Hg concentrations have been positively correlated with TL in several species (Endo et al., 2008; McMeans et al., 2007), and although there appeared to be an increase in both muscle and liver total Hg with TL in this study, the relationships were not significant (Figure 5.1). The lack of a clear relationship between tissue Hg concentrations and size/age is likely because of a high degree of variability in tissue total Hg concentrations measured among the smaller sharks. The largest shark analyzed in this study, a 330-cm male captured off central Oregon, exhibited relatively low levels of total Hg in white muscle (1.36 μg/g), likely because of lower levels of exposure compared with the SCB. Acquiring tissue samples from larger, older sharks may help determine to what degree bioaccumulation is occurring in White Sharks caught in the SCB.

OC concentrations in the young White Sharks in our study were not significantly related to age class or TL (Figure 5.4), although there may be an increasing trend with age (Figure 5.5). Sharks at least 1 or 2 yrs. of age had higher liver concentrations of DDT and PCBs (Figure 5.5). However, one shark estimated to be 5 yrs. old had much lower levels than all other age classes (Figure 5.5). Although age has been shown to be significantly related to OC concentrations in a number of teleosts (Pandelova et al., 2008; Bodiguel et al., 2009) and pinnipeds (Lee et al., 1996; Blasius and Goodmanlowe, 2008), it has not been observed in other shark species (Fisk et al., 2002; Schlenk et al., 2005; Strid et al.,

Table 5.3 Tissue Concentrations (μg/g lipid weight) of PCBs, DDT, and Chlorinated Pesticides (CHL) from the Present Study and Previously Published Studies

Species	Location	Sample Year	*n*	Age Class Total Length (cm)	PCBs (lw) Mean Range	DDT (lw) Mean Range	CHL (lw) Mean Range	Source
Carcharodon carcharias	SCB	2005–2009	19	YOY to juvenile 137–283	17.53 6.05–43.49	72.37 7.99–286.43	1.42 0.49–2.87	Present study
Sphyrna tiburo	Florida Gulf and Atlantic coast	N.R.	20	Juvenile to adult 54–86§	1.50 0.21–3.54	0.21 0.10–0.32	0.18 0.07–0.37	Gelsleichter et al. (2008)
Dasyatis Sabina	St. John's River and Tampa Bay, FL	N.R.	24	Adult N.R.	N.R.	0.03 0.01–0.08	0.05 0.03–0.09	Gelsleichter et al. (2006)
Chiloscyllium plagiosum	Hong Kong	2003–2004	20	N.R.	0.002* 0.001–0.005	0.001* 0.0006–0.023	<0.001*	Cornish et al. (2007)
Carcharodon carcharias	East coast of South Africa	2003	3	Juvenile 232–282	N.R.	1.50 0.60–2.63	N.R.	Schlenk et al. (2005)
Somniosus microcephalus	Iceland	2001–2003	10	Adult 350–480	4.40 0.99–10.0	N.R.	N.R.	Strid et al. (2007)
Sphyrna zygaena	Ionian Sea	2001	4	Adult 277–303	4.26* 0–17.39	N.R.	N.R.	Storelli et al. (2003)
Sphyrna tiburo	Florida estuaries	1998–2001	62	N.R.	<0.40	0.68† 0.12–0.60	0.33 0.15–0.63	Gelsleichter et al. (2005)
Somniosus microcephalus	Canadian Arctic	1999	15	Adult 283.6	3.44 N.R.	7.16 N.R.	1.82 N.R.	Fisk et al. (2002)
Centrophorus granulosus	South Adriatic Sea	1999	25	N.R. 80–90	2.25‡ 1.53–3.32	5.78‡ 4.32–7.79	N.R.	Storelli and Marcotrigiano (2001)
Squalus blainvillei	South Adriatic Sea	1999	20	N.R. 60–74	1.70‡ 0.65–3.35	2.89‡ 2.24–4.34	N.R.	Storelli and Marcotrigiano (2001)
Centroscymnus coelolepis	Canary Islands	1994–1995	13	Juvenile to adult 80.5–111	3.05§ N.R.	N.R.	N.R.	Serrano et al. (2000)

* Only wet weight reported
† $n = 46$
‡ Lipid weight concentration calculated off of reported average lipid weight
§ Averaged males and females
N.R., information not reported for a study.

2007). The lack of correlation between OC concentrations and age or length for the White Sharks reported here may be due to the limited age range, the small sample size, or possibly variability in the diets of YOY and juvenile White Sharks or where sharks are feeding (Fisk et al., 2002).

Geographical Trends

Geographical location will likely have a large effect on observed concentrations with sharks from contamination hot spots likely to exhibit elevated tissue concentrations of contaminants. White Sharks from the SCB exhibited elevated muscle levels of total Hg, second only to adult Smooth Hammerheads, Gulper Sharks (*Centrophorus granulosus*), Longnose Spurdog (*Squalus blainvilii*), and Kitefin Sharks (*Dalatias licha*) from the Ionian Sea. The Mediterranean Sea is recognized as a global hot spot for Hg contamination of marine wildlife because of urban and industrial input as well as abundant natural sources (Pirrone et al., 2003; Storelli et al., 2003). The exceptionally high levels of total Hg in both muscle and liver tissue of adult Smooth Hammerheads are likely the result of bioaccumulation in a region with high baseline levels of Hg. The SCB nursery area for young White Sharks is also impacted by industrial sources and urban runoff that have contributed trace elements and heavy metals (Schiff et al., 2000); however, environmental levels of Hg in this region are low compared with those of the Mediterranean Sea. Although regional differences are potentially important indicators of Hg concentrations, lifestyle and habitat likely determine exposure levels, with more coastal demersal species exhibiting elevated tissue concentrations compared to deep water benthic species (Table 5.2). Pelagic Blue Sharks from western Africa exhibited a wide range of liver total Hg concentrations (0.2–2.2 µg/g ww; Branco et al., 2007), whereas benthic deepwater Sleeper Sharks (*S. microcephalus* and *Somniosus pacificus*) from northern Canada exhibited relatively low liver total Hg concentrations [*S. pacificus*: 0.122 µg/g ww; *S. somniosus*: 0.492 µg/g ww (McMeans et al., 2007)].

Elevated concentrations of DDT and PCBs have been documented in a number of shark species worldwide (Storelli and Marcotrigiano, 2001; Fisk et al., 2002; Storelli et al., 2003; Schlenk et al., 2005; Gelsleichter et al., 2008); however, these levels are dwarfed in comparison with the SCB White Sharks. South Africa resumed the use of DDT in 2000 to combat malaria (van den Berg, 2009), and although the use of DDT has been banned in the United States since 1972, YOY and juvenile White Sharks from the SCB have 50 times higher levels of DDT than juvenile White Sharks from South Africa (Schlenk et al., 2005). These greatly elevated levels in SCB White Sharks are likely the results of exposure to food sources closely associated with DDT and PCBs dumped offshore in large quantities prior to the ban (Schiff et al., 2000). The Mediterranean Sea, with industrialized nations to the north and highly developed agricultural countries around the basin, is also highly impacted by OCs, specifically PCBs. Despite this high level of impact, young SCB White Sharks exhibited PCB liver concentrations 4, 7, and 10 times higher than adult Smooth Hammerheads (Storelli et al., 2003), Gulper Sharks, and Longnose Spurdog Sharks, respectively (Storelli and Macrotrigiano, 2001). The observed differences in Smooth Hammerhead, Gulper Shark, and Longnose Spurdog liver PCB concentrations are likely due to dietary differences, with the two latter squalomorph species feeding at a lower trophic level than Smooth Hammerhead Sharks. In addition, young SCB White Sharks exhibited DDT liver concentrations 12–25-fold higher than Gulper Sharks and Longnose Spurdogs, respectively (Storelli and Macrotrigiano, 2001).

Physiological Impairment

It has been suggested that Se, an essential element, can be upregulated to help offset the toxic effects of Hg, through conversion of mercuric selenide (McMeans et al., 2007). Previous studies have suggested that under "normal" conditions, Hg:Se ratios will be much greater than 1, but as Hg concentrations reach a species specific threshold of toxicity, Hg:Se ratios will approach 1:1 because of increased selenium uptake (McMeans et al., 2007). Most liver tissue exhibited relatively low levels of Hg and high Hg:Se ratios (Figure 5.2a). Muscle tissue exhibited relatively high levels of the Hg concentration and relatively low Hg:Se ratios (Figure 5.2b). This pattern suggests a physiological response to elevated tissue levels of Hg in the muscle and potentially a threshold level of ~1 μg/g ww for mercury toxicity in the White Shark, although this requires further testing to validate. The muscle total Hg concentrations of several sharks from our study exceed the threshold for sublethal effects seen in adult teleosts (~6–20 μg/g ww) (Wiener et al., 2003). Sublethal effects at these muscle-tissue concentrations include behavioral alterations, emaciation, cerebral lesions, and impaired gonadal development (Wiener et al., 2003).

Organochlorine Ratio

The extremely high concentrations of DDT deposited off the Palos Verdes Shelf provide an interesting potential marker of where young White Sharks may be feeding. Because of the elevated concentrations of DDT to PCBs off the Palos Verdes Peninsula, the ratio of these contaminants in White Sharks suggests that some individuals are acquiring food from this particular location in the SCB. Seven out of the 17 sharks examined had DDT:PCB ratios exceeding 4, an indicator of the Palos Verdes contaminant signature. This is very similar to high ratios of DDT:PCB concentrations found in the tissue of Southern California pinnipeds (Blasius and Goodmanlowe, 2008).

Conclusions

Overall, young White Sharks from the SCB exhibit high tissue levels of nonessential elemental and organic contaminants. Despite the overall high levels of contaminants, there is a high degree of variability among individuals, especially in younger sharks, and further testing will be necessary to determine the ultimate source of contaminants.

Regardless of how juvenile White Sharks are accumulating contaminants, observed levels suggest the possibility of physiological impairment, and Hg:Se analysis suggests an induced physiological response to elevated Hg concentrations. The magnitude and composition of contaminants is highly indicative of feeding within the SCB. Because of the regionally specific distribution of these contaminants, there is the potential that they could be used as indirect "environmental tags" to infer the feeding behavior and movement patterns of young sharks. Nevertheless, it is still unclear whether these high contaminant loads are causing deleterious physiological impacts that result in lower survival or future reproductive impairment, all of which may impact population maintenance.

ACKNOWLEDGMENTS

We thank Monterey Bay Aquarium for helping coordinate collections, California State University Long Beach Shark Lab, and the Institute for Integrated Research in Materials Environments and Society for use of their analytical equipment. We thank C. Winkler and local fishers for providing us with specimens and the California State University Long Beach Shark Lab White Shark rapid response team for assistance in collections and dissections of specimens.

REFERENCES

Blasius, M.E., Goodmanlowe, G.D. 2008. Contaminants still high in top-level carnivores in the Southern California Bight: Levels of DDT and PCBs in resident and transient pinnipeds. *Marine Pollution Bulletin* 56: 1973–1982.

Bligh, E.G., Dyer, W.J. 1959. A rapid method of total lipid extraction and purification. *Canadian Journal of Wildlife Diseases* 33: 438–449.

Bodiguel, X., Loizeau, V., Le Guellec, A., Roupsard, F., Philippon, X., Mellon-Duval, C. 2009. Influence of sex, maturity and reproduction on PCB an p,p′DDE concentration and repartitions in the European hake (*Merluccius merluccius* L.) from the Gulf of Lions (N.W. Mediterranean). *Science of the Total Environment* 408: 304–311.

Branco, V., Vale, C., Canário, J., Neves dos Santos, M. 2007. Mercury and selenium in blue shark (*Prionace glauca*, L. 1758) and swordfish (*Xiphias gladius*, L. 1758) from two areas of the Atlantic Ocean. *Environmental Pollution* 150: 373–380.

Bush, A., Holland, K. 2002. Food limitation in a nursery area: Daily ration in juvenile scalloped hammerhead sharks, *Sphyrna lewini*, in Kāne'ohe Bay, O'ahu, Hawai'i. *Journal of Experimental Marine Biology and Ecology* 278: 155–176.

Cailliet, G.M., Natanson, L.J., Welden, B.A., Ebert, D.A. 1985. Preliminary studies on the age and growth of the white shark, *Carcharodon carcharias*, using vertebral bands. *Memoirs of the Southern California Academy of Sciences* 9: 49–60.

Cochran, J.W., Frame G.M. 1999. Recent developments in the high-resolution gas chromatography of polychlorinated biphenyls. *Journal of Chromatography* 843: 323–368.

Connolly, J.P., Glaser, D. 2002. *p,p′*-DDE bioaccumulation in female sea lions of the California Channel Islands. *Continental Shelf Research* 22: 1059–1078.

Cornish, A.S, Ng, W.C., Ho, V.C.M., Wong, H.L., Lam, J.C.W., Lam, P. K.S., Leung, K.M.Y. 2007. Trace metals and organochlorines in the bamboo shark *Chiloscyllium plagiosum* from the southern waters of Hong Kong, China. *Science of the Total Environment* 376: 335–345.

Cross, J.N. 1988. Evidence for impaired reproduction in white croaker (*Genyonemus lineatus*) from contaminated areas off Southern California. *Marine Environmental Research* 24: 185–188.

Endo, T., Hisamichi, Y., Haraguchi, K., Kato, Y., Ohta, C., Koga, N. 2008. Hg, Zn and Cu levels in the muscle and liver of tiger sharks (*Galeocerdo cuvier*) from the coast of Ishigaki Island, Japan: Relationship between metal concentrations and body length. *Marine Pollution Bulletin* 56: 1774–1780.

Fisk, A.T., Tittlemier, S.A., Pranschke, J.L., Norstrom, R.J. 2002. Using anthropogenic contaminants and stable isotopes to assess the feeding ecology of Greenland sharks. *Ecology* 83: 2162–2172.

Gelsleichter, J., Walker, C.J. 2010. Pollutant exposure and effects in sharks and their relatives. In *Sharks and their relatives II: Biodiversity, adaptive physiology, and conservation* (pp. 491–537). Carrier, J.C., Musick, J.A., Heithaus, M.R. (eds.). Boca Raton, FL: CRC Press.

Gelsleichter, J., Manire, C.A., Szabo, N.J., Carlson, J., Lombardi-Carlson, L. 2005. Organochlorine concentrations in bonnethead sharks (*Sphyrna tiburo*) from four Florida estuaries. *Archives of Environmental Contamination and Toxicology* 48: 474–483.

Gelsleichter, J., Walsh, C.J., Szabo, N.J., Rasmussen, L.E.L. 2006. Organochlorine concentrations, physiology, and immune function in unique populations of freshwater Atlantic stingrays (*Dasyatis sabina*) from Florida's St. John's river. *Chemosphere* 63: 1506–1522.

Gelsleichter, J., Szabo, N.J., Belcher, C.N., Ulrich, G.F. 2008. Organochlorine contaminants in bonnethead sharks (*Sphyrna tiburo*) from Atlantic and Gulf estuaries on the US east coast. *Marine Pollution Bulletin* 56: 348–379.

Hose, J.E., Cross, J.N., Smith, S.G., Diehl, D. 1989. Reproductive impairment in a fish inhabiting a contaminated coastal environment off Southern California. *Environmental Pollution* 57: 139–148.

Hussey, N.E., Witner, S.P., Dudley, S.F.J., Cliff, G., Cocks, D.T., MacNeil, M.A. 2009. Maternal investment and size-specific reproductive output in carcharhinid sharks. *Journal of Animal Ecology* 79: 184–193.

Jarvis, E., Schiff, K., Sabin, L., Allen, M.J. 2007. Chlorinated hydrocarbons in pelagic forage fishes and squid of the Southern California Bight. *Environmental Toxicology and Chemistry* 26: 2290–2298.

Jorgensen, S.J., Reeb, C.K, Chapple, T.K., Anderson, S., Perle, C., Van Sommeran, S.R., Fritz-Cope, C., Brown, A.C., Klimley, A.P., Block, B.A. 2010. Philopatry and migration of pacific white sharks. *Proceedings of the Royal Society B, Biological Sciences* 277: 679–688.

Klimley, A. 1985. The areal distribution and autoecology of the white shark, *Carcharodon carcharias*, off the west coast of North America. *Memoirs of the Southern California Academy of Sciences* 9: 15–40.

Lee, J.S., Tanabe, S., Umino, H., Tatsukawa, R., Loughlin, T.R., Calkins, D.C. 1996. Persistent organochlorines in Steller sea lion (*Eumetopias jubatus*) from the bulk of Alaska and the Bering Sea, 1976–1981. *Marine Pollution Bulletin* 32: 535–544.

Lee, H.J., Sherwood, D.E., Drake, D.E., Edwards, B.D., Wong, F., Hamer, M. 2002. Spatial and temporal distribution of contaminated, effluent-affected sediment on the Palos Verdes Margin, Southern California. *Continental Shelf Research* 22: 859–880.

Lowe, C.G. 2002. Bioenergetics of free-ranging juvenile scalloped hammerhead sharks (*Sphryna lewini*) in Kaneohe Bay, Hawaii. *Journal of Experimental Marine Biology and Ecology* 278: 139–154.

McMeans, B.C., Borga, K., Bechtol, W.R., Higginbotham, D., Fisk, A.T. 2007. Essential and non-essential element concentrations in two sleeper shark species collected in arctic waters. *Environmental Pollution* 148: 281–290.

Muir, D., Morita, M. 2003. Use of analytical methods in environmental monitoring and surveillance: A critical appraisal. In *Proceedings of the Scientific and Technical Advisory Panel Workshop on the Use of Bioindicators, Biomarkers, and Analytical Methods for the Analysis of POPs in Developing Countries*, December 2003, http://www.unep.org/stap/LinkClick.aspx?fileticket=sSEnKbRwLy0%3D&tabid=3080&language=en-US.

Murray, C.J., Lee, H.J., Hampton, M.A. 2002. Geostatistical mapping of effluent-affected sediment distribution on the Palos Verdes Shelf. *Continental Shelf Research* 22: 881–897.

Pandelova, M., Henklemann, B., Simm, M., Jarv, L., Benfenati, E., Schramm, K.-W. 2008. Levels of PCDD/F and dioxin-like PCB in Baltic fish of different age and gender. *Chemospere* 71: 369–378.

Pirrone, N., Ferrara, R., Hedgecock, I.M., Kallos, G., Mamane, Y., Munthe, J., Pacyna, J.M., Pytharoulis, I., Srovieri, F., Voudouri, A., Wangberg, I. 2003. Dynamic processes of mercury over the Mediterranean region: Results from the Mediterranean Atmospheric Cycle System (MAMCS) project. *Atmospheric Environment* 37: 21–39.

Schiff, K. C., Allen, M.J., Zeng, E.Y., Bay, S.M. 2000. Southern California. *Marine Pollution Bulletin* 41: 76–93.

Schlenk, D., Sapozhnikova, Y., Cliff, G. 2005. Incidence of organochlorine pesticides in muscle and liver tissues of South African great white sharks *Carcharodon carcharias*. *Marine Pollution Bulletin* 50: 208–211.

Schulz, D.E., Petrick, G., Duinker, J.C. 1989. Complete characterization of polychlorinated biphenyl congeners in commercial arochlor and clophen mixtures by multidimensional gas chromatography-electron capture detection. *Environmental Science Technology* 23: 852–859.

Serrano, R., Fernandez, M., Rabanal, R., Hernandez, M., Gonzalez, M. 2000. Congener-specific determination of polychlorinated biphenyls in shark and grouper livers from the northwest African Atlantic Ocean. *Archives of Environmental Contamination and Toxicology* 38: 217–224.

Soffientino, B., Nacci, D.E, Specker, J.L. 2010. Effects of the dioxin-like PCB 126 on larval summer flounder (*Paralichthys dentatus*). *Comparative Biochemistry and Physiology, Part C* 152: 9–17.

Storelli, M.M., Marcotrigiano, G.O. 2001. Persistent organochlorine residues and toxic evaluation of polychlorinated biphenyls in sharks from the Mediterranean Sea (Italy). *Marine Pollution Bulletin* 42: 1323–1329.

Storelli, M.M., Ceci, E., Storelli, A., Marcotrigiano, G.O. 2003. Polychlorinated biphenyl, heavy metal and methylmercury residues in hammerhead sharks: Contaminant status and assessment. *Marine Pollution Bulletin* 46: 1035–1039.

Strid, A., Jorunsddottir, H., Papke, O., Svavrsson, J., Bergman, A. 2007. Dioxins and PCBs in Greenland shark (*Somniosus microcephalus*) from the north-east Atlantic. *Marine Pollution Bulletin* 54: 1514–1522.

Sun, Y. Takeda, M., Matsumoto, T, Oshita, K. 2004. Application of microwave-assisted extraction to the analysis of PCBs and CBzs in fly ash from municipal solid waste incinerators. *Special Instrumental Techniques and Pattern Recognition* 66: 321–327.

U.S. Environmental Protection Agency. 2006. Method 8260c: Volatile organic compounds by gas chromatography/mass pectrometry (GC/MS), August 2006, http://www.epa.gov/osw/hazard/testmethods/pdfs/8260c.pdf.

van den Berg, H. 2009. Global status of DDT and its alternatives for use in vector control to prevent disease. *Environmental Health Perspectives* 117: 1656–1663.

Weng, K. C., O'Sullivan, J.B., Lowe, C.G., Winkler, C.E., Dewar, H., Block, B.A. 2007. Movements, behavior and habitat preferences of juvenile white sharks *Carcharodon carcharias* in the eastern Pacific. *Marine Ecology-Progress Series* 338: 211–224.

Wiener, J.G., Krabbenhoft, D.P., Heinz, G.H., Scheuhammer, A.M. 2003. Ecotoxicology of mercury. In *Hanbook of ecotoxicology, second edition* (pp. 409–463). Hoffman, D.J., Rattner, B.A., Burton, G.A. Jr., Cairns, J., Jr. (eds.). Boca Raton, FL: CRC Press.

CHAPTER 6

Boat-Strike Wound Healing in *Carcharodon carcharias*

Alison Towner*
Dyer Island Conservation Trust and University of Cape Town

Malcolm J. Smale
Port Elizabeth Museum at Bayworld and Nelson Mandela Metropolitan University

Oliver Jewell
Oceans Research

ABSTRACT

Wound healing of a 2.3-m male Great White Shark is documented from Dyer Island, a known White Shark aggregation hot spot off South Africa, where several cage-diving companies operate. The injury was caused by a boat strike and resulted in a severe laceration estimated to measure 25 cm long, 30 cm wide, and 8.5 cm deep on the dorsum above the gill slits in front of the first dorsal fin. The wound was photographed over a 2-month period, while the shark remained in the vicinity of Dyer Island. The shark moved away from Dyer Island after 2 months but returned approximately 9 months later, when it was identified by its dorsal fin and body patterning by comparison with images in our extensive photographic database. The wound had closed, and a pigmented scar remained. It was subsequently sighted again in October 2010, and the healed wounds were visible but not pronounced; there was little evidence of long-term damage in that the animal swam and behaved in the normal manner for this species. Boat-strike incidents and wound healing in other species are discussed to provide context to this incident.

INTRODUCTION

The process of wound healing in sharks has received very little scientific attention, although their natural ability to heal quickly from various types of damage has been previously noted (Bird,

1978; Domeier and Nasby-Lucas, 2007). Captive elasmobranches have been observed to overcome serious injuries with relative freedom from infection (Bird, 1978; M. J. S., unpublished data).

Mechanical damage inflicted on marine animals, particularly by boat propellers, is problematic in various parts of the world. The ability to heal such injuries varies significantly between species. Studies on Manatees (*Trichechus manatus latirostris*) in Florida and Right Whales (*Eubalaena glacialis*) in the North Atlantic revealed both suffer substantially from anthropogenic mortalities. An estimated 24% of all of Florida's dead Manatees between 1993 and 2003 were killed by boat injuries (Lightsey et al., 2006). The positioning of these marine mammals' major organs along with the arrangement of major blood vessels within the subcutaneous layers of the epidermis makes cetaceans more vulnerable to fatality from boating injuries than sharks (Campbell-Malone et al., 2008). The need to understand wound healing capabilities in large marine animals is important and to date has not been documented in detail amongst Great White Sharks. This short communication describes observations of a free-ranging juvenile White Shark from serial photographs over a period of more than 2 yrs. and describes the prime response stages of healing and tissue regeneration of a serious anthropogenic wound.

MATERIALS AND METHODS

The study shark, "Prop," was hit by a recreational whale-watching vessel at approximately 15.50 p.m. on November 27, 2008 at a position of 34° 38.220 min. S, 19° 25.803 min. E in the vicinity of Gansbaai near Cape Agulhas on the south cost of South Africa (see map in Figure 6.1). The vessel was a Gemini inflatable mono-hull measuring 10.8 m in length, powered by two 300-horsepower Suzuki outboard motors. The incident was immediately reported to other operators and boaters in the area. After the strike, photographs were taken each time the shark was encountered. Detailed images of the shark and the wound were collected using a Canon 350D camera and a 32–75-mm lens. In addition, complementary subsurface footage was obtained using a Panasonic video camera in a custom-made underwater housing. Wound sizes were estimated from photographs by comparing the dimensions with adjacent features on the shark such as the height of the first dorsal fin and known reference points on the dive cage when the shark was close to it.

RESULTS

The shark was observed on nine separate occasions between December 12, 2008 and January 30, 2009; between November 1, 2009 and February 2010; and once again on October 17, 2010. The

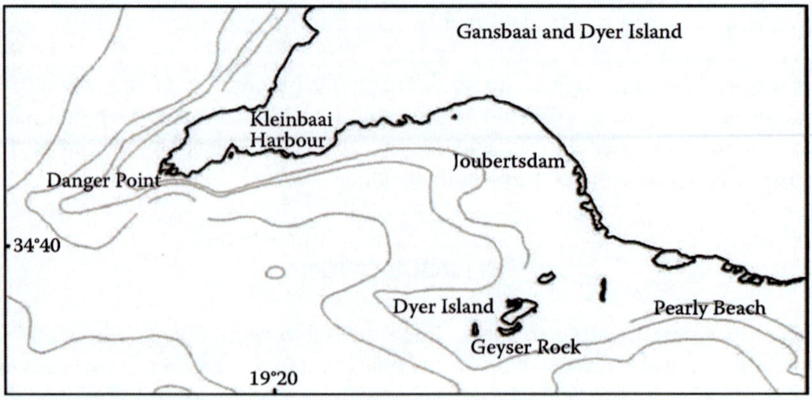

Figure 6.1 Map showing Gansbaai and Dyer Island.

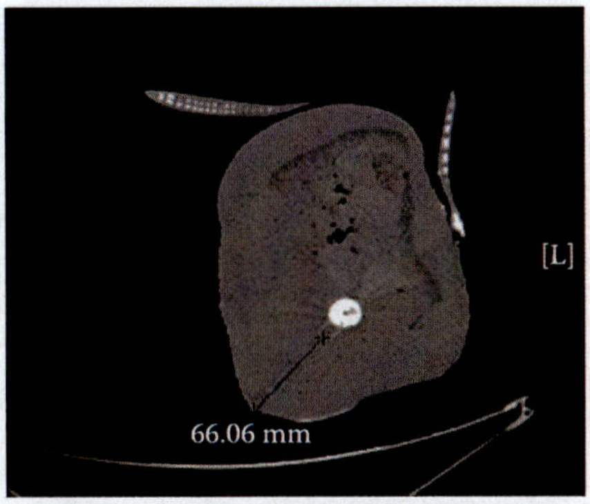

Figure 6.2 Cross-sectional scan of juvenile GWS from the Eastern Cape. The distance from neural arch to back is 6.6 cm.

boat strike occurred on the dorsum of the shark, approximately 2–3 cm in front of the first dorsal fin. Although the wound depth could not be measured directly, we estimated the size of the wound and its depth by comparing it with CAT scan images of smaller individuals revealing the distance from the spinal vertebrae to the dermis. The wound was estimated to be approximately 8.5–8.7 cm deep. The distance from the skin to the top of the neural arch of the vertebral column of a White Shark of 255 cm total length (TL) measured 9 cm at a distance 15 cm anterior to the first dorsal. A small White Shark of 145 cm had a predorsal muscle thickness of 6.6 cm (Figure 6.2).

Direct observations of the wound showed that the wound was initially narrow but deep and that it gaped after approximately a week, probably as a result of the damage to both muscle and skin. The chronology of healing is illustrated in Figure 6.3.

DISCUSSION

The observations of the shark's injury over the first month after strike suggested that the wound, which was initially a cut, went through a stage of gaping. The boat propeller clearly cut deep into the muscle, shearing its connective, tissue and disputed the skin's elasticity properties and its ability to hold together. Shark skins are intimately connected to their internal propulsion muscles through myosepta (Wainwright, 1983) and are classed in the category of "active skins." Shark skin, like that of most vertebrates, is made up of an epidermis and a dermis. In a detailed study of the mechanical behavior of shark skin by Naresh et al. (1996), the elastin distribution in the epidermis recorded in subject sharks indicated that collagen fibers are thicker in the anterior region, whereas finer fibers are more prevalent in the posterior region. The structural organization of shark skin is believed to act as a whole body "exotendon" to withstand contractions of muscles through various swimming speeds.

When any animal is injured, the wound healing process is a complex series of events that begins from the moment of injury and can continue for months to years. The physiology of the normal wound-healing trajectory has been categorized into four phases: hemostasis, inflammation, granulation, and maturation/remodeling (Naresh et al., 1996).

An observation of the White Shark made 15 days poststrike suggested that the wound was most likely in the inflammatory phase where hemostasis or blood clotting and the migration of inflammatory cells engulf the wound (Padgett et al., 2007). To accurately determine the exact stage of healing, more detailed images along with microscopy would have been required. By January 30, 2009 (63 days after the collision), we suggest the wound was most likely on its way to reaching its final remodelling phase, although the formation of repair scales was not observed in this study. Repair-scale formation most likely occurred during the sharks 9-month absence from the area when

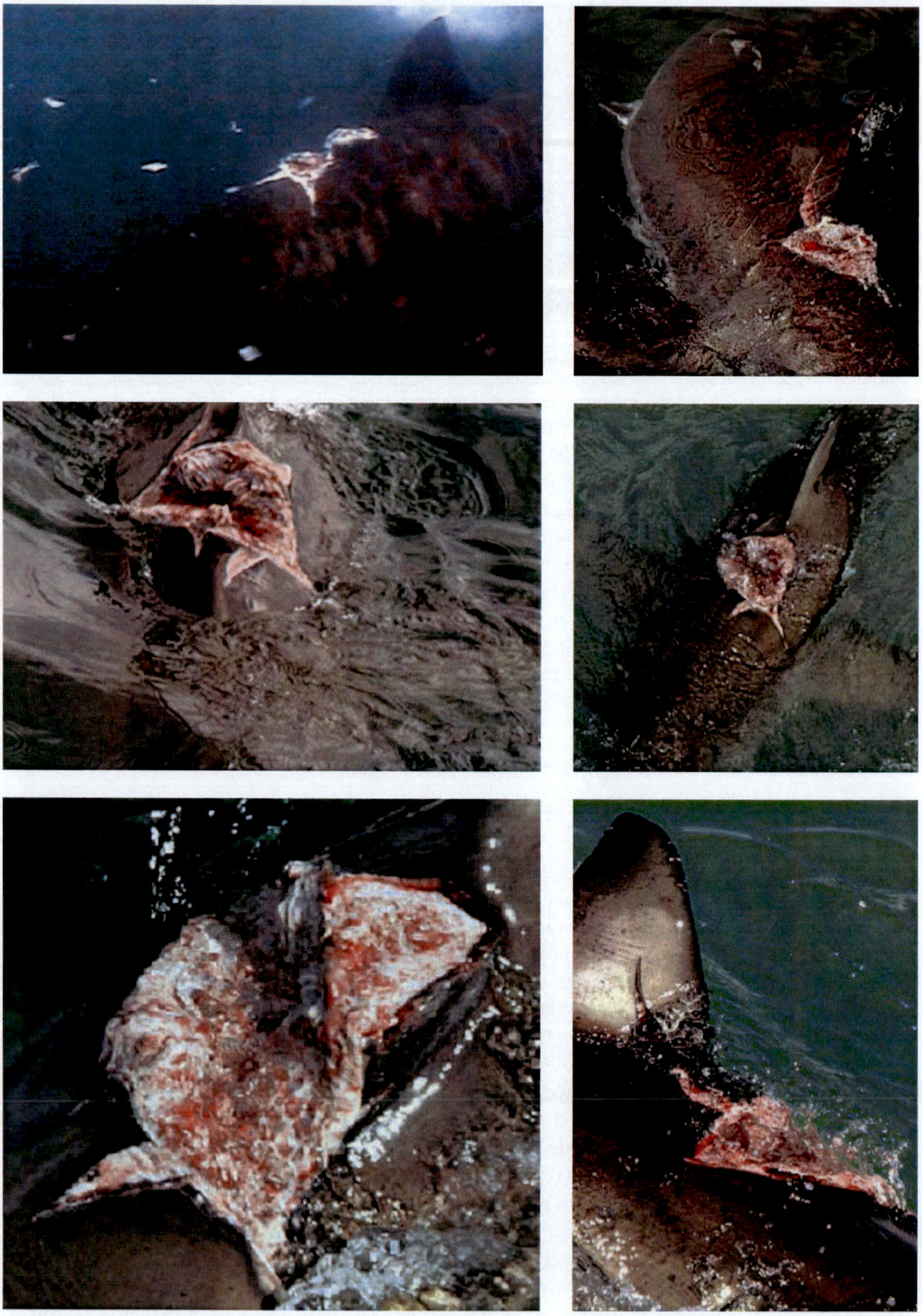

Figure 6.3 Photos illustrating the chronology of healing.

Figure 6.3 (Continued)

it presumably went about a migration. Studies on captive shark species surgically wounded reveal that scales are pulled into the wound area in the final stage of healing and larger repair scales erupt to fill in gaps at polar angles to the remaining denticles (Reif, 1978).

The behavior of the injured White Shark in this study revealed that, despite the depth and severity of the injury, there appeared to be little constraint on the shark's natural movements. The wound flexed and compressed with no observable influence on the shark's movement. This accords with previous studies because the anterior region of muscle on sharks seems to be the most elastic and damage-tolerant area (Naresh et al., 1996).

The effects of ensnaring on sharks has been described in three wild species *Carcharinus leucus* (244 cm TL), *Galcerdio cuvier* (324 cm TL), and *Carcharinus obscurus* (318 cm TL) that were caught up in plastic leashes off the coast of Florida. Wound healing of the anterior part of the body appeared to be taking place at a very rapid rate (Bird, 1978). Wounds on all three sharks started to heal over the straps within a short time period. Similarly, 43 entangled shark species were examined at the KwaZulu-Natal Sharks board labotorary off the east coast of South Africa. Plastic leashes had cut deeply into the flesh of various large species, including *C. obscurus*, *C. leucas*, *C. carcharias*, and *Carcharias taurus*. The same process of tissue healing over the straps was observed even with the presence of goose barnacles and fouling on the straps. Superficial wounds in sharks have been found to heal quickly, and foreign bodies are isolated by connective tissue. Sharks are known to have natural substances within them to enhance their immune system, making them very resistant without the complication of infection (Bird, 1978).

Livers may contribute to a substantial part of the weight of large sharks and may be used as an index of metabolic health (Springer, 1960). Small liver size was documented in all ensnared shark studies. The KZN study of 43 ensnared sharks found that the livers were approximately 19% less than expected, and the liver masses of 16 ensnared dusky sharks were 6% of body mass rather than the expected 10% for normal sharks. Even though a liver inspection was not possible in this study, it was immediately obvious on the return of the White Shark 9 months later in the November 2009 that it had decreased in body mass and appeared thinner. This would suggest that although the shark managed to survive and the wound healed, there was a biologically significant cost incurred. It may have spent more energy avoiding predators or may not have been as effective in predation as normal during its absence.

The varying degrees of response depend on the extent of the injury or the degree of damage. The propeller injury in this White Shark penetrated much deeper into the anterior muscle and resulted in a significantly larger gaping wound compared with the ensnared and surgical wounds referenced above and therefore would likely take longer to repair and heal.

CONCLUSION

This first detailed observation of a boat strike on a White Shark showed that the shark survived and was apparently little impeded by the strike. Nevertheless, the cost was apparently sufficient for it to have a lower body condition 12 months after the strike. The shark appeared thinner between the time it left Gansbaai and its return 9 months later, which was likely a consequence of the energy demands of the wound-healing process. However, the injury was close to the vertebral column, and had the neural arch been severed, the shark would clearly not have survived. The process of wound healing appeared to follow that described in other studies for other species and shows that these animals can survive despite apparently very severe injuries. Furthermore, the estimated rate at which the Great White Shark recovered from injury seems to be relatively rapid compared with other taxa. Wound-healing rates and models have been developed and applied to terrestrial animals (Alizadeh et al., 2009).

When sighted again in October 2010, the shark appeared to have normal body conformation, suggesting that it had recovered completely from the strike and was healthy. It is important to understand the wound-healing process and possibilities of survival in large marine species, particularly with increasing marine traffic in most ocean systems. Furthermore, in the scientific field, these observations stand to support the White Shark's ability to cope well overall with lacerations to their bodies, which can occur from the implication of satellite transmitters and tissue sampling.

ACKNOWLEDGMENTS

We thank the following: the Dyer Island Trust and namely Wilfred Chivell for all the funding and for allowing A. T. to attend the conference; Dr. Sandra Basson for her expertise in taking the CT scan in Figure 6.2; and Dr. Sheldon Dudley of KZN Sharks Board for providing measurements of juvenile White Sharks. Thanks to Lukas, Jaco, and Tracey Fourie of Fasttrax Marine for supplying some underwater imagery and illustration assistance. Thanks also to Steven Lamberth from Oceans and Coats for sourcing some helpful literature on the surgical healing process. Finally thanks to Professor Les Underhill at the University of Cape Town for helpful guidance.

REFERENCES

Alizadeh, A., Malihe, P., Amorallah, A., et al. 2009. The study of the effect of HESA-A on the wound healing process in rats. *Medical Journal of Islamic World Academy of Sciences* 17: 17–22.

Bird, P.M. 1978. Tissue regeneration in three carcharanid sharks encircled by embedded straps. *Copeia* 1978: 345–349.

Campbell-Malone, R., Barcom, S., Daoust, P. 2008. Gross and histologic evidence of sharp and blunt trauma in north Atlantic right whales *Eubaelena glacialis* killed by vessels. *Journal of Zoology and Wildlife Medicine* 39: 37–55.

Domeier, M., Nasby-Lucas, N. 2007. Annual re-sightings of photographically identified white sharks (*Carcharodon carcharias*) at the Eastern Pacific aggregation site (Guadeloupe, Mexico). *Marine Biology* 150: 977–984.

Lightsey, J.D., Senitel, A., Rommel, M.S., et al. 2006. Methods used during gross necropsy to determine water craft related mortality in the Florida manatee *Trichehcus manatus latirostris*. *Journal of Zoo and Wildlife Medicine* 37: 262–275.

Naresh, M.D, Arumgam, V., Sanjeevi, R. 1996. Mechanical behaviour of shark skin. *Journal of Bioscience* 22: 431–437.

Padgett, D.A., Phillip, T.M, Sheridan, J.F. 1997. Restraint stress slows subcutaneous wound healing in mice. *Brain Behaviour and Immunity* 12: 64–73.

Reif, W.E. 1978. Wound healing in sharks. *Zoomorphologie* 90: 101–111.

Springer, S. 1960. Natural history of the sandbar shark, *Eulamia milberti*. *Fishery Bulletin of the Fish and Wildlife Service* 61: 1–38.

Wainwright, S.A. 1983. To bend a fish. In Webb, P. W., Weighs, D., eds., *Fish Biomechanics* (pp. 68–91). New York: Praeger.

CHAPTER 7

A Summary of Observations on the Maximum Size Attained by the White Shark, *Carcharodon carcharias*

José I. Castro*
National Oceanic and Atmospheric Administration

Because of its large size and its record of attacks on humans, the White Shark, *Carcharodon carcharias* (Linneus, 1758), is usually described in awe-inspiring or exaggerated terms. Linné [Linneus] (1806) described it as "the most dreadful and voracious of all animals; preys on everything that comes its way, even its own tribe, and has been known to swallow a man whole." Smith (1833: 83) stated, "Of all others, the White Shark is the most terrific, and therefore regarded by mariners with peculiar dread." Jordan (1907: 195) described it as "the most voracious of fish-like animals," whereas J. L. B. Smith (1949: 49) wrote that "this swift, voracious and ferocious Shark is a terror to all who venture on or in the water."

Because of its reputation, it was predictable that the White Shark would be described in the literature, popular and scientific, as much larger than it really is. In his catalogue of the fishes in the British Museum, Günther (1870: 392) reported, under the genus *Carcharodon*, "Jaws from specimens 36 ½ feet long" presumably from "Port Fairey [Port Fairy?], Australia." Later, Jordan and Evermann (1896: 50) reported "One caught near Soquel, California, was about 30 feet long," although apparently the specimen was not measured, and there is no extant evidence of its size. For many years, a size of 36.5 ft. (11.17 m) was reported as the maximum size of the White Shark based on Günther's report. Bigelow and Schroeder (1948: 137) referred to the 36.5-ft. Australian specimen and the 30-ft. California specimen as "giants of their kind," because, as they noted, the next three largest specimens were 21, 17, and 19 ft. in length, although 20–25-footers had been reported several times.

It would not be until late in the twentieth century that the myth of gigantic White Sharks would be dispelled. Perry Gilbert was the first to question the accuracy of the 36.5-ft. size attributed to the British Museum specimen, after examining the jaws of a 5-m specimen from Durban, South Africa (Gilbert and Gilbert, 1973). Noting that the jaws of the two animals were the same size, Gilbert concluded that the 36.5 ft. (11.17 m) length attributed to the British Museum specimen was probably a printer's error, and that the correct figure should have been 16.5 ft. (5 m). Randall (1973) also examined the jaws of the British Museum specimen, comparing it with the jaws of specimens of known lengths. After comparing tooth enamel height measurements, Randall also concluded that the specimen was about 5 m long.

Once the "giants of their kind" were relegated to the mythology realm, the next largest specimen reported was the 6.4 m (21 ft) specimen from Cojímar, Cuba, taken around 1943. The measurement comes from Bigelow and Schroeder (1948), who had received it as a personal communication from Luis Howell-Rivero from Havana, Cuba. This maximum size has found its way into countless articles and books. However, this measurement also appears to be anecdotal or even erroneous.

Randall (1987) analyzed the enamel height (44 mm) of a tooth from the Cojímar specimen illustrated in Guitart and Milera (1974), and the diameter (80 mm) of a vertebra from the same shark, and determined that teeth and vertebrae of such sizes would belong to a 5-m shark. I obtained photographs of the fabled Cojímar shark from persons present at the shark's landing. These photographs depict the same shark as depicted in Guitart and Milera (1974) but with different people in the background. These photographs and the inscriptions on the photographs cast doubt on the reported size. One of the photographs is shown in Figure 7.1. A handwritten note on the back of one photograph (Figure 7.2) reads (translated from Spanish) that the "monster" weighed 7,125 lb., its liver weighed 1005 lb., and it measured 19.5 ft. (5.9 m), thus 2 ft. less than the Howell-Rivero length. The weights are the same as those given by Guitart and Milera (1974). The Randall analysis, these photographs, and the lack of specimens between the Cojímar shark and the next largest authenticated specimens do cast doubt on Howell-Rivero's measurement. A similar exaggeration occurred with a tiger shark report by Howell-Rivero published by Bigelow and Schroeder (1948: 269).

Several factors complicate the use of teeth and vertebrae comparisons in determining the size of White Sharks. Gordon Hubbell (1996), who has probably examined more White Shark jaws than anyone, stated that tooth structure is highly variable among sexually mature White Sharks. Some individuals have relatively long narrow teeth, while others have short, wide teeth (G. Hubbell, personal communication, January 2001). Vertebrae also vary greatly according to their position along the spinal column and among individuals. Thus, length calculations based on teeth or vertebrae are not sufficient to establish the size of White Sharks with certainty.

The many myths about the White Shark are difficult to dispel. Although there are many cases of White Sharks attacking and even consuming people under a variety of circumstances, the myth of

Figure 7.1 Photograph of the fabled "Cojímar monster," reputed to be 6.4 m in total length. (Courtesy of Jorge Rey.)

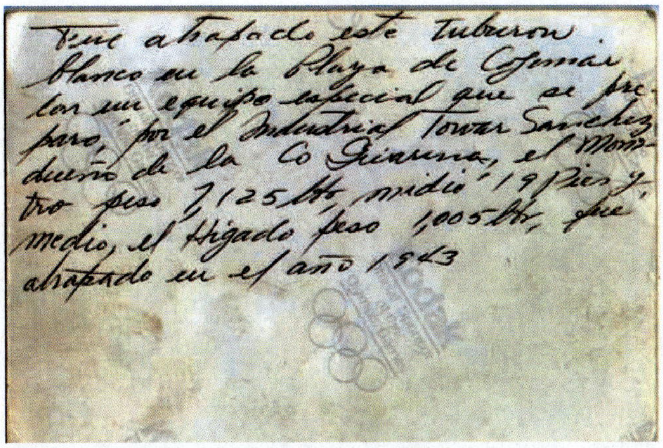

Figure 7.2 Inscription from the back of the photograph, stating that the shark measured 19.5 ft.

its ferocity is slowly being dispelled. Today there are numerous films and accounts of divers swimming with White Sharks, something that would have been considered a form of suicide a couple of decades ago. The myth of its gigantic size has proven more difficult to debunk, and the maximum size of 6.1 m is commonly cited in the literature, despite the 1987 article by Randall, perhaps because of the uncertainties cited in the preceding paragraph.

In western Mexico, White Sharks are taken incidentally in gillnet fisheries. In the fishing camps of Sonora and Baja California, I have seen dried jaws of large juvenile White Sharks (~3–4 m total length) hanging in fishermen's huts and countless heads of small juvenile White Sharks (<2 m total length) in the refuse piles that surround each camp. Unfortunately, these White Sharks are seldom examined by biologists.

The migrant fishermen (Chiapanecos) based at El Barril, Baja California, Mexico, took a very large female White Shark the summer of 2004 (Figures 7.3, 7.4, and 7.5). Unfortunately, the obviously large shark was not measured.* This specimen is of interest because it is an adult female. Although juvenile sharks are common in the Gulf of California, large adult White Sharks have not been reported from the area. The shark was quickly cut up, and its teeth and vertebrae were distributed around the local fishing community. Photographs of the shark were taken by Rancho El Barril personnel and given to me by Clark Crabbe. Several of the photographs reveal what appear to be mating bites around the pectoral fins. If these marks are indeed mating/courtship bites, it may indicate areas where White Sharks mate. One of the fishermen retained two of the shark's vertebrae. I examined and photographed both vertebrae (Figure 7.6). Both had a maximum diameter of 80 mm. Using the regression provided by Cailliet et al. (1985), this shark should have been about 5 meters long, assuming they are anterior vertebrae. If they had been tail vertebrae, the animal would have been a truly gigantic shark, much larger than indicated by the pictures.

It is unfortunate that these rare catches of large White Sharks are not reported or recorded. Posting a reward for documentation of such catch would be controversial because some could claim that it would encourage fishermen to take such sharks to claim the reward. However, dispensing plastic or corrosion-resistant measuring tapes and some nice hats or tee shirts in some of the fishing camps of Baja California could induce fishermen to measure some of their large White Shark catches, save some teeth and vertebrae, and yield some evidence of the sizes of these large sharks. Large adult White Sharks also inhabit the coast of southern California. Because the species is

* A measuring tape is not something usually found in a fishing camp. Unlike biologists, fishermen have no use for fish-length measurements. Fishermen are interested in the *weights* of the fishes they catch, as that determines what they will be paid for them.

Figure 7.3 Large female White Shark taken off El Barril, Baja California in 2004.

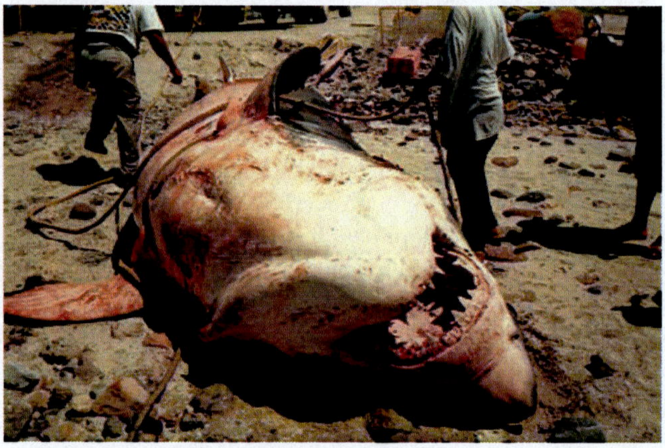

Figure 7.4 Large female White Shark taken off El Barril, Baja California in 2004. Frontal view. Note circular marks around the base of the left pectoral fins.

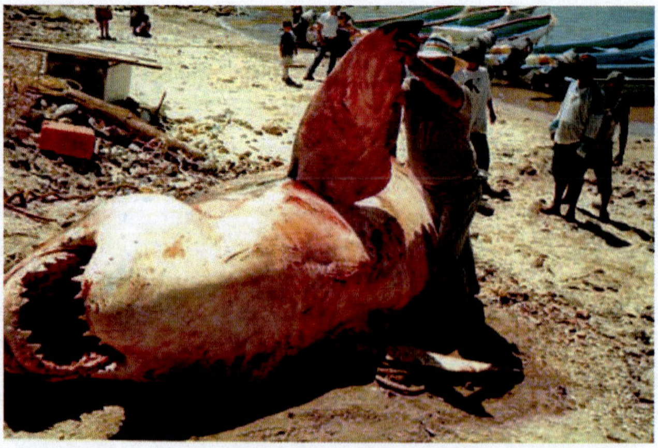

Figure 7.5 Large female White Shark taken off El Barril, Baja California in 2004. Note large size.

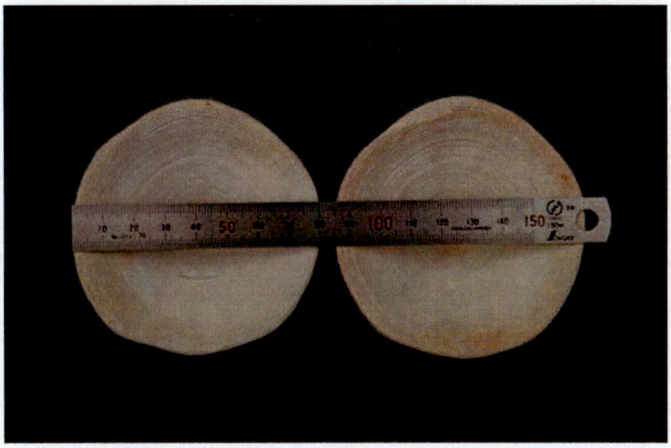

Figure 7.6 Vertebrae from the White Shark illustrated previously.

protected in California waters, it is unlikely that a large specimen is going to be taken or accurately measured there.

The largest White Shark believed to have been measured reliably is a 6-m specimen from Ledge Point, Western Australia, caught on March 22, 1984 (Randall, 1987). The jaws of this specimen are currently in Gordon Hubbell's collection in Gainesville, Florida. As Randall (1987: 168) said: "Undoubtedly *Carcharodon carcharias* exceeds 6.1 m in length, but as yet there is no authenticated record of such a size." If White Sharks measuring 6.1 m exist today, they must be specimens at the maximum of their allotted life spans. The chances of such shark surviving in coastal waters or oceans for decades, without encountering the ubiquitous gillnets or longlines, must be very low. We still await some reliable observer who can attest that he or she measured a White Shark at 6 m or larger and provide sufficient proof of the measurement.

ACKNOWLEDGMENTS

I thank Clark and Diana Crabbe for their hospitality at Rancho El Barril and for the photographs of the large shark taken there. Thanks also to Manny Puig for introducing me to Jorge Rey and for his interest in this project and to Jorge Rey for the use of the photographs of the Cojímar shark.

REFERENCES

Bigelow, H. B., and W. C. Schroeder. 1948. *Fishes of the Western North Atlantic, Lancelets, Cyclostomes, and Sharks*. Memoir No. 1, Part 1. New Haven: Sears Foundation for Marine Research, Yale University.

Cailliet, G. M., L. J. Natanson, B. A. Welden, and D. A. Ebert. 1985. Preliminary studies on the age and growth of the white shark, *Carcharodon carcharias*, using vertebral bands. *Memoirs of the Southern California Academy of Sciences* 9: 49–60.

Gilbert, P. W., and C. Gilbert. 1973. Sharks and shark deterrents. *Underwater Journal* 5: 69–79.

Guitart, D., and J. Fernandez Milera. 1974. El monstruo marino de Cojímar. *Mar y Pesca* 104: 10–11.

Günther, A. 1870. *Catalogue of the Fishes of the British Museum. Catalogue of the Physostomi, Containing the Families Gymnotidae, Symbranchidae, Muraenidae, Pegasidae, and of the Lophobranchii, Plectognathi, Dipnoi, Ganoidei, Chondropterygii, Cyclostomata, Leptocardii, in the British Museum*. Vol. 8. London: The Trustees of the British Museum.

Hubbell, G. 1996. Using tooth structure to determine the evolutionary history of the white shark. In *Great White Sharks: The Biology of Carcharodon carcharias* (pp. 9–18), edited by A. P. Klimley and D. G. Ainley. San Diego: Academic Press.

Jordan, D. S. 1907. *Fishes*. New York: Henry Holt and Company.

Jordan, D. S., and B. W. Evermann. 1896. The fishes of North and Middle America: A descriptive catalogue of the species of fish-like vertebrates found in the waters of North America, north of the Isthmus of Panama. *Bulletin of the United States National Museum*, No. 47. Washington, DC: Government Printing Office. 4 vols.

Linné [Linneus], C. 1806. *A General System of Nature*. London: Lackington, Allen, and Co. 7 vols.

Randall, J. E. 1973. Size of the great white shark (*Carcharodon*). *Science* 181: 169–170.

Randall, J. E. 1987. Refutation of lengths of 11.3, 9.0 and 6.4 m attributed to the white shark, *Carcharodon carcharias*. *California Fish and Game* 73: 163–168.

Smith, J. L. B. 1949. *The Sea Fishes of Southern Africa*. South Africa: Central News Agency, Ltd.

Smith, J. V. C. 1833. *Natural History of the Fishes of Massachusetts, Embracing a Practical Essay on Angling*. Facsimile, New York: Freshet Press, Inc., 1970.

CHAPTER 8

Investigatory Behavior toward Surface Objects and Nonconsumptive Strikes on Seabirds by White Sharks, *Carcharodon carcharias*, at Seal Island, South Africa (1997–2010)

Neil Hammerschlag*
University of Miami

R. Aidan Martin
University of British Columbia

Chris Fallows
Apex Expeditions

Ralph S. Collier
Shark Research Committee

Rob Lawrence
African Shark Eco-Charters

ABSTRACT

Investigation of any unfamiliar object carries inherent risks but can also provide significant benefits, including the discovery of new food sources. White Sharks (*Carcharodon carcharias*) combine numerous characteristics (e.g., large body, developed senses, broad prey spectrum, and few natural enemies) that should foster investigation of novel objects. Over a 13-year period (1997–2010), we observed White Shark investigatory behavior toward 26 floating objects and 61 strikes on seabirds at Seal Island in False Bay, South Africa. Attacks on seabirds were typically forceful and could have resulted from mistaken identity or been stimulated by contest competition from conspecifics. Nonconsumption of captured seabirds may also represent food rejection because of unpalatability or low fat content; however, we provide evidence that rejection of low-fat foods by White Sharks is inconsistent with quantitative studies of this species' diet and with available data on dietary thermogenesis. The adaptive value of White Sharks' observed behavior of biting objects that they do not consume is examined. We also present evidence that White Shark teeth function as mechanosensory structures, fostering tactile investigation.

INTRODUCTION

Investigatory behavior may be classified as extrinsic, as in searching for food or other resources, or intrinsic, as in examining novel objects toward no obvious goal (Berlyne, 1963). Most forms of food selection include some degree of tactile investigation (McFarland, 1987). Thus, it can be difficult to determine whether a predator's tactile exploration of objects or animals is motivated by extrinsic or intrinsic factors. Investigatory behavior is largely characterized by motivational conflict between curiosity and fear, often resulting in vacillation between approach and withdrawal, combined with tentative attempts to manipulate, "sniff," or bite the object of interest (Baerends et al., 1975; McFarland, 1987). The investigation of any unfamiliar object or creature carries inherent risks, including potential injury and (if the object proves inedible) lost foraging time; these risks increase with relative proximity and duration, respectively. However, investigation may also bear significant benefits, including the discovery of new food sources or the ability to recognize and respond appropriately to dangerous or inedible objects/creatures on subsequent encounters (Lima and Dill, 1990).

Investigatory behavior has been documented in a variety of taxa (Verbeek et al., 1994, 1996; Marchetti and Drent, 2000; Mettke-Hoffman et al., 2002; Van Oers et al., 2004), including White Sharks (*Carcharodon carcharias*) (Tricas and McCosker, 1984; Strong, 1996). White sharks are responsive to a wide variety of visual, chemical, mechanical, olfactory, and electrical stimuli, all of which may play a role in mediating investigatory behavior (Strong, 1996; Hueter et al., 2004; Yopak et al., 2007; Kajiura et al., 2010; Gardiner and Atema, 2007, 2010).

White Sharks feed on a wide variety of prey items (Compagno, 2001; Martin, 2003). However, White Sharks frequently mouth or bite objects they do not consume, such as penguins (Randall et al., 1988), sea otters (Ames and Morejohn, 1980; Ames et al., 1996), and other inedible objects (Anderson et al., 1996; Collier et al., 1996). Additionally, individual White Sharks often directly approach submerged and surface-borne objects (Miller and Collier, 1981; Tricas and McCosker, 1984; Strong, 1996; Collier, 2003), but how and why White Sharks investigate novel objects in their environment are not fully understood.

The present paper describes observations of White Shark investigatory behaviors toward various floating objects, including nonconsumptive strikes on six species of seabirds at Seal Island in False Bay, South Africa, over a 13-year period. Observations of nonconsumptive strikes on seabirds are compared with those of predatory strikes on Cape Fur Seals (*Arctocephalus pusillus pusillus*) at this site. The adaptive value of White Sharks' observed behavior of biting objects they do not consume is discussed. The relative costs and benefits to a White Shark of biting edible and nonedible objects is examined, and evidence is presented that White Shark teeth function as mechanosensory structures, fostering tactile investigation.

MATERIALS AND METHODS

Seal Island is situated at the foot of False Bay, with its long axis oriented roughly north-south (Figure 8.1). The islet measures about 400 × 50 m and is centered at latitude 34.1374°S, longitude 18.5825°E. Seal Island is inhabited by about 64,000 Cape fur seals (*Arctocephalus pusillus pusillus*), which feed 12–30+ km offshore, outside the 25-km-wide mouth of False Bay, and return to the

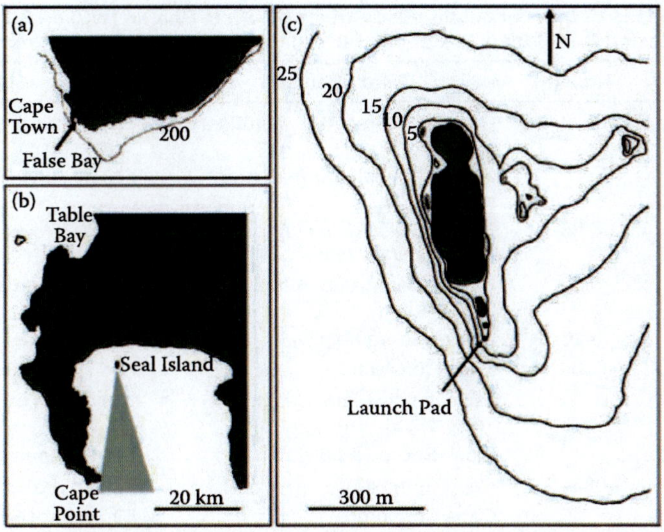

Figure 8.1 Study sites. (a) Location of False Bay, South Africa, with the 200-m depth contour indicated. (b) Location of Seal Island within False Bay, showing the primary path of pinniped movement leaving and returning to the island (gray triangle). (c) Depth contours (meters) around Seal Island.

island at irregular intervals (Martin et al., 2005). The primary pinniped entry/exit point, called the "Launch Pad," is a small craggy outcrop located off the south side of the island (Martin et al., 2005, 2009). At Seal Island, White Shark predatory attacks on Cape fur seals occur primarily between May and August and are concentrated within 2 h of sunrise, which occurs at about 7:30 a.m. during these months (Hammerschlag et al., 2006). For further site details, see Martin et al. (2005).

Between 1997 and 2010, we scanned the waters surrounding Seal Island for White Shark strikes on live, resting seabirds. When a strike on a seabird was observed, at least two observers from the research team were dedicated to guiding our observation vessel to the location of the initial strike, where we recovered and examined bite wounds on the bird. The data recorded included species of bird, date, time, and whether the bird escaped, was killed and consumed, injured and not consumed, or killed and not consumed. Seabird identification and common names follow Sinclair et al. (1993).

Whenever observed, White Shark tactile explorations of inanimate surface-borne objects were recorded. Observational data collected included type of object, date, time, and whether the object was consumed or not.

For the purpose of examining White Shark tooth morphology, we towed behind our vessel a seal-shaped decoy made of soft rubber to elicit a strike by a shark and potentially obtain a shedding tooth. Tooth terminology follows Kent (1994) and Shimada (2002). Tooth measurements follow Chandler (1995).

RESULTS

During the study period, we documented 61 nonconsumptive strikes on six species of seabirds (Table 8.1 and Figure 8.2a and b). Species (and number of times observed in parentheses) of birds attacked but not consumed were as follows: African Penguin, *Spheniscus demersus* (3); Black-Backed Kelp Gull, *Laurus dominicanus vetula* (16); Cape Cormorant, *Phalacrocorax capensis* (17); Sooty Shearwater, *Puffinus griseus* (1); Subantarctic Skua, *Catharacta Antarctica* (5); Cape Gannet, *Morus capensis* (15); unidentified bird (1); and White Breasted Cormorant, *Phalacrocorax*

Table 8.1 Descriptions of 61 Nonconsumptive Strikes by White Sharks (WS) on Six Species of Seabirds and One Consumption of a Penguin at Seal Island, South Africa (1997–2010)

Date	Time (h)	Bird Species	Remarks
July 1, 1997	0918	Cape Cormorant (*Phalacrocorax capensis*)	Shark polaris breach; bird escaped shark strike
July 7, 1999		African Penguin (*Spheniscus demersus*)	120 m; penguin killed
July 18, 1999	1235	Black-Backed Kelp Gull (*Larus dominicanus vetula*)	100 m; bird escaped strike
May 24, 2000	1050	Sooty Shearwater (*Puffinus griseus*)	Bird escaped strike
June 5, 2000	1420	Black-Backed Kelp Gull	150 m; bird escaped strike
June 21, 2000	0930	Cape Cormorant	Bird killed by shark
July 17, 2000		Subantarctic Skua (*Catharacta antarctica*)	100 m; bird escaped strike
September 3, 2000		Black-Backed Kelp Gull	Bird escaped strike
August 14, 2001	0854	Cape Cormorant	150 m; bird escaped strike
April 5, 2002		Cape Cormorant	Bird escaped strike
May 5, 2002		Black-Backed Kelp Gull	Bird escaped strike
July 21, 2002	0731	African Penguin	Kill, carcass found floating on very calm day with WS bite; shark not seen
August 20, 2002		African Penguin	150 m; bird killed and consumed (whole carcass); male shark
June 12, 2003	0927	Cape Cormorant	150 m; bird killed; carcass found with WS bite
June 13, 2003	1200	Cape Cormorant	60 m; bird escaped; shark lunged at bird at surface
June 27, 2003	903	Cape Cormorant	100 m; bird escaped; shark performed polaris breach
July 1, 2003	0923	Unidentified bird	750 m; bird escaped
July 17, 2003		Black-Backed Kelp Gull	Bird escaped; shark made single strike with upper anterior teeth; calm water, no wind, 8–10-m visibility
July 23, 2003	0959	Cape Cormorant	Dead bird found with WS bite; calm day, great visibility
July 23, 2003	1257	Black-Backed Kelp Gull	Bird escaped
July 29, 2003	1047	Cape Cormorant	Dead bird found with WS tooth marks; calm water, 16-m visibility
June 23, 2004	1044	White Breasted Cormorant (*Phalacrocorax lucidus*)	200 m; bird escaped
June 30, 2004	0902	Black-Backed Kelp Gull	200 m; shark breach; bird escaped
September 14, 2004	0932	Black-Backed Kelp Gull	250 m; shark breach; bird escaped
April 21, 2004	0732	Black-Backed Kelp Gull	200 m; shark killed bird
May 24, 2005	0732	Black-Backed Kelp Gull	Bird killed
June 16, 2005	0825	Black-Backed Kelp Gull	80 m; bird killed
June 23, 2005		White Breasted Cormorant	80 m; shark breach; bird escaped
August 23, 2005	0930	Black-Backed Kelp Gull	Bird killed by WS
September 2, 2005	0748	Cape Cormorant	100 m; shark breach; bird escaped
June 2, 2006		Cape Gannets (*Morus capensis*) (×8)	300 m; shark breach on eight birds sitting on surface; birds escaped strike
June 18, 2006	0726	Cape Gannet	600 m; polaris strike on bird

Table 8.1 (Continued) Descriptions of 61 Nonconsumptive Strikes by White Sharks (WS) on Six Species of Seabirds and One Consumption of a Penguin at Seal Island, South Africa (1997–2010)

Date	Time (h)	Bird Species	Remarks
June 23, 2006	0732	Cape Gannet	800 m; strike on bird (survived)
July 4, 2006		Cape Gannet	800 m; bird found with fatal WS bite
June, 2, 2007		Cape Cormorant group	150 m; shark surface attack on group; birds escaped
July 10, 2007		Cape Cormorant	100 m; shark breach; bird escaped
July 24, 2007	1218	Cape Cormorant	150 m; bird found with fatal WS bite
August 4, 2007		White Breasted Cormorant	100 m; dead bird found floating with fatal WS bite
August 12, 2007	0758	African Penguin	1800 m; WS strike on penguin; escaped
August 19, 2007	0722	Black-Backed Kelp Gull	50 m; WS shark fatal attack on bird
August 20, 2007	0742	Black-Backed Kelp Gull	450 m; WS fatal strike on bird
May 3, 2008		Cape Gannet	WS fatal strike on bird
May 13, 2008	0810	Cape Cormorant	100 m; WS breach; bird escaped
May 18, 2008	0830	Cape Cormorant	100 m; fatal strike
July 12, 2008	0845	Cape Cormorant	600 m; WS breach; bird escaped
July 13, 2008	0830	Subantartic Skua (×4)	400 m; WS breach; birds escaped
July 19, 2008	0823	Black-Backed Kelp Gull	30 m; WS fatal strike
July 28, 2008	0804	Black-Backed Kelp Gull	150 m; WS breach (outcome unknown)
June 29, 2009	1604	Cape Gannet	80 m; bird found with WS bite
July 21, 2009	1544	Cape Gannet	250 m; bird found with fatal WS bite
August 2, 2010	0845	Cape Cormorant	Bird found with nonfatal shark bite

The distance from the island (in meters) is noted when possible.

lucidus (3). Nonconsumptive strikes on seabirds occurred predominantly within an hour of sunrise. We also observed a single consumption of an African Penguin (Figure 8.2c).

Twenty-six instances of White Sharks investigating floating novel objects, resulting in contact, were also observed (Table 8.2). These objects ranged in shape from spherical to conical to rectangular to irregular and in length from 20 to 250 cm. Object colors included white, yellow, blue, green, and brown. These items were composed of kayaks, apple cores, oranges, cuttlebones, foam mats, brown kelp (*Laminaria* sp.), plastic bags, mesh bags, packets, towed cameras, floats, and buoys (Table 8.2 and Figure 8.2d).

Throughout the course of the study, we towed a seal decoy on 121 occasions around the Island. On September 7, 2000 at 8:56 a.m., a 3.2-m White Shark attacked our decoy, leaving a lower left anterior tooth embedded in it. The tooth was damaged, the distal blade bearing a spall fracture that was not caused by the relatively soft material of the decoy. The tooth measured 42.4 mm in slant height and 22.9 mm in enamel height. A large nutrient pore was apparent near the centre of the lingual surface of the root, around which were scattered four smaller foramina (Figure 8.3). The root of the fresh tooth bore a distinct pink blush and, from two of the small foramina extended about 4 mm of tissue that did not resemble a blood vessel in general aspect, most notably in its apparently solid cross-section and pale yellow color. Examination of cleaned and dried White Shark teeth revealed a common pattern of a large, centrally located nutritive pores surrounded by two to five scattered small foramina (Figure 8.3). We suggest that the yellow soft material removed from the root of the tooth recovered from the decoy was nerve tissue and that the small foramina scattered near the nutritive pore in each White Shark tooth are the openings through which tiny nerves pass into the living root.

Figure 8.2 Photographic examples of (a) a White Shark performing tactile investigation of a dead seabird, which it killed but did not consume; (b) recovered carcass of a Cape Cormorant and (c) African Penguin following a nonconsumptive strike by a White Shark; and (d) a White Shark tactile investigation of Brown Kelp at Seal Island, South Africa. (Courtesy of Chris Fallows, http://www.apexpredators.com.)

DISCUSSION

Why White Sharks attack animals that they do not eat is perplexing. Heithaus (2004) suggested that for ambush predators such as White Sharks, attacking a nonprey item at the surface allows sharks to gather further clues as to its identity. The large number (>5,000) of natural predations by White Sharks on Cape Fur Seals at Seal Island recorded during the study period (1997–2010) provides a useful baseline for comparison against strikes on seabirds (Martin et al., 2005, 2009; Hammerschlag et al., 2006; Chapter 9, this book). Initial strikes by White Sharks on seals were typically sudden, forceful, and devastating, lasting <2 min from initial strike to full consumption (Figure 8.4). Our observed initial strikes on seabirds were similarly forceful and brief. However, during the study period, only one seabird (an African Penguin) was observed to be consumed after the initial strike, compared with 61 nonconsumptive strikes (Figure 8.2).

Klimley (1994) and Klimley et al. (1996) hypothesized that manipulation of a potential food item in its jaws enables a White Shark to assess its fat content and proposed that this species preferentially consumes blubber-rich prey, such as juvenile Northern Elephant Seals (*Mirounga angustirostris*), to fuel its elevated metabolic needs. They further proposed that nonconsumptive strikes on sea otters, seabirds, inedible objects, and humans may represent food rejection because of inadequate energy content. In support of this claim, Pratt et al. (1982) found that White Sharks appeared to feed selectively on the blubber but not the underlying muscle layers of the floating carcass of a mysticete whale. Klimley (1987) found that White Sharks consumed offered seal carcasses but rejected sheep carcasses. Similarly, Polar Bears (*Ursus maritimus*) and Leopard Seals (*Hydrurga leptonyx*) often selectively consume only the fatty layer from phocids (D. Siniff, cited in Klimley et al., 1996). The theory that White Sharks selectively consume fat-rich prey and reject low-fat prey has been cited in numerous popular and scientific works (e.g. Burgess and Callahan, 1996; Bright, 1999; Benchley, 2000).

Table 8.2 Description of 26 Investigations of Floating Objects by White Sharks (WS) at Seal Island, South Africa (1997–2010)

Date	Time (h)	Object	Remarks
June 15, 2001		20-cm white cuttlebone	Twice approached on surface and circled object before biting
June, 27, 2001	1115	Large potato chip packet	Surfaced on object before biting
July 12, 2001		Plastic packet	Approached object on surface followed by bite
July 16, 2001	1345	Blue mat (1.5 × 2.5 m)	Left bait, resurfaced 120 m away; approached, snout poked, bit, and nibbled object
July 17, 2001	1025	Buoy	Tactile investigation
August 11, 2001		Brown Kelp	Approached, bumped, and nipped object
May 30, 2002		Mesh bag (empty)	WS surfaced and bit object
July 11, 2002		Two kayaks	Approached objects, circled, bit
July 12, 2002		Yellow buoy	Approached object, poked with snout, bit
July 18, 2002		Brown Kelp	Gaped at object
July 21, 2002		Two white apple cores, orange	WS consumed apple cores; snout poked orange
June 18, 2003	1555	Plastic packet	Approached and bit object in presence of bait
July 25, 2003	0827	Brown Kelp	Broached on and bit object; may have mistaken it for group of seals moving over object at time of strike
July 25, 2003	1121	Brown Kelp	Surfaced and bit on object
August 3, 2003	1134	Yellow buoy	Surfaced and bit on object
July 7, 2004	1256	Brown Kelp	Shark surfaced on kelp
August 16, 2004	1053	Tow camera	Shark followed camera for 15 min
August 22, 2004	1240	Tow camera	Shark followed camera
July 8, 2005	1330	Brown Kelp	Visual and tactile investigation floating kelp; two more sharks followed within the hour
July 31, 2005	0825	Brown Kelp	200 m; shark circled and contacted surface kelp
June 20, 2005		Yellow buoy	Visually inspected and consumed by 3.7-m female WS
July 7, 2009	1037	White foam float	3.2-m female WS visually inspects and consumes float
June 30, 2010		Yellow buoy	4-m female consumed float
July 4, 2010	1145	Potato chip packet	3.5-m female visually inspected and nibbled package
July 12, 2010	1057	Brown Kelp	Shark visually inspected and contacted kelp
July 15, 2010	1515–1715	Floating black plastic square (1.5 × 1.5 m)	Three sharks visually inspected and bit plastic; one shark exhibited a startle response when plastic folded because of swell

The distance from the island (in meters) is noted when possible.

However, very large White Sharks routinely consume low-fat food items, such as a 4.4-m male specimen captured off Washington state that contained, in addition to a few seal remnants, teleost remains and 150 cancrid crabs (Le Mier, 1951). In the diets of three South African White Sharks examined, Hussey et al. (Chapter 3, this book) found 300, 300, and 477 Sardines (*Sardinops sagax*) in each and suggested that these feeding events were linked to the annual sardine run off the South

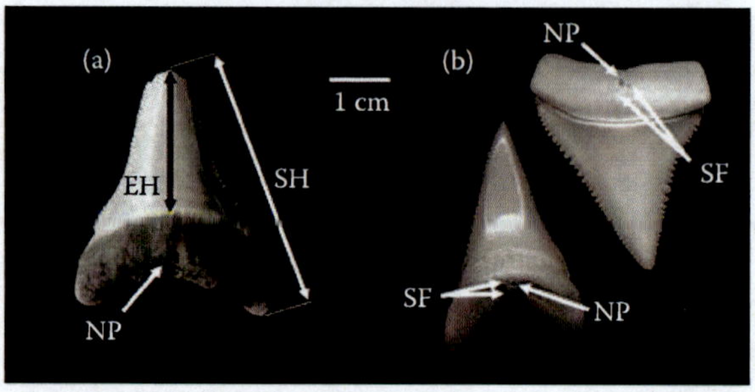

Figure 8.3 Lingual surfaces of representative White Shark teeth. (a) Lower left anterior tooth removed from a seal-shaped decoy towed at Seal Island, South Africa. (b) Upper and lower teeth from a ±4-m TL White Shark captured in Natal Sharks Board mesh nets set off Natal, South Africa in June 1986. EH, enamel height; NP, nutritive pore; SF, small foramina; SH, slant height; O, foramina from which putative nervous tissue was removed.

Figure 8.4 Attacks by White Sharks. (a and b) White Shark breach attack on a Cape Fur Seal at Seal Island in False Bay, South Africa. (c and d) Attacks by White Sharks on seals were forceful and devastating, lasting <2 min. from initial strike to full consumption. (Courtesy of Chris Fallows, http://www.apexpredators.com.)

African coast (Cliff et al., 1996; Dudley and Cliff, 2010). They also found squid, loligo squid, chiroteuthid squid, and cuttlefish in the stomachs of both small and large White Sharks. Smale and Cliff (Chapter 4, this book) repeatedly found large numbers of cephalopod beaks in the stomachs of African White Sharks.

Pop-up satellite archival transmitting tags have revealed that adult White Sharks of the northeastern Pacific occupy an entirely pelagic habitat, called the Shared Offshore Foraging Area (SOFA), for 4–6 months of the year where they appear to exhibit vertical diving behavior indicative of active foraging (Weng et al., 2007; Domeier and Nasby-Lucas, 2008; Nasby-Lucas et al., 2009;

Chapter 12, this book). However, small cetaceans and pinnipeds are markedly absent in the SOFA, suggesting that White Sharks likely feed upon nonmammalian (low-fat) prey while occupying this habitat (Chapter 12, this book).

Based on fourteen quantitative studies of White Shark feeding biology ($n = 259$ stomachs, representing all growth stages), Cortés (1999) found that almost 75% of this species' diet consists of low-fat food categories (teleosts, cartilaginous fishes, and cephalopods), whereas only slightly more than 20% consists of marine mammals. Similarly, White Shark diet studies from South Africa (Cliff et al., 1989) and South Australia (Bruce, 1992) indicate that the majority of the diet at both localities consisted of fishes. Based on stomach content analysis of 255 South African White Sharks, Hussey et al. (Chapter 3, this book) found White Sharks consumed a large number of prey items from several functional prey categories. At Gudalupe Island, Mexico, Domeier (2009) documented a White Shark preferentially feeding on tuna bait over an adjacent seal carcass. Additionally, while conducting feeding studies on captive juvenile White Sharks, Ezcurra et al. (Chapter 1, this book) observed one individual consistently feeding on energy-poor items from an array of energy-rich prey from which the shark could have selected.

So do White Sharks at Seal Island refrain from consuming seabirds that they have already killed because the birds lack the fat needed to fuel the shark's elevated body temperatures? In nutrition experiments, fat contains 9 kilocalories per gram, whereas protein contains only 4 kilocalories per gram (Whitney and Hamilton, 1984). Thus, on a per unit mass basis, fat-rich, protein-poor foods would seem more than twice as energy-rich as protein-rich, fat-poor foods. However, animals are not bomb calorimeters. Because of recent research into the claimed weight-loss benefits of high-protein, low-fat diets, precise measurements of physiological responses in humans to both food types are available (Halton and Hu, 2004). Fat and protein servings of equivalent energy content differ greatly in postprandial thermogenesis (specific dynamic action), with protein generating 200% more heat than fat within 2.5 h of consumption (Johnston et al., 2002). Another study allowing *ad libitum* food intake found that dietary thermogenesis of protein was 17% more than that of fat 5 h after consumption (Raben et al., 2003). Thus, protein is significantly more effective than fat at maintaining elevated body temperature in humans. Gut secretion and digestion in elasmobranchs is similar to that of other vertebrates, including mammals (Holmgren and Nilsson, 1999). White Shark stomach temperature is relatively constant at about 26°C (Lowe and Goldman, 2001), and intestinal valve temperatures of other lamnid sharks are 40–56% higher than their stomach temperatures (Carey et al., 1985). Because gut physiology and operating temperatures of White Sharks are similar to those of humans, it is likely that dietary thermogenesis of fats versus proteins is also similar. Therefore, we suggest that the theory that White Sharks selectively reject low-fat foods and consume fat-rich foods to facilitate maintaining elevated body temperature is inconsistent with quantitative studies of this species' diet and with the available data on dietary thermogenesis.

We suggest that most studies of White Shark feeding are biased toward over-reportage of marine mammal remains because of (1) the relative ease of sampling White Sharks near pinniped colonies or floating whale carcasses and (2) the greater durability in the gut of marine-mammal bones and fur compared with fish bones, skin, scales, or invertebrate skeletons. Accordingly, White Sharks may consume an even higher percentage of low-fat food items than present data indicate. Thus, the diet of the White Shark may be characterized best as highly euryphagous and opportunistic locally (Chapter 3, this book). In his review of shark foraging behavior, Bres (1993) found that numerous shark species shift their diet to whatever prey is locally most abundant.

In their discussion of nonconsumptive attacks by White Sharks on African Penguins, Randall et al. (1988) refrained from speculating as to why the birds were attacked or why they were not consumed. Most of the nonconsumptive seabird strikes reported in the present study occurred during winter months under conditions of moderate to high sea chop and low ambient-light levels, including periods of intense competition among White Sharks for seals (Hammerschlag et al., 2006; Martin et al., 2009). Under these environmental conditions, seabirds may have simply been attacked by White Sharks because they were mistaken for pinnipeds. It is also plausible that under high levels of competition, attacks on seabirds may have occurred to prevent conspecifics from usurping feeding opportunities. Unpalatable chemical (uropygial secretions) or tactile (feathers) cues exuded by seabirds may also lead White Sharks to reject and not consume seabird prey. Dense mats of feathers are likely also difficult to digest such that cost of consumption outweighs the benefits of a small meal. Similarly, Sea Otters (*Enhydra lutris*) have dense fur and secrete pungent oil from their anal glands (C. Harvey-Clark, personal communication). It has been suggested that White Sharks may reject sea otters as food because of their strong smell as well as their dense fur and its batting-like texture (Limbaugh, 1963).

White sharks at Seal Island investigated a wide variety of floating objects (Table 8.2 and Figure 8.4). These objects ranged in shape from spherical to rectangular to irregular; they ranged in size from 20 to 250 cm in length and were white, yellow, blue, and brown in color. These parameters do not closely resemble Cape Fur Seals or other prey known from South African White Sharks. In at least one case, a White Shark left bait apparently to investigate a novel object. At 1:45 p.m. on July 16, 2001, a 2.8-m White Shark had been orbiting the tuna bait at the stern of our observation vessel, when a sudden gust of wind blew from the bow a 1.2 × 2.5-m pale-blue foam-rubber mat onto the water surface some 120 m away. The shark immediately dived and then reappeared a minute or so later at the surface next to the mat, where it began to repeatedly bite the mat. This lasted for 25 min. until we retrieved the mat.

So why do White Sharks frequently bite novel floating objects or creatures? Investigation of unfamiliar objects by animals can result in injury, which may negatively impact future foraging and/or reproductive success (Vermeij, 1987). Thus, animals must weigh potential benefits against potential risks when deciding whether to investigate a novel object or not. White Sharks are exceptionally large among elasmobranchs and have very few natural predators (Taylor et al., 1983; Compagno, 2001). Further, White Sharks are fast, maneuverable, well armed, and protected by a tough, armored hide that heals remarkably quickly (Reif, 1978; Kemp, 1999; Compagno, 2001; Chapter 6, this book). Thus, for White Sharks, the risks of exploring novel objects or creatures may be low. Thus, we propose that if energy cost and risk of injury are minimal, a White Shark's best strategy is to bite any novel object or creature, regardless of whether or not it proves edible.

Although much work has been done on mechanoreception in sharks, including the acousticolateralis system and cutaneous mechanosensitivity (Bleckmann and Hofmann, 1999), relatively less attention has been given to studying mechanoreception in shark jaws. Roberts and Witkovsky (1975) used microelectrodes to record electrical activity in peripheral branches of the trigeminal (V) nerves caused by direct mechanical deflection of the teeth and gums in two species of carcharhinoid shark (*Scyliorhinus canicula* and *Mustelus canis*). They concluded that sensory endings located in the jaws of these sharks behave as high threshold mechanoreceptors. White Sharks are members of the order Lamniformes, which is believed to be the primitive sister taxon to the carcharhinoids (Compagno, 1999). Thus, the mechanoreceptive abilities of the teeth and gums of the carcharhinoid sharks tested by Roberts and Witkovsky (1975) are likely shared by White Sharks.

White Sharks are well known to be able to protrude their jaws with astonishing dexterity (Tricas and McCosker, 1984; Martin, 2003). Moreover, our findings suggest that nerves penetrate the roots of living White Shark teeth until they are shed (Figure 8.3). The combination of likely mechanorecptive abilities of White Shark teeth and gums, jaw protrusibility, and the innervation of the teeth and gums may provide White Sharks with tactile information from objects held in their jaws. Thus, based on our observations of White Sharks biting floating objects, we propose that teeth and gums are important sensory structures for this species. With this being said, future studies are needed to fully understand the sensory modalities employed, as well as the adaptive costs and benefits of White Shark investigation. However, White Shark jaws can do far more than gouge flesh from prey too large to swallow whole. In addition to enabling feeding, a White Shark's jaws may function as remarkably dexterous tactile sensory and manipulative structures used in investigation.

ACKNOWLEDGMENTS

For their invaluable assistance and ongoing support both in and out of the field, we thank M. Fallows, P. Jacobs, M. Bright, K. Cooper, and A. Barron. Thanks to M. Domeier, A. Gallagher, C. M. Hammerschlag, and the anonymous referees whose comments helped strengthen this chapter. For sharing data and ideas, thanks to B. Bruce of the Commonwealth Scientific and Industrial Research Organisation (CSIRO) of Australia and C. Harvey-Clark of the University of British Columbia. Thanks also to A. E. Martin, who prepared the line figures.

REFERENCES

Ames, J. A., and Morejohn, G. V. (1980). Evidence of white shark, *Carcharodon carcharias*, attacks on sea otters, *Enhydra lutris*. *California Fish and Game* 66: 196–209.

Ames, J. A., Geibel, J. J., Wendell, F. E., and Pattison, C. A. (1996). White shark-inflicted wounds of sea otters in California, 1968–1992. In *Great White Sharks: The Biology of Carcharodon carcharias* (A. P. Klimley and D. G. Ainley, eds.). San Diego, CA: Academic Press. pp. 309–316.

Anderson, S. D., Henderson, R. P., Pyle, P., and Ainley, D. G. (1996). White shark reactions to unbaited decoys. In *Great White Sharks: The Biology of Carcharodon carcharias* (A. P. Klimley and D. G. Ainley, eds.). San Diego, CA: Academic Press. pp. 223–228.

Baerends, G., Beer, C. and Manning, A. (1975). *Function and Evolution in Behavior: Essays in Honor of Professor Niko Tinbergen*. Oxford, UK: Clarendon Press.

Benchley, P. (2000). Great white sharks. *National Geographic* 197: 2–29.

Berlyne, D. E. (1963). Motivational problems raised by exploratory and epistemic behaviour. In *Psychology: A Study of a Science, Vol. 5*. (S. Koch, ed.). New York, NY: McGraw-Hill. pp. 284–364.

Bleckmann, H., and Hofmann, M. H. (1999). Special senses. In *Sharks, Skates, and Rays: The Biology of Elasmobranch Fishes* (W. C. Hamlett, ed.). Baltimore, MD: Johns Hopkins University Press. pp. 300–328.

Bres, M. (1993). The behaviour of sharks. *Reviews in Fish Biology and Fishes* 3: 133–159.

Bright, M. (1999). *The Private Life of Sharks: The Truth Behind the Myth*. London, UK: Robson Books.

Bruce, B. D. (1992). Preliminary observations on the biology of the white shark, *Carcharodon carcharias*, in South Australian waters. *Australian Journal of Marine and Freshwater Research* 43: 1–11.

Burgess, G. H., and Callahan, M. (1996). Worldwide patterns of white shark attacks on humans. In *Great White Sharks: The Biology of Carcharodon carcharias* (A. P. Klimley and D. G. Ainley, eds.). San Diego, CA: Academic Press. pp. 457–469.

Carey, F. G., Casey, J. G., Pratt, H. L., Jr., Urquhart, D., and McCosker, J. E. (1985). Temperature, heat production, and heat exchange in lamnid sharks. *Memoirs of the Southern California Academy of Sciences* 9: 92–108.

Chandler, R. (1995). *Cretaceous and Paleogene Fossils of North Carolina: A Field Guide*. Durham, NC: North Carolina Fossil Club.

Cliff, G., Dudley, S. F. J., and Davis, B. (1989). Sharks caught in the protective nets off Natal, South Africa: 2. The great white shark *Carcharodon carcharias* (Linnaeus). *South African Journal of Marine Science* 8: 131–144.

Cliff, G., Dudley, S. F. J., and Fry, M. R. (1996). Catches of white sharks in KwaZulu-Natal, South Africa and environmental influences. In *Great White Sharks: The Biology of Carcharodon carcharias* (A. P. Klimley and D. G. Ainley, eds.). San Diego, CA: Academic Press. pp. 351–362.

Collier, R. S. (2003). *Shark Attacks of the Twentieth Century from the Pacific Coast of North America*. Chatsworth, CA: Scientia Publishing.

Collier, R. S., Marks, M., and Warner, R. W. (1996). White shark attacks on inanimate objects along the Pacific Coast of North America. In *Great White Sharks: The Biology of Carcharodon carcharias* (A. P. Klimley and D. G. Ainley, eds.). San Diego, CA: Academic Press. pp. 217–222.

Compagno, L. J. V. (1999). Systematics and body form. In *Sharks, Skates, and Rays: The Biology of Elasmobranch Fishes* (W. C. Hamlett, ed.). Baltimore, MD: Johns Hopkins University Press. pp. 1–42.

Compagno, L. J. V. (2001). *Sharks of the World: An Annotated and Illustrated Catalogue of Shark Species Known to Date: Volume 2. Bullhead, mackerel, and Carpet Sharks (Heterodontiformes, Lamniformes and Orectolobiformes).* FAO species catalogue for fishery purposes, No. 1, Vol. 2. Rome: FAO.

Cortés, E. (1999). Standardized diet compositions and trophic levels of sharks. *ICES Journal of Marine Science* 56: 707–717.

Domeier, M. L. (2009). Experimental scavenging preference for the adult white shark, *Carcharodon carcharias. California Fish and Game* 95: 140–145.

Domeier, M. L., and Nasby-Lucas, N. (2008). Migration patterns of white sharks *Carcharodon carcharias* tagged at Guadalupe Island, Mexico, and identification of an eastern Pacific shared offshore foraging area. *Marine Ecology Progress Series* 370: 221–237.

Dudley, S. F. J., and Cliff, G. (2010). Influence of the annual sardine run on catches of large sharks in the protective gillnets off KwaZulu-Natal, South Africa, and the occurrence of sardine in shark diet. *African Journal of Marine Science* 32: 383–397.

Gardiner, J. M., and Atema, J. (2007). Sharks need the lateral line to locate odor sources: Rheotaxis and eddy chemotaxis. *Journal of Experimental Biology* 210: 1925–1934.

Gardiner, J. M., and Atema J. (2010). The function of bilateral odor arrival time differences in olfactory orientation of sharks. *Current Biology* 20: 1187–1191.

Halton, T. L., and Hu, F. B. (2004). The effects of high protein diets on thermogenesis, satiety and weight loss: A critical review. *Journal of the American College of Nutrition* 23: 373–385.

Hammerschlag, N., Martin, R. A., and Fallows, C. (2006). Effects of environmental conditions on predator-prey interactions between white sharks (*Carcharodon carcharias*) and Cape fur seals (*Arctocephalus pusillus pusillus*) at Seal Island, South Africa. *Environmental Biology of Fishes* 76: 341–350.

Heithaus, M. R. (2004). Predator-prey interactions. In *Biology of Sharks and Their Relatives* (J. C. Carrier, J. A. Musick, and M. R. Heithaus, eds.). Boca Raton, FL: CRC Press. pp. 487–521.

Holmgren, S., and Nilsson, S. (1999). Digestive system. In *Sharks, Skates, and Rays: The Biology of Elasmobranch Fishes* (W. C. Hamlett, ed.). Baltimore, MD: Johns Hopkins University Press. pp. 144–173.

Hueter, R. E., Mann, D. A., Maruska, K. P., Sisneros, J. A., and Demski, L. S. (2004). Sensory biology of elasmobranchs. In *Biology of Sharks and Their Relatives* (J. C. Carrier, J. A. Musick, and M. P. Heithaus, eds.). Boca Raton, FL: CRC Press. pp. 325–368.

Johnston, C. S., Day, C. S., and Swan, P. D. (2002). Postprandial thermogenesis is increased 100% on a high-protein, low-fat diet versus a high-carbohydrate, low-fate diet in healthy, young women. *Journal of the American College of Nutrition* 21: 55–61.

Kajiura, S. M., Cornett, A. D, and Yopak, K. E. (2010). Sensory adaptations to the environment: Electroreceptors as a case study. In *Biology of Sharks and Their Relatives* (J. C. Carrier, J. A. Musick, and M. P. Heithaus, eds.). Boca Raton, FL: CRC Press. pp. 393–429.

Kemp, N. E. (1999). Integumentary system and teeth. In *Sharks, Skates, and Rays: The Biology of Elasmobranch Fishes* (W. C. Hamlett, ed.). Baltimore, MD: Johns Hopkins University Press. pp. 43–68.

Kent, B. W. (1994). *Fossil Sharks of the Chesapeake Bay Region.* Columbia, MD: Egan Rees & Boyer.

Klimley, A. P. (1987). Field studies of the white shark, *Carcharodon carcharias*, in the Gulf of the Farallones National Marine Sanctuary. In *Current Research Topics in the Marine Environment* (M. Croom, ed.). San Francisco, CA: Gulf of the Farallones National Marine Sanctuary. pp. 33–36.

Klimley, A. P. (1994). The predatory behavior of the white shark. *American Scientist* 82: 122–133.

Klimley, A. P., Pyle, P., and Anderson, S. D. (1996). The behavior of white sharks and their pinniped prey during predatory attacks. In *great white sharks: The Biology of Carcharodon carcharias* (A. P. Klimley and D. G. Ainley, eds.). San Diego, CA: Academic Press. pp. 175–191.

Le Mier, E. H. (1951). Recent records of the great white shark, *Carcharodon carcharias*, on the Washington coast. *Copeia* 1951: 259.

Lima, S. L., and Dill, L. M. (1990) Behavioral decisions made under the risk of predation: A review and prospectus. *Canadian Journal of Zoology* 68: 619–640.

Limbaugh, C. (1963). Field notes on sharks. In *Sharks and Survival* (P. W. Gilbert, ed.). Boston, MA: D.C. Heath. pp. 63–94.

Lowe, C. G., and Goldman, K. J. (2001). Thermal and bioenergetics of elasmobranchs: Bridging the gap. *Environmental Biology of Fishes* 60: 251–266.

Marchetti, C., and Drent, P. J. (2000). Individual differences in the use of social information in foraging by captive great tits. *Animal Behaviour* 60: 131–140.

Martin, R. A. (2003). *Field Guide to the Great White Shark.* Vancouver, Canada: ReefQuest Centre for Shark Research. Special Publication No. 1.

Martin, R. A., Hammerschlag, N., Collier, R. S., and Fallows, C. (2005). Predatory behavior of white sharks (*Carcharodon carcharias*) at Seal Island, South Africa. *Journal of the Marine Biological Association of the United Kingdom* 85: 1121–1135.

Martin, R. A., Rossmo, D. K., and Hammerschlag, N. (2009). Hunting patterns and geographic profiling of white shark predation. *Journal of Zoology* 279: 111–118.

McFarland, D. (1987). Exploratory behaviour. In *The Oxford Companion to Animal Behaviour* (D. McFarland, ed.). Oxford, UK: Oxford University Press. p. 159.

Mettke-Hoffman, C., Winkler, H., and Leisler, B. (2002). The significance of ecological factors for exploration and neophobia in parrots. *Ethology* 108: 249–272.

Miller, D. J., and Collier, R. S. (1981). Shark attacks in California and Oregon, 1926–1979. *California Fish and Game* 67: 76–104.

Nasby-Lucas, N., Dewar, H., Lam, C. H., Goldman, K. J., and Domeier, M. L. (2009). White shark offshore habitat: A behavioral and environmental characterization of the eastern Pacific shared offshore foraging area. *PLoS ONE* 4: e8163.

Pratt, H. L., Jr., Casey, J. G., and Conklin, R. B. (1982). Observations on large white sharks, *Carcharodon carcharias*, off Long Island, New York. *Fisheries Bulletin* 80: 153–156.

Raben, A., Agerholm-Larsen, L., Flint, A., Holst, J. J., and Astrup, A. (2003). Meals with similar energy densities but rich in protein, fat, carbohydrate, or alcohol have different effects on energy expenditure and substrate metabolism but not on appetite and energy intake. *American Journal of Clinical Nutrition* 77: 91–100.

Randall, B. M., Randall, R. M., and Compagno, L. J. V. (1988). Injuries to jackass penguins (*Spheniscus demersus*): Evidence for shark involvement. *Journal of the Zoological Society of London* 214: 589–599.

Reif, W.-E. (1978). Wound healing in sharks: Form and arrangement of repair scales. *Zoomorphologie* 90: 101–111.

Roberts, B. L., and Witkovsky, P. (1975). A functional analysis of the mesencephalic nucleus of the fifth nerve of the selachian brain. *Proceedings of the Royal Society of London – B* 190: 473–495.

Shimada, K. (2002). Dental homologies in lamniform sharks (Chondrichthyes: Elasmobranchii). *Journal of Morphology* 251: 38–72.

Sinclair, I., Hockey, P., and Tarboton, P. (1993). *Birds of Southern Africa.* Cape Town, South Africa: Struik.

Strong, W. R., Jr. (1996). Shape discrimination and visual predatory tactics in white sharks. In *Great White Sharks: The Biology of Carcharodon carcharias* (A. P. Klimley and D. G. Ainley, eds.). Academic Press: San Diego, CA. pp. 229–240.

Taylor, L. R., Compagno, L. J. V., and Struhsaker, P. J. (1983). Megamouth: A new species, genus, and family of lamnoid shark (*Megachasma pelagios*, family Megachasmidae) from the Hawaiian Islands. *Proceedings of the California Academy of Sciences* 43: 87–110.

Tricas, T. C., and McCosker, J. E. (1984). Predatory behavior of the white shark (*Carcharodon carcharias*), with notes on its biology. *Proceedings of the California Academy of Science* 43: 221–238.

van Oers, K., Drent, P. J., de Goede, P., and van Noordwijk, A. J. (2004). Realized heritability and repeatability of risk-taking behaviour in relation to avian personalities. *Proceedings of the Royal Society of London – B* 271: 65–73.

Verbeek, M. E. M., Drent, P. J., and Wiepkema, P. R. (1994). Consistent individual differences in early exploratory behaviour of male great tits. *Animal Behaviour* 48: 1113–1121.

Verbeek, M. E. M., Boon, A., and Drent, P. J. (1996). Exploration, aggressive behavior and dominance in pair-wise confrontations of juvenile male great tits. *Behaviour* 133: 945–963.

Vermeij, G. J. (1987). *Evolution and Escalation: An Ecological History of Life.* Princeton, NJ: Princeton University Press.

Weng, K. C., Boustany, A. M., Pyle, P., Anderson, S. D., Brown, A., and Block, B. A. (2007). Migration and habitat of white sharks (*Carcharodon carcharias*) in the eastern Pacific Ocean. *Marine Biology* 152: 877–894.

Whitney, E. N., and Hamilton, E. M. N. (1984). *Understanding Nutrition* (3rd edition). St. Paul, MN: West Publishing.

Yopak, K. E., Lisney, T. J., Collin, S. P., and Montgomery, J. C. (2007). Variation in brain organization and cerebellar foliation in chondrichthyans: Sharks and holocephalans. *Brain Behavior and Evolution* 69: 280–300.

CHAPTER 9

Comparisons between White Shark–Pinniped Interactions at Seal Island (South Africa) with Other Sites in California

Chris Fallows
Apex Expeditions

R. Aidan Martin
University of British Columbia

Neil Hammerschlag*
University of Miami

ABSTRACT

The spatiotemporal patterns of nearly 2,600 predatory interactions between White Sharks and Cape Fur Seals at Seal Island, South Africa were studied. These data, in addition to previously reported data from this site, were compared against published patterns of White Shark–pinniped interactions at sites in California. In the present study, similarities between Seal Island and Californian sites identified included pinniped age class (juvenile), highest attack frequency versus distance from island (<400 m), tidal state (high), and depth range (5–50 m). However, many factors significantly affecting White Shark frequency and success rates of attack differed between Seal Island versus California sites, including seasonal distribution (winter versus autumn), mean frequency of attacks (6.7 per day versus 0.7 per day), prey capture success rate (48% versus 64%), size of attacking sharks (2.1–4.5 m versus 3.5–5.9 m), prey mass (low versus high), timing of attacks (frequency greatest early morning versus all day), wind direction (Northerly versus not significant), and light levels (low versus indiscriminate). The factors affecting White Shark predatory success have been reported from Seal Island, but not from the California sites. These similarities and differences are discussed in terms of prey characteristics and environmental factors at each site. Our results suggest that factors affecting White Shark predatory behavior and success rate are likely prey- and site-specific.

INTRODUCTION

Predator-prey interactions are of central importance in ecology, with important implications for population dynamics, management, and conservation (Walters, 1997; Musick, 1999). However, natural predation is rarely documented, in part because of the inherent difficulties of observing such behavior in the wild, which is especially true for predation in the marine environment (Busse, 1980; Lind and Cresswell, 1995). Thus, how marine predators select a specific prey animal from a group of apparently similar prospects is not fully known (Strong, 1991). Understanding prey selection processes of predators incorporates the antipredatory tactics of their prey. This aspect of predator-prey interaction involving elasmobranchs with mobile prey remains poorly understood (Heithaus, 2004). White Sharks (*Carcharodon carcharias*) aggregate seasonally at pinniped colonies, where they can be observed attacking seals and sea lions at the surface (Compagno, 2001). One such site is Seal Island, in False Bay, South Africa, where White Sharks actively prey upon Cape Fur Seals (*Arctocephalus pusillus pusillus*) during winter months, with up to 42 attacks recorded on a single day (Martin et al., 2005; Hammerschlag et al., 2006). The high frequency of predatory attacks on Cape Fur Seals at Seal Island provides an exceptional opportunity to examine predator-prey interactions between an apex predator and its prey.

White Shark predation on pinnipeds has been studied extensively off California, at the Farallon Islands (Klimley et al., 1992, 1996; Pyle et al., 1996) and at Año Nuevo Island (Le Boeuf and Crocker, 1996; Klimley et al., 2001). Juvenile Northern Elephant Seals (*Mirounga angustirostris*) are the principal pinniped prey at the Farallon Islands (Ainley et al., 1981, 1985) and at Año Nuevo Island (Le Boeuf and Crocker, 1996; Klimley et al., 2001). Klimley et al. (1992) found that attacks on pinnipeds at the South Farallon Islands occur during autumn (late August to early December) at similar times and locations on consecutive days. They also found that attacks were equally distributed throughout daylight hours in a "high-risk zone," concentrated near pinniped entry and exit points, between 25 and 450 m offshore at depths of 5–50 m, with a decrease in attack frequency with increasing depth. Significantly more White Shark attacks were on juvenile pinnipeds along the coast of central California (Long et al., 1996). Frequency of White Shark attacks on Northern Elephant Seals at South Farallon Islands was greatest during high tides, possibly because competition for reduced space on the island forced concentrations of seals into the water (Anderson et al., 1996). Pyle et al. (1996) found that wind direction, air temperature, barometric pressure, swell direction, and sea surface salinity showed no significant correlation with frequency of White Shark predatory attacks at South Farallon Islands; however, attack frequency increased significantly with swell height and decreased with water clarity, factors that likely affect the ability of a pinniped to detect and respond to a stalking White Shark. Pyle et al. (1996) found that the duration of predatory attacks on pinnipeds lasted more than 5 min. White Sharks at California pinniped colonies appear to spend most of their time swimming near the bottom (Goldman and Anderson, 1999). Thus, factors affecting White Shark predatory behavior have been identified based on the Farallon Island studies, but the importance of these factors have not been compared with other sites.

White Sharks are known to attack and consume a wide variety of pinnipeds (Compagno, 2001), but these powerful carnivorous prey are far from defenseless. It is not uncommon to see White Shark snouts and heads bearing patterns of four punctures or four parallel gashes, corresponding to the canine teeth or flipper nails, respectively, of pinnipeds (Martin et al., 2005; Domeier and Nasby-Lucas, 2007; see Figure 9.5). Active antipredator defense by pinnipeds has been reported in the form of juvenile and adult female and male Galapagos Sea Lions (*Zalophus californianus wollebaeki*) mobbing sharks (Barlow, 1972; Eibl-Eibesfeldt, 1984; Trillmich, 1996). Similar mobbing of White Sharks by adult male Fur Seals has been reported in Australian Fur Seals (*Arctocephalus pusillus doriferus*) and Cape Fur Seals (Stewardson and Brett, 2000; Kirkwood and Dickie, 2005). In theory, mobbing may drive off a hunting shark or alert other group members to it as a potential predator (Stewardson and Brett, 2000). Grouping can also reduce the probability that an individual pinniped is attacked, through the "selfish herd" ("dilution" or "swamping") effect, as has been described for other animals (Hamilton, 1971; Ims, 1990). Group living may also confer predator-avoidance advantage through increased overall vigilance (Pulliam and Caraco, 1984). For example, Terhune (1985) and da Silva and Terhune (1998) reported a reduction in scanning or vigilance behavior of individual Harbor Seals (*Phoca vitulina concolor*) in large groups compared with those in small groups, but overall group vigilance increased in larger groups. Le Boeuf and Crocker (1996) suggested that Northern Elephant Seals (*Mirounga angustirostris*) may reduce their vulnerability to attack by White Sharks by minimizing surface intervals.

This chapter presents spatiotemporal patterns of more than 2,600 natural predations by White Sharks on Cape Fur Seals studied at Seal Island in False Bay, South Africa and compares these patterns with those reported from California sites. Specific attention has been paid to the significance of attack locations and prey behaviors on frequency and success rate of White Shark predation on Cape Fur Seals at Seal Island. In addition to data collected in the present study, we also incorporate data from published reports at this site. Although some similarities between White Shark–pinniped interactions at Seal Island and California sites exist, factors affecting White Shark predatory behavior and success rate appear to be prey- and site-specific.

MATERIALS AND METHODS

Predator-prey interactions between White Sharks (*Carcharodon carcharias*) and Cape Fur Seals (*Arctocephalus pusillus pusillus*) were studied between 1997 and 2004 at Seal Island in False Bay, South Africa, and the site descriptions and methodologies described below are paraphrased from Martin et al. (2005) and Hammerschlag et al. (2006). The island is an elongated rocky islet at the foot of False Bay, centered at latitude 34.1374°S, longitude 18.5825°E, with its south end facing the 25-km-wide mouth of False Bay (Figure 9.1a). The bottom topography of Seal Island features a

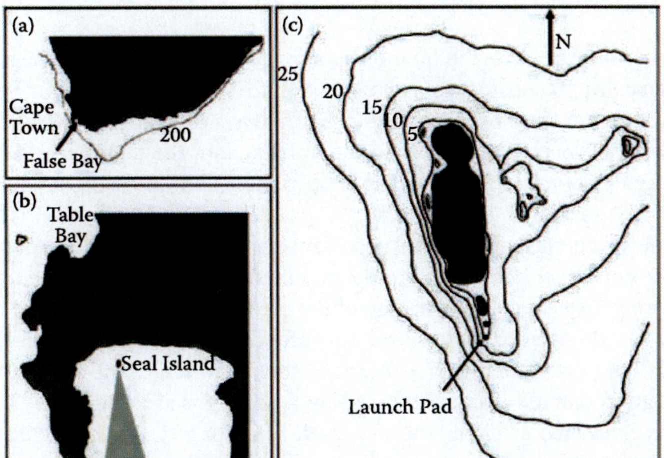

Figure 9.1 Study sites. (a) Location of False Bay, South Africa, with the 200-m depth contour displayed. (b) Location of Seal Island within False Bay, showing the main path of pinniped movement leaving and returning to the island (gray triangle). (c) Depth contours (meters) with location of the Launch Pad indicated.

steep drop-off along most of the western side of the islet, where the water depth reaches 20 m within 50 m of the island and a wide, shallow shelf along the northeastern side, where the water does not reach similar depths until 400 m or more from the island (Hammerschlag et al., 2006). Seal Island is inhabited by over 60,000 Cape Fur Seals, which typically leave the island in coordinated groups of 5–20 to feed in False Bay or up to 12–30 km or more offshore, outside of the bay, returning to the island at irregular intervals as solitary individuals or in groups of two or three (Figure 9.1b). The primary pinniped entry/exit point, called the "Launch Pad," is a distinguishable spot seaward of a small craggy outcrop situated off the south end of the island (Figure 9.1c). Seals travel to and from the island via porpoising at the surface, where they are attacked.

Observations were made at Seal Island during every month, weather permitting, by groups of two to eight from a single 8-m Butt Cat powered by two 90-horsepower outboards from 1997 to 2000 and by teams of 4–16 from two such boats from 2001 to 2003. At least part of the research team averaged some 200 days per year surveying the island, although most predatory activity occurs from late May to late August (winter). Arrival at Seal Island was before sunrise at about 7:30 a.m., sea conditions permitting. Predations occurred at the surface on porpoising seals and were focused on the southern end of the island, close to shore (within 2 km) and primarily when seals are traveling to and from the Launch Pad (Hammerschlag et al., 2006; Martin et al., 2009). By stationing at either end of the island, a single vessel can survey approximately 270° uninterrupted to a distance of at least 3.5 km. With two vessels, nearly all the water surrounding Seal Island could be surveyed to a similar distance.

Predatory events were detected at the surface by one or more of the following:

1. White Shark breach with a seal in its jaws or a seal leaping away from its mouth (Figure 9.2a and b)
2. An abrupt change in the travel behavior of seals, switching from porpoising at the surface to zigzag evasive movements with a shark in pursuit (Figure 9.2c and d)
3. A large splash accompanied by a blood stain, oil slick, distinctive odor, and any of the following secondary indicators such as a floating seal head, excised heart and/or lungs, entrails floating on the surface or trailing from the gills of a White Shark in the vicinity (Figure 9.2e), and/or focused plunge-diving Black-Backed Kelp Gulls (*Larus dominicanis vetula*) picking up seal entrails (Figure 9.2f)

Any subsurface kills could be detected by the appearance of a blood stain at the surface and floating seal entrails. Evidence of 10 subsurface strikes by White Sharks on Cape Fur Seals were recorded during the study period but were not included in the analysis of nearly 2,600 surface attacks. When a predatory event was observed, at least two researchers from the team were dedicated to guiding our vessel to the site of the initial strike to the best of their abilities using an onboard GPS (generally within 15 m).

Predatory events were classified as unsuccessful, in which the seal escaped, or successful, in which the seal was consumed. Where it could be identified, the following data recorded at each event: time of the predatory event and duration of the predatory event scored and defined as the time from initial White Shark strike to either seal consumption or escape. For the few cases in which a seal was killed but not consumed ($n = 10$), end of the predation event was calculated as the time after which the floating carcass was unattended by a shark for at least 120 s. The size of the seal group attacked was estimated as either solitary, 2–4, 5–10, or >11, and their direction of travel was categorized as either outgoing from or incoming to the island. Cape Fur Seals were divided into four classes that combine their size, maturity state, and sex:

1. Class 1: neonates [black pelt, <70 cm total length (TL)]
2. Class 2: young of the year (YOY, 70–100 cm TL)
3. Class 3: adult females and subadult males (lack of sagittal crest, 1.1–1.5 m TL)
4. Class 4: adult males (pronounced sagittal crest, >2 m TL)

Figure 9.2 Predatory events were detected by the following: (a and b) White Shark breach with a seal in its mouth or a seal leaping away from its mouth; (c and d) a sudden change in the movements of seals, switching from directional porpoising to zigzag evasive manoeuvres with a shark in pursuit; (e) a large splash in combination with a blood stain, oil slick, distinctive odor, and a floating seal head, excised heart and/or lungs, or seal entrails floating on the surface or trailing from the gills of sharks in the immediate vicinity; and (f) highly focused plunge-diving Black-Backed Kelp Gulls (*Larus dominicanis vetula*) picking up entrails.

Although we are within close proximity to seals both porpoising and during attacks, our age-class categories are based on a combination of size and distinct characteristics so that we can assign seals to these categories under field conditions. During or after many predatory events, sharks were close enough (<3 m) for sufficient time to permit estimation of size. Estimations were made independently by two trained observers. Shark lengths were estimated to the nearest 0.5 m against known dimensions of the vessel and placed into one of three size categories: small (<3 m), medium (3–3.5 m), or large (>3.5 m). Shark length was only recorded when it could be estimated with confidence by both trained observers. The time and number of seal groups leaving from and returning to Seal Island per day was recorded over 21 days between 7:30 a.m. and 1:30 p.m. Seal swimming behaviors and interactions with sharks were recorded topside by videography and photography, underwater via pole camera, recorded on field data sheets, and documented via Dictaphone.

Statistical analysis was conducted following Hammerschlag et al. (2006), chi-squared analysis was used to compare the frequency of observed attacks versus expected if attacks were random with respect to seal availability. The expected values were calculated by multiplying the proportion of seals available by the total number of attacks on them. For example, the expected number of attacks on different seal age classes if random with respect to their availability was calculated by multiplying the proportion of seals of the different age classes by the total number of attacks on all seals. Data on White Shark attack success with respect to month, time of day, seal group size, age class, seal movement relative to the island, and shark size were compared using a contingency table and chi-squared analysis. Each variable under investigation was recorded only when it could be clearly identified. Because of a lack of all factors being identified at each predatory event, interactions could not be analyzed using applications of multivariate analysis and/or generalized linear models for examining probability of success.

Published data by Hammerschlag et al. (2006) on environmental factors affecting White Shark attack frequency and success rate at Seal Island were incorporated in the discussion of the this paper. These factors included: wind direction and speed, tidal height, ocean depth, month of the year, and light intensity.

RESULTS

Between 1997 and 2004, a total of 2,546 predator-prey interactions between White Sharks (*Carcharodon carcharias*) and Cape Fur Seals (*Arctocephalus pusillus pusillus*) were documented. Predations peaked from June through August; success rates were not significantly different between months ($n = 2507$, $p > 0.284$; Figure 9.3a). Attacks occurred during all daylight hours, but frequency of predations at different times were nonrandom and were influenced by seal availability ($n = 282$, $p < 0.001$; Figure 9.3b). The frequency of attacks was significantly higher within an hour of sunrise, before 8:30 a.m. The second highest frequency of attack occurred between 8:30 and 9:30 a.m., followed by between 9:30 and 10:30 a.m. ($n = 1948$, $p < 0.001$). The success rate did not differ with time of day ($n = 2482$, $p > 0.213$). The group size of seals attacked ranged from 1 to at least 15, but were nonrandom, with most attacks on solitary seals ($n = 436$, $p < 0.001$; Figure 9.4a). Although not significant ($n = 436$, $p < 0.22$), the success rate showed a decreasing trend with increasing seal group size. Seals representing all size classes except Class 1 were attacked, but predations occurring on

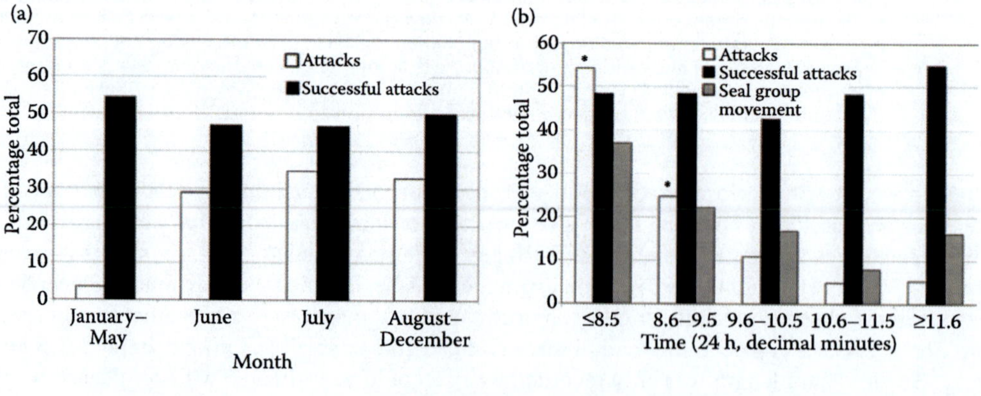

Figure 9.3 (a) Percentage of total ($n = 2507$) and successful ($n = 1207$) predatory attacks by White Sharks on Cape Fur Seals versus months. (b) Percentage of total ($n = 2482$) and successful ($n = 1194$) attacks by White Sharks on Cape Fur Seals and percentage of total seal group movement about the island ($n = 644$) versus time of day. Significance is indicated with an asterisk; see "Results" for further details.

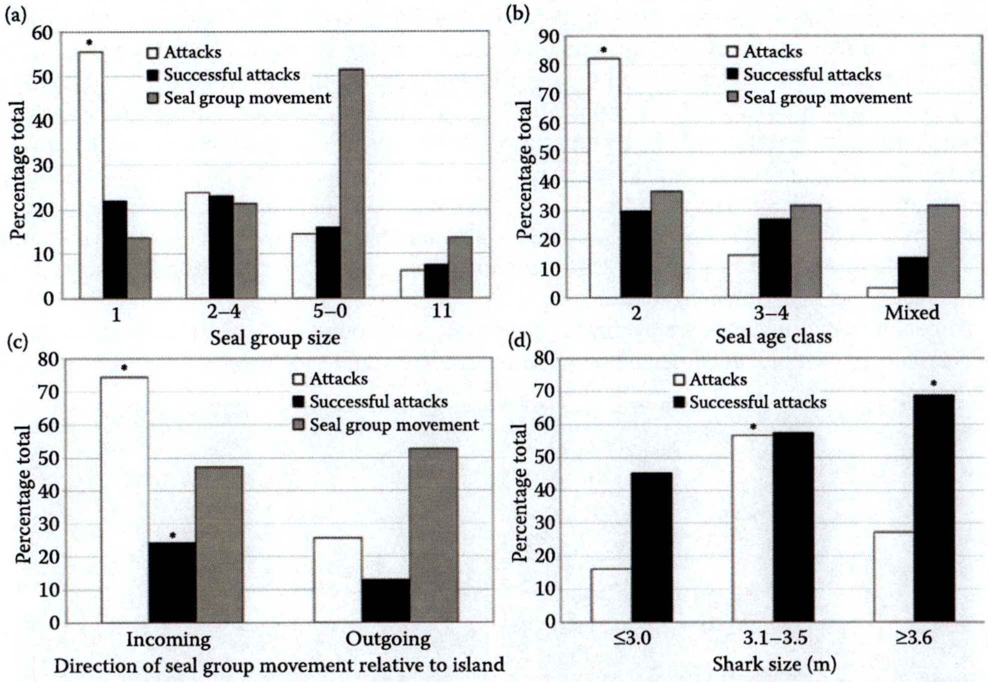

Figure 9.4 (a) Percentage of total (n = 436) and successful (n = 89) predatory attacks by White Sharks on Cape Fur Seals and percentage of total seal group movement about the island (n = 103) versus seal group size. (b) Percentage of total (n = 1132) and successful (n = 326) attacks by White Sharks on Cape Fur Seals and percentage of total seal group movement about the island (n = 41) versus seal age class. (c) Percentage of total (n = 390) and successful (n = 83) attacks by White Sharks on Cape Fur Seals and percentage of total seal group movement about the island (n = 645) versus seal travel direction. (d) Percentage of total (n = 669) and successful (n = 392) predatory attacks by White Sharks on Cape Fur Seals versus size of attacking shark. Significance is indicated with an asterisk; see "Results" for further details.

different seal age classes was not random ($n = 1132$, $p < 0.001$) with significantly more attacks on Class 2 seals than any other size class ($n = 1008$, $p < 0.001$; Figure 9.3b). Although not significant, the success rate showed a decreasing trend with attacks occurring on increasing seal age class ($n = 1132$, $p < 0.085$). Although both incoming and outgoing seals were attacked, the frequency of predations occurring on incoming versus outgoing seal groups was significantly different from random with respect to their availability ($n = 390$, $p < 0.001$; Figure 9.4c). Frequency ($n = 287$, $p < 0.001$) and success rate ($n = 390$, $p < 0.019$) were significantly higher on incoming seals (Figure 9.4c). Size of attacking sharks ranged from 2.1 to 4.5 m; however, 3.1–3.5-m sharks were responsible for significantly more attacks than any other shark size category ($n = 632$, $p < 0.001$), whereas sharks larger than 3.6 m possessed significantly higher success rates ($n = 669$, $p < 0.001$; Figure 9.4d).

DISCUSSION

White Shark predation at Seal Island occurs during the winter, peaking from late June to early August, at a time in which YOY Cape Fur Seals have just finished being weaned, have a thick layer of insulating blubber but limited energy stores, and are starting to forage away from Seal Island for the first time (Martin et al., 2005). The seals are likely targeted by White Sharks because they are small, inexperienced, have undeveloped swimming and diving abilities, and are probably easier to overpower and consume than older, larger seals (Le Boeuf et al., 1982; Brodie and Beck, 1983; Ainley et al., 1985; Le Boeuf and Crocker, 1996). Seals moving about Seal Island range in size and weight from yearlings (70 cm TL and 4.5 kg) to adult bulls (>200 cm TL and 360 kg) (Jefferson et al., 1993; E. Keith, personal communication). Cape Fur Seals possess strong claws and teeth; in many predatory events, seals bite and scratch sharks, even if grasped within the latter's jaws (Figure 9.5a–c). Large seals possess significant mass and represent a challenge, if not a danger, to any pursuing shark. Investigations of shark predation on Cape Fur Seals from the Eastern Cape of South Africa by Stewardson (1999) showed that a large number of adults, as compared with juvenile seals, had old bite wounds, suggesting that predatory attacks on older, more experienced, seals may be less successful than on younger seals.

Although Cape Fur Seals are available in the waters surrounding Seal Island from dawn to at least early afternoon, the documented increase in frequency and success of White Shark predation

Figure 9.5 (a and b) Cape Fur Seal biting a White Shark. (c) Cape Fur Seal clawing a White Shark. Injuring may cause shark to release seal or otherwise abort attack. (d) Swimming in large groups provides shared vigilance responsibilities, which increases probability of predator detection at reduced costs. (e) A seal swimming upside down within a porpoising group to maintain subsurface shared vigilance. (f) Groups of seals leave the island in coordinated groups traveling via rapid low porpoising (3+ body lengths/second). [(a–e) Courtesy of Chris Fallows, http://www.apexpredators.com; (f) courtesy of Neil Hammerschlag, http://www.neilhammer.com.]

during the early mornings, within an hour of sunrise, is likely due to low light conditions (Martin et al., 2005). Hammerschlag et al. (2006) concluded that during scotopic conditions at Seal Island, a seal's ability to detect a hunting shark is compromised, whereas White Sharks are at a visual and tactical advantage and as such benefit from higher success rates. Preliminary research indicates a smaller secondary daily peak in predation frequency occurs during dusk. Interestingly, on days that are overcast, predation continues throughout the day, rather than attenuating at midday (Fallows et al., unpublished data). For example, on August 8, 2003, 12 predatory attacks were recorded between 9:30 a.m. and 1:30 p.m., while there was little available ambient light all day because of an overcast sky. At the Farallones, attacks occur during all daylight hours; however, Pyle et al. (1996) reported a significant increase in nonpredatory sighting (including surfacing and breaches) of sharks at Southeast Farallon Island when cloud cover was high.

White Sharks appear to target solitary seals, which are unable to share vigilance or defensive duties and thus may be more vulnerable to ambush attacks by White Sharks than larger groups maintaining shared vigilance (Figure 9.5d and e). Large seal groups experience fewer attacks than solitary individuals, probably because grouping provides early warning of approaching danger and confuses sharks trying to select and attack a particular group member (Caro and Fitzgibbon, 1992; Savino and Stein, 1982). There are no reports from the Farallones showing that single or small groups of seals travelling about the islands are attacked more frequently than large groups of seals (Le Boeuf and Crocker, 1996).

Differences in swimming behavior, physical condition, and vigilance capabilities of seals returning to versus leaving from the island likely resulted in the documented difference in attack frequency and success. When leaving the island, Cape Fur Seals travel in large coordinated groups, locomoting via low porpoising at high speeds in the direction of the mouth of the bay, increasing shared subsurface vigilance (Figure 9.5f). Cape Fur Seals spend an average of 2.3 days at sea foraging (David and Rand, 1986), and YOY seals often return to the island as lone individuals or in small groups that are unable to benefit from shared vigilance and are likely also tired and affected by postprandial torpor, which may further decrease vigilance and make them more susceptible to attack (Martin et al., 2005). No comparison of attacks on incoming versus outgoing seals has been reported from the Farallones. Although White Sharks appear to target individual or small groups of seals returning to Seal Island, it should be noted that periods occur during which most seals return to the island in larger groups, and attack frequency on such groups increases, which is likely a reflection of prey availability.

Between 1998 and 2003, length range of attacking sharks at Seal Island ranged from 2.2 to 4.6 m, whereas at the Farallon Islands, sharks observed feeding between 1988 and 1992 ranged from 3.46 to 5.86 m. The occurrence of larger sharks feeding on seals in California sites, compared with Seal Island may be related to differences in pinniped prey size and a sharks' subsequent ability to incapacitate and subjugate prey accordingly. Northern Elephant Seals at the Farallones are massive (30–2200 kg). In comparison, Cape Fur Seals at Seal Island are relatively small (4.5–360 kg) (Jefferson et al., 1993; E. Keith, personal communication). It is logical that small sharks (<4 m) probably cannot subjugate large Elephant Seals; however, it stands to reason that larger sharks (>5 m) should still visit Seal Island to predate on relatively small Cape Fur Seals. The reason that latter is not observed may be related to locomotor performance of both predators and prey. Our data suggest that a 1.1-m-long Cape Fur Seals can reverse direction in about the same amount of time as a 3.5-m-long White Shark but in only 10–35% the distance. Thus, during secondary pursuit of a Cape Fur Seal by a White Shark, superior agility favors the former (Fallows et al., unpublished data). Thus, the lack of large (>5 m) White Sharks observed hunting at Seal Island may be because it is both difficult and energy-inefficient to leap out of the water in pursuit of an agile small prey. Further evidence to support this hypothesis comes from data collected when whale carcasses have been found at Seal Island. During the peak of the hunting season (May–August), the average shark size at Seal Island is 3.5 m, with sharks in excess of 4 m being a relatively rare occurrence, comprising 11% of the population observed (Fallows et al., unpublished data). However, during the three occasions that a whale carcass was found at Seal Island, in less than 24 h, we have observed over two dozen White Sharks feeding on the carcasses, mostly animals exceeding 4 m, including individuals over 5 m. The speed at which large White Sharks came to feed on the carcasses leads us to hypothesize that larger sharks are relatively nearby but choose not to hunt Cape Fur Seals. Over the course of 13 yrs., we have documented numerous individuals returning to feed at Seal Island on an annual basis, but when they grow in excess of 4 m, they disappear. For example, we first documented a female White Shark in 1997, measuring 2.5 m. This individual returned to the island annually, until 2004, when the shark measured 4.2 m in length. This shark has yet to return over the past 6 yrs. (Fallows et al., unpublished data). One slight exception to this general pattern was a large female that was first observed at 2.9 m, which grew to over 5 m in a period of 10 yrs. Upon reaching 4.2 m, she only returned to Seal Island every second year up until 2008, after which she never seen again at the island. With this being said, we cannot discount the possibility that size of sharks at these sites may be independent from the size of their prey. Particularly, emerging evidence suggests that large, sexually mature White Sharks in the eastern Pacific are aggregating at seal rookeries to mate (Chapter 11, this book).

Similarities and differences in White Shark–pinniped interactions at Seal Island and the Farallon Islands are summarized in Table 9.1. Similarities between the two sites include most frequently

Table 9.1 A Comparison of Factors Affecting Predation at Seal Island and the Farallon Islands

Factors	Seal Island	Farallon Islands	Similar
Frequency of attacks	High (6.68 per day)	Low (0.7 per day) (Klimley et al., 1992)	No
Mean predatory success rate	48%	64%*	Yes
Time of year in which attack frequency is highest	Winter (June to August)	Autumn (August to December) (Klimley et al., 1992)	No?
Time of day in which attack frequency is highest	Early morning (7:30–10:30 a.m.)	Frequency is high all day (Klimley et al., 1992)	No
Prey age class on which attack frequency is highest	Juvenile	Juvenile (Ainley et al., 1981, 1985; Long et al., 1996)	Yes
Prey mass	Low (4.5–360 kg)	High (30–2200 kg)	No
Location of attack relative to the island in which attack frequency is highest	Entry and exit points on southern side of Seal Island	All around island (Klimley et al., 1992)	No
Distance from island where attack frequency is highest	Near island (<400 m) (Martin et al., 2005; Hammerschlag et al., 2006)	Near island (<400 m) (Klimley et al., 1992)	Yes
Size of shark responsible for the highest frequency of attacks	Small (3.1–3.5 m TL)	Large (3.46–5.86 m TL)	No
Tidal height during which attack frequency is highest	High (>1500 cm above mean sea level) (Hammerschlag et al., 2006)	High (Anderson et al., 1996)	Yes
Depth range of in which attacks take place	Deep (5–46 m) (Hammerschlag et al., 2006)	Deep (5–50 m) (Klimley et al., 1992)	Yes
Wind direction in which attack frequency is highest	Northerly (Hammerschlag et al., 2006)	Not significant (Pyle et al., 1996)	No
Affect of light levels on attack frequency	Highest during low light levels (0–200 μE) (Hammerschlag et al., 2006)	Occur during all light levels	No

* Methodology may result in an under-representation of failed predation by White Sharks on juvenile elephant seals. Klimley et al. (1992) observed predatory events from a tower 102 m above sea level, where attention was first drawn to predatory attack by the appearance of blood, an oil slick, or an explosive splash where the initial bite was not usually seen (Ainley et al., 1981, 1985; Klimley, 1994; Klimley et al., 1996), and thus they were more likely to notice successful predations.

attacked prey age class (juvenile), most frequent attack distance from the island (<400 m), most frequent attack depth range (<50 m), and most frequent attack tide state (high). Juvenile prey are naïve and have less-developed defensive capabilities (diving and fighting) and thus may require less effort for a predator to subjugate. Distance from the island and depth range are inherently linked, and tide state also affects depth at a given distance from shore. These variables may reflect prey availability, which is greatest close to the island and when haul-out space is reduced (Anderson et al., 1996; Hammerschlag et al., 2006). Differences in such interactions at Seal Island and the Farallones include attack frequency (high versus low, respectively), mean predatory success rate (48% versus 64%), most frequent attack time of day and light levels (dawn versus all day), peak attack season (winter versus autumn), most frequent prey mass and shark length (small versus large, for both), and most frequent attack wind direction (westerly versus no significant difference). Differences in attack frequency may be attributable to concentration of prey near the Launch Pad at Seal Island compared with lack of such an aggregating point at the Farallones. Differences in success rate may not be significant or may reflect a logistically imposed bias toward conspicuous successful attacks at the Farallones, where predations events are typically detected by the appearance of blood, an oil slick, or an explosive splash and are spotted from atop a 102-m-tall observation point (Klimley et al., 1992). Time of day and light levels are causally related. Differences between Seal Island and Farallones may reflect relative heights of these islands, the former being low and the latter being high, thus casting a directional shadow that may prolong shark crypsis within its influence. Further data from the Farallon Islands are needed to test this idea. Differences in prey mass and attacking-shark size may be related, because it is unlikely that a 2-m White Shark could subjugate a 1.5-m Northern Elephant Seal pup as reliably as it could a 60-cm Cape Fur Seal pup. Differences in the significance or nonsignificance of wind direction may reflect the fact that Seal Island is located at the foot of False Bay, so northerly winds transport surface-borne scents seaward, whereas the Farallones are isolated offshore islands, essentially surrounded by open ocean on all sides.

The present study suggests that predation at Seal Island is significantly affected by factors that influence a shark's ability to encounter, ambush, and subjugate its pinniped prey and its prey's ability to detect, avoid, and injure its predator as proposed by Hammerschlag et al. (2006). Specific factors affecting the frequency of White Shark predation at Seal Island (prey age class, distance from island, tidal height, and depths at which attacks range) are similar to those identified at the Farallon Islands, off California; however, most (time of year, time of day, prey group size, prey mass, location of attacks relative to the island, distance from island, depth, wind direction, light intensity, and shark size) are different or have not been reported. Factors significantly affecting predatory success rate were identified at Seal Island (location of attacks relative to island, distance from island, seal direction of travel, tidal height, light intensity, and shark size); factors significantly affecting White Shark predatory success have not been reported from the Farallones. Therefore, factors affecting White Shark predatory behavior and success rate are likely prey- and site-specific.

ACKNOWLEDGMENTS

We would like to thank M. Fallows for her ongoing support, tremendous help, skill in field observations, data recording, and sensitive boat handling, without which most of this project could not have been accomplished. We would also like to thank R. Lawrence for his ongoing efforts to help in data collection and also for his respect of the area as a sensitive ecosystem. Thanks also to R. Strong for helping in the initial phases of setting up methodology for data collection as well as many of the film crews, including the BBC, Discovery, National Geographic, and many others who have allowed us to review footage shot with us and confirm observed behavior. We thank our corporate sponsors Canon South Africa and Fuji South Africa, YSI Environmental, Oregon Scientific, Li-Cor Biosciences, and Ben Meadows for generous donation of environmental sensor equipment. Thanks to all volunteer research assistants during the study period. Thank you to R. Farre, S. Gildenhuys, and the South African Navy: Hydrographic Office via the Institute for Maritime Technology for providing hydrographical data.

REFERENCES

Ainley DG, Henderson RP, Huber HR, Boekelheide RJ, Allen SG, McElroy TL (1985) Dynamics of white shark/pinniped interactions in the Gulf of the Farallones. *Mem South Calif Acad Sci* 9: 109–122.

Anderson SD, Klimley AP, Pyle P, Henderson RP (1996) Tidal height and white shark predation at the Farallon Islands, California. In: Klimley AP, Ainley DG (eds), *Great white sharks: The biology of Carcharodon carcharias*. Academic Press, San Diego. pp. 275–279.

Ainley DG, Strong CS, Huber HR, Lewis TJ, Morrell SJ (1981) Predation by sharks on pinnipeds at the Farallon Islands. *Fish Bull* 78: 941–945.

Barlow GW (1972) A paternal role for bulls of the Galapagos sea lion. *Evolution* 26: 307–310.

Brodie P, Beck B (1983) Predation by sharks on the grey seal (*Halichoerus grypus*) in eastern Canada. *Can J Fish Aquat Sci* 40: 267–271.

Busse C (1980) Leopard and lion predation upon chacma baboon living in the Moremi Wildlife Reserve. *Botswana Notes Rec* 12: 15–21.

Caro TM, Fitzgibbon CD (1992) Large carnivores and their prey: The quick and the dead. In: Crawley MJ (ed), *Natural enemies: The population biology of predators, parasites and diseases*. Blackwell Scientific Publications, London. pp. 117–142.

Compagno LJV (2001) Sharks of the world. *FAO species catalogue for fishery purposes*, No. 1, Vol. 2. Bullhead, mackerel and carpet sharks (Heterodontiformes, Lamniformes, and Orectolobiformes). FAO, Rome.

da Silva J, Terhune JM (1988) Harbour seal grouping as an anti-predator strategy. *Anim Behav* 36: 1309–1316.

David JHM, Rand RW (1986) Attendance behavior of South African fur seals. In: Gentry RL, Kooyman GL (eds), *Fur seals: Maternal strategies on land and at sea*. Princeton University Press, Princeton, NJ. pp. 126–141.

Domeier ML, Nasby-Lucas N (2007) Annual re-sightings of photographically identified white sharks (*Carcharodon carcharias*) at an eastern Pacific aggregation site (Guadalupe Island, Mexico). *Mar Biol* 150: 977–984.

Eibl-Eibesfeldt I (1984) The Galapagos seals: Part 1. Natural history of the Galapagos sea lion (*Zalophus californianus wollebaek*, Sivertsen) In: Perry R (ed), *Key environments: Galapagos*. Permagon Press, Oxford, UK. pp. 207–214.

Goldman KJ, Anderson SD (1999) Space utilization and swimming depth of white sharks, *Carcharodon carcharias*, at the South Farallon Islands, central California. *Environ Biol Fish* 56: 351–364.

Hamilton WD (1971) Geometry of the selfish herd. *J Theor Biol* 31: 295–311.

Hammerschlag N, Martin RA, Fallows C (2006) Effects of environmental conditions on predator-prey interactions between white sharks (*Carcharodon carcharias*) and Cape fur seals (*Arctocephalus pusillus pusillus*) at Seal Island, South Africa. *Environ Biol Fish* 76: 341–350.

Heithaus MR (2004) Predator-prey interactions. In: Carrier JC, Musick JA, Heithaus MR (eds), *Biology of sharks and their relatives*. CRC Press, Boca Raton, FL. pp. 488–512.

Ims RA (1990) On the adaptive value of reproductive synchrony as a predator swamping strategy. *Am Nat* 136: 485–498.

Jefferson TA, Leatherwood S, Webber MA (1993) *Marine mammals of the world*. FAO species identification guide. FAO, Rome.

Kirkwood R, Dickie J (2005) Mobbing of a great white shark (*Carcharodon carcharias*) by adult male Australian fur seals (*Arctocephalus pusillus doriferus*). *Mar Mamm Sci* 21: 336–339.

Klimley AP (1994) The predatory behaviour of the white shark. *Am Sci* 82: 122–133.

Klimley AP, Anderson SD, Pyle P, Henderson RP (1992) Spatiotemporal patterns of white shark (*Carcharodon carcharias*) predation at the South Farallon Islands, California. *Copeia* 1992: 680–690.

Klimley AP, Le Boeuf BJ, Cantara KM, Richert JE, Davis SF, Sommeran SV, Kelly JT (2001) The hunting strategy of white sharks (*Carcharodon carcharias*) near a seal colony. *Marine Biology* 138: 617–636.

Klimley AP, Pyle P, Anderson SD (1996) Residency patterns of white sharks at the South Farallon Islands, California. In: Klimley AP, Ainley DG (eds), *Great white sharks: The biology of Carcharodon carcharias*. Academic Press, San Diego. pp. 365–373.

Le Boeuf BJ, Crocker DE (1996) Diving behavior of elephant seals: Implications for predator avoidance. In: Klimley AP, Ainley DG (eds), *Great white sharks: The biology of Carcharodon carcharias*. Academic Press, San Diego. pp. 193–205.

Le Boeuf BJ, Riedman M, Keyes RS (1982) White shark predation on pinnipeds in California coastal waters. *Fish Bull* 80: 891–895.

Lind J, Cresswell W (2005) Determining the fitness consequences of antipredation behavior. *Behav Ecol* 16: 945–956.

Long DJ, Hanni KD, Pyle P, Roletto J, Jones RE, Bandar R (1996) White shark predation on four pinniped species in central California waters: Geographic and temporal patterns inferred from wounded carcasses. In: Klimley AP, Ainley DG (eds), *Great white sharks: The biology of Carcharodon carcharias*. Academic Press, San Diego. pp. 263–274.

Martin RA, Hammerschlag N, Collier RS, Fallows C (2005) Predatory behaviour of white sharks (*Carcharodon carcharias*) at Seal Island, South Africa. *J Mar Biol Ass UK* 85: 1121–1135.

Martin RA, Rossmo DK, Hammerschlag N (2009). Hunting patterns and geographic profiling of white shark predation. *J Zool* 279: 111–118.

Musick JA (1999) Ecology and conservation of long-lived marine animals. *Am Fish Soc Symp* 23: 1–10.

Pulliam HR, Caraco T (1984) Living in groups: Is there an optimal group size? In: Krebs R, Davies NB (eds), *Behavioural ecology*, 2nd ed. Blackwell Scientific, Oxford, UK. pp. 122–147.

Pyle P, Anderson SD, Klimley AP, Henderson RP (1996) Environmental factors affecting the occurrence and behavior of white sharks at the Farallon Islands, California. In: Klimley AP, Ainley DG (eds), *Great white sharks: The biology of Carcharodon carcharias*. Academic Press, San Diego. pp. 281–291.

Savino JF, Stein RA (1982) Predator-prey interaction between largemouth bass and bluegills as influenced by simulated submersed vegetation. *Trans Am Fish Soc* 11: 255–266.

Stewardson CL (1999) Preliminary investigations of shark predation on Cape Fur seals *Arctocephalus pusillus pusillus* from the Eastern Cape coast of South Africa. *Trans R Soc S Afr* 54: 91–203.

Stewardson CL, Brett M (2000) Aggressive behaviour of an adult male Cape fur seal (*Arctocephalus pusillus pusillus*) towards a great white shark (*Carcharodon carcharias*). *Afr Zool* 35: 147–150.

Strong WR (1991) Instruments of natural selection: How important are sharks. In: Gruber SH (ed), *Discovering sharks*. American Littoral Society, Highlands, NJ. pp. 70–73.

Terhune JM (1985) Scanning behavior of harbor seals on haul-out sites. *J Mamm* 66: 392–395.

Trillmich F (1996) Parental investment in pinnipeds. *Adv Study Behav* 25: 533–577.

Walters C (1997) Challenges in adaptive management of riparian and coastal ecosystems. *Conserv Ecol* 1:1 (http://www.consecol.org/vol1/iss2/art1).

SECTION II

Migratory Patterns and Habitat Use

CHAPTER 10

Fine-Scale Habitat Use by White Sharks at Guadalupe Island, Mexico

Michael L. Domeier*
Marine Conservation Science Institute

Nicole Nasby-Lucas
Marine Conservation Science Institute

Chi H. Lam
University of Southern California

ABSTRACT

Pop-up archival transmitting (PAT) tags and satellite-linked radio transmitting (SLRT) tags were deployed on subadult and adult White Sharks (*Carcharodon carcharias*) at Guadalupe Island, Mexico from 2002 to 2009. PAT tags provided high-resolution archival data, as well as summarized data for swimming depth and temperature. SLRT tags provided fine-scale position data that was used to identify spatial preferences of White Sharks at Guadalupe Island. PAT tag data sets indicated average maximum daily dive depths from 161 to 282 m, with more time at depths deeper than 40 m noted in late summer and autumn. Changes in seasonal distribution of White Sharks around the island were noted with shark activity spread out over the eastern half of the island in August through October and then moving to regions on the northeast and northwest beginning in November through March. This corresponded with the temporal and spatial differences in island habitat use by Guadalupe Fur Seals (*Arctocephalus townsendi*) and Northern Elephant Seals (*Mirounga angustirostris*).

INTRODUCTION

Guadalupe Island (GI), Mexico is one of only two known aggregation sites for subadult and adult White Sharks (*Carcharodon carcharias*) in the northeastern Pacific Ocean. White sharks are present at GI during autumn and winter months (Domeier and Nasby-Lucas, 2008; Chapters 11 and

25, this book) but spend spring and summer within a pelagic habitat called the Shared Offshore Foraging Area (SOFA), centered between the Hawaiian Islands and Baja California, Mexico, with a few individuals traveling as far as Hawaii (Domeier and Nasby-Lucas, 2008; Nasby-Lucas et al., 2009; Chapter 16, this book). Sharks then return to GI in late summer (Domeier and Nasby-Lucas, 2008, 2011; Chapters 16 and 25, this book).

Guadalupe Island is located 240 km off the west coast of Baja California and is a volcanic island that stretches 35 km north to south and 10 km east to west. The island has a narrow shelf that drops more than 3,000 m in as little as 12 km from the coast (Figure 10.1). The subtidal coastal area is composed of basaltic features (e.g. dikes, caverns, pillars, and blocks) and sand pebbles (Pierson, 1987) that support a diversity of benthic algae species (Gallo-Reynoso et al., 2005a). Deep canyons are found on the northeastern section and in the middle portion of the island. The island forms a barrier against the California current, which generates upwelling and eddies, bringing nutrient-rich cold water to the surface around the island. The eastern side of GI is somewhat protected from the prevailing northwesterly winds. Water clarity is often high, with a Secchi depth of 25–30 m (Gallo-Reynoso et al., 2005a).

The island serves as a haul-out and pupping site for three pinniped species:

1. The Northern Elephant Seal, *Mirounga angustirostris*
2. Guadalupe Fur Seal, *Arctocephalus townsendi*
3. California Sea Lion, *Zalophus californianus* (Figure 10.1)

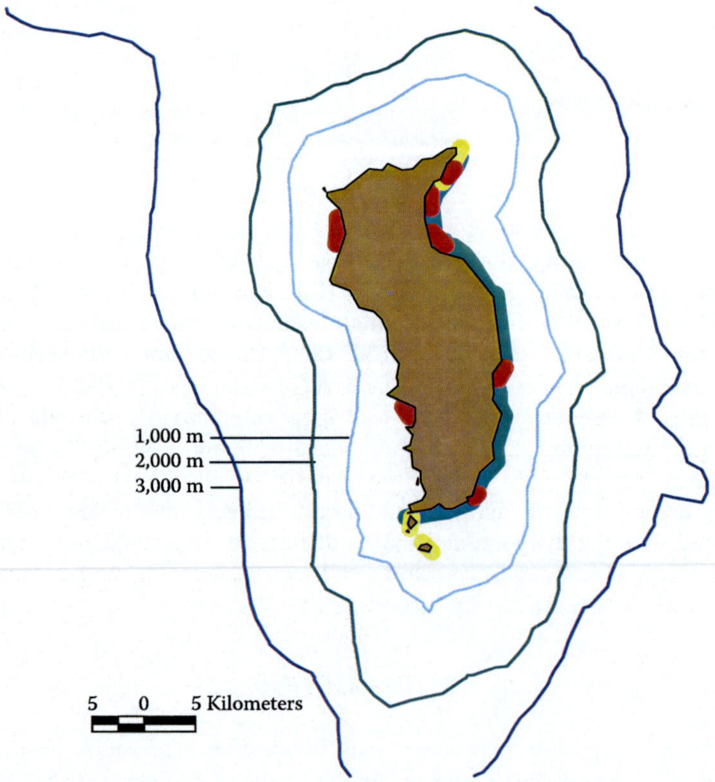

Figure 10.1 Map of Guadalupe Island showing bathymetry contours and locations of pinnipeds around the island. Green indicates the distribution of Guadalupe Fur Seals, yellow is California Sea Lions, and red indicates areas with Northern Elephant Seals.

Northern Elephant Seals are present at GI throughout the year with the highest numbers, around 13,000, during the breeding season in late January (Gallo-Reynoso et al., 2005b). The number of Guadalupe Fur Seals is estimated to be no more than 12,000 (Gallo-Reynoso et al., 2005b; Figueroa-Carranza, 1994) and is composed predominantly of adult females and juveniles. California Sea Lions have a peak population size of less than 1,200, with the largest number of individuals sighted from January to April (Gallo-Reynoso et al., 2005b).

Guadalupe Island has an abundance of marine life, including more than 126 species of fish; 17 cetacean species, including Bottlenose Dolphin (*Tursiops truncates*) and Baird's Beaked Whale (*Berardius bairdii*) (Gallo-Reynoso and Figueroa-Carranza, 2005); and several squid species (*Onychoteuthis banksii*, *Onychoteuthis borealijaponica*, and *Dosidicus gigas*) (Gallo-Reynoso, 1994). More than 10 other shark species are also observed around GI, which includes the Leopard Shark (*Triakis semifasciata*), Horned Shark (*Heterodontus francisci*), Blue Shark (*Prionace glauca*), Mako Shark (*Isurus oxyrinchus*), and Hammerhead Shark (*Sphyrna* sp.) (Gallo-Reynoso et al., 2005a).

Fine-scale horizontal (Chapter 11, this book) and vertical (Nasby-Lucas et al., 2009) movement patterns of White Sharks have been documented for the time these sharks are in the SOFA, but similar studies are lacking for the period that the sharks aggregate at GI. Combining vertical movement data from PAT tags with high-resolution position data from SLRT tags has allowed us to present this first assessment of fine-scale habitat use by White Sharks at GI.

MATERIALS AND METHODS

Satellite-Linked Radio Transmitting Tag

Adult and subadult White Sharks were captured at GI and fitted with SLRT tags (SPOT5; Wildlife Computers, Redmond, WA) (see Chapter 11, this book for detailed methods). Each time a tagged shark's dorsal fin was out of the water, a wet/dry switch activated the transmitter. If the tag remained out of the water long enough for an Argos satellite to receive four consecutive transmissions, the Doppler shift-induced frequency change allowed the calculation of the tag's location (Taillade, 1992). The Argos tracking and environmental monitoring system assigns a location class value (LC) that corresponds to one standard deviation of the estimated location error. Argos states that the estimated error for the location data is <250 m for LC 3, between 250 and 500 m for LC 2, and between 500 and 1500 m for LC 1. Argos does not provide estimates for LC A, B, and 0, but estimates have been published in the literature, showing an estimated location error for A of approximately 1 km and approximately 8 km for B (Hays et al., 2001).

A subset of position data was created by filtering out all positions that placed a tagged shark more than 5 km from GI. To avoid the bias introduced by variation in sampling frequency, we regularized the filtered tracks to a frequency of 1 position per day. Monthly fixed-kernel analyses (Worton, 1989) were performed using the home range extension (Rodgers and Carr, 1998) for ArcView 3.2 (Environmental Systems Research Institute, Redlands, CA). The 25% probability contour indicates the regions of the highest utilization by the sharks, the 50% probability contour shows areas with moderate utilization, and the 95% contour regions of low utilization. Location data for February and March were combined because there were fewer position points available for this period as many sharks begin their offshore migration by February (Domeier and Nasby-Lucas, 2008).

Pop-Up Archival Transmitting Tags

White Sharks were tagged at GI, Mexico using PAT tags manufactured by Wildlife Computers (WC) (PAT and MK10-PAT; Redmond, WA) and Microwave Telemetry (MT) (PTT-100; Columbia, MD). Tuna carcasses were used to lure sharks alongside the research vessel before tagging with a hand-held tagging pole. Tags were deployed by inserting a nylon tag head into the dorsal musculature at the base of the dorsal fin. Tags were rigged with an umbrella-style dart (Domeier et al., 2005) and either 136-kg test Sufix Superior monofilament (Yao I Fabric) or 113-kg test nylon-coated braided stainless steel leader (Sevenstrand). Sex and visual length estimates were recorded when possible. Sexual maturity was inferred from length estimates using published size at maturity data; males were considered mature when >3.6 m, and females were considered mature when >4.5 m (Francis, 1996; Pratt, 1996).

In total, 75 PATs (16 MT and 59 WC) were deployed on White Sharks between 2000 and 2007 (Domeier and Nasby-Lucas, 2008). Tags stayed attached to the individual shark until a preprogrammed date and time when an electrolytic release mechanism was activated, and the tag floated to the surface and transmitted data to the Argos satellite system. MT tags transmitted depth or temperature readings at a sampling rate of one per h, and data from recovered MT tags were available at the same sampling frequency. WC tags were programmed to collect data every 2 min. and then transmit summarized depth and temperature data for each 24-h period. Recovered WC tags provided high-resolution (2-min. sampling frequency) archival data sets of temperature, depth, and relative light level. Depth values in one WC tag [tag number 16 on a male shark (16M)] drifted systematically over time and were adjusted with a linear correction. The data in the summarized form were used to analyze the seasonal shifts in the use of the water column; this included data from all WC tags as well as hourly data from MT tags. Data that were not associated with GI were excluded by determining the dates of each shark's departure and return to the island. Changes in longitude along with characteristic surface travel (Boustany et al., 2002; Weng et al., 2007; Domeier and Nasby-Lucas, 2008) were used to determine the arrival and departure dates for each shark. Data from the first day after tagging and the last day before tag detachment were excluded to reduce potential influences caused by tagging effects.

Oceanographic Data

Monthly climatological values of dissolved oxygen (DO) were obtained from the National Oceanographic Data Center World Ocean Atlas 2005 (Garcia et al., 2006). Based on the extracted data, an oxygen value was assigned for each 2-min. data point for archival data sets. Monthly mixed-layer depth values were estimated from daily depth-temperature profile in the recovered data and averaged among all tags.

RESULTS

A total of 10 male and 7 female White Sharks were SLRT-tagged at Guadalupe Island between 2007 and 2009. Because of problems in early tag design, poor tag performance, and variation in individual shark behavior, some tags resulted in only Z class location estimates or very few location estimates. Data from 12 tagged sharks provided location data within 5 km of GI and were analyzed in this study. Location data at GI were available from August 1 through March 16. This included data from one subadult male, eight adult males, and four adult females.

Out of the 56 PAT tags reported in Domeier and Nasby-Lucas (2008), 35 WC and five MT tags were included in this study. Tags were not used if there were a large number of missing records because of poor data transmission or if multiple tags on the same individual created duplicate data sets, in which case the tag set containing more days of data was used. Thirteen of the tags were deployed in August, one in November, 26 in December, and one in January. Twenty tags stayed on until the shark began the seasonal offshore migration with an average departure date of February 10 (earliest December 21 and latest April 24). The greatest amounts of transmitted data were available in the months of December and January. The months of May through July were excluded because of little to no tagged White Shark presence at Guadalupe Island in this time frame.

Table 10.1 Tagging Data for Recovered PAT Tags from White Sharks Tagged at Guadalupe Island, Mexico

Tag Number and Sex	Photo Identification Number[a]	Sexual Maturity	Tagging Date	Date Left GI	Date Returned	Pop-Up Date	Days at GI
11U		No	8/12/2001			8/27/2001	15
16M	17	No	8/12/2002			11/17/2002	97
18M	19	Yes	12/5/2002	2/2/2003	7/28/2003	8/8/2003	70
30M	10	No	12/5/2003	5/3/2004	8/4/2004	8/6/2004	152
37U		Unknown	12/12/2004			12/28/2004	16
40M	10	No	12/12/2004	4/4/2005	7/15/2005	8/8/2005	137
41F	64	Unknown	12/7/2005	2/7/2006		5/18/2006	62

[a] Photo identification number from Chapter 25, this book.

Among the analyzed tags, seven WC tags were physically recovered after deployment periods of 0.5 to 8.2 months. One tag was recovered floating at GI, four of the tags were physically removed from the sharks while at GI, one was found on Laysan Island in the northwestern Hawaiian Islands, and another was recovered on the island of Hawaii. Recovered tags provided a total of 549 days of 2-min. archival data detailing shark behavior while at GI (Table 10.1). Two of the recovered tags were deployed in August and five in December. The period from September to November was represented by only one shark, which lost its tag in mid-November.

Fine-Scale Habitat Use of White Sharks at Guadalupe Island

Fixed-kernel analysis of SLRT tag data demonstrated that the northeastern corner of GI is generally the most important White Shark habitat; however, seasonal changes in the distribution of White Sharks around GI were noted (Figure 10.2). In August and September, which is the early part of the

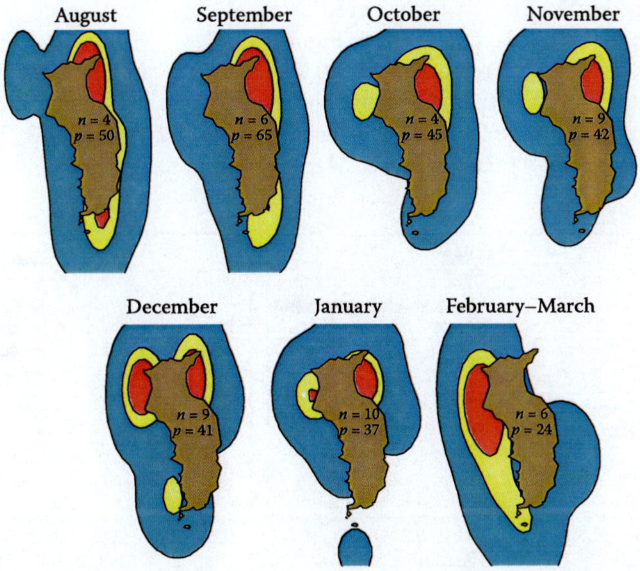

Figure 10.2 Fixed-kernel density contours of SLRT tag location data by month for tagged sharks around Guadalupe Island (red = 25%, yellow = 50%, blue = 95%; *n* is the number of sharks detected for the specified time period; and *p* is the number of position points).

White Shark season characterized predominantly by the presence of males, the 25 and 50% density contours were situated entirely on the eastern side of GI, with concentrations in the northeastern corner and southwest tip of the island. Starting in October, the 50% density contour no longer included the southeastern tip but began to include the northwestern portion of GI, and by February the 25 and 50% density contours were situated entirely on the west side of the island (data combined for February through March because of sharks departing for the SOFA, resulting in less data).

Vertical Movement Patterns

Archival records of depth provided a detailed look at the vertical movements of tagged White Sharks while at GI, indicating a strong diel pattern characterized by deep dives during the day and shallow dives within the mixed layer at night (Figure 10.3a). Maximum depths recorded by all tags while at GI ranged from 318 to 496 m. All tags recorded significantly different average maximum depths between day and night (161–258 m during the day and 89–128 m at night; Wilcoxon rank sum test, $p < 0.001$) (Table 10.2). Data from recovered tags indicated that sharks spent an average of 22% of their time below 50 m (15% at night and 27% during the day) at GI, differing from the 65% time below 50 m while in the SOFA (62% at night and 66% during the day; Figure 10.3). Although the tagged sharks spent 90% of the nighttime in the mixed layer (average mixed-layer depth, 60 m) while at GI, only 10% was spent near the surface (0–5 m); in comparison, sharks in the SOFA spent 65% of the night in the mixed layer (average mixed-layer depth 126 m) with 20% near the surface (Figure 10.3).

Archival tag data described tremendous variation in vertical movement patterns between individual sharks within the same time frame. In general, oscillatory bounce diving was the most prevalent vertical behavior. Although dive patterns between individuals may look similar at a temporal scale of several months (Figure 10.4), examining these same data at higher resolution reveals distinct differences. For example, 18M and 41F exhibited differing daytime vertical excursion patterns; in December, 18M made oscillatory dives between the mixed layer and 300 m, whereas 41F conducted longer dives at depth with deeper oscillations between 200 and 350 m (Figure 10.5a and b). In addition to deep oscillatory dives, recovered archival data also showed periods when sharks rarely dove beneath the mixed layer (e.g. 18M; Figure 10.5c).

Summarized 24-h time-at-depth data from all PAT tags indicated a change in vertical behavior by season (Figure 10.6). Sharks spent the most time above 40 m throughout their time at GI; however, more time was spent deeper than 40 m from August to December (44.8, 36.3, 26.4, 28.0, and 26.8% by month, respectively) than January to March (20.8, 14.8, and 15.1% by month, respectively)

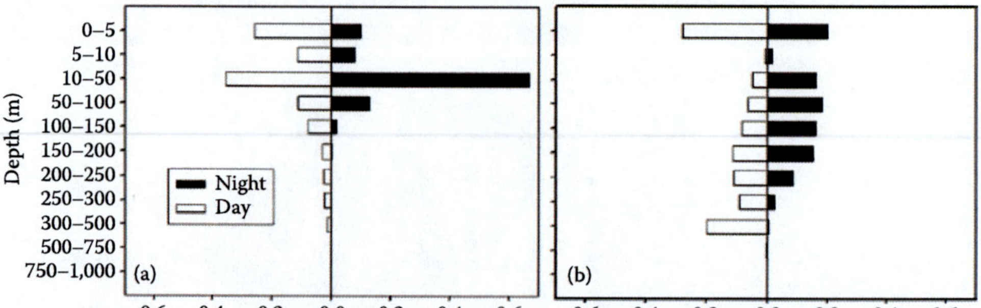

Figure 10.3 Day and night depth distribution (archival PAT data, tags 18M, 30M, and 40M) for tagged White Sharks while near Guadalupe Island, Mexico (a) and while in the Shared Offshore Foraging Area (b).

Table 10.2 Depth and Dissolved-Oxygen Levels Experienced by Tagged White Sharks while at Guadalupe Island

Tag Number and Sex	Max depth (m)	Day Average Maximum Depth (m)	Night Average Maximum Depth (m)	% Time below 50 m	Average Daily Minimum O_2 (ml L^{-1})	Minimum O_2 (ml L^{-1})
11U	358	207	114	39	2.9	1.4
16M	326	167	89	32	3.5	1.3
18M	461	282	128	21	2.3	0.7
30M	496	258	125	21	2.6	0.8
37U	318	185	104	10	3.5	1.5
40M	396	161	95	9	3.9	1.4
41F	414	215	113	26	3	0.7

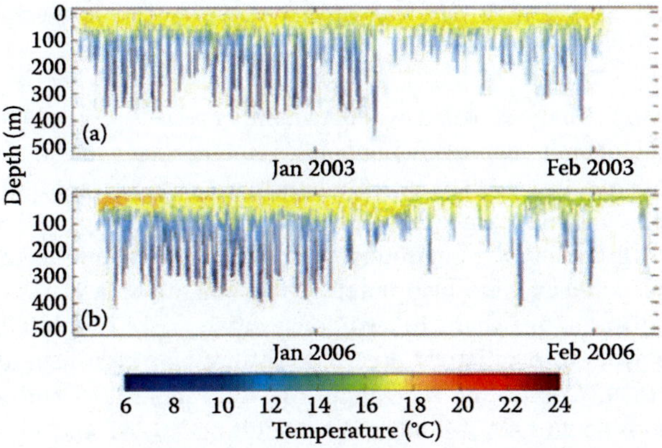

Figure 10.4 Ambient temperature-depth profiles displaying 2-min. archival data at Guadalupe Island for 18M (a) and 41F (b). Temperature is indicated by color.

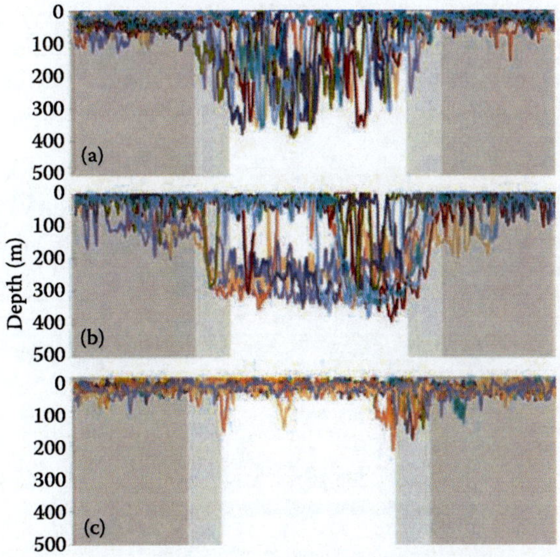

Figure 10.5 Depth profiles for 10 days overlaid in one 24-h period for 18M (a) and 41F (b) from December 20–29 and 18M from January 8–17 (c). The different colors indicate individual days. Dark gray indicates night, and light gray indicates twilight.

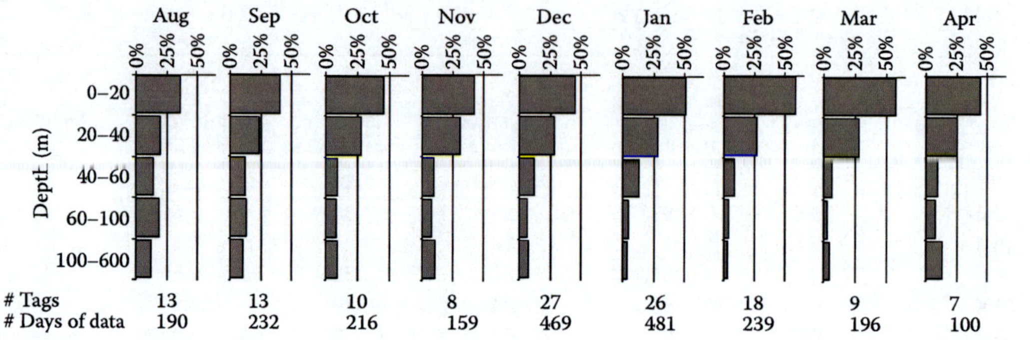

Figure 10.6 Bar graphs indicating percentage of time at depth by month for sharks around GI using all PAT tag data. These data show that a significantly higher percentage of time was spent at depths deeper than 40 m from August to December compared with January to March. The period from May to July was excluded because of very little data.

(Student's *t* test, p = 0.02). Analysis of archival data confirmed this seasonal depth change; 18M and 41F both shifted from a deeper vertical diving pattern in December to a shallow vertical pattern in January (Figure 10.5). This shift was statistically significant with average daily maximum depths of 338 m in December and 178 m in January (Wilcoxon rank sum test, $p < 10^{-11}$). Both 18M and 41F began their offshore migration in the beginning of February following this transition (Table 10.1).

Unique behavioral patterns were also noted for a subadult shark. Tags 30M and 40M were deployed on the same immature shark in two consecutive years (Table 10.1). In both years the immature shark displayed an oscillatory dive pattern that was significantly shallower than that of the mature male 18M. Comparing the month of December, 18M had a significantly deeper average daily maximum depth (326 m) than the subadult male (237 and 167 m for 30M and 40M, respectively) (Wilcoxon rank sum test, $p < 10^{-8}$). Although the immature shark exhibited periods of shallow diving within the mixed layer between January and March (in both years), it did not make a transition from a pattern of deep vertical dives to shallower vertical dives (Figure 10.7) (average December daily maximum depths of 208 and 195 m in January) (Wilcoxon rank sum test, $p < 0.46$). A period of vertical oscillatory activity with increased maximum depth was observed in both years beginning in late March. The percentage of time above 50 m for 30M from December to March was 82% and decreased to 61% in April (Wilcoxon rank sum test, $p < 10^{-8}$). In both years,

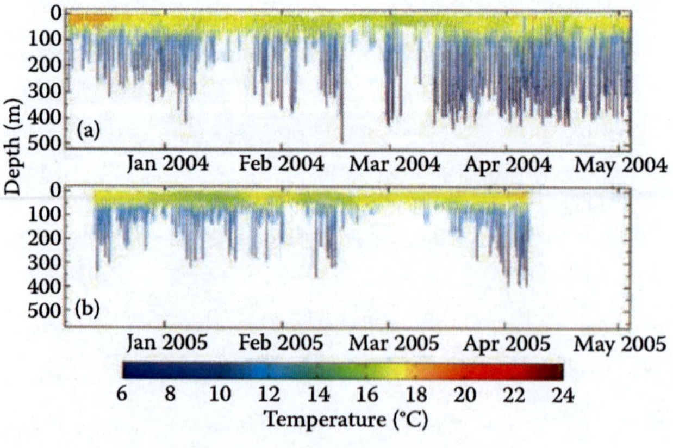

Figure 10.7 Ambient temperature-depth profiles displaying 2-min. archival data from recovered PAT tags at Guadalupe Island for 30M (a) and 40M (b). Both of these tags were deployed on the same shark in two consecutive years. Temperature is indicated by color.

the shark began the offshore migration in April or May (Table 10.1). An analysis of all PAT tag data from sharks estimated as subadult (Domeier and Nasby-Lucas, 2008) gave an average departure date of April 7, compared with those estimated as mature, which departed on an average date of January 31.

Mixed-Layer Depth and Dissolved Oxygen

During the time the sharks were present at GI, the mixed layer depth ranged from 30 to 90 m and was deeper in late winter and early spring than in summer and fall. Depths with low DO levels at GI were relatively shallow. Climatological data implied that the average depth of the oxygen-minimum zone (defined as <1.5 ml L^{-1}) ranged between 225 and 300 m. Matching the archival data sets with the World Ocean Atlas data (Garcia et al., 2006) indicated that the lowest estimated DO experienced for each shark was between 0.7 and 1.5 ml L^{-1} (Table 10.2). The maximum percentage of time in a 24-h period below 1.5 ml L^{-1} DO was 18.8% for 41F; this occurred in the month of December with a single dive having a maximum depth of 357 m and minimum depth of 216 m over a period of 9 h 36 min. The average DO during this dive was 1.6 ml L^{-1} and involved DO exposures less than 1.5 ml L^{-1} for periods of approximately 30 min. Male sharks generally spent less time below the oxygen-minimum zone with maximum exposures in a 24-h period ranging from 0.7 to 11.7% (tags 16M, 18M, 30M, and 40M). Average daily minimum DO for all tags ranged from 2.3 to 3.9 ml L^{-1} (Table 10.2).

DISCUSSION

Although detailed descriptions of horizontal and vertical movement patterns for northeastern Pacific White Sharks in the SOFA and in central California have been published (Goldman and Anderson, 1999; Jorgenson et al., 2010; Nasby-Lucas et al., 2009; Chapter 11, this book), these are the first comprehensive analyses of fine-scale habitat use by White Sharks at GI. There are notable differences and similarities between the habitat preferences of White Sharks at GI to the habitat preferences of White Sharks in other areas, with some apparent seasonal variation. Archival data sets of depth and temperature provide a first look at White Shark activity around GI, but it should be noted that the records of high-resolution data presented here constitute a small sample size, and precise behaviors such as foraging and mating cannot be determined from tag data.

SLRT tag data indicated seasonal changes in the spatial distribution of White Sharks around GI; not surprisingly, the spatial distribution of pinnipeds around the island appears to be impacting the distribution of White Shark activity. Although Fur Seals and Elephant Seals can be found at the island year round, each has seasonal peaks in abundance relative to pupping, mating, and molting. Guadalupe Fur Seals are found throughout the east side of the island (Figure 10.1) and have a pupping and mating season from mid-June through August, after which the males depart but the females remain until the pups are weaned 9–11 months later (Aurioles and Trillmich, 2008). The Northern Elephant Seals are most abundant in two locations at GI: one is a series of small beaches interspersed along the northeast corner of the island, and the other is a large beach on the northwest side of the island (Figure 10.1). Elephant Seal presence peaks during both the breeding season in late January and the molting season of the females and juveniles in the spring (Gallo-Reynoso et al., 2005b).

Shark activity was confined and spread out over the eastern half of the island in August and September, correlating with the presence of Guadalupe Fur Seals. The hot spots of activity from October through March match the two most important Elephant Seal habitats, particularly in December and January. Although the match between pinniped and White Shark spatial distribution confirms the importance of these prey species to White Sharks, it is noteworthy that the seasonal presence of White Sharks misses the June–July peak in Fur Seal presence as well as the latter half of the Elephant Seal peak and times when elephant seals are returning to molt. This is also the case in central California, with sharks departing on an average date of January 7 (Weng et al., 2007) and missing the peak breeding and molting season for Elephant Seals (Le Boeuf and Laws, 1994).

We noted a shift in tagged White Shark diving behavior from August to March, with more time at depths deeper than 40 m noted August through December. This shift corresponded with seasonal distribution of White Sharks around the island, as well as the abundance and distribution of pinnipeds, and may be an indication of the differences between hunting strategies for Fur Seal and Elephant Seal. Bruce et al. (2006) demonstrated plasticity in the vertical swimming behavior of a 3.0-m White Shark tracked along different regions in South Australia and suggested that such changes reflect prey-specific hunting strategies. The shift in vertical behavior to a shallow pattern was noted in January at the peak of the Northern Elephant Seal population. This may indicate foraging on Elephant Seals and might also explain why the subadult sharks (30M and 40M) did not exhibit the same transition in January and left the island later in the season, because smaller sharks may be exploiting different prey.

Shifts in diving behavior could also be related to the arrival of females in September and October (Chapters 11 and 25, this book). Aggression and perhaps courtship and mating (Chapter 16, this book) could alter the sharks' pattern of vertical habitat use as the season progresses. An alternative simple explanation could involve differences in bathymetry at the preferred GI habitats, which changes with prey availability. All of the above hypotheses require considerable additional research in order to determine the true nature of theses movement patterns.

Vertical Movement Patterns

Vertical movement patterns for White Sharks at GI differed from those reported for White Sharks near pinniped colonies in central California and New Zealand. White Sharks in central California (Weng et al., 2007) and New Zealand (Chapter 22, this book) rarely ventured below 50 m (1 and 4% of their time, respectively), whereas sharks at GI were found to spend 22% of their time in this depth range. The simplest explanation for this difference is that the bathymetry surrounding the central California and New Zealand study sites is very shallow compared with GI. The deepest waters within 5 km of southeast Farallon Island are approximately 100 m, with waters immediately adjacent to this site only 40 m deep, whereas the bathymetry at GI drops quickly to depths as deep as 1,900 m at 5 km. GI also has much clearer water than central California, which may allow sharks to visually stalk pinnipeds from greater depths. It is possible that the discrepancy between the number of pinniped predation events observed at GI, compared with southeast Farallon Island and Seal Island, South Africa, is also due to this difference in bathymetry. The steep drop in depth surrounding GI may give the seals an advantage by allowing them to rapidly drop into deep, darker water while remaining very close to the substrate.

It is also not surprising that White Sharks exhibit markedly different diving behavior at GI when compared with that exhibited while in the SOFA, because they are undoubtedly pursuing different prey. Repetitive oscillatory vertical movements were observed in both regions, but average daily maximum depths in the SOFA averaged 469 m (Nasby-Lucas et al., 2009) and were presumed to be associated with foraging within the deep scattering layer. In comparison, average daily maximum depths while at GI averaged 211 m with periods of shallow vertical movements, potentially associated with foraging and the seasonal cycles of pinnipeds.

Oxygen is likely a factor that impacts the depth to which White Sharks can dive throughout their range. The oxygen-minimum zone is considerably shallower in the eastern Pacific than in the central and western Pacific, effectively compressing the vertical habitat (Prince and Goodyear, 2007). Although the precise minimum oxygen limit for White Sharks is unknown, vertical movements from both GI and the SOFA may be constrained by approximately 1.5 ml L^{-1} DO, because average daily maximum diving depths were near the shallow boundary of the oxygen minimum layer at both GI (258 m) and the SOFA (493 m) (Nasby-Lucas et al., 2009). The maximum percentage of time spent below the oxygen-minimum zone (<1.5 ml/L DO) in a 24-h period was less at GI when compared with the SOFA (37.5% for a single tagged female and 13.5–23.6% for tagged males (Nasby-Lucas et al., 2010), suggesting that perhaps DO is not limiting vertical excursions at GI. Little information is available on hypoxia tolerance in sharks, but compared with other fish species, the levels reported are lower than those for Yellowfin Tuna (*Thunnus albacares*) and Skipjack Tuna (*Katsuwonus pelamis*) but comparable with those for Bigeye Tuna (*Thunnus obesus*) and Swordfish (*Xiphias gladius*) (Carey and Robison, 1981; Hanamoto, 1987; Lowe et al., 2000).

Both SLRT and PAT tags noted behavioral differences between small and large sharks at GI. This study, along with previous satellite tagging data (Domeier and Nasby-Lucas, 2008), showed that subadult males begin their offshore migration months after mature sharks. In addition, vertical excursions by smaller sharks tended to be shallower and could be an indication of different feeding strategies because of the onotenogenic shift in diet or the fact that smaller sharks may not be able to spend as much time in cold temperatures as larger individuals. Similarly, there are hints of sex-specific differences in diving behavior that future studies with more data should examine closely. Finally, the difficulty in assigning motivation to behaviors detected in depth and temperature records will hopefully inspire new forms of tags, as technology evolves, that can archive photographic and/or video records to compliment the physical data.

ACKNOWLEDGMENTS

We gratefully acknowledge the financial support of C. Fischer, Guy Harvey Ocean Foundation, the George T. Pfleger Foundation, and the Offield Family Foundation. We also thank T. Pfleger for his help with the PAT tagging and C. Fishcer and B. McBride, as well as the crew of M/V *Ocean* for their assistance with the SLRT tagging. We thank O. Sosa-Nishizaki, F. Galván Magaña, and M. Hoyos-Padilla for the assistance in obtaining Mexican permits. Research was conducted in accordance with permits through Secretaría de Medio Ambiente y Recursos Naturales and Comisión Nacional de Áreas Naturales Protegidas. C. H. Lam was supported by NASA Headquarters under Earth and Space Science Fellowship Program Grant NNX08AU86H.

REFERENCES

Aurioles, D., and F. Trillmich. 2008. *Arctocephalus townsendi*. In *IUCN Red List of Threatened Species*. Version 2010.4. http://www.iucnredlist.org.

Boustany, A. M., S. F. Davis, P. Pyle, S. D. Anderson, B. J. Le Boeuf, and B. A. Block. 2002. Expanded niche for white sharks. *Nature* 415: 35–36.

Bruce, B. D., J. D. Stevens, and H. Malcolm. 2006. Movements and swimming behaviour of white sharks (*Carcharodon carcharias*) in Australian waters. *Marine Biology* 150: 161–172.

Carey F. G., and B. H. Robison. 1981. Daily patterns in the activities of swordfish, *Xiphias gladius*, observed by acoustic telemetry. *Fishery Bulletin* 79: 277–292.

Domeier, M. L., and N. Nasby-Lucas. 2008. Migration patterns of white sharks *Carcharodon carcharias* tagged at Guadalupe Island, Mexico, and identification of an eastern Pacific shared offshore foraging area. *Marine Ecology-Progress Series* 370: 221–237.

Domeier, M. L., D. Kiefer, N. Nasby-Lucas, A. Wagschal, and F. O'Brien F. 2005. Tracking Pacific bluefin tuna (*Thunnus thynnus orientalis*) in the northeastern Pacific with an automated algorithm that estimates latitude by matching sea-surface-temperature data from satellites with temperature data from tags on fish. *Fishery Bulletin* 103: 292–306.

Figueroa-Carranza, A. L. 1994. Early lactation and attendance behavior of Guadalupe fur seal females (*Arctocephalus townsendi*). MA thesis, University of California Santa Cruz.

Francis, M. P. 1996. Observations on a pregnant white shark with a review of reproductive biology. In *Great White Sharks: The Biology of Carcharodon Carcharias* (pp. 157–172). Edited by A. P. Klimley and D. G. Ainley. San Diego, CA: Academic Press.

Gallo-Reynoso, J. P. 1994. Factors affecting the population status of Guadalupe fur seal, *Arctocephalus townsendi* (Merriam, 1897) at Isla de Guadalupe, Baja California, Mexico. PhD thesis, University of California Santa Cruz.

Gallo-Reynoso, J. P., and A. L. Figueroa-Carranza. 2005. Los cetáceos de la Isla de Guadalupe. In *Isla Guadalupe: Restauración y Conservación* (pp. 203–218). Edited by K. Santos del Prado and E. Peters. Mexico: Instituto Nacional de Ecología, Secretaría de Medio Ambiente y Recursos Naturales.

Gallo-Reynoso, J. P., A. L. Figueroa-Carranza, and M. P. Blanco-Parra. 2005a. Los tiburones de la Isla de Guadalupe. In *Isla Guadalupe: Restauración y Conservación* (pp. 143–170). Edited by K. Santos del Prado and E. Peters. Mexico: Instituto Nacional de Ecología, Secretaría de Medio Ambiente y Recursos Naturales.

Gallo-Reynoso, J. P., B. J. Le Boeuf, A. L. Figueroa-Carranza, and M. O. Maravilla-Chávez 2005b. Los pinnípedos de la Isla Guadalupe. In *Isla Guadalupe: Restauración y Conservación* (pp. 171–202). Edited by K. Santos del Prado and E. Peters. Mexico City, Mexico: Instituto Nacional de Ecología, Secretaría de Medio Ambiente y Recursos Naturales.

Garcia, H. E., R. A. Locarnini, T. P. Boyer, and J. I. Antonov. 2006. World ocean atlas 2005. Volume 3: Dissolved oxygen, apparent oxygen utilization, and oxygen saturation. In *NOAA Atlas NESDIS*, Vol. 63. Edited by S. Levitus. Washington, DC: U.S. Government Printing Office.

Goldman, K. J., and S. D. Anderson. 1999. Space utilization and swimming depth of white sharks, *Carcharodon carcharias*, at the South Farallon Islands, central California. *Environmental Biology of Fishes* 56: 351–364.

Hanamoto, E. 1987. Effect of oceanographic environment on distribution. *Bulletin of the Japanese Society of Fisheries Oceanography* 51: 203–216.

Hays, G. C., S. Akesson, B. J. Godley, P. Luschi, and P. Santidrian. 2001. The implications of location accuracy for the interpretation of satellite-tracking data. *Animal Behavior* 61: 1035–1040.

Jorgensen, S., C. A. Reeb, T. K. Chapple, S. Anderson, C. Perle, S. Van Sommeran, et al. 2010. Philopatry and migration of Pacific white sharks. *Proceedings of the Royal Society B Biological Sciences* 277: 679–688.

Le Boeuf, B. J., and R. M. Laws, editors. 1994. *Elephant Seals: Population Ecology, Behavior and Physiology.* Berkeley, CA: University of California Press.

Lowe, T. E., R. W. Brill, and K. L. Cousins. 2000. Blood oxygen-binding characteristics of bigeye tuna (*Thunnus obesus*), a high-energy-demand teleost that is tolerant of low ambient oxygen. *Marine Biology* 136: 1087–1098.

Nasby-Lucas, N., H. Dewar, C. H. Lam, K. J. Goldman, and M. L. Domeier. 2009. White shark offshore habitat: A behavioral and environmental characterization of the eastern Pacific shared offshore foraging area. *PLoS One* 4: e8163.

Pratt, H. L. Jr. 1996. Reproduction in the male white shark. In *Great White Sharks: The Biology of Carcharodon Carcharias* (pp. 131–138). Edited by A. P. Klimley and D. G. Ainley. San Diego, CA: Academic Press.

Pierson, M. O. 1987. Breeding behavior of the Guadalupe fur seal, *Arctocephalus townsendi.* In *Status, Biology, and Ecology of Fur Seals* (pp. 83–94). Edited by J. P. Croxall and R. L. Gentry. NOAA Technical Report NMFS 51. Washington, DC: NOAA.

Prince, E. C., and C. P. Goodyear. 2007. Hypoxia-based habitat compression of tropical pelagic fishes. *Fisheries Oceanography* 15: 451–464.

Rodgers, A. R., and A. P. Carr. 1998. *HRE: The Home Range Extension for ArcView*. Thunder Bay, Canada: Ontario Ministry of Natural Resources, Centre for Northern Forest Ecosystem Research.

Taillade, M. 1992. Animal tracking by satellite. In *Wildlife Telemetry: Remote Monitoring and Tracking of Animals* (pp. 149–160). Edited by I. G. Priede and S. M. Swift. London: Ellis Horwood.

Weng, K. C., J. B. O'Sullivan, C. G. Lowe, C. E. Winkler, H. Dewar, and B. A. Block. 2007. Movements, behavior and habitat preferences of juvenile white sharks *Carcharodon carcharias* in the eastern Pacific. *Marine Ecology-Progress Series* 338: 211–224.

Worton, B. J. 1989. Kernel methods for estimating the utilization distribution in home-range studies. *Ecology* 70: 164–168.

CHAPTER 11

Sex-Specific Migration Patterns and Sexual Segregation of Adult White Sharks, *Carcharodon carcharias*, in the Northeastern Pacific

Michael L. Domeier*
Marine Conservation Science Institute

Nicole Nasby-Lucas
Marine Conservation Science Institute

ABSTRACT

Attaching satellite-linked radio transmitter tags to adult White Sharks provided long-term movement data, leading to the discovery of sex-specific migration patterns in the northeastern Pacific. While males annually consistently migrated to a region known as the Shared Offshore Foraging Area (SOFA), females unpredictably roamed over a much larger pelagic area and were more likely to be found in the SOFA during the autumn when the males were absent. Males exhibited a 1-year migration pattern with up to 9 months spent offshore, whereas females were capable of remaining offshore for up to 15 months as part of a biennial migration pattern. One female returned to the adult aggregation site the year after tagging but then remained offshore during the subsequent aggregation period, indicating that females may rest between reproductive cycles. The spatial distribution of mature male and female White Sharks may only overlap for 90–120 d over a 2-year span.

INTRODUCTION

Sexual segregation has been defined as the separation of members of a species such that males and females live apart, either individually or in single-sex groups (Wearmouth and Sims, 2008). The underlying factors that influence sexual segregation are seemingly diverse; therefore, a single unifying hypothesis to explain the benefits of such segregation is lacking. Some species split into single-sex social groups, whereas others segregate into different habitats but live a solitary lifestyle; these two situations are fundamentally different from a functional perspective. Sexual dimorphism with respect to body size is another confounding factor; different body sizes likely present different energy budgets and nutritional requirements to each sex, driving the occupation of separate ecological niches during the nonreproductive season (Conradt, 1998; Ruckstuhl and Neuhaus, 2002).

Sexual segregation is particularly difficult to observe and study in the marine environment because direct behavioral observations are precluded for species that do not inhabit shallow clear habitats (Sundstrom et al., 2001) and because most marine species cannot be visually sexed. Elasmobranchs are readily sexed via external characters, and thus this group is best represented in the marine literature with respect to sexual segregation [e.g. Scalloped Hammerhead *Sphyrna lewini* (Klimley, 1987); Grey Reef Shark, *Carcharhinus amblyrhynchos* (Economakis and Lobel, 1998); White Shark, *Carcharodon carcharias* (Robbins, 2007; Chapter 20, this book); and Small-Spotted Catshark, *Scyliorhinus canicula* (Jacoby et al., 2010)]. Sharks are of particular interest with respect to segregation, because the cited examples above encompass sex and size segregation, in addition to both habitat and social sexual segregation. This suggests that multiple factors have likely influenced the evolution of sexual segregation among sharks.

Sex-specific seasonal visitation patterns of individual male and female White Sharks to northeastern Pacific aggregation sites (Guadalupe Island, Mexico, and central California, United States) have suggested that sexual segregation may be occurring. For example, photo-identification studies of adult White Sharks have demonstrated that males have a 1-year migration cycle, whereas females may have a 2-year migration cycle (Anderson and Pyle, 2003; Domeier and Nasby-Lucas, 2007; Chapter 25, this book), presumably associated with a gestation period in excess of 1 yr. (Francis, 1996; Mollet et al., 2000). White Sharks tagged with pop-up archival transmitting (PAT) tags at both Guadalupe Island and central California have been shown to undertake seasonal offshore migrations to an area referred to as the shared offshore foraging area (SOFA) (Boustany et al., 2002; Weng et al., 2007, Domeier and Nasby-Lucas, 2008; Chapter 12, this book). Although PAT tags have revealed a consistent offshore/inshore migration pattern for males (Domeier and Nasby-Lucas, 2008; Weng et al., 2007), this tag technology has a limited deployment duration, which precludes the full description of the multiyear female migration cycle. Furthermore, PAT tag data do not allow for fine scale analyses of movement because of the level of error associated with light-based geolocation estimates (Welch and Eveson, 1999; Musyl et al., 2001). Here we employ linked radio transmitter (SLRT) tags, capable of providing long-term, high-resolution tracking data, to describe multiyear migration cycles of adult female White Sharks and evidence of sexual segregation in the northeastern Pacific.

MATERIALS AND METHODS

Satellite-Linked Radio Transmitter Tagging

Adult White Sharks were tagged at Guadalupe Island (GI), Mexico (29.12°N, 118.27°W), and Southeast Farallon Island (SEFI), United States (36.69°N, 122.99°W). Sharks were attracted to the research vessel (126-ft. decommissioned Bering Sea crab boat) by baiting a circle hook that was 33.0 cm long by 17.8 cm wide, with a 12.7-cm gap between the point and the shank (custom hook; Mustad, Gjövik, Norway) with a tuna or salvaged marine mammal carcass. The baited hook was suspended behind the vessel via a plastic float. The circle hook was designed to increase the probability of jaw hooking, thereby minimizing the risk of serious injury to the shark. The hook was rigged with 5 m of 63-mm braided stainless-steel cable as a leader, which in turn was attached to approximately 75 m of coiled nylon rope; all but 6–10 m of rope remained on the deck of the research vessel until a shark was hooked. Four to six large plastic floats (22 kg of flotation each) were evenly spaced along the line to keep the shark near the surface while providing drag. Once a shark was hooked, a smaller boat was used to follow the shark, bring the animal to the surface by shortening the distance between the floats and the shark, and guide the shark onto a large submerged platform that was affixed to the larger research vessel. Once on the platform, the shark was hydraulically raised above the waterline such that tagging could be safely performed. An irrigation hose was immediately placed in the mouth of the shark to flush seawater over the gills, the hook was removed, and a wet towel was placed over the head to protect the eyes and calm the animal. The shark was measured and the sex was recorded prior to release. Determination of sexual maturity was based on total length (TL) of the shark [males ≥ 3.6 m TL, females ≥ 4.5 m TL, following Francis (1996) and Pratt (1996)].

SLRT tags (SPOT5; Wildlife Computers, Redmond, WA) were affixed to the apex of the shark's first dorsal fin by drilling four small holes through the fin and securing the tag with plastic bolts. Each time a tagged shark's dorsal fin was out of the water, a wet/dry switch activated the transmitter. If the tag remained out of the water long enough for an Argos satellite to receive four consecutive transmissions, the Doppler shift-induced frequency change allowed the calculation of the tag's location (Taillade, 1992) with associated location error.

Argos assigns a location class value (LC) that corresponds to one standard deviation of the estimated location error. Argos states that the estimated error for the location data is <250 m for LC 3, between 250 and 500 m for LC 2, between 500 and 1500 m for LC 1, and >1500 m for LC 0. However, the precision of these locations may be better than stated by Argos (Vincent et al., 2002). If only two or three messages are received by the satellite, an LC of A or B is assigned, and a location is calculated, but an error value is not provided. If only one message is received, an LC of Z is assigned, and location data are not available.

Nearly real-time movements of the tagged sharks were tracked using the SLRT tag location data transmitted through the Argos system. A utilization distribution analysis was performed on location data for sharks while in the SOFA using the fixed-kernel method (Worton, 1989) with the home-range extension (Rodgers and Carr, 1998) for ArcView 3.2 (Environmental Systems Research Institute, Redlands, CA). To avoid the bias introduced by variation in sampling frequency, we normalized the position estimates to a frequency of one per day. Position data during the period when the male sharks tagged at GI and SEFI were at their respective aggregation sites were analyzed separately, because they represented two distinct unimodal utilization distributions, but were combined when males from both regions were present in the SOFA. The results were displayed as density contours indicating core regions of occupancy. The 25% density contour indicates the regions of the highest utilization by the sharks, the 50% density contour shows major focal regions, and the 95% density contour encompasses nearly all regions of utilization. The centers of fixed-kernel density contours were determined using the real centroid generator extension for ArcView.

Shark locations, as transmitted by the SLRT tags, were divided into biologically meaningful, temporal subsets. For fixed-kernel density analyses, shark locations were grouped into the following sets: transition from offshore to male-only presence at the adult aggregation sites (July through August); male and female adult aggregation period (September through December); transition from adult aggregation sites to offshore (January through March); and the offshore period (April through June). Because the timing of departure from the adult aggregation sites was staggered rather than coordinated, the transition period contains shark locations at both the aggregation sites and the offshore region.

Sea surface temperature (SST) data were transmitted from the SLRT tags and plotted by month for both male and female sharks. The thermistor on the SLRT tags used in this study was embedded within the body of the tag, resulting in an unknown lag between the time the shark enters water of a different temperature and the time the sensor will record the new temperature. Because White Sharks undergo oscillatory diving that exposes them to widely varying temperatures, the transmitted SST data will contain error if the tag transmits immediately after a deep dive. Furthermore, if the tagged shark fins at the surface for a prolonged time period, the tag can warm above the ambient SST due to solar irradiance. To reduce these errors, we filtered the temperature data for obvious spurious data (greater than five degrees difference within a 24-h period). The resulting SST data were analyzed for average monthly temperature by sex to further characterize the environments inhabited by males and females throughout the year.

RESULTS

A total of 11 adult male and 6 adult female White Sharks were SLRT-tagged at northeastern Pacific adult aggregation sites. Fifteen were tagged at Guadalupe Island between 2007 and 2009, and two were tagged at Southeast Farallon Island in 2009. Because of problems in early tag design, poor tag performance, and variation in individual shark behavior, some tags resulted in only Z class location estimates or very few location estimates. Only tags that produced 60 or more location estimates with a location class of B or better were used for analyses, leaving a total of eight adult males and four adult females (Table 11.1). At the time of this publication, these 12 sharks produced a total of 6,199 d of tracking data (males = 4,040 d, females = 2,159 d), with individual tracks spanning 356–722 d (mean, 517 d; SD, 151 d) (Table 11.1).

Males

Adult male White Sharks had a predictable migratory pattern that involved a residency at their respective adult aggregation sites, beginning as early as July 26 and ending as late as April 2, followed by an offshore migration to the SOFA (Table 11.1 and Figure 11.1). Male residency in the SOFA ranged from 126 to 269 d (mean, 189 d; SD, 52 d), with the SEFI individuals having longer mean offshore residency periods (247 d) than those from GI (170 d) (sample size not robust enough for statistical comparison) (Table 11.1 and Figure 11.1). Males tagged at SEFI began their offshore migration earlier than GI males (mean departure from SEFI, November 23; mean departure from GI, February 7), but they returned to SEFI at the same approximate time as males returning to the GI aggregation site (mean return to SEFI, July 30; mean return to GI, August 7) (Table 11.1). No males from either aggregation site migrated beyond the SOFA to Hawaii, and males from both GI and SEFI returned to their respective aggregation site.

Position data from 2009 and 2010 revealed that adult males from both GI and SEFI were traveling to the same core region of the SOFA. A fixed-kernel analysis was performed using only locations from when adult male White Sharks were offshore and provided boundaries for a pelagic White Shark habitat that is roughly circular in shape with a geographic center at 136.9°N, 23.5°W. The 95% density contour has an approximate diameter of 1,000 km, covering an area over 785,000 km^2 (Figure 11.2). The 50% density contour for combined years was a rough circle covering nearly 200,000 km^2. SLRT tag locations for 2009 and 2010 demonstrated interyear temporal and spatial stability of the SOFA; kernel analyses produced a roughly circular 50% density contour for each year with centers only 53 km apart (Figure 11.2). Analyzing the position and size of the SOFA by month revealed a temporal constriction and spatial shift of the 50% contour during the months of June and July. Adult male White Sharks concentrated activity 300 km northeast of the overall SOFA center, occupying an elliptical 50% density contour centered at 134.61°N, 25.35°W; this ellipse stretched 495 km east-west and 165 km north-south occupying an area of 64,150 km^2 (Figure 11.2). This movement was observed by three males in 2009 (7M, 14M, and 33M) and three males in 2010 (7M, 14M, and 101M). Males left this constricted SOFA region for their respective adult aggregation sites in July.

Females

Four adult female White Sharks were tracked from Guadalupe Island, with two tracks originating in 2008 and two more in 2009. After tagging, the females began an offshore migration as early as January 24 and as late as February 22 (mean, February 5). During their months in the offshore environment, the females were not closely associated with the offshore region occupied by the

Table 11.1 Individual White Sharks Tagged with SLRT Tags off Guadalupe Island, Mexico and Southeast Farallon Island, United States, in 2008 and 2009

Shark Number[a]/ Sex	Tagging Date	Tagging Location	TL (m)	Span of Days for Location Data[b]	Years Previously Sighted[a]	Began Offshore Migration	Returned to Coastal Aggregation	Days Offshore
14M	12/04/2008	GI	4.42	633	2002–2007	3/21/2009, 4/2/2010	8/10/2009, 8/7/2010	142, 127
33M	12/06/2008	GI	4.47	713	2003–2005	3/30/2009, unknown	8/3/2009, unknown	126, NA
7M	12/07/2008	GI	4.57	722	2002–2007	1/13/2009, 1/24/2010, 2/4/2011	8/1/2009, 8/22/2010, still offshoreb	200, 210, NA
FI1M	10/29/2009	SEFI	4.27	410	NA	12/13/2009, 11/29/2010	7/26/2010, still offshoreb	225, NA
FI2M	11/02/2009	SEFI	4.52	415	NA	11/8/2009, 11/12/2010	8/4/2010, still offshoreb	269, NA
19M	11/19/2009	GI	4.39	421	2001–2007	12/4/2009, 2/4/2011	7/30/2010, still offshoreb	238, NA
101M	11/20/2009	GI	4.72	370	NA	3/5/2010	7/29/2010	146
102M	11/20/2009	GI	5.41	356	NA	unknown	8/19/2010	NA
77F	12/03/2008	GI	5.08	755	2004 and 2006	2/1/2009, 1/24/2010	9/17/2009, still offshoreb	228, >365
98F	12/09/2008	GI	4.98	565	NA	1/28/2009	4/16/2010c	443
6F	11/19/2009	GI	4.62	415	2003	2/22/2010	Still offshoreb	>336
100F	11/20/2009	GI	4.62	424	NA	2/16/2010	Still offshoreb	>342

SEFI, Southeast Farallon Island; NA, not applicable.

[a] From Guadalupe Island photo-identification database (see Chapter 25, this book).

[b] At time of publication.

[c] Date of first location point within the Sea of Cortez.

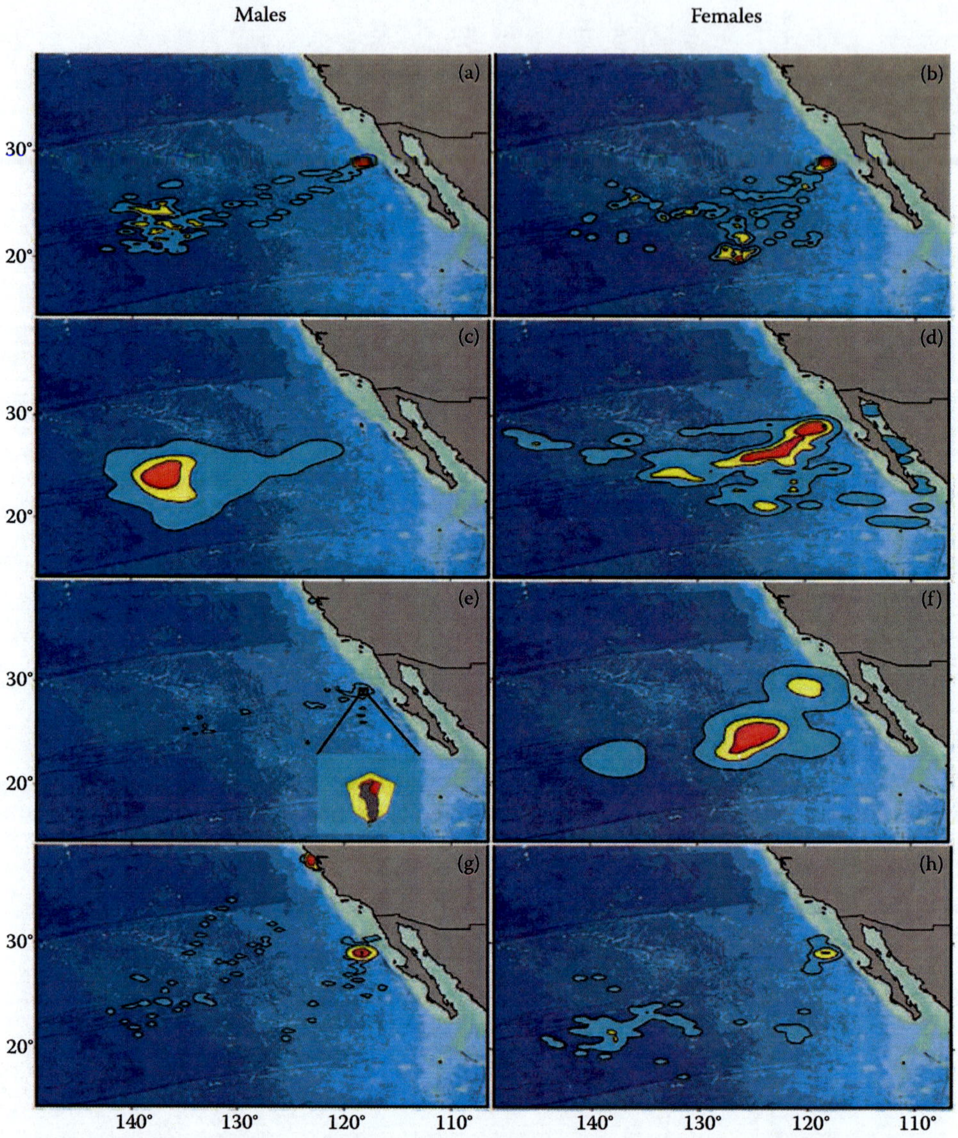

Figure 11.1 Kernel density contours for SLRT-tagged White Sharks from December 2008 to February 2011 for males (a) and females (b) January through March; males (c) and females (d) April through June; males (e) and females (f) July and August; and males (g) and females (h) September through December. The colors indicate core areas of habitat utilization, where red = 25%, yellow = 50%, and blue = 95%.

males (Figures 11.1 and 11.3). Whereas the males occupied the SOFA, the females roamed a very large expanse of ocean between GI and the eastern edge of the SOFA, rarely passing through the 50% density contour of the SOFA (Figures 11.3 and 11.4). There was considerable interyear variation in female spatial distribution while offshore; during the transition phase and the male-only adult aggregation phase (January through August) of 2009, tracked females (98F and 77F) never traveled west of 130.7°W longitude, but females (6F, 77F, 98F, and 100F) traveled as far west as 147.2°W during this same time in 2010 (Figure 11.3).

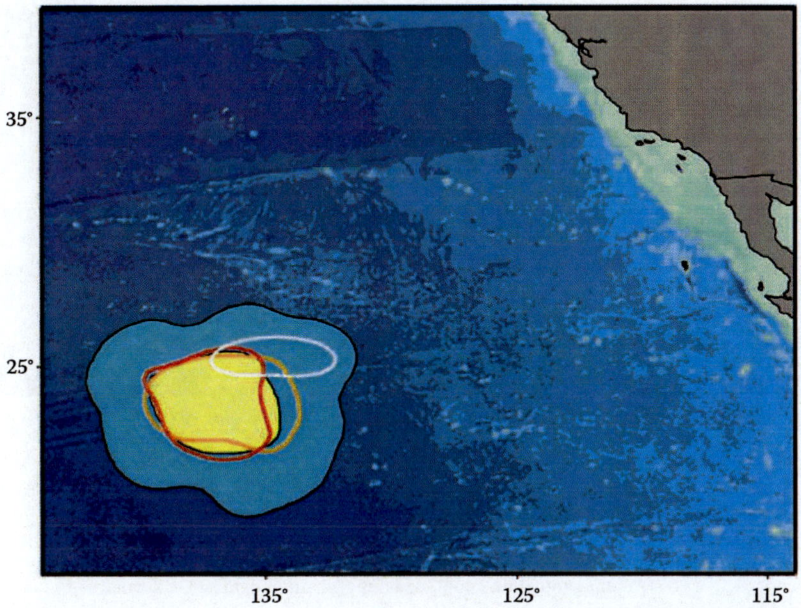

Figure 11.2 Fixed kernel density contours for tagged adult male White Sharks during the periods they occupied the SOFA 2009 through 2011. The blue- and yellow-shaded regions are the 95 and 50% density contours for all offshore data; the orange and red circles are the 50% contours from 2009 and 2010 data, respectively; and the white ellipse represents the seasonal June through July shift and constriction of the SOFA (50% density contour for all June and July data).

Although the SLRT-tagged females were not detected precisely at GI between February and August, three individuals migrated very close to the island between May and July. There were four such visits to the vicinity of GI (Figure 11.5). Female 77F came within 11 km of GI on May 5, 2009 and immediately turned southeast, heading back into the open ocean. From May 22 to 27, 2009, 77F returned to the vicinity of GI, meandering 40 km west and north of the island, eventually passing within 10 km of GI on June 1 and again immediately heading out to sea on the same course as her previous exit. On May 12, 2010, female 100F left the eastern edge of the SOFA's 95% density contour, heading east until she was within 50 km of GI on May 30. Upon reaching GI, 100F immediately reversed course, reaching the edge of the SOFA on June 26, a 3,350-km roundtrip lasting 45 d with an average speed of 74.4 km/d. Female 6F migrated within 30 km of GI on July 17, 2010, departing to the southeast for 300 km before reversing and traveling within 25 km of GI on 14 July; 5 d later she headed back into the open ocean. Regardless of the direction from which these females approached GI, they all departed in a southeasterly direction.

Three females did not return to GI for the adult aggregation season the year following tagging (6F, 98F, and 100F). In contrast, female 77F returned to GI on September 17, 2009, 228 d after her departure from the island; she migrated back into the open ocean on January 24, 2010 and did not return to GI in 2011. After the 2009–2010 aggregation season at GI, females 6F, 77F, and 100F all remained offshore in excess of 1 yr., where they remain at the time of this writing. Female 98F, tagged in 2008, spent 443 d continuously at sea before migrating into the Sea of Cortez in April 2010 (first detection on April 14 was near Loreto, Mexico, 325 km north of the entrance to the sea). She ventured to the northern tip of the Gulf of California before reversing and eventually exiting the Sea of Cortez on June 20, 2010. Her foray into the Sea of Cortez/Gulf of California was in excess of 2 months. We received no signals from her tag after June 2010.

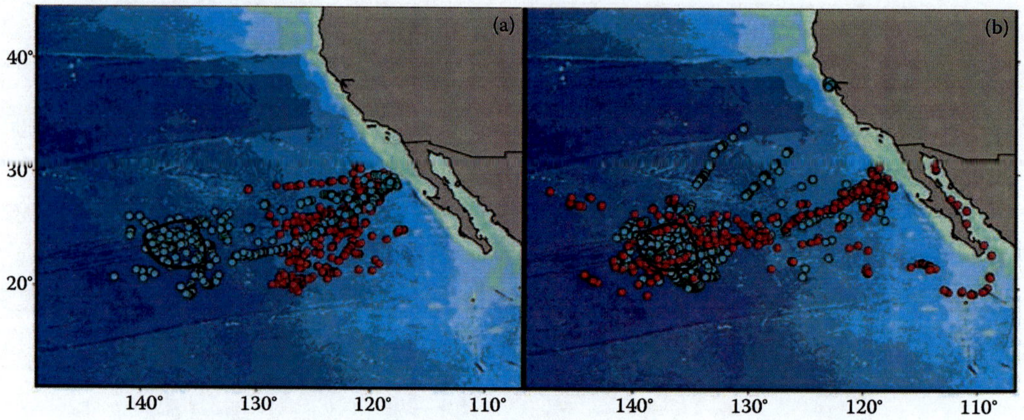

Figure 11.3 Position data for tagged male (blue) and female (pink) White Sharks from December 2008 through August 2009 (a) and September 2009 through August 2010 (b). The black circle indicates the 50% density contour for the adult males.

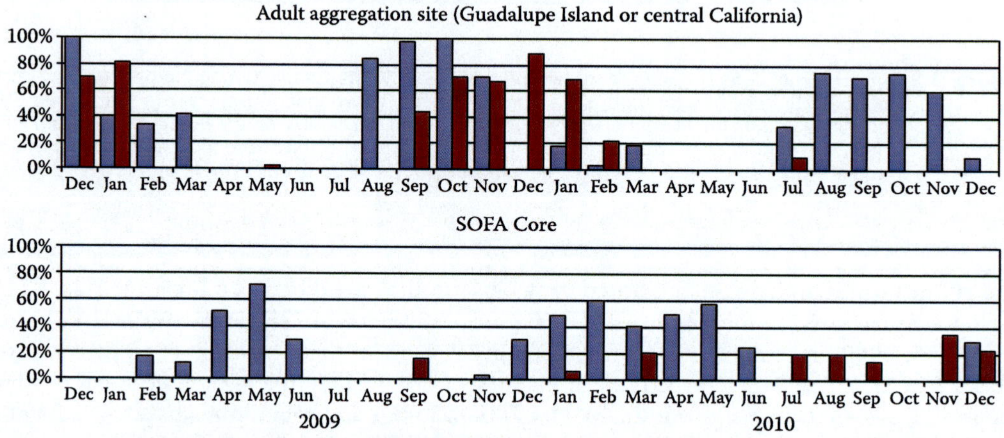

Figure 11.4 Percentage of hits for males and females that were at Guadalupe Island or within the 50% fixed contour (core) of the SOFA, calculated by month for each year.

Sexual Segregation

The percentage of SLRT tag positions in given areas over time provides insight into the spatial distribution of males and females. We analyzed SLRT tag data in this manner to describe the spatial proximity of males and females at GI and within the 95 and 50% density contours of the SOFA. The SLRT tag results reveal that males and females are rarely in close proximity unless they are visiting an adult aggregation site (Figures 11.1 and 11.4), a fact supported by a reanalysis of previously published PAT tag data (from Domeier and Nasby-Lucas, 2008) (Figure 11.6). Because females have a migration cycle of at least 2 yrs., some portion of the adult female population is found offshore year round, with females having a bimodal spatial distribution during the time adults aggregate at GI and SEFI. The females that remained offshore during the adult coastal aggregation season were found within the male-defined SOFA region and even further west (Figure 11.1).

The highest incidence of female hits within the SOFA occurred during the months the males were entirely absent from the SOFA (September through November). The greatest degree of male/

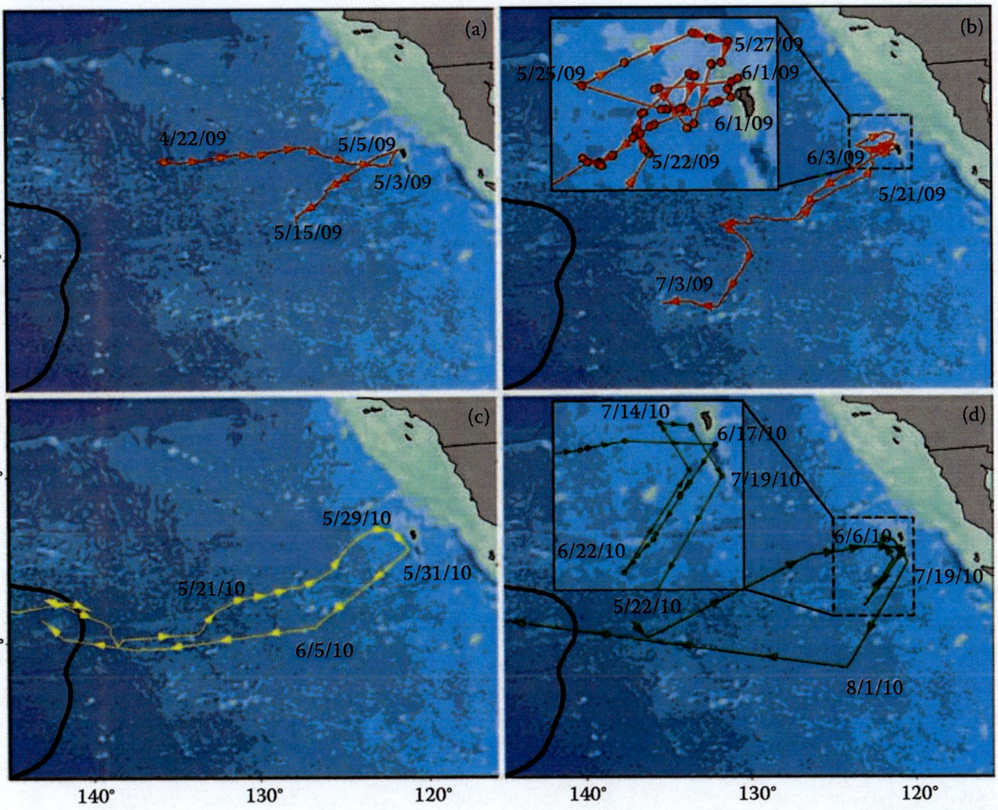

Figure 11.5 Position data for three adult tagged female White Sharks demonstrating short excursions to the vicinity of Guadalupe Island in May or June. Data show tracks from 77F in early-May 2009 (a) and in late-May and early-June 2009 (b) and 100F in late-May 2010 (c) and 6F in mid-June and mid-July 2010 (d). The black outline on the left indicates the 95% density contour for the adult males in the SOFA.

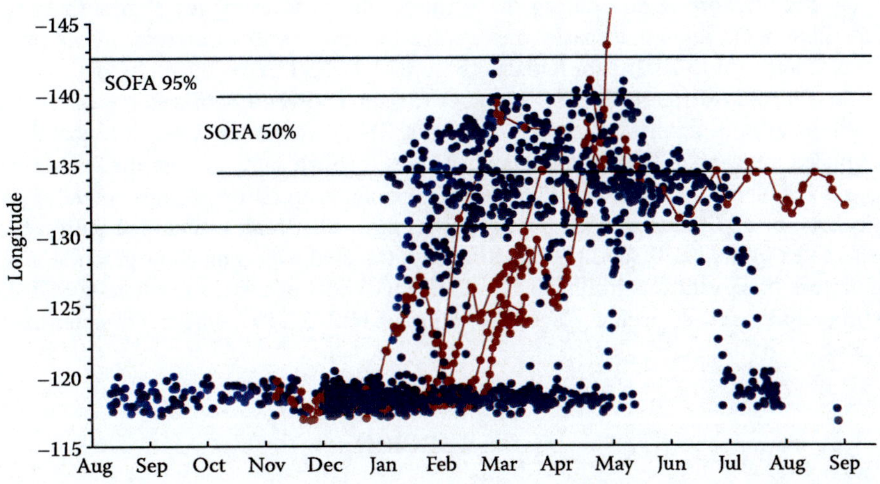

Figure 11.6 PAT tag data from 2001 to 2005 showing offshore movement patterns for males (blue) and females (red). The presence of females east of the SOFA agrees with new SLRT tag data presented here.

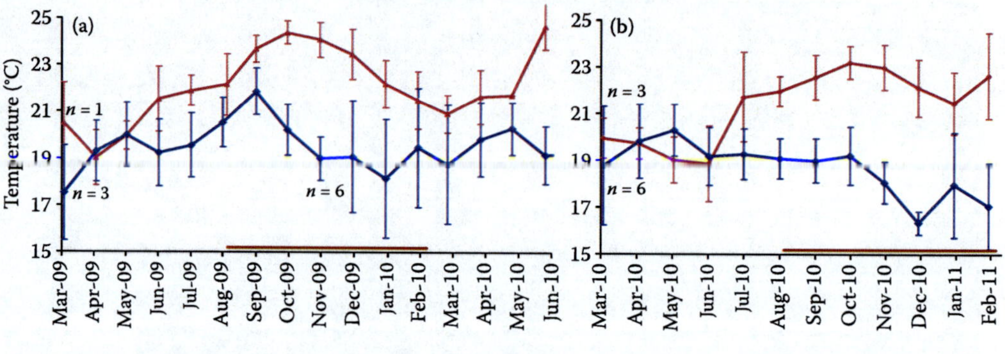

Figure 11.7 Monthly average transmitted SST for male (blue) and female (purple) White Sharks tagged at Guadalupe Island. Displayed are females that stayed offshore in 2009–2010 (a) and in 2010–2011 (b). The brown bar at the bottom indicates the months that the male sharks are at Guadalupe Island.

female offshore overlap occurred in March, but at relatively low levels: 26 and 7% of SLRT locations for males and females, respectively, in the 50% density contour, and 47 and 24% for males and females in the 95% density contour (Figure 11.4). During the peak season of male presence in the SOFA, the females spend much of their time in the space between the SOFA and GI (Figures 11.1, 11.4, and 11.5). None of the tracked females crossed the SOFA's eastern boundary (95% density contour) in 2009, but in 2010 females did cross this eastern boundary, but only rarely were they detected near the periphery of the SOFA's 50% density contour.

Temperature

Females that did not return to GI for the annual adult aggregation remained in the pelagic environment (Figure 11.1h). This large spatial separation between males and females resulted not only in sexual segregation but also habitat segregation. One measurable difference in the offshore and inshore habitats was temperature; tagged females experienced higher sea surface temperatures in the offshore region during the autumn months than tagged sharks that returned to Guadalupe Island. On an annual basis, the average monthly temperatures for males ranged from 16.4 to 21.9°C and 16.1 to 24.7°C for females (Figure 11.7). SST experienced for males and females were not significantly different from September to January when both sexes were at Guadalupe Island (Wilcoxon rank sum tests, $p = 0.1–1.0$), but SST differences were significantly different when the males returned to Guadalupe Island (August through January) and the females remained offshore (ranges of 15.2–23.9°C and 17.6–26.0°C for males and females, respectively) (Wilcoxon rank sum tests, $p < 0.001$) (Figure 11.7). From March to July, males remained within the SOFA, whereas females individually (and not concurrently) traveled over a much larger area, resulting in no fixed pattern of significant similarities or differences between male and female SST data; during this time, males experienced average daily SSTs of 16.2–22.5°C, whereas females experienced 17.0–23.5°C.

DISCUSSION

An admittedly small sample size of SLRT-tagged adult White Sharks revealed consistent sex-specific migration patterns that resulted in extended periods of sexual segregation. The highest incidence of coordinated location data for tagged male and female sharks indicated that individual females may only occupy the same space as males for 90–120 d over a 2-year span, with females spending 15 months or more in the pelagic environment. Although the number of SLRT-tagged individuals is small, the results are supported by patterns observed in previously published studies using PAT tags and photo-identification methods. The multiyear migration pattern of adult females has been suggested by photo-identification studies that found adult females may not return to adult aggregation sites in consecutive years (Anderson and Pyle, 2003; Domeier and Nasby-Lucas, 2007; Chapter 25, this book). Tagging with PATs also revealed that White Sharks are widely dispersed while in the SOFA (Domeier and Nasby-Lucas, 2008), and the distribution of males and females while offshore differs (Jorgensen et al., 2010).

Hypotheses to explain the underlying causes of sexual segregation are many, and several logically apply to the case of White Sharks. Female White Sharks reach a considerably larger size than males, a phenomenon common among sharks (Pratt and Casey, 1983; Hoenig and Gruber, 1990), and therefore the differing energy budgets may cause the sexes to choose different habitats to meet their specific needs. Although both sexes exploit pinniped prey while at adult aggregation sites, differences in spatiotemporal prey availability in the pelagic environment may explain the variation in their distribution patterns. SLRT tagging determined that the SOFA (Domeier and Nasby-Lucas, 2008; Nasby-Lucas et al., 2009), also termed Offshore Focal Area (Weng et al., 2007) and White Shark Café (Jorgensen et al., 2010), is actually a region defined by the presence of males. This area was found to be devoid of marine mammals with the exception of Sperm Whales (Chapter 12, this book); a noted cetacean researcher refers to this part of the Pacific as a "cetacean dead zone" (R. Pitman, personal communication). Whereas male sharks meander within the SOFA pursuing unknown prey, females roam a much larger area in relatively straight, east-west lines, sometimes traveling to or near GI, or into the Sea of Cortez. The region between the SOFA and the mainland is occupied by a number of small species of pelagic cetaceans (Chapter 12, this book), and pinnipeds are found at GI and the Sea of Cortez, suggesting that females may be targeting marine mammals much of the year. However, females do occupy the male-defined SOFA region and areas further west while males are visiting adult aggregation sites, indicating that whatever the food source in the SOFA, females likely exploit it as well.

Our tagging revealed evidence of habitat segregation, but some of the data could also be interpreted as supporting social segregation. In many cases, social segregation involves species that occur in social groups that exhibit intrasexual affinity, but it can also describe situations where intersexual asynchrony or aversion exists (Wearmouth and Sims, 2008). In the case of adult White Sharks, we saw some evidence that females may actively avoid males. During the peak season of male residency in the SOFA (April through June), none of our tagged females traveled within the 50% density contour for adult males, and only rarely did they travel through the 95% density contour. Once the males left the SOFA for the nearshore adult aggregation sites, females that remained offshore frequented the SOFA, departing again when the males returned.

The thermal niche hypothesis may also explain our observation of sexual segregation in adult White Sharks. The observation that female Grey Reef Sharks occupy shallower, warmer water than males led to the assumption that females may select warmer water to facilitate the optimum growth of their developing embryos (Economakis and Lobel, 1998). This hypothesis was also proposed to explain differences in the sex ratio of adult White Sharks at the Neptune Islands, Australia, where months with warmer sea surface temperature resulted in a higher incidence of females (Robbins, 2007; Chapter 20, this book). Because the thermistor in our SLRT tags was embedded in the body of the tag, the SST data certainly contained error that impacted the magnitude of the standard deviation. Despite this known weakness, the data were valuable. We have demonstrated that adult female White Sharks experience significantly warmer SSTs when they do not return to the adult aggregation site and remain offshore. Although this observation is consistent with the thermal-niche hypothesis, our data currently do not permit us to evaluate this. Furthermore, adult females have been shown to spend considerable time at depth while in the offshore habitat (Nasby-Lucas et al., 2009), thus undergoing prolonged exposure to cold temperatures and possibly nullifying the advantage that developing embryos might otherwise experience.

Although sexual segregation in northeastern Pacific White Sharks is evident, the motivation for this separation is still unclear. We have strong data to describe the areas utilized by the adult males but still lack a full understanding of the preferred habitat for the females. Evidence of segregation based on a foraging strategy, social interaction, or temperature cannot be ruled out. Further, we lack data from sexually immature female White Sharks for comparison with data for adults presented here.

Our SLRT tag data provided results that complemented previous studies of northeastern Pacific White Sharks from PAT tag data. Two additional studies conducted off South Australia produced conflicting results with respect to sexual segregation. For example, a female-biased sex ratio of adult White Sharks at Dangerous Reef, South Australia (Bruce, 1992), was later contradicted by a more detailed study that found an overall male bias at the nearby Neptune Islands (Chapter 20, this book). The fine-scale temporal data of Chapter 20, this book also revealed changes in sex ratios correlated with changes in water temperature. It is possible that the earlier study of Bruce (1992) did not have enough temporal coverage to generate the data necessary to observe patterns of sexual segregation noted by Robbins. Another possibility is that adult White Sharks may not be visiting the Neptune Islands and Dangerous Reef during the same portion of their annual (and biennial) life cycle as their conspecifics at GI and SEFI. Longer-term observations will eventually determine whether the South Australia site has a consistent or changing pattern of sexual segregation. Currently, it is not possible to conclude that White Sharks universally exhibit the patterns of offshore migration and sexual segregation that we observed using SLRT tags. We propose that applying SLRT tags to sharks in other regions may help address some of the gaps in our knowledge with respect to the fine-scale spatiotemporal residency patterns and habitat use of White Sharks.

ACKNOWLEDGMENTS

We gratefully acknowledge the financial support of C. Fischer, Guy Harvey Ocean Foundation, the George T. Pfleger Foundation, and the Offield Family Foundation. We also thank C. Fischer, Brett McBride and the crew of M/V *Ocean* for their assistance in SLRT tagging adult White Sharks. We thank Oscar Sosa-Nishizaki, Felipe Galván-Magaña, and Mauricio Hoyos-Padilla for assistance in obtaining Mexican permits. Research was conducted in accordance with permits through Secretaría de Medio Ambiente y Recursos Naturales, Comisión Nacional de Áreas Naturales Protegidas, California Department of Fish and Game, and the National Oceanic and Atmospheric Administration Office of National Marine Sanctuaries Program. We thank Liz Wheeler and Emily Andrews of the Marine Mammal Center for their assistance in the collection of salvaged whale blubber.

REFERENCES

Anderson, S. D., and P. Pyle. 2003. A temporal, sex-specific occurrence pattern among white sharks (*Carcharadon carcharias*) at the south Farallon Islands, California. *California Fish and Game* 89: 96–101.

Boustany A. M., S. F. Davis, P. Pyle, S. D. Anderson, B. J. Le Boeuf, and B. A. Block. 2002. Expanded niche for white sharks. *Nature* 415: 35–36.

Bruce, B. D. 1992. Preliminary observations on the biology of the white shark, *Carcharodon carcharias*, in South Australian waters. *Australian Journal of Marine and Freshwater Research* 43: 1–11.

Conradt, L. 1998. Could asynchrony in activity between the sexes cause intersexual social segregation in ruminants? *Proceedings of the Royal Society B Biological Sciences* 265: 1359–1363.

Domeier, M. L., and N. Nasby-Lucas. 2007. Annual re-sightings of photographically identified white sharks (*Carcharodon carcharias*) at an eastern Pacific aggregation site (Guadalupe Island, Mexico). *Marine Biology* 150: 970–984.

Domeier, M. L., and N. Nasby-Lucas. 2008. Migration patterns of white sharks *Carcharodon carcharias* tagged at Guadalupe Island, Mexico, and identification of an eastern Pacific shared offshore foraging area. *Marine Ecology Progress Series* 370: 221–237.

Economakis, A. E., and Lobel, P. S. (1998). Aggregation behavior of the grey reef shark, *Carcharhinus amblyrhychos*, at Johnston Atoll, central Pacific Ocean. *Environmental Biology of Fishes* 51: 129–139.

Francis, M. P. 1996. Observations on a pregnant white shark with a review of reproductive biology. In *Great White Sharks: The biology of Carcharodon carcharias*. Edited by A. P. Klimley and D. G. Ainley, pp. 157–172. San Diego, CA: Academic Press.

Hoenig, J.M., and S. H. Gruber. 1990. Life-history patterns in the elasmobranchs: Implications for fisheries management. In *Elasmobranchs as living resources: Advances in biology, ecology, systematics and status of the fisheries*. Edited by H. L. Pratt, Jr., S. H. Gruber, and T. Taniuchi, pp. 1–16. NOAA Report NMFS 90. Washington, DC: NOAA.

Jacoby, D. M. P., D. S. Busawon, and D. W. Sims. 2010. Sex and social networking: The influence of male presence on social structure of female shark groups. *Behavioral Ecology* 21: 808–818.

Jorgensen, S., C. A. Reeb, T. K. Chapple, S. Anderson, C. Perle, S. Van Sommeran, et al. 2009. Philopatry and migration of Pacific white sharks. *Proceedings of the Royal Society B Biological Sciences* 277: 679–688.

Kimber, J. A., D. W. Sims, P. H. Bellamy, and A. B. Gill. 2009. Male-female interactions affect foraging behaviour within groups of small-spotted catshark, *Scyliorhinus canicul. Animal Behavior* 77: 1435–1440.

Klimley, A. P. 1987. The determinants of sexual segregation in the scalloped hammerhead *Sphyrna lewini*. *Environmental Biology of Fishes* 18: 27–40.

Mollet, H. F., G. Cliff, H. L. Pratt, Jr., and J. D. Stevens. 2000. Reproductive biology of the female shortfin mako, *Isurus oxyrinchus Raninesque*, 1810, with comments on the embryonic development of lamnoids. *Fisheries Bulletin* 98: 299–318.

Musyl, M., R. Brill, D. Curran, J. S. Gunn, J. R. Hartog, R. D. Hill, et al. 2001. Ability of archival tags to provide estimates of geographical position based on light intensity. In *Electronic tagging and tracking in marine fisheries*. Edited by J. R. Sibert and J. L. Nielsen, pp. 343–368. Dordrecht, the Netherlands: Kluwer Academic Press.

Nasby-Lucas, N., H. Dewar, C. H. Lam, K. J. Goldman, and M. L. Domeier. 2009. White shark offshore habitat: A behavioral and environmental characterization of the eastern Pacific shared offshore foraging area. *PLoS ONE 4: e8163*.

Pratt, H. L., Jr. 1996. Reproduction in the male white shark. In *Great white sharks: The biology of Carcharodon carcharias*. Edited by A. P. Klimley and D. G. Ainley, pp. 131–138. San Diego, CA: Academic Press.

Pratt, H. L., Jr., and J. G. Casey. 1983. Age and growth of the shortfin mako, *Isurus oxyrinchus*, using four methods. *Canadian Journal of Fisheries and Aquatic Sciences* 40: 1944–1957.

Robbins, R. L. 2007. Environmental variables affecting the sexual segregation of great white sharks *Carcharodon carcharias* at the Neptune Islands South Australia. *Journal of Fish Biology* 70: 1350–1364.

Rodgers, A. R., and A. P. Carr. 1998. HRE: The home range extension for ArcView. Thunder Bay, Canada: Ontario Ministry of Natural Resources, Centre for Northern Forest Ecosystem Research.

Ruckstuhl, K. E., and P. Neuhaus. 2002. Sexual segregation in ungulates: A comparative test of three hypotheses. *Biological Reviews* 77: 77–96.

Sundstrom, L. F., S. H. Gruber, S. M. Clermont, J. P. S. Correia, J. R. C. de Marignac, J. F. Morrissey, et al. 2001. Review of elasmobranch behavioural studies using ultrasonic telemetry with special reference to the lemon shark, *Negaprion brevirostris*, around Bimini Islands, Bahamas. *Environmental Biology of Fishes* 60: 225–250.

Taillade, M. 1992. Animal tracking by satellite. In *Wildlife telemetry: Remote monitoring and tracking of animals*. Edited by I. G. Priede and S. M. Swift, pp. 149–160. London: Ellis Horwood.

Vincent, C., B. B. McConnell, V. Ridoux, and M. Fedak. 2002. Assessment of Argos location accuracy from satellite tags deployed on captive gray seals. *Marine Mammal Science* 18: 156–166.

Wearmouth, V. J., and D. W. Sims. 2008. Sexual segregation behaviour of marine fish, reptiles, birds and mammals: Patterns, mechanisms and conservation implications. *Advances in Marine Biology* 54: 107–170.

Welch, D. W., and J. P. Eveson. 1999. An assessment of light-based geoposition estimates from archival tags. *Canada Journal of Aquatic Sciences* 56: 1317–1327.

Weng, K. C., A. M. Boustany, P. Pyle, S. D. Anderson, A. Brown, and B. A. Block. 2007. Migration and habitat of white sharks (*Carcharodon carcharias*) in the eastern Pacific Ocean. *Marine Biology* 152: 877–894.

Worton, B. J. 1989. Kernel methods for estimating the utilization distribution in home-range studies. *Ecology* 70: 164–168.

CHAPTER 12

The Northeastern Pacific White Shark Shared Offshore Foraging Area (SOFA)

A First Examination and Description from Ship Observations and Remote Sensing

Michael L. Domeier*
Marine Conservation Science Institute

Nicole Nasby-Lucas
Marine Conservation Science Institute

Daniel M. Palacios
University of Hawaii at Manoa and National Oceanic and Atmospheric Administration/
National Marine Fisheries Service/Southwest Fisheries Science Center

ABSTRACT

Previous studies have shown that adult White Sharks (*Carcharodon carcharias*) migrate from aggregation sites near the coast to pelagic habitats situated between North America and Hawaii. Studies using satellite-linked radio transmitting (SLRT) tags have allowed for a better spatial description of this region while also delineating sex-specific offshore-habitat partitioning. Although females roam widely in the pelagic habitat, males occupy a more defined region called the Shared Offshore Foraging Area (SOFA). Here we report the first direct observations of the SOFA, made from a ship in June 2009, and compare these observations with oceanographic remote sensing data to characterize the SOFA habitat. White sharks tagged with SLRT tags provided real-time tracking positions to guide the research vessel toward offshore regions used by White Sharks. The timing of the expedition coincided with a seasonal contraction of the SOFA. Ship observations in the region where the sharks were detected revealed an absence of epipelagic fish and small cetaceans, but Sperm Whales (*Physeter macrocephalus*) and three species of spawning squid (*Architeuthis* and two species of Ommastrephid squids) were present. Oceanographic remote sensing data showed the region to be an epipelagic cold spot, broadly characterized by downwelling conditions, reduced horizontal motions or other dynamical processes, no major temperature fronts, and very low phytoplankton biomass. However, the presence of apex predators like White Sharks, Sperm Whales, and squid suggests that the SOFA ecosystem supports a considerable mesopelagic biomass. Oceanic hot spots are currently described primarily by epipelagic and surface observations; this study demonstrates the need to incorporate mesopelagic observations into the characterization of hot and cold spots.

INTRODUCTION

Pop-up satellite archival transmitting and satellite-linked radio transmitting (SLRT) tags have revealed that adult White Sharks (*Carcharodon carcharias*) of the northeastern Pacific occupy an entirely pelagic habitat for at least 3 months and as much as 15 months of the year (Boustany et al., 2002; Weng et al., 2007; Domeier and Nasby-Lucas, 2008, 2011; Nasby-Lucas et al., 2009; Jorgensen et al., 2010; Chapter 11, this book), raising many questions regarding the ecosystem that sustains them during this time. Individual White Sharks from both central California and Guadalupe Island, Mexico, migrate to the Shared Offshore Foraging Area (SOFA) centered between the Hawaiian Islands and Baja California (Boustany et al., 2002; Weng et al., 2007; Domeier and Nasby-Lucas, 2008; Nasby-Lucas et al., 2009; Jorgensen et al., 2010; Chapter 11, this book). Although the use of SLRT tags allowed for the SOFA to be better defined spatially (Chapter 11, this book), our understanding of the ecosystem and environmental characteristics of this seemingly vast open-ocean habitat is in its infancy.

The factors that motivate White Sharks to make these round-trip offshore migrations are unknown, but hypotheses that propose foraging and/or mating have been suggested. The length of time the sharks spend in the pelagic realm, coupled with very active vertical diving behavior (Nasby-Lucas et al., 2009), suggests that the area is used for foraging (Domeier and Nasby-Lucas, 2008). Alternatively, Jorgensen et al. (2010) proposed that the timing of the arrival of young-of-the-year White Sharks into coastal nursery areas, together with the presumed length of gestation (Francis, 1996; Mollet et al., 2000), suggests that the SOFA may function primarily as a mating area. However, evidence presented in this book demonstrated that males and females are spatially segregated while offshore, with the SOFA being primarily occupied by males (Chapter 11, this book). This apparent sexual segregation makes the offshore mating hypothesis unlikely (Chapter 16, this book).

Whether or not mating is occurring offshore, it is accepted that the sharks must forage during their prolonged pelagic residence (Domeier and Nasby-Lucas, 2008; Jorgensen et al., 2010). It is not known what prey species form the diet of White Sharks in the SOFA, but archival tag data from both Guadalupe Island and central California sharks show that White Sharks regularly dive to great depths while in the SOFA, recording mean maximum daily swimming depths of 400–500 m (Nasby-Lucas et al., 2009; Jorgensen et al., 2010). Furthermore, White Sharks in the SOFA exhibit diel diving behavior, suggesting that these sharks may be exploiting the deep scattering layer (Nasby-Lucas et al., 2009). Given the remote location, White Sharks have never been observed feeding while in the SOFA.

The prolonged residence of adult White Sharks in the SOFA leads us to question whether the SOFA may be an oceanic hot spot; hot spots are defined as regions of energetic and dynamic oceanographic processes induced by winds, fronts, current instabilities, mesoscale eddies, or complex topography that lead to enhanced biological productivity. Hot spots can have high fisheries yields and are known regions of apex predator aggregation (Palacios et al., 2006). Direct and indirect measurements of the physical and biological parameters that persist in the SOFA are necessary to properly describe the ecosystem. Biological productivity in the pelagic realm can be both temporally and spatially variable. Therefore, a detailed study of White Shark habitat in the offshore region requires more precision with respect to the time and location that the sharks occupy while in this region, precision that is afforded by the use of SLRT-tagged White Sharks.

The primary objective of this study was to describe the SOFA via remote sensing and direct observations made from a research vessel. Analyses performed on location data from adult White Sharks fitted with SLRT tags indicated that adult male White Sharks occupy the same offshore region each year, whereas females were found over a broader and more temporally and spatially unpredictable region (Chapter 11, this book). Furthermore, it has been noted that males occupy a significantly smaller pelagic space during the months of June and July, resulting in a predictable but temporary constriction of the SOFA (Chapter 11, this book). The real-time, precise locations of the tagged adult male White Sharks were used as a reference to guide the research vessel into this constricted space during the month of June, allowing the first direct examination of the SOFA environment. For the purpose of this paper, the core of the SOFA is defined as the 50% density contour identified by Domeier and Nasby-Lucas (Chapter 11, this book) for the June/July period of spatial constriction.

MATERIALS AND METHODS

Data from of eight SLRT (SPOT5; Wildlife Computers, Redmond, Washington) tagged males (see Chapter 11, this book for deployment methods) were used as a reference before and after the research expedition to the SOFA. Six of the eight males were tagged at Guadalupe Island in 2008 and 2009 (sharks 7M, 14M, 19M, 33M, 101M, and 102M)), and two were tagged at the Farallon Islands in 2009 (sharks FI1M and FI2M). A research cruise was planned in the month of June 2009 to the northwestern region of the SOFA using near real-time position data from the tagged sharks at the time of the cruise as a guide for the ship's track. The research vessel departed San Diego on June 8, 2009 and returned June 25, 2009 (Figure 12.1).

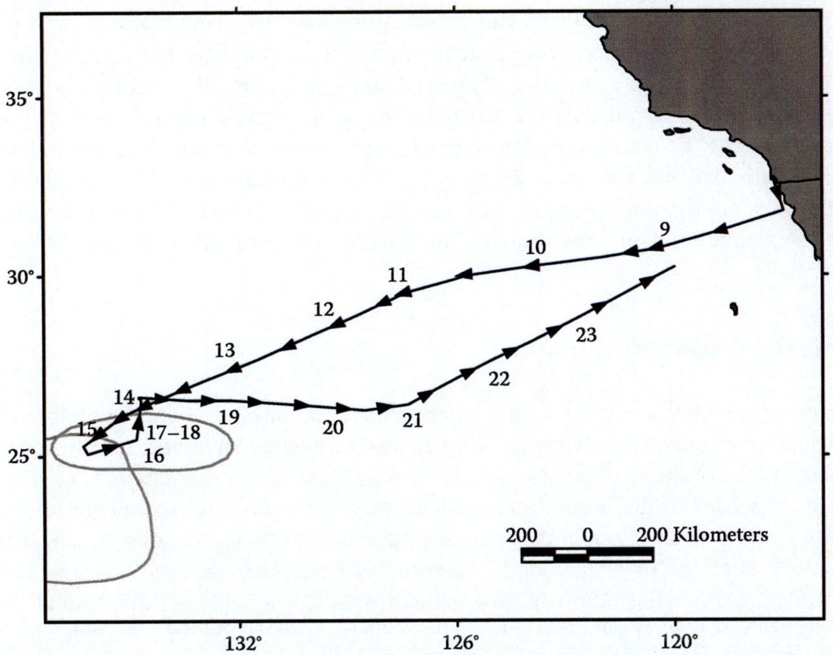

Figure 12.1 The ship's actual track for the SOFA cruise in June 2009. Each number indicates the position along the track of midday for the specified date in June 2009 and corresponds to marine mammal observations along the ship's tracks by date in Table 12.2. The gray contours represent the core utilization areas for sexually mature male White Sharks, with the smaller oval being the 50% density contour for the months of June and July (Chapter 11, this book).

Visual Strip Transects of Biological Diversity within the SOFA

Strip transects were conducted, following the methods described by Olson et al. (2001). Observations were made from the flying bridge, with eye level at 7 m above the waterline. In summary, all organisms that could be observed and counted, at or above the surface out to 300 m from the vessel, were recorded and identified to species if possible. Floating marine debris were also identified and quantified. Distance to the observed animal or object was estimated using a fixed-interval range finder (Heinemann, 1981), and binoculars were used to assist in species identification. Counts took place on only one side of the ship (whichever side had better visibility), from the bow through a 90° arc that ended perpendicular to the vessel at observer's position on the flying bridge for 1 h. The exact position, bearing, and speed of the ship were recorded at the beginning and end of each transect so that the total area of the surveyed strip could be calculated. If the bearing of the ship was altered during a transect, that particular transect was discontinued. There were not enough qualified personnel on the vessel to maintain continuous surveys; instead, such surveys were conducted daily when time and weather permitted. All of the transects were conducted by the same person (M. L. D.).

The counts of organisms and debris were converted to density (#/km^2) by dividing by the area covered during the 1-h transect. On a few occasions, a specific organism was encountered in such large numbers that it was impossible to count; in these instances an entry saying "too many to be counted" was entered to document the event.

Marine Mammal Surveys

Similar surveys were conducted for marine mammals, but in this case the surveys were conducted to the horizon on both sides of the vessel. Whenever a marine mammal was sighted, the course of the vessel was changed to approach, count, and identify the species. Marine mammals were spotted both via active searching with gyro-stabilized binoculars and with the naked eye. Searching occurred during all daylight hours by the ship's crew on watch, and they alerted the expedition leader (M. L. D.) of any sightings so that he could attempt to count and identify the species. The methods were not rigorous enough to calculate densities. For example, larger cetaceans could be spotted from a greater distance than small cetaceans so a precise search radius could not be calculated. Thus, raw marine mammal counts provided a rough index of relative abundance.

Environmental Characterization

The research vessel was not fitted with oceanographic sampling gear, and therefore in situ environmental data were not collected from the ship. Instead, we relied on remote sensing data to provide a general description of the survey region, with the important caveat that satellites only measure the near-surface properties of the ocean, and consequently, potentially important subsurface features are not detected. We obtained monthly composites for June 2009 for various remote sensing products available through the Environmental Research Division's Data Access Program (ERDDAP), a data pass-through service hosted by the CoastWatch West Coast Node of the National Oceanic and Atmospheric Administration (NOAA) (http://coastwatch.pfeg.noaa.gov/erddap/index.html).

The primary products obtained from ERDDAP were sea surface temperature (SST), phytoplankton chlorophyll-*a* concentration (CHL), and sea surface height deviation (SSH). SST is a 0.1° (~11 km) spatial-resolution product blending observations from several satellite platforms for optimal coverage Moderate Resolution Imaging Spectroradiometer (MODIS), on the National Aeronautics and Space Administration (NASA) Aqua satellite; Advanced Very High Resolution Radiometer on

NOAA's Polar Operational Environmental Satellites; Imager on NOAA's Geostationary Operational Environmental Satellites; and Advanced Microwave Scanning Radiometer on Aqua. CHL is a 0.05° (~5.5 km) spatial resolution product from MODIS/Aqua. Finally, SSH is a 0.25° (~25 km) spatial resolution product that merges observations from multiple satellite altimeters and that is provided by the Archiving, Validation and Interpretation of Satellite Oceanographic data program and Collect Localisation Satellites (CLS) in France.

SSH measures the ocean's surface topography, from which geostrophic current vectors (u and v for the zonal and meridional components, respectively) can be derived. A related product, geostrophic current anomaly (u' and v', respectively), which measures current deviations caused by mesoscale processes like eddies, was also obtained in order to compute eddy kinetic energy [EKE; where $EKE = 0.5 * (u'^2 + v'^2)$], a more direct description of mesoscale motions.

RESULTS

Strip Transect Results

Strip transects began on June 9, 2009, after the ship's track had moved west of the continental shelf. Twenty-two 1-h strip transects were performed along the ship's track between June 9 and 22, 2009 (Figure 12.1). On June 17, high sea-state conditions caused the cessation of all research activities, preventing all but one transect from being sampled on the return voyage (June 22).

Seabirds encountered during strip surveys included Leach's Storm Petrel (*Oceanodroma leucorhoa*; n = 60), Laysan Albatross (*Phoebastria immutabilis*; n = 1), Black-Footed Albatross (*Phoebastria nigripes*; n = 2), Red-Tailed Tropicbird (*Phaethon rubricauda*; n = 3), unidentified petrel (*Ptetodroma* sp.; n = 1), and an unidentified shearwater (*Puffinus* sp.; n = 1). Combined seabird densities ranged from 0 to 8.4/km^2 and were highest near the continental shelf. Eleven of the twenty-two strip transects encountered no seabirds of any kind. Leach's Storm Petrel was the most abundant species observed, but only one individual was seen west of 127.3°W longitude. Seabirds were very sparse in the core of the SOFA, with only two Red-Tailed Tropicbirds and one unidentified petrel counted during strip transects. However, several Black-Footed Albatross (n = 8) were seen within the SOFA during times when strip transects were not underway (Table 12.1 and Figure 12.2).

Flying Fish densities were rarely above 0.0, but small peaks were found at 120°, 127°, and 132°W longitude. Between 122.7° and 123.3°W longitude, *Velella velella* (By-The-Wind-Sailor) were encountered in such high numbers that they were impossible to count. A small debris field (debris density = 0.5/km^2) was observed at approximately 123°W longitude, and a much larger debris field (peak density = 3.63/km^2) was encountered between 131° and 135°W longitude (Table 12.1 and Figure 12.2). The most abundant identifiable objects in these fields were plastic fishing floats, but bits of rope, wood, and other plastics were also observed.

Marine Mammal Surveys

Short-Beaked Common Dolphin (*Delphinus delphis*), Striped Dolphin (*Stenella coeruleoalba*), Pantropical Spotted Dolphin (*Stenella attenuata*), unidentified beaked whales (family Ziphiidae), unidentified large rorqual (*Balaenoptera sp.*), Risso's Dolphin (*Grampus griseus*), and Sperm Whales (*Physeter macrocephalus*) were observed (Table 12.2 and Figure 12.3). With the exception of Risso's Dolphin (near 130°W), no small cetaceans were found west of about 127°W. The only cetaceans found in the core SOFA region were Sperm Whales. Note that only animals at the surface were used to estimate numbers, and no allowance was made for those members of the pod underwater.

Table 12.1 **Observations of Seabirds, Flying Fish, Hydrozoans, and Floating Debris in Density per Square Kilometer from Visual Transects**

Transect	Date	Start Time	Distance (km)	LSP	BFA	LA	SW	P	TB	FF	V	FD
1	6/9/2009	7:10	14.1	2.12	0	0.24	0	0	0.24	0	0	0
2	6/9/2009	11:46	14.2	8.44	0	0	0	0	0	0.23	0	0.23
3	6/10/2009	7:57	15.2	0.44	0	0	0	0	0	0	TNC	0
4	6/10/2009	8:59	16.3	0	0	0	0	0	0	0	TNC	0
5	6/10/2009	11:30	14.2	0	0	0	0	0	0	0	TNC	0.47
6	6/10/2009	15:15	15.6	0	0	0	0	0	0	0	0	0
7	6/11/2009	8:00	14.5	0.46	0	0	0	0	0	0	0	0
8	6/11/2009	12:45	7.0	2.38	0	0	0.48	0	0	0	0	0
9	6/11/2009	16:34	13.6	1.23	0	0	0	0	0	0.74	0	0
10	6/12/2009	8:15	13.7	0	0.24	0	0	0	0	0	0	0
11	6/12/2009	14:05	14.1	0	0	0	0	0	0	0	0	0
12	6/12/2009	16:00	14.5	0	0	0	0	0	0	0	0	0
13	6/12/2009	19:12	14.6	0	0	0	0	0	0	0	0	0
14	6/13/2009	9:15	15.3	0.22	0	0	0	0	0	0.66	0	0.87
15	6/13/2009	11:00	14.6	0	0.23	0	0	0	0	0.23	0	1.14
16	6/13/2009	14:18	13.7	0	0	0	0	0	0	0.24	0	1.22
17	6/14/2009	12:35	11.9	0	0	0	0	0.28	0	0	0	3.63
18	6/15/2009	9:45	15.9	0	0	0	0	0	0.42	0	0	0
19	6/15/2009	13:28	15.0	0	0	0	0	0	0	0	0	0
20	6/16/2009	8:30	12.3	0	0	0	0	0	0	0	0	0.27
21	6/16/2009	12:17	11.5	0	0	0	0	0	0	0	0	0
22	6/22/2009	13:00	12.4	0	0	0	0	0	0	0	0	0

LSP, Leach's Storm Petrel; BFA, Black-Footed Albatross; LA, Laysan Albatross; SW, shearwater; P, unidentified petrel; TB, tropicbird; FF, Flying Fish; V, *Velella velella*; FD, floating debris; TNC, too numerous to count.

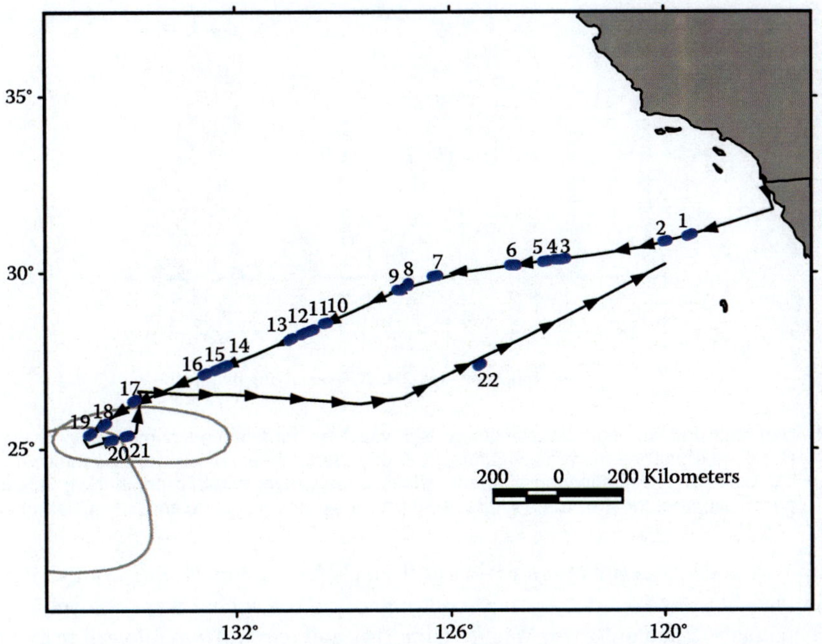

Figure 12.2 The ship's actual track for the SOFA cruise in June 2009. The thick blue segments indicate 1-h strip transects and are labeled by transect number. The gray contours represent the core utilization area for sexually mature males, with the smaller oval being the 50% density contour for adult male White Sharks for the months of June and July (Chapter 11, this book).

Table 12.2 Marine Mammal Sightings

Date[a]	Species	Number	Latitude (°N)	Longitude (°W)
6/9/2009	*Delphinus delphis*	29	30.9	119.9
6/10/2009	*Stenella attenuata*	15	30.3	123.4
6/10/2009	*Balaenoptera* sp.	3	30.2	124.3
6/11/2009	*Delphinus delphis*	100	29.5	127.5
6/11/2009	Unidentified beaked whale	4	28.4	129.8
6/11/2009	*Stenella delphis*	100	28.3	129.8
6/11/2009	*Delphinus delphis*	40	28.3	129.9
6/12/2009	*Grampus griseus*	50	28.1	130.4
6/14/2009	*Physeter macrocephalus*	35	26.0	135.5
6/15/2009	*Physeter macrocephalus*	20	25.2	136.4
6/15/2009	*Physeter macrocephalus*	20	25.1	136.5

[a] Dates correspond to midday locations from Figure 12.2.

Anecdotal Observations in the SOFA

On June 14, the dorsal fin of a shark was spotted at the surface (26.4°N 134.8°W) in the SOFA core. A maximum of 10 cm of fin broke the surface with a flat-calm sea state, as the shark repeatedly surfaced and submerged, allowing it to be tracked visually at intervals for 60 min. The tip of the caudal fin was never visible, and the body of the shark was not seen well enough to estimate its length. The shape and color of the dorsal fin were consistent with that of a White Shark, but without more visual evidence, it was impossible to assign any species-level identification to this sighting. Attempts to attract and catch the shark using bait failed.

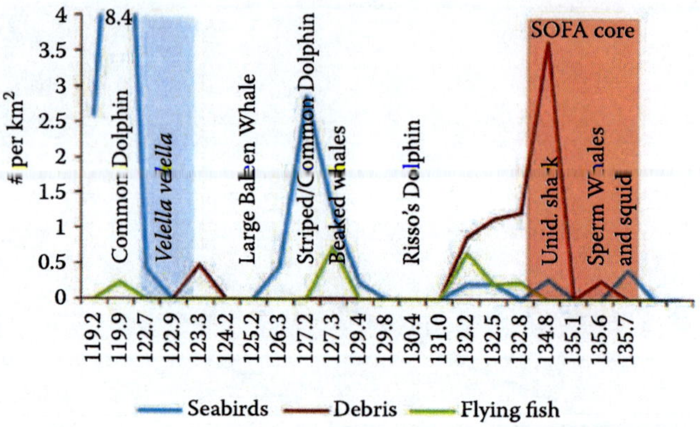

Figure 12.3 Sightings and concentrations of seabirds, Flying Fish, and debris along this ship's track (longitude). Positions of marine mammal sightings and anecdotal observations are also indicated. Blue shading indicates the region where *Velella velella* was sighted, and the red shading indicates the 50% density contour for adult male White Sharks for the months of June and July (Chapter 11, this book).

The ship was in the June/July core region of the SOFA (Chapter 11, this book) from June 14 to 17. Although the ship was underway almost continuously, a sea anchor was deployed in the vicinity of the first encountered pod of Sperm Whales (June 14), and the ship was allowed to drift overnight. Lights were deployed during the nighttime drifts, which attracted two species of Ommastrephid squid (Red Squid [*Ommastrephes bartramii*] and Purpleback Flying Squid [*Sthenoteuthis oualaniensis*]) (Figure 12.4); for both species, males were running ripe, and females were gravid. On the

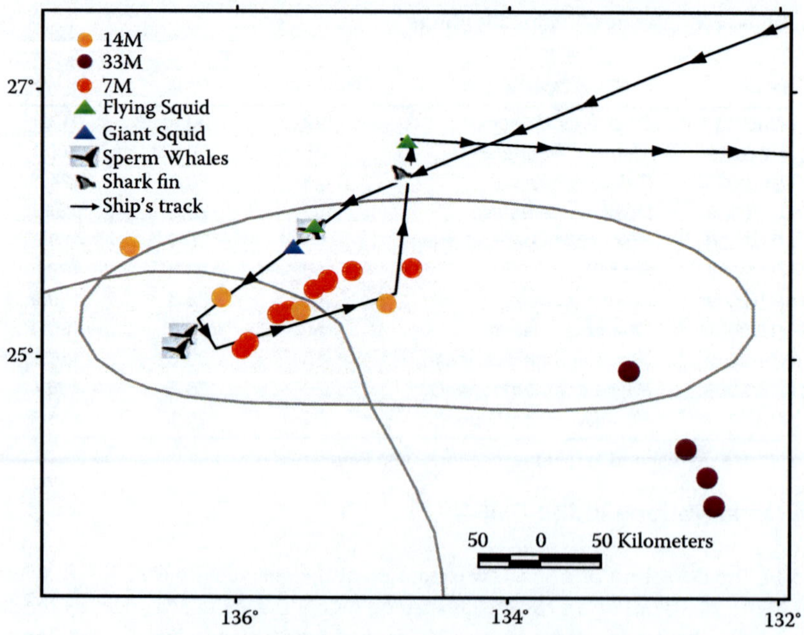

Figure 12.4 Sperm Whale, Flying Squid, Giant Squid, unidentified shark fin, and SLRT-tagged sharks detected between June 14 and 19, 2009. The upper gray contour is the fixed-kernel 50% density contour indicating the major focal area of sexually mature male White Sharks during the months of June and July, and the lower gray contour is the fixed-kernel 50% density contour for all offshore location data from SLRT-tagged sexually mature males (Chapter 11, this book).

morning of June 14, a freshly dead Giant Squid (*Architeuthis dux*) was found floating at the surface near the first Sperm Whale pod sighting. The Giant Squid was missing all but the bases of its arms and tentacles, as well as its eyes. Dissection revealed it to be a running-ripe male. Sperm Whales were again sighted on June 15 (Figure 12.4). From June 17 to 18, the ship returned to the region where the shark fin was originally sighted, making more observations while drifting for 24 h. SLRT tags on sharks 7M, 14M, and 33M all reported positions, between June 14 and 19, 2009, in close proximity to the ship's track while in the SOFA core, also placing the White Sharks in close spatial and temporal proximity to both Sperm Whales, squid, and the unidentified shark (Figure 12.4).

Environmental Description

Based on the satellite remote sensing observations for June 2009, oceanographic conditions at the SOFA were characterized by SSTs that were intermediate between the cold, upwelled waters typical off the North American coast and the warm waters around Hawaii and to the south (Figure 12.5a), but no major temperature gradient or front occurred within this region. In contrast to the very high CHL concentrations typical of the coastal upwelling ecosystem off the North American coast, the SOFA occurred in an area with very low phytoplankton abundance, characteristic of the oligotrophic North Pacific Gyre (Figure 12.5b). SSH throughout much of this region was generally below the mean sea level in June 2009, and at the SOFA it was punctuated by a few cyclonic (i.e., downwelling) mesoscale eddies (Figure 12.6a). In contrast to the high EKE values off the North

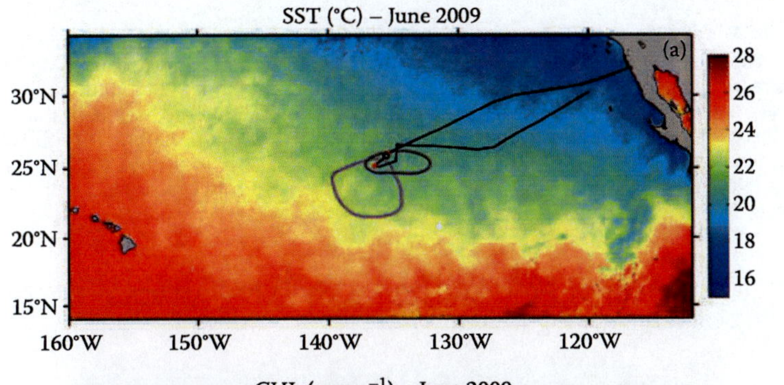

Figure 12.5 Monthly satellite composites of SST (a) and CHL (b) for the month of June 2009 in the region of the SOFA. The upper gray contour is the fixed-kernel 50% density contour indicating the major focal area of sexually mature male White Sharks during the months of June and July, and the lower gray contour is the fixed-kernel 50% density contour for all offshore location data from SLRT-tagged sexually mature males (Chapter 11, this book). The ship's track is indicated as a black line. Red circles represent Sperm Whale sightings, and the white circle with black edge is the location where the dead Giant Squid was found.

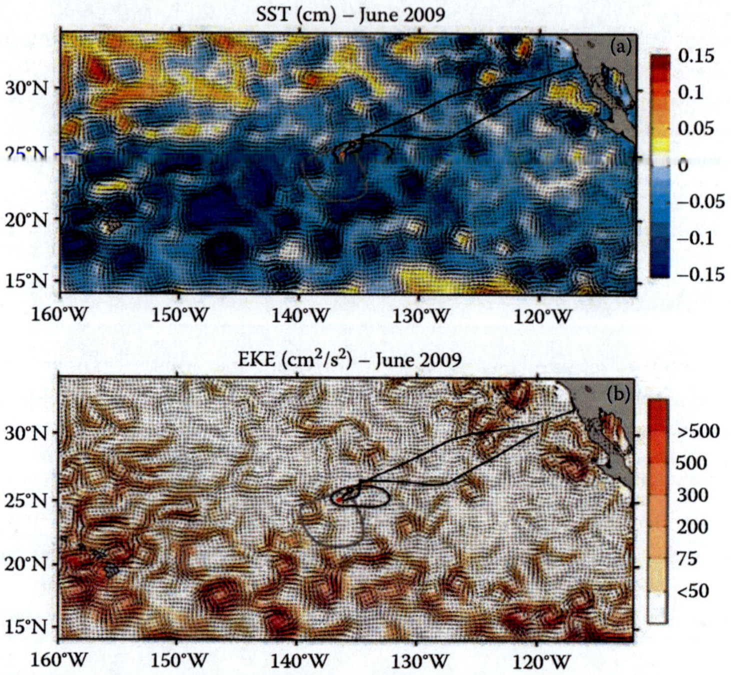

Figure 12.6 Monthly satellite composites of SSH (a) and EKE (b) for the month of June 2009 in the region of the SOFA. In both images the geostrophic current anomalies are shown as black vectors. The upper gray contour is the fixed-kernel 50% density contour indicating the major focal area of sexually mature male White Sharks during the months of June and July, and the lower gray contour is the fixed-kernel 50% density contour for all offshore location data from SLRT-tagged sexually mature males (Chapter 11, this book). The ship's track is indicated as a black line. Red circles represent Sperm Whale sightings, and the white circle with black edge is the location where the dead Giant Squid was found.

American coast, near the Hawaiian islands, and along a latitudinal band centered at about 17°N, EKE at the SOFA was very low (Figure 12.6b).

DISCUSSION

Contrary to our expectation that the SOFA may occur within an oceanographic hot spot, the region was broadly characterized by downwelling conditions, reduced horizontal motions, no major temperature fronts, and very low phytoplankton biomass. Very stable water columns, a strong thermocline, and depauperate conditions at the surface are typical of these regions because they have a tightly coupled production cycle from primary to tertiary producers, resulting in negligible accumulation of surplus biomass (Longhurst, 2007). However, the remote sensing data used in this characterization cannot detect potentially important subsurface features or dynamic processes in these regions. Therefore, although the SOFA may appear as a "cold spot," relatively devoid of epipelagic life, the presence of apex predators like White Sharks, Sperm Whales, and squid indicates that the SOFA ecosystem supports considerable mesopelagic biomass. This suggests that in oligotrophic regions where there is a strong coupling between production and consumption, in situ and water column observations may be necessary in addition to remotely sensed measurements to more completely describe these ecosystems. Certainly more measurements of the mesopelagic faunal composition and habitat characteristics within the SOFA are needed.

Sperm Whales and three species of spawning squid were found within the core of the SOFA. Red Squid in the North Pacific are known to undertake extensive migrations from subarctic to subtropical waters to spawn in winter and summer (Bower and Ichii, 2005), whereas Purpleback Flying Squid are believed to continuously reside and spawn in subtropical waters (Nesis, 1993). It is possible that a spawning aggregation of squid, predictable in time and space, is driving the seasonal constriction of the SOFA and subsequent concentration of White Sharks. It is unclear why such an event would attract only male White Sharks, although sexual segregation in sharks is common (Sims, 2005; Wearmouth and Sims, 2008).

It was previously thought that adult White Sharks feed largely on marine mammals (Tricas and McCosker, 1984; Casey and Pratt, 1985; Klimley, 1985), whereas juveniles feed primarily on invertebrates, demersal teleosts, and elasmobranchs. The absence of small cetaceans and pinnipeds observations during our SOFA expedition, coupled with the fact that White Sharks remain in the SOFA for extended periods, suggests that White Sharks may rely upon nonmammal prey during this portion of their annual life cycle. White sharks may be directly preying upon squid while in the SOFA or preying upon other species that feed on squid. In previous studies, pelagic cephalopod species have been shown to be an important part of the diet for other shark species (Smale and Cliff, 1998; Chapter 4, this book), although the stable isotope signature has been found to drop off for large White Sharks (Chapter 3, this book). Even though no small cetaceans were observed in the SOFA, it remains possible that species such as Pygmy Sperm Whales (family Kogiidae) and Risso's Dolphin occur in low abundance but provide some predation opportunities for White Sharks.

ACKNOWLEDGMENTS

We gratefully acknowledge the financial support and fieldwork of Chris Fischer, the Guy Harvey Ocean Foundation, and the Offield Family Foundation. We also thank B. McBride and the crew of *M/V Ocean* for their assistance in this project. We thank O. Sosa-Nishizaki, F. Galván-Magaña, and M. Hoyos-Padilla for assistance in obtaining Mexican permits. Research was conducted in accordance with permits through Secretaría de Medio Ambiente y Recursos Naturales, Comisión Nacional de Áreas Naturales Protegidas, California Department of Fish and Game, and NOAA Office of National Marine Sanctuaries Program. D. M. P. was supported by funding from the Gordon and Betty Moore Foundation and from the NASA Applied Sciences Program, Earth Science Division, through a grant provided by Research Announcement NNH07ZDA001N, Research Opportunities in Space and Earth Sciences (2007), Program Element A.20: Decision Support through Earth Science Research Results. Environmental satellite data were provided courtesy of NOAA, CoastWatch, NASA, and CLS.

REFERENCES

Boustany, A. M., S. F. Davis, P. Pyle, S. D. Anderson, B. J. Le Boeuf, and B. A. Block. 2002. Expanded niche for white sharks. *Nature* 415: 35–36.

Bower, J. R., and T. Ichii. 2005. The red flying squid (*Ommastrephes bartramii*): A review of recent research and the fishery in Japan. *Fisheries Research* 76: 39–55.

Casey, J. G., and H. L. Pratt, Jr. 1985. Distribution of the white shark, *Carcharodon carcharias*, in the western North Atlantic. *Memoirs of the Southern California Academy of Sciences* 9: 2–14.

Domeier, M. L., and N. Nasby-Lucas. 2008. Migration patterns of white sharks *Carcharodon carcharias* tagged at Guadalupe Island, Mexico, and identification of an eastern Pacific shared offshore foraging area. *Marine Ecology Progress Series* 370: 221–237.

Francis, M. P. 1996. Observations on a pregnant white shark with a review of reproductive biology. In *Great White Sharks: The biology of Carcharodon carcharias*. Edited by A. P. Klimley and D. G. Ainley, pp. 157–172. San Diego, CA: Academic Press.

Heinemann, D. 1981. A range finder for pelagic bird censusing. *Journal of Wildlife Management* 45: 489–493.

Jorgensen, S., C. A. Reeb, T. K. Chapple, S. Anderson, C. Perle, and S. Van Sommeran. 2010. Philopatry and migration of Pacific white sharks. *Proceedings of the Royal Society B Biological Sciences* 277: 679–688.

Klimley, A. P. 1985. The areal distribution and autecology of the white shark (*Carcharodon carcharias*), off the west coast of North America. *Memoirs of the Southern California Academy of Sciences* 9: 15–40.

Longhurst, A. R. 2007. *Ecological geography of the sea*, 2nd ed. San Diego, CA: Academic Press.

Mollet, H. F., G. Cliff, H. L. Pratt, Jr., and J. D. Stevens. 2000. Reproductive biology of the female shortfin mako, *Isurus oxyrinchus* Raninesque, 1810, with comments on the embryonic development of lamnoids. *Fisheries Bulletin* 98: 299–318.

Nasby-Lucas, N., H. Dewar, C. H. Lam, K. J. Goldman, and M. L. Domeier. 2009. White shark offshore habitat: A behavioral and environmental characterization of the eastern Pacific shared offshore foraging area. *PLoS ONE* 4: e8163.

Nesis, K. N. 1993. Population structure of oceanic ommastrephids, with particular reference to *Sthenoteuthis oualaniensis*: A review. In *Recent advances in cephalopod fishery biology*. Edited by T. Okutani, R. K. O'Dor, and T. Kubodera, pp. 375–383. Tokyo: Tokai University Press.

Olson, P., R. Pitman, L. Ballance, K. Hough, P. Dutton, and S. Reilly. 2001. Summary of seabird, marine turtle, and surface fauna data collected during a survey in the eastern tropical Pacific Ocean, July 28–December 9, 2000. *NOAA Technical Memorandum* NOAA-TM-NMFS-SWFSC-304. Washington, DC: NOAA.

Palacios, D. M., S. J. Bograd, D. G. Foley, and F. B. Schwing. 2006. Oceanographic characteristics of biological hot spots in the North Pacific: A remote sensing perspective. *Deep-Sea Research II* 53: 250–269.

Sims, D. W. 2005. Differences in habitat selection and reproductive strategies of male and female sharks. In *Sexual segregation in vertebrates: Ecology of the two sexes*. Edited by K. E. Ruckstuhl and P. Neuhaus, pp. 127–147. Cambridge, UK: Cambridge University Press

Smale, M. J., and G. Cliff, 1998. Cephalopods in the diets of four shark species (*Galeocerdo cuvier*, *Sphyrna lewini*, *S. zygaena*, and *S. mokarran*) from Kwazulu-Natal, South Africa. *South African Journal of Marine Science* 20: 241–253.

Tricas, T. C., and J. E. McCosker. 1984. Predatory behavior of the white shark (*Carcharodon carcharias*), with notes on its biology. *Proceedings of the California Academy of Sciences* 43: 221–238.

Wearmouth, V. J., and D. W. Sims. 2008. Sexual segregation behaviour of marine fish, reptiles, birds and mammals: Patterns, mechanisms and conservation implications. *Advances in Marine Biology* 54: 107–170.

Weng, K. C., A. M. Boustany, P. Pyle, S. D. Anderson, A. Brown, and B. A. Block. 2007. Migration and habitat of white sharks (*Carcharodon carcharias*) in the eastern Pacific Ocean. *Marine Biology* 152: 877–894.

CHAPTER 13

Connectivity among White Shark Coastal Aggregation Areas in the Northeastern Pacific

Salvador J. Jorgensen*
Stanford University and Monterey Bay Aquarium

Taylor K. Chapple
University of California, Davis

Scot Anderson
Inverness, California

Mauricio Hoyos
Centro Interdisciplinario de Ciencias Marinas

Carol Reeb
Stanford University

Barbara A. Block
Stanford University

ABSTRACT

White Sharks (*Carcharodon carcharias*) in the northeastern Pacific comprise a genetically distinct population, demographically isolated from other populations in Australia/New Zealand and South Africa. Within the northeastern Pacific, mature and subadult White Sharks show strong fidelity to aggregation sites in Central California and Guadalupe Island, Mexico, respectively.

Individuals tagged at both sites migrate offshore to common pelagic habitats but routinely return to the coastal site where they were tagged. To date, evidence for the connectivity of these coastal sites has been lacking, raising the question whether they comprise one continuous or two distinct populations. A third coastal area, comprising the near-shore waters from Point Conception, California, and extending southward through the Southern California Bight to Sebastian Vizcaino Bay, Mexico, has been recognized as a White Shark nursery area. Although incidental catches in the Southern California Bight comprise mostly neonates and large females, little evidence exists demonstrating the movement of mature females to this purported pupping area, leaving open the possibility that parturition might occur elsewhere followed by immigration of neonates into the nursery area. We illustrate how rare events in a large tagging data set provide important information about the connectivity of aggregation areas within the northeastern Pacific. Acoustic tagging revealed the rare movement of one single female out of 140 deployments, from Central California to Guadalupe Island; the first result to demonstrate movement connecting these aggregation sites. Of 97 pop-up archival transmitting (PAT) tags deployed in Central California, only one deployment was on a large female with healing mating scars indicative of prior mating activity. This female was tracked offshore and eventually into the Southern California Bight where the tag popped up 362 d later off Santa Catalina Island. The movement of this female into the Southern California Bight coincided with the peak period for neonates in the area.

INTRODUCTION

In the past, many viewed the sea as a habitat largely free of physical barriers, a place where globally distributed species could roam unrestricted subsequently exchanging genes along the way as one very large population (Palumbi, 1994). Indeed, early genetic studies on marine fishes were consistent with this view. Most species showed little genetic divergence across very large spatial scales (Avise, 2000). However, electronic tagging studies have provided new insight into the complicated movements of highly migratory marine fish species, including the White Shark. These new data for sharks now show high levels of site fidelity and homing, setting the stage for fine-scale genetic population structure that was, until recently, unexpected.

White Sharks in the northeastern Pacific (NEP) comprise a distinct phylogenetic clade based on mitochondrial DNA markers (Jorgensen et al., 2010). To date, published genetic data collected for the NEP clade only included sharks in Central California (CC). Nonetheless, electronic tagging data indicate that the genetic isolation of NEP sharks is potentially maintained through strong site fidelity to very specific coastal and pelagic habitats (Jorgensen et al., 2010).

Tagging has shown that site fidelity also occurs at more localized scales in the NEP; White Sharks tagged near seal rookeries in CC regularly return to these same rookeries following visits to offshore locations, including Hawaii and the White Shark "Café" (Jorgensen et al., 2010). Similarly, White Sharks tagged at the Guadalupe Island (GI) Northern Elephant Seal rookery also visited the same offshore areas (also referred to as Hawaii and the Shared Offshore Foraging Area or SOFA) (Domeier and Nasby-Lucas, 2008; Chapters 11 and 12, this book) and consistently returned to GI. Although sharks from these coastal aggregation sites in the NEP overlap in a pelagic setting 1,500–4,000 km offshore, there has been no evidence of individuals moving between these coastal sites, leaving open the possibility of further population structure within the NEP. It is unknown whether the groups at these two aggregation sites comprise one continuous or two distinct populations. Understanding the degree of connectivity of these two NEP groups separated by the Mexico-U.S. border is vital for successfully determining their population status, the degree of intermixing, and management options.

Management of White Sharks in the NEP also requires information on the juvenile life stages. The Southern California Bight (SCB) has been hypothesized as a nursery area based on the concentration of neonate White Sharks occurring there in summer and fall (Klimley, 1985; Chapters 14 and 16, this book). This nursery area likely extends from Point Conception, California, southward along the coast across the Mexico-U.S. border to Sebastian Vizcaino Bay (Chapters 14, 15, and 16, this book). Recent reporting on capture records, as well as tracking data from an adult female (Chapter 16, this book), indicate that the nursery may extend further south to Cabo San Lucas and into the Gulf of California, although the smallest size classes (~150 cm; neonates) did not appear in records from the Gulf of California (Galvan-Magana et al., 2010). Neonates are either born in the SCB or are born elsewhere then immigrate into this nursery area. Klimley (1985) showed captures of White Sharks within the SCB consisted primarily of neonates and large females in summer and fall, whereas north of Point Conception captures mainly comprised males and females ≥2 m, in summer, fall, and winter. The seasonal co-occurrence of large females and neonates suggests that late-term females migrate into the nursery area for parturition. However, none of the large females caught were pregnant (five females of >4.5 m total length), and neonates were usually captured closer to shore than the mature females, raising some doubt about conclusions reached from capture data alone.

We used electronic tagging data to determine connectivity within the NEP between White Sharks in CC and GI and between CC and the SCB nursery area. The specific goals of this study were to (1) track potential movement patterns between CC and GI; (2) describe the movement of a mature female from CC that moved into the White Shark nursery habitat in the summer/fall pupping season; and (3) further elucidate the life history, population structure, and connectivity of White Sharks within the NEP.

MATERIALS AND METHODS

Ninety-seven pop-up archival transmitting (PAT) tags (PAT 2.0, 3.0, and 4.0 and Mk10-PAT; Wildlife Computers, Redmond, WA) were deployed on White Sharks in CC. Deployment and position estimation methods are described in detail in Jorgensen et al. (2010). To determine the rate of travel during migration from satellite tag tracks, we calculated the net displacement distance over time between position estimates connecting the coastal area and the Café.

Additionally, we deployed 110 individually coded acoustic transmitter tags (V16-4H; Vemco, Halifax, Nova Scotia) in CC [described in detail by Jorgensen et al. (2010)] and 31 of the same tag model on White Sharks at Guadalupe Island (see Figure 13.1 for deployment date information). Six VR-3 acoustic listening stations were placed at aggregations sites in CC from October 16, 2006 through December 2, 2009. Three VR-2 acoustic receivers were actively stationed around GI from August 2008 through October 30, 2009. Acoustic data were analyzed using MATLAB® (The MathWorks, v.7.9.0.529). Total length was estimated as each shark swam alongside a research vessel of known length. Video and photography were used to help determine sex and individual identification.

RESULTS

Of the 141 acoustic tag deployments, 110 at CC and 31 at GI, a single individual was detected at both locations (Figure 13.1a). The 366-cm female, S153, was tagged on September 14, 2008 near Tomales Point in CC and remained locally during the following month until October 13, 2008. During this 29-d period the shark was detected 528 times with a mean time between detections of 28.9 min. (minimum = 1.1 min.; maximum = 5.4 d). The shark was next detected at GI, 1,000 km to the southeast, 29 d later on November 12, 2008. The minimum mean velocity the shark would have to travel from CC to GI was 34.5 km/d. S153 then remained at GI until March 7, 2009. During this 115-d period, the shark was detected 15,365 times with a mean time between detections of 11.0 min. (minimum = 1.1 min.; maximum = 7.9 d). On March 7, 2009, the shark apparently departed, coincident

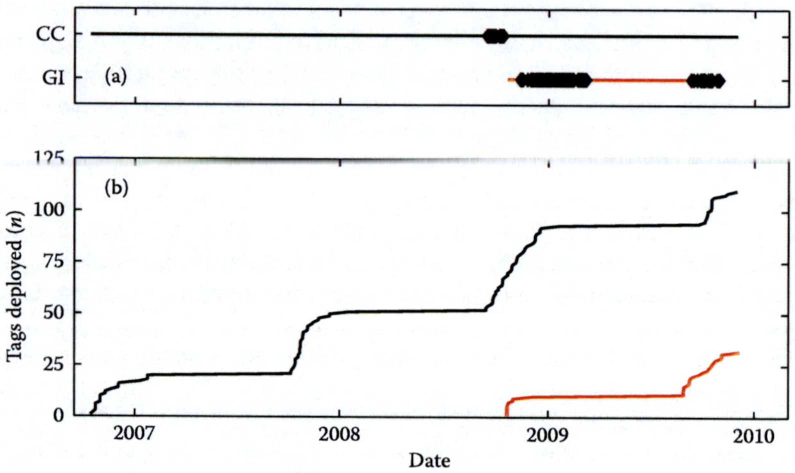

Figure 13.1 Movement of *Carcharodon carcharias* between CC and GI determined from acoustic tagging. (a) Receivers were actively deployed at CC from October 15, 2006 and December 14, 2009 (black line) and from October 22, 2008 to October 30, 2009 at Guadalupe Island (red line). White Shark S153 was tagged in Central California on October 13, 2008 and was detected (black diamonds) locally for 29 d before leaving the area. The shark was subsequently detected at GI on November 12, 2008. S153 remained near GI for 115 d until March 7, 2009. Then, after a 185-d absence, the shark returned to GI on September 10, 2009 and remained there for an additional 50 d. (b) The cumulative number of acoustic tags deployed at CC (black line) and GI (red line) over time totaled 110 in CC and 31 at GI. The movement of only a single individual suggests that such exchanges are relatively rare.

with the expected offshore migration season, and was not detected again for 185 d. On September 10, 2009, S153 returned to GI, where it was detected for another 50 d through the remainder of the data collection period, ending October 30, 2009. During these 50 d, the shark was detected 386 times with a mean period between detections of 3.1 h (minimum = 1.1 min.; maximum = 8.5 d).

A total of 68 White Sharks (23 females, 32 males, and 13 unsexed) were successfully tracked using satellite tags. The mean deployment duration was 199 d. The tagging dates, size estimates, deployment durations, and gender information are summarized in Supplemental Table S1 in Jorgensen et al. (2010). S70, a female tagged with a PAT tag on October 9, 2006 at the South East Farallon Island in CC, was tracked for 362 d, the longest deployment. Following tagging, the shark remained in the CC vicinity for more than 3 months until heading offshore near the middle of January (Figure 13.2). The shark travelled toward the Café region at a mean displacement rate of 87 km/d. Once offshore, the shark remained around the Café region until April and then made one relatively direct movement toward SCB (mean displacement rate = 60 km/d) only to return offshore (68 km/d) again. The "turnaround" occurred apparently before reaching the continental shelf, at which time the tag records showed a minimum sea-surface temperature of 13.6°C reached on April 16 and 17. In June, the shark travelled to the SCB (50 km/d), arrived at the beginning of July, and remained there until at least October 5, 2007, when the tag released according to the preprogrammed date and was first detected near Two Harbors on Santa Catalina Island (Figure 13.2). The estimated total length for S70 when tagged was 488 cm. Notably, females of >450 cm total length are believed to be reproductively mature (Francis, 1996), and S70 was the only female bearing evidence of relatively recent mating activity (Pratt and Carrier, 2001) at the time of tagging (Figure 13.3). No other females or males with satellite tags moved into the SCB during the period of tracking. Of the 23 females tagged with PAT tags, nine individuals were >450 cm (mature) and spanned a tracking period past May 16, the earliest recorded calendar date of neonate (≤150 cm) capture (Klimley, 1985). Of these nine records, two individuals returned to Central California in August [shark 50 and 59 (Supplemental Table S1 in Jorgensen et al., 2010)], and six tags popped

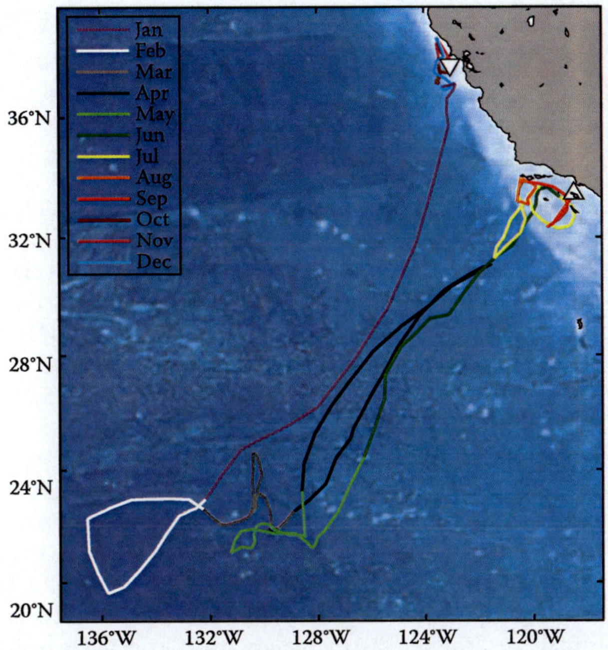

Figure 13.2 The movement of a female White Shark from Central California to the Southern California Bight. S70 was tagged on October 9, 2006 (∇) near the Farallon Islands in CC, and the tag popped off, according to schedule, October 5, 2007 (Δ) near Fisherman's Cove off the northeast side of Santa Catalina Island in the SCB. The track line describes a five-point moving average of geolocation positions, and color coincides with the month.

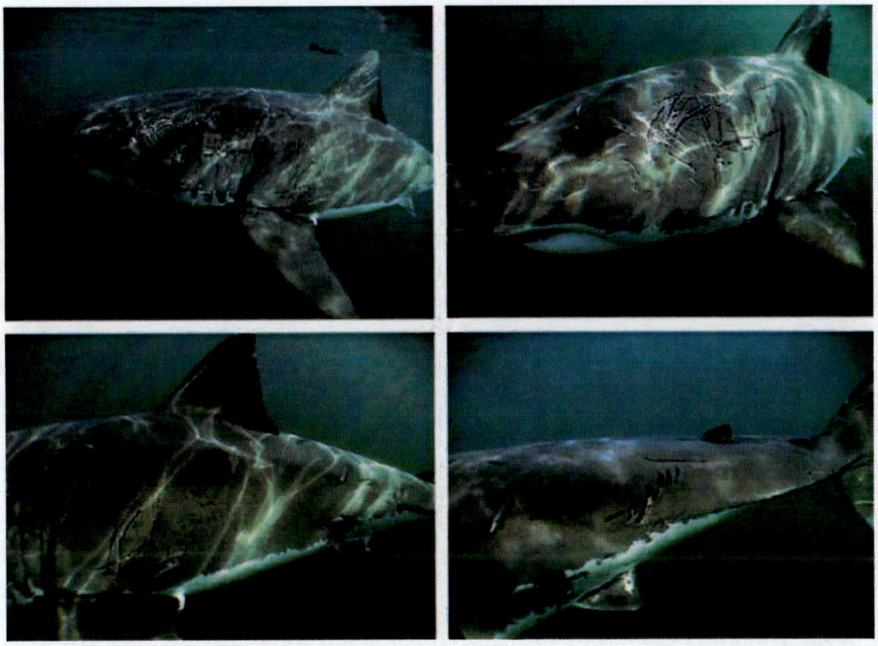

Figure 13.3 Shark S70 just prior to tagging on October 8, 2006, showing evidence of partially healed presumed mating-induced injury.

up in the Café region [two each in July, August, and September; sharks 8, 91, 35, 42, 46, and 137, respectively (Jorgensen et al., 2010)].

DISCUSSION

The predictable migratory patterns in White Sharks frequenting GI compared with those showing fidelity to seal rookeries in CC provide a compelling possibility of population structure within the NEP. Indeed, previous tagging results suggested no movement between the two aggregation sites (Domeier and Nasby-Lucas, 2008; Jorgensen et al., 2010; Chapters 11 and 16, this book). The documentation here of only one single individual moving between CC and GI despite the large number of tags suggests that such movements are relatively rare occurrences. Nonetheless, following tagging, S153 resided in CC for 1 month then moved to GI, revealing for the first time connectivity between these two NEP aggregation sites.

Given the apparent segregation between White Sharks frequenting CC and GI, the discussion of the potential genetic exchange between the groups has focused on the possibility of intermating offshore in a shared area, the White Shark Café (Jorgensen et al., 2010), or mixing in a common pool of offspring, a single nursery area in the SCB from which mixed juveniles could recruit to either GI or CC. However, this study provides the first empirical documentation of exchange through direct migration between the two groups. The continual use of acoustic receivers and tags (which have a >3-year life span), will help determine how frequently these exchanges occur, provided receivers are maintained at both locations.

Evidence of these movements might be further elucidated using individual photo-identification methods. Anderson et al. (2011) described nearly 200 individuals over 22 yrs. using dorsal fin photographs in CC. Fin morphology was shown to be conserved over long time periods (Anderson et al., 2011) and should allow comparison of sharks between CC and GI sites for years to come. At GI, 113 unique individuals were identified over 9 yrs. (Chapter 25, this book) using unique body pigmentation markers. Chapple et al. (2011) estimated the current population of mature and subadult White Sharks off CC to be 219 individuals (between 130 and 275; 95% credible intervals). At GI, an open mark-recapture framework of the 9-year photo-identification collection suggests an index of abundance near 120 adults and subadults (Chapter 26, this book); however, this may be an underestimate. Nonetheless, potentially a high percentage of adults and a fair number of subadults are currently identified in the NEP, raising the likelihood of detecting exchanges between CC and GI in the future. Acquiring body pigmentation photos in CC is problematic because of poor water clarity, whereas no limitation on acquiring dorsal fin photos exists at GI. Therefore, efforts should be made to standardize individual identification methods using fin morphology at both locations.

Further molecular testing will determine whether GI and CC adults could be genetically divergent. The answer is important for White Shark conservation, because estimates of the number of individuals in both groups is alarmingly low. If these are in fact separate gene pools, separate management measures will need to be developed. However, answering this question conclusively will be challenging. First, mitochondrial DNA has a slow mutation rate in sharks (Martin et al., 1992). Very few polymorphic sites were observed in the DNA sequence that could distinguish these putative populations from each other. In fact, over 60% of sampled CC individuals belonged to two main female lineages (Jorgensen et al., 2010), which means many more samples will be needed to test for statistical significance. Second, even low levels of mixture between CC and GI, either by gene flow or spatial overlap, can inhibit sensitivity of the genetic analyses. The use of faster evolving markers, such as microsatellite loci, may offer a better test for GI and CC adults. These markers also provide a means to genetically fingerprint individuals (Gubili et al., 2009) and quantify measures of kinship (Feldheim et al., 2004). Matching adults to their offspring in the SCB is possible given a small population size and should help discriminate between potential mixing mechanisms.

To quantify mixing between the two sites, the use of acoustic and satellite tags, photo identification and genetics are important complementary tools. The extent and patterns of mixing will be important to understand whether this population functions as a single mixed population in the NEP or as separate subpopulations with little exchange. Additionally, this will help determine to what extent management policy should be coordinated across the U.S.-Mexico boundary.

As with mixing between CC and GI, evidence that female sharks move from CC to the purported nursery grounds is rare. S70 is the first tagging record showing an adult White Shark traveling from the CC aggregation site into the vicinity of the SCB nursery area. The only other published record is from a White Shark that was photographed at the South Farallon Islands on November 7, 1988, and then photographed again 6 yrs. later on October 18, 1994, halfway between Santa Catalina Island and Newport Beach, California in the SCB (Anderson and Goldman, 1996). Interestingly, the location where this shark was subsequently photographed in the SCB was less than 30 km from where S70's tag popped up. Of the nine mature females with records extending past May 16, S70 was the only tagging record to show this movement pattern; however, this was also the only female with evidence of prior mating activity to be tagged. Two of the nine sharks returned to Central California within 1 yr. of tagging. However, the possibility that any of the remaining six individuals moved to the SCB during the parturition period after the tags popped up (in the Café region) could not be ruled out because the tags popped up between July and September. Observations of female White Sharks with relatively recent mating marks are rare in CC. During fieldwork conducted in CC over the past two decades, 395 encounters with White Sharks were photographically documented (Anderson et al., 2011), yet only five instances of apparent mating-induced trauma occurred, and of these only S70 was tagged. All of the cases were similar in that the wounds were partially healed. That mating marks are infrequently observed on White Sharks in CC suggests either that mating events are rare or that females generally remain offshore following mating during the period when such marks would be healing. Also, the absence of observations of fresh mating marks, despite such extensive observation, is inconsistent with the hypothesis that mating occurs with any regularity at coastal foraging sites in CC. Healing rates are high in White Sharks (Chapter 6, this book). Nonetheless, the possibility that the wounds on S70 occurred offshore at the Café prior to arriving at the Farallones cannot be ruled out. Despite where mating may actually occur, the presence of the partially healed scars and the subsequent movement of S70 into the SCB support the hypothesis that gravid females from the CC group migrate to the SCB to birth (Klimley, 1985).

The gestation period for White Sharks is thought to be between 12 and 18 months (Francis, 1996). Assuming this is accurate, and given the mean calendar date of neonate (≤150 cm; Francis, 1996) captures in the SCB is in August (Klimley, 1985), fertilization should occur between February and August of the previous year. If the mating wounds on S70 occurred approximately >45 d prior to the date photographed (Figure 13.3; October 9), then this observation would fit the prediction. However, the precise gestation period for White Sharks remains unknown, and the possibility of sperm storage also has not been determined for the species.

CONCLUSION

Previously, extensive, satellite tagging data sets had suggested isolation between the CC and GI aggregation sites within the NEP. White Sharks showed strong fidelity to these sites with no indication of movement between them, despite habitat overlap and seasonal mixing at offshore areas. Additionally, electronic tracking data had yet to confirm the movement of females (or males) from adult sites to the SCB. Through satellite and acoustic tagging, photo identification, and data sharing between CC and GI study sites, we revealed rare but important observations illustrating the connectivity of these three coastal areas vital to the understanding and management of NEP White Shark population. Internationally, White Sharks (*Carcharodon carcharias*) have been categorized as vulnerable to extinction on the World Conservation Union "red list" (IUCN 2010-04-27, Category VU A2cd+2cd) and are listed under Appendix II of the Convention on International Trade in Endangered Species. This study provides an update to the spatial dynamics of the NEP White Shark population offering a better understanding of a complex species by providing further information on the reproductive and migratory ecology of the species and underscoring the importance of U.S.-Mexico collaboration in research and management.

ACKNOWLEDGMENTS

This project was funded by the Sloan, Moore and Packard Foundations as a part of the Tagging of Pacific Pelagics program of the Census of Marine Life. The Monterey Bay Aquarium Foundation, Shark Diver, The Guadalupe Island Conservation Fund, Pfleger Institute of Environmental Research, Instituto Politecnico Nacional (COFAA, EDI), Iemanya Oceanica, International Community Foundation, WWF-Telcel Alliance, and Grupo Bursatil Mexicano also provided financial support. We thank J. P. Kanive, J. Barlow, C. Logan, R. Elliot, J. Cornelius, T. O'Leary, P. Douglas, G. Grivetto, K. Neff, B. Cornapple, J. Schaeffer, J. Fitzgerald, B. Becker, R. Theiss, C. Farwell, A. Carlisle, S. McAfee, J. O'Sullivan, M. Castleton, J. Ganong, L. Rodriguez, and A. Swithenbank for assistance with field work, lab work, data processing, and editing. We thank CICIMAR, the Mexican Navy, Comision de Areas Naturals Protegidas, Island Conservation, and local fishermen from Guadalupe Island for assistance. We are grateful to SeaLife Conservation, R. Repass, E. Homer and the crew of the *R.S.V. Derek M. Bayliss*, Horizon Charters, Islander Charters, *Nautilus Explorer*, *Sea Escape*, *Solmar V*, and *M/V Andrea Lynn* for vessel support. The project was conducted with U.S. permits from CDFG, MBNMS, GFNMS, NMFS, NPS, and under Stanford University animal subjects Protocol 10765. In Mexico, work was conducted with permits from Secretaria del Medio Ambiente y Recursos Naturales (SGPA/DGVS/04053/07), Comisión Natural de Áreas Naturales Protegidas (F00.RNO.RBIG.-154/07 and F00.RNO.RBIG-166/07), and Secretaria de Gobernación con Oficio (DICOPPU/211/1827/07).

REFERENCES

Anderson, S. D., and K. J. Goldman. 1996. Photographic evidence of white shark movements in California water. *California Fish and Game* 82: 182–186.

Anderson, S. D., T. K. Chapple, S. J. Jorgensen, A. P. Klimley, and B. A. Block. 2011. Long-term individual identification and site fidelity of white sharks, *Carcharodon carcharias*, off California using dorsal fins. *Marine Biology* doi:10.1007/s00227-011-1643-5.

Avise, J. C. 2000. *Phylogeography: The history and formation of species*. Cambridge, MA: Harvard University Press.

Chapple, T. K., S. J. Jorgensen, S. D. Anderson, P. E. Kanive, A. P. Klimley, L. W. Botsford, and B. A. Block. 2011. A first estimate of white shark, *Carcharodon carcharias*, abundance off Central California. *Biology Letters* 7: 581–583.

Domeier, M. L., and N. Nasby-Lucas. 2008. Migration patterns of white sharks *Carcharodon carcharias* tagged at Guadalupe Island, Mexico, and identification of an eastern Pacific shared offshore foraging area. *Marine Ecology-Progress Series* 370: 221–237.

Feldheim, K. A., S. H. Gruber, and M. V. Ashley. 2004. Reconstruction of parental microsatellite genotypes reveals female polyandry and philopatry in the lemon shark, *Negaprion brevirostris*. *Evolution* 58: 2332–2342.

Francis, M. P. 1996. Observations on a pregnant white shark with a review of reproductive biology. *Great white sharks: The biology of Carcharodon carcharias*. San Diego, CA: Academic Press. pp. 157–172.

Galvan-Magana, F., E. M. Hoyos-Padilla, C. J. Navarro-Serment, and F. Marquez-Farias. 2010. Records of white shark, *Carcharodon carcharias*, in the Gulf of California, Mexico. *Marine Biodiversity Records* doi:10.1017/S1755267210000977.

Gubili, C., R. Johnson, E. Gennari, W. Oosthuizen, D. Kotze, M. Meÿer, D. Sims, C. Jones, and L. Noble. 2009. Concordance of genetic and fin photo identification in the great white shark, *Carcharodon carcharias*, off Mossel Bay, South Africa. *Marine Biology* 156: 2199–2207.

Jorgensen, S. J., C. A. Reeb, T. K. Chapple, S. Anderson, C. Perle, S. R. Van Sommeran, C. Fritz-Cope, A. C. Brown, A. P. Klimley, and B. A. Block. 2010. Philopatry and migration of Pacific white sharks. *Proceedings of the Royal Society B Biological Sciences* 277: 679–688.

Klimley, A. P. 1985. The areal distribution and autoecology of the white shark, *Carcharodon carcharias*, off the west coast of North America. *Memoirs of the Southern California Academy of Sciences* 9: 15–40.

Martin, A. P., G. J. P. Naylor, and S. R. Palumbi. 1992. Rates of mitochondrial DNA evolution in sharks are slow compared with mammals. *Nature* 357: 153–155.

Palumbi, S. R. 1994. Genetic divergence, reproductive isolation, and marine speciation. *Annual Review of Ecology and Systematics* 25: 547–572.

Pratt, H. L., and J. C. Carrier. 2001. A review of elasmobranch reproductive behavior with a case study on the nurse shark, *Ginglymostoma cirratum*. *Environmental Biology of Fishes* 60: 157–188.

CHAPTER 14

Historic Fishery Interactions with White Sharks in the Southern California Bight

Christopher G. Lowe*
California State University, Long Beach

Mary E. Blasius
California State University, Long Beach

Erica T. Jarvis
California State University, Long Beach

Tom J. Mason
California State University, Long Beach

Gwen D. Goodmanlowe
California State University, Long Beach

John B. O'Sullivan
Monterey Bay Aquarium

ABSTRACT

The degree to which White Sharks (*Carcharodon carcharias*) have interacted with various fisheries in Southern California is unknown, despite their high public interest and recent protection under state, federal, and international regulations. Data on White Shark fishery interactions in Southern California were mined from news reports, state and federal management agencies, fisher logbooks, and research institutions. Of the 369 records of reported White Shark catch between 1936 and 2009, 39% were categorized as young of the year (YOY), 21% as juveniles, 5% as subadult/

adults, and the remaining 35% were of unreported size. YOY sharks were caught in nearshore waters (<50-m depth) more often than adult sharks, which were mainly caught in offshore waters (>50-m depth). In addition, entangling net fisheries (e.g., trammel nets, set and drift gillnets) caught more White Sharks (81%) than other fisheries (purse seine, trawl, set line, hook-and-line, harpoon, etc.). Incidental reported catch rates of YOY and juvenile White Sharks have increased in Southern California since the California nearshore-gillnet ban in 1994 and regulation of the offshore drift gillnet fishery, despite a significant decrease in overall gillnet fishing effort since the mid-1990s. This suggests that the White Shark population off California may be increasing because of reduced nearshore gillnet fishing effort and harvest protection in state and federal waters.

INTRODUCTION

Although there have been extensive commercial and recreational fisheries for several species of the nearshore and pelagic sharks off California (e.g., Thresher Sharks, *Alopias* sp.; Shortfin Mako Shark, *Isurus oxyrinchus*; Pacific Angel Shark, *Squatina californica*; Tope, *Galeorhinus galeus*; Basking Shark, *Cetorhinus maximus*; Leopard Shark, *Triakis semifasciata*; and Spiny Dogfish, *Squalus acanthias*), there have been no directed fisheries for White Sharks (*Carcharodon carcharias*) (Holts, 1988). Although attempts were made in the 1970s and 1980s to develop a market for White Shark meat in the United States (Holts, 1988), their low relative abundance and reputation for attacks on humans reduced their marketability and motivation for a directed fishery. Nevertheless, there is a fairly substantial record of incidental and periodically targeted catch of White Sharks over the last 100 yrs., particularly in Southern California.

Klimley (1985) used fishery catch data, observational reports, and museum records from 1935 to 1984 to determine the spatial distribution and ecology of White Sharks along the west coast of North America. Based on size distribution and seasonality from catch records and reports, Klimley (1985) theorized that Southern California was a pupping and nursery ground for White Sharks, because adults were most frequently seen and caught in Central and Northern California in fall and winter months, whereas young-of-the-year (YOY)-sized sharks were only caught south of Point Conception in the Southern California Bight (SCB) during summer months. Stomach content data from museum specimens and fisher reports, along with field observations of marine mammal predation, suggested that YOY and juveniles [<3 m total length (TL)] fed heavily on benthic fishes in nearshore habitats, whereas adults (>3 m TL) fed heavily on marine mammals, which were concentrated at rookeries in central California and offshore islands in Southern California. Interestingly, Klimley (1985) found that over 62% of the reports of White Shark captures came from Southern California; however, there was no correlation between report occurrence and catch per unit effort (CPUE) along the coast of California.

Although previous analysis of historic records has greatly furthered our knowledge of the distribution, life history, and behavior of the White Shark along the west coast of the United States, some of which has subsequently been supported by fishery independent data (Boustany et al., 2002; Dewar et al., 2004; Weng et al., 2007), fisheries in Southern California have changed considerably since 1984 because of overharvesting concerns and increased fisheries regulations. Marine commercial and sport fisheries in Southern California constitute a large portion of the total fishing effort and production for the state of California (Oliphant et al., 1990; Dotson and Charter, 2003). The northern portion of the SCB (coastline including the Channel Islands) constitutes one of the greatest expanses of continental shelf habitat in California. In addition to a large sport fishing community, which reported over 2.7 million angler trips in 2008 (CDFG, 2009), Southern California has supported a wide variety of commercial fisheries over the last 100 years, including trawl, set line, purse seine, trap, and entangling net in both nearshore and offshore habitats (Love, 2006). Entangling net fishing (e.g., trammel nets, gillnets) rapidly increased in popularity in the 1970s in Southern

California with the development of monofilament, where nearshore (<3 nm from shore, <50-m depth) set-gillnet (SGN) fishers targeted California Halibut (*Paralichthys californicus*), Pacific Angel Shark, and White Seabass (*Atractoscion nobilis*) (Barsky, 1990; Richards, 2001; Pondella and Allen, 2008). An offshore (>3 nm from shore, >50-m depth) drift-gillnet (DGN) fishery also developed in the 1970s for Thresher Shark (*Alopias* spp.) and Swordfish (*Xiphias gladius*) (Hanan et al., 1993; Holts, 1988; Holts et al., 1998). These entangling net fisheries enabled commercial fishers to fish virtually the entire SCB; however, large-scale stock declines in the 1980s and 1990s resulted in increased regulations of these fisheries (Holts, 1988; Holts et al., 1998; Pondella and Allen, 2008). Significant declines in nearshore fish species and bird and marine mammal interactions resulted in a ban of nearshore gillnets in California state waters in 1994. The SGN fishery still exists outside state waters (>3 nm offshore and >1 nm of offshore islands) for these nearshore species; however, there are only a few locations in Southern California with sufficient shallow habitat to permit effective fishing for California Halibut, Pacific Angel Shark, and White Seabass. The nearshore SGN ban and increased regulation of the offshore DGN fishery significantly reduced the overall entangling net effort in Southern California after 1994 (Holts et al., 1998; Pondella and Allen, 2008).

In addition to changes in fishery trends in Southern California, the White Shark was protected and could no longer be landed in California waters in 1994 (Heneman and Glazer, 1996) and was listed as a prohibited species in U.S. federal waters in 2005. Therefore, a re-examination of White Shark catch records since 1984 may provide insight into the degree to which nondirected fisheries may have interacted with or impacted White Shark populations in Southern California.

MATERIALS AND METHODS

Data Acquisition

Reports of White Sharks captured in the Southern California Bight came from four primary sources (Table 14.1):

Table 14.1 Sources of White Shark Capture Reports in the Southern California Bight

Source	*n*	Year(s)	Overlap (*n*)
Resource agencies			
CDFG DGN/SGN fishery logbook data	22	1981–2008	19
CDFG shark tagging program	3	1983–2001	
CDFG game wardens	2		
NOAA Fisheries DGN/SGN fishery logbook data	187		78
NOAA Fisheries DGN/SGN fishery observer data	9	1981–2008	1
NOAA Fisheries-*Los Angeles Times* database	1	1990–2008	1
PacFIN	91	1959–1998	29
MBA Juvenile White Shark Tagging Program	77	1979–2005	30
M. Domeier White Shark Tagging Study	3	2002–2009	
Scientific collections		2009	
Natural History Museum of Los Angeles County	33		11
California Academy of Sciences	8		4
Scripps Institution of Oceanography	17		5
Newspaper and scientific journal articles	13		1

Some sources have specific date ranges for when they started recording their data, whereas other sources have no consistent date ranges. *n* represents the number of records obtained from each source. Overlap (*n*) represents the number of reports found in duplication in other source databases.

1. Records/landings from resource agencies California Department of Fish and Game (CDFG) and the National Oceanic and Atmospheric Administration National Marine Fisheries Service (NOAA Fisheries)
2. Data from tagged or dead sharks collected by the authors and associates of the Monterey Bay Aquarium (MBA) Juvenile White Shark Tagging Program
3. Scientific collections
4. Newspaper articles

Commercial-fishing records of White Sharks were obtained from a variety of agencies/programs. CDFG and NOAA Fisheries provided data from logbooks of the commercial DGN and SGN fisheries and from commercial market receipts (landing tickets) reporting landings of catch for a variety of gear types, including entangling nets, trawl, purse seine, harpoon, set line, and trap. Reports of White Shark captures were also obtained from data provided by the NOAA Fisheries Southwest Region Drift and Set Gillnet Observer Program. The observer program for this fishery was initiated in July 1990 (to present) and has provided approximately 21% observer coverage. We also extracted additional reports from commercial market receipt landings made available on the Pacific Coast Fisheries Information Network (PacFIN), a database of Pacific coast-wide commercial-fishing landings obtained from market receipts at the time of landing.

Data collected by the authors and associates of the MBA Juvenile White Shark Tagging Program (initiated in 2002) were primarily from sharks incidentally caught by commercial fishermen, whereas others were caught by MBA personnel and other researchers (M. Domeier) using either longline gear, rod and reel, and purse seine. Data were also obtained from the CDFG shark-tagging program (1983–2001) based on reports from recreational anglers and/or commercial fishermen who voluntarily tagged and released sharks throughout Southern California waters. After 2001, these records were obtained from the CDFG and NOAA Fisheries joint shark longlining and tagging cruises. There were two reports provided by game wardens at the CDFG, where cases had been filed against recreational anglers who illegally landed White Sharks. Additional recreational fishing reports of White Sharks were obtained from the NOAA-*Los Angeles Times* database, which is a daily report of marine-fish landings by commercial-passenger fishing vessels in Southern California between 1959 and 1998. Additional reports were obtained from several museum collections or records, including the Natural History Museum of Los Angeles County (J. Seigel, personal communication), California Academy of Sciences (J. McCosker, personal communication), and Scripps Institution of Oceanography Fish Collection (H. J. Walker, personal communication). Confirmed White Shark captures reported in Los Angeles and Orange County newspaper articles were also included in our analysis.

The following data were tabulated for each verified White Shark record: date, location description, latitude and longitude, TL, weight, sex, capture method, fishing gear, target species, source, and general remarks. Sharks were assigned age classes defined as YOY, individuals <1.75 m TL; juveniles, 1.75–3.00 m TL; and subadults/adults, >3.00 m TL (McCosker, 1985; Francis, 1996; Mollet and Cailliet, 1996; Wintner and Cliff, 1999; Malcolm et al., 2001; Chapter 17, this book). When total length was not available, but a whole weight was given, age class was estimated from total length derived from length-weight relationship developed for White Sharks (Mollet and Cailliet, 1996). When only landing weights were available (dressed weight), whole weight was estimated from a conversion factor (round weight = dressed weight * 1.45) based on a formula for Shortfin Mako Sharks (California Code of Regulations, Title 14, Section 187), which was then converted to a length and age class estimate.

The type of fishery was categorized as recreational or commercial, and gear type was categorized as entangling net (e.g., trammel, SGN, DGN), harpoon, long line, purse seine, trawl, trap, and recreational (hook and line).

Data Analysis

All of the records were cross-referenced, and duplicate records were removed. Capture locations were reported as CDFG fishing blocks (10 min. × 10 min.), site names, landmarks, or global positioning system (GPS) coordinates. For analysis, all capture locations were assigned to their respective CDFG fishing blocks.

Records of reported White Shark captures were analyzed by age class, capture month, capture season, fishing gear, and spatially, by CDFG fishing block for all categories. Temporal trends in YOY captures were plotted against trends in fishing effort in the DGN and SGN fisheries from 1981 to 2008. Fishing effort was not available for 2009. We used a chi-squared test of independence to determine whether the age-class distribution of reported White Sharks in entangling net fisheries was the same before (1980–1993) and after (1994–2008) the nearshore-gillnet ban. We also tested for differences in average annual fishing effort (number of net sets) in the DGN and SGN fisheries between the two time periods using an independent two-sample t test (assuming unequal variances). Similarly, we used an independent two-sample t test to test for differences in average annual YOY White Shark catch per unit effort (CPUE = YOY captures/1,000 sets) in the SGN (assuming unequal variances) and DGN (assuming equal variances) fisheries between the two time periods. YOY CPUE values were square root transformed; however, reported means and 95% confidence limits were back-transformed.

RESULTS

Frequency and Distribution of Reported White Shark Captures

We obtained 369 fishery-dependent records of reported White Shark captures occurring in the Southern California Bight from 1936 to 2009. Of the 369 records analyzed, 35% of the reports provided no indication of size. Of the remaining records, young-of-the-year White Sharks comprised 60% of the reports, followed by juveniles (32%) and subadult/adults (8%). Reports of White Shark captures were sporadic throughout the early and mid-twentieth century but increased from the 1980s through the early 1990s and peaked in 1985 and 2009 (Figure 14.1). Reported captures

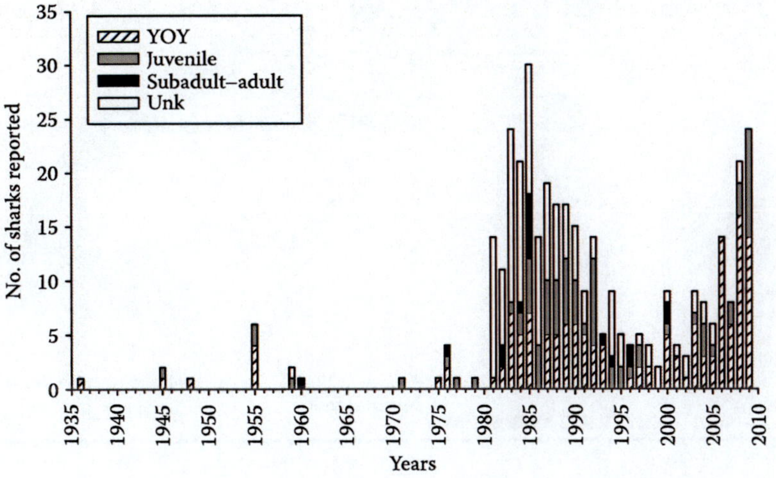

Figure 14.1 Temporal trends in reported Southern California White Shark captures by age class, 1935–2009. YOY, <175 cm TL; juveniles, 176–300 cm TL; subadult/adult, >300 cm TL; and Unk, size unknown.

of juveniles and adults were less frequent, with the majority of reports occurring during the 1980s and 1990s (Figure 14.1). Examination of reports by capture location indicated that YOY were reported along the entire coast of Southern California from Ventura to San Diego, with occasional captures in offshore waters (Figure 14.2a). YOY were most frequently captured along coastal fishing blocks near the ports of Ventura, Los Angeles/Long Beach, Danas Point/San Onofre, and San Diego. Reported captures of juvenile White Sharks were less frequent but showed a similar pattern in distribution and concentration (Figure 14.2b). Adult White Shark captures were even less commonly reported, and although the overall distribution was similar, captures were highest in the Santa Barbara and San Pedro channels (Figure 14.2c). Reports of White Shark captures were highest during the spring and summer months, occurring in the northern and central coastal fishing blocks of the SCB in the spring and extending further south in the summer (Figure 14.3a and b). During fall months, the distribution of reported captures of White Sharks was relatively less frequent than during spring and summer, with fewer sharks appearing off the central coastal areas of the SCB and more appearing just offshore (Figure 14.3c). Reported captures of White Sharks were least common during winter months, and those captures were evenly distributed along the coast and at the islands (Figure 14.3d).

Fishery Interactions

Commercial entangling nets accounted for 81% of all reported White Shark captures in the SCB, followed by recreational hook-and-line fishing (8%; Figure 14.4). YOY were primarily caught in entangling nets, and juvenile captures were dominant among hook-and-line captures. White Shark captures by entangling nets primarily occurred in coastal fishing blocks, with the highest

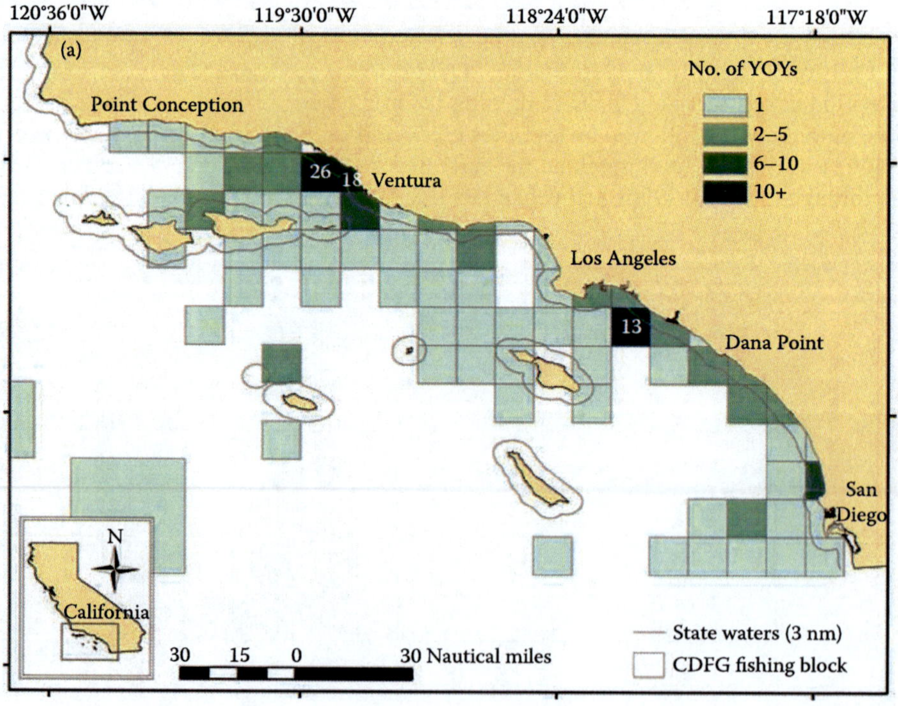

Figure 14.2 Spatial distribution of reported YOY (a), juvenile (b), and adult (c) White Shark captures occurring in Southern California, 1935–2009. The numbers represent sample sizes for fishing blocks (10 min. × 10 min.) with greater than 10 sharks.

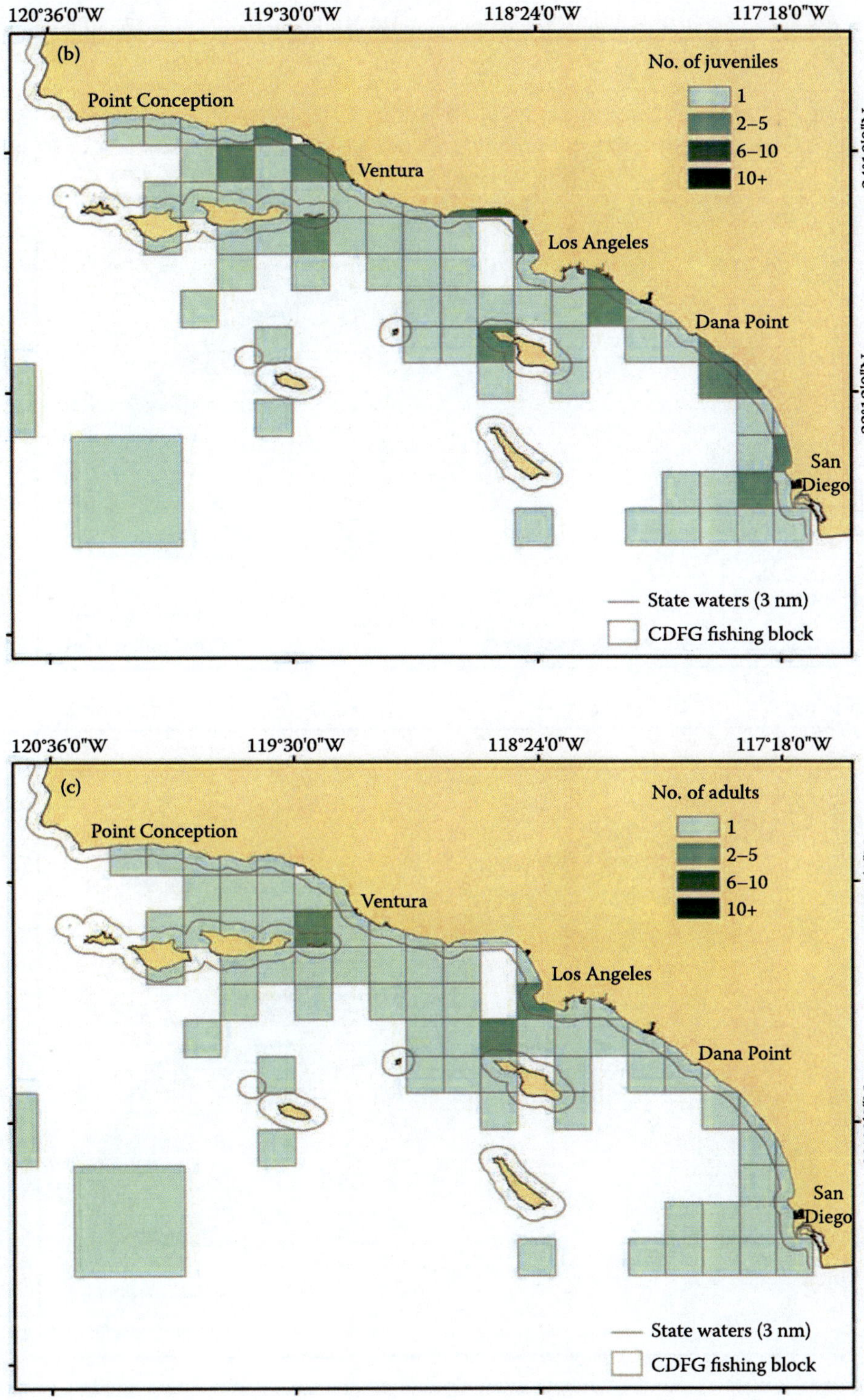

Figure 14.2 (Continued)

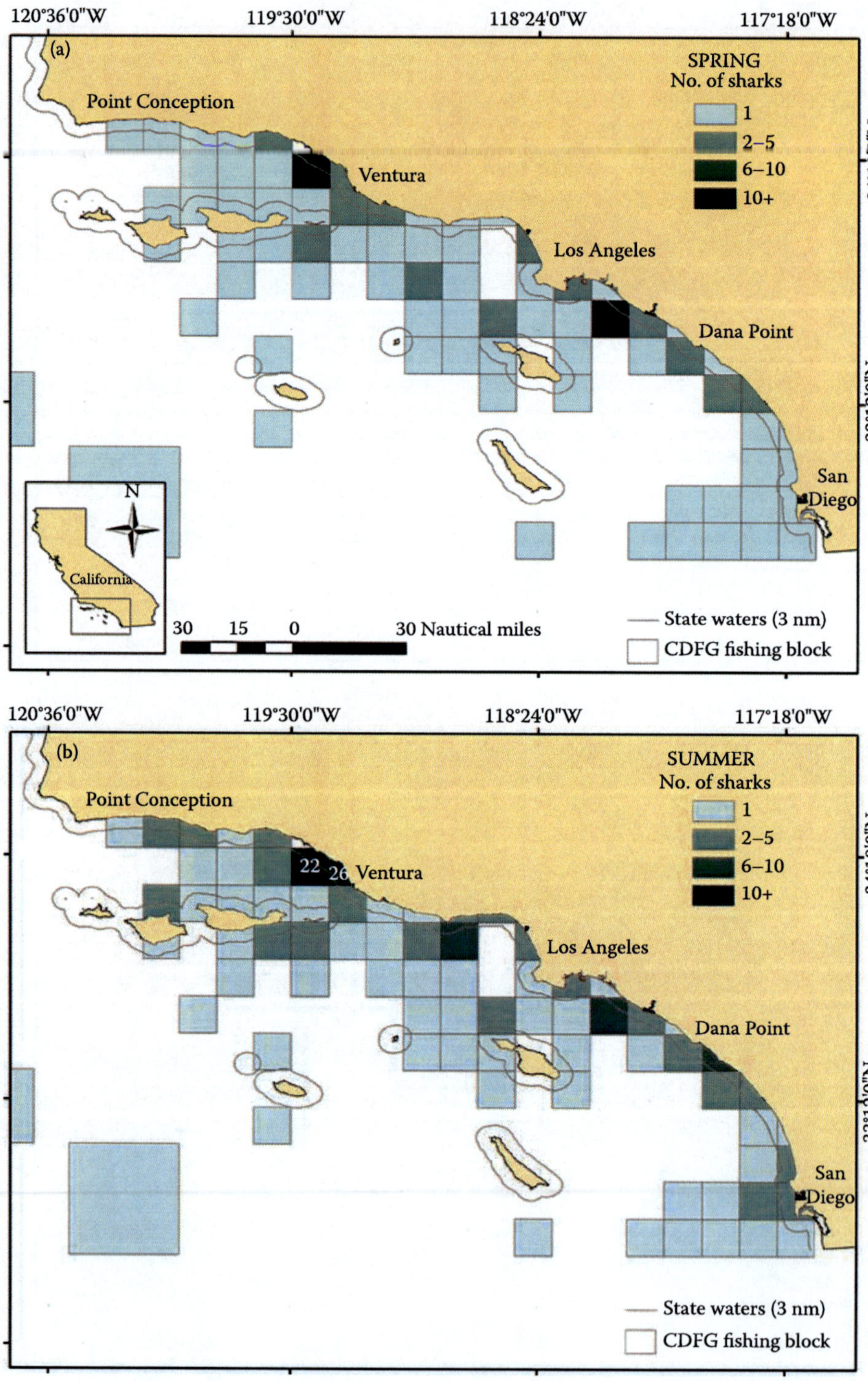

Figure 14.3 Spatial distribution of reported White Shark captures occurring in Southern California during spring (April to June, a), summer (July to September, b), fall (October to December, c), and winter (January to March, d), 1935–2009. The numbers represent sample sizes for fishing blocks with greater than 10 sharks.

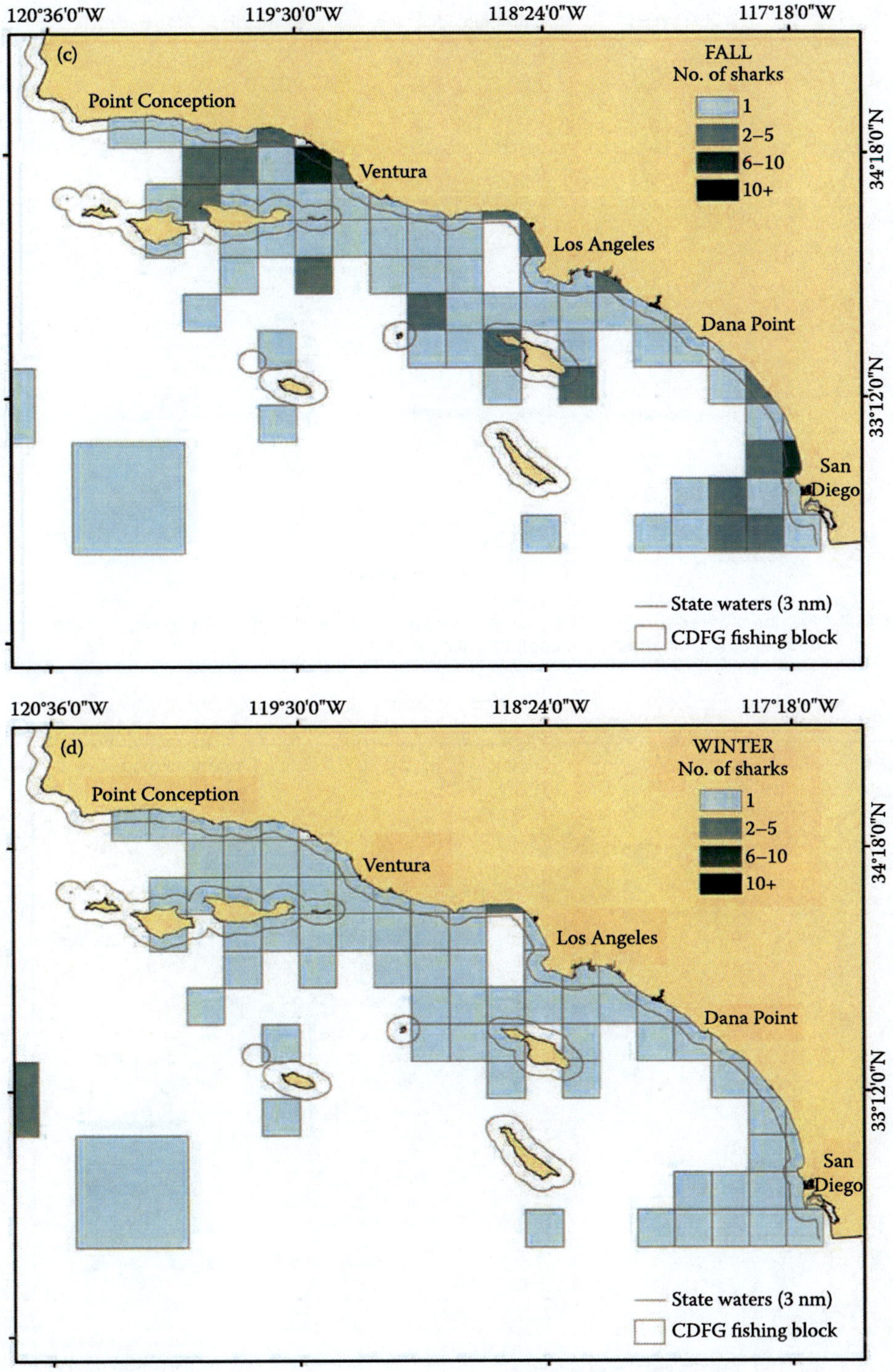

Figure 14.3 (Continued)

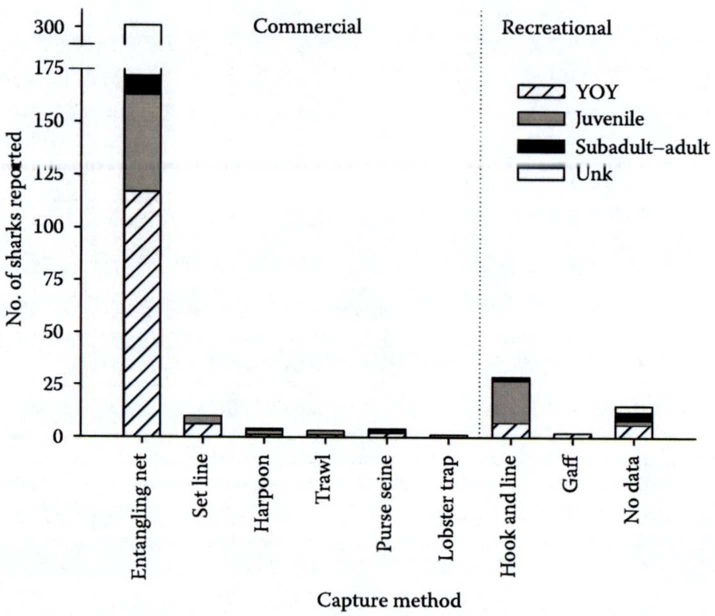

Figure 14.4 Numbers of reported White Shark captures occurring in Southern California by capture method, 1935–2009. Unk, method of capture unknown.

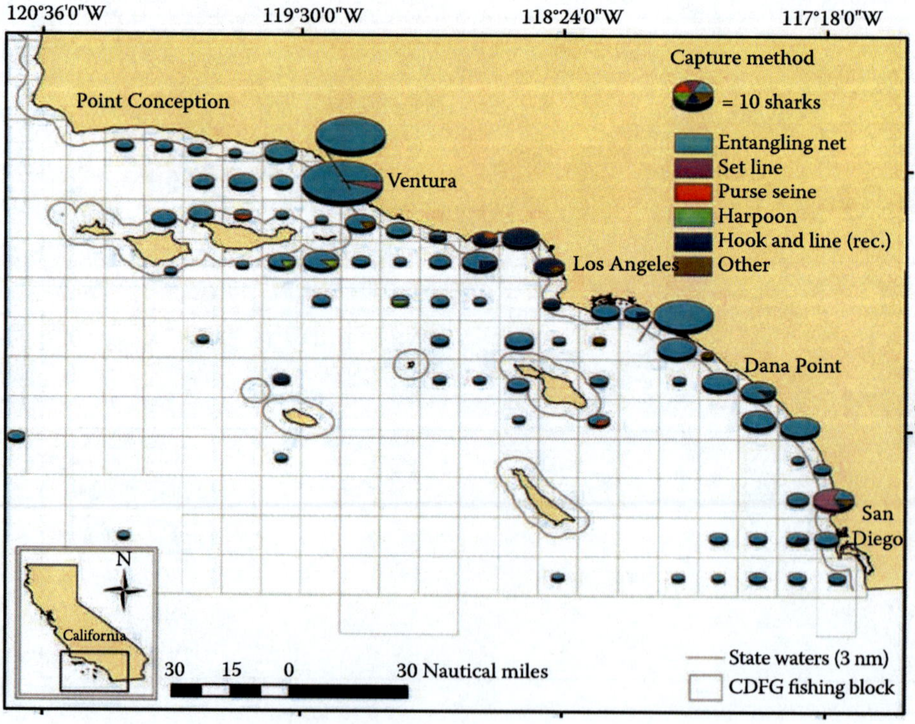

Figure 14.5 Spatial distribution of reported White Shark captures occurring in Southern California by capture method, 1935–2009. Capture methods with three or fewer capture records (e.g. trawl, lobster trap, and gaff) were pooled into the "Other" category.

reports off Ventura Flats and the San Pedro Shelf near the Los Angeles/Long Beach harbor (Figure 14.5). Recreational hook-and-line captures occurred mainly within Santa Monica Bay.

Of the 300 entangling net records, 62% of the reported captures occurred in the set-gillnet fishery targeting California Halibut, Pacific Angel Shark, and/or White Seabass, and 32% occurred in the DGN fishery targeting pelagic sharks and Swordfish; 6% of the reports provided no indication of entangling net fishery type. White Sharks captured in the SGN and DGN gillnet fisheries were reported across all seasons, with a prominent peak in set-gillnet captures occurring from May through July (Figure 14.6). The frequency of reported White Shark captures in the DGN fishery was more evenly distributed from April through November than that of the SGN fishery (Figure 14.6). Reports of White Sharks in entangling net fisheries were dominated by YOY and began in the late 1970s, peaked in 1985, decreased into the late 1990s, and began an upward trend in 2006 (Figure 14.7a).

Reported captures of YOY mirrored temporal trends in fishing effort (number of sets) in the SGN fisheries from 1981 to 2005 (Figure 14.7b). Fishing effort remained relatively stable in both fisheries from the mid-1990s to 2008; however, following 2005, the incidence of reported White Shark captures steadily increased. During the period prior to the nearshore-gillnet ban (1981–1993), the average (±SD) number of sets per year in the SGN fishery was 10,882 ± 3,964 compared with 5,821 ± 3,251 in the DGN fishery. After the nearshore-gillnet ban (1994–2008), the average number of sets decreased significantly in both fisheries (set gillnets: $\overline{X} \pm \text{SD} = 2{,}905 \pm 936$, $t = 2.16$, $p < 0.0001$; drift gillnets: $\overline{X} = 1{,}123 \pm 893$, $t = 2.14$, $p < 0.0001$). We found a significant difference in the age-class distribution of reported White Sharks captured by gillnets before and after the nearshore-gillnet ban in state waters ($X^2 = 6.19$, $p = 0.045$). Prior to the nearshore-gillnet ban, YOY White Sharks accounted for 61% of reported captures in gillnets, whereas after the ban this proportion increased by 16% to 77%. The average reported YOY CPUE (YOY/1,000 sets) in the SGN fishery was significantly higher after the closure ($\overline{X} = 0.93$, 95% confidence interval: 0.22 to 2.11) than before ($\overline{X} = 0.23$, 95% confidence interval: 0.05 to 0.55; $t = 2.12$, $p = 0.02$).

Prior to the nearshore-gillnet ban in state waters, SGN effort was primarily concentrated along the mainland coast (<3 nm from shore) from Point Conception to San Diego (Figure 14.8a). Other areas with concentrated effort included the north and south sides of Santa Rosa Island. Higher numbers of set-gillnet YOY captures appeared to coincide with higher fishing

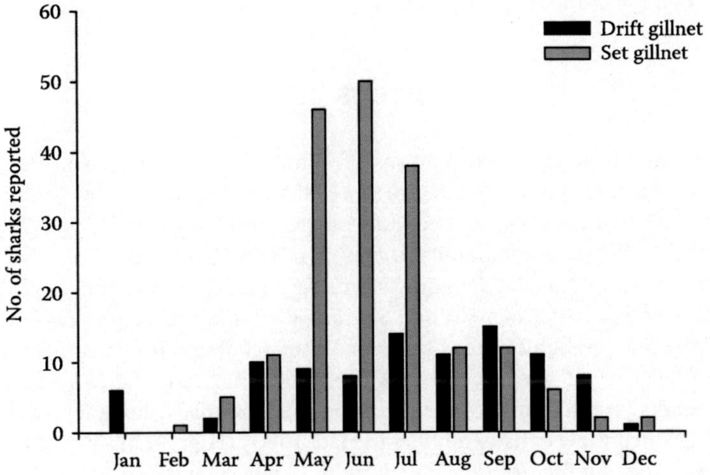

Figure 14.6 Total number of White Shark captures reported per month for the set- and drift-gillnet fisheries in Southern California, 1981–2009.

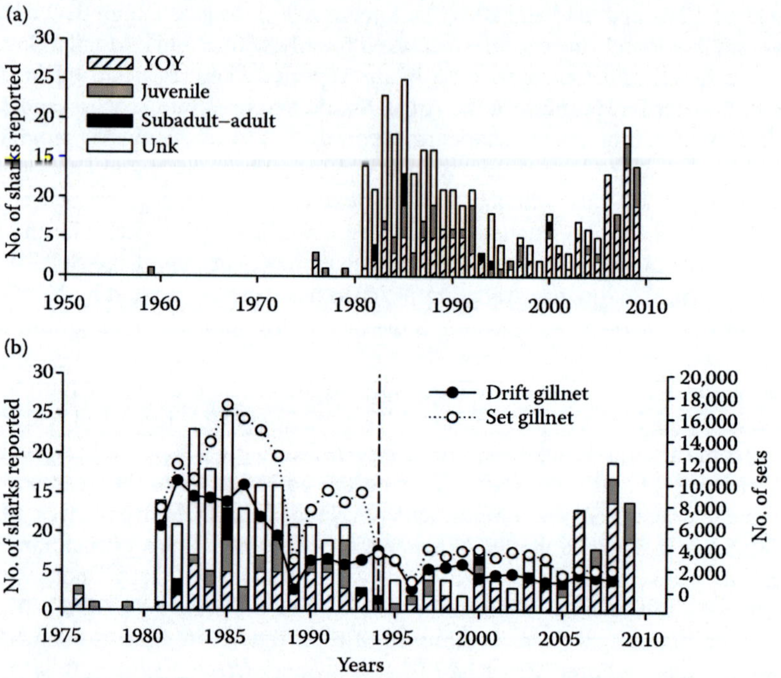

Figure 14.7 Temporal trends in reported White Sharks captured in entangling nets by stage class from a historical perspective, 1950–2009 (a), and relative to fishing effort in the set- and drift-gillnet fisheries, 1981–2008 (b).

effort along the coast of La Jolla, San Onofre, and just to the north off Newport Beach (Figure 14.8a). Following the nearshore-gillnet ban, the distribution of SGN effort changed dramatically (Figure 14.8b). Only four mainland areas (Ventura, Los Angeles/Long Beach, San Onofre, and San Diego) still contained concentrated effort, with slight offshore shifts in effort beyond state waters (Figure 14.8b). Coastal fishing blocks with high SGN effort following the ban coincided with higher reports of entangling net YOY captures occurring off Ventura and Los Angeles/Long Beach harbor (Figure 14.8a and b).

DISCUSSION

Spatial- and temporal-distribution patterns of White Sharks captured in Southern California fisheries follow patterns observed by Klimley (1985) from earlier catch records (1935–1984). However, changes in catch patterns, adjusted for changes in fishing activity and pressure after 1984, may reflect changes in White Shark populations in Southern California.

Despite intensive recreational hook-and-line fishing pressure in Southern California over the last 100 yrs., it is surprising that more White Sharks (8%) were not caught via this fishing method. However, it is likely that the lighter gear used in most nearshore recreational fisheries would reduce the probability of successfully landing even a YOY White Shark and therefore may under-represent the numbers of sharks interacting with the nearshore recreational fishery. The pelagic recreational fishery for sharks, which grew considerably in the late 1990s (Holts et al., 1998), typically uses steel leaders and larger hooks and therefore is much more likely to land White Sharks. However, as the spatial distribution in White Shark catch records suggests, it is also less likely that recreational fishers would encounter YOY or juvenile White Sharks in these pelagic habitats.

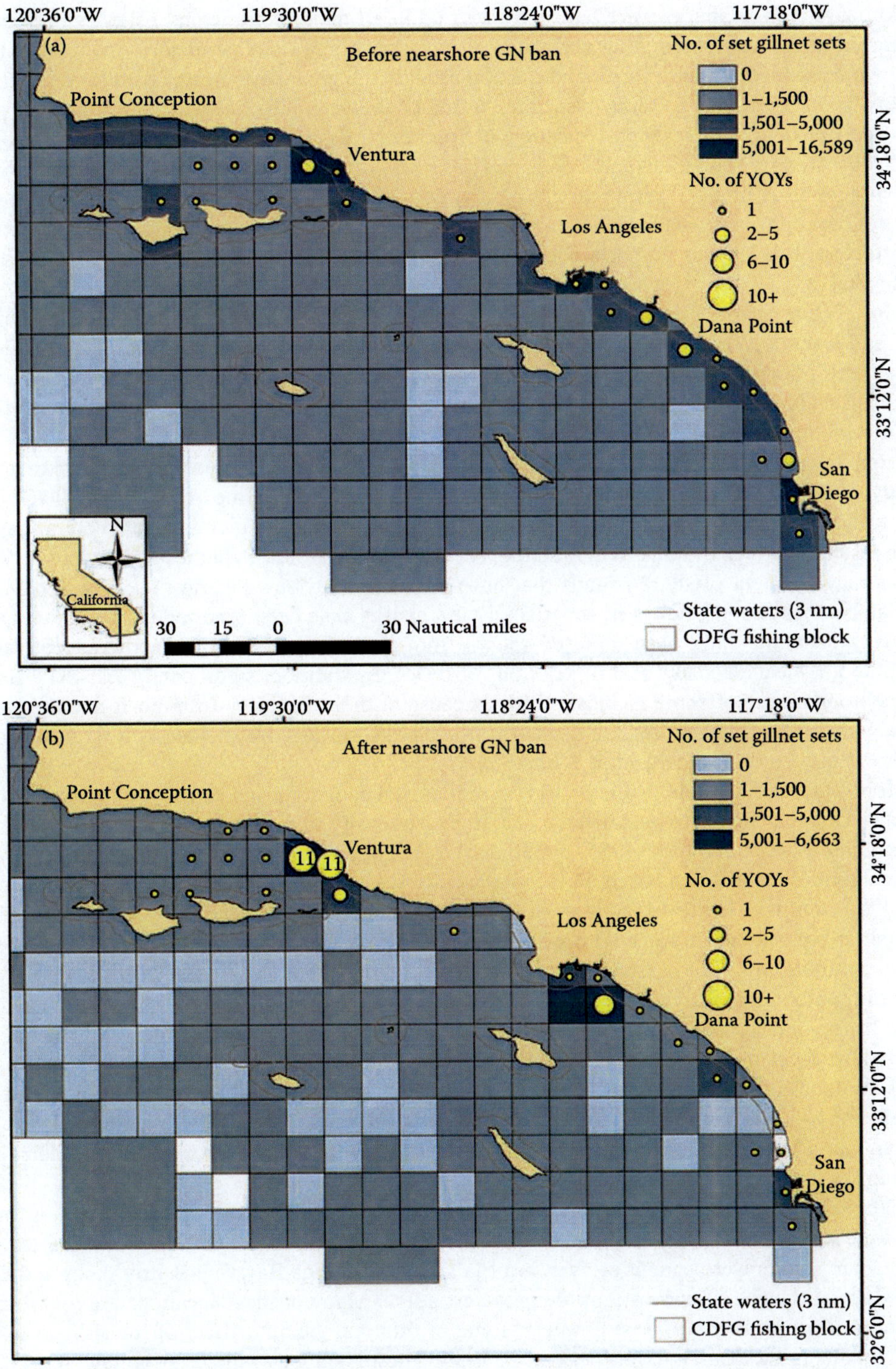

Figure 14.8 Spatial distribution of set-gillnet effort and associated YOY White Shark captures in Southern California before (1981–1993, a) and after (1994–2008, b) the nearshore-gillnet ban. The numbers represent sample sizes for fishing blocks with greater than 10 sharks. GN, gillnet.

Although White Sharks were reported captured in all major commercial fisheries in Southern California (e.g., trawl, entangling net, set line, harpoon, trap), the vast majority of reports (81%) came from entangling net fisheries and thereby provide the greatest source of data for determining White Shark distribution patterns over time. Young-of-the-year White Sharks constituted the greatest proportion of age classes reported caught in Southern California. This is similar to what Klimley (1985) observed from previous catch data, and it is likely that this pattern is influenced by the higher gear susceptibility of YOY White Sharks compared with the larger sharks. Larger sharks are more likely to break through gear (gillnets and set line) and avoid capture and thus be under-reported. Nevertheless, this further supports Klimley's (1985) theory that the SCB is a nursery ground for White Sharks.

More White Sharks were reported caught in nearshore SGN than in offshore DGN. In addition, YOY and juvenile White Sharks were most frequently reported caught in nearshore fishing blocks, particularly prior to 1994. Although this is where a majority of the nearshore SGN fishery for California Halibut, Pacific Angel Shark, and White Seabass took place prior to the nearshore-gillnet ban in 1994, recent aerial observations (J. O'Sullivan, personal observations) and satellite telemetry data from tagged sharks (Weng et al., 2007; J. O'Sullivan et al., unpublished data) indicate that YOY and juvenile White Sharks frequently swim within 100 m of the shoreline. Adult White Sharks were most frequently reported caught offshore or near offshore islands, which is also where a majority of the visual sightings of adult White Sharks occur. Klimley (1985) hypothesized that adult White Sharks may frequent offshore islands that have seal and sea lion rookeries, such as San Miguel and San Nicolas Islands; however, surprisingly few sharks have been reported caught around those islands. Most adults have been reported caught from around Santa Catalina and Anacapa Island, which do not have extensive seal or sea lion rookeries. In addition, significantly less SGN fishing occurs around these offshore islands, mostly because of their distances from ports and the lack of available fishing habitat more than 1 nm from island shorelines. Nevertheless, it is unlikely that SGN gear may catch and hold adult White Sharks.

More YOY and juvenile White Sharks were reported caught during summer months than other seasons. This may be partly attributed to the higher intensity of nearshore SGN fishing effort during summer months. Klimley (1985) hypothesized that juvenile White Sharks were most common during summer months and based on the presence of yolk scars, may indicate that pupping occurs during the summer. However, adults were thought to move north in the fall and winter months. This pattern of seasonality was not observed in adult White Sharks because they were reported as being captured during all seasons. Recent satellite-tagging data have suggested that YOY White Sharks migrate to Mexican waters during winter months (Weng et al., 2007; Chapter 28, this book). These fishery-independent data further support findings that YOY and juvenile White Sharks are primarily present in Southern California during the summer months. A review of northeastern Pacific White Shark life history (Chapter 16, this book) analyzed a subset of the fishery data presented here, as well as other data, to suggest that the Southern California Bight is indeed a pupping ground, and adults are relatively rare in this region, except for the presence of females during the pupping season.

Although White Sharks were reportedly caught along the entire coastline of Southern California, there were several areas where more White Sharks were reported caught more frequently than other areas. More sharks were caught off Ventura Flats, Santa Monica Bay, San Pedro Shelf, and Dana Point Canyon than at other locations along the coastline. Many of these locations are within close proximity to major fishing ports; however, some of this may be an artifact of the shift in fishing effort following the nearshore-gillnet ban in 1994. These locations represent the few areas along the Southern California coastline and offshore islands where gillnet fishers can effectively fish for California Halibut, Pacific Angel Shark, and White Seabass outside state waters. Satellite-tagging data for YOY and juvenile White Sharks in Southern California indicate that sharks may be using these areas (Ventura Flats, Santa Monica Bay, San Pedro Shelf, and Dana Point Canyon) more than other areas along the coastline, particularly during the summer months (Dewar et al., 2004; Weng et al., 2007; J. O'Sullivan et al., unpublished data).

Overall, there were very few and sporadic reports of White Shark captures prior to the 1980s in Southern California. This level of reporting prior to the 1980s could have been attributed to a lack of interest in White Sharks, lower fishery interaction, or a lower White Shark population. Klimley (1985) hypothesized that significantly reduced marine mammal populations (sea otters, pinnipeds, and cetaceans) along the California coast because of overharvesting could have reduced adult White Shark food availability and thereby lowered the White Shark population. However, two prominent peaks were observed in reported White Shark captures after the federal protection of marine mammals in 1972: one in 1985 and another larger peak in 2009. Klimley (1985) attributed the rise in White Shark reportings in the 1980s to an increased awareness of White Sharks following the movie *Jaws* in 1975 and an interest from Sea World in attempting to display White Sharks in public aquariums. Although there is little doubt that both of these factors increased awareness by commercial fishers and the public of White Sharks in California, entangling net-based commercial fisheries also peaked in the 1980s, which showed peak effort and landings for many targeted species (Methot, 1983; Holts, 1988; Holts et al., 1998; Pondella and Allen, 2008). It is likely that the increased effort of entangling net fisheries (both nearshore SGN and offshore DGN) in the 1980s and early 1990s increased fishery interactions with YOY and juvenile White Sharks in Southern California, thereby increasing the mortality rates of these age classes. Prior to state and federal protection, White Sharks were being landed and sold in California, albeit at low levels compared with other shark species (Holts, 1988; Holts et al., 1998). Although it is unknown to what degree this fishery interaction reduced the White Shark population in Southern California, there was a marked decrease in reported White Shark catch from the peak in 1985 to 1995. Interestingly, this decline coincides with a series of entangling net gear and season restrictions that began shortly after 1985.

Coincidentally, the ban of nearshore gillnets and California protection of White Sharks both began in 1994. Concerns regarding overfishing of Thresher and Mako Sharks and the incidence of marine mammal interactions in the offshore drift-net fishery resulted in substantial restrictions placed on this fishery beginning in the late 1980s (Holts, 1988; Carretta et al., 2004). These fishery restrictions resulted in a significant reduction in nearshore and offshore gillnet fishing effort in California and prohibited landing of any incidentally caught White Sharks. Since the mid-1990s, fishing effort has remained fairly stable or declined in both the SGN and DGN fisheries; however, reported White Shark captures have steadily increased since 2005. One explanation for this increased trend in reported White Shark captures in Southern California may be attributed to the initiation of the Monterey Bay Aquarium White Shark Program, where MBA worked with commercial fishers to coordinate tag and release of incidentally caught sharks. Fishers were paid for their participation in this program, because this process is allowed under the current White Shark protection regulations in the State of California. It is possible that incidentally caught sharks may have been simply released and not reported in logbooks prior to the initiation of this program.

On the other hand, it is also possible that this increase in reported White Shark catch since 2005 is the result of the recovery of the White Shark population following impacts from previous YOY and juvenile mortality in gillnet fisheries prior to the nearshore-gillnet ban and changes in offshore gillnet regulations. Because a majority of the "unknown" age class reports ($n = 131$) came from the entangling net records, and the majority of the sharks caught in the entangling net fishery were reported as YOY and juveniles, it is likely the data presented under-represent the magnitude of the fishery interaction with these age classes, especially prior to gillnet-fishing restrictions (Figure 14.7). Recovery of other large nearshore-fish populations has also been documented as the result of the nearshore-gillnet ban. Pondella and Allen (2008) found fishery-dependent and fishery-independent evidence of increased abundances of Giant Black Sea Bass (*Stereolepis gigas*), Tope, Leopard Sharks, and White Seabass in Southern California following the nearshore gillnet ban. They also noted that these abundance trends were not correlated with changes in Pacific Decadal

Oscillation or El Niño/Southern Oscillation events. Although there are no fishery-independent data available to corroborate this pattern in White Sharks, there has been a steady increase in reported predation events on Sea Otters (*Enhydra lutris nereis*) off Central and Northern California since 2003. Necropsy results of recovered Sea Otters suggest juvenile White Sharks may be responsible for these mortalities and that the frequency of these mortalities has doubled since 1992 (Kreuder et al., 2003; Ames et al., 1996; M. Harris, personal communication). This pattern may be indicative of a growing number of juvenile sharks, relieved from fishing mortality since the mid-1990s.

Preliminary data and observations from the MBA Juvenile White Shark Tagging Program have indicated that YOY and juvenile White Sharks exhibit fairly high postrelease survival after being caught in gillnet gear and, in some cases, being held in on-board fish totes for periods of up to several hours (Chapter 28, this book). This indicates that despite continued fishery interactions, release of incidentally caught White Sharks has likely reduced YOY and juvenile mortality of sharks and that postcapture release can be considered an effective conservation tool. Although there are numerous problems with using fishery-dependent data for estimating population trends, future fishery-independent methods such as telemetry tagging, aerial surveys, and better observer-based records could help substantiate some of the hypotheses put forth in this study. Nevertheless, increases in marine mammal populations, protection of YOY and juvenile food base, and reduced fishing mortality has probably enabled the White Shark population to increase in California in recent years.

ACKNOWLEDGMENTS

We would like to thank the following people for their help in obtaining records: CDFG: Sandy Owen, Steve Wertz, Rebecca Hartman, Rod Buckler, and Valerie Taylor; National Marine Fisheries Services: Al Coan, Darlene Ramon, Lyle Enriquez, Suzy Kohin, and Rand Rasmussen; scientific collections: Jeff Siegel, John McCosker, and H. J. Walker. We thank Chuck Winkler and the Southern California commercial fishers for their information on incidental White Shark catches and data, the CSULB Shark Lab Rapid Response Team (Chris Martin, Kate Jirik, Bonnie Rogers, Thomas Farrugia, Mario Espinoza, Kerri Loke, Kim Anthony, Haley Zemel, Chris Mull, Megan McKinzie, Carrie Espasandin, Erika Fox, and Carlos Mireles) for their detailed notes, and Monterey Bay Aquarium for their support of the Southern California Juvenile White Shark Tagging Program.

REFERENCES

Ames, J.A., J.J. Geibel, F.E. Wendell, and C.A. Pattison. 1996. White shark-inflicted wounds of sea otters in California, 1968–1992. In: *Great white sharks: The biology of Carcharodon carcharias*, edited by A.P. Klimley and D.G. Ainley, pp. 309–316. San Diego, CA: Academic Press.

Barsky, K.C. 1990. History of the commercial California halibut fishery. *Cal. Dept. Fish. & Game Fish. Bull.* 174: 217–228.

Boustany, A.M., S.F Davis, P. Pyle, S.D. Anderson, B.J. LeBoeuf, and B.A. Block. 2002. Satellite tagging: Expanding niche for white sharks. *Nature* 415: 35–36.

Carretta, J.V., T. Price, D. Petersen, and R. Read. 2004. Estimates of marine mammal, sea turtle, and seabird mortality in the California drift gillnet fishery for swordfish and thresher shark, 1996–2002. *Mar. Fish. Rev.* 66: 21–25.

CDFG. 2009. *Annual fisheries forum fisheries report*. California Department of Fish and Game Marine Region 2009 (http://nrm.dfg.ca.gov/FileHandler.ashx?DocumentID=36309)

Dewar, H., M. Domeier, and N. Nasby-Lucas. 2004. Insights into young of the year white shark, *Carcharodon carcharias*, behavior in the Southern California Bight. *Envir. Biol. Fish.* 70: 133–143.

Dotson, R.C., and R.L. Charter. 2003. Trends in the southern California sport fishery. *CalCOFI Reports* 44: 94–106.

Francis, M. 1996. Observation on a pregnant white shark with a review of reproductive biology. In: *Great white sharks: The biology of Carcharodon carcharias*, edited by A.P. Klimley and D.G. Ainley, pp. 157–172. San Diego, CA: Academic Press.

Hanan, D.A., D.B. Holts, and A.L. Coan, Jr. 1993. The California drift gill net fishery for sharks and swordfish, 1981–82 through 1990–1991. *Cal. Dept. Fish & Game Fish Bull.* 175: 1–90.

Heneman, B., and M. Glazer. 1996. More rare than dangerous: A case study of white shark conservation in California. In: *Great white sharks: The biology of Carcharodon carcharias*, edited by A.P. Klimley and D.G. Ainley, pp. 481–491. San Diego, CA: Academic Press.

Holts, D.B. 1988. Review of US west coast commercial shark fisheries. *Mar. Fish. Rev.* 50: 1–8.

Holts, D.B., A. Julian, O. Sosa-Nishizaki, and N.W. Bartoo. 1998. Pelagic shark fisheries along the west coast of the United States and Baja California, Mexico. *Fish. Res.* 39: 115–125.

Klimley, A.P. 1985. The aerial distribution and autecology of the white shark (*Carcharodon carcharias*), off the west coast of North America. *Mem. South. Calif. Acad. Sci.* 9: 15–40.

Kreuder, C., M.A. Miller, D.A. Jessup, L.J. Lowenstine, M.D. Harris, J.A. Ames, T.E. Carpenter, P.A. Conrad, and J.A.K. Mazet. 2003. Patterns of mortality in southern sea otters (*Enhydra lutris nereis*) from 1998 to 2001. *J. Wildlife Dis.* 39: 495–509.

Love, M.S. 2006. Subsistence, commercial, and recreational fisheries. In: *The ecology of marine fishes: California and adjacent waters*, edited by L.G. Allen, D.J. Pondella, and M.H. Horn, pp. 567–594. Berkeley, CA: University of California Press.

Malcolm, H., Bruce, B. D., and Stevens, J. D. 2001. *A review of the biology and status of white sharks in Australian waters.* Hobart, Australia: CSIRO Marine Research.

McCosker, J. 1985. White shark attack behavior: Observations of and speculations about predator and prey strategies. *Mem. South. Calif. Acad. Sci.* 9: 123–135.

Methot, R.D., Jr. 1983. Management of California's nearshore fishes. *Mar. Rec. Fish.* 8: 161–172.

Mollet, H.F., and G.M. Cailliet. 1996. Using allometry to predict body mass from linear measurements of the white shark. In: *Great white sharks: The biology of Carcharodon carcharias*, edited by A.P. Klimley and D.G. Ainley, pp. 81–89. San Diego, CA: Academic Press.

Oliphant, M.S., P.A. Gregory, B.J. Ingle, and R. Madrid. 1990. California marine fish landings for 1977–1986. *Calif. Dept. Fish & Game, Fish. Bull.* 173: 1–52.

Pondella, D.J., and L.G. Allen. 2008. The decline and recovery of four predatory fishes from the Southern California Bight. *Mar. Biol.* 154: 307–313.

Richards, J.B. 2001. Pacific angel shark. In: *California's marine living resources: A status report*, edited by W.S. Leet, C.M. Dewees, R. Klingbeil, and E.J. Larson, pp. 248–250. Berkeley, CA: University of California Press.

Weng, K.C., J.B. O'Sullivan, C.G. Lowe, C.E. Winkler, H. Dewar, and B.A. Block. 2007. Movements, behavior, and habitat preferences of juvenile white sharks *Carcharodon carcharias* in the eastern Pacific. *Mar. Ecol. Prog. Ser.* 338: 211–224.

Wintner, S.P., and G. Cliff. 1999. Age and growth determination of the white shark, *Carcharodon carcharias*, from the east coast of South Africa. *Fish Bull.* 1999: 153–169.

CHAPTER 15

Incidental Catch and Ecological Observations of Juvenile White Sharks, *Carcharodon carcharias*, in Western Baja California, Mexico

Conservation Implications

Omar Santana-Morales
Centro de Investigación Científica y de Educación Superior de Ensenada

Oscar Sosa-Nishizaki*
Centro de Investigación Científica y de Educación Superior de Ensenada

Miguel A. Escobedo-Olvera
Centro de Investigación Científica y de Educación Superior de Ensenada

Erick C. Oñate-González
Centro de Investigación Científica y de Educación Superior de Ensenada

John B. O'Sullivan
Monterey Bay Aquarium

Daniel Cartamil
University of California, San Diego

ABSTRACT

Between 1999 and 2010, we documented incidental catch of 111 juvenile White Sharks (*Carcharodon carcharias*) during surveys of the artisanal and commercial fisheries along the Pacific coast of Baja California, Mexico and opportunistically collected stomach samples. Sharks ranged in size from 123 to 274 cm total length (TL). Most (79.8%) were young of the year (YOY) (≤175 cm TL), and the remainder were juvenile White Sharks (JWS) (>175–300 cm TL). Of the four documented fishing-gear types, bottom gillnets represented 74.7% of the catch, followed by drift gillnet (18.0%), artisanal seine net (4.5%), and surface longline (2.7%). Catch data indicate that the continental shelf is an important habitat for YOY and JWS in the region, with a possible core nursery area in Bahía Sebastián Vizcaíno, where 66 YOY specimens were documented. Stomach content analysis for 14 JWS agree with previous studies showing that the species feeds primarily on bony fish and demersal elasmobranchs. National (NOM-059 and NOM-029) and international [Convention on International Trade of Endangered Species (CITES)] regulations ban the capture and trade of this species. However, monitoring of all White Shark catches is complicated by the remote nature of the fishing grounds, as well as socioeconomic considerations. Our data suggest that future White Shark conservation must include cooperative research and management plans between Mexico and the United States.

INTRODUCTION

White Sharks (*Carcharodon carcharias*) are taken as bycatch in several fisheries worldwide, with gear that includes longline, gillnet, trawl, handline, and rod-and-reel fisheries, and they are also opportunistically fished for jaws and other curios (Bruce, 2008). Although widely distributed, the White Shark is an apex predator with a naturally low abundance, low natural mortality, and low reproductive potential (Bruce, 2008). Because of these life-history characteristics and the perceived declining trend in some aggregation sites and sport fisheries (mainly in Australia and South Africa), the White Shark was listed as "vulnerable" by the International Union for the Conservation of Nature (Anonymous, 2002; Fergusson et al., 2005). In addition, this species was included in Appendix II of the 2004 CITES of fauna and flora to monitor trade in its products (fins, meat, jaws, and teeth) and ensure that such trade does not contribute further to its population decline (UNEP-WCMC, 2010).

An important component of conservation efforts is an analysis of White Shark habitat utilization, and the fisheries that may operate in those habitats, to assess the potential for fishery interactions and develop strategies to minimize fishery-induced mortality. Juvenile habitat may be particularly important in this regard, because low juvenile natural mortality could make fishing mortality a more significant factor in population dynamics (Kinney and Simpendorfer, 2008). Thus, restricted fishing in juvenile habitats may be warranted as a management tool for this species (Mollet and Cailliet, 2002). However, the distribution of juvenile White Sharks (JWS) is not as well known as that of adults. Electronic tagging studies in the eastern Pacific have shown that juveniles occur mainly in nearshore waters (Dewar et al., 2004; Weng et al., 2007b; Chapters 14, 16, and 28, this book), whereas adults show a seasonal affinity for both nearshore and pelagic habitats, including deep sea habitats, and exhibit wide-ranging travels (Boustany et al., 2002; Weng et al., 2007a; Domeier and Nasby-Lucas, 2008; Jorgensen et al., 2010; Chapters 11 and 16, this book). In the northeastern Pacific, JWS of <300 cm total length (TL) (Cailliet et al., 1985; Mollet et al., 1996) are associated with coastal waters over the continental shelf (Klimley, 1985; Dewar et al., 2004; Weng et al., 2007b; Chapters 14 and 16, this book), and their range extends southward into Mexican waters. The distribution of newborn individuals (120–175 cm TL; Francis, 1996) suggests that nursery grounds for the species include the west coast of the Baja California peninsula and into the Gulf of California (Klimley, 1985; Compagno, 2002; Weng et al., 2007a; Galván-Magaña et al., 2010; Chapters 14 and 16, this book).

Concerns for the conservation of White Sharks in Mexican waters are becoming more apparent, because fisheries there have historically not been well regulated. The White Shark is now recognized as a "threatened" species in Mexico by the Natural Resources Agency (SEMARNAT) (DOF, 2002), and in 2007, regulations in the form of the NOM-029-PESCA-2006 were published to regulate elasmobranch fisheries and prohibit fisheries that target White Shark, Whale Shark (*Rhincodon typus*), Basking Shark (*Cetorhinus maximus*), and other highly vulnerable elasmobranchs (DOF, 2007). Nevertheless, capture of White Sharks persists in Mexican waters, and recent reports have shown that both juveniles and adults are caught incidentally on the western coast of the Baja California peninsula and in the Gulf of California (Escobedo-Olvera, 2009; Galván-Magaña et al., 2010; Cartamil et al., 2011).

Artisanal fisheries account for up to 80% of elasmobranch fishing activity in Mexican waters, and an extensive artisanal fishery is known to operate on the Pacific coast of Baja California (Cartamil et al., 2011). Artisanal fishing is conducted from "pangas," which are small (7–10 m long), outboard motor-powered open boats with a crew of (typically) three or fewer undertaking 1–2-d fishing trips, setting and hauling the gear by hand. Depending on the region and season, fishing gear can include small bottom longlines (around 400 hooks), pelagic longlines (around 500 hooks), small drift gillnets, and bottom fixed gillnets (Sosa-Nishizaki et al., 2008). Elasmobranchs and other fishes in the catch are brought back to a base camp on the beach where they are processed. Because much of the fishing activity takes place in nearshore waters, interactions of the artisanal fleet with White Sharks have historically occurred, but because many of these are remote locations, historic landings have not been quantified.

Commercial fisheries also have high potential for interactions with White Sharks. Historically, a commercial drift-gillnet fishery targeting Swordfish (*Xiphias gladius*) and sharks (principally Thresher Shark, *Alopias vulpinus*, and Blue Shark, *Prionace glauca*) operated mainly off the west coast of the Baja California peninsula (Holts et al., 1998). Most of the drift-gillnet fleet, which at its height included 34 fishing vessels, was medium-sized boats (10–24 m long) based in the port city of Ensenada, Baja California. Although commercial drift-gillnet fishing was banned by the NOM-029-PESCA-2006, log book records exist that can be used to obtain data on White Shark landings.

In order to understand White Shark population trends, Cailliet (1996) suggested several approaches, including assessment of the temporal component of the catch, and the use of catch per unit effort when possible. Analysis of White Shark incidental catch in California fisheries since 1935 has contributed greatly to our knowledge of White Shark life history and informed management strategies for this species on the west coast of the United States (Klimley, 1985; Holts, 1988; Hanan et al., 1993; Chapter 14, this book). In this chapter, we summarize records of JWS caught incidentally by artisanal and commercial fisheries operating along the Pacific coast of the Mexican state of Baja California and discuss the options for conservation of the species in this region.

MATERIALS AND METHODS

Data Acquisition

White shark records were obtained from four primary sources:

1. Surveys of artisanal-fishing camps
2. Carcass discard site surveys
3. Log book from one commercial drift-gillnet vessel
4. Observations onboard the commercial drift-gillnet vessel from Source 3

Surveys of artisanal-fishing camps along the Pacific coast of Baja California (Figure 15.1) were conducted during the period 2006–2010 [see Cartamil et al. (2011) for a complete description of the camps] to assess elasmobranch capture in the region. When White Sharks were encountered, TL was measured to the nearest 0.1 cm, sex was determined, and stomach content samples were collected when possible. In some cases, we had access only to heads of sharks, which were identified

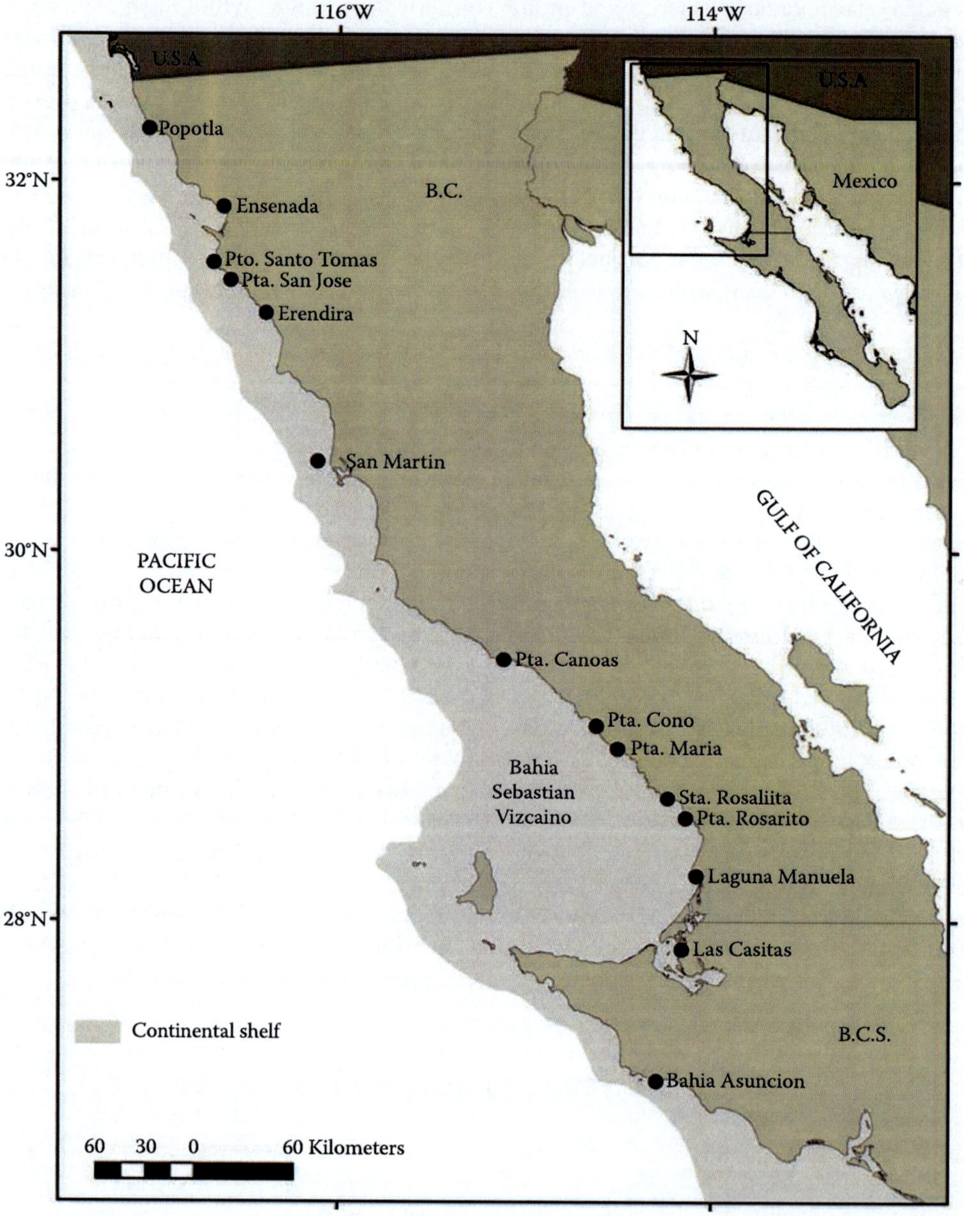

Figure 15.1 Location of the fishing camps with juvenile White Shark records on the Pacific coast of Baja California, Mexico. The shaded area indicates the continental shelf up to 200-m depth. The inset shows the location of the study area along the northwest coast of Mexico's border with the United States.

to species using physical characteristics such as chondrocranium, jaws, and teeth (Santana-Morales, 2008). For these specimens, TL was calculated based on the equation:

$$TL = (0.1206*CH) - ((0.0284)*100) \tag{15.1}$$

where CH is the crown height (measured with calipers with accuracy of 1 mm) of the second upper anterior tooth in right side of the jaw. The equation was generated using reported data of TL and CH for 51 White Sharks in Mollet et al. (1996) and from 13 White Sharks measured during this study (size range, 130–252 cm TL).

Cartamil et al. (2011) also assessed catch composition by surveying carcass discard sites at the Laguna Manuela fishing camp (Figure 15.1) during the period 2006–2008. At these sites, the high heat and low humidity of the desert had desiccated elasmobranch remains and preserved them in a state resembling mummification (Figure 15.2). Because most of the mummified carcasses consisted only of heads, TL was estimated based on the tooth-size equation above.

We analyzed and included log book data from one commercial drift-gillnet vessel based in Ensenada, which operated along the west coast of the Baja California peninsula from April 1999 to December 2009. The log book contained detailed information on numbers of all specimens captured during 1,333 sets targeting swordfish and thresher shark but did not include length or weight data. Additional log book data considered in our analyses include set location, target species, and sea-surface temperature. An observer was placed on this vessel on four fishing trips between July 2006 and February 2008 and observed 47 sets targeting thresher shark. White sharks captured during these trips were measured (TL, with accuracy of 0.5 cm) and sexed, and whenever possible stomach content samples were obtained.

Stomach Content Analysis

Stomach samples were placed in plastic bags, labeled, and stored in ice coolers or frozen. In the laboratory, each stomach sample was separated, thawed, rinsed to remove excess water, and weighed full. The percentage of fullness (for the esophageal stomach) was estimated visually, and contents were removed and weighed. Materials and slurry were rinsed and sorted with screen sieves with mesh size of 425 and 560 μm. Prey items were separated and identified to the lowest taxonomic level possible (Preti et al., 2001). Fish crystalline lenses were counted in pairs, with the highest count representing the minimum number of fish present in the stomach.

Figure 15.2 Mummified White Shark heads sampled at a carcass discard site at Laguna Manuela, Baja California, Mexico.

RESULTS

We obtained 111 records of White Sharks caught incidentally during commercial and artisanal fishing operations on the continental shelf of the west coast of Baja California from 1999 to 2010. From the Laguna Manuela carcass discard site, we recorded 64 sharks (57.7%), whereas 27 White Sharks were sampled at beach landing sites (24.3%), 17 were recorded from log books (15.3%, number of sets = 1,333), and 3 were recorded from observer efforts (2.7%, number of sets = 46). White Sharks were recorded at 14 of the 44 artisanal-fishing camps visited (Figure 15.1 and Table 15.1). The actual or estimated sizes of 84 (75.7%) sharks ranged from 123 to 274.5 cm TL, with shark sizes more frequent in the 140–150- and 160–170-cm TL intervals (Figure 15.3). Sixty-seven White Sharks (79.8%) were young of the year (YOY) (≤1.75 m TL; Francis, 1996; Chapter 17, this book), and the remaining 17 were juveniles. Sex was determined for 21 sharks, with a sex ratio of 1.3:1 (12 females and 9 males).

Most YOY White Sharks were caught inside Bahia Sebastian Vizcaino (BSV) and landed at Laguna Manuela (60), Las Casitas (5), and Punta Canoas (1) fishing camps (Figure 15.1 and Table 15.1). The northernmost YOY (151 cm TL, female) was recorded at Popotla. In two of the YOY White Sharks, the umbilical scar was still visible (one female with 150 cm TL and one male with 163 cm TL), indicating that they were newborn. Excluding the carcass discard site records ($n = 64$) without associated date of catch, most sharks were caught during the summer months ($n = 28$), although high frequency of capture was also recorded in January ($n = 9$). Catch of White Sharks was not observed during March, April, and December in this study (Figure 15.4).

Highest incidental catch of White Sharks was recorded in the artisanal bottom-set gillnet, with 83 (74.7%) sharks caught using this fishing gear and landed in the Popotla, Erendira, Santa Rosaliita, and Laguna Manuela camps where fishing occurs all year (Figure 15.5 and Table 15.1). This was followed by drift gillnet with 20 (18%) White Sharks caught during fishing operations at the edge of or over the continental shelf, in waters with sea-surface temperature ranging between 15.7 and 23.1°C. Five White Sharks (4.5%) were caught in the artisanal seine-net fishery that operates inside Laguna Ojo de Liebre and lands at the Las Casitas fishing camp. Artisanal surface-longline fishing at Bahia Sebastian Vizcaino accounted for only three (2.7%) White Sharks, all landed at the Laguna Manuela fishing camp (Table 15.1).

Table 15.1 Catch Information for Juvenile White Sharks by Fishing Site

Fishing Site	N Latitude	W Longitude	*n*	Fishing Gear	Source
Popotla	32.282	117.035	3	DG, BG	LbD, BS
Ensenada	31.861	116.628	6	DG	LbD
Punta San Jose	31.463	116.590	1	DG	LbD
Puerto Santo Tomas	31.559	116.681	3	DG	LbD
Erendira	31.286	116.398	1	BG	BS
San Martin Island	30.489	116.114	2	DG	LbD
Punta Canoas	29.420	115.120	3	DG	ObS
Punta Cono	28.977	114.580	1	DG	LbD
Punta Maria	28.936	114.507	1	DG	LbD
Santa Rosaliita	28.667	114.239	3	BG	BS
Punta Rosarito	28.560	114.142	1	DG	LbD
Laguna Manuela	28.247	114.083	80	BG, SL	DS, BS
Las Casitas	27.850	114.160	5	SN	BS
Bahia Asuncion	27.138	114.296	1	DG	LbD

This table shows catch information for juvenile White Sharks by fishing site (including geographical position), number of organisms, fishing gear (DG, drift gillnet; BG, bottom gillnet; SL, surface longline; and SN, artisanal seine net), and source (LbD, log book data; BS, beach landing sites; ObS, onboard surveys; and DS, discard sites).

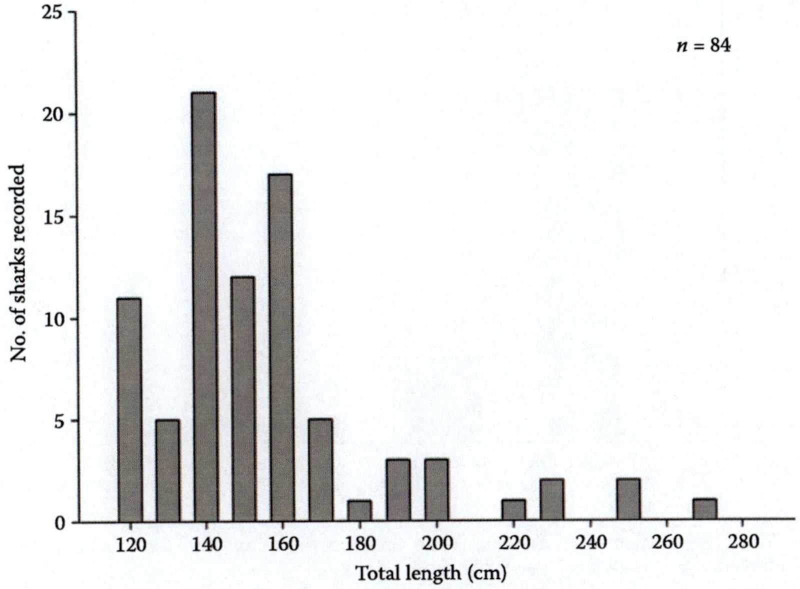

Figure 15.3 Size-frequency distribution of juvenile White Sharks caught on the Pacific coast of Baja California, Mexico. $n = 84$, including shark TL measured or estimated.

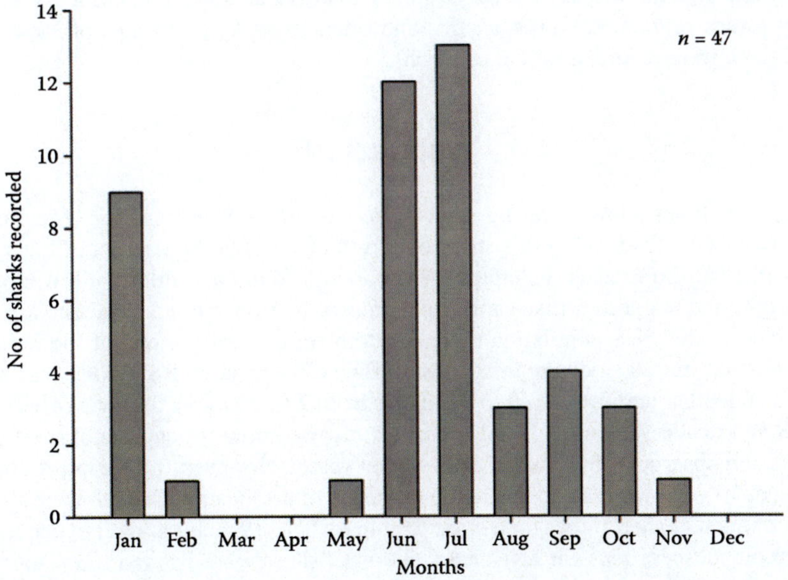

Figure 15.4 Temporal catch-frequency distribution of juvenile White Sharks on the Pacific coast of Baja California, Mexico. $n = 47$, excluding mummified carcasses without date of catch.

Stomach Content Description

A total of 14 stomach contents of White Sharks (134–207 cm TL) were analyzed. Twelve of these were caught in BSV, whereas two were sampled at the Popotla landing site. The esophageal stomach weight range was 500–3,350 g, with a fullness index between 20 and 60%. The overall

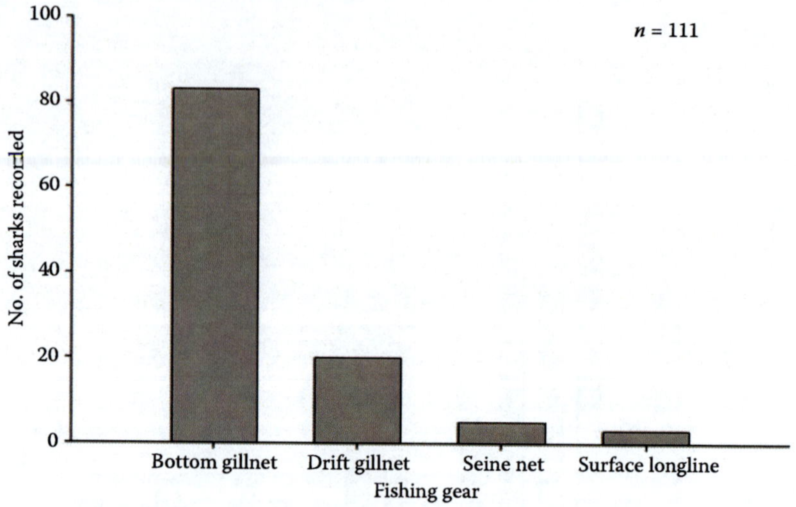

Figure 15.5 Fishing gear involved in juvenile White Shark captures on the Pacific coast of Baja California, Mexico. $n = 111$, including all White Shark data.

weight of contents was 1–2,750 g. Three stomachs were empty, and in all cases, the pyloric stomach content was slurry. Identified prey items included specimens of the taxa: *Thunnus* spp., *Scomber japonicus*, *Atractosion nobilis*, *Myliobatis californica*, *Mustelus* spp., unidentified Scombridae spp., cephalopods (order *Teuthoidea*), crustaceans, and other items (egg capsules of *Raja* spp., and the shark's own teeth were recorded in one stomach).

DISCUSSION

Different gear types show differing selectivities to White Shark capture. Most of the White Shark incidental catch observed in this study occurred in artisanal bottom-set gillnets during summer months in BSV. Lowe et al. (Chapter 14, this book) found a similar pattern in the Southern California Bight and suggested that California's recent ban on gillnets may explain an apparent increasing trend in the JWS population there. Bottom-set gillnets are one of the most commonly used gear types year-round along the west coast of Baja California. In BSV, bottom gillnets are usually operated at depths ranging from 9 to 32 m and from 2 to 8 km off the coast. Drift gillnets also capture JWS, but to a lesser extent. In the present study, we noted the capture of twenty JWS in the log book data and observer effort on one drift-gillnet vessel. However, these captures occurred over a total period of 10 yrs. This may be due to the influence of set locations, which were typically at the edge of the continental shelf rather than close to shore where JWS are known to occur.

Although our observations for artisanal seine-net fisheries are limited, they suggest that this fishery may have a high capacity to capture JWS. For example, five JWS (size range, 130–148 cm TL) were captured inside the coastal lagoon of Ojo de Liebre (Baja California) using this gear type. These coastal lagoons are used by fishermen for the extraction of commercially important clams and octopus. In addition, fishermen use artisanal seines to catch elasmobranchs and bony fishes primarily for local consumption. This is the first record of a White Shark being landed within a coastal lagoon in the northeastern Pacific and suggests that YOY and JWS may use these habitats as a refuge and feeding area. Artisanal surface longlines do not appear to have a high selectivity for JWS capture, probably because they target Blue (*Prionace glauca*) and Shortfin Mako Sharks (*Isurus oxyrinchus*) and thus fish mostly offshore of the continental shelf. The three JWS captured in surface longlines in this study were located in the central part of BSV over the shelf and landed at Laguna Manuela camp, where the fishermen commented that capture of this species was an atypical event for this gear type. It must be noted that gear types used by the artisanal fishery may not be of adequate strength to retain larger White Sharks, and thus the capture of primarily JWS may not reflect the actual demography of the general population.

The young of many shark species spend the first few years of life in nursery areas that are discrete from the areas occupied by adult sharks (Cartamil et al., 2010). Our study results indicate that BSV is an important component of JWS habitat and may constitute a core part of their nursery area. However, most of our sampling effort was concentrated in BSV (Cartamil et al., 2011), and this may bias these results to some extent.

BSV is a large embayment (35,678 km^2) that at its southern extreme (Punta Eugenia) represents the southern limit of the distribution of many fish taxa common to the San Diegan Province, and the northern limit of distribution for many fish species common to the tropical Panamic Province (Cartamil et al., 2011). Most of the JWS reported in this study were captured at BSV, where the continental shelf extends to as far as 140 km offshore (Figure 15.1), whereas in the northern part of the study region the continental shelf is more narrow, extending only from 5 to 20 km offshore (Cartamil et al., 2011). This extensive shelf area may provide JWS with a larger forage base, because it is also important habitat for many teleost and elasmobranch species upon which the JWS feed.

Heupel et al. (2007) proposed three testable criteria for a shark nursery area:

1. Sharks are more commonly encountered in the nursery than in other areas.
2. Sharks have a tendency to remain or return for extended period.
3. The habitat is repeatedly used across years.

If we take the nursery area to be the continental shelf, then JWS are more commonly encountered here than in offshore areas. Our results also show that JWS can be caught across years at years, such as BSV. However, it is unclear whether JWS remain on the shelf of the Pacific coast of Baja California year-round. This is complicated by the fact that fishing effort exhibits seasonal variability in effort, with the greatest effort during summer and greatly reduced effort in winter. Weng et al. (2007b) suggested that this habitat might represent an expansion of the Southern California Bight nursery during autumn months. However, our results have shown that JWS are present in summer in BSV. Further research, including electronic tagging and genetics studies (Jorgensen et al., 2010) are needed in order to clarify the role of this region as a nursery for JWS.

Potential prey items inhabit continental shelf waters, no doubt influencing JWS geographical distribution (Tricas and McCosker, 1984). Dewar et al. (2004) suggested that JWS foraging may occur in the mixed layer and near the surface at night, whereas during the day feeding occurs at or near the bottom of the continental shelf on demersal and benthic species. BSV is considered a center of high biological activity and has high productivity because of physical process dynamics and local upwelling (Hernández-Rivas et al., 2000). The stomach contents of 14 White Sharks closely resemble prey composition previously reported for JWS by Klimley (1985). The finding of *Raja* sp. eggs corroborates that their feeding is coastal. A 149.5-cm-TL White Shark was found with 12 of its own teeth in its stomach. The size range of these teeth (between 3 and 6.5 mm crown height) corresponds to its embryonic stage (Francis, 1996; Kabasakal and Özgür-Gedikoğlu, 2008), suggesting that the shark was recently born in summer in coastal Baja California waters.

Socioeconomic Aspects

Cartamil et al. (2011) recorded approximately 518 artisanal vessels fishing on elasmobranchs and bony fishes in the west coast of Baja California. These boats typically have a crew of two, with each crew member supporting a family. Thus, we can estimate that at least 1,000 families in the region depend on this activity. Although this situation does not justify unregulated fishing on elasmobranchs, it highlights the fact that management of artisanal fisheries must address social issues related to the poorest social sectors of Mexico. This is coupled with the fact that there are strong monetary incentives to capture and process White Sharks. Although White Shark meat does not command a high ex-vessel price (e.g., approximately US$1/kg^{-1}, relative to US$2.2/kg^{-1} for Shortfin Mako Shark, and US$4/kg^{-1} for California Halibut, *Paralichthys californicus*), the price for 1 kg of dry fins is about US$100, and the jaws (according to their size) can command up to US$500.

The remoteness of many artisanal camps in western Baja California is one of the main reasons that monitoring of fishing activity is almost nonexistent and that landings occur despite regulations against commercialization of the species, sometimes because of a lack of knowledge of the regulations by the fishermen of these localities. In addition, common names for the White Shark used by fishers are varied and confusing. We found that even in the most organized fishing camps (e.g., Erendira fishing cooperative), elasmobranchs are classified into at most eight groups for marketing, thus losing important taxonomic data. In addition, most artisanal fishers come from different Mexican states and use the fish common names of their home state. In the case of the White Shark, we recorded four different common names (tiburon blanco, mako barroso, bonita barrosa, and lobero) in the study area. We suggest that standardization of common names used by fishers would clarify landings data and thus facilitate management.

Standards and Regulations

Worldwide, the White Shark is protected by the CITES of wild fauna and flora, which restricts its exploitation based upon the species' low fecundity (UNEP-WCMC, 2010). On the other hand, the International Union for the Conservation of Nature considers the species as "vulnerable" because relatively little is known about its biology (Fergusson et al., 2005). In addition to international protections, the Mexican Official Norm 059 (DOF, 2002) ranked the White Shark as an endangered species in 2002, and the Mexican Fishery Gazette (DOF, 2004) prohibited the finning of the species in 2004. Since May 2007, the Mexican Official Norm 029 (DOF, 2007) limits artisanal fisheries outside of a coastal strip 18–13 km and industrial fisheries outside of 37.02 km. This regulation prohibits the catch, retention, or marketing of any part of the shark. With these regulations, Mexico has made advances in the protection of many elasmobranchs species, including the White Shark. We believe that additional regulations are not necessarily needed but that stronger efforts must be made to enforce existing regulations, and the education level of the fishermen community must be improved. Efforts to promote the release of live White Sharks by fishers should also be encouraged, especially in BSV. Lowe et al. (Chapter 14, this book) found a high survival rate of tagged and released JWS caught by several California fisheries. We also recommend studies of population structure, habitat utilization, movements, abundance, and fishery interactions in Mexican waters. Finally, international management and research plans must be coordinated by the United States and Mexico, because both countries contain important JWS habitat, and fishing impacts in both countries contribute to conservation concerns for the species.

ACKNOWLEDGMENTS

This research was supported by California Sea Grant, UC MEXUS-CONACYT, the Moore Family Foundation, the Tinker Foundation, the Save our Seas Foundation, Monterey Bay Aquarium, and CICESE. Carmen Rodriguez assisted with laboratory work, and Jonathan Sandoval, Luis Malpica, Alfonso Medellin, Alfonsina Romo, Jennyfer Chong, Francisco Lopez, Francisco Martinez, Mario Navarrete, and Cesar Guerrero assisted with field data collection. Thanks to Mariela Brito, Melissa Mayorga, and Emiliano Garcia for their help on graphic materials. We thank

Kevin Weng and anonymous reviewers for their recommendations and improvement of an early version of this chapter. We especially acknowledge all of the fishermen in Baja California for the help and information provided during the development of this work.

REFERENCES

Anonymous. 2002. *White shark (Carcharodon carcharias) recovery plan*. Marine Conservation Branch, Environment, Commonwealth of Australia.

Boustany, A. M., S. F Davis, P. Pyle, S. D. Anderson, B. J. LeBoeuf, and B. A. Block. 2002. Satellite tagging: Expanding niche for white sharks. *Nature* 415: 35–36.

Bruce, B. D. 2008. The biology and ecology of the white shark, *Carcharodon carcharias*. In *Sharks of the open ocean: Biology, fisheries and conservation*, edited by M. D. Camhi, E. K. Pikitch, and E. A. Babcock, pp. 69–81. Oxford, UK: Blackwell Publishing.

Cailliet, G. M. 1996. An evaluation of methodologies to study the population biology of white sharks. In *Great white sharks: The biology of Carcharodon carcharias*, edited by A. P. Klimley and D. G. Ainley, pp. 415–416. Orlando, FL: Academic Press.

Cailliet, G. M., L. J. Natanson, B. A. Weldon, and D. A. Ebert. 1985. Preliminary studies on the age and growth of the white shark, *Carcharodon carcharias*, using vertebral bands. *Memoirs of the Southern California Academy of Science* 9: 49–60.

Cartamil, D., N. C. Wegner, D. Kacev, N. Ben-aderet, S. Kohin, and J. B. Graham. 2010. Movement patterns and nursery habitat of juvenile thresher sharks *Alopias vulpinus* in the Southern California Bight. *Marine Ecology Progress Series* 404: 249–258

Cartamil, D., O. Santana-Morales, M. Escobedo-Olvera, D. Kacev, L. Castillo-Geniz, J. B. Graham, R. D. Rubin, and O. Sosa-Nishizaki. 2011. The artisanal elasmobranch fishery of the Pacific coast of Baja California, Mexico. *Fisheries Research* 108: 393–403.

Compagno, L. J. V. 2002. *Sharks of the world: An annotated and illustrated catalogue of shark species known to date*. Rome: FAO.

Dewar, H., M. Domeier, and N. Nasby-Lucas. 2004. Insights into young of the year white shark, *Carcharodon carcharias*, behavior in the Southern California Bight. *Environmental Biology of Fishes* 70: 133–143.

DOF. 2002. Norma Oficial Mexicana NOM-059-ECOL-2001, Protección ambiental–Especies nativas de México de flora y fauna silvestres–Categorías de riesgo y especificaciones para su inclusión, exclusión o cambio–Lista de especies en riesgo. SEMARNAT. *Diario Oficial de la Federación*, 6 de marzo de 2002, segunda sección.

DOF. 2004. Carta Nacional Pesquera. SAGARPA. *Diario Oficial de la Federación*, 15 de marzo de 2004, quinta sección.

DOF. 2007. Norma Oficial mexicana NOM-029-PESC-2006, Pesca responsable de tiburones y rayas: Especificaciones para su aprovechamiento. SAGARPA. *Diario Oficial de la Federación*, 14 de febrero de 2007, primera sección.

Domeier, M. L., and N. Nasby-Lucas. 2008. Migration patterns of white sharks (*Carcharodon carcharias*) tagged at Guadalupe Island, Mexico, and identification of an eastern pacific shared offshore foraging area. *Marine Ecology Progress Series* 370: 221–237.

Escobedo-Olvera, M. A. 2009. Análisis biológico pesquero de la pesquería con red agallera de deriva en la península de Baja California durante el periodo 1999–2008. MA thesis. CICESE. Ensenada, Baja California, Mexico.

Fergusson, I., L. J. V. Compagno, and M. Marks. 2005. *Carcharodon carcharias*. In *IUCN Red List of Threatened Species*. Version 2010.1. http://www.iucnredlist.org.

Francis, M. P. 1996. Observations on a pregnant white shark with a review of reproductive biology. In *Great white sharks: The biology of Carcharodon carcharias*, edited by A. P. Klimley and D. G. Ainley, pp. 157–172. Orlando, FL: Academic Press.

Galván-Magaña, F., E. M. Hoyos-Padilla, C. J. Navarro-Serment, and F. Márquez-Farías. 2010. Records of white shark, *Carcharodon carcharias*, in the Gulf of California, Mexico. *Marine Biodiversity Records* 3: 1–6.

Hanan, D. A., D. B. Holts, and A. L. Coan, Jr. 1993. The California drift gill net fishery for sharks and swordfish, 1981–82 through 1990–91. *Fish Bulletin* 175: 89–95.

Hernández-Rivas, M., S. P. Jiménez-Rosenberg, R. Funes-Rodríguez, and R. J. Saldierna-Martínez. 2000. El centro de actividad biológica de la Bahía Sebastián Vizcaíno, una primera aproximación. In *BAC Centros de actividad biológica del Pacífico mexicano*, edited by D. Lluch-Belda, J. Elourduy-Garay, S. E. Lluch-Cota, and G. Ponce-Díaz, pp. 65–85. La Paz, Mexico: CIBNOR.

Heupel, M. R., J. K. Carlson, and C. A. Simpfendorfer. 2007. Shark nursery areas: Concepts, definition, characterization and assumptions. *Marine Ecology Progress Series* 337: 287–297.

Holts, D. B. 1988. Review of U.S. west coast commercial shark fisheries. *Marine Fisheries Review* 50: 1–8.

Holts, D. B., A. Julian, O. Sosa-Nishizaki, and N. Bartoo. 1998. Pelagic shark fisheries along the west coast of the United States and Baja California, Mexico. *Fisheries Research* 39: 115–125.

Jorgensen, S. J., C. A. Reeb, T. K. Chapple, S. Anderson, C. Perle, S. R. Van Sommeran, C. Fritz-Cope, A. C. Brown, A. P. Klimley, and B. A. Block. 2010. Philopatry and migration of Pacific white sharks. *Proceedings of the Royal Society – B* 277: 679–688.

Kabasakal, H., and S. Özgür-Gedikoğlu. 2008. Two new-born great white sharks, *Carcharodon carcharias* (Linnaeus, 1758) (Lamniformes; Lamnidae) from Turkish waters of the north Aegean Sea. *Acta Adriatica* 49: 125–135.

Kinney, M. J., and C. A. Simpfendorfer. 2008. Reassessing the value of nursery areas to shark conservation and management. *Conservation Letters* 2: 53–60.

Klimley, A. P. 1985. The areal distribution and autoecology of the white shark, *Carcharodon carcharias*, off the west coast of North America. *Memoirs of the Southern California Academy of Science* 9: 15–40.

Mollet, H. F., and G. M. Cailliet. 2002. Comparative population demography of elasmobranchs using life history tables, Leslie matrices and stage-based matrix models. *Marine and Freshwater Research* 53: 503–516.

Mollet, H. F., G. M. Cailliet, A P. Klimley, D. A. Ebert, A. T. Testi, and L. J. V. Compagno. 1996. A review of length validation methods for large white sharks. In *Great white sharks: The biology of Carcharodon carcharias*, edited by A. P. Klimley and D. G. Ainley, pp. 91–108. Orlando, FL: Academic Press.

Preti, A., S. E. Smith, and D. A. Ramon. 2001. Feeding habits of the common thresher shark (*Alopias vulpinus*) sampled for the California-based drift gill net fishery, 1998–1999. *CalCOFI Reports* 42: 145–152.

Santana-Morales, O. 2008. Composición especifica de elasmobranquios capturados por la pesca artesanal en Bahía Vizcaíno, B.C., México. Análisis de un registro histórico. MA thesis. CICESE. Ensenada, Baja California, Mexico.

Sosa-Nishizaki, O., J. F. Márquez-Farías, and C. J. Villavivencio-Garayzar. 2008. Case study: Pelagic shark fisheries along the west coast of Mexico. In *Sharks of the open ocean: Biology, fisheries and conservation*, edited by M. D. Camhi, E. K. Pikitch, and E. A. Babcock, pp. 275–282. Oxford, UK: Blackwell Publishing.

Tricas, T. C., and J. E. McCosker. 1984. Predatory behavior of the white shark (*Carcharodon carcharias*), with notes on its biology. *Proceedings of the California Academy of Science* 43: 221–238.

UNEP-WCMC. April 20, 2010. *UNEP-WCMC species database: CITES-listed species*. http://www.unep-wcmc.org/isdb/CITES/Taxonomy/tax-species-result.cfm/isdb/CITES/Taxonomy/tax-species-result.cfm?Genus=Carcharodon&Species=carcharias&source=animals&tabname=all.

Weng, K. C., A. M. Boustany, P. Pyle, S. D. Anderson, A. Brown, and B. A. Block. 2007a. Migration and habitat of white sharks (*Carcharodon carcharias*) in the eastern Pacific Ocean. *Marine Biology* 152: 877–894.

Weng, K. C., J. B. O'Sullivan, C. G. Lowe, C. E. Winkler, H. Dewar, and B. A. Block. 2007b. Movements, behavior and habitat preferences of juvenile white sharks *Carcharodon carcharias* in the eastern Pacific. *Marine Ecology Progress Series* 338: 211–224.

CHAPTER 16

A New Life-History Hypothesis for White Sharks, *Carcharodon carcharias*, in the Northeastern Pacific

Michael L. Domeier*
Marine Conservation Science Institute

ABSTRACT

A review of studies focused on White Sharks (*Carcharodon carcharias*) in the northeastern Pacific, together with new data, has allowed for a revised regional hypothesis for the life history of this species. Current evidence suggests that White Sharks in the northeastern Pacific begin their life in the nearshore, shallow waters of the Southern California Bight and Baja California. During their first winter, they may migrate to the warmer waters of coastal Baja California. As the juveniles grow, they gain the ability to remain in cooler water, exploiting deeper depths and migrating north of Point Conception. As they continue to grow, their diet changes from one dominated by fish and invertebrates to one dominated by marine mammals. As White Sharks approach maturity, they begin annual (in the case of males) and bi-annual (in the case of females) migrations between offshore habitats and two known adult aggregation sites, one located off central California in the United States and the other off Guadalupe Island, Mexico. Males from both aggregation sites travel to a Shared Offshore Foraging Area, whereas females roam a much larger and less defined offshore habitat. Both males and females may go to the Hawaiian Islands. Indirect evidence suggests that mating occurs at the adult aggregation sites. After more than 15 months at sea and an approximately 18-month gestation period, pregnant females return to coastal regions off Southern California, United States, and Baja California, Mexico, to give birth. Parturition occurs primarily between May and August, after which females return to one of the two adult aggregation sites. The understanding of this species' migration patterns and reproductive biology was advanced through capture and satellite-linked radio transmitting tagging of adults, as well as the direct examination of claspers and the screening of blood samples for the presence of circulating reproductive hormones.

INTRODUCTION

Until recently, White Sharks (*Carcharodon carcharias*) were a species about which very little was known because of their naturally low abundance and great size. In the past 10 yrs., northeastern

Pacific White Sharks have been the focus of several electronic tagging programs (Boustany et al., 2002; Weng et al., 2007a; Domeier and Nasby-Lucas, 2008; Nasby-Lucas et al., 2009; Jorgensen et al., 2010; Chapters 11 and 12, this book), as well as detailed photo-identification studies (Anderson and Pyle, 2003; Domeier and Nasby-Lucas, 2007; Chapter 25, this book), rapidly advancing our understanding of their seasonal migration patterns, habitat-preferences and behavior. White Sharks have proven ideal for electronic tagging studies because they retain tags for long periods and exhibit philopatry for certain aggregation sites, allowing for recovery of tags upon return to the tagging site. What was once thought to be a species restricted to coastal, temperate regions is now known to spend considerable time in deep-ocean habitats, thousands of kilometers from shore (Boustany et al., 2002; Bonfil et al., 2005; Weng et al., 2007a; Domeier and Nasby-Lucas, 2008; Nasby-Lucas et al., 2009; Jorgensen et al., 2010; Chapters 11, 12, and 21, this book).

Earlier published works on the life history of White Sharks were based upon very small sample sizes, few observations, and anecdotes, a testament to the extreme difficulty of conducting controlled studies on such a rare and large fish. Klimley (1985a) compiled records of incidentally captured specimens and records of White Shark/human and White Shark/marine mammal interactions to synthesize a life-history hypothesis for White Sharks in the northeastern Pacific. In summary, Klimley's hypothesis suggested the following:

1. Adult females give birth around the offshore islands of Southern California, United States during the late summer and early fall. Young-of-the-year (YOY) White Sharks remain in the nearshore regions of the Southern California Bight.
2. As the pups reach a size of 2 m, they migrate north of Point Conception, where they inhabit both inshore and offshore island habitats. This migration is thought to coincide with a shift in diet from fish to marine mammals.
3. Mature White Sharks feed primarily on marine mammals off the coast of central California, but pregnant females migrate south of Point Conception in the summer and fall to give birth in waters around offshore islands, closing the life-cycle loop.

Despite the fact that electronic tagging has revealed that White Sharks undergo long migrations to spend considerable time in oceanic habitats, no alternative to Klimley's hypothesis has been suggested. This is largely due to remaining gaps in some of the most basic life-history parameters: age, growth, and reproductive biology. Without such data, the life-history pattern of White Sharks is difficult to accurately describe. White Sharks have not been historically targeted by any commercial fisheries, so access to samples has been opportunistic and infrequent (Chapter 14, this book). Fishery-independent studies are impractical because of the difficulty in targeting White Sharks, their large size, and most recently, a 2005 listing in Appendix II of the Convention of the International Trade in Endangered Species of wild fauna and flora. Some studies promoting the basic understanding of White Shark age, growth, and reproductive biology have been published (Cailliet et al., 1985; Pratt, 1996; Francis, 1996; Uchida et al., 1996; Mollet et al., 2000), but these studies were based on very small sample sizes. Adequate, systematic, lethal sampling of White Sharks will never be acceptable, forcing the use of opportunistic sampling, anecdotal information, and satellite tagging to describe their life-history pattern.

The number of publications, old and new, that relate to White Sharks of the northeastern Pacific outnumber those that represent White Sharks from other regions. These papers, together with new observations and data presented here, allow for the synthesis of a new regional life-history hypothesis for White Sharks. Innovative sampling of blood hormones, sperm samples, and new long-term satellite-linked radio transmitting (SLRT) tag data, presented here, mark a significant step forward in our interpretation of past and present data. As with all White Shark studies, conclusions must rely somewhat on speculation, but proposing a new life-history pattern for this species updates outdated theories and helps direct future studies.

MATERIALS AND METHODS

Previously published studies and the new material presented in the various chapters of this book were reviewed to provide a framework to which new observations, described below, could be applied to synthesize the life-history hypothesis.

Seasonal Abundance of Adult and Subadult White Sharks

Since adult and subadult White Sharks occasionally interact with humans, these encounters can be used as a proxy for the seasonal presence of relatively large White Sharks (Klimley, 1985a). The relative frequency of White Shark/human encounters in the northeastern Pacific were compiled by month from 1926 to 2010 [data from McCosker and Lea (1996); data from the Shark Research Committee website: http://www.sharkresearchcommittee.com/2000.htm; 2009 California Department of Fish and Game (CDFG) data]. These encounters were also split into two geographic groupings: those that occurred south of Point Conception and those that occurred north of Point Conception.

YOY White Sharks

Although White Sharks have been protected in California since 1994, the law allows commercial fishers to land incidentally caught White Sharks. Juvenile White Sharks are occasionally captured by commercial gillnets targeting nearshore species such as California Halibut (*Paralichthys californicus*), Pacific Angel Shark (*Squatina californica*), and White Seabass (*Atractoscion nobilis*) (Heneman and Glazer, 1996; Chapter 14, this book). These nets are made of light monofilament and are not normally capable of capturing large sharks, which destroy the net and escape. Both documented White Shark specimens (from the literature) and CDFG White Shark landing data were gathered to analyze the timing and occurrence of YOY White Sharks. The data used here represent a subset of data analyzed by Lowe et al. (Chapter 14, this book) for a description of fisheries interactions with White Sharks in Southern California.

YOY White Sharks are found primarily in Southern California (Klimley, 1985a; Weng et al., 2007b; Chapter 14, this book), and therefore White Shark landing data were obtained from CDFG for the Southern California ports of Santa Barbara (catches from Ventura/Santa Barbara region), Los Angeles (Los Angeles/Orange County region), and San Diego (San Diego County region). CDFG data were obtained in table format, with the number of landing receipts and weight of the landings summed by month and year. Because tables of individual catches are not correlated with individual landing receipts, they cannot normally be matched; however, White Shark catches are typically zero for any given month and rarely exceed one, allowing them to be manually matched. Furthermore, individual landings typically consisted of a single fish (obvious from weights of landings). Because YOY White Sharks are estimated to obtain 44.4 kg in their first year (Cailliet et al., 1985; Mollet and Cailliet, 1996), eliminating landing receipts and landings (kg) that contained sharks larger than 45 kg excluded sharks more than 1 yr. old. The remaining data were analyzed and plotted for seasonal occurrence of YOY. CDFG set-gillnet effort was also plotted by month to determine whether effort was biasing seasonal landings.

Seasonality of Reproductive Activity

The capture and restraint of adult White Sharks for the purpose of SLRT tagging presented the opportunity to measure and examine the sharks prior to their release (Chapter 11, this book). Claspers were examined for calcification and the presence of seminal fluid and spermatophores in the clasper groove. Blood samples were drawn and analyzed for the presence of reproductive hormones (Chapter 30, this book). Under the assumption that male White Sharks grasp females with their jaws during copulation, as has been documented in other shark species (Pratt and Carrier, 2001), images from the photo-identification database (Domeier and Nasby-Lucas, 2007; Chapter 11, this book) were examined for the presence of bite wounds on females.

Satellite-Linked Radio Transmitting Tags Studies

Several chapters of this book detail recent results of SLRT tagging of adult and juvenile White Sharks. The methods of capture and attachment of these tags are detailed by Domeier and Nasby-Lucas in Chapter 11, this book. A selection of tracks from White Sharks tagged at Guadalupe Island (GI), Mexico is presented here to illustrate new findings that both address life-history questions as well as raise new questions. Sharks are referred by their number in the GI photo-identification database (see Chapter 25, this book).

RESULTS

Early Life History

Fifty-one published records of known YOY were compiled from the literature; only records that included a total length (TL) and date of capture were included. Klimley (1985a) contained 28 records of YOY White Sharks, 25 of which had associated dates of capture and TL (two records did not have a date, and one record had an impossibly small TL). Seventeen records came from Santana-Morales et al. (Chapter 15, this book), and a combined total of nine additional records were taken from Dewar et al. (2004) and Weng et al. (2007b; Chapter 28, this book). When the total lengths of these 51 YOY sharks are plotted, there is a trend for increasing length from May to November (Figure 16.1). The timing of the occurrence of the smallest individual YOY provided the most insight into the possible timing of parturition. The size at birth likely varies depending upon the size and condition of the female, but it has been estimated that neonates fall within the range of 120–150 cm (Francis, 1996). The number of YOY within the neonate size range were divided by the total YOY catch for each month; the resulting percentage of neonates collected per month falls from 100% in April, 75% in June, 81% in July, ~40% August-September, 25% in October, and 0%

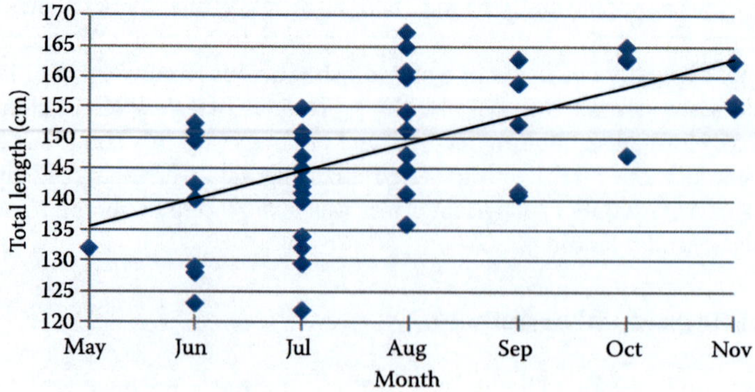

Figure 16.1 Total length (centimeters) at capture of known YOY White Sharks ($n = 51$) with trend line ($r^2 = 0.30$) indicating that parturition occurs in the early months of this season.

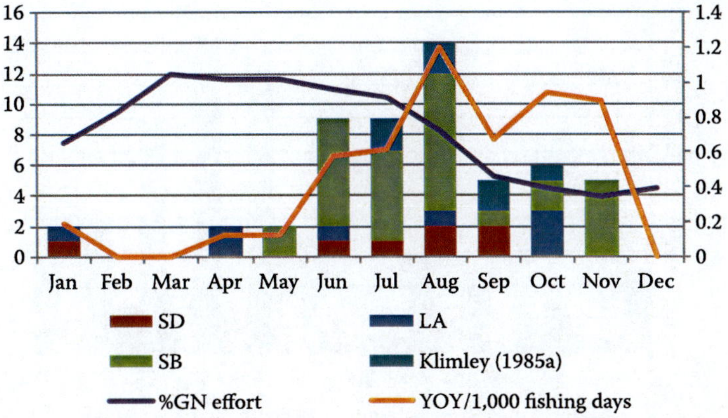

Figure 16.2 Seasonal pulse of young-of-the-year White Sharks incidentally captured in the set-gillnet fishery. Most landings occurred at the port of Santa Barbara (green), followed by similar numbers landed in Los Angeles and San Diego (blue and red, respectively). Light blue indicates set-gillnet specimens added to the California Department of Fish and Game records (from Klimley, 1985a).

in November. July contributed the largest number of neonates in any given month ($n = 16$), and only one YOY was less than 140 cm after July. All of these YOY sharks were captured south of Point Conception, California.

Examination of CDFG set-gillnet data (1980–2000) resulted in forty-seven additional records of YOY White Sharks in Southern California. The majority ($n = 32$) were landed in the port of Santa Barbara, with much fewer landed in Los Angeles and San Diego ($n = 8$ and 7, respectively). Seven of the records in Klimley (1985a) were caught by set gillnet, so these were added to the CDFG data, but capture location data were missing. These 64 YOY specimens were collected between the months of April through November, and two were collected in January. The first appearance of YOY in the fishery occurs in April, with a steep increase in June followed by a peak frequency of occurrence in August (Figure 16.2). Gillnet-fishing effort peaked during the months of lowest YOY occurrence (Figure 16.2); therefore, fishing effort did not bias the timing of the seasonal peak in YOY captures, but the reduced effort during and just after the observed peak could indicate that relatively more YOY would be captured during these times if effort were the same. Plotting the capture rate of YOY White Sharks (number of YOY/effort) gives some indication of the impact of effort on the number of fish landed (Figure 16.2).

Seasonal Abundance of Adult and Subadult White Sharks along the Coast of California

Adult and subadult White Shark encounters with humans in the northeastern Pacific were combined from McCosker and Lea (1996), Lea and McCosker (2008 online CDFG database: http://www.dfg.ca.gov/marine/whiteshark.asp), and Shark Research Committee (online database: http://www.sharkresearchcommittee.com). The seasonal distribution of these encounters was used as a proxy for the occurrence of subadult and adult White Sharks along the west coast of North America (1926–2010). White shark encounters occurred in every month of the year, with the lowest occurrence in March and a peak in September (Figure 16.3). The temporal distribution of shark attacks differed between regions north and south of Point Conception, California. July was the peak month for shark attacks south of Point Conception, whereas September was the peak north of Point Conception (Figure 16.3). Overall, regions north of Point Conception experienced far more shark attacks on humans.

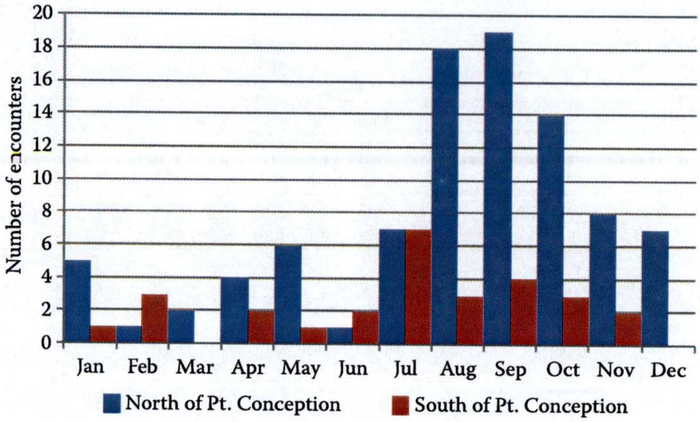

Figure 16.3 Seasonal occurrence of White Shark attacks on humans (1926–2010).

Indirect Evidence of Reproductive Activity

The claspers of 11 adult male White Sharks (nine from GI and two from Southeast Farallon Island) were examined between December 2007 and December 2009. Six of the 11 males (~55%) were found to have seminal fluid in the groove of at least one clasper, and three of these fish had spermatophores present within the seminal fluid (Table 16.1 and Figure 16.4). Seminal fluid was microscopically examined from one male that had no visible spermatophores; spermatophores were microscopically examined from two individuals. Very few sperm cells were present in the seminal fluid, but the spermatophores consisted of a dense matrix of closely packed sperm cells (Figure 16.5).

It is presumed that male White Sharks must grasp females with their jaws during copulation and that this act would leave bite marks on the posterior flank, head, pectoral fin, or gill region. Fresh conspecific bite marks on their flanks, head, or gill region were observed on 17 mature female White Sharks on 19 occasions at GI (Table 16.2 and Figure 16.6); on two occasions the same individual was seen with bite marks in different years. No individual female observed with such bite marks was ever sighted at the GI aggregation the following year.

Table 16.1 Results of Clasper Exams from 11 Mature Adult Male White Sharks

Date	Location	Total Length (cm)	Clasper Condition	Reproductive Status	Seminal Fluid	Spermatophores
12/6/2007	Guadalupe Island	368	Calcified	Mature	Yes	No
12/4/2008	Guadalupe Island	457	Calcified	Mature	No	No
12/4/2008	Guadalupe Island	442	Calcified	Mature	Yes	No
12/6/2008	Guadalupe Island	447	Calcified	Mature	No	No
12/7/2008	Guadalupe Island	457	Calcified	Mature	No	No
10/29/2009	Southeast Farallon Island	427	Calcified	Mature	Yes	Yes
11/2/2009	Southeast Farallon Island	452	Calcified	Mature	No	No
11/19/2009	Guadalupe Island	439	Calcified	Mature	Yes	No
11/20/2009	Guadalupe Island	462	Calcified	Mature	No	No
11/20/2009	Guadalupe Island	472	Calcified	Mature	Yes	Yes
11/20/2009	Guadalupe Island	541	Calcified	Mature	Yes	Yes

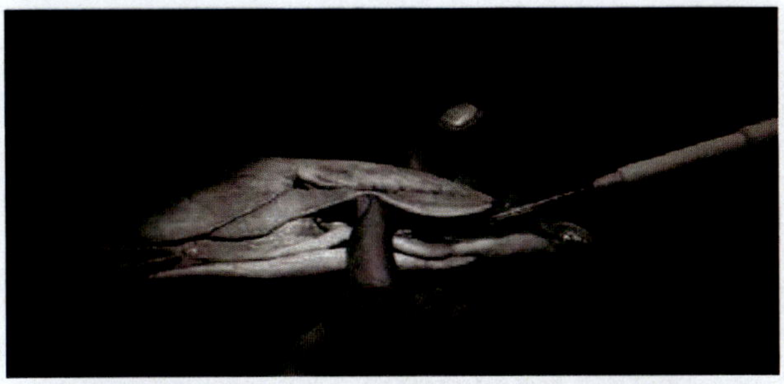

Figure 16.4 Sampling a spermatophore from the clasper groove. The spermatophore, resembling a grain of rice, can be seen just under the tip of the pipette.

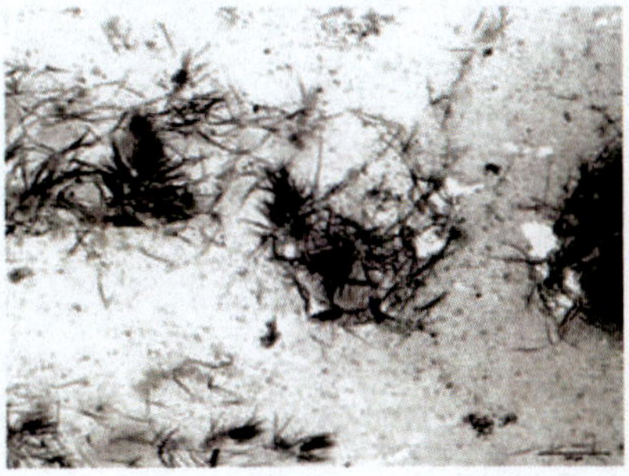

Figure 16.5 Teased apart spermatophore revealing dense sperm aggregation.

Table 16.2 Long-Term Pattern of Presumed Mating Bites on Female White Sharks at GI

Shark ID	2002	2003	2004	2005	2006	2007	2008	2009
2	gray	gray	gray	yellow				
4	gray	gray	yellow		gray			yellow
20		gray	gray	gray	gray		yellow	
30			gray	gray	gray	gray		yellow
36	gray	gray	yellow		yellow		gray	
38			yellow			gray		
39			gray		yellow		gray	gray
43	gray		yellow					
44			gray	yellow				
53	yellow							
54	gray	gray		yellow		gray		
57		yellow						
64				yellow				
67		gray	gray	yellow		gray		
77			gray		yellow		gray	gray
84						gray		yellow
108				gray		gray		yellow

The gray cells indicate years each individual was sighted; the yellow cells indicate years bite wounds were present.

Figure 16.6 Conspecific bite wounds on female White Sharks photographed at Guadalupe Island during the seasonal adult aggregation.

Novel Satellite-Linked Radio Transmitting Tag Tracks

First Long-Term Track of a Subadult

During efforts to capture and SLRT tag adult White Sharks at Guadalupe Island, a single male subadult was tagged and subsequently tracked (shark 85M, 3.15 m TL). The shark was tagged on December 6, 2007 with data still transmitting at the present time, over 1,100 d after tagging (Figure 16.7). The shark departed GI on April 20, 2008 for his first tracked offshore migration, considerably later than the average adult male departure from GI of February 7 (Chapter 11, this book). Shark 85M traveled westward, crisscrossing the Clarion Fracture to approximately 145°W longitude before turning northwest until he hit the Murray Fracture. He briefly moved west along the fracture,

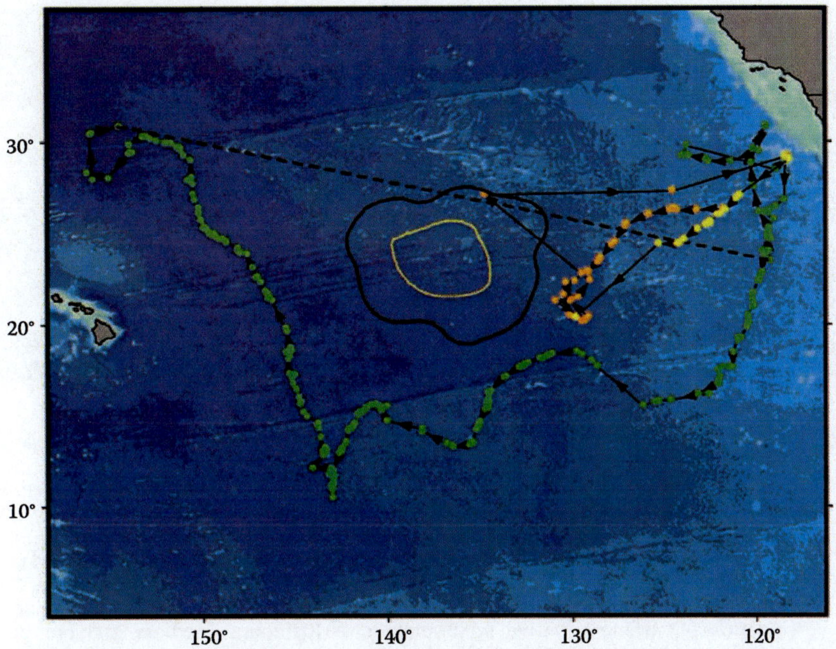

Figure 16.7 Multiyear SLRT tag track of a subadult male White Shark tagged at Guadalupe Island. His first offshore migration occurred in 2008 (green), followed by migrations in 2009 (orange) and 2010 (yellow). The black and yellow circles are the 95 and 50% density contours (fixed-kernel analyses) of the SOFA, defined by all adult male offshore data (Chapter 11, this book).

at which point he was 1100 km due north of the island of Hawaii. Shark 85M began moving eastward and then ceased transmitting on July 17 until October 7 when he was 615 km southwest of GI; he arrived at GI on October 29, 2008. On April 29, 2009, he again departed GI for an offshore migration that took him to a region approximately 800 km southwest of the center of the Shared Offshore Foraging Area (SOFA), as defined in Chapter 11, this book. Considerably less position data were received during the time he was offshore in 2009. He returned to GI on October 11, 2009. His 2010 offshore migration began on April 4, when the few positions we received produced a track very similar to that of 2009. Not enough position data were received to determine when shark 85M returned to GI in 2010.

First Long-Term Track of Adult Females

On December 9, 2008, a mature female (shark 98F, 4.98 m TL) was fitted with a SLRT tag at GI. She departed GI on January 28 and remained offshore for 443 d before entering the Gulf of California on April 16, 2010 (Figure 16.8a). For the first 222 d of her offshore migration, she remained a minimum of 940 km east of the central point of the core SOFA. She moved west for the next 205 d before traveling into the Gulf of California. She remained within the Gulf of California for 67 d.

Another mature female was tagged on December 3, 2008 (shark 77F, 5.1 m TL). This shark departed GI on February 1 and remained offshore for 228 d before returning to Guadalupe Island on September 17. She then departed the island on January 24 and is still offshore at the time of this manuscript (>402 d) (Figure 16.8b). During her 2009 offshore migration, 77F remained at least 840 km east of the central point of the core SOFA, traveling back and forth between the offshore region and GI.

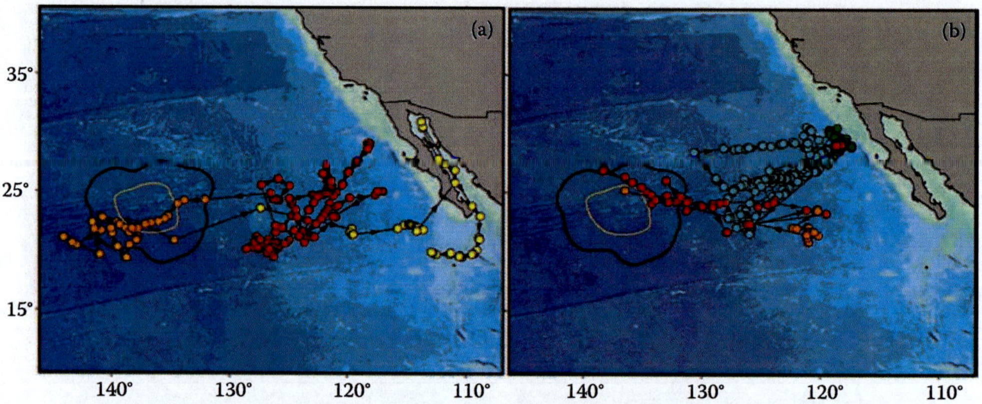

Figure 16.8 Multiyear SLRT tag tracks of two adult female White Sharks tagged at Guadalupe Island in December 2008. (a) Shark 98F migrated offshore where she spent December 2008 through August 2009 between the continental shelf and the SOFA (red). September 2009 to March 2010 was spent primarily west of the SOFA (orange) before migrating to the Gulf of California for the parturition season (April to June 2010). (b) In contrast, female 77F also migrated offshore (December 2008 through August 2009; blue), but she returned to Guadalupe Island for the subsequent adult aggregation season. She again departed for offshore waters December 2009 through August 2010 (red) but then remained offshore during the 2010 aggregation season, where she remains as of the time of this writing (September 2010 through February 2011; orange). The black and yellow circles are the 95 and 50% density contours (fixed-kernel analyses) for all adult male offshore data (Chapter 11, this book).

Atypical Adult Male Movements

On November 20, 2009, the largest male White Shark ever measured anywhere in the world (shark 102M, 5.41 m TL, 4.95 m fork length, and 3.20-m girth) was captured and SLRT-tagged at GI. This male had never been previously sighted at GI. The relatively small number of positions received from 102M did not allow for the determination of when he departed GI, but he did migrate to the SOFA during the offshore season of 2010 and then returned to GI (date unknown). After returning to GI, shark 102M made at least five forays away from the island, traveling 200 km northeast, 435 km south, 170 km north, and 133 km southwest during the months of September and October (Figure 16.9a). On one occasion, he was detected 70 km from the Pacific coast of Baja California, Mexico. For comparison, the track of shark 19M (4.4 m TL) illustrates a typical adult

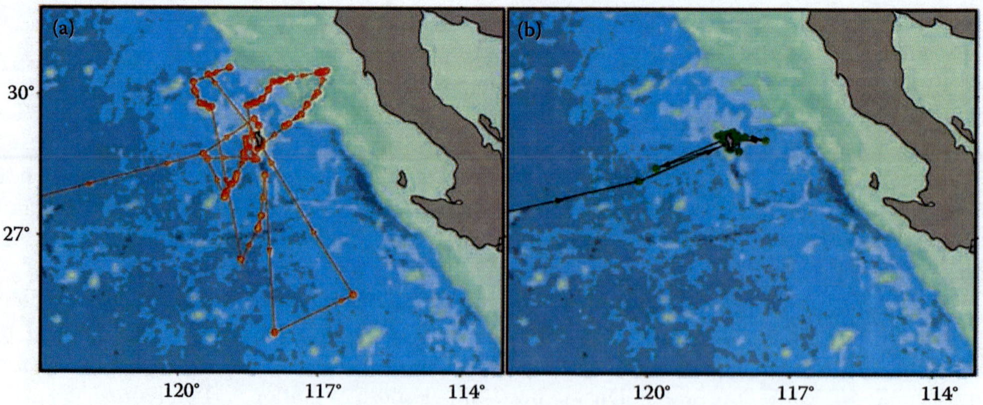

Figure 16.9 Atypical (a) and typical (b) adult male movement patterns during the seasonal adult aggregation at Guadalupe Island.

male pattern of seasonal GI residency, with a very tight pattern of positions in the vicinity of the island (Figure 16.9b).

DISCUSSION

Early Life-History Stage

Both scientific collections and commercial fishery bycatch data indicate an April through November seasonal presence of YOY White Sharks in the nearshore regions of Southern California and Baja California, Mexico (Figures 16.1 and 16.2) (Chapters 14 and 15, this book). The YOY length over time trend (Figure 16.1) suggests that parturition begins in May and ends in October; parturition appears to peak in July and is largely complete by the end of August. Uchida et al. (1996) reported two northern hemisphere female White Sharks, collected in Japan, with full-term litters in their uteri; both were captured in May, consistent with the conclusions presented here. If future studies decrease or increase the maximum size of neonates, then this estimation of the parturition season will shift accordingly. The timing of parturition is important relative to discussions of the timing of mating and the length of gestation.

The location of the gillnet fishery that interacts with YOY White Sharks indicates that relatively shallow coastal waters (15–60 m) over flat sandy substrate are an important nursery habitat (Chapter 14, this book), but because this fishery does not target reef and kelp habitats, the relative importance of these three habitats is unknown. YOY and juvenile White Sharks are often observed in close proximity to the shoreline, within or just outside of the surf zone (M. L. Domeier, personal observations; Chapter 17, this book). Anecdotal but consistent localized sightings of juvenile White Sharks over multiple years suggest that juveniles may have very localized preferred sites/hot spots within the Southern California Bight (Chapter 14, this book), and a large number of YOY incidental captures suggests another hot spot in Vizcaino Bay, Mexico (Chapter 15, this book). These observations parallel new findings from eastern Australia that have identified a relatively small number of preferred habitat areas for juvenile White Sharks over multiple years (Chapter 17, this book). The discovery of juvenile White Shark hot spots has research and conservation implications that could ease population monitoring and specific spatial protections.

Young-of-the-year White Sharks tracked with pop-up archival transmitting (PAT) and SLRT tags have remained in coastal regions south of Point Conception through summer and fall before migrating to the warmer coastal waters of Baja California, Mexico, where they remain during the autumn and winter (Figure 16.10) (Dewar et al., 2004; Weng et al., 2007b; Chapter 28, this book). The newly reported records of juveniles and YOY from the Pacific coast of Baja California (Chapter 15, this book) confirms Klimley's (1985a) suggestion that Mexico is also an important nursery and juvenile region for northeastern Pacific White Sharks.

The limited tracking of juvenile White Sharks at least 1 yr. old suggests they may remain south of Point Conception for the first year or two and then begin exploiting habitats north of Point Conception by the third year (Figure 16.10) (Weng et al., 2007b). During this time, juvenile White Sharks are feeding on a variety of fishes (teleost and elasmobranch), crustaceans, and perhaps cephalopods (Klimley, 1985a; Dewar et al., 2004; Weng et al., 2007b; Chapter 4, this book). Evidence suggests that as the sharks get larger they are capable of exploiting colder water and therefore forage deeper and further north (Figure 16.10) (Weng et al., 2007b).

Subadult Life-History Stage

The subadult stage of White Sharks is the least known life-history stage in the northeastern Pacific. The transition from preying on fish and invertebrates to a diet that also includes marine

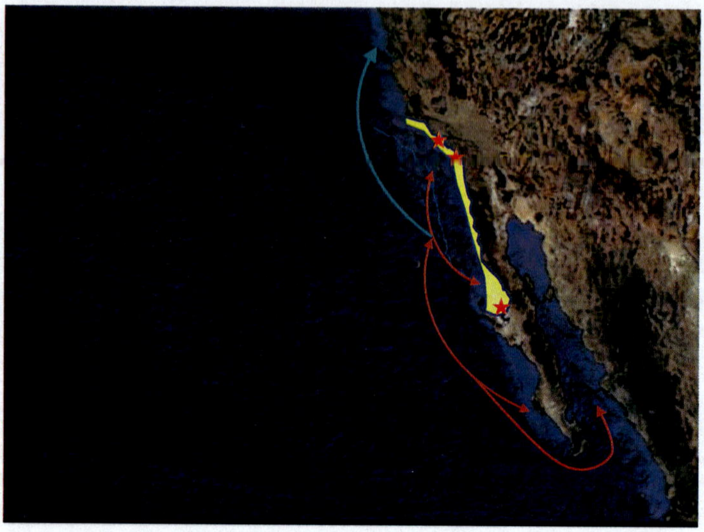

Figure 16.10 Early life-history migratory patterns. Adult females come from pelagic habitats to give birth in coastal nursery areas. Young-of-the-year White Sharks appear in the nursery region (yellow) May through August, where they remain until the water cools in the winter. They migrate south to coastal Baja California, Mexico and return to more northern nursery areas in the spring (pink arrows). Concentrations of juveniles are found within preferred habitat or hot spots (red stars). Older juveniles and subadults begin to exploit habitats north of Point Conception (blue arrow). (Courtesy of Google Earth, Data SIO, NOAA, U.S. Navy, NGA, GEBCO, Image USDA Farm Service Agency. Image © 2011 DigitalGlobe and © 2011 Cnes/Spot Image.)

mammals is likely gradual and opportunistic, with juvenile sharks first preying on small marine mammals and scavenging carcasses (Dicken, 2008; Chapter 14, this book) before acquiring the size and skills to take larger prey. Tooth morphology changes during the subadult stage, from a narrow grasping tooth to a broad cutting tooth, when the sharks reach approximately 3 m (Tricas and McCosker, 1984). Attacks on humans have occurred year-round along the coast of California, and because adults White Sharks are known to be offshore from late winter through early summer, attacks during this time (Figure 16.3) likely indicate a year-round coastal presence of subadults (Figure 16.11). Larger subadults are found primarily north of Point Conception (Klimley, 1985a), and marine mammals are an important prey source, but records from Baja California indicate the presence of subadults on both the Pacific and Gulf of California coastlines as well (Galván-Magaña et al., 2010; Chapter 15, this book).

It is not known how or when subadult White Sharks recruit to adult aggregation sites (Figure 16.11) or how they learn to make seasonal migrations to the SOFA. The track of subadult male 85M is intriguing because this was one of the smallest individuals ever observed at the GI adult aggregation site. This male had recently recruited to the aggregation, and it is possible that his 2008 offshore migration was one of his first such migrations, if not the first. This individual's offshore migrations have been atypical in that they are not temporally or spatially synched with those of adult males; shark 85M arrived and departed GI later than adult males, and his SLRT tag has not transmitted any positions from the core SOFA region in his first or two subsequent offshore migrations. This subadult male's migration patterns are not constrained by reproductive needs but are likely a balance between the seasonal availability of prey and minimizing competition and aggression from much larger individuals.

PAT-tag and SLRT-tag data have revealed that White Sharks travel near the surface in relatively straight lines when they are moving between habitats of high importance (Domeier and

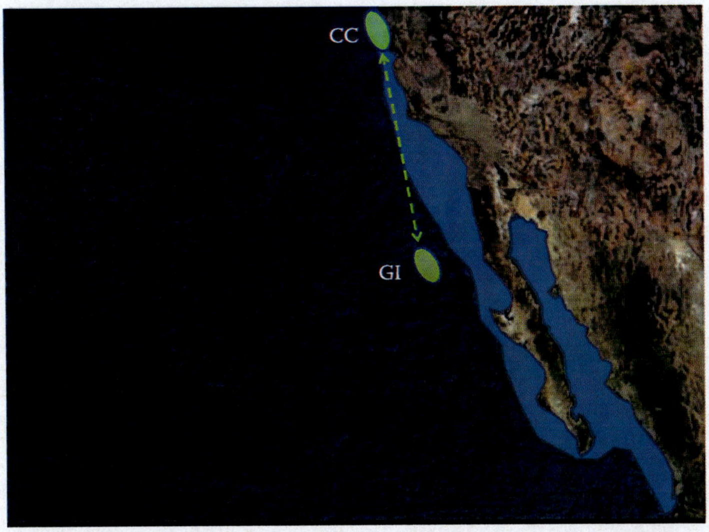

Figure 16.11 Coastal habitats of subadult and adult White Sharks of the northeastern Pacific. Subadult White Sharks range widely over the continental shelf of the northeaster Pacific (blue-shaded region). Subadults recruit to adult aggregation sites off central California (northern green oval) and Guadalupe Island (southern green oval). Adults demonstrate strong seasonal philopatry to their respective aggregation site, with movement between these sites being very rare (dashed green line). (Courtesy of Google Earth, Data SIO, NOAA, U.S. Navy, NGA, GEBCO, Image USDA Farm Service Agency. Image © 2011 DigitalGlobe and © 2011 Cnes/Spot Image.)

Nasby-Lucas, 2008; Chapter 11, this book). Shark 85M's first offshore track consisted of regular SLRT positions until he reached an area 1,100 km north of the island of Hawaii, after which transmissions were few until he was detected 615 km southeast of GI, heading toward the island. The outbound portion of his track suggests a period of travel and searching for desirable prey and/or habitat, and interestingly, his path caused him to completely miss the SOFA, where all adult males reside during their offshore season. Over the next two offshore migrations, shark 85M made much shorter tracks, traveling to a region southeast of the SOFA. The difference between the first and two subsequent tracks leads to the question of whether preferred offshore habitats are discovered via searching and then these locations referenced for the subsequent migrations (Papastamatiou et al., 2011).

It is not known what stimulates White Sharks to drastically change a life-history pattern that keeps them near the continental shelf as juveniles to one that seasonally takes them into a deep pelagic habitat. Nor is it known how sharks can navigate across ocean basins with precision, but it is assumed that they are referencing the earth's magnetic field (Klimley, 1993; Montgomery and Walker, 2001; Meyer et al., 2005). Perhaps it is coincidental, but shark 85M's movement along both the Clarion and Murray Fracture Zones is interesting. Other PAT- and SLRT-tagged White Sharks have moved along the Molokai Fracture, suggesting that these fracture zones may be a reference for east-west movements.

Adult Life-History Stage

Adult White Shark Aggregations

Adult White Sharks seasonally aggregate at two known sites along the Pacific coast of North America; subadults recruit to these aggregations prior to maturity, but these aggregations form as

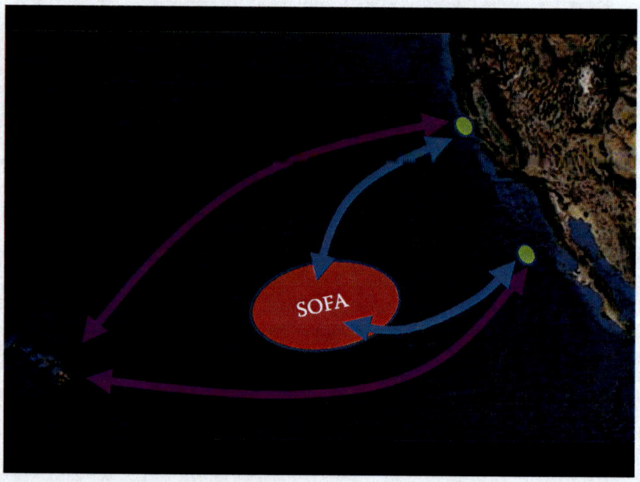

Figure 16.12 Adult male offshore migration pattern. Adult males (blue arrows) annually migrate between their adult aggregation site and the Shared Offshore Foraging Area (pink oval). Some males may migrate to the Hawaiian Islands. (Courtesy of Google Earth, Data SIO, NOAA, U.S. Navy, NGA, GEBCO, Image USDA Farm Service Agency. Image © 2011 DigitalGlobe and © 2011 Cnes/Spot Image.)

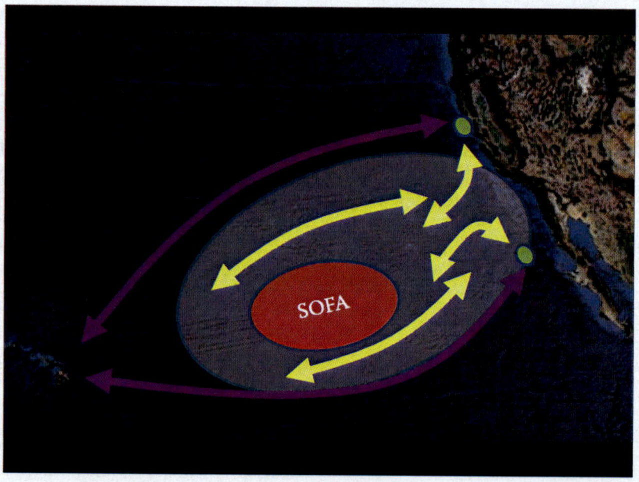

Figure 16.13 Adult female offshore migration pattern. Adult females (yellow arrows) biennially migrate between the adult aggregation sites and a much more diffuse and expansive offshore region (large pale oval). Pregnant females will migrate to coastal nursery grounds to give birth before returning to the adult aggregation sites. Some females migrate to the Hawaiian Islands. (Courtesy of Google Earth, Data SIO, NOAA, U.S. Navy, NGA, GEBCO, Image USDA Farm Service Agency. Image © 2011 DigitalGlobe and © 2011 Cnes/Spot Image.)

an important aspect of the adult life-history stage. One aggregation site is off central California (CC) in the United States (Boustany et al., 2002; Weng et al., 2007a; Jorgensen et al., 2010), and the other is GI (Figures 16.12 and 16.13) (Domeier and Nasby-Lucas, 2007, 2008; Chapters 10 and 11, this book). GI is a large island situated 255 km from the coast of Baja California, Mexico; White Sharks that frequent GI tend to remain at the island. The CC consists of a network of hot spots within a triangular-shaped region centered off the coast of San Francisco, California. Both CC and

GI support important breeding colonies of pinnipeds; Northern Elephant Seals and California Sea Lions are present at both aggregation sites, but GI also supports a large number of Guadalupe Fur Seals, and the CC site supports Northern Fur Seals, Stellar Sea Lions, and Pacific Harbor Seals. The most important pinniped sites within the overall CC region are two small islands: Southeast Farallon Island and Año Nuevo Island.

Pinnipeds have been demonstrated to be important White Shark prey at the CC site (Tricas and McCosker, 1984; Ainley et al., 1985; Klimley et al., 1992, 1996; Pyle et al., 1996), but similar evidence has been lacking from GI. Domeier and Nasby-Lucas (2007) reported they had never observed a White Shark prey upon a seal at GI; subsequently, I have witnessed a White Shark prey upon an Elephant Seal at GI (M. L. Domeier, unpublished personal observation), but these incidents are not observed with the same frequency as those reported from South Africa (Martin et al., 2005; Hammerschlag et al., 2006) and CC (Klimley et al., 1992). Domeier et al. (Chapter 10, this book) hypothesize that the differences in bathymetry between GI and these other sites may be responsible for the lower frequency of observed predation events. The steep drop in water depth around GI could make seals more elusive to White Sharks or result in small seals being entirely consumed underwater and preventing the event from being observed from the surface. Interestingly, White Sharks at GI routinely feed on Yellowfin Tuna hooked by recreational anglers (Domeier and Nasby-Lucas, 2007), and they showed a preference for fresh tuna (a food source not available off CC) over a rotting California Sea Lion carcass (Domeier, 2009). Whether White Sharks can capture free swimming tuna is unknown, and until proven otherwise, one must assume that pinnipeds are the most important prey species for White Sharks at GI.

The timing of adult arrival to the aggregation sites differs between males and females (Domeier and Nasby-Lucas, 2007; Chapters 11 and 25, this book). Males generally arrive at GI and CC at approximately the same time, over a span of a few weeks from late July through August. The males remain at the aggregation sites until they begin their offshore migration to the SOFA, but unlike their arrival, the departure is not coordinated and can begin as early as November (in the case of central California males) and as late as April. Most males begin their offshore migration between January and March. Females appear at the aggregation sites several weeks after the males have returned; although a female arrival has been observed as early as August at GI, most arrive in October (Chapter 25, this book). The observed sex ratio at GI is closest to parity in November (similar data not published for CC); however, observational effort at both GI and CC significantly fall off after November. Given the movement data provided by satellite tags, it is likely that December and January are also important months of male/female overlap at the aggregation sites, particularly at GI. Adult males are known to visit these aggregation sites on an annual basis, but photo-ID and SLRT tag studies have demonstrated that mature females generally visit the aggregation sites every other year (Anderson and Pyle, 2003; Domeier and Nasby-Lucas, 2007; Chapters 11 and 25, this book).

The atypical SLRT track of shark 102M (Figure 16.9a) raises interesting questions regarding the recruitment of individuals to adult aggregation sites. First, his extraordinary size makes it unlikely that he had been annually visiting the GI aggregation site, because he would very likely have been spotted in at least one of the previous 10 yrs. of observations (Chapter 11, this book). This leads to the speculation that he may have recently recruited to the aggregation site as an adult, and his track shows that he did return in the following year. His track also demonstrated atypical forays to and from the aggregation site during the aggregation season, and such behavior could lead an individual to discover a new preferred aggregation site over time; perhaps a site like San Benito Island (260 km south-southeast of GI), which supports an Elephant Seal colony in the apparent absence of a White Shark aggregation. These atypical forays could also explain why this shark has not been sighted previously, since he spends less time at GI than his conspecifics, and therefore the likelihood of sighting him at the island is less.

Male Offshore Migration Pattern and the Shared Offshore Foraging Area

Adult male White Sharks annually migrate between their respective aggregation site and an offshore habitat approximately halfway between the mainland and the Hawaiian Islands. This region has also been referred to as the Offshore Focal Area (Weng et al., 2007a), the Shared Offshore Foraging Area (SOFA) (Domeier and Nasby-Lucas, 2008), and the White Shark Café (Jorgensen et al., 2010). The use of the word "focal" insinuates a convergence of sharks within this area, the use of SOFA defines it as a region where both CC and GI sharks forage during their offshore season, and the authors of the term White Shark Café have suggested that adults are gathering offshore to mate (Jorgensen et al., 2010). Recent SLRT-tag data have allowed for a much more accurate geographic description of the area, as well as the discovery that it is only males that are routinely occupying this region (Chapter 11, this book). Although females will travel through the region, preliminary data suggest they may actively avoid the area while males are exploiting the forage base that is found there.

Although an expedition to the SOFA found the presence of sperm whales, the area was devoid of smaller marine mammals that could support White Sharks (Chapter 12, this book). In fact, the SOFA was an epipelagic "cold spot" with very little epipelagic productivity. However, vertically migrating mesopelagic squids (*Ommastrephes bartramii*, *Sthenoteuthis oualaniensis*, and *Architeuths dux*) were present (Figure 16.14), leading to the supposition that White Sharks may be directly feeding on squid or feeding on species that target squid (e.g. other sharks, swordfish, etc.). Squid are capable of supporting large cetaceans and have been identified as prey for White Sharks (Chapter 4, this book), so it is not unrealistic that adult White Sharks could survive on a diet of squid. Because squid are devoid of fat, this hypothesis is contrary to the historical view that adult White Sharks actively select fat-rich prey (Klimley et al., 1996). However, observations made while these sharks are occupying coastal habitats cannot be generalized to describe their pelagic life-history strategies.

The SOFA is a vast area centered at 136.9°N, 23.5°W, encompassing 785,000 km^2 (95% density contour) with a general core region of 200,000 km^2 (50% density contour). During the months of June and July, the offshore distribution of males shifts 300 km northeast, occupying considerably less space (50% density contour: 64,150 km^2) (Chapter 11, this book). Using current estimates of the combined adult and subadult White Shark populations from GI (132; Chapter 16, this book) and CC (219; Chapple et al., 2011) and assuming a 1:1 sex ratio, male White Sharks are occurring at a density of 1/1140 km^2 while in the SOFA, increasing to 1/366 km^2 in June/July. These density estimates assume that all male sharks are within the 50% density contour, which certainly is not the case. The SOFA cannot be considered a "convergence" of White Sharks, particularly when compared with

Figure 16.14 Ommastrephid squid collected during an expedition to the SOFA. (Courtesy of the author.)

their density while concentrated around the adult aggregation sites, but it is an area where individuals from CC and GI are most likely to interact.

The June/July constriction and shift of the SOFA is an interesting phenomenon that could be driven by a predictable, seasonal pulse of prey availability. All the squid species sampled in the SOFA were in spawning condition (Chapter 12, this book), and squid spawn in spatially and temporally predictable aggregations. The recent discovery that myctophids form massive seasonal spawning aggregations (Australian Institute of Marine Science press release, 2011) could also produce a temporary spatiotemporal concentration of the food chain. Although there is no current evidence that myctophids are spawning in the SOFA, I did find a myctophid in the beak of a sampled ommastrephid squid while in the SOFA (personal observation).

The Hawaiian Islands

Both male and female White Sharks tagged at both CC and GI have made return migrations to the Hawaiian Islands (Figures 16.12 and 16.13). In reviewing published records from both GI and CC, it is clear that a higher proportion of sharks from CC migrate to the Hawaiian Islands than sharks from GI [22% versus 5% of PAT tagged sharks, respectively (Jorgensen et al., 2010; Domeier and Nasby-Lucas, 2008)]. I consider this to be a separate region from the SOFA, and sharks that make this migration may be targeting prey that is not found in the SOFA.

Stomach content analyses have indicated that small odontocetes are an important White Shark prey item in some regions (Arnold, 1972; Cliff et al., 1989; Bruce, 1992), but it has not been possible to determine whether the dolphins were moribund or dead prior to being consumed. A predatory attack on a bottlenose dolphin adjacent to island of Lanai, Hawaii (P. Morton and C. Blair, personal communication), is the first evidence that White Sharks possess the speed and agility to pursue and capture live, seemingly healthy Bottlenose Dolphin. Small cetaceans are abundant around the Hawaiian Islands and may be the target prey for White Sharks that migrate to the islands; however, similar prey can be found near the continental shelf of North America (Figure 16.15), so the motivation for such a long migration must be more complex.

Female Offshore Migratory Pattern

SLRT tags have provided the first multiyear tracks of mature female White Sharks, leading to the discovery that they are primarily a pelagic animal (Chapter 11, this book). Photo-identification studies previously noted that mature females do not annually return to adult aggregation sites, and

Figure 16.15 Small cetaceans, such as the Common Dolphin, were absent from the SOFA, but they are numerous closer to the continental shelf and near the Hawaiian Islands. (Courtesy of the author.)

SLRT tags revealed that, with the exception of moving into parturition grounds to give birth, the females are continuous residents of the pelagic environment during this absence. Although males predictably occupy the SOFA while they are offshore, the spatial use of offshore habitat by females is much less predictable (Chapter 11, this book). SLRT-tagged females range widely from hundreds of kilometers west of the SOFA east to the vicinity of GI, and PAT-tagged females have ranged as far west as Hawaii (Figure 16.13) (Weng et al., 2007a; Domeier and Nasby-Lucas, 2008; Jorgensen et al., 2010; Chapter 11, this book) (Figure 16.11). SLRT-tagged females have been observed making repeated east/west migrations between the eastern edge of the SOFA to within a few hundred kilometers of the continental shelf of Baja California (Figure 16.8). Females may be found in the SOFA, but typically only when the males have returned to GI and CC. Because females are at times coming much closer to the continental shelf while they are offshore, they will encounter small cetaceans (Figure 16.15) (Chapter 12, this book), leading to the possibility that their preferred prey is different from that of adult males. Chapter 11, this book also notes that females spend more time in warmer water than males, something that could be tied to the optimum temperature for fetal development.

Insight into the Reproductive Cycle

Readily available prey could explain the seasonal presence of White Sharks off the Pacific coast of North America, but it does not explain why White Sharks would aggregate in unusual density at just two relatively concentrated areas. Both male and female White Sharks at the aggregation sites are routinely observed with conspecific bite wounds, indicating a high level of intraspecific aggression. CC and GI are unique in that they both support large Northern Elephant Seal colonies, but additional Elephant Seal colonies exist between these two sites, in the apparent absence of large numbers of adult White Sharks. Other prey species for White Sharks abound throughout the coasts of California and Mexico, so it is not logical that adult White Sharks would subject themselves to the risk of serious injury simply to feed at relatively confined sites.

Although White Sharks gather at sites that offer an abundance of prey, it is feasible that these aggregations form primarily for the purpose of mating. White sharks have never been observed in the mating act, anywhere in the world. A single anecdotal account of two sharks locked together at the surface off New Zealand (Francis, 1996) may have been a mating act, but the observer was not a trained shark biologist, and there is no mention of the sex of the participating sharks. Also, the presumed mating observation occurred over a 40-min. period entirely at the surface; the fact that sharks will sink once they begin to copulate makes this report less credible. Although no direct evidence of mating exists from either GI or CC, a substantial amount of circumstantial and indirect evidence has accumulated.

Mating can only occur when the sexes are in close proximity, and it has been demonstrated that the sexes are widely segregated except when they are at the aggregation sites (Domeier and Nasby-Lucas, 2008; Jorgensen et al., 2010; Chapter 11, this book). Individual mature females are typically absent from adult aggregation sites every other year (Anderson and Pyle, 2003; Domeier and Nasby-Lucas, 2007; Chapters 11 and 25, this book); if the aggregations were solely for the purpose of feeding, one would expect the females to return each year. The suspected 18-month gestation period (Mollet et al., 2000) would place females on a 2-year reproductive cycle, a periodicity that matches their typical visitation to adult aggregation sites. However, the female visitation interval to aggregation sites can deviate from the expected; both photo identification (Domeier and Nasby-Lucas, 2007; Chapter 25, this book) and SLRT tagging (Figure 16.8) (Chapter 11, this book) have documented mature females returning to GI in consecutive years, suggesting that females may not always successfully mate and/or that they sometimes rest for a year between pregnancies.

The presence of running-ripe males (Table 16.1 and Figures 16.4 and 16.5) and females with conspecific bite marks (Figure 16.6) (Chapter 13, this book) is indirect evidence that mating has occurred (Gilmore, 1993) at both aggregation sites. It cannot be a coincidence that every individual female observed with a presumed mating bite wound was not sighted at the adult aggregation the following year (Table 16.2). Also, a small sample size of mature females from GI were analyzed for the presence of circulating reproductive hormones (Chapter 30, this book): all had low levels of progesterone, indicating they were not pregnant (Chapter 30, this book) [although in Atlantic stingrays progesterone was found to peak only at ovulation and parturition (Tricas et al., 2000)], and one individual had a high level of 17-b-estradiol, suggesting impending ovulation. If females are mating in October through January, and parturition is occurring in May through August (Figures 16.1 and 16.2), then a back-calculation estimates gestation to be 16–22 months, a temporal window that overlaps with gestation estimates of 12–18 months (Francis, 1996; Mollet et al., 2000). Unfortunately, current estimates of gestation are guesses based upon very little empirical data; furthermore, we do not know how long females may store sperm between mating and ovulating, so our current level of understanding does not allow the exact determination of the mating season or gestation period.

At a first glance, the circulating blood testosterone levels in the adult male White Sharks sampled at the aggregation sites seem contradictory to the hypothesis that these sharks are reproductively active while at the aggregation (n = 8; Chapter 30, this book). Testosterone levels were lower than what has been observed during the mating season in other elasmobranchs; however, some of these males were found to have spermatophores in their clasper grooves, undeniably indicating that these individuals were capable of impregnating females (Table 16.1 and Figures 16.4 and 16.5). Both male and female elasmobranchs are capable of storing sperm for an undetermined amount of time, meaning males can undergo a peak of testosterone and spermatogenesis and then store the sperm for mating sometime in the future (Pratt, 1993; Maruska et al., 1996; Mull et al., 2008). If individual females store sperm until environmental factors are suitable for early embryonic development, parturition may not be precisely synchronous and occur over a protracted time frame, a condition for which we see some evidence (Figures 16.1 and 16.2). These factors cloud the interpretation of circulating-blood hormone levels, but it is an intriguing field that requires far more sampling before it may eventually prove to be a valuable tool in determining the reproductive cycle of this protected species.

It has been suggested that White Sharks are mating while in the SOFA (Weng et al., 2007a; Jorgensen et al., 2010), but no direct or indirect evidence exists to support that hypothesis. Alone, none of the data presented here are conclusive that White Sharks are aggregating for the purpose of mating, but taken together, the evidence makes a compelling argument. Aggregations of other shark species have been implicated to serve a reproductive function (Klimley, 1985b; McKibben and Nelson, 1986; Carrier et al., 1994; Economakis and Lobel, 1998; Harvey-Clarke et al., 1999; Whitney et al., 2004), and White Sharks are another likely example. For many shark species, the female will mate immediately after giving birth and then store sperm for up to 12 months before ovulating and fertilizing the next batch of eggs (Pratt, 1979). In the case of northeastern Pacific White Sharks, there are probably very few or no adult males in the coastal nursery/parturition areas (they are offshore in the SOFA) at the time of parturition, and the next likely mating opportunity would be at the adult aggregation sites. It will take much more research to determine the full reproductive biology of this species, because there are so many plastic variables that can lead to unexpected realities. It is not unfathomable that both sexes can store sperm, allowing for mating to take place at any opportunity, and even more startling are observations of parthenogenesis in two species of captive shark (Chapman et al., 2007, 2008), leading to the conclusion that female White Sharks could possibly give birth in the complete absence of mating.

Parturition Grounds

Because no pregnant female White Sharks have ever been landed by commercial fishing fleets in California or Mexico, our best available evidence of parturition grounds comes from studies of

YOY White Sharks. The data presented here and in Chapters 14, 15, and 28, this book, all indicate that coastal areas of Southern California and Baja Mexico are important YOY nursery areas and therefore are likely the sites of parturition. The geographic and seasonal pattern of shark attacks, combined with satellite-tag data, provide further clues. The sudden August–December spike in White Shark attacks on humans (Figure 16.3) aligns with the arrival of adults, returning to their seasonal adult aggregation site from the SOFA. There is a relatively low incidence of shark attacks south of Point Conception, and it is noteworthy that the peak month of White Shark attacks in Southern California is July, a month adults are typically offshore. The July spike in shark attacks, which coincides with the peak of the parturition season, could be a signal of adult females traveling to the nursery area to give birth.

April is the month shark 98F, the first mature female ever tracked over her entire multiyear migration, entered the Gulf of California after 443 d offshore. She moved to the northern tip of the Gulf of California before reversing and exiting in early June. Her absence from GI in 2009 and the length of time she remained offshore strongly indicate she moved into the Gulf of California to give birth sometime during the months of April and June. Larger YOY and juvenile White Sharks have been recorded from the Gulf of California (Galván-Magaña et al., 2010; Domeier, personal observations), and YOY have been tracked from California to the Gulf of California (Chapter 28, this book), but very small neonate-sized YOY have not been documented from this region. The remote nature of the region presents challenges when attempting to obtain samples for the purpose of documenting the size and seasonal presence of White Sharks in the Gulf of California, so whether or not neonates occur in Gulf of California is not known. Unfortunately, the SLRT tag on shark 98F never transmitted again shortly after she exited the Gulf of California, so it is not known whether she visited other coastal regions or if she rejoined the GI aggregation later in 2010. If female had continued on her last known course, she would have returned to the offshore habitat.

Jorgensen et al. (Chapter 13, this book) proposes the migration of a CC PAT-tagged female (estimated TL, 4.88 m) as first evidence of a female parturition in Southern California. There are many problems with this assumption. First, this female returned to CC in 2007, just 1 yr. after being tagged, which is not consistent with the multiyear migration pattern of mature females from CC (Anderson and Pyle, 2003) or GI (Domeier and Nasby-Lucas, 2007; Chapters 11 and 25, this book). SLRT tagging at GI documented a similar case of a mature female returning to GI in consecutive years, after which she remained offshore during the subsequent GI aggregation (Chapter 11, this book). The most reasonable explanation for consecutive visits of mature females to an adult aggregation is that they are resting a year between pregnancies. In the case described by Jorgensen et al. (Chapter 13, this book), it is also possible that the female in question was not mature, because the visual estimate of length was close to the minimum estimated size of maturity; visual estimates are not precise, and the size of maturity is based upon relatively few samples. Finally, the CC female was reported to have mating wounds at the time of tagging, but to fit their hypothesis, these wounds would have to be 45 d old (Chapter 13, this book). White sharks heal very rapidly (Chapter 6, this book), and it is not plausible that these wounds would remain open that long. A more reasonable explanation is that these wounds were a result of more recent intraspecific aggression or a mating attempt that occurred at CC.

Why then, did the female described by Jorgensen et al. (Chapter 13, this book) travel to the vicinity of Southern California on two occasions? This is precisely the type of migratory behavior exhibited by SLRT-tagged females from GI; whereas the males remained within the SOFA, the females made repeated east/west migrations, sometimes coming very near the continental shelf. The presence of YOY in Southern California predicts that pregnant females from one or both adult aggregation sites give birth in the Southern California Bight, but I do not believe this has yet been directly documented via active tracking. The pattern of GI females migrating east/west near the coast of Baja and CC females migrating east/west near the coast of California raises the possibility that individuals from these aggregation sites maintain discrete parturition areas.

Summary

Parturition and Young of the Year

Parturition peaks between the months of June and August, likely over the shallow coastal nursery habitats from Point Conception south into Baja California and perhaps the Gulf of California. YOY spend their first summer and fall in these nursery habitats, as far north as Point Conception, before migrating to warmer waters in Baja California for their first winter. YOY prey upon a variety of fish and invertebrates.

Juvenile White Sharks

Juvenile White Sharks migrate north and south along the coast of California and Baja California, with some individuals entering the Gulf of California. They are consistently found at local regions of preferred habitat that create juvenile hot spots. Juveniles prey upon a variety of fish and invertebrates and opportunistically scavenge marine mammal carcasses.

Subadult White Sharks

The movement patterns of subadults are not well understood. They range widely and are frequently found from the Gulf of California to Oregon. Subadults begin to visit adult aggregation sites, but it is not known how these aggregation sites are located. The subadult stage may be a source of mixing between the two metapopulations of adults. Subadults also begin to make inshore/offshore migrations, but it is not known precisely when or how this migration pattern begins. Subadults begin to actively prey upon marine mammals.

Adult White Sharks

Adult White Sharks seasonally aggregate at two known aggregation sites, Guadalupe Island, Mexico, and central California, United States. Male sharks arrive at the aggregation sites in August, and females arrive late September through November. Adults prey upon pinnipeds, particularly Northern Elephant Seals, while they frequent the aggregation sites. Indirect evidence suggests that adults are likely mating at these aggregation sites. Males visit the aggregation sites every year, whereas adult females typically visit every other year. Both sexes depart the aggregation sites for pelagic habitats between December and April. Males spend their time in a region called the Shared Offshore Foraging Area, an area with little epipelagic productivity, no pinnipeds, and very few cetaceans. The SOFA does contain several species of squid and Sperm Whales. Adult females are primarily pelagic, spending the vast majority of their time far offshore. Females range widely while offshore and typically do not inhabit the same space as males; they may be targeting the same or different prey as the males.

Vulnerable Life-History Stages

YOY and juvenile White Sharks are susceptible to incidental capture by gillnet and longline fisheries. Adults are less vulnerable to commercial fisheries because their large size allows them to escape most fishing gears; occasionally they can get entangled in gillnet or longline gear. Very little commercial-fishing activity currently takes place in the SOFA, so the offshore region acts as a refuge, but there are no international laws in place to protect White Sharks in international waters. Females migrating through parturition regions are likely more vulnerable to incidental fishery related mortality during that time (female 98F is a likely example). Adults are extremely vulnerable to exploitation while they occupy adult aggregation sites, but these sites have protective laws forbidding the take of White Sharks. All life-history stages may be vulnerable to high body burdens of anthropogenic toxins (Chapter 5, this book); how these may impact the population is not known.

Suggestions for Future Research

Electronic tagging, photo identification, blood sampling, and spermatophore sampling have all contributed to a much clearer life-history hypothesis for northeastern Pacific White Sharks (Figure 16.11). Lines of research that will increase our understanding of the differences between the SOFA, the Hawaiian Islands, and the offshore region used by mature females will be valuable contributions to our knowledge of this species. Further work on correlating circulating-blood hormone levels with reproductive states could provide a future means of validating the reproductive biology of sharks without the need of sacrificing individuals. There is a diminished need to study adult male White Sharks in the northeastern Pacific, but the atypical movements of male 102M indicate that not all is known about male migratory patterns. Effort should be spent to locate, tag, and track adult individuals, during the adult aggregation season, that are not at the aggregation (perhaps along the coast of Southern California or Baja Mexico). Population estimates assume that all of the subadult and adult sharks are visiting the aggregation sites, but this may not be the case. The least understood life-history stage is that of the subadult; developing appropriate long term satellite tags for subadults should be a high priority. The study of subadults will allow much more insight into the connectivity of adult aggregation sites and the recruitment of sharks to the offshore environment, likely leading to some novel findings relative to learned versus instinctive migratory behaviors.

ACKNOWLEDGMENTS

N. Nasby-Lucas provided long-term, invaluable assistance to many aspects of this research project. The passion and field assistance of T. Pfleger was the spark for this long-term program. C. Fischer and the crew of the *Ocean* assisted in the capture and tagging of adult White Sharks. L. Laughlin, J. Robertson, and B. Read from the California Department of Fish and Game provided the fishery-dependent data necessary for many of the analyses. P. Morton and C. Blair provided a detailed description of the White Shark preying upon a dolphin. O. Sosa, F. Galvan, and M. Hoyos provided invaluable assistance in obtaining Mexican research permits (SEMARNAT permit GAMSX0300311). This chapter is the result of many years of White Shark research, and over this time funding was provided by the George T. Pfleger Foundation, the Offield Family Foundation, the Guadalupe Island Conservation Fund, the Guy Harvey Ocean Foundation, and Chris Fischer.

REFERENCES

Ainley, D. G., R. P. Henderson, H. R. Huber, R. J. Boekelheide, S. G. Allen, and T. L. McElroy. 1985. Dynamics of white shark/pinniped interactions in the Gulf of the Farallones. *Memoirs of the Southern California Academy of Sciences* 9: 109–122.

Anderson, S. D., and P. Pyle. 2003. A temporal, sex-specific occurrence pattern among white sharks (*Carcharodon carcharias*) at the South Farallon Islands, California. *California Fish and Game* 89: 96–101.

Arnold, P. W. 1972. Predation on harbour porpoise, *Phocena phocena*, by a white shark, *Carcharodon carcharias*. *Journal of the Fisheries Research Board of Canada* 29: 1213–1214.

Bonfil, R., M. Meyer, M. C. Scholl, R. Johnson, S. O'Brien, H. Oosthuizen, S. Swanson, D. Kotze, and M. Paterson. 2005. Transoceanic migration, spatial dynamics, and population linkages of white sharks. *Science* 310: 100–103.

Boustany, A. M., S. F. Davis, P. Pyle, S. D. Anderson, B. J. Le Boeuf, and B. A. Block. 2002. Expanded niche for white sharks. *Nature* 415: 35–36.

Bruce, B. D. 1992. Preliminary observations on the biology of the white shark, *Carcharodon carcharias*, in South Australian waters. *Australian Journal of Marine and Freshwater Research* 43: 1–11.

Cailliet, G., L. Natanson, B. Weldon, and D. Ebert. 1985. Preliminary studies on the age and growth of the white shark *Carcharodon carcharias*, using vertebral bands. *Memoirs of the Southern California Academy of Sciences* 9: 49–60.

Carrier, J. C., H. L. Pratt, Jr., and L. K. Martin. 1994. Group reproductive behaviors in free-living nurse sharks, *Ginglymostoma cirratum*. *Copeia* 1994: 646–656.

Chapman, D., M. Shivji, E. Louis, J. Sommer, H. Fletcher, and P. Prodöhl. 2007. Virgin birth in a hammerhead shark. *Biology Letters* 3: 425–427.

Chapman, D., B. Firchau, and M. Shivji. 2008. Parthenogenesis in a large-bodied requiem shark, the blacktip *Carcharhinus limbatus*. *Journal of Fish Biology* 73: 1473.

Chapple, T. K., S. J. Jorgensen, S. D. Anderson, P. E. Kanive, A. P. Klimley, L. W. Botsford, and B. A. Block. 2011. A first estimate of white shark, *Carcharodon carcharias*, abundance off central California. *Biology Letters* 7: 581–583.

Cliff, G., S. F. J. Dudley, and B. Davis. 1989. Sharks caught in the protective gill nets off Natal, South Africa: 2. The great white shark *Carcharodon carcharias* (Linnaeus). *South African Journal of Marine Science* 8: 131–144.

Dewar, H., M. L. Domeier, and N. Nasby-Lucas. 2004. Insights into young of the year white shark, *Carcharodon carcharias*, behavior in the southern California bight. *Environmental Biology of Fishes* 70: 133–143.

Dicken, M. L. 2008. First observations of young of the year and juvenile great white sharks (*Carcharodon carcharias*) scavenging from a whale carcass. *Marine and Freshwater Research* 59: 596–602.

Domeier, M. L. 2009. Experimental scavenging preference for the adult white shark, *Carcharodon carcharias*. *California Fish & Game Bulletin* 95:140.

Domeier, M., and N. Nasby-Lucas. 2007. Annual re-sightings of photographically identified white sharks (*Carcharodon carcharias*) at an eastern Pacific aggregation site (Guadalupe Island, Mexico). *Marine Biology* 150: 970–984.

Domeier, M., and N. Nasby-Lucas. 2008. Migration patterns of white sharks *Carcharodon carcharias* tagged at Guadalupe Island, Mexico, and identification of an eastern Pacific shared offshore foraging area. *Marine Ecology Progress Series* 370: 221–237.

Economakis, A. E., and P. S. Lobel. 1998. Aggregation behavior of the grey reef shark, *Carcharhinus amblyrhynchos*, at Johnston Atoll, central Pacific Ocean. *Environmental Biology of Fishes* 51: 129–139.

Francis, M. P. 1996. Observations on a pregnant white shark with a review of reproductive biology. In *Great white sharks: The biology of Carcharodon carcharias*, edited by A. P. Klimley and D. G. Ainley, pp. 157–172. San Diego, CA: Academic Press.

Galván-Magaña, F., E. Hoyos-Padilla, C. Navarro-Serment, and F. Márquez-Farías. 2010. Records of white shark, *Carcharodon carcharias*, in the Gulf of California, Mexico. *Marine Biodiversity Records* 3: e111 doi:10.1017/S1755267210000977.

Gilmore, R. G. 1993. Reproductive biology of lamnoid sharks. *Environmental Biology of Fishes* 38: 95–114.

Hammerschlag, N., R. A. Martin, and C. Fallows. 2006. Effects of environmental conditions on predator-prey interactions between white sharks (*Carcharodon carcharias*) and Cape fur seals (*Arctocephalus pusillus pusillus*) at Seal Island, South Africa. *Environmental Biology of Fishes* 76: 341–350.

Harvey-Clark, C. J., W. T. Strobo, E. Helle, and M. Mattson. 1999. Putative mating behaviour in basking sharks off the Nova Scotia coast. *Copeia* 3: 780–782.

Heneman, B., and M. Glazer. 1996. More rare than dangerous: A case study of white shark conservation in California. In: *Great white sharks: The biology of Carcharodon carcharias*, edited by A. P. Klimley and D. G. Ainley, pp. 481–491. San Diego, CA: Academic Press.

Jorgensen, S. J., C. A. Reeb, T. K. Chapple, S. Anderson, C. Perle, S. R. Van Sommeran, C. Fritz-Cope, A. C. Brown, A. P. Klimley, and B. A. Block. 2010. Philopatry and migration of Pacifc white sharks. *Proceedings of the Royal Society B Biological Sciences* 277: 679–688.

Klimley, A. P. 1985a. The areal distribution and autoecology of the white shark, *Carcharodon carcharias*, off the West Coast of North America. *Memoirs of the Southern California Academy of Sciences* 9: 15–40.

Klimley, A. P. 1985b. Schooling in *Sphyrna lewini*, a species with low risk of predation: A non-egalitarian state. *Zeitschrift für Tierpsychologie* 7: 297–319.

Klimley, A. P. 1993. Highly directional swimming by scalloped hammerhead sharks, *Sphyrna lewini*, and sub surface irradiance, temperature, bathymetry, and geomagnetic field. *Marine Biology* 117: 1–22.

Klimley, A., S. Anderson, P. Pyle, and R. Henderson. 1992. Spatiotemporal patterns of white shark (*Carcharadon carcharias*) predation at the South Farallon Islands, California. *Copeia* 1992: 680–690.

Klimley, A., P. Pyle, and S. Anderson. 1996. The behavior of white sharks and their pinniped prey during predatory attacks. In *Great white sharks: The biology of Carcharodon carcharias*, edited by A. P. Klimley and D. G. Ainley, pp. 175–191. San Diego, CA: Academic Press.

Martin, R., N. Hammerschlag, R. Collier, and C. Fallows. 2005. Predatory behavior of white sharks (*Carcharodon carcharias*) at Seal Island, South Africa. *Journal of the Marine Biological Association of the United Kingdom* 85: 1121–1135.

Maruska, K. P., E. G. Cowie, and T. C. Tricas. 1996. Periodic gonadal activity and protracted mating in elasmobranch fishes. *Journal of Experimental Zoology* 276: 219–232.

McCosker, J. E., and R. N. Lea. 1996. White shark attacks in the eastern Pacific Ocean: An update and analysis. In *Great white sharks: The biology of Carcharodon carcharias*, edited by A. P. Klimley and D. G. Ainley, pp. 419–434. San Diego, CA: Academic Press.

McKibben, J. R., and D. H. Nelson. 1986. Patterns of movements and grouping of grey reef sharks, *Carcharhinus amblyrhynchos*, at Enewetak, Marshal Islands. *Bulletin of Marine Science* 38: 89–110.

Meyer, C. G., K. N Holland, and Y. P. Papastamatiou. 2005. Sharks can detect changes in the geomagnetic field. *Journal of the Royal Society Interface* 2: 129–130.

Mollet, H. F., and G. M. Cailliet. 1996. Using allometry to predict body mass from linear measurements of the white shark. In *Great white sharks: The biology of Carcharodon carcharias*, edited by A. P. Klimley and D. G. Ainley, pp. 81–89. San Diego, CA: Academic Press.

Mollet, H., G. Cliff, H. Pratt, Jr., and J. Stevens. 2000. Reproductive biology of the female shortfin mako, *Isurus oxrtinchus Rafinesque*, 1810, with comments on the embryonic development of lamnoids. *Fishery Bulletin* 98: 299–318.

Montgomery, J. C., and M. M. Walker. 2001 Orientation and navigation in elasmobranchs: Which way forward? *Environmental Biology of Fishes* 60: 109–116.

Mull, C. G., C. G. Lowe, and K. A. Young. 2008. Photoperiod and water temperature regulation of seasonal reproduction in male round stingrays (*Urobatis halleri*). *Journal of Comparative Biochemistry and Physiology: Part A* 151: 717–715.

Nasby-Lucas, N., H. Dewar, C. H. Lam, K. J. Goldman, and M. L. Domeier. 2009. White shark offshore habitat: A behavioral and environmental characterization of the eastern Pacific Shared Offshore Foraging Area. *PLoS ONE* 4: e8163.

Papastamatiou, Y. P, D. P. Cartamil, C. G. Lowe, C. G. Meyer, B. M. Wetherbee, and K. N. Holland. 2011. Scales of orientation, directed walks and movement path structure in sharks. *Journal of Animal Ecology* 80: 864–874.

Pratt, H. L. 1979. Reproduction in the blue shark, *Prionace glauca*. *Fishery Bulletin* 77: 445–470.

Pratt, H. L. 1993. The storage of spermatozoa in the oviducal glands of western North Atlantic sharks. *Environmental Biology of Fishes* 38: 139–149.

Pratt, H. 1996. Reproduction in the male white shark. In *Great white sharks: The biology of Carcharodon carcharias*, edited by A. P. Klimley and D. G. Ainley, pp. 131–138. San Diego, CA: Academic Press.

Pratt, H., and J. Carrier. 2001. A review of elasmobranch reproductive behavior with a case study on the nurse shark, *Ginglymostoma cirratum*. *Environmental Biology of Fishes* 60: 158–188.

Pyle, P., S. D. Anderson, A. P. Klimley, and R. P. Henderson. 1996. Environmental factors affecting the occurrence and behavior of white sharks at the Farallon Islands, California. In *Great white sharks: The biology of Carcharodon carcharias*, edited by A. P. Klimley and D. G. Ainley, pp. 281–291. San Diego, CA: Academic Press.

Tricas, T., and J. McCosker. 1984. Predatory behavior of the white shark (*Carcharodon carcharias*), with notes on its biology. *Proceedings of the California Academy of Sciences* 43: 221–238.

Tricas, T., K. P. Maruska, and L. E. L. Rasmussen. 2000. Annual cycles of steroid hormone production, gonad development, and reproductive behavior in the Atlantic stingray. *General and Comparative Endocrinology* 118: 209–225.

Uchida, U., M. Toda, K. Teshima, and K. Yano. 1996. Pregnant white sharks and full-term embryos from Japan. In *Great white sharks: The biology of Carcharodon carcharias*, edited by A. P. Klimley and D. G. Ainley, pp. 139–155. San Diego, CA: Academic Press.

Weng, K., A. Boustany, P. Pyle, S. Anderson, A. Brown, and B. Block. 2007a. Migration and habitat of white sharks (*Carcharodon carcharias*) in the eastern Pacific Ocean. *Marine Biology* 152: 877–894.

Weng, K., J. O'Sullivan, C. Lowe, C. Winkler, H. Dewar, and B. Block. 2007b. Movements, behavior and habitat preferences of juvenile white sharks *Carcharodon carcharias* in the eastern Pacific. *Marine Ecology Progress Series* 338: 211–224.

Whitney, N. M., H. L. Pratt, and J. C. Carrier. 2004. Group courtship, mating behaviour and siphon sac function in the whitetip reef shark, *Triaenodon obesus. Animal Behavior* 68: 1435–1442.

CHAPTER 17

Habitat Use and Spatial Dynamics of Juvenile White Sharks, *Carcharodon carcharias*, in Eastern Australia

Barry D. Bruce*
CSIRO Marine and Atmospheric Research

Russell W. Bradford
CSIRO Marine and Atmospheric Research

ABSTRACT

The movements and behavior of 22 juvenile White Sharks, *Carcharodon carcharias* [1.75–2.6 m total length (TL)], were monitored in eastern Australia using satellite tracking and pop-up archival tags, providing a total of 1,941 d of data. Broad-scale movements were primarily coastal, were highly directional with high angular concentration values r (mean $r = 0.81$; SD = 0.16), and ranged over 2,000 km of the Australian coastline from eastern Tasmania to southern Queensland. One 2.1-m male also traveled 3,600 km across the Tasman Sea to New Zealand, through Cook Strait, and east onto the Chatham Rise. Directional movement was interspersed with periods of temporary residency denoted by significantly lower r values (mean $r = 0.22$; SD = 0.17) and a highly restricted geographic footprint. Residency sites were limited in number, and season of occupancy suggesting that juvenile White Sharks have a relatively small number of interconnected preferred habitat areas in eastern Australia. The consistent occupancy of two residency sites over multiple years, the Corner Inlet-Ninety Mile Beach region off eastern Victoria and the Port Stephens region of New South Wales, defined these as seasonal nursery areas. On-shelf shark activity was primarily shoreward of the 120-m depth contour in nursery areas. Sharks moving between nursery areas showed a propensity to travel over a corridor spanning bottom depths of 60–120 m, with 55% of all ARGOS positions located over this zone. This depth zone includes relic coastline structures in eastern Australia, which may offer cues for navigation or enhanced foraging opportunities by providing focal areas for prey species. However, sharks also made occasional dives to 300–500 m when traveling and when off-shore. The maximum depth recorded by a 1.9-m female was 984 m. Sharks showed considerable plasticity in swimming behavior, recording most patterns previously identified in juvenile, subadult, and adult stages of this species. The similarity in the dominant temperature range experienced by juvenile White Sharks between our data and other studies suggests a propensity to occupy waters of 18–20°C. Nursery areas in eastern Australia are coincident in time and space with aggregations of various finfish species that provide suitable prey and coincided with areas of seasonal upwelling and subsequent nutrient enrichment. The smaller sizes of juveniles recorded from the Corner Inlet region of eastern Victoria suggests that pups may initially recruit to this nursery area prior to establishing a seasonal migratory pathway along the east coast to the Port Stephens nursery area of central New South Wales. The potential exists for juvenile White Sharks to be taken as bycatch in nearshore, shelf, and slope fishing activities in the vicinity of these nursery areas. Forays seaward of the shelf break, offshore diving behavior, and open-ocean travel may also expose juvenile White Sharks to incidental capture in offshore pelagic fisheries previously considered not to pose such a threat.

INTRODUCTION

Determining the spatial and temporal scale of movements for species of conservation concern and their habitat preferences are important for identifying areas of high interactions with human activities. Quantifying behavior in such areas is important for assessing in which localities those species may require enhanced protection (Wilson et al., 2007). For many species, preferred habitats and spatial dynamics vary over their life history, and these differences may have ramifications for threat assessment and conservation actions during different life-history stages. Worldwide research on White Sharks (*Carcharodon carcharias*) has primarily focused on subadult and adult sharks >3.0 m in length because these are the stages that are commonly observed in the vicinity of pinniped colonies where most research has been undertaken (Strong et al., 1996; Pyle et al., 1996; Goldman and Anderson, 1999; Malcolm et al., 2001; Martin et al., 2005; Bruce et al., 2005a, 2005b; Robbins, 2007; Domeier and Nasby-Lucas, 2007; Chapters 8, 9, 10, 11, 12, 13, 16, 21, 22, 25, 26, 27, and 30, this book). Research on movement patterns based at such localities using electronic tags have identified that White Sharks >3.0 m in length can make extensive coastal migrations coupled with periods of temporary residency; may show a propensity for returning to specific sites, sometimes on a seasonal basis; and can make excursions into the open ocean (Boustany et al., 2002; Bruce et al., 2006; Weng et al., 2007a; Domeier and Nasby-Lucas, 2008; Bonfil et al., 2010; Chapters 11, 16, and 25, this book) or cross ocean basins (Bonfil et al., 2005). These data have shed new light on the habitat preferences, spatial dynamics, and connectivity between populations of White Sharks (see Chapters 10, 11, and 16, this book). However, despite these advances, comparatively little effort has been directed at determining the movements and habitat preferences of juvenile (≤3.0 m) White Sharks prior to their appearance at pinniped colonies (Klimley et al., 2002; Dewar et al., 2004; Weng et al., 2007b). This paucity of data is due to sharks in these earlier life-history stages being less reliably located in sufficient numbers to facilitate focused research efforts. However, available data suggest that sharks in these stages can be encountered close to shore in areas that make them vulnerable to bycatch in commercial and recreational fisheries (Casey and Pratt, 1985; Klimley, 1985; Malcolm et al., 2001; Weng et al., 2007b; Chapters 14, 15, and 16, this book). Residency in some of these areas also potentially increases other chances of human-shark interactions. These range from changed recreational use of an area because of the increased risk of attack (either real or perceived), to "nuisance" sharks interfering with recreational or commercial fishing by taking hooked fish or baits, to pressure from sectors wishing to pursue an ecotourism type experience through viewing or baiting sharks in areas close to other users or populated centers.

White sharks are a listed threatened species in several regions of the world, including the United States, South Africa, Malta, Namibia, New Zealand, and Australia, as well as being listed on Appendix II of the Convention for International Trade in Endangered Species and under the Convention on

Migratory Species [reviewed by Bruce (2008)]. In Australian waters, White Sharks have been protected since 1996 (Malcolm et al., 2001), and a National Recovery Plan was established in 2002 (Environment Australia, 2002). The primary objective of this plan is to recover White Shark numbers in Australian waters to a level that will see the species removed from the threatened species schedules under Australian legislation. However, there are no current metrics from which to monitor the population status of White Sharks and thus no way of adequately determining either the efficacy of the plan or when this recovery point is reached. This is also hampered by a lack of long-term population indices. Most recovery actions are directed toward habitat preferences and the risk of interaction with White Sharks >3.0 m in length, because of the bias in available data for these life-history stages. However, the proximity to shore, and thus various human-based impacts, suggests that juvenile stages <3.0 m may require additional considerations with respect to threat identification and mitigation, as well as resulting interactions that may require some balance between threats to the sharks and public safety.

Because of their low fecundity and low natural mortality (Francis, 1997), a useful index of population status may be the abundance of juvenile sharks recruiting to the population. Recruitment monitoring programs are used to assess current, or predict the future, status of various commercial species, including finfish and invertebrates, e.g., Southern Bluefin Tuna, *Thunnus maccoyii* (Cowling, 1999), and Western Rock Lobster, *Panulirus cygnus* (Phillips et al., 2000). Prerequisites for utilizing juvenile abundance in such programs include knowledge of the location and extent of juvenile habitat, the time period during which juveniles occupy these habitats, the influence of behavior on the ability to observe and count juveniles, and the ability to correlate estimates of juvenile abundance with population trends or size. There are currently few data available about these aspects of White Shark life history.

Anecdotal reports suggest that sightings of juvenile White Sharks have escalated in recent years at certain beaches on the central coast of New South Wales (NSW) off eastern Australia during the late winter to early summer period. However, it is not clear whether such sightings represent an increase in reporting, a redistribution of sharks in response to variations in local environmental or biological parameters, or an increase in the number of juveniles recruiting to the population in response to the species protection. In addition, the connectivity between this and other Australian regions where juvenile White Sharks have also been recorded is unclear. Such data are also a vital part of identifying the number of populations of the species in Australian waters.

In this study, we have used satellite tracking to identify the movement patterns, habitat, and behavior of 22 juvenile White Sharks of 1.75–2.6 m total length. Our overall objectives were to identify regions of important habitat that may constitute nursery areas for juvenile White Sharks and to examine the connectivity via their movements between such areas. This information is essential for assessing the importance of different habitats to the species, linkages between areas where juvenile White Sharks have been encountered, and potential sources of threat and, in turn, for facilitating ways to adequately protect juvenile life-history stages. Nursery areas may also provide opportunities for monitoring juvenile recruitment to complement other indices of population status and provide additional clues to the location of pupping grounds in Australian waters.

MATERIALS AND METHODS

Life-History Stage Definitions

We use the following life-history definitions for White Sharks:

1. Young-of-the-year (YOY) sharks (≤ 1.75 m total length): Based on the size of full-term pups, the smallest free-swimming White Sharks, and estimated growth rates from the literature (Cailliet et al., 1985; Francis, 1996; Wintner and Cliff, 1999; Malcolm et al., 2001), we define White Sharks in their first year, young of the year, to be ≤1.75 m TL.

2. Juvenile White Sharks (JWS) (>1.75–3.0 m TL): Juvenile White Sharks are those with sizes greater than YOY sharks but prior to the smallest sizes commonly observed visiting pinniped colonies (>3.0 m). At these sizes, sharks feed primarily on teleosts and other elasmobranchs (Malcolm et al., 2001).
3. Subadult sharks (>3.0–3.6 m TL for males and >3.0–4.8 m TL for females). Male and female White Sharks mature to different sizes (Francis, 1996; Pratt, 1996). Subadult sharks are those between the size when first visiting pinniped colonies and the size on reaching maturity. Arrival at pinniped colonies defines a new predatory regime and ensuing predatory strategies in White Sharks via the addition of marine mammal to their fish diet (Martin et al., 2005). The smallest size commonly observed at Australian pinniped colonies is 3.0 m (Bruce et al., 2006).
4. Adult sharks (>3.6 m TL for males and >4.8 m TL for females). Adult White Sharks are those that have reached sexual maturity as defined by Pratt (1996) and Francis (1996). We use 4.8 m for maturity in females based on data for Australian sharks from Bruce (1992) and Malcolm et al. (2001).

Unless otherwise stated, all lengths refer to TL as defined by Mollet and Cailliet (1996). Other length measurements (e.g., fork length) have been converted to TL for comparison using equations derived by these authors. All of the lengths and depths are specified in meters (m), including depth units on all figures.

Study Sites

Juvenile White Sharks were tagged in two regions of eastern Australia: the Corner Inlet–Ninety Mile Beach region (referred to as "Corner Inlet" herein) in southeast Victoria and off the Port Stephens region in central NSW (Figure 17.1). Corner Inlet is a coastal lagoon-estuary system at the southern end of a 150-km southeast-facing beach. Port Stephens is a large estuarine system with a narrow and shallow entrance. At Port Stephens, tagging of sharks was undertaken along open coast beaches immediately to the north of the estuary (Hawks Nest) and to the south (Stockton Beach) (Figures 17.2–17.6). The area is popular for commercial and recreational fishing, as well as surfing, and Stockton Beach is backed by one of the largest continuous mobile sand-dune systems in eastern Australia. The southern end of Stockton Beach is meshed as part of the NSW shark control program (SCP) and has a history of taking JWS (Reid and Krogh, 1992). Hawks Nest, immediately north of the Port Stephens estuary, is part of a 44-km beach (including Mungo Beach) running north to the prominent headland of Seal Rocks and has a similar southeasterly aspect (Figure 17.6). Seal Rocks represents a significant change in trajectory of this section of the NSW coast from a southeasterly aspect to north-south and is the general location where the East Australian Current (EAC) separates from the coast into the Tasman Sea (Tranter et al., 1986).

Beaches and coastal reefs in each of these regions are seasonally frequented by schools of commercial and recreational finfish and other shark species, including Australian Salmon (*Arripis trutta*), Australasian Snapper (*Pagrus auratus*), Mulloway (*Argyrosomus hololepidotus*), Striped Mullet (*Mugil cephalus*), and Gummy Shark (*Mustelus antarcticus*), as well as various other elasmobranchs (e.g., Eagle Rays *Myliobatis australis*), all of which are common prey for JWS. Both areas have a history of White Shark interactions with commercial and recreational fisheries (Malcolm et al., 2001).

Tagging

Various combinations of satellite tracking tags (SIRTRACK KiwiSat 101, Wildlife Computers SPLASH), pop-up satellite archival tag (PSAT) tags (Wildlife Computers Mk10), and coded acoustic tags (Vemco-Amirix R64k-V16) were fitted to 22 White Sharks ranging in size from 1.75 to 2.6 m total length (Table 17.1). Two sharks (S1 and S2) were tagged in 2000 and 2001,

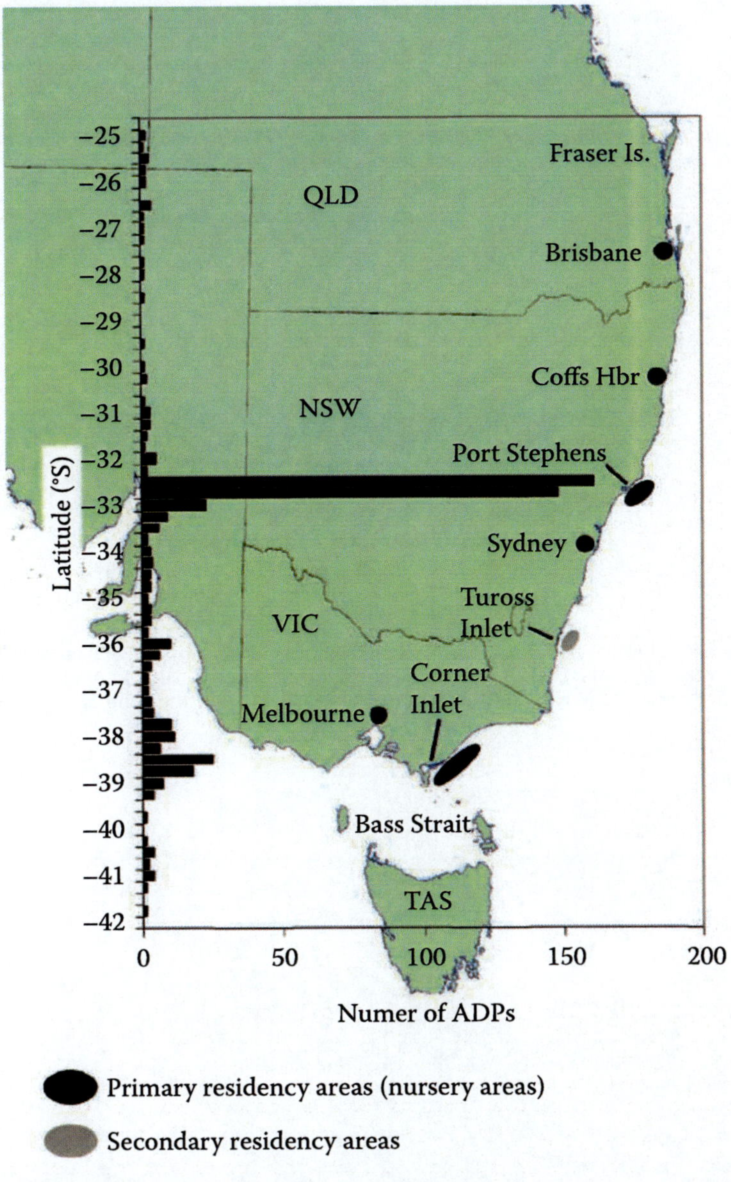

Figure 17.1 The latitudinal distribution of ADPs of 20 juvenile White Sharks tracked in eastern Australia. The data are binned by 0.25º latitude.

respectively, in the vicinity of Corner Inlet; 10 sharks (S5–S14) were tagged at Stockton Beach in October 2007; 2 (S15 and S16) at Hawks Nest Beach in October 2008, and 8 (S17–S24) were tagged at Hawks Nest in October 2009. Sharks S1 and S2 were captured on baited set lines and fitted with position only tags each utilizing a KiwiSat 101 PTT [see Bruce et al. (2006) for details]. Sharks S5–S23 were captured using a baited line presented to each shark after they were individually located from a vessel searching immediately seaward of the beach front. These nineteen sharks were fitted with Wildlife Computer SPLASH tags. Three of these sharks (S6, S9, and S17) and an additional shark (S24) were fitted with Wildlife Computer Mk10 PSAT tags (Table 17.1). Sharks S5–S14 were fitted with external acoustic tags (Vemco-Aimirix R64k-V16 coded tags);

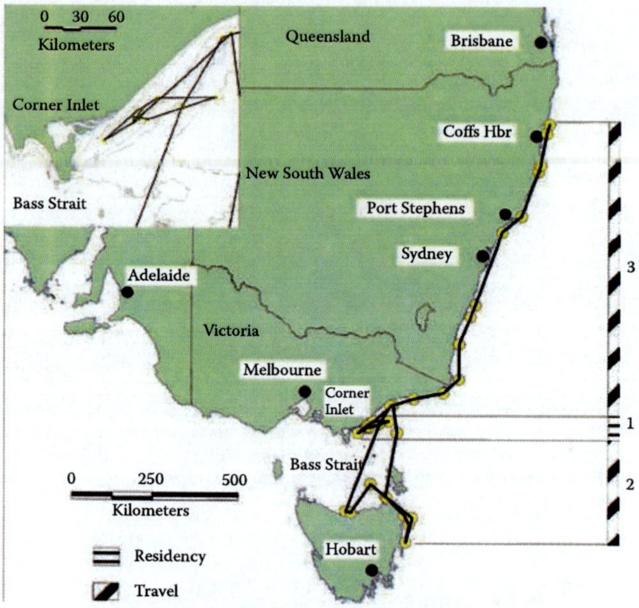

Figure 17.2 Patterns of travel and residency for shark S2 tagged at Corner Inlet based on the angular concentration (*r*) and RCM over its track. Period 1 was March 2, 2001 to April 20, 2001; period 2 was April 22, 2001 to May 8, 2001; and period 3 was May 9, 2001 to June 22, 2001.

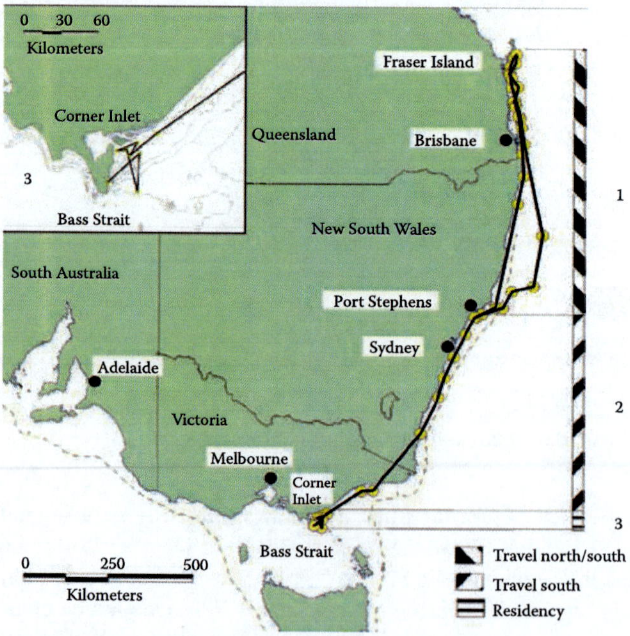

Figure 17.3 Patterns of travel and residency for shark S6 tagged at Port Stephens based on the angular concentration (*r*) and RCM over its track. Period 1 includes travel north to Fraser Island (October 12, 2007 to November 11, 2007) and return south via Port Stephens (November 12, 2007 to December 1, 2007); period 2 was December 2, 2007 to December 13, 2007; and period 3 was December 15, 2007 to January 17, 2008.

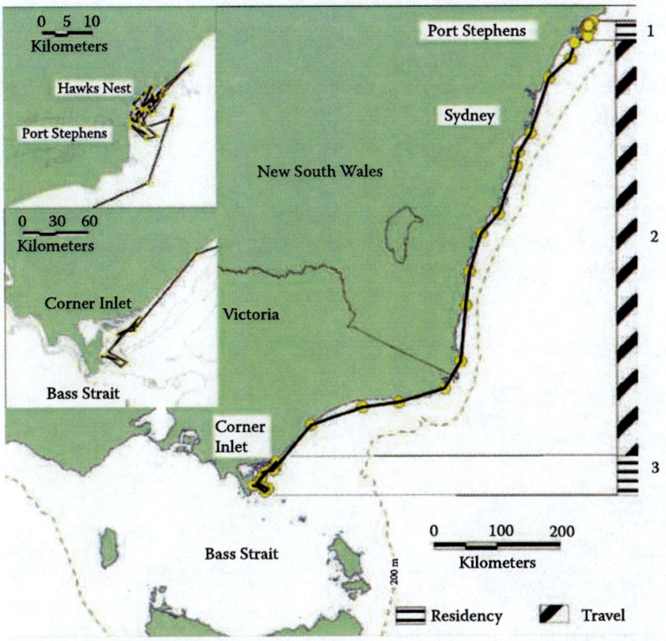

Figure 17.4 Patterns of travel and residency for shark S18 tagged at Port Stephens identified based on the angular concentration (*r*) and RCM over its track. Period 1 was October 28, 2009 to January 5, 2010; period 2 was January 6, 2010 to January 27, 2010; and period 3 was January 29, 2010 to April 2, 1010.

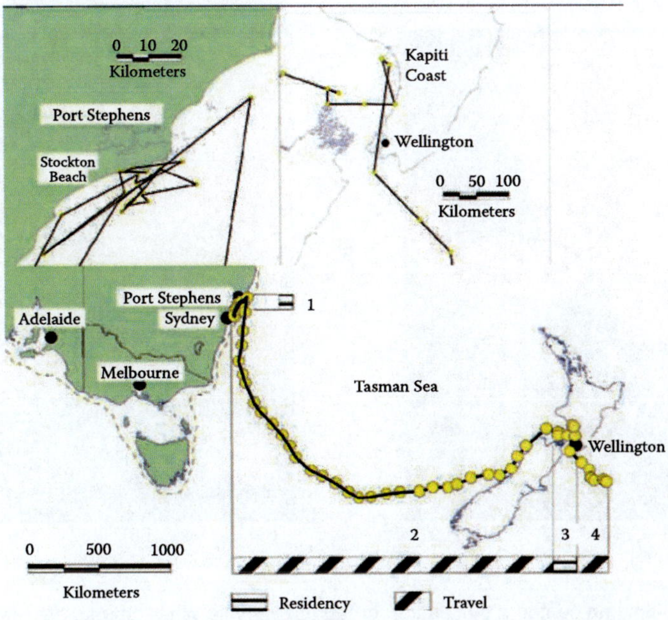

Figure 17.5 Patterns of travel and residency for shark S5 tagged at Port Stephens identified based on the angular concentration (*r*) and RCM over its track. Period 1 was October 9, 2007 to November 20, 2007); period 2 was December 1, 2007 to January 1, 2008; period 3 was January 3, 2008 to January 10, 2008; and period 4 was January 11, 2008 to January 20, 2008.

sharks S15–S24 were fitted with surgically implanted acoustic tags. Detections of acoustic tags are not reported herein.

Satellite tracking tags were positioned on the first dorsal fin of each shark and attached using the methods described by Bruce et al. (2006). SPLASH tags were programmed to collect and archive water temperature and swim depth data at 10-sec intervals, and these data were compressed into 6-h bins for transmission. PSAT tags were attached by 15-cm tethers of braided 200-kg stainless steel multistrand wire to a 32- × 8-mm stainless steel dart. The darts were inserted in the dorsal musculature below the trailing edge of the first dorsal fin such that the tags towed clear of the fin and body. Where possible, a secondary tag head attached to a nylon monofilament loop secured the

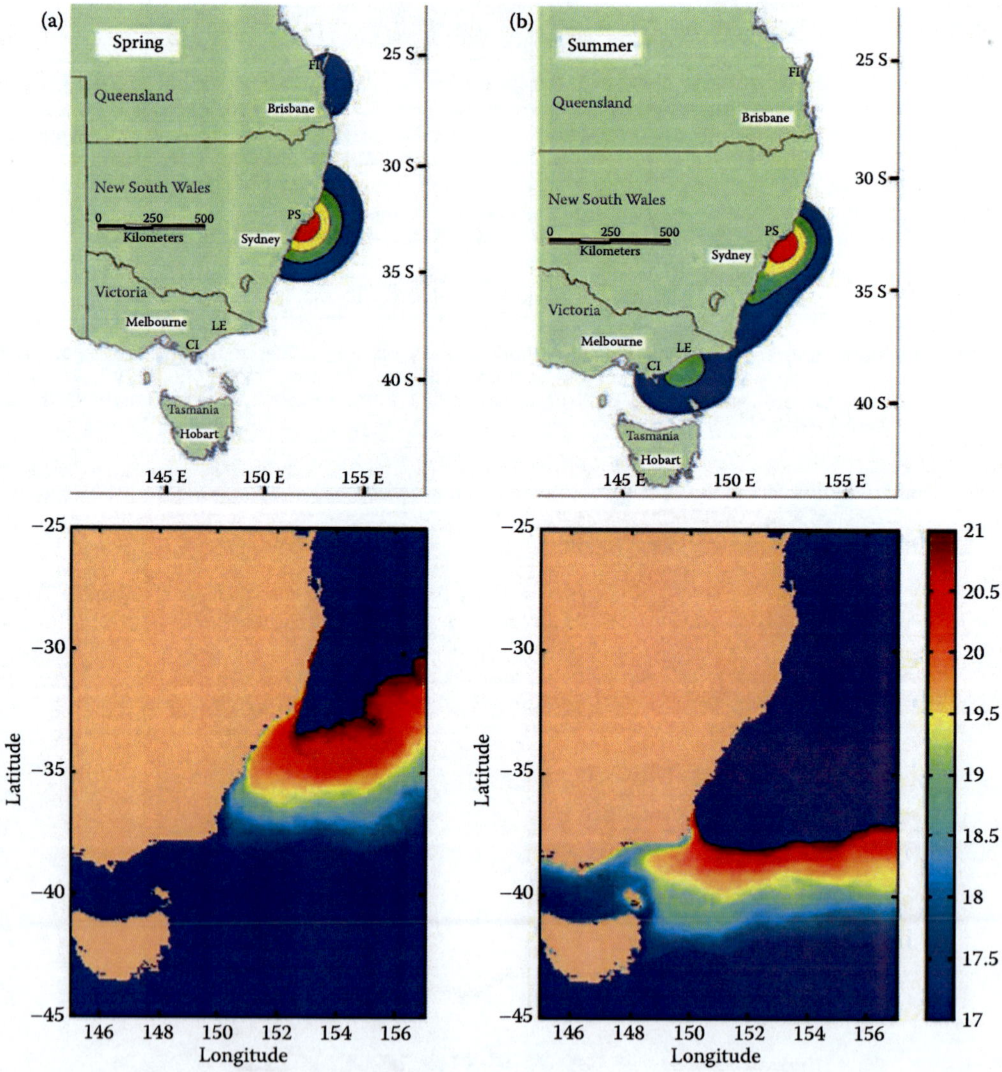

Figure 17.6 Kernel-volume contours by season for tagged juvenile White Sharks off eastern Australia. Red, 25%; yellow, 50%; green, 75%; blue, 95% of all average daily positions. Note that the winter data were restricted to one shark only following a consistent track north along the coast. The data are plotted as ADPs. Images below give the corresponding sea surface temperature climatology 2005–2007 showing the seasonal location of the 18–20°C temperature zone off eastern Australia. (a) Spring. (b) Summer. (c) Autumn. (d) Winter. The temperature scale is in °C.

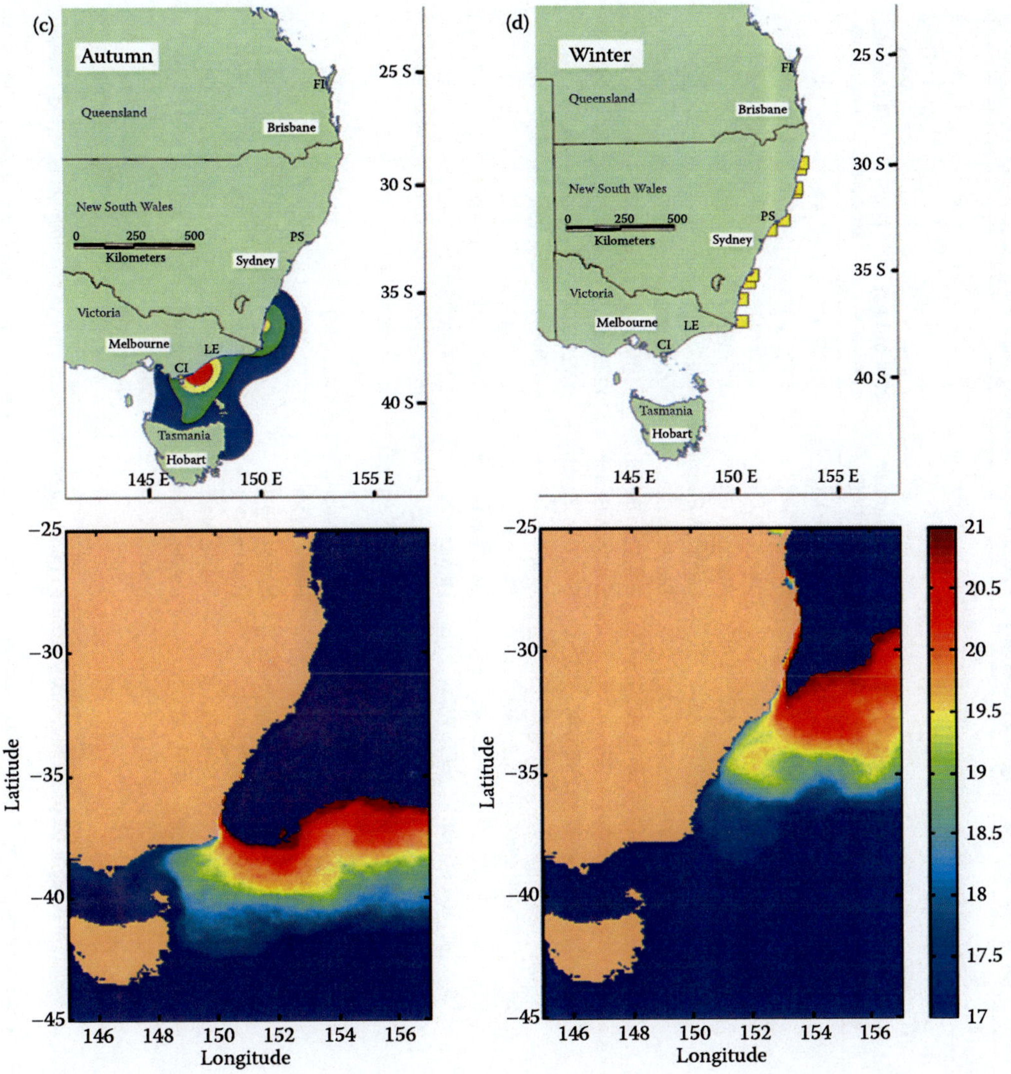

Figure 17.6 (Continued)

body of each PSAT tag in place so as to stabilize the tag on the shark. PSAT tags were programmed to archive data at 10-sec intervals and detach from the sharks after 90–120 d.

Data Handling and Processing

Transmissions from the satellite tracking tags were detected by the ARGOS data collection and location system, which is adequately described elsewhere (Eckert and Stewart, 2001). We followed the methods of Bruce et al. (2006) in accepting or rejecting positions based on ARGOS-derived location classes and a swim-speed filter. Positions registered on land were rejected from analyses. SPLASH and PSAT tag temperature and depth data were decoded via Wildlife Computers propriety software.

The seasonal distribution of sharks was identified based on the fixed-kernel method (Worton, 1989) using accepted ARGOS locations, and analyses were performed using Biotas 1.03.1

Table 17.1 Estimated Length, Sex, and Tagging Details for White Sharks Tagged with Archival and Satellite Tracking Tags

			Age Estimates (yrs.)								
Shark[a]	Sex	Length (m)	After Cailliet et al. (1985)	After Malcolm et al. (2001)	After Wintner and Cliff (1999)	Date Tagged	Tag Types[b]	Location of Tagging	Date of Last Position	Days Tracked	No. of Days with Accepted Positions
S1	F	1.8	1.1	2.2	1.8	3/1/2000	CS	Corner Inlet	4/18/2000	49	42
S2	M	2.4	3.0	4.0	4.6	3/2/2001	CS	Corner Inlet	6/23/2001	129	71
S5	M	2.1	2.0	3.1	3.1	10/8/2007	SPL, AC	Stockton Beach	1/20/2008	104	78
S6	F	1.9	1.4	2.5	2.2	10/8/2007	SPL, AC, PSAT	Stockton Beach	1/17/2008	98	40
S7	F	2.4	3.0	4	4.6	10/10/2007	SPL, AC	Stockton Beach	11/22/2007	44	19
S8	M	2.6	3.6	4.7	5.6	10/10/2007	SPL, AC	Stockton Beach	1/20/2008	102	46
S9	M	1.9	1.4	2.5	2.2	10/11/2007	SPL, AC, PSAT	Stockton Beach	4/1/2008	177	13
S10	F	2.5	3.3	4.4	5.1	10/11/2007	SPL, AC	Stockton Beach	12/30/2007	81	60
S11	F	2.0	1.7	2.8	2.7	10/11/2007	SPL, AC	Stockton Beach	11/2/2007	22	5
S12	F	2.0	1.7	2.8	2.7	10/11/2007	SPL, AC	Stockton Beach	12/29/2007	80	17
S13	M	2.1	2.0	3.1	3.1	10/11/2007	SPL, AC	Stockton Beach	11/1/2007	22	11
S14	F	2.4	3.0	4.0	4.6	10/11/2007	SPL, AC	Stockton Beach	11/22/2007	43	9
S15	F	1.75	1.0	2.0	1.6	10/28/2008	SPL, AC	Hawks Nest	6/10/2009	225	61
S16	F	2.3	2.6	3.7	4.1	10/28/2008	SPL, AC	Hawks Nest	12/30/2008	63	23
S17	M	2.0	1.7	2.8	2.7	10/28/2009	SPL, AC PSAT,	Hawks Nest	No positions transmitted	0	0
S18	M	2.1	2.0	3.1	3.1	28/10/2009	SPL, AC	Hawks Nest	4/2/2010	156	81
S19	F	2.4	3.0	4.0	4.6	10/28/2009	SPL, AC	Hawks Nest	3/16/2010	139	25
S20	M	2.3	2.6	3.7	4.1	10/29/2009	SPL, AC	Hawks Nest	1/17/2010	80	24
S21	F	2.1	2.0	3.1	3.1	10/30/2009	SPL, AC	Hawks Nest	1/3/2010	65	13
S22	F	2.4	3.0	4.0	4.6	10/30/2009	SPL, AC	Hawks Nest	4/22/2010	174	9
S23	F	2.2	2.3	3.4	3.6	10/30/2009	SPL, AC	Hawks Nest	1/26/2010	88	15
S24	F	2.3	2.6	3.7	4.1	10/28/2009	PSAT, AC	Hawks Nest	Wash up on Ninety Mile Beach		

[a] Shark numbers are consecutive based on sharks tagged as part of a larger program than solely reported here. Sharks S3 and S4 are larger animals reported on elsewhere [see Bruce et al. (2006) for details].

[b] SPL, Wildlife Computers SPLASH tag; AC, Vemco/Amirix coded acoustic tag; PSAT, Wildlife Computers Mk 10 pop-off archival tag; CS, CSIRO designed position-only tag [see Bruce et al. (2006) for details].

(Ecological Software Solutions LLC, Hegymagas, Hungary). The results were displayed as volume contours (25%, 50%, 75%, and 95%) to identify core areas of distribution. To avoid biasing caused by days where sharks transmitted more frequently than on others, the position of the each shark was reduced to an average daily position (ADP) for these analyses. ADP was calculated by averaging all accepted ARGOS-derived positions for a shark received on a single calendar day (0000–2359 local time). Transmissions for individual sharks were not received for each day of tracking. Thus some data gaps occurred in all tracks. Such gaps can create biases in kernel-density plots, particularly if they occur in a systematic fashion where, for example, sharks do not surface and transmit their location while undertaking a specific behavior. Various methods have previously been used to interpolate between transmitted positions to reconstruct tracks and reduce such biases (Tremblay et al., 2006). However, interpolation can create its own biases (Aebischer et al., 1993), particularly if the gaps between positions are large and thus does not always provide an effective solution. In this study, because there were no observed systematic patterns in when gaps in data occurred, and large gaps were relatively few compared to consecutive days of data, we have chosen to use ADPs without reconstructing missing data via interpolation for the analysis of sharks' seasonal distribution.

Tracks of individual sharks were compared for similarity in their rate, timing, linearity, and geographic scale of movement to provide insight into habitat use. Rate of movement (ROM) was calculated by dividing the distance between successive accepted ARGOS positions by the sampling interval between positions. Angular concentration (*r*) was calculated for track segments using the Biotas 1.03.1 (ESS) movement module. Angular concentration is a relative measure of linearity in movement over a series of chronological points on a scale of 0 to 1, with 1 representing a perfectly linear trajectory (Batschelet, 1981). We used Rayleigh's *z*-test, following the methods of Cartamil and Lowe (2004), to test for significant linearity in movement. We combined these results with the scale of geographic movement to identify areas of residency and periods of traveling. Significant linearity in the trajectory of a track over multiple days may occur during traveling when a shark is consistently moving along a path of similar bearings. However, significant linearity may also occur when intermittent ARGOS positions happen to fall in a linear pattern despite the shark following a highly irregular and/or a geographically confined track (e.g., Shark 1; Table 17.2). Thus it is important to include both the spatial and temporal scale of movement over the tracking period when interpreting values of *r* as indicators of traveling and residency behavior. The spatial and temporal scale of track sections was assessed by defining a measure of rate of coastal movement (RCM), which is defined as follows:

$$\text{RCM} = \text{LCD}/\text{d} \qquad (17.1)$$

Linear coastal distance (LCD) was the minimum straight line distance parallel to the coast that bounded all ARGOS positions over the number of days for a defined period of an individual shark's track. RCM values were pooled for track sections that showed significant linearity in their *r* values and those with nonsignificant *r* values, respectively, and the two groups were compared using a Mann-Whitney U test.

RESULTS

Transmissions were received from all twenty-one of the satellite tracking tags over periods ranging from 22 to 225 d (Table 17.1). However, one tag only transmitted a very limited data set and failed to register positions via the ARGOS system, although a number of location class "z" transmissions were registered. The remaining 20 satellite tracking tags provided a cumulative period of 1,941 tracking days during which transmissions were received on 662 separate days. Gaps between transmitted (and accepted) positions ranged from a single day to 116 d (mean = 3.12; SD = 8.47)

Table 17.2 Track Details for 20 Juvenile White Sharks off Eastern Australia

Shark	Track Period	Number of Days	*r*	*p*	LCD	RCM	Behavior Mode	Description of Track
S1	3/1/2000 to 3/22/2000	21	**0.667**	**0.008**	65	3.1	R[a]	Off Corner Inlet
	3/24/2000 to 4/7/2000	14	**0.773**	**<0.001**	430	30.7	T	Travel north to Tuross Inlet
	4/8/2000 to 4/15/2000	7	0.227	0.697	18	2.6	R	Off Tuross Inlet
S2	3/2/2001 to 4/20/2001	49	0.23	0.558	60	1.2	R	Off Corner Inlet
	4/22/2001 to 5/8/2001	16	**0.890**	**0.042**	420	26.3	T	Travel south to NE Tasmania
	5/9/2001 to 6/22/2001	42	**0.664**	**<0.001**	1480	35.2	T	Travel north to northern NSW
S5	10/9/2007 to 11/20/2007	42	0.138	0.658	43	1.0	R	Off Port Stephens
	12/1/2007 to 1/1/2008	31	**0.752**	**<0.001**	2010	64.8	T	Travel to New Zealand
	1/3/2008 to 1/10/2008	7	0.514	0.267	110	15.7	R	In Cook Strait
	1/11/2008 to 1/20/2008	9	**0.719**	**0.01**	370	41.1	T	Travel east to Chatham Island
S6	10/12/2007 to 11/11/2007	30	**0.834**	**<0.001**	880	29.3	T	Travel north to Fraser Island
	11/12/2007 to 12/13/2007	31	**0.842**	**<0.001**	1800	58.1	T	Travel south to Corner Inlet
	12/15/2007 to 1/17/2008	33	0.048	0.993	40	1.2	R	Off Corner Inlet
S7	10/10/2007 to 11/22/2007	43	0.013	0.998	28	0.7	R	Off Port Stephens
S8	10/11/2007 to 12/15/2007	65	0.093	0.813	41	0.6	R	Off Port Stephens
	12/17/2007 to 12/12/2007	8	**0.983**	**0.021**	720	90.0	T	Travel south to Corner Inlet
	12/26/2007 to 1/17/2008	22	0.282	0.384	98	4.5	R	Off Corner Inlet
S9	10/11/2007 to 1/25/2008	106	0.054	0.98	43	0.4	R	Off Port Stephens
S10	10/11/2007 to 10/18/2007	7	**0.804**	**0.11**	250	35.7	T	Travel north to South West Rocks
	10/19/2007 to 10/24/2007	5	**0.92**	**0.015**	250	50.0	T	Travel south West Rocks to Port Stephens
	10/25/2007 to 12/15/2007	51	0.088	0.779	33	0.6	R	Off Port Stephens
	12/18/2007 to 12/24/2007	6	0.347	0.486	243	40.5	T	Travel to south to Jervis Bay
	12/25/2007 to 12/30/2007	5	**0.994**	**0.002**	250	50.0	T	Travel north Jervis Bay to Port Stephens
S11	10/12/2007 to 10/28/2007	16	0.55	0.298	30	1.9	R	Off Port Stephens
S12	10/12/2007 to 10/28/2007	16	0.2	0.618	22	1.4	R	Off Port Stephens
	12/19/2007 to 12/29/2007	10	—	—	—	—	—	Travel south to Corner Inlet

Table 17.2 (Continued) Track Details for 20 Juvenile White Sharks off Eastern Australia

Shark	Track Period	Number of Days	*r*	*p*	LCD	RCM	Behavior Mode	Description of Track
S13	10/11/2007 to 10/30/2007	19	0.328	0.342	69	3.6	R	Off Port Stephens
S14	10/12/2007 to 10/29/2007	17	0.372	0.331	20	0.5	R	Off Port Stephens
S15	10/28/2008 to 12/12/2008	45	0.039	0.949	89	2.0	R	Off Port Stephens
	12/14/2008 to 12/29/2008	15	**0.932**	**<0.001**	820	54.7	T	Travel south to Corner Inlet
	5/16/09 to 6/10/2009	25	**0.751**	**0.006**	730	29.2	R	Travel north to Port Stephens
S16	10/28/2009 to 12/25/2009	58	0.083	0.871	20	0.3	R	Off Port Stephens
S18	10/28/2009 to 1/5/2010	69	0.058	0.863	18	0.3	R	Off Port Stephens
	1/6/2010 to 1/27/2010	21	**0.909**	**<0.001**	890	42.4	T	Travel south to Corner Inlet
	1/29/2010 to 4/2/2010	63	0.296	0.27	48	0.8	R	Off Corner Inlet
S19	11/1/2009 to 2/3/2010	94	0.128	0.821	15	0.2		Off Port Stephens
S20	10/29/2009 to 1/17/2010	80	0.264	0.216	56	0.7		Off Port Stephens
S21	10/30/2009 to 1/3/2010	65	0.227	0.568	13	0.2		Off Port Stephens
S22	10/29/2009 to 2/28/2010	122	0.103	0.938	135	1.1		Off Port Stephens
S23	11/7/2009 to 12/25/2009	48	0.182	0.938	29	0.6		Off Port Stephens
	1/21/2010 to 1/26/2010	5	—	—	—	—		Travel south to Corner Inlet

[a] Despite significant linearity, the low RCM over this 21-day section of the track suggests temporary residency. Examination of data identified intermittent ARGOS-based positions following a linear pattern in a geographically confined area.

r, angular concentration; *p*, probability statistic, Rayleigh's *z*-test. Bold text denotes values that differ significantly from random ($p < 0.05$) ; —, insufficient positions over track period to reliably assess angular concentration. LCD, linear coastal distance; RCM, rate of coastal movement (see definitions in text). Behavior mode *R* = residency; *T* = traveling.

with 63% of records consisting of data over consecutive days (no gaps). In addition to these data, three PSAT tags were returned, and their archived data were downloaded. On November 9, 2007, a PSAT tag was located on Noosa Beach, southern Queensland (26.3228°S, 153.0573°E) after prematurely detaching from shark S6. The PSAT tag from shark S24 was recovered from just north of Corner Inlet (38.1243°S, 147.5245°E) on February 25, 2009. The PSAT tags deployed on sharks S9 and S17 failed to transmit on, or after, the scheduled release date. However, the tag from shark S9 was found on Hawks Nest Beach (32.6879°S, 152.1925°E) and returned on June 1, 2009, 15 months after deployment.

Movement Patterns

Movements of JWS covered an extensive range from the site of tagging at Stockton Beach in NSW, north 850 km to Fraser Island in southern Queensland, south 1,100 km to northern Tasmania, and east 3,600 km to the Chatham Rise in New Zealand waters (Figures 17.2–17.5). Sharks tagged off Corner Inlet moved as far south as eastern Tasmania (400 km) and as far north as Coffs Harbour in NSW (1,200 km). In all, sharks ranged along 2,000 km of the east Australian coast from eastern Tasmania to southern Queensland and across the Tasman Sea to New Zealand.

All twenty sharks showed extended periods of residency where their tracks retained low values of angular concentration (mean $r = 0.22$; SD = 0.17), and movements were contained along restricted areas of the coast (mean LCD = 47.6; SD = 32.5). Most sharks also showed periods of traveling, where tracks showed significantly nonrandom r values approaching 1 (mean $r = 0.81$; SD = 0.16) and where movements covered distances hundreds to thousands of kilometers (mean LCD = 769.5; SD = 574.6) or both (Table 17.2). RCMs were significantly higher during periods of travel (median RCM = 41.1 km d^{-1}; SD 16.9) than periods of residency (median RCM = 0.9 km d^{-1}; SD = 3.2), Mann-Whitney U test, $p < 0.0001$. Eastern Australian residency sites were not randomly distributed but seasonally focused at two sites based on the distribution of ADPs and from fixed kernel density analyses (Figures 17.1 and 17.6). Sharks were resident in the Corner Inlet region of eastern Victoria from mid-summer through the autumn, and sharks were resident in the Port Stephens region of central NSW from spring to early- to mid-summer. Periods of residency in these two sites ranged from 21 to 122 d (mean = 51.5; SD = 30.4). These were the only areas in eastern Australia where multiple JWS showed residency behavior, and we termed these "primary residency regions." Shark S1 remained resident for a period of 7 d in the vicinity of Tuross Inlet, southern NSW (36.1023°S, 150.1987°E) after traveling from the Corner Inlet–Lakes Entrance primary residency region. No other tracked sharks recorded similar behavior in this area, although several traveled past. We termed this a "secondary residency region." Shark S6 spent a period of 10 d in the vicinity of Fraser Island southern Queensland (25.2189°S, 151.3133°E), denoted by a period of reduced RCM (10.7), which also suggests that this as an area of residency. However, the path described by ARGOS positions was highly linear and appeared to comprise a slow patrol north and then south along the eastern side of Fraser Island over this period before the shark resumed a more rapid, consistent, and linear path south, returning past Port Stephens to Corner Inlet.

The date of departure of sharks from the Port Stephens region varied from October 16 (within a week of tagging) to January 25 (Figure 17.7). However, all departures prior to December 1 involved northward travel of 100–850 km. In each of these northerly movements, sharks subsequently

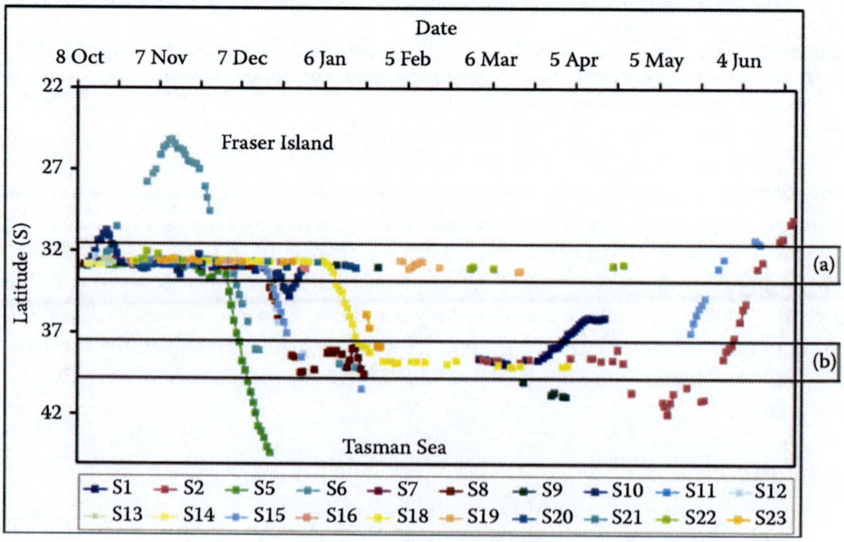

Figure 17.7 Timing of residency and movement for juvenile White Sharks along the easy coast of Australia. (a) Latitude of the Port Stephens residency region. (b) Latitude of the Corner Inlet residency region.

returned to the Port Stephens region by December 1. Sharks departing the Port Stephens region after December 1, with the exception of shark S5, which crossed the Tasman Sea to New Zealand, traveled consistently south. Eight sharks were tracked from the Port Stephens area south approximately 850 kilometers to Bass Strait. Four of these sharks (S6, S8, S15, and S18) recorded extended periods of residency at the second primary residency region off Corner Inlet. These were coincident with the residency areas registered by sharks (S1 and S2) tagged in the Corner Inlet region in 2000 and 2001. Two additional sharks (S12 and S23) were recorded traveling to the Corner Inlet region from Port Stephens but failed to register sufficient positions to establish whether they spent extended periods in region. One shark (S9) registered positions in eastern Bass Strait after a gap in transmissions of 61 d, passing Flinders Island and traveling to northern Tasmanian waters. The PSAT tag from shark (S24), also tagged at Port Stephens, was recovered from the Corner Inlet–Lakes Entrance area after washing ashore within a few days of releasing, indicating the shark's presence in the same area. When traveling outside of the Port Stephens and Corner Inlet residency areas, sharks showed a propensity for movement over a corridor spanning the 60–120-m depth contours, with 55% of all accepted ARGOS positions within this zone (Figure 17.8).

Not all sharks tagged in the Port Stephens region were recorded moving away from that area. Ten sharks recorded their last positions from the Port Stephens region. In eight of these cases, track duration was short with the final position registered between November 1 and January 17. Thus movement away or continued residency in the region could not be confirmed. However, two sharks (S19 and S22) continued to transmit positions from the Port Stephens region until their last recorded positions on March 16 and April 22, respectively, suggesting continuous residency in the region for up to 5–6 months, although gaps in transmissions of up to 43 d precluded confirmation.

While resident in the Port Stephens region, several sharks showed fidelity to particular areas of within the geographic footprint of the nursery area (Figure 17.9). This was marked by a high degree of consistency in registered positions (ARGOS Location Classes 1-3) for individual sharks either along or immediately offshore from the northern half of Stockton Beach, along the beach off Hawks Nest, or off Mungo Beach. In some cases, individual sharks returned to the same beaches after temporary departures. Temporary departures ranged from 2 to 18 d and included sharks traveling between the nearshore zone and the mid- to outer shelf and upper slope waters or coastal excursions over return longshore distances of 60–480 km.

Despite some gaps between recorded positions, the time periods between the date of departure from the Port Stephens area and arrival at the Corner Inlet region generally indicated rates of

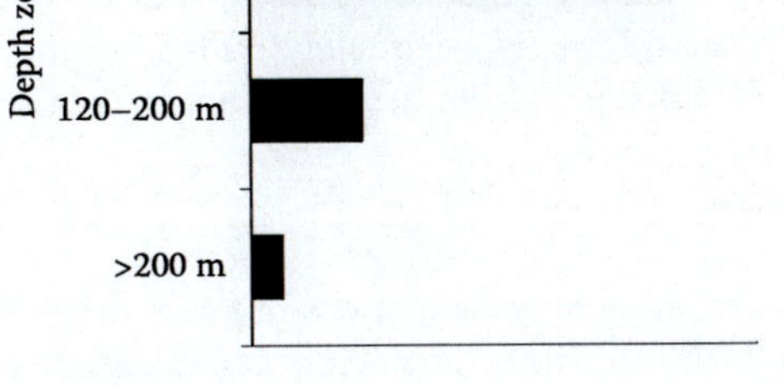

Figure 17.8 The location of ARGOS-based positions over cross-shelf depth zones during periods of travel by juvenile White Sharks between the Port Stephens and Corner Inlet nursery areas.

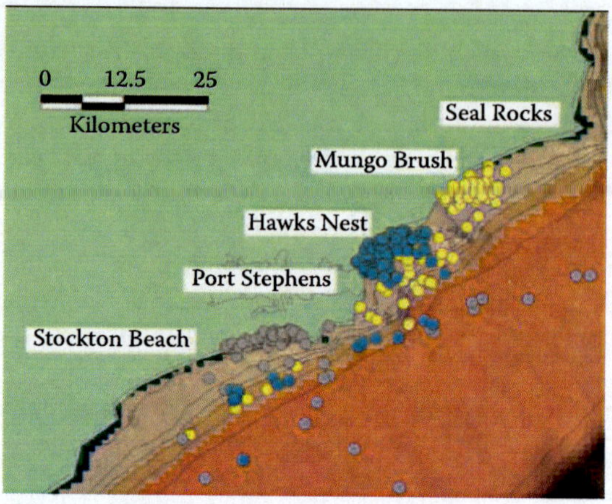

Figure 17.9 Registered positions (ARGOS location class 1, 2, or 3) for three sharks (S10, gray; S15, blue; and S18, yellow) showing site-area fidelity to three separate beaches in Port Stephens region of central NSW.

movement within the range previously reported for White Sharks (2–4 km/h), suggesting direct travel between these regions with little evidence of intervening residency periods.

The two sharks (S1 and S2) tagged in the Corner Inlet region remained resident for periods of 24 and 52 d, departing this region on March 24 and April 22, respectively. Shark S1 headed steadily north to a secondary residency site off Tuross Inlet on the southern NSW coast before transmissions ceased; shark S2 initially traveled south to northern and eastern Tasmania before passing back via the Corner Inlet area on June 11 and then tracking 1,100 km to northern NSW where transmissions ceased on June 22. Shark S15 recommenced transmitted positions from southern NSW on May 16 after being recorded off Corner Inlet on January 19. This shark also progressed steadily north, passing the Port Stephens region on May 26 and then tracking 150 km further north to the Port Macquarie region on June 10, where transmissions ceased.

Tasman Sea Crossing

Movements and areas where sharks spent periods of temporary residency were not confined to Australian waters. Shark S5 departed the continental shelf off the Port Stephens area on December 1, moving steadily southeast to 44.3°S, 160.12°E by December 17, approximately 1,000 km east of southern Tasmania, before changing trajectory east and crossing onto the New Zealand shelf on December 28. The mean ROM for this crossing was 3.61 km/h (SD = 0.80) and thus consistent with other reported ROM for this species. Shark S5 then tracked north along the west coast of New Zealand's South Island and entered Cook Strait on January 1, 2008. It then headed east where it spent 7 d in the vicinity of the Kapiti Coast (40.127°S, 175.129°E) northwest of Wellington. The shark departed this area on January 11, heading east, out of Cook Strait where it subsequently took up residency on the Chatham Rise approximately 300–340 km southeast of Wellington and approximately 3,600 km tracked distance from the Port Stephens region (Figure 17.5).

Depth Swimming Behavior

Overall, sharks showed a strongly bimodal pattern of swim depth with peaks in activity at the surface (0–5 m) and in the 60–100-m depth zone. Time at the surface dominated in all individuals (44% of overall time at depth) with 92% of records confined to depths less than 100 m. This pattern was similar between periods of traveling and residency and between day and night (Figures 17.10

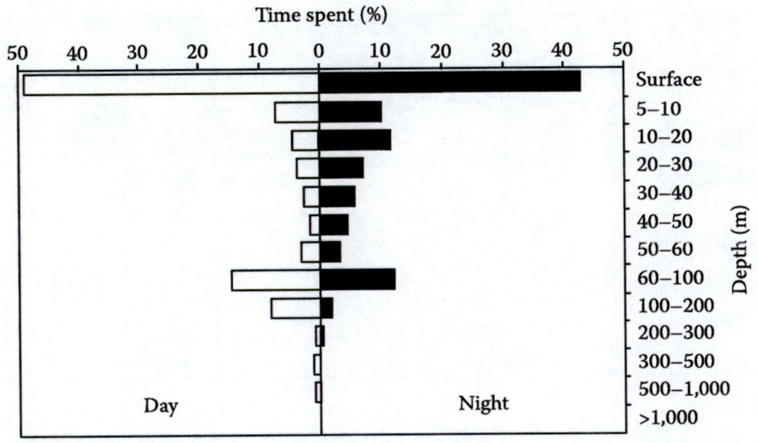

Figure 17.10 Time spent at depth during day and night periods based on SPLASH tag data (all juvenile White Sharks combined).

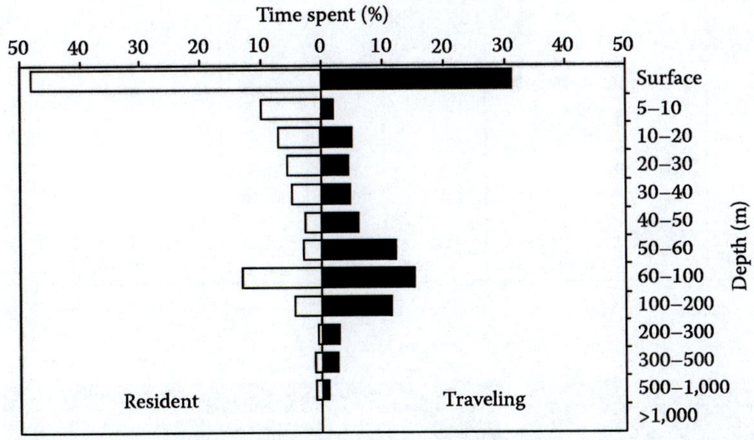

Figure 17.11 Time spent at depth during residency and traveling periods based on SPLASH tag data (all juvenile White Sharks combined).

and 17.11). There was, however, a tendency for sharks to occupy shallower depths during the night, particularly when seaward of the continental shelf. Dive behavior over individual days was highly variable, even when sharks were consistently traveling along the coast or resident in either the Port Stephens or Corner Inlet regions. Consequently, combined analyses of behavior for all sharks, or for the same individual, over the period of a track were ambiguous. When resident, sharks showed either prolonged periods of time in the 0–10-m zone, bimodal behavior in the surface and 60–100-m zones or, on occasions, evidence of cross-shelf and upper-slope forays to 300–500 m (Figure 17.12a). When traveling, sharks showed the same surface and 60–100-m bimodal pattern as well as slope forays, again to 300–500 m (Figure 17.12b). Shark S5 provided the only prolonged exception to depth pattern during the 28-day period of traveling across the Tasman Sea. During this period, S5 showed a highly bimodal pattern with activity peaks at the surface and at 300–500 m (Figure 17.12c).

The retrieval of the PSAT tags from sharks S6 and S24 provided high resolution data for these sharks' tracks (Figure 17.13). ARGOS-derived positions from the SPLASH tag on shark S6 were used to extract bottom depths from bathymetric charts for comparison with the archived PSAT depth data. PSAT data confirmed the rapid switching between different swimming behaviors during

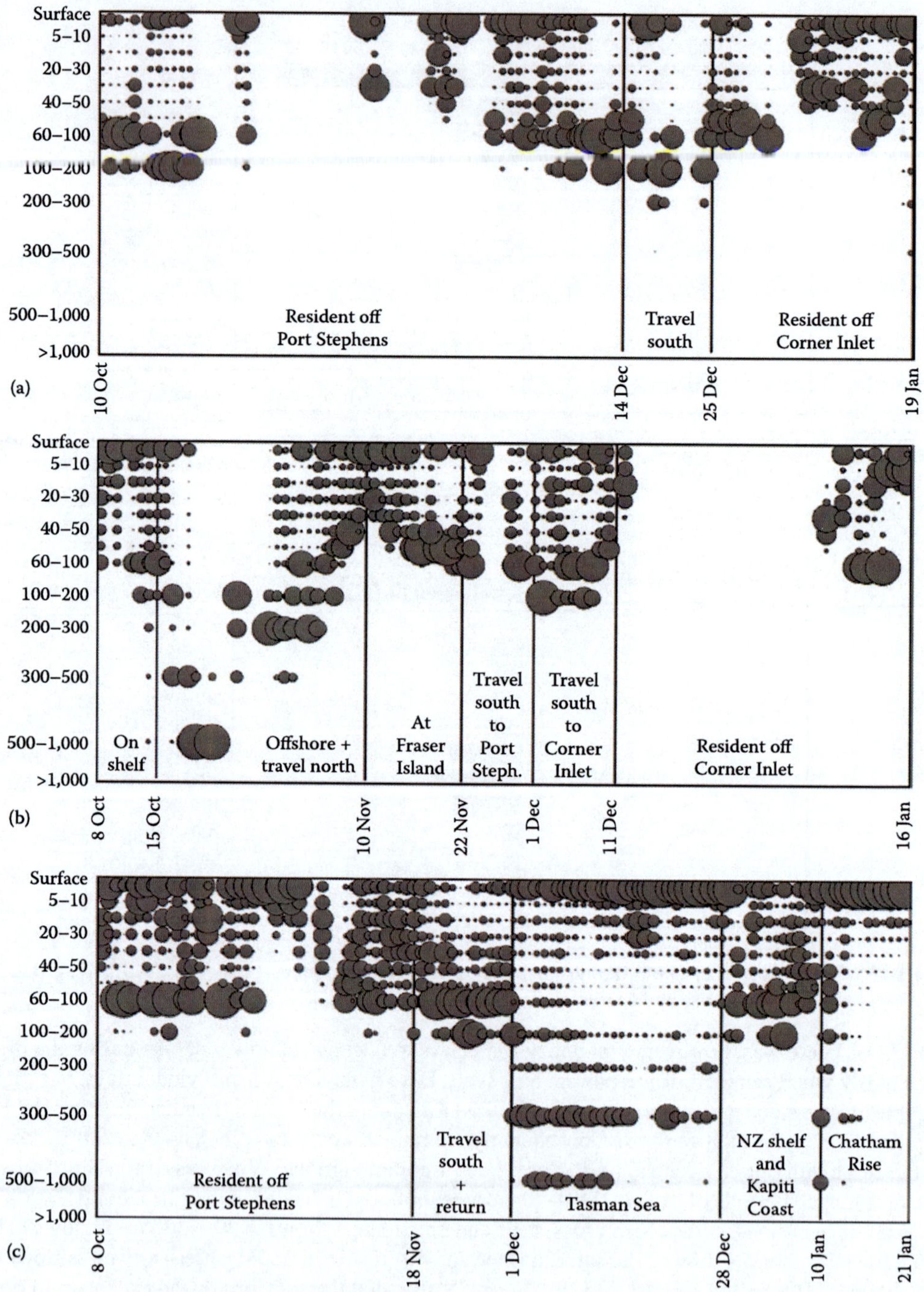

Figure 17.12 Daily depth summaries. Circles represent daily percentage time at depth. (a) Shark S8 showing periods of bimodal behavior between the surface and 60–100 or 100–200 m. (b) Shark S6 showing periods of off-shelf diving activity and bimodal behavior between the surface and 60–100 or 100–200 m. (c) S5 showing bimodal behavior between the surface and 60–100 m in shelf waters, as well as deep-diving and bimodal behavior between the surface and 300–1,000-m depths during its Tasman Sea crossing. Circles represent daily percentage time at depth.

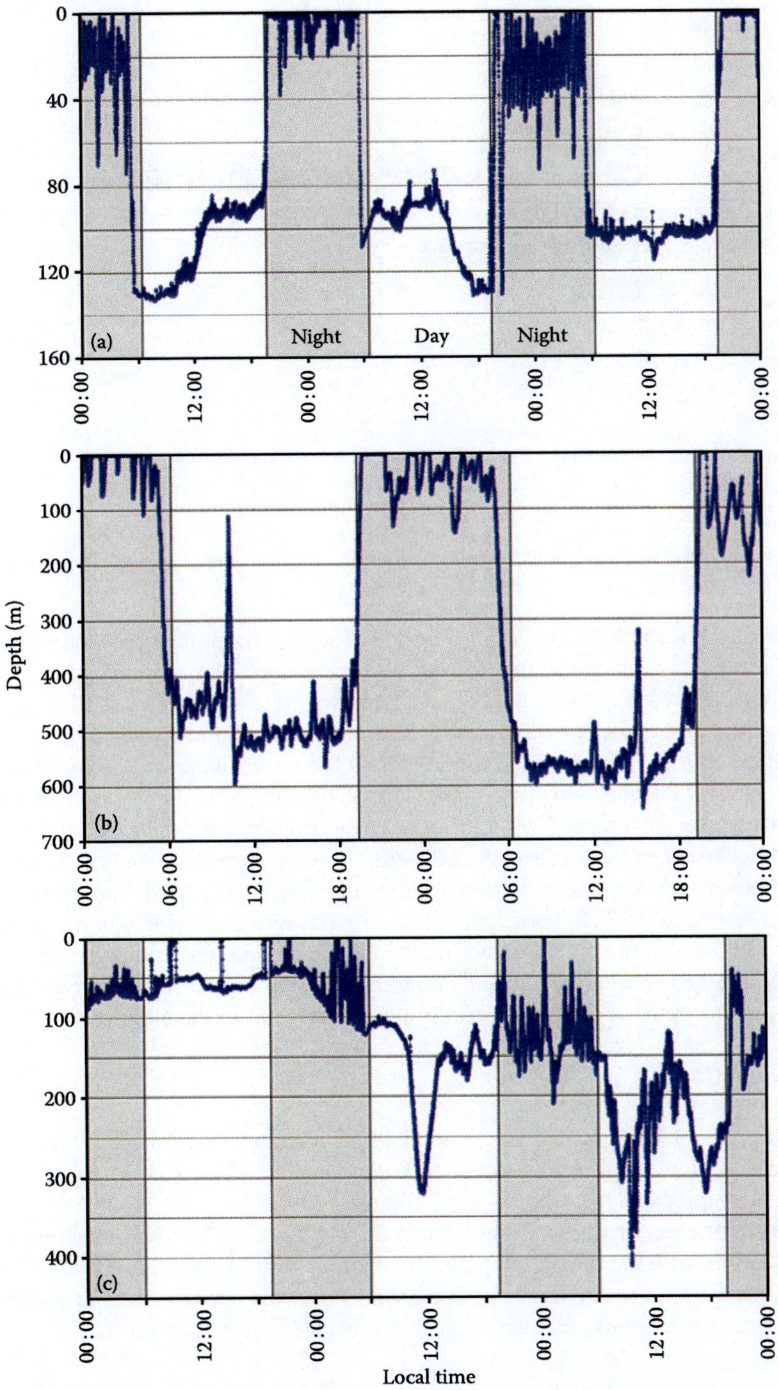

Figure 17.13 Swimming behavior of shark S6. (a) Archival record of diel behavior when on shelf in vicinity of Stockton Beach. ARGOS positions suggest that the shark was moving steadily north in the vicinity of the 100-m contour. (b) Archival record of diel-dive behavior when seaward of the continental shelf off central NSW. The data are consistent with the shark swimming at depths that maintained consistent light levels. (c) Archival record showing a lack of pattern of diel diving behavior during northwards travel over shelf and shelf edge waters off northern NSW.

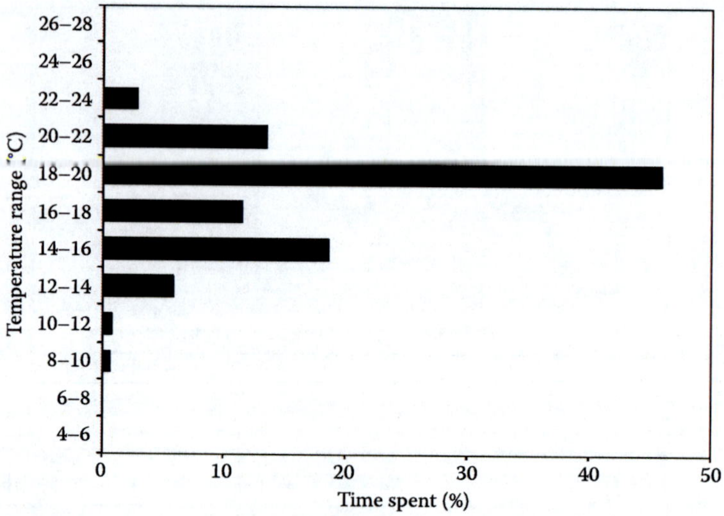

Figure 17.14 Time spent at temperature based on SPLASH tag data (all juvenile White Sharks combined).

the northward travel of shark S6, including a period of several days spent seaward of the continental shelf where the shark made regular dives to depths exceeding 500 m and on one occasion recorded a maximum depth of 984 m. Given that the tag's sampling interval was set at 10 s, the resolution of the tag's depth sensor was rated at ±0.5 m to a depth of 1,000 m, and consecutive records spanning 1 min. either side of this reading ranged from 978 to 981.5 m, the depth of 984 m is likely very close to the maximum depth achieved by this shark. This shark showed examples of highly diel-oriented behavior between the surface and 80–130 m when on the shelf near Port Stephens, highly diel-oriented behavior between the surface and 400–600 m when over slope and offshore waters north-east of Port Stephens, and periods where there was a complete lack of diel behavior over shelf and slope waters in northern NSW. Several behavioral subsets were also evident in these data, including prolonged periods swimming at the surface and along the bottom, dawn and dusk dives, oscillatory diving both within and through the surface mixed layer and the thermocline, rapid deep descents and ascents, and examples of extended periods at the surface after deep dives.

Temperature Data

SPLASH tags recorded temperatures in binned ranges from 6–8 to 24–26°C, with sharks spending the majority of time (89.4%) in temperatures ranging from 14–22°C. The most common temperature range occupied (45.9%) was 18–20°C (Figure 17.14).

The data from the returned PSAT tags (sharks S6 and S24) revealed similar overall temperature ranges (6.8–23.3°C), with the lowest temperature recorded during a dive in offshore waters off central NSW by shark S6 to 984 m. Temperatures below 14°C only occurred on days where the depths registered by sharks were consistent with dives seaward off the continental shelf.

DISCUSSION

Juvenile White Sharks off eastern Australia show broad-scale patterns of movement ranging from southern Queensland to Tasmanian waters. Our data also expand the niche for JWS to include offshore waters, deep-diving, and open-ocean travel similar to that recorded for subadult and adults

(Boustany et al., 2002; Weng et al., 2007a; Domeier and Nasby-Lucas, 2008, Bonfil et al., 2010). Broad-scale movements were interspersed with periods of temporary residency. However, residency sites were limited in both number and geographic extent, suggesting that JWS may have a relatively small number of interconnected preferred habitat areas in eastern Australia. Two geographically restricted, primary residency regions were identified where sharks spent extended periods of time and at which abundances, based on satellite tracking, were above that of the surrounding areas: one in central NSW (Port Stephens) and one, 850 kilometers south, in eastern Victoria (Corner Inlet). The data demonstrate that these areas were utilized repeatedly on a seasonal basis across different years and are thus consistent with the definition of shark nursery areas as applied by Heupel et al. (2007). Tagged JWS utilized the Port Stephens nursery area from spring to mid-summer and the Corner Inlet nursery area from summer through autumn. Supporting evidence for limited preferred habitat areas in eastern Australia is also apparent in data from the NSW SCP, where approximately 20% of all White Sharks captured from 51 beaches over an 18-yr. period (1990–2007) were recorded from the southern end of Stockton Beach just south of the Port Stephens nursery area (NSW Industry and Investment, unpublished data; Chapter 19, this book). These patterns of site-specific nearshore nursery areas combined with broadscale geographic movements, including occasional offhore excursions, are also consistent with the behavior of YOY and JWS off California and Mexico (Chapter 28, this book).

Travel between the Port Stephens and Corner Inlet nursery areas was directed and relatively rapid, with sharks showing a propensity for travel over depths between the 60- and 120-m contours. A marked change in bottom profile occurs in this depth zone, marking the change from a progressively sloping inner shelf to an expansive area of low relief substrate extending to the outer shelf. This depth zone shows evidence of relic coastline structures in eastern Australia, which may offer cues for navigation or enhanced foraging opportunities for JWS by providing focal areas for prey species.

The NSW nursery area was restricted to a 65-km section of the coast centered on the Port Stephens estuarine system and encompassing three beaches: the northern end of Stockton Beach, Hawks Nest, and off Mungo Brush. When resident, sharks often showed fidelity to individual beaches. Sharks that returned after periods of traveling along the coast generally did so to the same areas of the particular beach that they had left. However, both transmitted and archived depth data confirmed that even while resident, sharks made forays out to mid-shelf depths (to the 60–120-m depth zone) before returning to the same areas of the beach, a distance of 25–30 km. This further suggests that the 60–120-m depth zone provides sharks some cue for orientation or foraging opportunities. Site fidelity, based on satellite tracking data, is consistent with our in situ observations during field operations where sharks were often resighted in the same vicinity on separate observation runs along the beach. Such site-area fidelity in White Sharks has previously been associated with feeding or predatory success. Goldman and Anderson (1999) and Klimley and Anderson (1996) reported that White Sharks >4.0 m occupied smaller activity areas than sharks <4.0 m, which patrolled the islands more extensively. These authors considered that this behavior was the result of larger sharks having learned to occupy areas where they had been successful in hunting. Martin et al. (2005) similarly found differences in the distribution of sharks <3.5 m and >3.5 m at Seal Island, South Africa, and observed that sharks in the larger category were more successful in attacks. They suggested that these differences may be due to accumulated experience in older, larger sharks. No territorial behavior was observed in either of these studies that would suggest that larger White Sharks excluded smaller sharks from their preferred areas. An individual annual pattern of preferred areas has also been reported by Domeier and Nasby-Lucas (2008) for White Sharks visiting a shared offshore foraging area in the central eastern Pacific, although the areal scale was somewhat larger than the previous two studies. We similarly observed no evidence of agonistic displays or actions that would suggest territorial behavior in JWS in the Port Stephens region, despite observing sharks in close proximity to each other on some occasions. It would, however, be difficult to observe subtle displays, if they occurred, from above the water.

Behavioral modification through learning (conditioning) is well established in elasmobranchs in both captive and field situations (Gruber and Myrberg, 1977; Bres, 1993) and was suggested as the underlying mechanism for adaptive behavioral responses in foraging by Dill (1983). We hypothesize that JWS in the Port Stephens area may also adaptively respond to areas of individual feeding success. Australian Salmon aggregate in large schools and move along the surf zones of beaches in the Port Stephens region. Successful foraging in a particular area (e.g., on schools of salmon) may encourage a shark to occupy and return to that particular section of the beach during its overall period of residency. However, only on one occasion was a shark sighted with a school of salmon, despite our estimate of sighting 60 individual JWS in surveys in 3 consecutive years. On that occasion, other sharks were present within 500–1000 m north and south of the same school. In all other cases, there were no prey sighted near to a patrolling shark, and on one occasion schools of salmon were sighted with no attending sharks despite sharks being present along the same section of the beach. These observations suggest that sharks may not actively follow schools of salmon but prey on passing schools while maintaining their preferred areas of the beach. Such a strategy may serve to maximize predatory efficiency and minimize energy expenditure for sharks where prey is aggregated into schools that are not uniformly distributed but follow predictable pathways along the beach front. Sharks cycling between the surf zone and the 60–120-m depth region may also assist in reducing competition between individuals in the absence of territorial behavior, as well as providing an opportunity to search for prey in deeper coastal waters seaward of the beach.

The Port Stephens region is a populated area and a noted tourist destination. Site-area fidelity in areas of human activity poses an increased risk of multiple encounters with the same individual shark. Increased encounter frequency may either lead to habituation where sharks ignore swimmers or may increase the risk of an encounter, resulting in an investigatory response by the shark. White sharks of the size tagged in this study have been implicated in approximately 40% of all White Shark attacks in Australian waters since 1990 (Chapter 31, this book). However, despite the high abundance of juvenile White Sharks in the Port Stephens region, there have been no recorded attacks leading to injury at the three beaches involved, despite numerous reported encounters (J. West Taronga Zoo, Australian Shark Attack File, Surf Life Saving Australia, personal communication). Similarly, juvenile White Shark hotspots have been identified in the northeastern Pacific (Chapters 14, 15, and 16, this book), and these sites have not been implicated as high-risk areas relative to shark attacks. This suggests that the presence of juvenile White Sharks alone is a poor indicator of attack risk.

The Corner Inlet nursery area is greater in geographic scale extending along approximately 200 km of coast from Wilsons Promontory to Lakes Entrance and to a distance of 25–30 km from the coast. Sharks ranged over a wider area in this region, with little sign of the highly restricted spatial scale of behavior shown at the Port Stephens nursery area. Activity was focused along the 20-m contour, which marks a suite of low-profile reef systems in the area where finfish (e.g., Snapper, *P. auratus*) aggregate during the summer autumn period, and it is likely that sharks are focusing on this species. Commercial fishers targeting snapper in the area take juvenile White Sharks as bycatch (Malcolm et al., 2001).

Open-Ocean Behavior

Although subadult and adult White Sharks have been recorded undertaking open-ocean forays and cross-ocean basin excursions (Boustany et al., 2002; Bonfil et al., 2005; Weng et al., 2007a; Domeier and Nasby-Lucas, 2008; Bonfil et al., 2010; Chapters 11, 16, and 21, this book), JWS have previously been considered to be coastal life-history stages. Our data now confirm that White Sharks 1.9–2.1 m similarly make offshore forays and open-ocean excursions. The trans-Tasman Sea crossing by a 2.1-m male White Shark was made at an average ROM of 3.61 km/h, which is consistent with other ROM values recorded during this study for sharks traveling between nursery areas and is consistent with ROM values recorded for White Sharks in other studies (Klimley et al., 2002; Strong et al., 1996; Bruce et al., 2006; Domeier and Nasby-Lucas, 2008). The consistency of this shark's progress suggests that it did not stop in any area during the crossing. This, combined with the corridor of coastal movements by the other sharks tracked in this study, provides further evidence for the hot spot highway style of movement recorded for adult and subadult White Sharks in Australian waters by Bruce et al. (2006) and across other areas of the species' range (Bonfil et al., 2005; Weng et al., 2007a; Domeier and Nasby-Lucas, 2008; Chapters 14, 15, and 16, this book). Overall, these data support the connectivity between White Shark habitats and populations both in eastern Australian waters and across the Tasman Sea to New Zealand as suggested in genetic analyses (Pardini et al., 2001) and conventional tagging data (Malcolm et al., 2001).

Dive Behavior

Our data expand the recognized habitat niche for JWS to include slope and offshore waters, including deep-diving behavior over a similar depth range to that reported for subadult and adult White Sharks (Boustany et al., 2002; Domeier and Nasby-Lucas, 2008; Bonfil et al., 2010; Chapter 20, this book). Similar slope-water and offshore excursions of JWS recorded off California (Chapter 28, this book) suggest that these habitats may be frequented by these life-history stages across their range. The depths reported here are the deepest that JWS have been reported diving to. Previously reported dive depths for JWS range from 240 to 400 m (Dewar et al., 2004; Weng et al., 2007b; Chapter 28, this book). Four of the sharks tracked during this study made regular slope and off-shelf excursions, particularly during periods of travel, recording dives to a maximum of 984 m. Juvenile White Sharks have also been recorded by baited underwater video units in depths of 300 m in southern Australia (CSIRO, unpublished data).

Unlike previous studies on YOY and JWS (Dewar et al., 2004; Weng et al., 2007b), combined analyses of diel-depth preferences during this study failed to identify significant overall patterns. However, individual tracks include examples of short periods (days) of common dive behavior, including highly diel-oriented patterns (e.g., surface at night, bottom-oriented behavior during the day, and the diel targeting of constant ambient-light levels), multiple short duration dives, extended periods at the surface, extended periods in shallow (nearshore) depths, nondiel behaviors, dawn-dusk dives, and extended bottom-oriented swimming behavior. These behaviors were not well correlated with location, and rapid switching of behavior from one mode to another was observed even when sharks were in the same location. Such plasticity makes descriptors of behavior that are integrated over the duration of a lengthy track ambiguous, and these results, as well as similar results for subadult and adult sharks in Australian waters (Bruce et al., 2006), suggest that some caution should be exercised in drawing conclusions if this is attempted.

Several authors have previously interpreted different swimming behaviors as evidence of foraging. Such behaviors include bottom-oriented swimming, vertical excursions during dusk or dawn, and the diel targeting of constant ambient-light levels (Dewar et al., 2004; Weng et al., 2007b; Patterson et al., 2008; Chapter 28, this book). Examples of these behaviors were more common during periods of travel than during periods of residency for JWS off eastern Australia. This seems counterintuitive from a foraging perspective because presumably food availability and hence foraging is more likely to drive patterns of temporary residency. The difficulty with such assumptions is that there are still few data on what sharks actually do during any of the observed periods of different swimming behavior. It is unlikely, for example, that sharks continually forage for prey when within the nursery area, and likewise it is unlikely that sharks continually travel without periods of foraging when moving between nursery areas. Foraging periods in both cases may only represent a small percentage of observed behaviors. It is also possible that some observed behaviors in areas of residency (e.g., slow surf-zone patrolling) may represent periods of rest between bouts of foraging similar to that suggested for White Sharks in coastal habitats in South Africa by Johnson et al. (2009). Interpreting dive and swim behaviors will come from the future incorporation of physiological sensors into satellite-based tag technology.

Temperature

Temperature ranges experienced by JWS in eastern Australian waters were similar to those previously reported by Dewar et al. (2004), Weng et al. (2007b), and Chapter 28, this book for both YOY and JWS stages. During deep dives, sharks experienced temperatures slightly lower (6.8°C) than the lowest temperatures reported by these studies (8.4–9.0°C) but only for short periods of time (typically 10–15 min./dive). The similarity in the dominant temperature range experienced by JWS between studies to date suggests an overall propensity for occupying waters of 18–20°C. The seasonal location of surface waters with this temperature signature off eastern Australia shows some similarity to movements of JWS (Figure 17.6). In summer and autumn, this temperature zone is located from central eastern Tasmania to southern NSW. In winter and spring, it is located further north from central southern NSW to Seal Rocks with a coastal band extending north to Noosa in southern Queensland. Whether this temperature zone represents a physiological optimum for JWS or defines the distribution of preferred prey is unclear.

The restricted spatial footprints of the two JWS nursery areas off eastern Australia and their cycle of occupancy suggest that their use has evolved around predictable seasonal events such as prey availability. Stomach content data were unavailable for either of these regions, but both areas are popular recreational and commercial fishing regions, and a variety of finfish and elasmobranchs are recorded in these areas during the periods when sharks are present. Snapper are a favored prey species for White Sharks in Australian waters (Malcolm et al., 2001; Bruce et al., 2006), and the summer autumn period of JWS residency coincides with the seasonal presence of snapper along coastal reefs off Corner Inlet. Chlorophyll a (1998–2009 SeaWifs data) concentrations peak in both areas during the period when JWS are present; during March–April in the Corner Inlet region and October–November in the Port Stephens region (Figure 17.15).

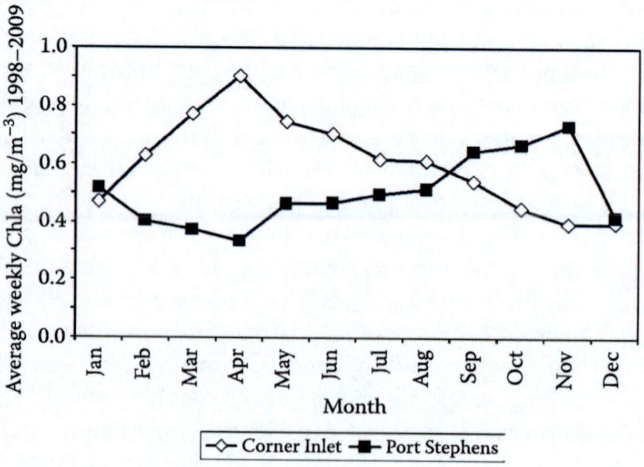

Figure 17.15 Seasonal cycle of chlorophyll in the Port Stephens and Corner Inlet regions based on weekly averaged SeaWiFS data (1998–2009).

Shelf waters in the Port Stephens region receive nutrient enrichment via upwelling generated by the interaction of the EAC and the local continental shelf topography (Oke and Middleton, 2001). These events are strongest during spring and early summer when the EAC flow is strongest (Wyrtki, 1960) and result in local chlorophyll peaks (Hallegraeff and Jeffrey, 1993). The marked change in trajectory of the coastline at Port Stephens results in back-eddy formation that enhances nearshore productivity in the area (Lee et al., 2007). This may be one of the factors contributing to the abundance of suitable prey in this area for JWS. Migrations of suitable prey during this period are not restricted to nearshore waters and include a large schooling Gempylid, *Rexea solandri*, which inhabits slope waters at depths of between 300 and 500 m. This species makes a late-winter spawning migration along the NSW coast from southern latitudes followed by a return south during spring (Rowling, 2001). Ichthyoplankton surveys have identified the Laurieton area, 100 km north of Port Stephens, as the primary spawning area for this species (Miskiewicz et al., 1987). This coincides with the depth range visited during slope forays by JWS at Port Stephens and along the NSW coast.

Pupping Grounds

All of the sharks observed or tagged at both Stockton Beach and the Corner Inlet regions, with the exception of the 1.75-m female tagged in 2008, were in excess of YOY size, suggesting that these areas represent juvenile habitat rather than pupping grounds for White Sharks. YOY-sized sharks have occasionally been reported by local fishers and have been taken at various beaches by NSW SCP operations, including Stockton Beach, but not in numbers sufficient at any one site to demonstrate the location of a specific pupping area in eastern Australia, and verification of measured sizes in these cases are lacking. Malcolm et al. (2001) reported that 50% of JWS taken as bycatch in a longline fishery off Corner Inlet were ≤2.0 m ($n = 30$). This compares with 28% in this size category for JWS tagged in the Port Stephens area as part of this study. White sharks are generally believed to pup during the spring to early summer period [reviewed by Bruce (2008)]. Recreational fisher's photographs confirm that neonate White Sharks have been caught in Bass Strait. We hypothesize that pupping occurs in Bass Strait and that Corner Inlet may represent the first summer's nursery area for YOY sharks in eastern Australia, followed by a migration north to the Port Stephens nursery area in the following spring. This is then followed by a cycling between these two nursery areas for up to at least 5 yrs. of age, which is the estimated ages of the larger sized sharks in both areas. However, confirmation of this will require further tracking and surveys of YOY and JWS sharks in the Corner Inlet and Bass Strait regions.

Duration of Tracking

Tracking periods ranged from 22 to 225 d (mean = 92), which were below those expected based on the battery configuration of the SPLASH tags. Hays et al. (2007) concluded that saltwater switch failure was a common cause of cessation of satellite-tag signals and that the simplest explanation for this failure was biofouling around saltwater switches, resulting in tags not turning on to transmit after surfacing. Photographs taken of one tagged shark, approximately 1 month after tagging, revealed barnacles already growing on areas of the tag that had not been antifouled. Thus it is likely that at least some tags on the sharks in this study were impacted by biofouling. The species composition and biomass of accumulated biofouling in offshore waters is generally less than in coastal waters (Thorstad et al., 2001; Yan and Yan, 2003), and this may explain why shark S5, which crossed the Tasman Sea, provided such a lengthy and reliable series of transmissions. This may also explain why satellite tags on sharks traveling in open-ocean waters have generally provided more lengthy tracks than those confined to shelf and coastal waters (Weng et al., 2005). Shark S6 traveled through the most extensive latitudinal and sea surface temperature range, and this behavior may also have resulted in limiting the impact of biofouling in any one habitat by the shark moving into habitats and temperature ranges that were unfavorable for the accumulated community. Similar conclusions were reached by Hays et al. (2007) regarding transmissions from tags attached to leatherback turtles when migrating between tropical and temperate waters and for Atlantic salmon returning to fresh water from the open sea (Thorstad et al., 2001).

Bycatch Potential

Juvenile White Sharks are more vulnerable to capture in commercial and recreational fisheries due to their size and are frequently taken in areas where their abundance inshore is high (Weng et al., 2007b; Chapters 14, 26, and 28, this book). The location of the two eastern Australian White Shark nursery areas renders sharks vulnerable to interactions with recreational fishers and nearshore commercial fishing operations. Captures of JWS have been reported by both sectors, and several sharks that we captured were trailing fishing gear. One tagged shark was also recaptured and released by a recreational fisher close to its original site of capture during the period of the study. The site-fidelity behavior shown by tagged sharks may also increase the risk of exposure of individual sharks to an area-specific threat, for example, the NSW SCP meshing activities at the southern end of Stockton Beach. The overall risks posed by the SCP for JWS in the region will depend on the extent to which sharks remain site-area specific during residency periods and the proportion of sharks temporarily resident to the nursery area whose individual site-area footprints coincide with the meshed section of the beach.

A commercial tuna and billfish longline fishery exists off eastern Australia. Forays into outer-shelf regions and movements seaward off the shelf break, including offshore diving behavior, may expose JWS to incidental capture in this fishery in areas adjacent to aggregation sites. The potential also exists for White Sharks to be taken as bycatch in shelf and slope trawling activities for prawns and finfish in the vicinity of aggregation sites, because examples of such bycatch have been reported in trawling operations in other Australian regions (Malcolm et al., 2001).

Our data suggest that these nursery areas, particularly the more spatially discrete Port Stephens area, offer considerable opportunities for monitoring the abundance of JWS and hence information on the status of the species in eastern Australia. Further information on the percentage of time individual JWS spend in the nearshore zone, where observation is possible, is required in order to interpret abundance data. The limited number of observed nursery sites in eastern Australia, and their connectivity, suggests that they provide important JWS habitat, and assessment of threats to the species should focus in these areas.

ACKNOWLEDGMENTS

This research was supported by a grant from the Department of the Environment, Water, Heritage and the Arts (Australia), a donation by Calypso Star Charters Pty. Ltd., and support from the Monterey Bay Aquarium and Sydney Aquarium Conservation Foundation. Specific resources crucial to the success of the project were supplied by NSW Industry & Investment and the Port Stephens-Great Lakes Marine Park. We specifically thank the following people for advice and assistance during the project: J. O'Sullivan, K. Weng, R. Forster, B. Talbot, P. Gallagher, C. Gallen, J. Sakker, D. Reid, M. Haste, D. Harasti, V. Peddemoors, K. Stannard, R. McAuley, J. Stevens, M. Heupel, A. Hobday, and G. Trinkler. This chapter was greatly enhanced by comments from K. Weng, J. O'Sullivan, J. Stevens, and K. Evans.

In memory of Georgia Clare Trinkler (December 15, 1995 to September 12, 2008).

REFERENCES

Aebischer, N. J., Robertson, P. A., and Kenward, R. E. 1993. Compositional analysis of habitat use from animal radio tracking data. *Ecology* 74: 1313–1325.

Batschelet, E. 1981. *Circular statistics in biology*. London, UK: Academic Press.

Bonfil, R., Meyer, M., Scholl, M. C., Johnson, R., O'Brien, S., Oosthuizen, H., Swanson, S., Kotze, D., and Paterson, M. 2005. Transoceanic migration, spatial dynamics, and population linkages of white sharks. *Science* 310: 100–103.

Bonfil, R., Francis, M. P., Duffy, C., Manning, M. J., and O'Brien, S. 2010. Large-scale tropical movements and diving behaviour of white sharks *Carcharodon carcharias* tagged off New Zealand. *Aquatic Biology* 8: 115–123.

Boustany, A. M., Davis, S. F., Pyle, P., Anderson, S. D., Le Boeuf, B. J., and Block, B. A. 2002. Expanded niche for white sharks. *Nature* 415: 35–36.

Bres, M. 1993. Behaviour in sharks. *Reviews in Fish Biology and Fisheries* 3: 133–159.

Bruce, B. D. 1992. Preliminary observations on the biology of the white shark, *Carcharodon carcharias*, in South Australian waters. *Australian Journal of Marine and Freshwater Research* 43: 1–11.

Bruce, B. D. 2008. White sharks: The biology and ecology of *Carcharodon carcharias*. In *Sharks of the open ocean*, edited by E. Pikitch and M. Camhi. London: Blackwell Scientific.

Bruce, B. D., Stevens, J. D., and Bradford, R. W. 2005a. *Site fidelity, residence times and home range patterns of white sharks around pinniped colonies*. Final report to Department of Environment and Heritage. Hobart, Australia: CSIRO.

Bruce, B. D., Stevens, J. D., and Bradford, R. W. 2005b. *Identifying movements and habitats of white sharks and grey nurse sharks*. Final report to Department of Environment and Heritage. Hobart, Australia: CSIRO.

Bruce, B. D., Stevens, J. D., and Malcolm, H. 2006. Movements and swimming behaviour of white sharks (*Carcharodon carcharias*) in Australian waters. *Marine Biology* 150: 161–172

Cailliet, G. M., Natanson, L. Welden, B., and Ebert, D. 1985. Preliminary studies on the age and growth of the white shark, *Carcharodon carcharias*, using vertebral bands. *Southern Californian Academy of Sciences Memoirs* 9: 49–60.

Cartamil, D. P., and Lowe, C. G. 2004. Diel movement patterns of ocean sunfish *Mola mola* off southern California. *Marine Ecology Progress Series* 266: 245–253.

Casey, J. G., and Pratt, H. L., Jr. 1985. Distribution of the white shark, *Carcharodon carcharias*, in the western North Atlantic. *Southern Californian Academy of Sciences Memoirs* 9: 2–14.

Cowling, A. 1999. Aerial surveys. In *Australian southern bluefin tuna industry-research-management workshop proceedings*, Port Lincoln, October 1999, pp. 12–16. Hobart, Australia: CSIRO Marine Research.

Dewar, H., Domeier, M., and Nasby-Lucas, N. 2004. Insights into young of the year white sharks (*Carcharodon carcharias*) behavior in the Southern California Bight. *Environmental Biology of Fishes* 70: 133–143.

Dill, L. M. 1983. Adaptive flexibility in the foraging behavior of fishes. *Canadian Journal of Fisheries and Aquatic Science* 40: 398–408.

Domeier, M. L., and Nasby-Lucas, N. 2007. Annual re-sightings of photographically identified white sharks (*Carcharodon carcharias*) at an eastern Pacific aggregation site (Guadalupe Island, Mexico). *Marine Biology* 150: 977–984.

Domeier, M. L., and Nasby-Lucas, N. 2008. Migration patterns of white sharks *Carcharodon carcharias* tagged at Guadalupe Island, Mexico, and identification of an eastern Pacific shared foraging area. *Marine Ecology Progress Series* 370: 221–237.

Eckert, S. A., and Stewart, B. S. 2001. Telemetry and satellite tracking of whale sharks, *Rhincodon typus*, in the Sea of Cortez, Mexico, and the North Pacific Ocean. *Environmental Biology of Fishes* 60: 299–308.

Environment Australia 2002. *White shark, Carcharodon carcharias, Recovery Plan*. Canberra, Australia: Environment Australia.

Francis, M. P. 1996. Observations on a pregnant white shark with a review of reproductive biology. In *Great white sharks: The biology of Carcharodon carcharias*, edited by A. P. Klimley and D. G. Ainley, pp. 157–172. San Diego, CA: Academic Press.

Francis, M. P. 1997. Reproductive strategy of white sharks, *Carcharodon carcharias*. *Shark News* 9: 8–9.

Goldman, K. J., and Anderson, S. D. 1999. Space utilization and swimming depth of white sharks, *Carcharodon carcharias*, at the South Farallon Islands, central California. *Environmental Biology of Fishes* 56: 351–364.

Gruber, S. H., and Myrberg, A. A. 1977. Approaches to the study of the behaviour of sharks. *American Zoologist* 17: 471–486.

Hallegraeff, G. M., and Jeffrey, S. W. 1993. Annually recurrent diatom blooms in spring along the New South Wales coast of Australia. *Australian Journal of Marine and Freshwater Research* 44: 325–334.

Hays, G. C., Bradshaw, C. J. A., James, M. C., Lovell, P., and Sims, D. W. 2007. Why do ARGOS satellite tags deployed on marine animals stop transmitting? *Journal of Experimental Marine Biology and Ecology* 349: 52–60.

Heupel, M. R., Carlson, J. K., and Simpfendorfer, C. A. 2007. Shark nursery areas: Concepts, definition, characterization and assumptions. *Marine Ecology Progress Series* 337: 287–297

Johnson, R., Bester, M. N., Dudley, S. F. J., Oosthuizen, W. H., Meyer, M., Hancke, L., and Gennari, E. 2009. Coastal swimming patterns of white sharks (*Carcharodon carcharias*) at Mossel Bay, South Africa. *Environmental Biology of Fishes* 85: 189–200.

Klimley, A. P. 1985. The areal distribution and autoecology of the white shark, *Carcharodon carcharias*, off the west coast of North America. *Southern Californian Academy of Science Memoirs* 9: 15–40.

Klimley, A. P., and Anderson S. D. 1996. Residency patterns of white sharks at the South Farallon Islands, California. In *Great white sharks: The biology of Carcharodon carcharias*, edited by A. P. Klimley and D. G. Ainley, pp. 365–373. San Diego, CA: Academic Press.

Klimley, A. P., Beavers, S. C., Curtis, T. H., and Jorgensen, S. J. 2002. Movements and swimming behaviour of three species of sharks in La Jolla Canyon, California. *Environmental Biology of Fishes* 63: 117–135.

Lee, R. S. Pritchard, T. R., Ajani, P. A., and Black, K. P. 2007. The influence of the East Australian Current eddy field on phytoplankton dynamics in the coastal zone. *Journal of Coastal Research* 50: 576–584.

Malcolm, H., Bruce, B. D., and Stevens, J. D. 2001. A review of the biology and status of white sharks in Australian waters. In *Report to Environment Australia, Marine Species Protection Program*. Hobart, Australia: CSIRO Marine Research.

Martin, R., Hammerschlag, N., Collier, R., and Fallows, C. 2005. Predatory behaviour of white sharks (*Carcharodon carcharias*) at Seal Island, South Africa. *Journal of the Marine Biology Association of the UK* 85: 1121–1135.

Miskiewicz, A. G., Gorman, T., and Graham, K. 1987. *The distribution of larvae of gemfish (Rexea solandri) along the New South Wales coast*. New South Wales Department of Agriculture, Fisheries Research Institute, Internal Report No. 25.

Mollet, H. F., and Cailliet, G. M. 1996. Using allometry to predict body mass from linear measurements of the white shark. In *Great white sharks: The biology of Carcharodon carcharias*, edited by A. P. Klimley and D. G. Ainley, pp. 81–89. San Diego, CA: Academic Press.

Oke, P. R., and Middleton, J. H. 2001. Nutrient enrichment off Port Stephens: The role of the East Australian Current. *Continental Shelf Research* 21: 587–606.

Pardini, A. T., Jones, C. S., Noble, L. R., Kreiser, B., Malcolm, H., Bruce, B. D., Stevens, J. D., Cliff, G., Scholl, M. C., Francis, M., Duffy, C., and Martin, A. P. 2001. Philopatric females and roving male great white sharks. *Nature* 412: 139–140.

Patterson, T. A., Evans, K., Carter, T. I., and Gunn, J. S. 2008. Movement and swimming behaviour of large southern bluefin tuna (*Thunnus maccoyii*) in the Australian region determined using pop-up satellite archival tags. *Fisheries Oceanography* 17: 352–367.

Phillips, B. F., Chubb, C. F., and Melville-Smith, R. 2000. The status of Australia's rock lobster fisheries. In *Spiny lobsters: Fisheries and culture*, edited by B. F. Phillips and J. Kittaka, pp. 45–77. Oxford, UK: Blackwell Science.

Pratt, H. L., Jr. 1996. Reproduction in the male white shark. In *Great white sharks: The biology of Carcharodon carcharias*, edited by A. P. Klimley and D. G. Ainley, pp. 131–138. San Diego, CA: Academic Press.

Pyle, P., Anderson, S. D., Ainley, D. G., and Henderson, R. P. 1996. Environmental factors affecting the occurrence and behaviour of white sharks at the Farallon Islands, California. In *Great white sharks: The biology of Carcharodon carcharias*, edited by A. P. Klimley and D. G. Ainley, pp. 281–291. San Diego, CA: Academic Press.

Reid, D. D., and Krogh, M. 1992. Assessment of catches from protective shark meshing off New South Wales beaches between 1950 and 1990. *Australian Journal of Marine and Freshwater Research* 43: 283–296.

Robbins, R. L. 2007. Environmental variables affecting the sexual segregation of great white sharks *Carcharodon carcharias* at the Neptune Islands South Australia. *Journal of Fish Biology* 70: 1350–1364.

Rowling, K. R. 2001. Comment on the "Spawning dynamics of the eastern gemfish (*Rexea solandri*) in relation to regional oceanography." *Marine and Freshwater Research* 52: 623–630.

Strong, W. R., Jr., Bruce, B. D., Nelson, D. R., and Murphy, R. C. 1996. Population dynamics of white sharks in Spencer Gulf, South Australia. In *Great white sharks: The biology of Carcharodon carcharias*, edited by A. P. Klimley and D. G. Ainley, pp. 401–414. San Diego, CA: Academic Press.

Thorstad, E. B., Okland, F., and Heggberget, T. G. 2001. Are long term negative effects from external tags underestimated? Fouling of an externally attached telemetry transmitter. *Journal of Fish Biology* 59: 1092–1094.

Tranter, D. J., Carpenter, D. J., and Leech, G. S. 1986. The coastal enrichment effect of the East Australian Current eddy field. *Deep Sea Research* 33: 1705–1728.

Tremblay, Y., Shaffer, S. A., Fowler, S. L., Kuhn, C. E., McDonald, B. I., Weise, M. J., Bost, C-A., Weimerskirch, H., Crocker, D. E., Goebel, M. E., and Costa, D. P. 2006. Interpolation of animal tracking data in a fluid environment. *Journal of Experimental Biology* 209: 128–140.

Weng, K. C., Castilho, P. C., Morrissette, J. M., Landeira-Fernandez, A. M., Holts, D. B., Schallert, R. J., Goldman, K. J., and Block, B. A. 2005. Satellite tagging and cardiac physiology reveal niche expansion in salmon sharks. *Science* 310: 104–106.

Weng, K. C., Boustany, A. M., Pyle, P., Anderson, S. D., Brown, A., and Block, B. A. 2007a. Migration and habitat of white sharks (*Carcharodon carcharias*) in the eastern Pacific Ocean. *Marine Biology* 152: 877–894.

Weng, K. C., O'Sullivan, J. B., Lowe, C. G., Winkler, C. E., Dewar, H., and Block, B. A. 2007b. Movements, behaviour and habitat preferences of juvenile white sharks *Carcharodon carcharias* in the eastern Pacific. *Marine Ecology Progress Series* 338: 211–224.

Wilson, R. P., Shepard, E. L. C., and Liebsch, N. 2007. Prying into the intimate details of animal lives: Use of a daily diary on animals. *Endangered Species Research* 3: 1–15.

Wintner, S. P., and Cliff, G. 1999. Age and growth determinations of the white shark *Carcharodon carcharias*, from the east coast of South Africa. *Fishery Bulletin* 97: 153–169.

Worton, B. J. 1989. Kernel methods for estimating the utilization distribution in home-range studies. *Ecology* 70: 164–168.

Wyrtki, K. 1960. *The surface circulation in the Coral and Tasman Seas*. CSIRO Australia, Division of Fisheries and Oceanography, Technical Paper No. 8.

Yan, T., and Yan, W. X. 2003. Fouling of offshore structures in China: A review. *Biofouling* 19: 1133–1138.

CHAPTER 18

Identifying Juvenile White Shark Behavior from Electronic Tag Data

Russell W. Bradford*
CSIRO Marine and Atmospheric Research

Alistair J. Hobday
CSIRO Marine and Atmospheric Research

Barry D. Bruce
CSIRO Marine and Atmospheric Research

ABSTRACT

Juvenile White Sharks (JWS) in Australian waters appear to occupy different coastal habitats during their early years. Two juvenile nursery areas are known in southeastern Australia, each with different resident periods. Juveniles are seasonally resident in the Port Stephens region of central New South Wales from mid-August to early January. Sharks commonly occur at the second known nursery area in coastal waters of eastern Bass Strait, 900 km south, from January to April. Movements between the two nursery areas occur in the nonresident periods. At the Port Stephens site, sharks are highly visible in the nearshore surf zone where interactions with recreational fishers and beach users are frequently reported. To examine potential interactions in these coastal zones, habitat use within and between the nursery areas is being studied with a range of electronic tags. Twenty-five JWS have been tagged in the Port Stephens region, seven of these with a pop-up archival transmitting (PAT) tag. Because six of the PAT tags were washed ashore and returned, the full archive data set was recovered. We used a novel mean first difference analysis, guided by supplemental location data from satellite and acoustic tags where available, to define behavioral modes and provide an indication of habitat use when sharks were resident at and when traveling between the Port Stephens and the Bass Strait nursery areas. These analyses identified a suite of behaviors for sharks ranging from shallow-water bottom-orientated swimming behavior consistent with surf-zone habitat use, diel behavior in deeper waters, continental slope (400–600 m), and open-ocean excursions with dives to nearly 1000 m. The transition phase between residency and traveling periods was characterized by multiple behavior modes indicative of short forays between the surf zone and shelf edge. Classification of the behavioral modes allows estimates of time when human interactions may occur, even without fine-scale position data.

INTRODUCTION

Developments in electronic tagging have delivered new insights into both fine-scale movements (Welch et al., 2004; Heupel et al., 2006) and large-scale oceanic migrations of marine animals (Nielson et al., 2009; Bonfil et al., 2005; Musyl et al., 2003; Teo et al., 2007; Walli et al., 2009). These discoveries have enriched the existing understanding of movement patterns established by more traditional release-recapture techniques (Graham and Roberts, 2007; Castro and Rosa, 2005; Klimley, 1996; Sims et al., 2000), in particular by uncovering the variation in individual fish behavior at a range of spatial and temporal scales. Coastal regions are where many anthropogenic pressures on marine species are greatest and where some species are most readily observed. Thus, resolving movements and behavior at appropriate spatial scales in these areas is important for understanding both these species' population dynamics and the potential impact of human activities.

Satellite-linked tags in particular have provided detailed knowledge of species movements (Gifford et al., 2007; Hsu et al., 2007) and, depending on the tag, can provide location at resolutions of hundreds of kilometers to hundreds of meters (ARGOS, 2008). When equipped with additional sensors, these tags can collect and archive continuous data on swimming behavior and a range of environmental parameters. However, realizing the full value of such data in a spatial context has been difficult because of the sporadic nature of position fixes generated by animals that can only be located when they surface [e.g. position-only tags: satellite-linked radio transmitting (SLRT) tags] or the low accuracy of light-based geolocation data [e.g. pop-up archival transmitting (PAT) tags]. Often the satellite receives only a summary of the archived data because of battery limits to power a satellite up-link (i.e. PAT), even though the full data set is collected on a tag. The deployment of satellite-linked tags on species that inhabit coastal waters has the advantage that tags bearing archival data can be washed ashore and returned if they are found. Given the pressures on the populations of many top predators, an understanding of habitat use and movement patterns are needed to provide spatially resolved management advice in order to reduce threats and interpret spatially and temporally variable patterns of observed abundance. Fully utilizing the continuous archived data that returned tags provide has the potential to make considerable advances in this understanding.

The White Shark is an iconic apex predator that occurs throughout the temperate waters of the world. Recent research has demonstrated that both juveniles and adults are capable of movement through the open ocean (Boustany et al., 2002; Bruce et al., 2006; Weng et al., 2007; Domeier and Nasby-Lucas, 2008; Chapters 10, 11, 16, and 21, this book) and even across ocean basins (Bonfil et al., 2005). White Sharks also make extensive coastal migrations; however, the drivers for transitions between coastal and oceanic movements are not yet understood. Understanding these drivers may allow greater predictive power regarding arrivals and departures from regions where interactions are expected. In many coastal regions, White Sharks are exposed to a range of coastal threats, including shark netting (Reid & Krogh, 1992; Cliff et al., 1996; Malcolm et al., 2001) and capture in commercial and recreational fisheries (Bruce, 1992; Pepperell, 1992; Malcolm et al., 2001; Chapter 14, this book), and in turn may represent a potential hazard for some beach users.

White Sharks are long-lived (at least 49–60 years; Cailliet et al., 1985; Wintner and Cliff, 1999; Bruce, 2008), have a relatively slow development, and have a low reproductive rate coupled with a long gestation period (estimated at up to 18 months; Francis, 1996; Mollet et al., 2000). These characteristics imply a low reproductive potential, reducing the rate at which depleted populations can recover and hampering the conservation of White Sharks. Despite a lack of reliable estimates on White Shark numbers (see Chapter 26, this book), either internationally or in Australian waters, there is evidence to indicate a population decline prior to the introduction of protective measures. Concerns over White Shark numbers resulted in the species being listed in Appendix II of the Convention for International Trade in Endangered Species (Bruce, 2008), and in Australian waters being listed as a threatened species under the Environmental Protection and Biodiversity Conservation Act 1999, which carries a legal requirement to develop and implement a recovery plan. One of the main aims of the White Shark recovery plan is: "To recover White Shark numbers in Australian waters, to a level that will see the species removed from the schedules of the Environmental Protection and Biodiversity Conservation Act 1999" (Environment Australia, 2002). Although the aim is to rebuild White Shark numbers, there is currently no metric to gauge the success of the conservation measures incorporated in the recovery plan. Estimating the number of sharks would be a first step in measuring success, and practically, this depends on the ability to survey these animals in coastal waters.

In Australia, White Sharks are distributed throughout southern waters from central Queensland to North West Cape, Western Australia (Last and Stevens, 2009). Tagging studies in Australia have identified two regions along the eastern seaboard of Australia: Stockton Bight, New South Wales (NSW) (32°52'S, 151°58'E) and the Corner Inlet region, Victoria (38°51'S, 146°45'E), where juvenile White Sharks (JWS) are seasonally resident (Bruce and Bradford, 2008). The JWS present in the Stockton Bight range from about 1.7 to 2.6 m total length and are approximately 2 to 6 years old (Bruce and Bradford, 2008). The Stockton Bight, and in particular Stockton Beach to the south and Hawks Nest to the north of the Port Stephens estuary, meet the definition of a nursery area defined by Heupel et al. (2007) in that JWS are more common within the area than outside of it; the juvenile sharks have a tendency to remain and/or return for extended periods, and the habitat is repeatedly used across years. Satellite tracking indicates that JWS are present in the area during the late Austral winter through to mid summer (Bruce and Bradford, 2008), after which the majority of sharks move south to the Corner Inlet region where they then reside until April or May.

The Stockton Bight is backed by one of the largest mobile sand dune systems in eastern Australia. The sand extends into the nearshore marine waters, resulting in a clear sandy, gently sloping bottom over which sharks are starkly contrasted and highly visible from land and the air. The unique characteristics of the region coupled with seasonal aggregations of JWS makes the Stockton Bight region an ideal location to conduct vessel-based or aerial surveys of abundance and trends across years. When resident in the Stockton Bight, JWS appear to move from the nearshore surf zone in depths as shallow as 0.5 m to mid-shelf depths of 120–140 m (Bruce and Bradford, 2008). However, it is only during periods of occupancy in the nearshore zone that JWS can be reliably observed and counted. Thus, an estimate of the time spent by White Sharks in the nearshore zone when resident in this nursery area is required in order to interpret abundance data from such surveys and provide estimates of encounter rates with nearshore based threats such as angling and terrestrial-based pollutants.

We have used archival data from returned PAT tags to characterize behavioral signatures at both coarse and fine scales and provide an indication of habitat use when sharks are resident in the nearshore environment or the offshore environment and when traveling between residency sites. Such results will lead to a better understanding of White Shark movement dynamics through the provision of quantitative output derived from archival tag data; in turn these results may be incorporated into models of JWS movement and abundance and into correction factors for abundance estimation (Hobday et al., 2009).

MATERIALS AND METHODS

Tagging relevant to this analysis was conducted between October 2007 and November 2009. To attach the tags, JWS were caught using a handline consisting of 12-mm soft, silver braided rope attached, with a swivel, to a short (1.2–1.4 m) stainless steel wire rope trace (2.5 mm diameter), and ending in a baited 12/0 steel hook with the barb filed down. A hook baited with dead mullet

(Mugilidae) was cast in front of an appropriate sized shark (<2.6 m total length) following sighting from a vessel. Once the shark was hooked, a small float (12-inch diameter) was attached close to the trace and the shark allowed to tire by pulling against the float. The shark was then guided into an in-water stretcher and an aerated (medical-grade oxygen) flow of seawater placed in the mouth. Sex was determined by presence/absence of claspers, and total length was measured. The shark was turned onto its dorsal surface and an acoustic tag (Vemco V16) surgically implanted into the peritoneal cavity. A total of 25 JWS ranging in length from 1.8 to 2.6 m (total length) were acoustically tagged in the Stockton Bight. The shark was then righted, and for twenty of the sharks, a satellite tag (SPLASH: Wildlife Computers) was attached to the first dorsal fin. Seven of the tagged sharks were also fitted with a PAT tag (Table 18.1). For all sharks, a single conventional tag was attached just posterior to the first dorsal fin. The entire tagging procedure was completed within 10 min. All of the tagging techniques are described in detail by Bruce and Bradford (2008) and Bradford et al. (2009).

Both SPLASH and PAT tags were programmed to record external temperature and pressure data at 10-s resolution; PAT tags were programmed to pop-up and transmit data to ARGOS satellites between 30 and 180 d after deployment. Sharks with acoustic tags could be detected at a number of acoustic receivers deployed in the Stockton Bight area and provide a reliable location to a distance of within 400 m (Hobday et al., 2009). SPLASH tags can provide a position estimate in real time with the most reliable positions having a root-mean-square error of <150 m. Location can be derived from PAT data using light-based geolocation methods but is generally coarse in resolution (~100 km) in eastern Australia using present processing algorithms (Patterson et al., 2008).

Because of bandwidth limitations inherent in the ARGOS satellite network and the battery limitations of tags, only binned data are transmitted from either the SPLASH or PAT tag. Such binned data form the basis of most SPLASH and PAT analyses. However, if the tag can be recovered, by washing ashore for example, a full set of data at the original sampling resolution can be obtained. Our behavioral analysis is based on archived data from returned PAT tags. Interpretation of the behavioral data has been assisted by using the more accurate location data derived from both SPLASH and acoustic tags, which were attached along with the PAT tags to the same animals.

Patterns in the archived PAT vertical data were examined using MATLAB® functions (MathWorks). Several techniques were used to classify the behavior of the sharks. First, overall behavioral patterns were examined using a Fourier transform to identify whether the depth time series contained behaviors at a particular frequency (e.g. diel pattern). Data sets that contained what appeared (subjectively) to be strong signals (frequencies) were analyzed further using spectral analysis to identify where in the time series such systematic behaviors were being expressed.

Second, fine-scale behavioral modes were investigated in the PAT depth data by using a new behavioral metric. Simple metrics, such as mean depth or variation in depth lose the temporal concordance between successive depth records and can result in several different behaviors that are visible in the raw data not being distinguished. Classification of animal behavior is difficult, and often human classification is more effective; however, for large time series and to improve "repeatability," algorithms may be preferred. After exploring a number of alternative metrics, including mean depth over a period of time and variance in depth over a period of time, we found that a differencing metric resolved a number of classes of depth-related diving behavior and showed good resolution between a range of behaviors that seem different when inspecting the depth trace. Here we define this metric as mean first difference (MFD): the average difference between successive depth measurements. MFD metrics calculated at full temporal resolution (10 s) were too noisy and did not produce coherent behaviors; therefore, we explored a range of filtering and smoothing of the data to obtain behavior metrics at a temporal resolution that produced more sensible results. Filtering was accomplished by subsampling the data at various time scales. Several levels of filtering, ranging from subsampling the archive data set at 30-s to 5-min. intervals, were examined. For example, the MFD for a 30-s interval was calculated between every third data point in the time series, whereas

Table 18.1 Pop-Up Satellite Archival Tag Deployment Details on White Sharks in Stockton Bight

Tag Identification	Shark Number[a]	Date Deployed	Days to Release			Sex	Total Length (m)	Summary Data	Archival Data	Notes
			Programmed	Actual	Difference					
65556	S6	10/8/2007	120	29	–91	M	1.9	Y	Y	PAT with tether intact found on beach
65554	S9	10/11/2007	120	10	–110	M	1.9	N	Y	Tag recovered May 2009, stopped recording 10/21/2007
81193	S25[b]	10/28/2008	180	114	–66	F	2.3	Y	Y	
65556	S26[b]	10/28/2008	90	111	21	F	2.3	Y	Y	
65555	S27[b]	10/30/2008	90	19	–71	F	2.2	N	Y	
65553	S28[b]	10/31/2008	120	dnt		M	2.4	N	N	
65555	S17	10/28/2009	31	31	0	M	2.0	N	Y	dnt; tag recovered May 2010, 31 d of data

[a] See Chapter 17, this book.
[b] Sharks not included in Chapter 17, this book.
dnt, data did not transmit.

the MFD for a 5-min. interval was calculated between every thirtieth data point. Smoothing using an unweighted moving average (replacing each element on the time series by the simple average of n surrounding elements) was also explored over a range of periods (30 min. to 48 h). Smoothing over short periods tended to result in erratic shifts in "behavioral mode," whereas smoothing over long time scales (e.g. >6 h) removed the ability to detect sensible short-period behaviors, including diurnal signals (Diggle, 2000). The final choices for filtering and smoothing were 1 min. of filtering (MFD calculated between every sixth data point) on a 2-h moving average (the MFD averaged over a 2-h window). These choices were based on subjective examination of a subset of the data that appeared to provide the greatest behavioral resolution and sensible classification compared with inspection of the data. Thus, behaviors that occurred at time periods shorter than 2 h were not detected. When sharks are moving at a constant depth, MFD will be small, whereas oscillating dive behaviors will result in larger values.

RESULTS

Six of the seven PAT tags were found and returned, allowing the archived data set to be recovered. All of the recovered PAT tags, except two, released prematurely. Two PAT tags never transmitted, but one was subsequently recovered. The average number of days of data collected was 52.3 ± 47.2 (mean ± 1 SD). One PAT tag (#65556, second deployment) exceeded the programmed deployment time by 21 d before it released and began transmitting. On the first deployment of PAT 65556, the tag was recovered close to a shark-netted beach before the tag could start transmitting. This tag was returned complete with tether, anchor, and burn pin intact, suggesting that it fell out of the shark. Where available, positions relative to nursery areas were ground-truthed from ARGOS positions derived from SPLASH tags or from detections of tagged sharks on moored acoustic receivers along Australia's east coast. Accompanying position information from SPLASH and acoustic tags was available for four of the six tags, covering a total of 41 d of archival data.

Fourier Analysis

Fourier transformation of the depth data showed several peaks in the frequency distribution (periodogram) (Figure 18.1) of all six data sets. In particular, all six archival data sets included a peak at 24 h, indicating a diel signal of varying strength in the behavior of the six JWS. This diurnal behavior was not consistently apparent from visual inspection of the raw dive data. Other peaks were also evident, with the majority corresponding to a factor of the 24-h signal (harmonics).

SPECTRAL ANALYSIS

To investigate the frequency with which the diel and other frequencies occurred in each data set, a spectrogram was calculated for each shark, showing the power in each frequency as a function of time. It was apparent from inspection of these plots that the diel behavior was not continuously present in any shark. Calculation of the "frequency of occurrence" of diel behavior was not possible, because the spectrograms were quite coarse and sparse; however, this remains a future research area.

Mean First Difference

The vertical behavior metric we developed, MFD, provides a continuous measure of depth differences between sequential (filtered to every minute) data points. Dividing this continuous

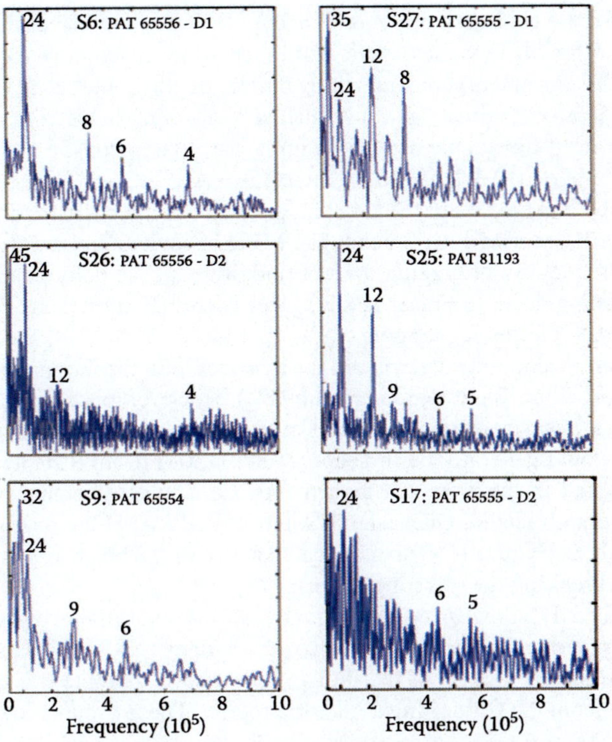

Figure 18.1 Periodograms of pop-up satellite archival tag data sets for six White Sharks indicating the dominant spectral peaks in hours. D1 and D2 denote first and second deployments of a PAT tag.

metric into categories allowed different "behaviors" to be identified. To determine whether natural breakpoints in the distribution of MFD values existed, we grouped all sharks together to determine whether modes of behavior occurred. For all sharks, behavior was classified into six categories representing "modes" (Figure 18.2). Mode 1 was characterized by very small changes between sequential data points indicating swimming within a constricted depth range. In contrast, Mode 6 was characterized by large changes between sequential data points indicating rapid changes in depth and greater use of the water column, perhaps a searching behavior. Modes 2, 3, and 6 were the most commonly expressed modes, representing 10.9, 55.3, and 25.4% of the time, respectively.

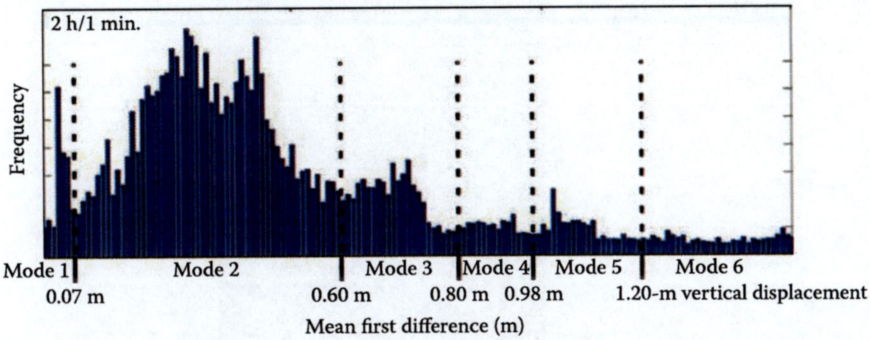

Figure 18.2 Combined frequency histogram for all archival data sets with the common behavioral boundaries indicated by vertical bars. Mean first difference is based on the moving average vertical displacement over a 2-h period between successive 1-min. intervals.

The expression of these six behavioral modes using MFD was evaluated in the archival data for individual sharks. Similar MFD values could still come from different shark vertical swimming behaviors. For example, movements constrained by bottom depth would result in small MFD values (small differences in depth over time), as would surface swimming in water of any depth or bottom swimming. Thus, we used the supplementary tagging data (i.e., acoustic and splash tag data) to provide additional insight on shark location for the behaviors.

Shark S6 was doubled tagged with a SPLASH tag which provided a more detailed picture of the movements of this shark against which to relate the MFD behaviors derived from the archival data set. S6 remained in the vicinity of tagging for a period of about 7 d before departing the Stockton Bight region and traveling north to Fraser Island, Queensland (Figure 18.3). The archival data set for this shark covered 29 d following tagging.

Using the common break points determined for all sharks, S6 displayed a mixture of behaviors while resident in the Stockton Bight region (Figure 18.4). However, once S6 departed the area, the behavior mode expressed was consistent with that of traveling (expressed as a higher MFD). When traveling, the sharks would generally be in deeper water (120–140 m) (Chapter 17, this book) and therefore less constrained in their vertical movements. On about November 4, 2007, S6 reached the northern tip of Moreton Island, Queensland, where the SPLASH tag positions formed a small cluster. The MFD analysis (Figure 18.4) also indicated a mixture of behaviors at that time consistent with the shark briefly breaking the traveling pattern.

Shark S25 was detected on the Australian Animal Tracking and Monitoring Systems (AATAMS) acoustic line running perpendicular to Bondi Beach, Sydney, NSW, on January 14, 2009. On February 17, 2009, premature release of the PAT tag resulted in satellite transmissions beginning on February 22, 2009 from the Corner Inlet, Victoria region. These events provided a georeference

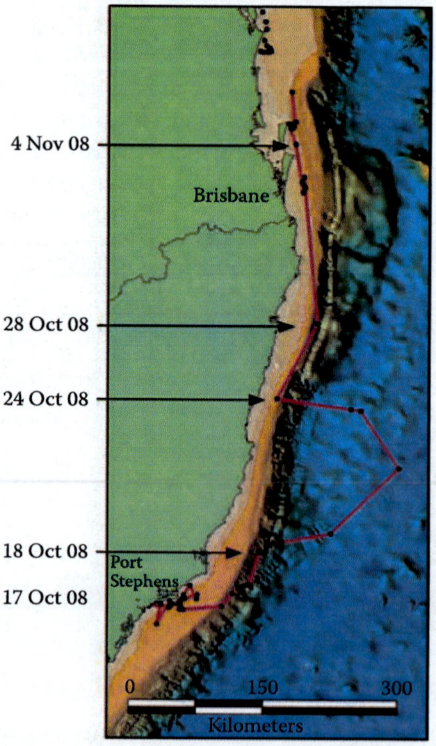

Figure 18.3 SPLASH track diagram for S6 (double-tagged with PAT 65556-D1).

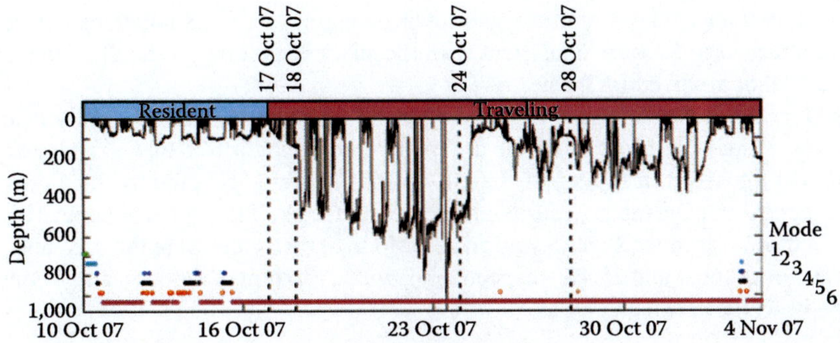

Figure 18.4 Plot of MFD analysis for S6 (PAT 65556 deployment 1) below the archival record of depth (at 10-s. resolution); resident and traveling modes derived from SPLASH tag data. MFD behavioral modes are color-coded with small vertical displacement indicated by green (Mode 1), grading to large vertical displacement indicated by purple (Mode 6).

for the archival data set (Figure 18.5). For the first 60 d following tagging, S25 exhibited behavior consistent with the shark being resident in the Stockton Bight in shallow waters. Following this period, a mixture of behaviors was exhibited until several days before the shark was detected on the AATAMS acoustic line at Bondi Beach. For about the next 10 d S25 exhibited behavior that was consistent with traveling before returning to a mixed behavior pattern when in the Corner Inlet region.

For shark S26, there were no accompanying data from either a SPLASH or acoustic tag to indicate where this shark was located during the period covered by the archival data set (although it was detected off Bondi, NSW, on January 22, 2009 after the PAT tag had been released). Sharks S9, S27, and S17 were double-tagged with a satellite tag; however, the satellite tags on S27 and S17 did not

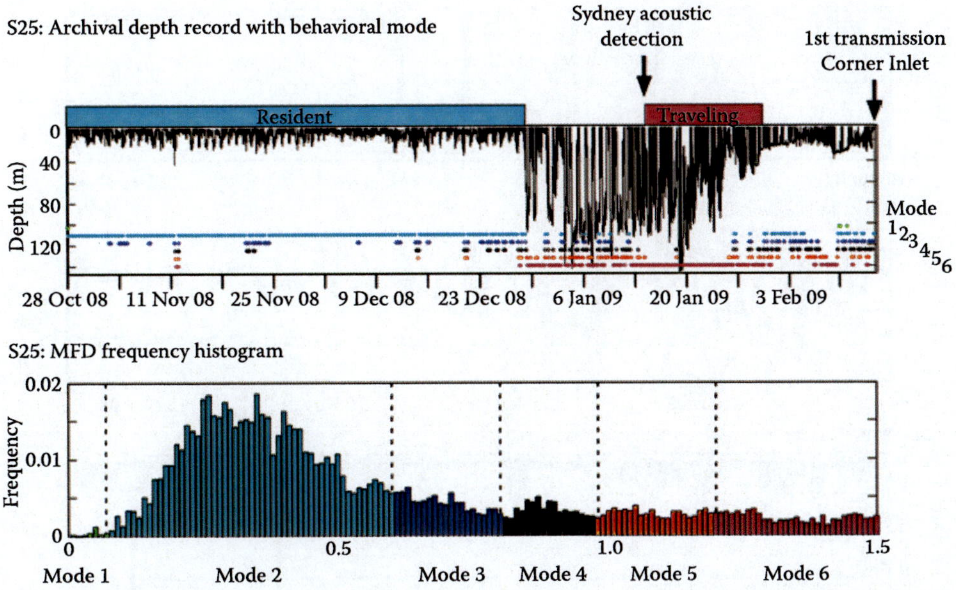

Figure 18.5 Plot of MFD analysis for shark S25 (PAT 81193) below the archival record of depth (at 10-s. resolution). (a) Behavioral modes are color-coded with small vertical displacement indicated by green (Mode 1), grading to large vertical displacement indicated by purple (Mode 6). (b) The frequency distribution of the MFD analysis with color coding for the behavioral modes. Acoustic detections were recorded on the AATAMS's line of receivers arranged perpendicular to Bondi Beach, Sydney.

transmit for more than 1 day immediately following tagging. ARGOS locations derived from the SPLASH tag attached to S9 were consistent with the shark remaining in the Stockton Bight region for the entire 10 d of archived PAT data.

The MFD analyses for S26, S9, and S27 (Figure 18.6) indicated predominantly nearshore residency behavior. Visual examination of the archival depth trace for S26 and S27 also suggests that these two sharks remained in water less than 40 m for the period covered by the archival data. S9 and S17 both appear to have made occasional forays over the shelf to depths of about 120 m. Overall the resident behaviors (Modes 2 and 3) and travel (Mode 6) behaviors were the two most commonly expressed categories (66.2 and 25.4%, respectively), with a mixture of modes 4 and 5 consistent with switching between these two categories.

DISCUSSION

Information regarding the movements and behavior of JWS are important for managing a range of potential threats to this species as well as shark-human interactions. When resident, JWS occupy a defined spatial footprint where direct interactions may occur between JWS, beach users, and

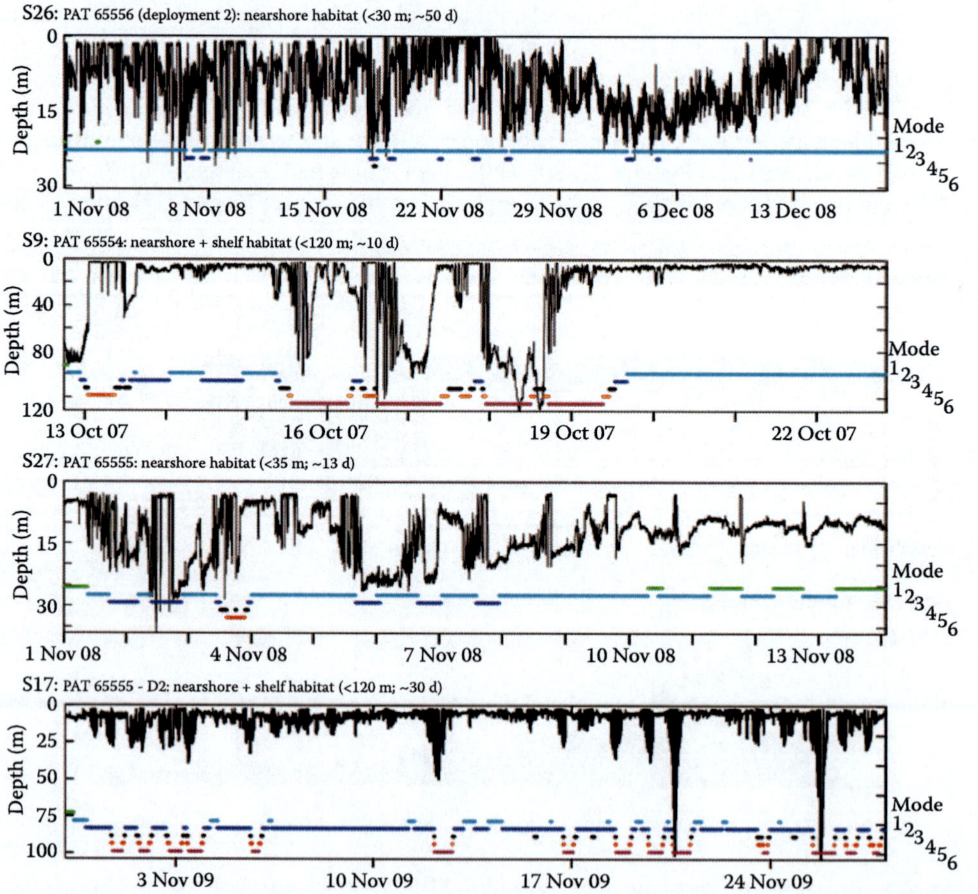

Figure 18.6 Plot of MFD analysis for sharks S26 (PAT 65556 deployment 2), S9 (PAT 65554), S27 (PAT 65555 deployment 1), and S17 (PAT 65555 deployment 2) below the archival record of depth (at 10-s. resolution). Behavioral modes are color-coded with small vertical displacement indicated by green, grading to large vertical displacement indicated by purple.

recreational fishers. When traveling, JWS show a tendency to move within a corridor bounded by the 60- and 120-m depth contours, with occasional forays over deeper water as well as in toward the shore (Chapter 17, this book), the former putting them at risk of interacting with commercial fishing operations, such as longlining. Through the analysis of archival tag data, we have been able to quantify the time spent in these two key behavioral modes.

Two high residency sites along the east coast of Australia have been identified that fit the criteria of a nursery area [as defined by Heupel et al. (2007)] for JWS: Stockton Bight, NSW, and Corner Inlet, Victoria (Figure 18.7) (Chapter 17, this book). While resident in either of these nursery areas, JWS remain within a relatively restricted zone extending from the shore to depths of about 120–140 m (Chapter 17, this book). These two areas are exploited on a seasonal basis, with travel to and from these areas occurring in a relatively direct manner. We use data from six archival data sets to examine patterns in depth use that may indicate changes in behavior characteristic of periods of residency or travel.

A visual examination of an archival data set can often reveal regular patterns. For example, diel changes in depth distribution are a common behavior exhibited by many marine animals (Bestley et al., 2009; Bertrand et al., 2004; Shepard et al., 2006; Nakano et al., 2003; Graham et al., 2006). Although not present throughout the archival data sets reported here, a diel pattern was visible for a portion of all six data sets. However, not all patterns are so visible. A Fourier transformation may be used with time-series data to identify systematic components, if present, in the time series that

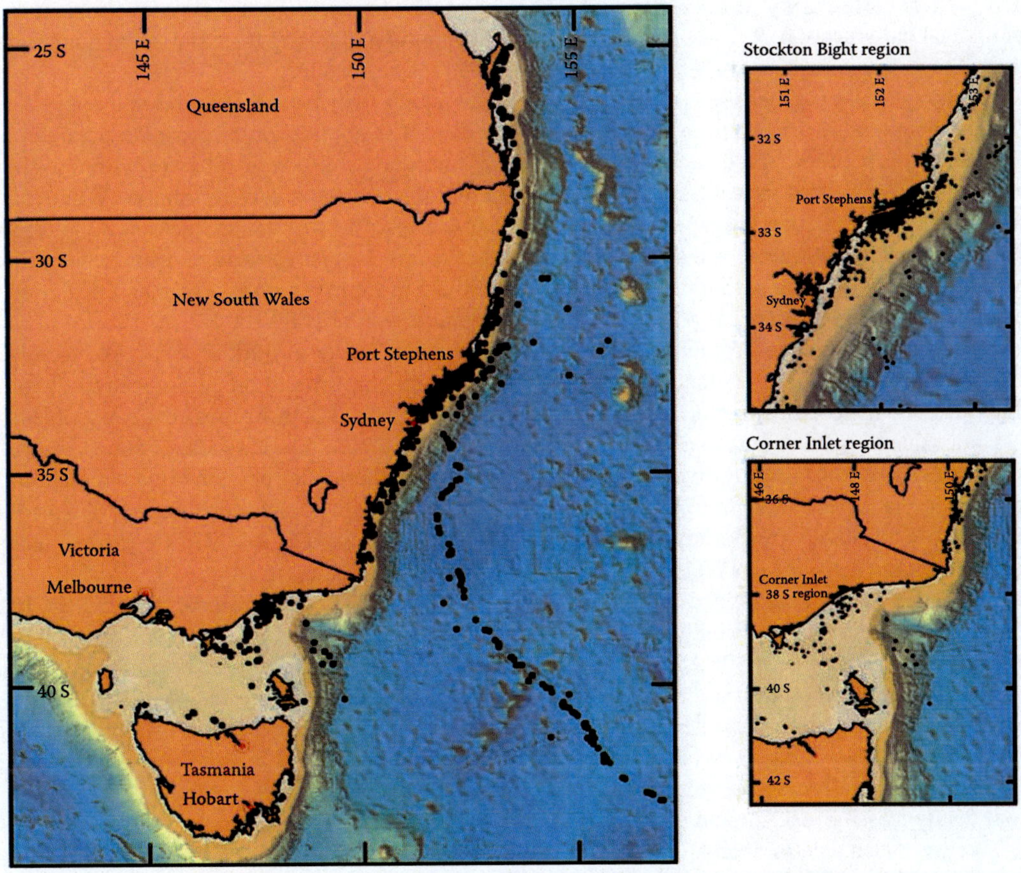

Figure 18.7 Location of juvenile White Shark temporary residency sites off eastern Australia.

have been masked by the overwhelming amount of noise inherent in time series data (Diggle, 2000). Of the six archival data sets examined, all contained a systematic component that occurred on a 24-h period, supporting the presence of a diel pattern. In addition, there were a number of other systematic components equivalent to factors of 24 h that may have been harmonics of the strong diel signal. The application of Fourier transformation yields no information on what time within the diel period particular behaviors occurred in the period of interest. Spectrogram analysis can be used to identify where periodic influences occur in the archival time series and are related to environmental or location data where known, and this analysis is planned for a future publication.

To investigate finer-scale JWS behaviors, in particular those representing resident or traveling behaviors, we used an analysis of changes in vertical distribution over time. To date, there have been many approaches in the literature to determine behavior through the analysis of fine-scale archival depth data. For example, the behavior of many marine animals has been examined through considering the mean depths occupied at day and night (Schaefer and Fuller, 2005), division into depth bins occupied during day and night (Willis and Hobday, 2007), classification of dive profiles based on maximum depth and dive duration (Kitagawa et al., 2004; Bestley et al., 2009), or a visual and/or statistical classification of dives based on simple geometry into U- and V-shaped dive profiles (Schreer and Testa, 1996; Baechler et al., 2002, Wilson and Block, 2009; Patterson et al., 2009). Our metric, MFD, representing the magnitude of change in successive depth readings, allows the sequential nature of movement to be retained (compared with "mean depth"). Linking the MFD-based behavior with the location of the JWS derived from satellite or acoustic tags concurrently deployed on these sharks allows new insight. This approach also overcomes some of the subjectivity of manual classification of dive profiles while at the same time allowing for a statistical quantification of time spent resident at a site or traveling between residency sites.

Although defining behavior modes in continuous MFD distribution is somewhat subjective, the resulting patterns in the time series were concordant. The application of common breakpoints for transitions between putative behaviors may mask fine-scale changes in behavior within a shark but, we argue, provides for a robust examination of common behaviors between sharks. Where data were available from other tag types to indicate a shark was resident within the Stockton Bight region (S6, S9, and S25), MFD showed a consistent pattern of small changes in vertical displacement (low MFD). Conversely, when a shark was outside of the Stockton Bight region, MFD analysis showed a consistent pattern of large changes in vertical displacement (high MFD).

Using the magnitude of change in archival depth data, Teo et al. (2007) were able to relate changes in the behavior of Atlantic Bluefin Tuna to transit behavior and breeding behavior within the Gulf of Mexico. In the North Atlantic, Wilson and Block (2009) examined dive profiles in Atlantic Bluefin Tuna and were able to assign putative behaviors to the two most common geometric shapes (V- and U-shaped dives). Restricted dive profiles that could not easily be assigned to a geometric shape were thought to be in response to thermal or bathymetric constraints (Wilson and Block, 2009). In the current study, restricted dive profiles, expressed as low MFD, were characteristic of resident behavior in JWS.

Shark S6 was double tagged with a SPLASH and a PAT tag. Archival data covering the first 29 d following the tagging event were recovered from the PAT tag. During those 29 d, the ARGOS positions derived from the SPLASH tag placed S6 in the Stockton Bight for approximately 1 week, before S6 departed and traveled north to the vicinity of Fraser Island, Queensland. Shark S6 traveled north in a very direct manner with only a single period at about the time the PAT tag detached where S6 showed a temporary break in travel. Examination of the archival data set alongside the reported positions for S6 provided us with an unprecedented opportunity to examine MFD and confidently classify resident and travel behavior.

We were also able to estimate positions to link the archival data set for S25 based on acoustic tag detections and the location where the PAT tag popped off. The MFD analysis of the archival data from shark S25 indicated a residency for about the first 2 months of the archival data set. This was followed by a period of mixed behaviors that we surmise was a series of short forays along the shelf to and from the Port Stephens area before the shark began directed travel toward Corner Inlet. Satellite tracks of several sharks (Chapter 17, this book) have shown a similar pattern of short forays over the shelf before leaving the Stockton Bight region.

The first confirmed location fix for S25 occurred on January 14, 2009, when it was detected on the AATAMS acoustic receiver line off Bondi Beach, NSW, and shortly after traveling behavior began. The presumed bottom depth during this period was about 120 m, consistent with the pattern of sharks traveling within the 60–120-m depth corridor reported from satellite tags (Chapter 17, this book). The bottom depth changed after about 10 d and coincided with a change in MFD to behavior more consistent with that expressed immediately prior to travel. Shortly thereafter, the PAT tag began transmissions from the Corner Inlet region and providing our second spatial anchor. Isolating the period of travel expressed in the MFD analysis, S25 traveled between Port Stephens and Corner Inlet (~850 km) at a speed of just over 3 km/h, which is consistent with cruising speeds reported elsewhere (Klimley et al., 2002; Strong et al., 1996; Bruce et al., 2006; Domeier and Nasby-Lucas, 2008; Chapter 17, this book). Once in the Corner Inlet region, shark S25 spent the majority of the time in water depths of about 40 m. Coinciding with the time White Sharks are in the Corner Inlet region, snapper aggregate along the low-profile reef system (Malcolm et al., 2001), and it is likely that S25 was focussed on this species (Chapter 17, this book).

For the period of the archival record available for S9, there is concurrent SPLASH tag data to indicate that S9 spent the whole period in the Stockton Bight region. During this period, S9 appeared to cycle between the shallow nearshore zone and deeper shelf waters. This pattern has been reflected in the MFD analysis, where S9 displayed a mixture of behaviors similar to that displayed by S25, indicating a period of shelf forays bracketed by periods of residency.

Based on the MFD analyses for S6, S9, and S25, we were confident that common behaviors could be assigned based on the rate of change between successive depth records. The analyses were extended to include the remaining two sharks, for which we had no concurrent spatial data to "ground truth" potential behavioral patterns expressed in the MFD. Sharks S26 and S27 displayed predominantly resident behavior for 50 and 13 d, respectively, following tagging. Satellite tracks indicate that sharks tended to remain in the Stockton Bight region during October/November (Bruce and Bradford, 2008; Chapter 17, this book), supporting the conclusion from MFD analyses that these sharks were likely resident during this period.

The analytical approach we present here has allowed us to uncouple the depth data from the spatial data to infer common behavioral patterns expressed in JWS. The MFD approach lends itself well to quantifying the amount of time individuals spend expressing certain behaviors. This is of particular importance for JWS, of which aerial surveys are being developed to provide an index of abundance. Aerial surveys will require information on the amount of time JWS spend resident in the Stockton Bight and in particular within the narrow visible zone over the nearshore sandy bottom. Using the fraction of time that sharks perform a "nearshore behavior" can be used to correct the abundance estimation based on sharks sighted. These behaviors can also be incorporated in models (e.g. individual-based models) of movement and population dynamics and, together with additional electronic tagging data, highlight times and regions of residency where JWS may be more vulnerable to human activity.

ACKNOWLEDGMENTS

This research was supported by a grant from the Department of Environment, Water, Heritage and the Arts (Australia) and support from the Monterey Bay Aquarium, Sydney Aquarium Conservation Foundation, and Calypso Star Charters (SA). All tagging conformed to animal ethics requirements under AEC Permit No. 5/2007-08 (Tasmanian Department of Primary Industries and

Water). We specifically thank the following people for their advice and assistance: John O'Sullivan, Kevin Weng, Chris Gallen, Bill Talbot, and Kent Stannard. We also thank the CSIRO Marine and Atmospheric workshop for the construction of a new in-water sling used during tagging in 2009.

REFERENCES

ARGOS. (2008). *Argos user's manual*. http://www.argos-system.org/manual.

Baechler, J., Beck, C. A., and Bowen, W. D. (2002). Dive shapes reveal temporal changes in the foraging behaviour of different age and sex classes of harbour seals (*Phoca vituline*). *Canadian Journal of Zoology* 80: 1569–1577.

Bestley, S., Gunn, J. S., and Hindell, M. A. (2009). Plasticity in vertical behaviour of migrating juvenile southern bluefin tuna (*Thunnus maccoyii*) in relation to oceanography of the south Indian Ocean. *Fisheries Oceanography* 18: 237–254.

Bertrand, A., Barbieri, M. A., Córdova, J., Hernández, C., Gómez, F., and Leiva, F. (2004). Diel vertical behaviour, predator-prey relationships, and occupation of space by jack mackerel (*Trachurus murphyi*) off Chile. *ICES Journal of Marine Science* 61: 1105–1112.

Bonfil, R., Meyer, M., Scholl, M. C., Johnson, R., O'Brien, S., Oosthuizen, H., Swanson, S., Kotze, D., and Paterson, M. (2005). Transoceanic migration, spatial dynamics, and population linkages of white sharks. *Science* 310: 100–103.

Boustany, A. M., Davis, S. F., Pyle, P., Anderson, S. D., Le Boeuf, B. J., and Block, B. A. (2002). Expanded niche for white sharks. *Nature* 415: 35–36.

Bradford, R. W., Hobday, A. J., Evans, K., and Lansdell, M. (2009). *CMAR code of practice for tagging marine animals*. CSIRO Marine and Atmospheric Research Paper 028. Hobart, Australia: CSIRO. http://www.cmar.csiro.au/e-print/open/2009/bradfordrw_a.pdf.

Bruce, B. D. (1992). Preliminary observations on the biology of the white shark, *Carcharodon carcharias*, in South Australian waters. *Australian Journal of Marine and Freshwater Research* 43: 1–11.

Bruce, B. D. (2008). White sharks: The biology and ecology of *Carcharodon carcharias*. In *Sharks of the open ocean*, edited by E. Pikitch and M. Camhi. Oxford, UK: Blackwell Science.

Bruce, B. D., and Bradford, R. W. (2008). *Spatial dynamics and habitat preferences of juvenile white sharks—identifying critical habitat and options for monitoring recruitment.* Final Report to the Department of Water, Heritage and the Arts. Hobart, Australia: CSIRO.

Bruce, B. D., Stevens, J. D., and Malcolm, H. (2006). Movements and swimming behaviour of white sharks (*Carcharodon carcharias*) in Australian waters. *Marine Biology* 150: 161–172.

Castro, A. L. F., and Rosa, R. S. (2005). Use of natural marks on population estimates of the nurse shark, *Ginglymostoma cirratum*, at Atol das Rocas Biological Reserve, Brazil. *Environmental Biology of Fishes* 72: 213–221.

Cliff, G., Dudley, S. F. J. and Jury, M. R. (1996). Catches of white sharks in KwaZulu-Natal, South Africa and environmental influences, pp. 351–362. In *Great white sharks: The biology of* Carcharodon carcharias, edited by A. P. Klimley and D. G. Ainley. San Diego, CA: Academic Press.

Diggle, P. J. (2000). *Time series: A biostatistical introduction*. Oxford, UK: Oxford Science Publications, Clarendon Press.

Domeier, M. L., and Nasby-Lucas, N. (2008). Migration patterns of white sharks *Carcharodon carcharias* tagged at Guadalupe Island, Mexico, and identification of an eastern Pacific shared foraging area. *Marine Ecology Progress Series* 370: 221–237.

Environment Australia (2002). *White shark, Carcharodon carcharias, recovery plan*. Canberra, Australia: Environment Australia.

Francis, M. P. (1996). Observations on a pregnant white shark with a review of reproductive biology, pp. 157–172. In *Great white sharks: The biology of Carcharodon carcharias,* edited by A. P. Klimley and D. G. Ainley. San Diego, CA: Academic Press.

Gifford, A., Compagno, L. J. V., Levine, M., and Antoniou, A. (2007). Satellite tracking of whale sharks using tethered tags. *Fisheries Research* 84: 17–24.

Graham, R. T., and Roberts, C. M. (2007). Assessing the size, growth rate and structure of a seasonal population of whale sharks (*Rhincodon typus* Smith 1828) using conventional tagging and photo identification. *Fisheries Research* 84: 71–80.

Graham, R. T., Roberts, C. M., and Smart, J. C. R. (2006). Diving behaviour of whale sharks in relation to a predictable food pulse. *Journal of the Royal Society Interface* 3: 109–116.

Heupel, M. R., Carlson, J. K., and Simpfendorfer, C. A. (2007). Shark nursery areas: Concepts, definition, characterization and assumptions. *Marine Ecology Progress Series* 337: 287–297.

Heupel, M. R., Simpfendorfer, C. A., Collins, A. B., and Tyminski, J. P. (2006). Residency and movement patterns of bonnethead sharks, *Spyrna tiburo,* in a large Florida estuary. *Environmental Biology of Fishes* 76: 47–67.

Hobday, A. J., Kawabe, R. Takao, Y. Miyashita, K., and Itoh, T. (2009). Correction of an abundance index using acoustic tag data for juvenile southern bluefin tuna in southern Western Australia: Tagging and tracking of marine animals with electronic devices II. *Methods and Technologies in Fish Biology and Fisheries* 9: 405–422.

Hsua, H.-H., Joung, S.-J., Liao, Y.-Y., and Liu, K.-M. (2007). Satellite tracking of juvenile whale sharks, *Rhincodon typus*, in the Northwestern Pacific. *Fisheries Research* 84: 25–31.

Kitagawa, T., Kimura, S., Nakata, H., and Yamada, H. (2004). Diving behaviour of immature, feeding Pacific bluefin tuna (*Thunnus thynnus orientalis*) in relation to season and area: The East China Sea and the Kuroshio-Oyashio transition region. *Fisheries Oceanography* 13: 161–180.

Klimley, A. P. (1996). Residency patterns of white sharks at the South Farallon Islands, California, pp. 365–374. In *Great white sharks: The biology of Carcharodon carcharias*, edited by A. P. Klimley and D. A. Ainley. San Diego, CA: Academic Press.

Klimley, A. P., Beavers, S. C., Curtis, T. H., and Jorgensen, S. J. (2002). Movements and swimming behaviour of three species of sharks in La Jolla Canyon, California. *Environmental Biology of Fishes* 63: 117–135.

Last, P. R., and Stevens, J. D. (2009). *Sharks and rays of Australia*, 2nd ed. Hobart, Australia: CSIRO.

Malcolm, H., Bruce, B. D., and Stevens, J. D. (2001). *A review of the biology and status of white sharks in Australian waters*. Report to Environment Australia, Marine Protection Program. Hobart, Australia: CSIRO.

Mollet, H. F., Cliff, G., Pratt, H. L., Jr., and Stevens, J. D. (2000). Reproductive biology of the female shortfin mako *Isurus oxyrinchus* with comments on the embryonic development of lamnoids. *Fishery Bulletin* 98: 299–318.

Musyl, M. K., Brill, R. W., Boggs, C. H., Curran, D. S., Kazama, T. K., and Seki, M. P. (2003). Vertical movements of bigeye tuna (*Thunnus obesus*) associated with islands, buoys, and seamounts near the main Hawaiian Islands from archival tagging data. *Fisheries Oceanography* 12: 152–169.

Nakano, H., Matsunaga, H., Okamoto, H., and Okazaki, M. (2003). Acoustic tracking of bigeye thresher shark *Alopias superciliousus* in the eastern Pacific Ocean. *Marine Ecology Progress Series* 265: 255–261.

Neilson, J. D., Smith, S., Royer, F., Paul, S. D., Porter, J. M., and Lutcavage, M. (2009). Investigations of horizontal movements of Atlantic swordfish using pop-up satellite archival tags: Tagging and tracking of marine animals with electronic devices II. *Methods and Technologies in Fish Biology and Fisheries* 9: 145–159.

Patterson, T. A., Evans, K, Carter, T. L., and Gunn, J. S. (2008). Movement and behaviour of large southern bluefin tuna (*Thunnus maccoyii*) in the Australian region determined using pop-up satellite archival tags. *Fisheries Oceanography* 17: 352–367.

Patterson, T. A., Basson, M., Bravington, M. V., and Gunn, J. S. (2009). Classifying movement behaviour in relation to environmental conditions using hidden Markov models. *Journal of Animal Ecology* 78: 1113–1123.

Pepperell, J. G. (1992). Trends in the distribution, species composition and size of sharks caught by gamefish anglers off south-eastern Australia, 1961–90. In *Sharks: Biology and fisheries*, edited by J. G. Pepperell. *Australian Journal of Marine and Freshwater Research* 43: 213–225.

Reid, D. D., and Krogh, M. (1992). Assessment of catches from protective shark meshing off New South Wales beaches between 1950 and 1990. In *Sharks: Biology and fisheries*, edited by J. G. Pepperell. *Australian Journal of Marine and Freshwater Research* 43: 283–296.

Schaefer, K. M., and Fuller, D. W. (2005). Behavior of bigeye (*Thunnus obesus*) and skipjack (*Katsuwonus pelamis*) tunas within aggregations associated with floating objects in the equatorial eastern Pacific. *Marine Biology* 146: 781–792.

Schreer, J. F., and Testa, J. W. (1996). Classification of Weddell seal diving behaviour. *Marine Mammal Science* 12: 227–250.

Shepard, E. L., Ahmed, M. Z., Southall, E. J., Witt, M. J., Metcalfe, J. D., and Sims, D. W. (2006). Diel and tidal rhythms in diving behaviour of pelagic sharks identified by signal processing of archival tagging data. *Marine Ecology Progress Series* 328: 205–213.

Sims, D. W., Speedie, C. D., and Fox, A. M. (2000). Movements and growth of a female basking shark re-sighted after a three year period. *Journal of the Marine Biological Association U.K.* 80: 1141–1142.

Strong, W. R., Jr., Bruce, B. D., Nelson, D. R., and Murphy, R. C. (1996). Population dynamics of white sharks in Spencer Gulf, South Australia, pp. 401–414. In *Great white sharks: The biology of Carcharodon carcharias*, edited by A. P. Klimley and D. G. Ainley. San Diego, CA: Academic Press.

Teo, S. L. H., Boustany, A., Dewar, H., Stokesbury, M. J. W., Weng, K. C., Beemer, S., Seitz, A. C., Farwell, C. J., Prince, E. D., and Block, B. A. (2007). Annual migrations, diving behaviour, and thermal biology of Atlantic bluefin, *Thunnus thynnus*, on their Gulf of Mexico breeding grounds. *Marine Biology* 151: 1–18.

Walli, A., Teo, S. L. H., Boustany, A., Farwell, C. J., Williams, T., Dewar, H., Prince, E., and Block, B. A. (2009). Seasonal movements, aggregations and diving behavior of Atlantic bluefin tuna (*Thunnus Thynnus*) revealed with archival tags. *PLoS One* 4: e6151.

Welch, D. W., Ward, B. R., and Batten, S. D. (2004). Early ocean survival and marine movements of hatchery and wild steelhead trout (*Oncorhynchus mykiss*) determined by an acoustic array: Queen Charlotte Strait, British Columbia. *Deep-Sea Research II* 51: 897–909.

Weng, K. C., Boustany, A. M., Pyle, P., Anderson, S. D., Brown, A., and Block, B. A. (2007). Migration and habitat of white sharks (*Carcharodon carcharias*) in the eastern Pacific Ocean. *Marine Biology* 152: 877–894.

Willis, J., and Hobday, A. J. (2007). Influence of upwelling on movement of southern bluefin tuna (*Thunnus maccoyii*) in the Great Australian Bight. *Marine and Freshwater Research* 58: 699–708.

Wilson, S. G., and Block, B. A. (2009). Habitat use in Atlantic bluefin *Thunnus thynnus* inferred from diving behaviour. *Endangered Species Research* 10: 355–367.

Wintner S. P., and Cliff, G. (1999). Age and growth determinations of the white shark *Carcharodon carcharias*, from the east coast of South Africa. *Fisheries Bulletin* 97: 153–169.

CHAPTER 19

Beach Areas Used by Juvenile White Shark, *Carcharodon carcharias*, in Eastern Australia

Jonathan M. Werry*
Australian Rivers Institute

Barry D. Bruce
CSIRO Marine and Atmospheric Research

Wayne Sumpton
Agri-Science Queensland

Dennis Reid
New South Wales Industry and Investment

David G. Mayer
Agri-Science Queensland

ABSTRACT

Juvenile White Sharks (JWS), *Carcharodon carcharias*, have been recorded in shark control programs (SCPs) since their inception in the late 1930s in New South Wales (NSW) and early 1960s in Queensland (Qld). Recent satellite tracking of JWS (1.75–2.6 m total length) indicates that they are seasonally resident, with extended periods in the nearshore surf zone, at a small number of interconnected areas along the east coast of Australia. The JWS also showed directed and consistent periods of travel, with infrequent coastal visits, when transiting between these areas. These differences in behavior may have important ramifications for interpreting long-term trends in catch-per-unit-effort (CPUE) from SCPs. We used generalized linear models to examine CPUE and size of JWS from 1993 to 2009 from both shark control programs. We further examined the movements of 20 satellite-tracked sharks in relation to SCP beach areas. Analyses indicated a concordance between beaches of higher CPUE and residency areas of tracked sharks, suggesting that data from these areas are more conducive to long-term monitoring. Significantly larger JWS were caught in Qld SCP gears. Seasonal peaks in SCP CPUE corresponded with movement of satellite-tracked JWS past SCP areas. Satellite tracking has previously indicated two nursery areas on the east coast: one in the Port Stephens region near Newcastle, NSW, and the other at Corner Inlet–Ninety Mile Beach region of eastern Victoria. A third region along the eastern side of Fraser Island, Qld, has also been suggested by satellite tracking, but data were comparatively few. CPUE data from the Qld SCP indicate that this area, and Bundaberg on the north of Fraser Island, may be an additional seasonally important area on the east Australian coast. Proximity to the 60–120-m contour, orientation of a beach relative to the directed migrations of JWS, and beach length may play a role in the frequency of encounter between SCP gear and JWS. Moon phase may also influence the onshore movement of JWS both at residency areas and over areas of transit, because the lowest SCP CPUE were recorded around the full moon.

INTRODUCTION

Insight into the behavior and large-scale movements of White Sharks (*Carcharodon carcharias*) between key habitats has advanced substantially over the last decade, largely through the application of satellite-tracking technologies. Recent studies have defined transoceanic migrations, diel and temporal movements in relation to physical parameters, tropical excursions, and seasonal migrations to specific pelagic and coastal areas (Bonfil et al., 2005, 2010; Bruce et al., 2006; Weng et al., 2007a, 2007b; Domeier and Nasby-Lucas, 2008; Jorgensen et al., 2009; Chapters 11, 12, 16, 17, 18, 21, and 22, this book). On the east coast of Australia, recent satellite tracking of juvenile White Sharks (JWS) (<3 m total length) suggest that they are seasonally resident at a relatively small number of interconnected areas where they spend extended periods in the nearshore surf zone (Chapter 17, this book) (Figure 19.1). Between these areas, they displayed directed and consistent periods of travel. The identification of areas that are consistently visited by White Sharks remains the subject of current research (Domeier and Nasby-Lucas, 2008; Jorgensen et al., 2009; Chapters 11 and 25, this book) as a result of concerns over the global status of the species (Baille et al., 2004).

Like many large shark species, White Sharks are widespread and have low natural abundances (Bruce, 2008). As such, long-term data on populations are scant. Throughout the twentieth century, however, several shark control programs (SCPs) designed to minimize adverse shark-human

Figure 19.1 Juvenile White Shark in the surf zone, Stockton beach, New South Wales (NSW) (2007). (Courtesy of Ben Doolan.)

interactions have captured White Sharks adjacent to beach bathing areas. These programs aimed to reduce the abundance of potentially dangerous shark species in the vicinity of popular bathing beaches. Historical catch records from the Kwala-Zulu SCP in South Africa, for example, have provided some insight into White Shark population characteristics and status along the East Africa Seaboard (Cliff et al., 1989; Dudley and Simpfendorfer, 2006). In Australian waters, the NSW SCP was implemented in the late 1930s, and the Queensland (Qld) SCP was implemented in the early 1960s (Reid and Krogh, 1992; Gribble et al., 1998). Long-term catch records from gillnets in both Australian SCPs and drumlines in the Qld SCP provide one of the few long-term data sets on coastal sharks in Australian waters. The documented long-term decline in the total catch per year of White Sharks in the NSW SCP (Krogh, 1994) was instrumental in the listing of White Sharks for protection in Australian waters under the application of the precautionary principle (Environment Australia, 2002).

Several studies have suggested that abiotic factors can influence temporal residency and movement of both juvenile and adult White Sharks (Pyle et al., 1996; Robbins, 2007; Weng et al., 2007b). Although most studies have focused on the movement and habitat of adult (>3 m total length) White Sharks, recent work (Chapter 17, this book) highlights the need for a more comprehensive understanding of habitats or areas important to JWS.

In this study we examined long-term catch-per-unit-effort (CPUE) records from Australian SCPs and the movement of satellite-tracked JWS in order to identify:

1. Beach areas and regions used by JWS
2. Seasonality, length, and relative abundance of JWS
3. Physical parameters that may influence CPUE of JWS in SCPs
4. SCP areas conducive to long-term monitoring of JWS population trends

These data are vital prerequisites for managing human impacts on White Sharks in the coastal zone, minimizing the risk of encounter between sharks and people, and establishing more effective long-term monitoring of shark abundance and population trends.

MATERIALS AND METHODS

Shark Control Programs

The east coast of mainland Australia extends from a tropical/subtropical climate in Qld to temperate waters in NSW and cool temperate waters in Victoria. The Qld SCP operates gillnets and drumlines at 10 locations between Cairns (17°S) and the Gold Coast (28°S) (Figure 19.2). Fishing effort differs considerably between locations, largely because of the varying number of beaches protected by the program (Table 19.1). JWS (≤3 m in total length) were captured throughout the region covered by the program, but few were caught in locations north of Capricorn Coast. Hence, analyses in this chapter were confined to catch data taken from the 400-km region from the Capricorn Coast to the Gold Coast (Figure 19.2). In NSW, the shark control program uses only gillnets, and these nets are removed from the water each year between May and August (Green et al., 2009). Nets are deployed at beaches from the southern end of Stockton beach near Newcastle, south to the Illawarra region near Wollongong, across a distance of approximately 200 km (Figure 19.2), although the maximum total length of nets in the water at any time is 7.5 km. Records of White Shark catch in the Qld SCP date from 1963. However, effort varied significantly across the area spanned by the program and was not fully documented prior to 1993. Thus, it was not feasible to accurately estimate CPUE. For this reason, we have chosen to restrict our analyses of the Qld SCP to the years 1993–2009, because this was a period of quantifiable and fairly consistent effort. For consistency in comparisons,

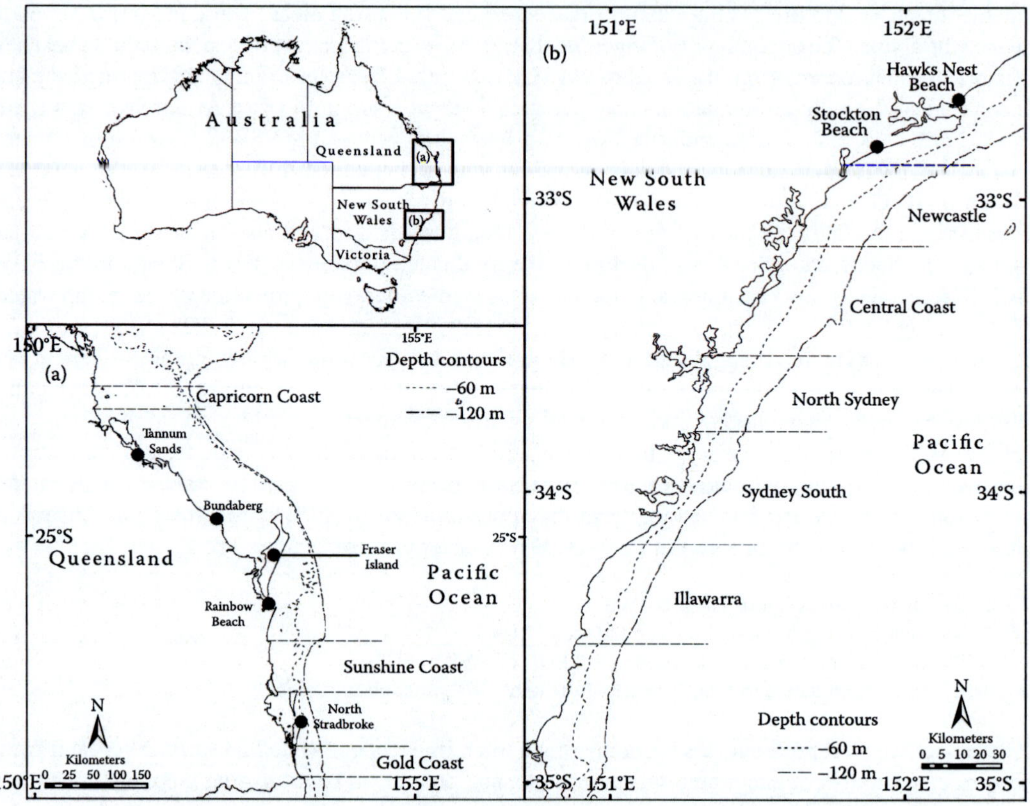

Figure 19.2 The study area in Queensland and New South Wales, Australia, showing the locations of the Queensland Shark Control Program (a) and the New South Wales Shark Meshing Program (b). The dots indicate beach areas, and lines show the extent of certain beach areas. Note that Fraser Island (a) is not an area of the Queensland Shark Control Program. Stockton beach and Hawks Nest beach (areas of most satellite tagging) are shown (b).

data from the NSW SCP were also analyzed over the same period. Prior analyses on JWS catches in the NSW SCP from October 1972 to December 1990 by Krogh (1994) are referred to in this study.

Satellite-Tagging Data

Satellite-tracking data for 20 JWS were assembled (Table 19.2) as part of two other studies [see the work of Bruce et al. (2006) and Chapter 17, this book for full details]. Sharks were tagged at Corner Inlet, Victoria, as well as Stockton and Hawks Nest beaches in NSW. Positions of sharks were detected via the ARGOS system and accepted based on location classes. Our analyses use average daily positions (ADPs) as described in Chapter 17, this book in order to avoid data biasing generated variations in the number of position fixes determined for each shark per day. By comparing the scale of movement between ADPs, Chapter 17, this book noted two types of primary behavior:

1. "Temporary residency" at certain beaches
2. "Travelling" or "transiting" behavior between these nearshore beach areas

Table 19.1 Beaches and Fishing Equipment Used by the Qld SCP and NSW SCP

Region	Location	Number of Beaches	Drumlines	Gillnets	Distance to 60-m Contour (km)	Distance to 120-m Contour (km)
Queensland	Capricorn Coast	9	54		117.3	132.99
	Tannum Sands	1	12		120.64	130.7
	Bundaberg	4	20		96.86	98.07
	Rainbow Beach	1	12	3	25.13	70.03
	Sunshine Coast	23	78	11	34.76	55.28
	Point Lookout	3	28		9.14	19.52
	Gold Coast	24	35	11	19.27	36.38
New South Wales	Newcastle	10		10	12.73	17.8
	Central Coast	11		11	12.97	27.18
	Sydney North	15		15	8.3	19.38
	Sydney South	10		10	3.06	9.8
	Illawara	5		5	10.98	19.232

The table shows the number of beaches and types of fishing equipment used by the Qld SCP and NSW SCP at 10 locations along the east coast of Queensland and five locations along the east coast of New South Wales. Distances to the 60- and 120-m contours were measured from a midpoint in each of the SCP locations.

We used these modes of behavior to compare with SCP CPUE data in subsequent analyses in this study. ADPs of satellite-tracked sharks were plotted according to SCP beach areas and seasonal occurrence in order to see whether patterns reflected SCP CPUE.

Statistical Analyses

Analyses were performed using Genstat (2008), SPSS (v17), and Sigmaplot (2000). The significance of all statistical tests was determined at the 0.05 probability level. Two-part conditional (MacNeil et al., 2009) generalized linear models (GLMs) (McCullagh and Nelder, 1989) were used to determine the patterns of CPUE in SCPs, combining a binomial and a conditional gamma model for the zero-truncated data. The significance of each term was largely taken from the binomial GLM, because this used all of the available data, whereas the gamma GLM only used a subset (the actual catches). CPUE is defined as the number of sharks caught per unit of gear deployed per day. An analysis of deviance was used to describe the fit of the data to the systematic components of the model. Separate models were run for gillnets and drumlines in the Qld SCP, because initial models indicated that the gears were not directly comparable within models. Gillnets were also analyzed separately for the Qld SCP and NSW SCP. GLMs tested the effects of location, year, and season on CPUE. Seasons were defined by summer (December, January, and February), autumn (March, April, and May), winter (June, July, and August), and spring (September, October, and November). Interactions were initially fitted but dropped as their level of importance was orders of magnitude less than the component main effects. To determine where differences among locations occurred within each data set, protected pair-wise *t* tests were used to compare the combined means from the two-part GLM. Daily physical measurements (regardless of capture of a White Shark) of rainfall (Bureau of Meteorology, http://www.bom.gov.au), fraction of the moon illuminated (lunar phase) (http://aa.usno.navy.mil/data/docs/MoonFraction.php), and sea-surface temperature, current direction, current velocity (m s^{-1}), and chlorophyll a (mg m^{-3}) (CSIRO, http://www.csiro.au) were also analyzed in relation to the CPUE using the conditional GLM. Independent variables used in the final models were location + year + season and physical

Table 19.2 Catch Metrics for Juvenile White Sharks Satellite-Tagged off the East Coast of Australia, 2001–2009

Shark Number	Bruce and Bradford Identifier (CSIRO Identifier)	Total Length (m)	Gender	Tagging				Days Tracked	Distance Moved (km)
				Date	Region	Location	Zone		
1	S1 (H)	1.8	F	3/1/2000	Vic	Corner Inlet	22	49	880
2	S2 (N)	2.4	M	3/2/2001	Vic	Corner Inlet	22	129	2,946
3	S5 (66011)	2.1	M	10/8/2007	NSW	Stockton Beach	12	104	4,180
4	S6 (36014)	1.9	F	10/8/2007	NSW	Stockton Beach	12	98	3,010
5	S7 (35973)	2.4	F	10/10/2007	NSW	Stockton Beach	12	44	100
6	S8 (66012)	2.6	M	10/10/2007	NSW	Stockton Beach	12	102	1,720
7	S9 (35970)	1.9	M	10/11/2007	NSW	Stockton Beach	12	177	1,170
8	S10 (35971)	0.0	F	10/11/2007	NSW	Stockton Beach	12	81	1,420
9	S11 (36029)	2.0	F	10/11/2007	NSW	Stockton Beach	12	22	100
10	S12 (36016)	2.0	F	10/11/2007	NSW	Stockton Beach	12	80	910
11	S13 (35972)	2.1	M	10/11/2007	NSW	Stockton Beach	12	22	200
12	S14 (35974)	2.4	F	10/11/2007	NSW	Stockton Beach	12	43	80
13	S15 (81191)	1.8	F	10/28/2008	NSW	Hawks Nest	12	225	2,570
14	S16 (81192)	2.3	F	10/28/2008	NSW	Hawks Nest	12	63	190
15	S18 (98038)	2.1	M	10/28/2009	NSW	Hawks Nest	12	85 (154)	990 (1,290)
16	S19 (98036)	2.4	F	10/28/2009	NSW	Hawks Nest	12	39	200
17	S20 (98037)	2.3	M	10/29/2009	NSW	Hawks Nest	12	76	260
18	S21 (97992)	2.1	F	10/30/2009	NSW	Hawks Nest	12	65	40
19	S22 (97993)	2.4	F	10/30/2009	NSW	Hawks Nest	12	16	360
20	S23 (97994)	2.2	F	10/30/2009	NSW	Hawks Nest	12	83	930

Refer to Chapter 17, this book for complete details on these sharks.

environmental variables. Total length (m) of JWS was also analyzed by a normal GLM testing for the effects of gear, gender, and state (Qld versus NSW).

RESULTS

The lengths of JWS differed significantly between gears ($\chi^2_{(1\text{ d.f.})} = 10.185$, $p < 0.001$), and between states ($\chi^2_{(1\text{ d.f.})} = 5.62$, $p < 0.001$) but not between genders ($\chi^2_{(2\text{ d.f.})} = 0.281$, $p = 0.423$) (df = 2 as we included individuals for which sex was unknown). Length distributions indicated larger JWS (2.50 ± 0.060 m) on Qld drumlines ($n = 49$) than Qld gillnets (2.22 ± 0.069 m), which, in turn, were larger than those caught in NSW gillnets (1.86 ± 0.044 m) (Figure 19.3). JWS caught in Qld SCP gillnets ranged from 1.6 to 3 m ($n = 34$), whereas those in NSW SCP ranged from <1.2 to 3 m ($n = 86$) with a peak in the 1.6–1.8-m size classes. Four potential neonates (<1.2 m) were captured in NSW SCP during 1993–2009, with one of these in the Newcastle region and the others on the Central Coast. The highest number of JWS caught were in the Newcastle region (Figure 19.4). Few JWS (≤2) were caught over the study period in the Capricorn Coast and Gladstone in Qld; however, 10 sharks were caught at Bundaberg with equal or greater numbers (for combined gears) in all Qld locations south of Bundaberg.

CPUE of SCPs in Qld varies significantly between seasons ($\chi^2_{(3\text{ d.f.})} = 9.65$, $p < 0.001$) and indicated that JWS occur in subtropical Qld during winter and spring with the highest catch in gillnets during this period (Figure 19.5). The largest catch in the NSW SCP occurred during autumn, although this program does not deploy nets during most winter months. Satellite tracking indicated that JWS migrate between distinct "residency" areas located along the temperate, and potentially the subtropical, east coasts of eastern Australia (Figure 19.6). For the satellite-tagged JWS, usage of these areas was seasonally dispersed between Qld and NSW during spring, NSW in summer and winter, and Victoria during autumn (Figure 19.5). Most JWS (mean length of 2.1 ± 0.56 m) sharks were tagged within the Stockton nursery area, and movement from this area varied between individuals (mean = 1113 km, with a range of 40 to 4180 km), largely because of the length of satellite transmission (79.9 ± 53.2 d) (Table 19.2). Seasonal variability in SCP CPUE and ADP of tracked sharks also showed a concordance, and Qld SCP gillnets indicated the largest CPUE occurs in spring in Qld (Figure 19.5).

The GLM indicated that CPUE varied significantly among SCP locations for drumlines ($\chi^2_{(6\text{ d.f.})} = 35.14$, $p < 0.001$) and NSW SCP gillnets ($\chi^2_{(4\text{ d.f.})} = 18.57$, $p < 0.001$), with significantly

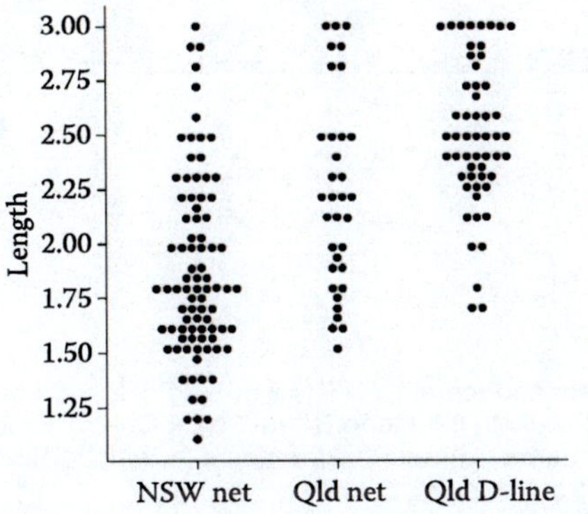

Figure 19.3 Length-frequency histogram for JWS (≤3 m in total length) caught over the period of the study (1993–2009) for drumlines and gillnets in the Queensland Shark Control Program and gillnets in the New South Wales Shark Meshing Program. Note the potential occurrence of neonatal White Sharks (≤1.2 m length) in NSW gillnets. One of these sharks was caught in Newcastle, and three were caught on the Central Coast, New South Wales.

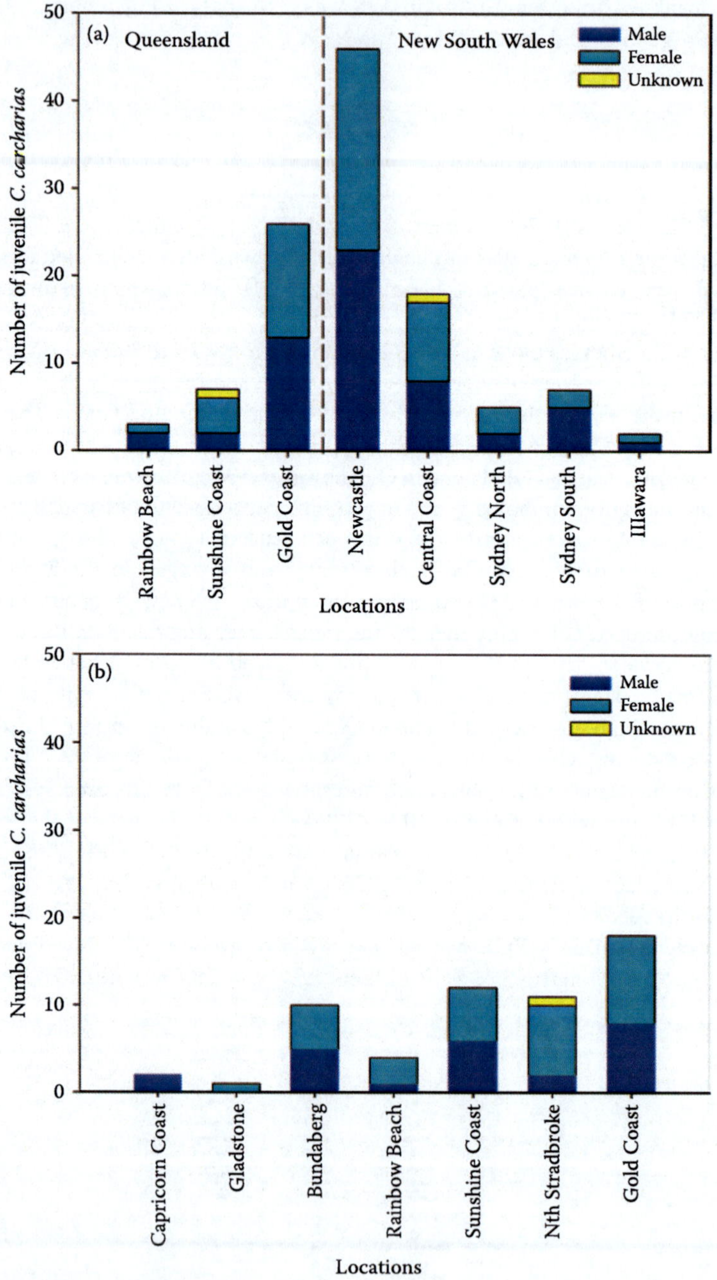

Figure 19.4 Total catch for male and female JWS (≤3 m in total length) caught over the period of the study (1993–2009) for gillnets in the Queensland Shark Control Program and in the New South Wales Shark Meshing Program (a) and for drumlines in the Qld SCP (b). Sharks for which no gender data was recorded are listed as unknown.

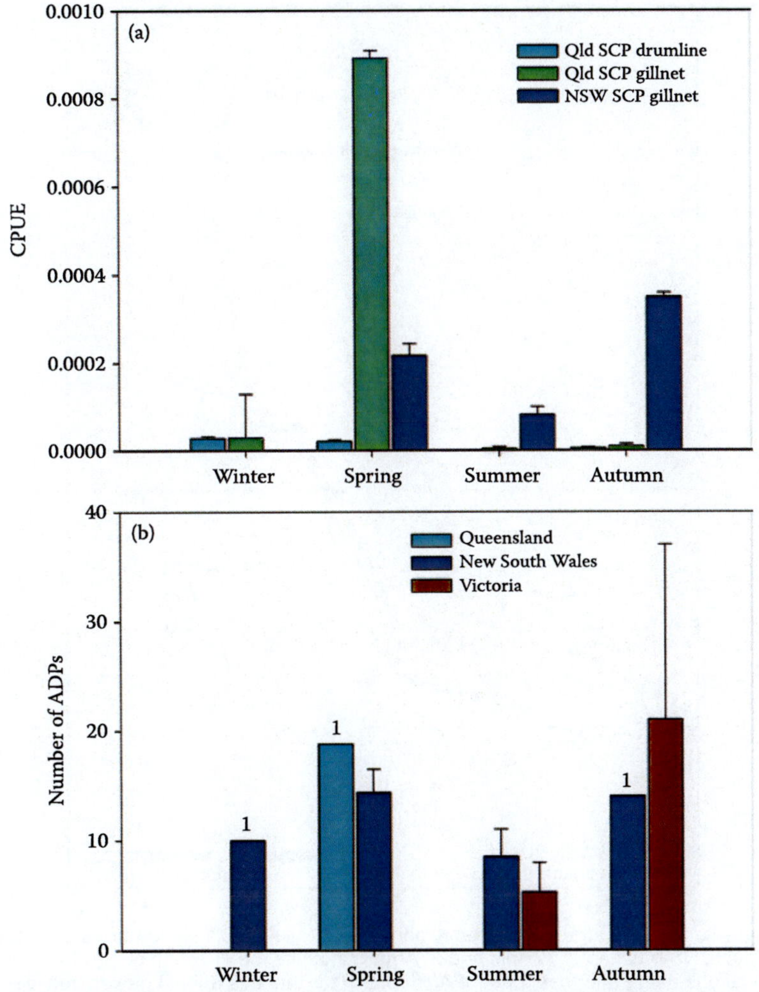

Figure 19.5 (a) Model-predicted seasonal variation in QLD SCP drumline and gillnet CPUE (number of sharks/number of gear/day) and NSW SCP (± SE) by season. Note that NSW SCP does not deploy gear during the winter period. (b) Seasonal distribution of number of ADPs of twenty satellite-tagged JWS in each state on the east coast of Australia (± SE). Columns with 1 above them indicate ADPs for a single satellite-tagged JWS.

higher CPUE at North Stradbroke in Qld and Newcastle and Sydney South regions in NSW (Figure 19.7). North Stradbroke in Qld and South Sydney in NSW were the two locations closest to the 60–120 m contours (Table 19.1). Rainbow Beach had the highest CPUE in Qld for areas with gillnets; however, there was high variability in the annual CPUE, and also given lower replication, it was not quite significantly different ($p = 1.61$) from the Sunshine and Gold Coast areas. These areas of large catches occurred with close proximity (several kilometers) to "temporal residency" areas of the satellite-tagged JWS. Drumline CPUE was also high at Bundaberg (Figure 19.7), which is close to an adult area of site fidelity confirmed with satellite tracking by Bruce et al. (2006) and Bonfil et al. (2010). SCP CPUE and ADPs also showed a concordance between SCP locations (Figure 19.7), and Table 19.3 shows that the only significant relationship between CPUE of JWS and the physical parameters tested was moon phase with drumline CPUEs ($\chi^2_{(1\ \mathrm{d.f.})} = 7.17$, $p = 0.007$). JWS catch decreased significantly from no moon (0% illumination) to full moon (100% illumination) on

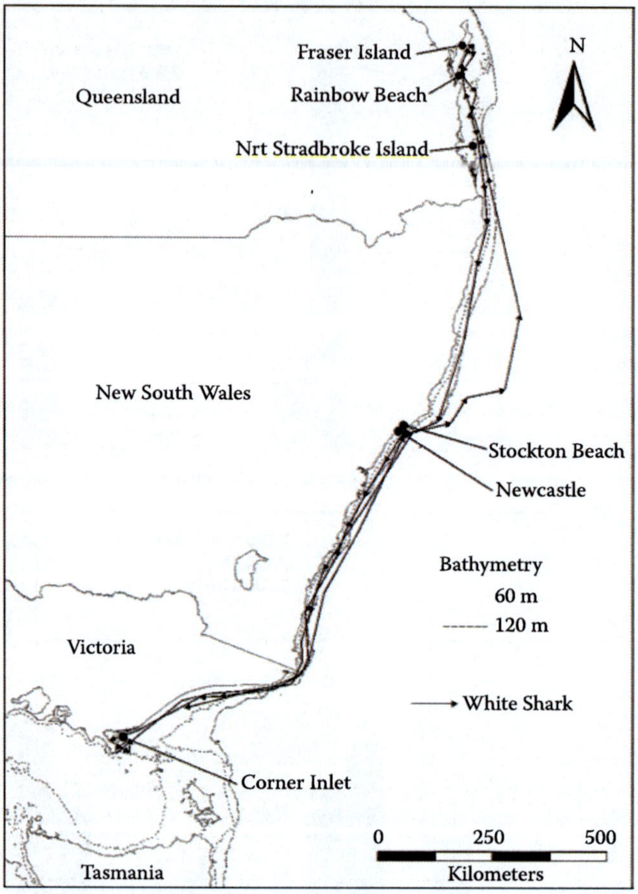

Figure 19.6 Satellite track illustrating the movement of a single JWS (S5; Table 19.2) tagged at Stockton Beach, NSW (Chapter 17, this book). Arrows indicate the direction of movement in coastal beach locations along the east coast of Australia. Note that this animal moves between the temperate waters of Victoria and NSW and into the subtropical waters of Queensland. Tagging locations for JWS (Stockton beach, NSW, and Corner Inlet, Victoria) are illustrated, with SCP beaches with high JWS CPUE also shown.

Qld SCP drumlines (Figure 19.8). Although the effect of moon phase on net captures in both NSW and Qld was not significant (Table 19.3), the fitted trends were similar to those in the drumline data, with the highest CPUEs occurring at no moon.

DISCUSSION

Defining key habitats or important areas used by a species is critical to their conservation because efforts to mitigate threats can be directed at such areas where appropriate (Hodson, 1999). The importance of shark nurseries has been an accepted concept in shark biology and behavioral studies for several decades, but a robust definition of such areas has only recently been proposed by Heupel et al. (2007). They defined shark nursery areas on the following standards:

1. Neonate and small juvenile sharks are more commonly encountered in the area than other areas.
2. Sharks have a tendency to remain or return for extended periods.
3. The area or habitat is repeatedly used across years.

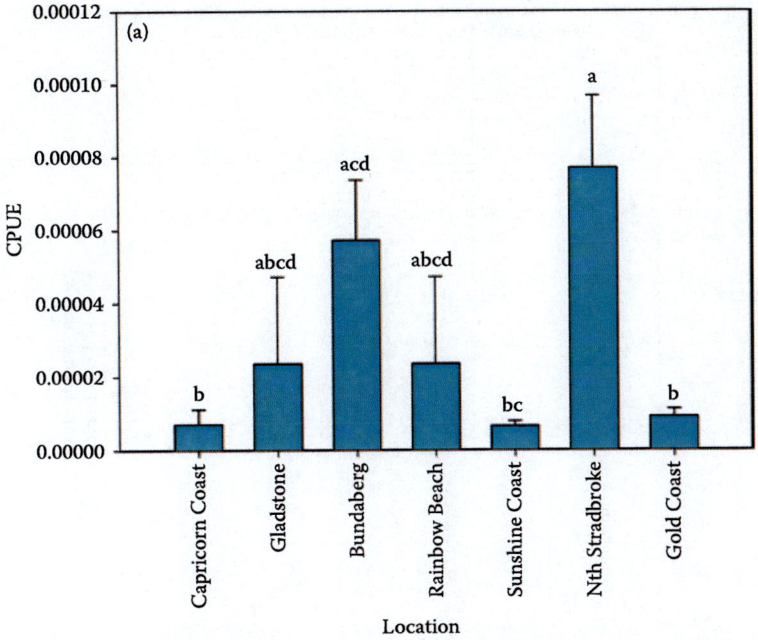

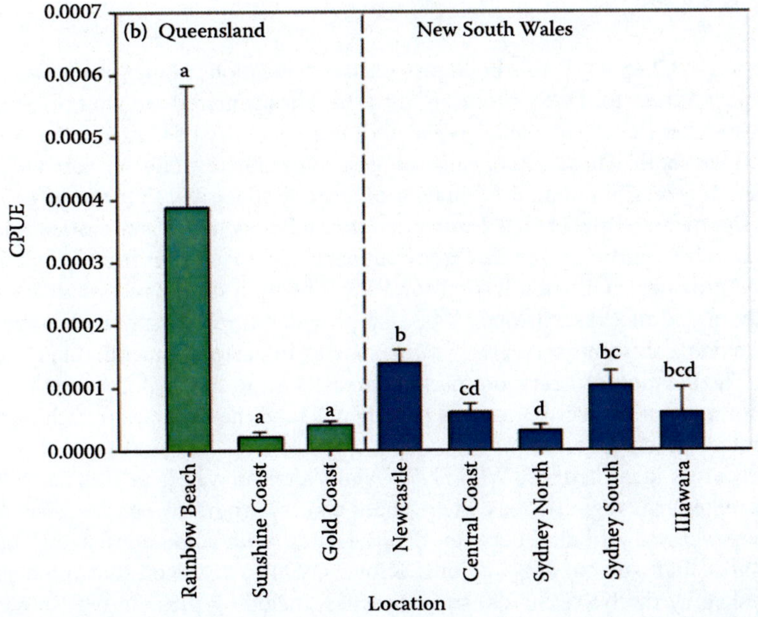

Figure 19.7 Model-predicted QLD SCP drumline (a) and gillnet for both SCPs (b) JWS CPUE (number of sharks/number of gear/day) (±SE) by location (1993–2009). The vertical bars are standard errors. It is interesting to note that Newcastle and Rainbow beach appear to be locations with highest overall predicted CPUE for JWS. Both of these locations are adjacent to suspected "residency" areas for JWS in these states. Locations with a common letter show no significant difference, from pair-wise *t* tests; a is allocated to the largest mean. (c) The number of ADPs (±SE) of 20 satellite-tracked JWS in SCP beach areas (gray bars). Dotted lines separate ADPs from each state on the east coast of Australia. Columns with 1 above them indicate ADPs for a single satellite-tagged JWS.

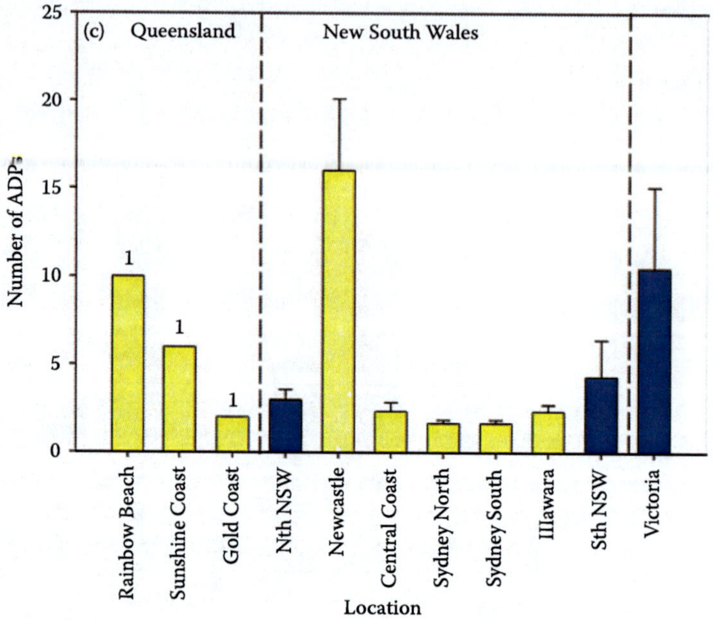

Figure 19.7 (Continued)

Using these criteria, Chapter 17, this book proposed two locations along the east coast of Australia as seasonal nursery areas for JWS: a series of three beaches centered around Port Stephens (including northern Stockton Beach) in central NSW and in the vicinity of Corner Inlet, Victoria (Figure 19.5) some 900 km south. These authors also suggest a possible third nursery area off Fraser Island, Queensland, but this was based on the behavior of a single shark and thus not sufficiently robust to confirm. The significantly higher CPUE recorded for SCP nets in the Newcastle region, and southern Stockton Beach in particular, in this study supports the conclusion from Chapter 17, this book that this region provides significant habitat for JWS. The high CPUE recorded at Rainbow Beach near the southern end of Fraser Island, Qld, also provides further support that the Fraser Island region may provide a third nursery area for the species in eastern Australia and warrants further investigation. The location of nearshore nursery areas close to major population centers increases the potential for interactions between sharks and people and the chances of such interactions leading to undesirable outcomes for both.

The nursery areas identified for JWS in NSW and Victorian waters fall outside (albeit just in the case of the Port Stephens region) areas where shark control programs operate. Even so, catch rates of JWS are not evenly distributed between SCP beaches, with some consistently taking a higher number of sharks than surrounding regions. Krogh (1994) also noted that more juvenile White Sharks were taken by the NSW SCP at some beaches, including those in the Newcastle region as well as Wattamolla and Garie beaches south of Sydney (Sydney South; Figure 19.2). The relatively high number of White Sharks caught at the latter two beaches were considered by Krogh (1994) to be an unusual result and difficult to explain other than the proximity to deeper water. Indeed, our CPUE analyses confirm the findings of Krogh (1994), but when combined with satellite tracking data, the significance of the proximity to deeper water is not that JWS are locally resident to these beaches but that these beaches are closer to the 60–120-m "in-transit" depth corridor for movement along the coast and thus possibly more likely to receive visits of passing sharks. Close proximity to this depth corridor could explain the significantly higher catches of JWS in drumlines by the Qld SCP at North Stradbroke, Qld, than at the surrounding beaches. However, this requires detailed

Table 19.3 **Probability Values for the Binomial GLM Analyzing the Effects of Various Factors on JWS Catch in Gillnets and Drumlines**

SCP	Gear	Rainfall	Sea-Surface Temperature	Chlorophyll a	Current Direction	Current Velocity	Lunar Phase
QLD	Drumline	0.066	0.551	0.840	0.661	0.132	**0.007**
	Gillnets	0.694	0.977	0.986	0.157	0.145	0.217
NSW	Gillnets	0.525	0.173	0.929	0.188	0.113	0.505

Significant values are highlighted in bold.

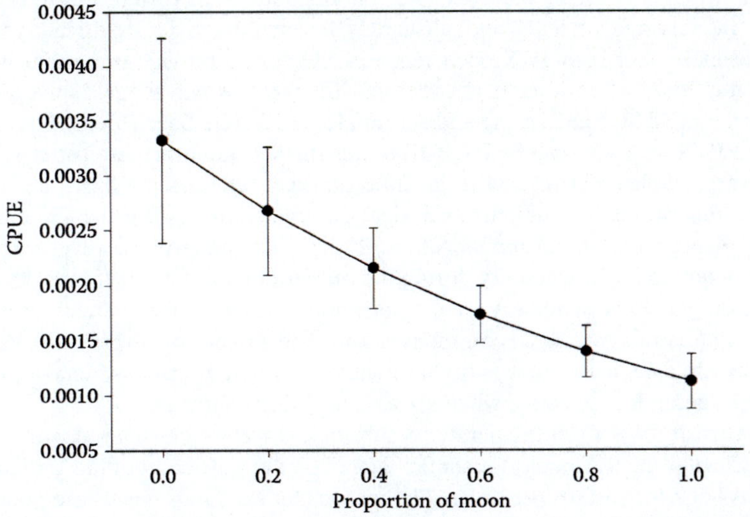

Figure 19.8 Model predicted trends in juvenile White Shark CPUE (number of sharks/number of gear/day) (±SE) in relation to moon phase from Qld SCP drumlines. 0 indicates no moon through to 1 indicating 100% moon illumination.

analyses but does support the hypothesis of a depth-movement corridor proposed in Chapter 17, this book.

The significant difference in the size of JWS between Qld and NSW SCPs may reflect the selectivity of the gears in each SCP or suggests that the seasonal movement into Qld may occur with larger juvenile size classes. Captures of JWS in NSW SCP and NSW recreation/game fishing captures are of the same size as those found in NSW nursery areas rather than those caught in Qld drumlines (Chan, 2001; Chapter 17, this book). Our data suggest that JWS move between important areas or temporary resident "hot spots" along the east coast, possibly moving into Qld waters during spring and winter from NSW. Hot spots are not uncommon in both adult and juvenile White Sharks; however, the characteristics of these hot spots are yet to be identified (Bruce et al., 2006; Bonfil et al., 2010; Domeier and Nasby-Lucas, 2008, Jorgenson et al., 2009). On the east coast of Australia, JWS may be more likely to be caught by SCP gears that are closer to the "in-transit" corridor, whereas beach areas with an orientation eastward toward the corridor possibly increase the likelihood of temporary residency by JWS. We hypothesise that JWS move from Corner Inlet to Stockton beach (Newcastle) and possibly stop at the south side of Fraser Island, which extends out across the shelf, on a seasonal basis before moving beyond Fraser Island and continuing northward to the hot spot identified by Bruce et al. (2006). As JWS begin to return south, the orientation of Fraser Island on the continental shelf may direct juveniles into the Bundaberg area at the north of Fraser Island, explaining the large drumline catch of White Sharks in this area. Stockton beach faces SSE (azimuth of the strike of the beach 65° relative to north), compared to an orientation of

ESE to E for most NSW beaches. The length of Stockton Beach is 31.8 km, compared to a mean length of NSW beaches of 1.4 km (Short, 2006). Beach length and orientation may play important roles in the temporary residency of JWS in these beach areas.

White Sharks have been implicated as one of the key shark species involved in attacks on humans in Australian and international waters, and JWS in particular have been identified in a number of recent Australian attacks (West, 1996; International Shark Attack File, 2010; Chapter 31, this book). On the east coast of Australia, White Shark attacks have occurred at beaches where JWS appear to be in transit rather than in residency or nursery areas. Behavior at residency areas often included extended periods within the nearshore surf-zone environment, whereas travel between areas involved infrequent coastal visits (Chapter 17, this book). The differences in behavior suggest that the risk of interaction with SCPs and, ultimately, beach users differ significantly between beach areas. The seasonal patterns in JWS occurrence in SCP gears provide important insight into the timing of potential risk to beach goers at different SCP areas, which suggest spring and winter are of most concern for Qld SCP and spring and autumn for NSW SCP beach areas. The concordance of satellite-tagged JWS with seasonal SCP CPUE trends further supports these patterns. No fatal JWS attacks, however, have been attributed to juvenile sharks. Many marine fishes display directional and long-range migrations in relation to physical parameters (e.g., sea-surface temperature) and seasonally abundant prey (Pittman and McApline, 2001). These patterns, however, do not necessarily equate to a reduction in feeding activity during the migrations. No fatal JWS attacks have occurred at Stockton Beach-Hawks Nest nursery area, despite numerous recorded interactions with swimmers (Surf Life Saving Australia, personal communication). The absence of adverse attacks on humans in nursery areas where JWS abundance is higher than areas where transitory behavior occurs may be a result of sharks being less selective when seeking prey during transit.

The migration of JWS from temperate to subtropical waters has been documented by Weng et al. (2007b; Chapter 28, this book) for sharks tagged on the west coast of the United States and in Chapter 17, this book for eastern Australia. The reasons for such movements are poorly understood but may presumably be driven by energetic requirements and the pursuit of migratory or seasonally abundant prey species. For example, several teleost species, that is, Striped Mullet, *Mugil cephalis*, and Tailor, *Pomatomus saltatrix*, migrate seasonally along the east coast of Australia to spawn in northern NSW or southern Qld waters (Miskiewicz et al., 1996; Kailola et al., 1993). JWS may time their migrations to feed on these species as they move north into subtropical areas. Long-term records in the Qld SCP indicate that the majority of JWS are caught in Qld waters in winter. This is coincident with the period when these fish species are most abundant in subtropical waters (Kailola et al., 1993). JWS have also been recorded during winter in the tropical waters of New Caledonia; at similar latitudes to southeast Qld (Bonfil et al., 2010; Chapters 21 and 23, this book), however, the reasons for these movements remain unknown. Nonetheless, subtropical migration appears to be an important component of the species ecology and life cycle in the Australasian region. The timing of and extent of movements of sharks away from subtropical Queensland is poorly known and thus a fertile area for further research.

Moon phase was the only variable having a statistically significant effect on shark-catch rates in either the NSW or Qld SCPs. Moon phase is a demonstrated cue for fish movement in both estuarine and shelf waters (Gaudreau and Boisclair, 2000; Naylor, 2001) and has been shown to play a role in the food availability and activity for example, in Lemon Sharks (Morrissey and Gruber, 1993). The movement of prey in relation to moon phase has been linked to CPUE in Blue Sharks, *Prionace glauca*, and Shortfin Mako, *Isurus oxyrinchus*, in pelagic waters (Lowry et al., 2007). Trillmich and Mohen (1981) suggested moonlight avoidance in the Galapogos Fur Seal, *Arctocephalus galapagoensis*, was partly related because of the influence of moonlight on the vertical distribution of prey. Weng et al. (2007b) noted a depth preference of JWS in relation to moon phase for sharks tracked on the west of the United States, possibly in relation to pursuit of prey. The relationship between drumline CPUEs and moon phase for JWS in the Qld SCP could be related to movement of fish species into beach areas in response to the lunar cycle or the avoidance of nets because of increased visibility at night. However, the testing of such a relationship will require additional data.

The status of JWS populations on the east coast of Australia remains unknown despite concerns regarding populations and their protection since 1996. This study supports the existence of a possible third important area located in southern Qld in the vicinity of Fraser Island and suggests on the basis of comparison between satellite tracking data and SCP catches that beaches within these programs with the highest White Shark catches are those in close proximity to defined important areas. Beaches close to the "in-transit" 60–120-m depth corridor may explain high CPUE between nursery areas, but this is an area requiring future research. Similarly beach orientation may be important in determining the probability of encountering a White Shark in beach areas. SCP data from beaches closest to nursery areas may provide the most robust data sets to monitor long-term trends in the species taken in these programs. A dedicated satellite-tagging program of JWS in subtropical Qld and southern Victorian waters would improve our understanding of the long-range movement between tropical and temperate areas of east Australian waters and provide greater insight into the patterns of movement observed in JWS tagged at Stockton/Hawks Nest beach.

ACKNOWLEDGMENTS

We gratefully Rocio Noriega and Ciaran Morris for ArcView production of maps. Ciaran Morris also provided assistance with the collation of physical parameter data for analyses. Tony Ham (Department of Employment, Economic Development and Innovation) for access to Qld SCP data. We thank Ben Doolan for the photo used in this paper.

REFERENCES

Baille, J.E.M., Hilton-Taylor, C., Stuart, S.N. (eds.). 2004. *2004 IUCN Red List of Threatened Species: A Global Species Assessment*, p. xxiv. Cambridge, UK: IUCN.

Bonfil, R., Meyer, M., Scholl, M.C., Johnson, R., O'Brien, S., Oosthuizen, H., Swanson, S., Kotze, D., Paterson, M. 2005. Transoceanic migration, spatial dynamics, and population linkages of white sharks. *Science* 310: 100–103.

Bonfil, R., Francis, M.P., Duffy, C., Manning, M.J., O'Brien, S. 2010. Large-scale tropical movements and diving behaviour of white shark *Carcharodon carcharias* tagged off New Zealand. *Aquatic Biology* 8: 115–123.

Bruce, B.D. 2008. White sharks: The biology and ecology of *Carcharodon carcharias*. In *Sharks of the Open Ocean*, edited by E. Pikitch and M. Camhi, pp. 69–81. Oxford: UK: Blackwell Scientific.

Bruce, B.D., Stevens, J.D., Malcolm, H. 2006. Movements and swimming behaviour of white sharks (*Carcharodon carcharias*) in Australian waters. *Marine Biology* 150: 161–172.

Chan, RW.K. 2001. *Biological Studies on Sharks Caught off the Coast of New South Wales*. Ph.D. thesis, University of New South Wales.

Cliff, G., Dudley, S.F.J., Davis, B. 1989. Sharks caught in the protective gill nets off Natal, South Africa: 2. The great white shark *Carcharodon carcharias* (Linnaeus). *South African Journal of Marine Science* 8: 131–144.

Domeier, M.L., Nasby-Lucas, N. 2008. Migration patterns of white shark *Carcharodon carcharias* tagged at Guadalupe Island, Mexico, and identification of an eastern Pacific shared offshore foraging area. *Marine Ecology Progress Series* 370: 221–237.

Dudley, S.F.J., Simpfendorfer, C.A. 2006. Population status of 14 shark species caught in the protective gill nets off KwaZulu-Natal beaches, South Africa, 1978 to 2003. *Marine and Freshwater Research* 57: 225–240.

Environment Australia. 2002. *White Shark (Carcharodon carcharias) Recovery Plan*. http://www.deh.gov.au/coasts/publications/gwshark-plan/pubs/greatwhiteshark.pdf.

Gaudreau, N., Boisclair, D. 2000. Influence of moon phase on acoustic estimates of the abundance of fish performing daily horizontal migration in a small oligotrophic lake. *Canadian Journal of Fisheries and Aquatic Sciences* 57: 581–590.

Green, M., Ganassin, C., Reid, D.D. 2009. *Report into the NSW Shark Meshing (Bather Protection) Program.* http://www.dpi.nsw.gov.au/data/assets/pdf_file/008/276029/Report-into-the-NSW-Shark-Meshing-Program.pdf.

Gribble, N.A., McPherson, G., Lane, B. 1998. Effect of the Queensland Shark Control Program on non-target species: Whale, dugong, turtle and dolphin: A review. *Marine and Freshwater Research* 49: 645–651.

Heupel, M.R., Carlson, J.K., Simpfendorfer, C.A. 2007. Shark nursery areas: Concepts, definition, characterization and assumptions. *Marine Ecology Progress Series* 337: 287–297.

Hodson, K.A. 1999. Tracing origins and migration of wildlife using stable isotopes: A review. *Oecologia* 120: 314–326.

International Shark Attack File. 2010. Florida Museum of Natural History. http://www.flmnh.ufl.edu/fish/sharks/isaf/access.htm.

Jorgensen, S.J., Reeb, C.A., Chapple, T.K., Anderson, S., Perle, C., Van Sommeran, S.R., Fritz-Cope, C., Brown, A.C., Klimley, A.P., Block, B.A. 2009. Philopatry and migration of Pacific white sharks. *Proceedings of the Royal Society B Biological Sciences* 270: 679–688.

Kailola, P.J., Williams, M.J., Stewart, P.C., Reichelt, R.E., McNee, A., Grieve, C. 1993. *Australian Fisheries Resources.* Canberra, Australia: Bureau of Resource Sciences, Department of Primary Industries and Energy, and the Fisheries Research and Development Corporation.

Krogh, M. 1994. Spatial, seasonal and biological analysis of sharks caught in the New South Wales protection beach meshing programme. *Australian Journal of Marine and Freshwater Research* 45: 1087–1106.

Lowry, M., Williams, D., Mettia, Y. 2007. Lunar landings: Relationship between lunar phase and catch rates for an Australian gamefish-tournament fishery. *Fisheries Research* 88: 15–23.

MacNeil, M.A., Carlson, J.K., Beerkircher, L.R. 2009. Shark depredation rates in pelagic longline fisheries: A case study from the Northwest Atlantic. *ICES Journal of Marine Science* 66: 708–719.

McCullagh, P., Nelder, J.A. 1989. *Generalized Linear Models* (2nd ed.). London, UK: Chapman and Hall.

Miskiewicz, A.G., Bruce, B.D., Dixon, P. 1996. Distribution and abundance of tailor larvae (*Pomatomus saltatrix*) off eastern Australia. *Marine and Freshwater Research* 47: 331–336.

Morrissey, J.F, Gruber, S.H. 1993. Habitat selection by juvenile lemon sharks, *Negaprion brevirostris. Environmental Biology of Fishes* 7: 341–347.

Naylor, E. 2001. Marine animal behaviour in relation to lunar phase. *Earth Moon Planets* 85–86: 291–302.

Pittman, S.J., McAlpine, C.A. 2001. Movements of marine fish and decapod crustaceans: Process, theory and application. *Advances in Marine Biology* 44: 205–294.

Pyle, P., Anderson, S.D., Klimley, A.P., Henderson, R.P. 1996. Environmental factors affecting the occurrence and behavior of white sharks at the Farallon Islands, California. In *Great White Sharks: The Biology of Carcharodon carcharias*, edited by A.P. Klimley and D.G. Ainley, pp. 111–120. San Diego, CA: Academic Press.

Reid, D.D., Krogh, M. 1992. Assessment of catches from protective shark meshing off NSW beaches between 1950 and 1990. *Marine and Freshwater Research* 43: 283–296.

Robbins, R.L. 2007. Environmental variables affecting the sexual segregation of great white sharks, *Carcharodon carcharias*, at the Neptune Islands South Australia. *Journal of Fish Biolo*gy 70: 1350–1364.

Short, A.D. 2006. *Beaches of the New South Wales Coast.* Australian Beach Safety and Management Program. Sydney, Australia: University Publishing Services, University of Sydney.

Trillmich, F., Mohren, W. 1981. Effects of the lunar cycle on the Galapagos fur seal, *Arctocephalus galagoensis. Oecologia* 48: 85–92.

Weng, K., Boustany, A., Pyle, P., Anderson, S., Brown, A., Block, B. 2007a. Migration and habitat of white sharks (*Carcharodon carcharias*) in the eastern Pacific Ocean. *Marine Biology* 152: 877–894.

Weng, K., O'Sullivan, J., Lowe, C., Winkler, C.H. 2007b. Movements, behavior and habitat preferences of juvenile white sharks *Carcharodon carcharias* in the eastern Pacific. *Marine Ecology Progress Series* 338: 211–224.

West, J. 1996. White shark attacks in Australian waters. In *Great White Sharks: The Biology of Carcharodon carcharias*, edited by A.P. Klimley and D.G. Ainley, pp. 449–455. San Diego, CA: Academic Press.

CHAPTER 20

Seasonal Sexual and Size Segregation of White Sharks, *Carcharodon carcharias*, at the Neptune Islands, South Australia

Rachel L. Robbins*
University of Technology and Fox Shark Research Foundation

David J. Booth
University of Technology

ABSTRACT

The seasonal temporal sexual and size segregation of White Sharks, *Carcharodon carcharias*, at the Neptune Islands, South Australia, is described. Under baited conditions, using chum to attract sharks to the vessel, the sex and size of White Sharks were recorded over a 3-yr. period from April 2001 to February 2004. Sexual and size segregation was observed in this species. There was a high degree of interannual variability in the data with the number of subadult males increasing and mature males decreasing over the 3 yrs. of the study. An overall trend was observed for larger White Sharks to be recorded in the winter/spring months (June to September) and smaller sharks in other months. Sea-surface temperature appeared to be related to sexual and size segregation in this species. It was found to be negatively related to subadult and mature male numbers, whereas no significant relationships with temperature were found for females of either size grouping, although female abundance tended to cluster toward the higher temperature ranges (15.7–18.1°C). The significantly lower temperatures in 2003 may explain the absence of females in this year. It is speculated that females may favor warmer waters to promote and accelerate embryonic development of young as a result of comparatively increased internal core body temperatures.

INTRODUCTION

Sexual segregation appears nearly universal in sharks (Bres, 1993), including the Grey Reef Shark, *Carcharhinus amblyrhynchos* (Economakis and Lobel, 1998); the Scalloped Hammerhead, *Sphyrna lewini* (Klimley, 1987); the Basking Shark, *Cetorhinus maximus*

(Watkins, 1958; Lien and Fawcett, 1986); the Salmon Shark, *Lamna ditropis* (Blagoderov, 1994; Goldman, 2000); the Blue Shark, *Prionace glauca* (Pratt, 1979); the Tiger Shark, *Galeocerdo cuvier* (Hiethaus, 2001); *Centrophorus cf. uyato* (McLaughlin and Morrissey, 2005); and the Nurse Shark, *Ginglyostoma cirratum* (Carrier et al., 1994). Sexual segregation in elasmobranchs occurs for a variety of reasons, including intraspecific competition, prey availability, refuge from mating, or reproductive strategies associated with pre- or postmating behaviors or through differing seasonal habitat and resource requirements (Sims et al., 2001). Shark populations are, at times, also segregated in groups consisting of juveniles, subadults, mature males, and mature females (Springer, 1967). Size segregation is widely reported in sharks (Simpfendorfer et al., 2005), including in the Blacknose Shark, *Carcharhinus acronotus* (Schwartz, 1984); Silky Shark, *Carcharhinus falciformis* (Branstetter, 1987); Tiger Shark, *Galeocerdo cuvier* (Hiethaus, 2001); Porbeagle, *Lamna nasus* (Department of Fisheries and Oceans, 1999); Whale Shark, *Rhincodon typus* (Norman, 1999; Eckert and Stewart, 2001); and White Shark, *Carcharodon carcharias* (Krogh, 1994). Previous studies have supported correlations between sea-surface temperatures and sexual and size segregation in other species (Springer, 1967; Bass et al., 1973; Compagno, 1984; Muñoz-Chápuli, 1984; Economakis and Lobel, 1998; Department of Fisheries and Oceans, 1999; Norman, 1999; Eckert and Stewart, 2001). Different body sizes at various life stages likely present different energy budgets and nutritional requirements to each sex (Ruckstuhl and Neuhaus, 2002).

Sexual segregation has been anecdotally cited in White Sharks (Bruce, 1992; Klimley et al., 1992; Goldman et al., 1996) and reported from Guadalupe, Mexico and South Australia (Robbins, 2007; Chapter 11, this book). Seasonal segregation in this species may be due to prey availability, differing physiological needs relating to abiotic variables such as sea-surface temperature, or reproductive habits. In-depth studies of northeastern Pacific White Sharks have shown clear evidence of sexual segregation. Females may actively avoid the male-dominated Shared Offshore Foraging Area (Chapter 12, this book) during the season the males occupy this region (Chapter 11, this book). Females remain offshore for nearly a year and a half, migrating to the coast only to give birth (Chapters 11 and 16, this book) and presumably to mate (Chapter 16, this book).

Size segregation on a spatial scale has been reported for White Sharks in some parts of their range (Klimley, 1985; Cliff et al., 1996) but not in other areas, for example, South Africa (Ferreira and Ferreira, 1996). In South Australian waters, spatial size segregation has been recorded in some studies (Bruce, 1992; Malcolm et al., 2001). Movements of mature and subadult White Sharks in populations are influenced by external changes in sea temperatures coupled with the potential changes in diet as a result of temperature variations (Ainley et al., 1985; Long and Jones, 1996; Krogh, 1994).

Seasonal variations in sea-surface temperature appear to play a role in the abundance and movements of White Sharks. Records from boreal waters of the eastern North Pacific show White Sharks to occur at sea-surface temperatures of 5–16°C, with most occurring where sea-surface temperatures were between 9 and 10°C (Martin, 2004). Other studies have shown that, on an annual basis, the average monthly temperatures for males ranged from 16.4 to 21.9°C and 16.1 to 24.7°C for females (Chapter 11, this book), suggesting a preferences for warmer waters in female sharks. In Florida, White Sharks seem only to be present between the months of January and April, during winter and early spring (Adams et al., 1994). At this time, water temperatures are at the lowest for this region and range between 18.7 and 21.6°C (Adams et al., 1994). White sharks have not been recorded at other times of year, suggesting that they move into these waters only when temperatures decrease sufficiently. Mediterranean White Sharks seemingly vanish from selected areas when temperatures exceed 25°C, although they may remain deep in the isothermal waters, only coming to the surface sporadically (Fergusson, 1994). Large White Sharks are found to occur seasonally at the Farallon Islands (Weng et al., 2007), exhibiting a strong peak in the autumn months (Ainley et al., 1985) and migrating to an offshore foraging area at other times

(Boustany et al., 2002; Weng et al., 2007; Chapters 11, 16, and 25, this book). Guadalupe Island has similarly been pinpointed as a seasonal aggregation area for large White Sharks (Domeier and Nasby-Lucas, 2007; Chapters 11, 16, and 25, this book). It has been suggested that mature adults migrate to water off Baja (Chapter 15, this book) and the Southern California Bight (Chapters 14 and 16, this book) from May to August (Chapter 16, this book) in order to pup, which is supported by capture data of young of the year (YOY) during the summer months (Chapter 14, this book). It has been concluded that Baja and Southern Californian waters are important nursery areas for YOY White Sharks (Chapters 14, 15, and 16, this book). This leads to sexual segregation, whereby mature females segregate from the rest of the population in order to travel to these areas to pup (Chapter 16, this book); the absence of males may reduce the risk of cannibalism of neonates (Springer, 1967). Adult White Sharks also segregate by sex in pelagic habitats (Chapter 11, this book).

Juveniles are noticeably absent from the Neptune Islands but are common around the western coast of South Australia and in the upper Spencer Gulf waters (Bruce, 1992), where waters are generally warmer than those of the offshore islands. Teleost and elasmobranch fishes are important food sources for White Sharks in Australian waters (Malcolm et al., 2001). Smaller White Sharks eat mainly teleosts and elasmobranchs, with an increase in mammalian prey and a corresponding decrease in these prey with increasing size of shark (Chapter 3, this book). Although whale and seal are the dominant prey in the diet of White Sharks >3 m (Ainley et al., 1981; Casey and Pratt, 1985; Chapter 3, this book), teleosts and smaller prey still encompass a portion of their diet (Chapter 3, this book). Chapter 4, this book suggests that cephalopods form an important component of the diet of large White Sharks in the pelagic environment.

Although the exact timing of reproduction is unknown, parturition (Uchida et al., 1996; Francis, 1996) and mating (Stevens, 1984; Francis, 1996) in Australian waters are thought to occur in the spring and summer based on landings of gravid females and neonates around this time (Francis, 1996). Pregnant White Sharks have been observed from the west coast waters of South Australia in October and November (Bruce, 1992), whereas captures of neonates are most common between December and May (Malcolm et al., 2001). The most complete investigation of White Shark life history has been conducted on the northeastern Pacific population. Current evidence suggests that White Sharks in the northeastern Pacific begin their life in the nearshore, shallow waters of the Southern California Bight and Baja California, and then, as they near maturity, they migrate between offshore habitats and aggregation sites (Boustany et al., 2002; Weng et al., 2007; Chapter 16, this book). Adult males are known to visit aggregation sites annually after up to 9 months spent offshore (Chapter 11 this book), but mature females generally only return to the inshore aggregation sites or parturition grounds every other year (Anderson and Pyle, 2003; Domeier and Nasby-Lucas, 2007; Chapters 16 and 25, this book), exhibiting a primarily pelagic zone residence (Chapters 11, 12, and 16, this book) and supporting a gestation period of more than one year (Francis, 1996; Mollet et al., 2000; Chapter 16, this book). Parturition occurs primarily between May and August (summer/autumn), after which females return to adult aggregation sites (Chapter 16, this book).

The reported annual incidental mortality of White Sharks in Australian waters is estimated to be around 200, although the number is probably much higher (Malcolm et al., 2001). If sexual and size segregation does occur throughout their range, the removal of White Sharks both accidentally and purposefully could result in a high proportion of the breeding population disappearing, or it may affect recruitment by removing neonates and juveniles from the population. The removal of large numbers of one sex or size grouping could have dramatic effects on the population. Thus, knowledge of segregation in population distribution in White Sharks would enable more effective implementation of current and future protection plans. The objective of this study was to ascertain how sea-surface temperature and season interact with sexual and size distribution of White Sharks at the Neptune Islands.

MATERIALS AND METHODS

Expeditions were conducted from April 2001 to February 2004, totalling 27 trips over 140 d to the Neptune Islands, South Australia. The Neptune Islands lie at the mouth of the Spencer Gulf off Port Lincoln, South Australia, and consist of two island groups: North Neptune Islands (35°14'S, 136°04'E) and South Neptune Islands (35°20'S, 136°06'E) (Figure 20.1). The Neptune Islands are the feeding and nationally significant breeding area for New Zealand Fur Seals (*Arctocephalus forsteri*) and Australian Sea Lions (*Neophoca cinerea*) (Robinson et al., 1996). They are also consequently an important feeding and aggregation ground for White Sharks.

The duration of each trip was between 4 and 11 d. Sharks were attracted to the boat by means of a standardized baiting method (Strong et al., 1992). A three-dimensional odor corridor was discharged, consisting of known amounts of unrefined fish oil, minced tuna, tuna blood, and seawater, delivered at a low rate. Baits of whole salmon or tuna chunks (weighing about 1 kg) were attached to a natural fiber bait line approximately 15 m in length and allowed to drift from the stern of the boat. Once a shark arrived and showed interest in the baits, the baits were drawn closer to the boat by crewmembers to allow tagging and identification to take place.

As each shark arrived at the vessel, its total length (TL) was estimated. This was done by visual comparison with known length vessel structures. For example, from pontoon to pontoon, the dive cages were 240 cm and, because the sharks often swam directly past or in line with the

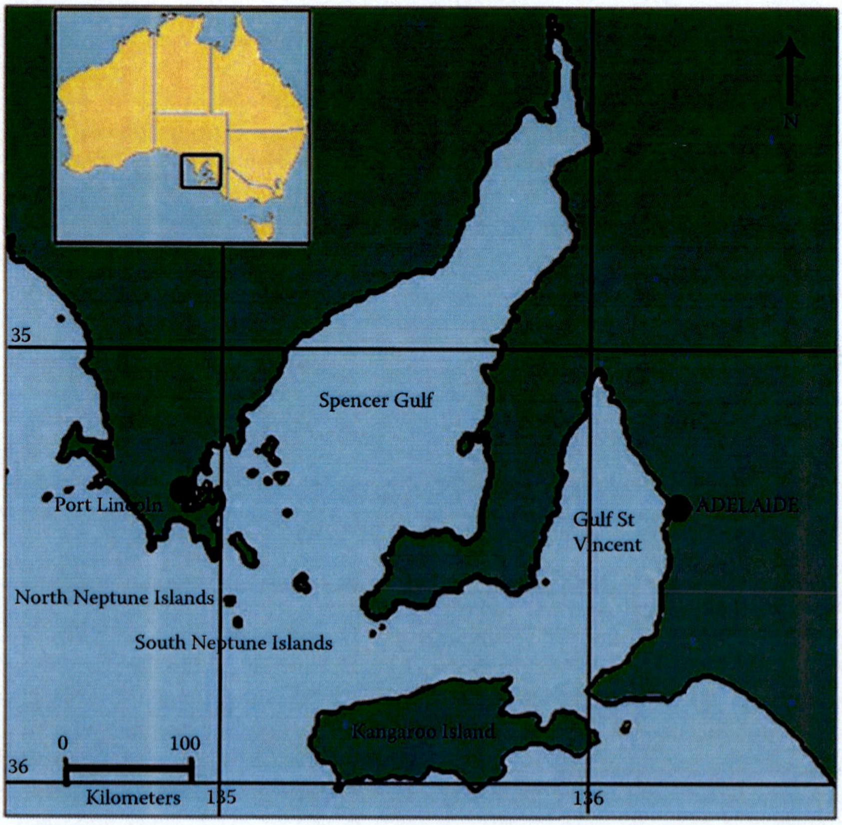

Figure 20.1 Location of the Neptune Islands in relation to well-known sites in South Australia. North Neptune Islands, 35°14'S, 136°03'E; South Neptune Islands, 35°20'S, 136°06'E.

cages, a length estimate could be made. This was also confirmed by photographic analysis of sharks alongside measured objects.

The sex of each shark was recorded by determining the presence or absence of claspers. White Sharks <300 cm TL were classified as juveniles, and White Sharks >300 cm TL were classified as subadults (Chapter 17, this book). Females were considered mature once they reached 480 cm TL (Chapter 17, this book). In males, sharks ≥360 cm TL were considered to be sexually mature (Francis, 1996; Pratt, 1996; Chapter 17, this book).

Sea-surface temperature was measured at a consistent location on the vessel using a calibrated temperature probe (CIE Testing Instruments, model 305; accuracy ±0.3%). It was ensured that readings were taken up-current from the chum dispenser so that the colder temperature of the chum mixture did not interfere with the true sea-surface temperature. Readings from the Neptune Islands, Spencer Gulf, Gulf St. Vincent, and Ceduna were also taken, using remote sensing satellite imagery (CSIRO Marine Research 2001–2003).

The sharks were categorized into groups, namely subadult females, mature females, subadult males, and mature males. A two-factor analysis of variance (ANOVA) was applied to determine the effects of life stage and year on the mean daily number of sharks. The data from the months of March and December were removed from ANOVA analyses because of the lack of replication seen in these months (only sampled in 2003 because of constraints on the expedition timetable). Regression analyses were utilized to determine relationships between mean daily numbers of subadult and mature sharks of each sex per month with sea-surface temperature. Multidimensional scaling was applied, using the environmental data, month, year, and the mean daily number of White Sharks as variables to construct an ordination of mature and subadult White Sharks of each sex. Similarities between counts of mature and subadult White Sharks were given by Bray-Curtis similarity coefficients (Clark and Warwick, 1994).

RESULTS

The overall sex ratio was biased toward males with a ratio of 3:1. For subadult sharks, the sex ratio was slightly biased toward males (ratio, 1.8:1), but for mature sharks, the male bias was much more obvious, with a sex ratio of 12:1. A chi-squared test confirmed that there was a highly significant departure from homogeneity in these sex rations ($\chi^2 = 8.42$, $p = 0.0037$).

Segregation of Mature and Subadult White Sharks

The population at the Neptune Islands over 2001–2003 consisted of 38.9% subadult male White Sharks, 25.4% subadult females, 23.8% mature males, and only 2.4% mature females (the remaining 9.5% of sharks were unidentified in terms of either sex or size and hence are excluded from analyses). Mature females were infrequently encountered compared with other life-history stages sampled (Figure 20.3). While subadult male numbers increased over the course of the study, the opposite was true for mature males and females.

Mature females were encountered only during autumn and early winter (May to July) of 2001, when temperatures were still relatively warm (Figure 20.2), and females were absent altogether in 2003, which had overall much colder temperatures (Figure 20.2). Subadult females were present throughout autumn and early to midwinter in 2001 and 2002 (Figure 20.4a), with their numbers decreasing as sea-surface temperatures dropped. Subadult males were present in all but one month sampled (April; Figure 20.4c). Mature males were recorded in all months sampled except March (Figure 20.4d). When data from all years were combined, a seasonal pattern was more easily discernible, whereby females were present in autumn and early winter before the influx of males in mid-winter to early spring (Figure 20.3). Males, in particular subadults, were also prevalent in February (austral summer) (Figure 20.3).

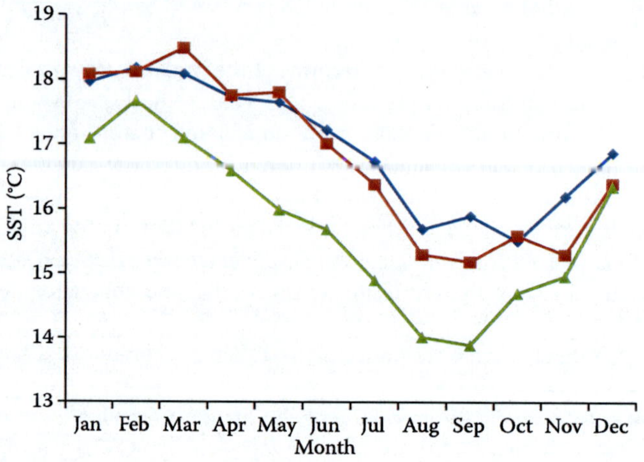

Figure 20.2 The differences between the monthly sea-surface temperatures at the Neptune Islands for each year of the study (n = 3 yrs.) (♦2001, ■ 2002, ▲ 2003).

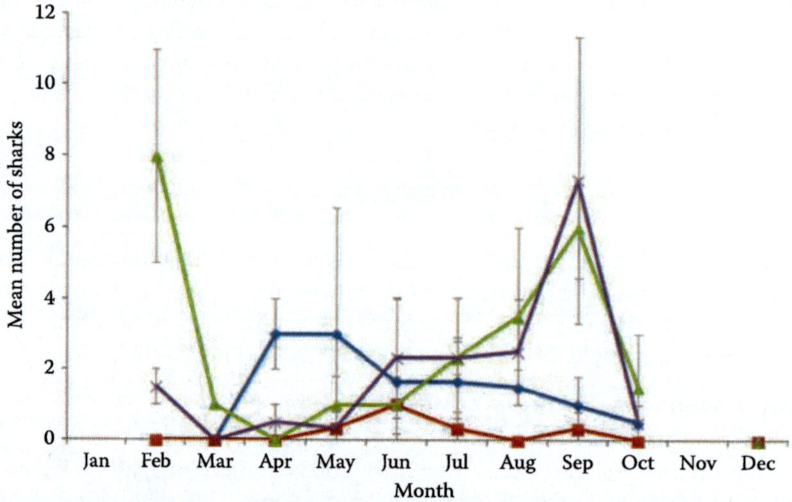

Figure 20.3 Mean daily number of White Sharks of each life history stage for each year of the study: ■ subadult females; ■ mature females; ■ subadult males; and ■ mature males.

Significant interactions existed between life-history stage and year (two-factor ANOVA, F = 2.552, p = 0.026), in relation to mean daily numbers of White Sharks (Table 20.1), with 2001 having a higher daily mean number of sharks than other years in all stages except subadult males. The mean daily numbers of females from both size groupings decreased from 2001 to 2002 (females were absent in 2003). A two-factor ANOVA also showed a significant interaction between life-history stage and month (two-factor ANOVA, F = 1.711, p = 0.048; Table 20.2) with females of both life-history stages being most abundant in late autumn to early winter (April to June), mature males being most abundant in late winter to early spring (August to October), and subadult males having peaks in both February and late winter (August and September). Figure 20.4 illustrates the differences in the abundances of each life-history stage in each month sampled and the significant differences between the years.

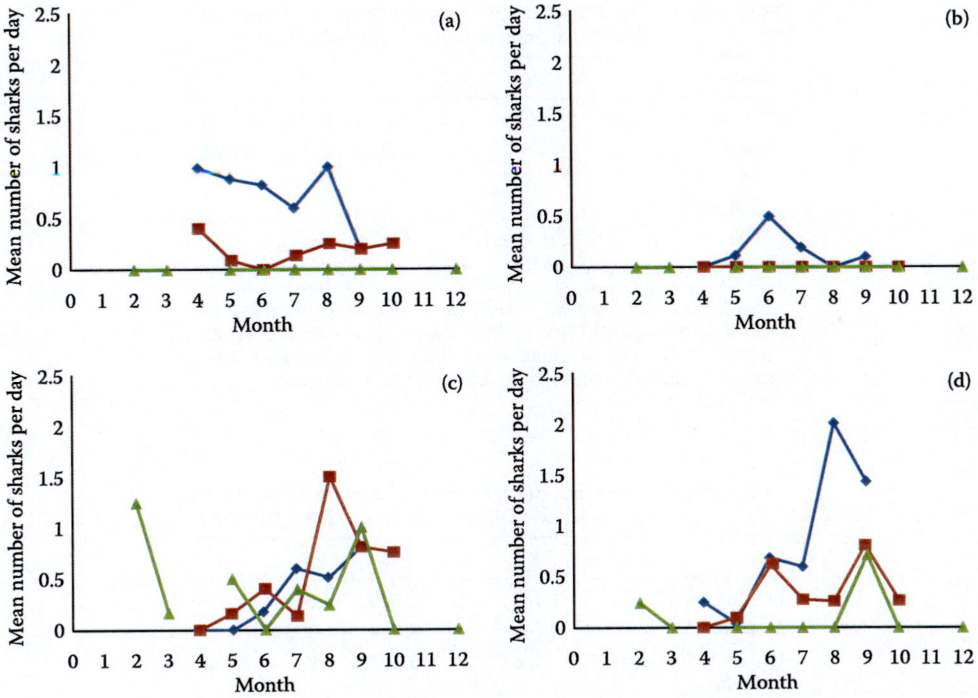

Figure 20.4 Mean daily numbers of subadult females (a), mature females (b), subadult males (c), and mature males (d) for each year of the study (n = 3 yrs.) (♦ 2001, ■ 2002, and ▲ 2003).

There was a negative relationship between sea-surface temperature and subadult male numbers (linear regression on log-transformed data, $R^2 = 0.55$, $p = 0.004$; Figure 20.5a) and with mature males (linear regression on log-transformed data, $R^2 = 0.45$, $p = 0.007$; Figure 20.5a). There was no difference between abundances of subadult and mature males with respect to sea-surface temperature (two-factor ANOVA, $F = 0.016$, $p = 0.899$). No linear relationships were found to exist between either of the female groups with sea-surface temperature, nor did a two-factor ANOVA reveal a significant relationship between maturity category and sea-surface temperature for female White Sharks. However, Figure 20.5b suggests that female abundance was skewed toward the higher temperatures, with no females encountered when temperatures dropped below around 16°C.

DISCUSSION

The population structure at the Neptune Islands was distinctly segregated spatially and temporally. The Neptune Islands are undoubtedly an important aggregation area for subadults of both sexes and for mature males, although mature females are less frequently encountered. Mature females were present from May to July, whereas mature males were recorded in every month sampled except March. Females at the Neptune Islands are thus shifted earlier than males, arriving and departing earlier in the season. This is in contrast to findings from Guadalupe Island, where adult males were found to arrive earlier in the season than females (Domeier and Nasby-Lucas, 2007; Chapters 11, 16, and 25, this book) and depart mostly between January and March (Chapter 25, this book). Females depart later than males at the Farallon Islands, which is suggested to be consistent with a higher-energy demand in females and the need to remain longer to provision for migration

Table 20.1 The Effect of Life-History Stage and Year on the Mean Daily Numbers of White Sharks

Source	SS	df	MS	*F*	*P*
Life-history stage	2.659	3	0.886	5.231	0.002
Year	1.746	2	0.873	5.151	0.008
Life -history stage × year	2.595	6	0.433	2.552	0.026
Error	13.558	80	0.169		
Total	29.057	92			

The table shows two-way ANOVAs investigating the effect of life-history stage and year on the mean daily numbers of White Sharks. SS, sum of squares; df, degrees of freedom; MS, mean square; *F*, ratio of mean squares; *P*, probability.

Table 20.2 The Effect of Life-History Stage and Month on the Mean Daily Numbers of White Sharks

Source	SS	df	MS	*F*	*P*
Life-history stage	2.147	3	0.716	4.643	0.006
Month	2.767	9	0.307	1.995	0.059
Life-history stage × month	7.118	27	0.264	1.711	0.048
Error	8.014	52	0.154		
Total	29.057	92			

The table shows two-way ANOVAs investigating the effect of life-history stage and month on the mean daily numbers of White Sharks. Note that n = 10 months (November and January were dropped because of missing data). SS, sum of squares; df, degrees of freedom; MS, mean square; *F*, ratio of mean squares; *P*, probability.

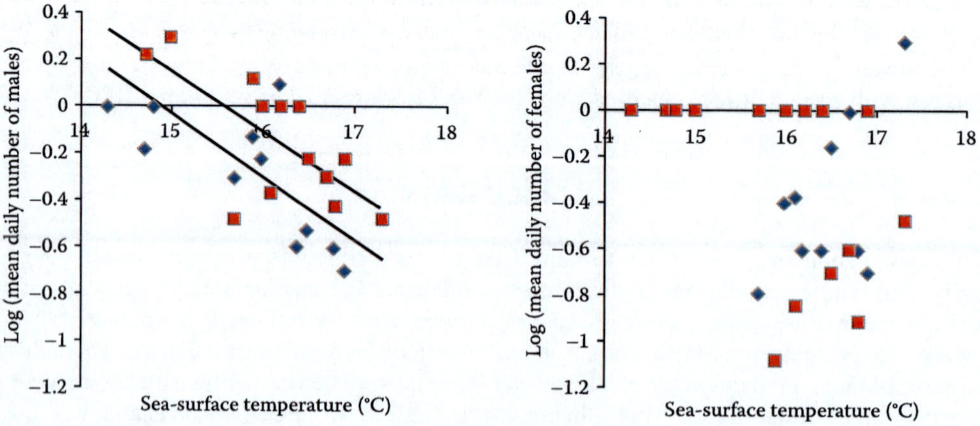

Figure 20.5 (a) Linear regression between logged mean daily number of subadult and mature male White Sharks and sea-surface temperature (R^2 = 0.55 and 0.45, respectively). (b) Logged mean daily numbers of subadult and mature female White Sharks with sea-surface temperature, at the Neptune Islands (♦ subadult sharks and ■ mature sharks).

(Weng at al., 2007). As sea-surface temperatures increased, the abundance of subadult and mature male White Sharks decreased linearly. Although no significant relationships were found to exist between females (mature or immature) and sea-surface temperature, female abundance was found to be skewed toward the mid- to high-temperature ranges. Statistical analyses for females were truncated because of the total lack of females from August to April in all of the years sampled.

The only sharks observed between October and April were males, and predominantly subadults, and in very much reduced numbers than seen at other times of the year (aside from a spike of subadult males in February 2004). It may be that mature males and females have departed for the breeding grounds at this time, because breeding and parturition is thought to occur in spring/summer, and hence only nonbreeding individuals would be encountered.

There was a strong bias toward males when comparing the number of mature White Sharks of each sex; a significant male bias was also reported at the aggregation site at Guadalupe Island (Chapter 25, this book), possibly resulting from the 2-yr. migratory pattern reported for mature females that results in a large portion of the female population being absent from the adult aggregation in any one year and less sampling effort during the time females are present (Chapters 11 and 16, this book). Males are classified as sexually mature at a smaller size than females (Francis, 1996; Pratt, 1996), and because growth rates are relatively slow in this species, this bias is to be expected. The significantly lower temperatures in 2003 may offer an explanation as to the absence of females during this year and provide further evidence for the temperature-related sex ratio differences observed (Robbins, 2007). Tagged females in the northeastern Pacific experienced higher sea-surface temperatures in the offshore region during the autumn months than tagged sharks that returned to Guadalupe Island (Chapter 11, this book). On an annual basis, the average monthly temperatures for males ranged from 16.4 to 21.9°C but was higher for females at 16.1 to 24.7°C (Chapter 11, this book). The seasonal size variation observed may be due to prey availability, sea-surface temperature, or reproductive habits, which stand to be closely related.

Juvenile White Sharks are piscivorous (Ainley et al., 1981; Casey and Pratt, 1985), with an increase in mammalian prey and a corresponding decrease in teleost and elasmobranch prey with increasing size of shark (Chapter 3, this book). The Neptune Islands may therefore be an unsuitable hunting site for juveniles, despite the relatively abundant teleost prey; not only would small White Sharks be competitively excluded and be at risk of injury from larger sharks, but other areas, such as the Spencer Gulf waters, are more species rich in fish prey items than the Neptunes (Malcolm et al., 2001).

Females were prevalent from April to June, which coincided with the time seal pups were beginning to enter the waters around the islands (Baylis et al., 2005) and would be easy prey compared with their adult counterparts, and so this time of year would be optimal provisioning periods for White Sharks. Male numbers were relatively low during these times, and those males that were present tended to be larger individuals.

During other months of the year, males predominated, with the population consisting of a mixture of mature males and subadults of each sex. Mature females were found to have departed this aggregation area, possibly to follow migratory paths to breeding and pupping grounds.

If prey availability were the main factor affecting size segregation, mature White Sharks of both sexes would be expected to be most abundant during times of increased prey availability (i.e., autumn and early winter). This was not the case, leading to the inference that another factor must play an important role in determining the segregations observed.

The distribution of White Sharks may be more related to temperature requirements associated with reproduction than to prey availability. As sea-surface temperature decreased, so the number of subadult and mature male White Sharks increased, whereas the opposite appeared true for females. Mature female White Sharks were only recorded from May to July. It has been suggested that, in some other shark species, females favor warmer waters in order to increase their growth rates and to achieve maturity size at a similar age to males (Klimley, 1987; Sims et al., 2001). Alternatively, mature pregnant females may be utilizing the warmer waters at the Neptune Islands in late summer/early autumn to increase their internal temperatures to accelerate the developmental rate of embryonic young. Chapter 11, this book also suggests a thermal niche hypothesis for females after describing sexual segregation in northeastern Pacific sharks, possibly relating to maintaining optimal temperature for fetal development, as the authors note that females spend more time in warmer water than males when they remain in the offshore habitat.

It should be noted that temperature may be an indirect controlling factor, and it could be that sharks are being influenced by one or more other variables that are closely associated with sea-surface temperature (SST). Sea-surface temperature and season would be highly related, and therefore any segregation occurring could be attributed to season rather than SST. However, if this were the case, a similar seasonal abundance would be expected from year to year regardless of SST, which was not observed in this study, with females being totally absent in 2003 when temperatures were lower than average. Other studies have shown female Grey Reef Sharks (Economakis and Lobel, 1998) and White Sharks to inhabit warmer waters than those experienced by males (Chapter 11, this book) and females to be absent from aggregation sites after intense cold water upwelling events (A. Towner, personal communication), which suggests that SST could be a controlling factor.

The results of this study could be implemented in management and conservation plans. Sexual and size segregation is a phenomenon in this species at the Neptune Islands; and their natural rarity, low natural mortality, and low reproductive rate result in White Sharks having a very low potential for population recovery, particularly if the composition of bycatch is heavily biased toward one sex or one size class of shark. The Neptune Islands appear to be an important feeding habitat for the development of White Sharks at various life-history stages. White Sharks would be particularly prone to exploitation and poaching at sites in which they aggregate and return to periodically; therefore, the importance of protection at the Neptune Islands is highlighted. The presence and timing of the segregations observed could be included into monitoring operations that plan to reduce the incidental mortality of this species in commercial fishing zones (Environment Australia, 2002).

There are many mysteries still surrounding adult White Shark life histories in Australian waters, and whether the emerging patterns from the northeastern Pacific will apply to the populations found here remain to be seen. Certainly, mature males return to this aggregation area annually, as seen in this species at Guadalupe Island (Chapters 16 and 25, this book), and additional annual data may show a similar pattern for mature females. In recent years, individual adult females have been photographically identified as visiting the Neptune Islands in nonconsecutive, and in one case consecutive, years (Robbins and Fox, unpublished data), again similar to findings from other aggregation areas (Chapter 25, this book). Further studies should focus on determining the local and long-term movements of mature and subadult sharks of both sexes once they leave the Neptune Islands, and whether the sexual and size segregation is observed throughout their range or only at feeding sites. The conclusions drawn here are based on the population at the Neptune Islands only, and so spatial segregation cannot be ascertained. A broader spatial segregation in Australian waters cannot be assumed without data from other sites; therefore further work should encompass other sites in their range. A long-term measure of abundance is needed that considers results from multiple years and decades to measure the true abundance, taking into account any fluctuations that may occur. Further data collection may also reveal alternate patterns, with recent anecdotal evidence suggesting that males may be present year-round (R. Robbins, personal observations). Interestingly, after recent increased research efforts at the South Neptune Islands, observations have shown adult females to be seasonally present and in the complete absence of males (which were concurrently present at the North Neptune Islands) (R. Robbins, personal observations). Further work should be done to confirm these distributions and to determine whether these patterns of segregation show consistency or plasticity. A number of satellite tagging projects have shown long and sometimes transoceanic migrations (Boustany et al., 2002; Bonfil et al., 2005; Chapters 11, 12, 16, and 21, this book). The use of additional satellite tags in Australian waters, particularly in the case of mature individuals, will advance our current level of understanding with respect to spatiotemporal residency patterns, habitat use, separation, and the location of mating and parturition areas, all of which are poorly understood in the Australian populations.

CONCLUSION

The marked sexual segregation and size segregation identified appeared to be interrelated. Mature females were encountered only during autumn and early winter, when temperatures were still relatively warm. Subadult females were present throughout autumn and early to midwinter, with their numbers decreasing with decreasing sea-surface temperatures. Mature and subadult male distributions were skewed toward the cooler SST experienced at the Neptune Islands. Both mean female and male size peaked in June, with mature females present only from May to July. It is hypothesized that mature females come to provision at the Neptune Islands from May to July, the time when New Zealand Fur Seal pups were beginning to enter the water in search of food, and then move away to reach the pupping or mating grounds in time for the breeding season. There was a trend for larger sharks in winter/spring months (June to September) and smaller sharks at other times. Sea-surface temperature had a negative linear relationship with subadult and mature male White Shark numbers. No relationships were found for females although, both maturity classes were skewed toward the higher temperatures. The proposed reasons for this population structure were centered around (1) a thermal niche hypothesis, with females favoring warmer waters to increase their internal temperatures to accelerate the developmental rate of embryonic young, and (2) differing physiological requirements associated with temperature.

REFERENCES

Adams, D.H., Mitchell, M.E., and Parsons, G.R. 1994. Seasonal occurrence of the white shark, *Carcharodon carcharias*, in waters off the Florida west coast, with notes on its life history. *Marine Fisheries Review* 56: 24.

Ainley, D.G., Strong, C.S., Huber, H.R., Lewis, T.J., and Morrell, S.H. 1981. Predation by sharks on pinnipeds at the Farallon Islands. *Fishery Bulletin* 78: 941–945.

Ainley, D.G., Henderson, R.P., Huber, H.R., Boekelheide, R.J., Allen, S.G., and McElroy, T.L. 1985. Dynamics of white shark/pinniped interactions in the Gulf of the Farallones. *Memoirs of the Southern California Academy of Sciences* 9: 109–122.

Anderson, S.D., and Goldman, K.J. 1996. Photographic evidence of white shark movements in California waters. *California Fish and Game* 82: 182–186.

Anderson, S.D., and Pyle, P. 2003. A temporal, sex-specific occurrence pattern among white sharks (*Carcharadon carcharias*) at the south Farallon Islands, California. *California Fish and Game* 89: 96–101.

Bass, A.J., D'Aubrey, D., and Kistnasamy, N. 1973. Sharks of the east coast of southern Africa. *International Oceanographic Research Institute* 33: 1–168.

Baylis, A.M.M., Page, B., Peters, K., McIntosh, R., Mckenzie, J., and Goldsworthy, S. 2005. The ontogeny of diving behaviour in New Zealand fur seal pups (*Arctocephalus forsteri*). *Canadian Journal of Zoology* 83: 1149–1161.

Blagoderov, A.I. 1994. Seasonal distribution and some notes on the biology of the salmon shark (*Lamna ditropis*) in the northwestern Pacific Ocean. *Journal of Ichthyology* 34: 115–121.

Bonfil, R., Meyer, M., Scholl, M.C., Johnson, R., O'Brien, S., Oosthuizen, H., Swanson, S., Kotze, D., and Paterson, M. 2005. Transoceanic migration, spatial dynamics, and population linkages of white sharks. *Science* 310: 100–103.

Boustany, A.M., Davis, S.F., Pyle, P., Anderson, S.D., Le Boeuf, B.J., and Block, B.A. 2002. Expanded niche for white sharks. *Nature* 415: 35–36.

Branstetter, S. 1987. Age, growth, and reproductive biology of the silky shark, *Carcharhinus falciformis*, and the scalloped hammerhead, *Sphyrna lewini*, from the Northwestern Gulf of Mexico. *Environmental Biology of Fishes* 19: 161–173.

Bres, M. 1993. The behaviour of sharks. *Reviews in Fish Biology and Fisheries* 3: 133–159.

Bruce, B.D. 1992. Preliminary observations on the biology of the white shark, *Carcharodon carcharias*, in South Australian waters. *Australian Journal of Marine and Freshwater Research* 43: 1–11.

Bruce, B.D., Stevens, J.D., and Malcolm, H. 2006. Movements and swimming behaviour of white sharks (*Carcharodon carcharias*) in Australian waters. *Marine Biology* 150: 161–172.

Carrier, J.C., Pratt, H.L., and Martin, L.K. 1994. Group reproductive behaviours in free-living nurse sharks, *Ginglyostoma cirratum*. *Copeia* 1994: 646–656.

Casey, J.G., and Pratt, H.L 1985. Distribution of the white shark, *Carcharodon carcharias*, in the western North Atlantic. *Memoirs of the Southern Californian Academy of Sciences* 9: 2–14.

Clark, K.R., and Warwick, R.M. 1994. *Change in Marine Communities: An Approach to Statistical Analysis and Interpretation*. Plymouth, MA: Plymouth Marine Laboratory.

Cliff, G., Dudley, S.F.J., and Jury, M.R. 1996. Catches of white sharks in KwaZulu-Natal, South Africa and environmental influences. In *Great White Sharks: The Biology of Carcharodon carcharias*, edited by A.P. Klimley and D.G. Ainley, pp. 351–362. New York: Academic Press.

Compagno, L.J.V. 1984. FAO species catalogue. Vol. 4. Sharks of the world: An annotated and illustrated catalogue of shark species known to date. Part 1: Hexanchiformes to Lamniformes. *FAO Fisheries Synopsis* 125: 1–655.

CSIRO Marine Research. 2001–2003. *CSIRO marine research remote sensing*. http://www.marine.csiro.au/~lband/web_point.

Department of Fisheries and Oceans. 1999. Porbeagle Shark in NAFO subareas 3–6. *DFO Scientific Stock Status Report*, B3–09.

Domeier, M.L., and Nasby-Lucas, N. 2007. Annual re-sightings of photographically identified white sharks (*Carcharodon carcharias*) at an eastern Pacific aggregation site (Guadalupe Island, Mexico). *Marine Biology* 150: 970–984.

Eckert, S.A., and Stewart, B.S. 2001. Telemetry and satellite tracking of whale sharks, *Rhincodon typus*, in the Sea of Cortez, Mexico, and the North Pacific Ocean. *Environmental Biology of Fishes* 60: 299–308.

Economakis, A.E., and Lobel, P.S. 1998. Aggregation behaviour of the grey reef shark, *Carcharhinus amblyrhynchos*, at Johnston Atoll, Central Pacific Ocean. *Environmental Biology of Fishes* 51: 129–139.

Environment Australia. 2002. *White Shark (Carcharodon carcharias) Recovery Plan*. Canberra, Australia: Environment Australia.

Fergusson, I.K. 1994. Notes on the shark fauna of the Sicilian Channel, with reference to future in-situ studies by means of tagging. In *Proceedings of the 2nd European Shark & Ray Workshop*, edited by R.C. Fowler and S.L. Earll, p. 83. British Museum of Natural History.

Ferreira, C.A., and Ferreira, T.P. 1996. Population dynamics of white sharks in South Africa. In *Great White Sharks: The Biology of Carcharodon carcharias*, edited by A.P. Klimley and D.G. Ainley, pp. 381–391. New York: Academic Press.

Francis, M.P. 1996. Observations on a pregnant white shark with a review of reproductive biology. In *Great White Sharks: The Biology of Carcharodon carcharias*, edited by A.P. Klimley and D.G. Ainley, pp. 157–172. New York: Academic Press.

Goldman, K. 2000. Distribution, segregation, and the potential for variable growth rates in the salmon shark, *Lamna ditropis*, in the western and eastern North Pacific. Annual Meeting of the American Elasmobranch Society, La Paz, Baja, Mexico.

Goldman, K.J., Anderson, S.D., McCosker, J.E., and Klimley, A.P. 1996. Temperature, swimming depth, and movements of a white shark at the South Farallon Islands, California. In *Great White Sharks: The Biology of Carcharodon carcharias*, edited by A.P. Klimley and D.G. Ainley, pp. 111–120. New York: Academic Press.

Hiethaus, M.R. 2001. Predator-prey and competitive interactions between sharks (order Selachii) and dolphins (suborder Odontoceti): A review. *Journal of Zoology* 253: 53–68.

Klimley, A.P. 1985. The areal distribution and autecology of the white shark, *Carcharodon carcharias*, off the west coast of North America. *Memoirs of the Southern California Academy of Sciences* 9: 15–40.

Klimley, A.P. 1987. The determinants of sexual segregation in the scalloped hammerhead shark, *Sphyrna lewini*. *Environmental Biology of Fishes* 18: 27–40.

Klimley, A.P., Anderson, S.D., Pyle, P., and Henderson, R.P. 1992. Spatiotemporal patterns of white shark, *Carcharodon carcharias*, predation at the South Farallon Islands California. *Copeia* 3: 680–690.

Krogh, M. 1994. Spatial, seasonal and biological analysis of sharks caught in the New South Wales protective beach meshing program. *Australian Journal of Marine and Freshwater Research* 45: 1087–1106.

Lien, J., and Fawcett, L. 1986. Distribution of basking sharks *Cetorhinus maximus* incidentally caught in inshore fishing gear in Newfoundland. *Canadian Field-Naturalist* 100: 246–252.

Long, D.J., and Jones, R.E. 1996. White shark predation and scavenging on cetaceans in the eastern North Pacific ocean. In *Great White Sharks: The Biology of Carcharodon carcharias*, edited by A.P. Klimley and D.G. Ainley, pp. 293–307. New York: Academic Press.

Malcolm, H., Bruce, B.D., and Stevens, J.D. 2001. *A Review of the Biology and Status of White Sharks in Australian Waters*. Hobart, Australia: CSIRO Marine Research.

Martin, R.A. 2004. Northerly distribution of white sharks, *Carcharodon carcharias*, in the eastern Pacific and relation to ENSO events. *Marine Fisheries Review* 66: 16–26.

McLaughlin, D.M., and Morrissey, J.F. 2005. Reproductive biology of *Centrophorus* cf. *uyato* from the Cayman Trench, Jamaica. *Journal of the Marine Biological Association of the United Kingdom* 85: 1185–1192.

Mollet, H.F., Cliff, G., Pratt, Jr., H.L., and Stevens, J.D. 2000. Reproductive biology of the female shortfin mako, *Isurus oxyrinchus Raninesque*, 1810, with comments on the embryonic development of lamnoids. *Fisheries Bulletin* 98: 299–318.

Muñoz-Chápuli, R. 1984. Ethologie de la reproduction chez quelques requins de l'Atlantique Nord-Est. *Cybium* 8: 1–14.

Norman, B.M. 1999. Aspects of the biology and ecotourism industry of the whale shark *Rhincodon typus* in north-western Australia. M.Phil. thesis, Murdoch University, Western Australia.

Pratt, H.L. 1979. Reproduction in the blue shark, *Prionace glauca. Fishery Bulletin* 77: 445–470.

Pratt, H.L., Jr. 1996. Reproduction in the male white shark. In *Great White Sharks: The Biology of Carcharodon carcharias*, edited by A.P. Klimley and D.G. Ainley, pp. 131–138. New York: Academic Press.

Robbins, R.L. 2007. Environmental variables affecting the sexual segregation of great white sharks, *Carcharodon carcharias*, at the Neptune Islands, South Australia. *Journal of Fish Biology* 70: 1350–1364.

Robinson, A.C., Canty, P.D., Mooney, P.A., and Ruddock, P.M. 1996. *South Australia's Offshore Islands*. Adelaide, Australia: South Australian Department of Environment and Natural Resources.

Ruckstuhl, K.E., and Neuhaus, P. 2002. Sexual segregation in ungulates: A comparative test of three hypotheses. *Biological Reviews* 77: 77–96.

Schwartz, F.J. 1984. Occurrence, abundance and biology of the blacknose shark, *Carcharhinus acronotus*, in North Carolina. *Northeast Gulf Science* 7: 29–47.

Simpfendorfer, C.A., Freitas, G.G., Wiley, T.R., and Heupel, M.R. 2005. Distribution and habitat partitioning of immature bull sharks (*Carcharhinus leucas*) in a southwest Florida estuary. *Estuaries and Coasts* 28: 78–85.

Sims, D.W., Nash, J.P., and Morritt, D. 2001. Movements and activity of male and female dogfish in a tidal sea lough: Alternative behavioural strategies and apparent sexual segregation. *Marine Biology* 139: 1165–1175.

Springer, S. 1967. Social organization of shark populations. In *Sharks, Skates and Rays*, edited by P.W. Gilbert, R.F. Mathewson, and D.P. Rall, pp. 149–174. Baltimore, MD: Johns Hopkins Press.

Stevens, J.D. 1984. Biological observations on sharks caught by sport fishermen off New South Wales. *Australian Journal of Marine and Freshwater Research* 35: 573–590.

Strong, W.R., Murphy, R.C., Bruce, B.D., and Nelson, D.R. 1992. Movements and associated observations of bait-attracted white sharks, *Carcharodon carcharias*: A preliminary report. *Australian Journal of Marine and Freshwater Research* 43: 13–20.

Uchida, S., Toda, M., Teshima, K., and Yano, K. 1996. Pregnant white sharks and full-term embryos from Japan. In *Great White Sharks: The Biology of Carcharodon carcharias*, edited by A.P. Klimley and D.G. Ainley, pp. 139–155. New York: Academic Press.

Watkins, A. 1958. *The Sea My Hunting Ground*. London, UK: Heinemann.

Weng, K.C., Boustany, A.M., Pyle, P., Anderson, S.D., Brown, A., and Block, B.A. 2007. Migration and habitat of white sharks (*Carcharodon carcharias*) in the eastern Pacific Ocean. *Marine Biology* 152: 877–894.

CHAPTER 21

Regional Population Connectivity, Oceanic Habitat, and Return Migration Revealed by Satellite Tagging of White Sharks, *Carcharodon carcharias*, at New Zealand Aggregation Sites

Clinton A. J. Duffy*
Department of Conservation and University of Auckland

Malcolm P. Francis
National Institute of Water and Atmospheric Research

Michael J. Manning[†]
National Institute of Water and Atmospheric Research

Ramon Bonfil
Brooklyn, New York

ABSTRACT

Twenty-five White Sharks (*Carcharodon carcharias*) were tagged with pop-up archival transmitting tags at Chatham Islands, Stewart Island, and one coastal location in New Zealand between April 2005 and September 2009. Nineteen tracks were successfully estimated using light-based geolocation and refined using remotely sensed sea-surface temperature data and unscented Kalman filtering. All but one shark made rapid directed movements north of New Zealand to subtropical or tropical regions, including eastern Australia, the Coral Sea, New Caledonia, Vanuatu, and Tonga during the austral autumn, winter, and spring. Four instances of return migration to New Zealand waters were observed, including three to the tagging location. Adult male and juvenile White Sharks occupied a variety of pelagic subtropical and tropical habitats for at least 5–7 months of the year. Sharks were resident in New Zealand waters for up to 5 months post-tagging, with only one shark, an immature female, overwintering in New Zealand. No shark moved between the Chatham

Islands and Stewart Island, but mixing of these sharks may occur off northeast North Island, eastern Australia, and Norfolk Island and in the Coral Sea. Effective conservation of White Sharks in the southwest Pacific is dependent upon ongoing international cooperation and the implementation of effective protection measures covering both national and international waters.

INTRODUCTION

Reported declines in the abundance of White Sharks (*Carcharodon carcharias*) in some parts of their range and their low resilience to fishery mortality have led to their protection in a number of countries and their inclusion in international conservation instruments such as the Convention on Migratory Species and the Convention on International Trade in Endangered Species of Wild Fauna and Flora (Reid and Krogh, 1992; Smith et al., 1998; Walker, 1998; Compagno, 2001; Natanson, 2002; Baum et al., 2003; McPherson and Myers, 2009). Despite these initiatives, White Sharks are subject to ongoing mortality throughout their range, principally as bycatch in commercial and artisanal fisheries (Compagno, 2001; Dudley and Simpfendorfer, 2006; McPherson and Myers, 2009). Effective conservation of White Sharks requires a comprehensive understanding of their spatiotemporal distribution, the linkages between critical habitats, the influence these factors have on population mixing, and the extent to which differences in behavior in different habitats affect fishery interactions. Until relatively recently, White Sharks were considered to inhabit mainly temperate coastal waters. We now know that juveniles and adults regularly undertake long-distance migrations between temperate and subtropical and tropical habitats and that in some parts of their range, they may be oceanic for considerable periods of time (Boustany et al., 2002; Bonfil et al., 2005, 2010; Bruce et al., 2006; Weng et al., 2007b; Domeier and Nasby-Lucas, 2008; Nasby-Lucas et al., 2009; Jorgensen et al., 2009; Chapters 11, 12, 13, 16, and 17, this book).

Significant regional variation in white shark migratory behavior and habitat use is apparent. In South Africa and Australia, White Sharks generally migrate between temperate and tropical waters along the shelf, often exhibiting periods of temporary residency at apparently favored sites along the migratory route (Bonfil et al., 2005; Bruce et al., 2006; Chapter 17, this book). In contrast, White Sharks in the northeast Pacific and New Zealand typically make rapid, directed movements across ocean basins (Boustany et al., 2002; Domeier and Nasby-Lucas, 2008; Nasby-Lucas et al., 2009; Bonfil et al., 2010; Jorgensen et al., 2009; Chapters 11, 12, 13, and 16, this book). Predominantly along-shelf migration in South Africa and Australia and transoceanic movement between New Zealand and Australia is consistent with what is known of the genetic population structure in the Southern Hemisphere (Pardini et al., 2001). However, Pardini et al. (2001) presented genetic evidence suggesting extensive dispersal of male White Sharks in the Southern Hemisphere, and Bonfil et al. (2005) subsequently documented return migration by a subadult female between South Africa and Australia. The finding of Bonfil et al. (2005) highlights the fact that movements of individuals between populations that do not result in reproduction may not be detected by genetic studies. If such movements are regular and involve return migration, they could be of significance to the management of the parent population.

To date, direct observations of transoceanic movements by White Sharks in the southwest Pacific Ocean are based on only four tagged sharks. Bruce et al. (2006) reported movement of a subadult female White Shark from South Australia to northwest North Island, New Zealand, and we (Bonfil et al., 2010) reported movement of three White Sharks from the Chatham Islands, New Zealand, to the Louisville Ridge, New Caledonia, and Vanuatu. The latter were tagged with pop-up satellite tags at Star Keys in April 2005. All exhibited a period of residency at the Chatham Islands followed by rapid, directed transoceanic movement. Although interesting, these observations raised questions about the return migration of White Sharks to New Zealand, the frequency and duration of transoceanic movements, and connectivity between White Shark populations at the Chatham Islands and mainland New Zealand, Australia, and the Pacific islands. Since 2005, we have deployed additional satellite tags on White Sharks at the Chatham Islands, as well as at two mainland New Zealand aggregation sites in order to address these questions. This has confirmed regular transoceanic movement of juvenile and adult White Sharks from New Zealand to tropical and subtropical habitats throughout the southwest Pacific and Coral Sea, as well as return migration to New Zealand. Interestingly, it has revealed that little or no mixing of individuals occurs between aggregation sites in New Zealand waters.

MATERIALS AND METHODS

White sharks were tagged at Star Keys and Hanson Bay (Te Awapatiki), Chatham Islands (44°00'S, 176°20'W) in 2005 and 2008; Titi/Muttonbird and Ruapuke Islands, near Stewart Island (46°55'S, 168°15'E) in 2007, 2008, and 2009; and at Ariel Bank, east coast North Island (38°43'S, 178°18'E) in 2009 (Figure 21.1 and Table 21.1). The Chatham Islands are located at the eastern end of the Chatham Rise, whereas the Muttonbird and Ruapuke Islands are located on the inner shelf close to Stewart Island. Both the Chatham and Stewart Island sites are located in or close to the Subtropical Front and support large breeding colonies of the New Zealand Fur Seal (*Arctocephalus forsteri*). Ariel Bank is a large submerged reef complex located on the outer shelf, close to the shelf break. It is influenced by the warm, southerly flowing East Cape Current and supports populations of large reef-associated fishes. Several winter Fur Seal haulouts occur along the adjacent coastline, and small numbers of Fur Seals have been observed foraging over the reef during tagging operations.

All but one shark were tagged free-swimming. White Sharks were attracted with chum (minced Albacore) and lured close enough with fish baits to tag. Tags were deployed using a hand-held tagging pole that inserted a nylon umbrella-style dart (Domeier et al., 2005) into the dorsal musculature of

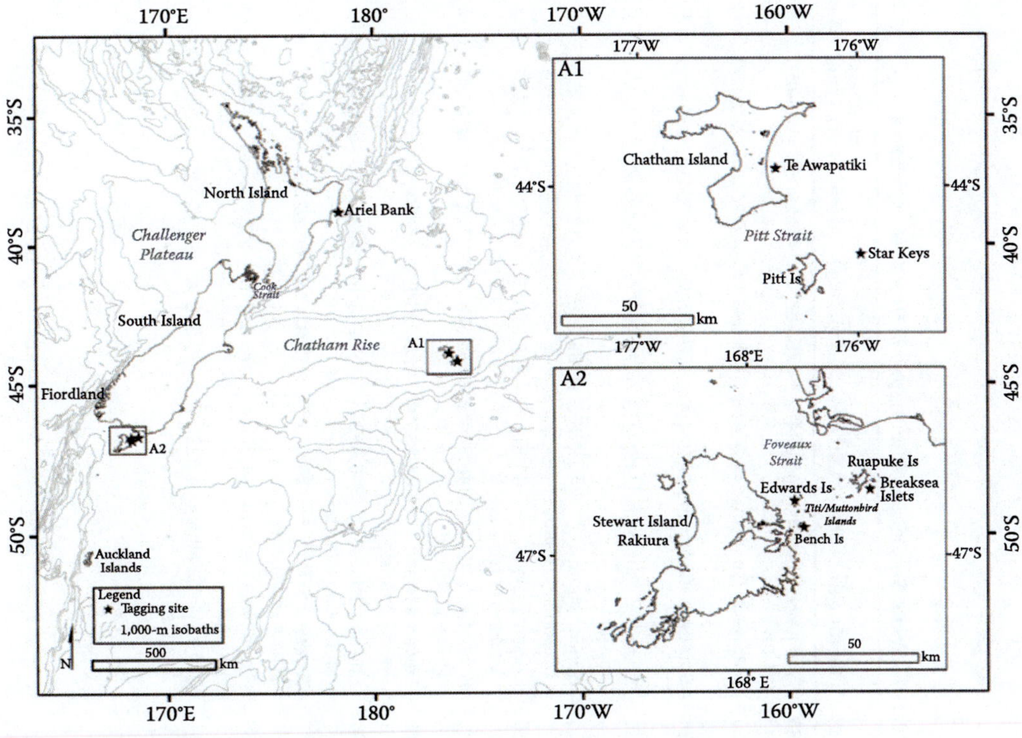

Figure 21.1 New Zealand White Shark (*Carcharodon carcharias*) tagging locations.

Table 21.1 White Sharks Satellite Tagged in New Zealand Waters

Shark	Tagging Location	Latitude	Longitude	Date	Estimated TL (cm)	Sex	Maturity	Days Deployed	Pop-up Location	Latitude	Longitude	Distance (km)
57033	Star Keys, Chatham Islands	44.224°S	175.986°W	4/8/2005	420–450	F	Immature	89	South Louisville Ridge	35.423°S	172.035°W	1035
57036	Star Keys, Chatham Islands	44.224°S	175.986°W	4/8/2005	320–350	M	Immature	181	West of Gemini Seamounts, Vanuatu	21.186°S	169.545 E	2883
57034	Star Keys, Chatham Islands	44.224°S	175.986°W	4/9/2005	380–400	F	Immature	5	Hanson Bay, Chatham Island, NZ	43.750°S	176.290°W	58
57035	Star Keys, Chatham Islands	44.224°S	175.986°W	4/9/2005	400	F	Immature	149	West of Isle of Pines, New Caledonia	22.755°S	167.155 E	2841
64036	Bench Island, Muttonbird Islands	46.908°S	168.255°E	3/18/2007	440	F	Immature	274	Swain Reefs, Australia	22.562°S	152.488°E	3053
64035	Edwards Island, Muttonbird Islands	46.829°S	168.228°E	3/20/2007	330	M	Immature	274	Mana Island, NZ	41.095°S	174.751°E	823
64033	Breaksea Islets, Ruapuke Island	46.794°S	168.558°E	2/27/2008	350–360	M	Mature	180	Swain Reefs, Australia	21.975°S	151.545°E	3151
64029	Edwards Island, Muttonbird Islands	46.835°S	168.225°E	3/5/2008	400	M	Mature	4	Foveaux Strait, NZ	45.569°S	171.281°E	274
64030	Edwards Island, Muttonbird Islands	46.830°S	168.213°E	3/13/2008	380–400	F	Immature	4	Foveaux Strait, NZ	46.816°S	168.527°E	32
64037	Te Awapatiki, Chatham Islands	43.943°S	176.374°W	4/12/2008	300	F	Immature	274	North Kermadec Trench, NZ	31.694°S	176.924°W	1363
78468	Star Keys, Chatham Islands	44.232°S	175.992°W	4/15/2008	400	M	Mature	202	Tongatapu	21.198°S	175.250°W	2562
64031	Star Keys, Chatham Islands	44.232°S	175.992°W	4/15/2008	350	F	Immature	271	Pitt Strait, Chatham Island, NZ	44.143°S	176.140°W	15

78470	Star Keys, Chatham Islands	44.232°S	175.992°W	4/15/2008	450	M	Mature	134	Northwest of Tongatapu	20.863°S	175.633°W	2599
64032	Te Awapatiki, Chatham Islands	43.942°S	176.374°W	4/19/2008	250–280	F	Immature	188	South Taranaki Bight, NZ	40.261°S	172.890°E	975
78469	Te Awapatiki, Chatham Islands	43.942°S	176.374°W	4/19/2008	450–480	F	Adolescent	180	East of Pelorus Reef, Tonga Ridge	22.621°S	175.678°W	2372
78471	Bench Island, Muttonbird Islands	46.912°S	168.257°E	3/24/2009	400	F	Immature	216	Oreti Beach, NZ	46.395°S	168.201°E	58
88235	Bench Island, Muttonbird Islands	46.912°S	168.257°E	3/24/2009	400	F	Immature	271	Kelso Bank, Coral Sea	24.121°S	159.327°E	2656
88234	Edwards Island, Muttonbird Islands	46.830°S	168.228°E	3/25/2009	480	M	Mature	109	Off Byron Bay, Australia	28.596°S	154.061°E	2372
78467	Bench Island, Muttonbird Islands	46.912°S	168.257°E	3/26/2009	280–300	M	Immature	116	south Kermadec Trench, NZ	37.858°S	179.768°E	1378
78473	Bench Island, Muttonbird Islands	46.912°S	168.257°E	3/26/2009	280	M	Immature	268	Southwest of Norfolk Island, Australia	31.448°S	164.422°E	1750
78474[a]	Bench Island, Muttonbird Islands	46.912°S	168.257°E	3/26/2009	380–400	F	Immature	107	south Kermadec Trench, NZ	36.957°S	179.550°W	1494
78475	Bench Island, Muttonbird Islands	46.912°S	168.257°E	3/27/2009	330	F	Immature	195	Northwest of Huon Atoll, New Caledonia	17.706°S	162.584°E	3289
78476	Bench Island, Muttonbird Islands	46.912°S	168.257°E	3/30/2009	290	M	Immature	130	East of Yaté Pass, New Caledonia	22.109°S	167.198°E	2760
78472	Edwards Island, Muttonbird Islands	46.832°S	168.228°E	4/1/2009	400	M	Mature	146	Bellona Plateau, Coral Sea	21.346°S	159.204°E	2949
78477	Ariel Bank, North Island	38.726°S	178.298°E	9/9/2009	300	F	Immature	17	North of Mahia Peninsula, NZ	39.025°S	177.959°E	44

Twenty-five White Sharks were satellite tagged in New Zealand waters. Maturity was assumed to be >350 cm TL for males and >470 cm TL for females (Francis, 1996; Pratt, 1996). Days deployed = the number of days the tag remained on the shark. Distance = the great circle distance that the tag was displaced from the tagging location.

[a] Shark 64030 was retagged.

the shark. Three sharks were tagged with Microwave Telemetry (MT) PTT100, and twenty-two were tagged with Wildlife Computers (WC) PAT4 and MK10 pop-up archival transmitting (PAT) tags. These were attached to the dart by a 23–25-cm leader of 2-mm diameter (181.4-kg test) monofilament nylon. One shark (64035), a 330 cm total length (TL) immature male, was caught using a baited line and secured in a cradle. This shark was double tagged with a MT PAT tag and a WC smart position and temperature (SPOT) satellite tag. The latter was bolted to the shark's first dorsal fin.

Estimated TL, sex, tag position, time of deployment, and sea-surface temperature (SST) at the tagging site (if known) were recorded for each shark. From 2008 onwards, all sharks tagged at the Muttonbird Islands were photographically identified using high-definition video according to Domeier and Nasby-Lucas (2007). PAT tags were deployed for periods between 3 and 11 months. WC PAT4 and MK10 tags archived light, temperature, and pressure (depth) data at 2–60-s intervals. Archived depth and temperature data transmitted by these tags via Argos satellites were summarised in 3- or 6-h bins (starting at 00:00 h local time). MT PTT100 tags recorded and transmitted point measurements of the same data at 15-, 30-, or 60-min intervals depending on the length of deployment (i.e., longer intervals for longer deployments). The full archival record was used to fit tracks from four recovered WC PAT tags.

Daily positions were estimated from ambient light data stored on WC PAT4 and Mk10 tags using WC-GPE: Global Position Estimator Program Suite software (www.wildlifecomputers.com). Dawn/dusk light-level data were extracted using WC-AMP, and daily longitudes and latitudes were estimated using WC-GPE. Daily records with poor dawn/dusk light level curves were excluded from the analyses. Microwave Telemetry provided estimated latitudes and longitudes for data from PTT 100 tags. The most probable tracks were fitted by matching tag-measured SST with remotely sensed SST data using unscented Kalman filtering (*UKFSST*) (Nielsen et al., 2006; Lam et al., 2008). Preliminary fits indicated that models incorporating longitude bias and SST bias always produced implausible tracks, so we omitted these parameters. For sharks whose tags popped up on the programmed date ($n = 8$), the *UKFSST* models incorporated parameters for latitude bias and solstice error variance (which accounts for greater error around the equinoxes), and the measurement error for the last position was set to zero. For sharks whose tags popped up prematurely or reported their first accurate location some time after pop-up ($N = 11$), the pop-up locations were not known accurately, so measurement error was estimated for the last position. For nine of these sharks, the *UKFSST* models incorporated parameters for latitude bias and solstice error variance; the remaining two sharks did not provide enough data to estimate the solstice error parameter, so uniform error was assumed. Dates for the commencement of oceanic movements away from the New Zealand shelf were estimated by examination of temporal changes in latitude, diving behavior, and temperature (Bonfil et al., 2010). Oceanic swimming speeds were estimated from point-to-point great circle distances and time spent travelling. Diving behavior is described in detail by Francis et al. (Chapter 22, this book).

RESULTS

Length, sex, and maturity data for tagged sharks are presented in Table 21.1. One subadult male, two adult males, and seven immature females were tagged at the Chatham Islands. Males ranged from 350 to 450 cm estimated TL, and females ranged from 250 to 480 cm TL. Five subadult males, three mature males, and six immature females were tagged near Stewart Island. The males ranged from 250 to 480 cm estimated TL, and females ranged from 300 to 440 cm TL. One 300 cm TL female was tagged at Ariel Reef in September 2009 (Table 21.1).

Twenty-one tags remained attached for more than 3 weeks, and nineteen of these provided useful data on shark movements (Figures 21.2 and 21.3). Track duration ranged from 89 to 274 d. Eighteen of these sharks made rapid long-distance movements north of New Zealand. Departure from the tagging site was not synchronized (Figure 21.4). Sharks left the Chatham Islands between

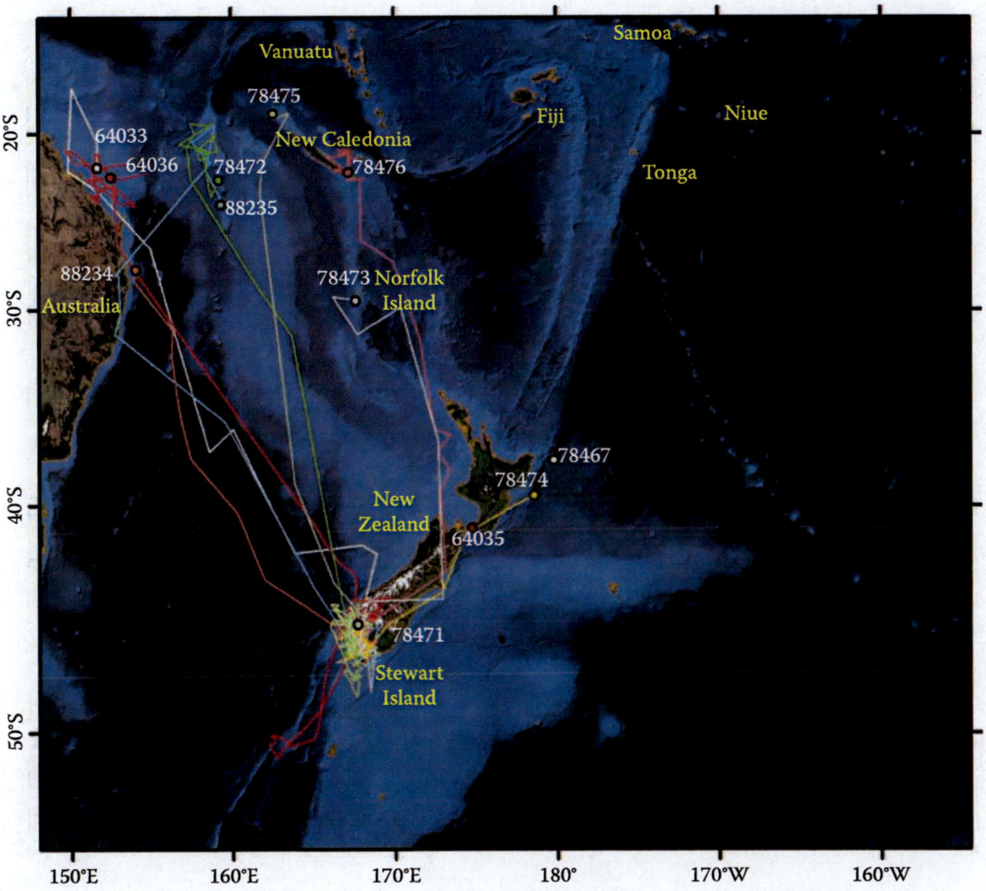

Figure 21.2 Long-distance movements of White Sharks tagged near Stewart Island, New Zealand.

mid-April and mid-September and left Stewart Island between early March and late August. At the Chatham Islands, departure dates of the three males coincided with those of the females, whereas at Stewart Island males left between early March and mid-June, and females left between early July and late August. There was no relationship between shark size and departure date. Although shark 64035's PAT tag did not transmit sufficient data to allow estimation of any daily positions, maximum transmitted temperatures for June 20, August 29, and September 17 were 23.5, 21.37, and 22.54°C respectively, indicating that it was north of New Zealand on those dates. The 20°C SST isotherm is usually located north of 30°S during these months (Chiswell, 1994). The SPOT5 tag attached to this shark stopped transmitting after only 23 d. The only shark that did not leave New Zealand waters was a 400 cm TL female (78471) tagged at Bench Island in March 2009 (Figures 21.1 and 21.2). A large conspecific bite, already in the process of healing, was noted on this shark's left flank. The estimated track, depth, and temperature data indicate that this shark remained over the shelf in the vicinity of Fiordland and Stewart Island for at least 216 d (Figure 21.2).

Sharks were resident at Chatham and Stewart Islands for between 4 d and 5 months post-tagging. No movement was observed between these aggregation sites, and only one shark (64036) moved south. That shark, a 440 cm TL female tagged at Bench Island in March 2007, remained near Stewart Island until July 26 and then traveled 628 km southwest to the southern end of the Puysegur Ridge, reaching 51°S on August 5. It spent about 10 d over this part of the ridge before swimming north over Solander Trough and Puysegur Bank to Fiordland. It reached the Fiordland coast (46°S)

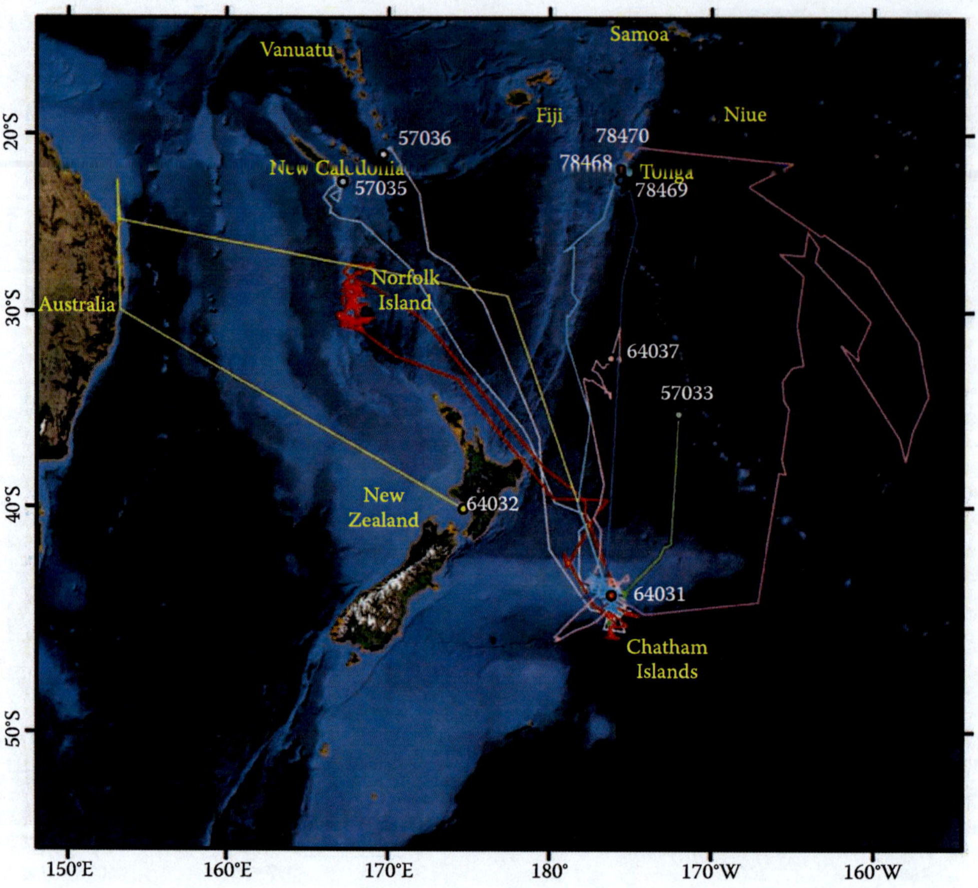

Figure 21.3 Long-distance movements of White Sharks tagged at Chatham Islands, New Zealand.

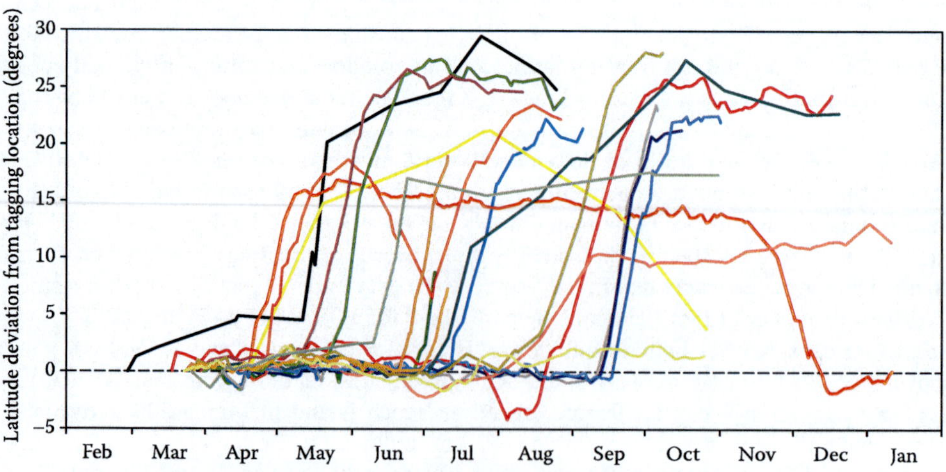

Figure 21.4 Seasonality of White Shark movements away from New Zealand tagging sites.

on or about August 22 and moved northeast, following the trend of the shelf beak before moving offshore on August 28 and crossing to the Swain Reefs, Great Barrier Reef, Australia (Figure 21.2). While over the southern Puysegur Ridge, this shark spent almost all of the time at the surface, making only four dives below 200-m depth. The deepest of these dives was to 317 m. Minimum charted depths in this area exceed 500 m. Ambient temperatures experienced over the Puysegur Ridge were between 2.7 and 8°C (Chapter 22, this book).

Five sharks tagged at the Muttonbird Islands moved to the Coral Sea (Swain Reefs, Kelso Bank, Bellona Plateau, Queensland Plateau, Huon Atoll, and northern New Caledonia). All left the New Zealand shelf off southwest Fiordland. Four, including 64036, appear to have crossed the Tasman Sea south of the Lord Howe Rise, whereas the remaining two appear to have followed the rise north. Two sharks tagged at the Muttonbird Islands moved north along the east coast of South Island before moving west through Cook Strait and then north to Norfolk Island and southern New Caledonia. Two other sharks (78474 and 78467) moved north along the east coasts of North and South Islands before their tags prematurely released near East Cape. Two sharks were resighted at the Muttonbird Islands in 3 consecutive yrs. Shark 64030, a 380–400 cm TL female, was tagged Edwards Island in March 2008, but the tag came off after only 4 d. It was retagged at Bench Island in March 2009, and its tag (78474) was released near East Cape in July. The shark was subsequently resighted at Edwards Island in February 2010. A 400 cm TL female originally photo-identified at Edwards Island with shark 64030 in 2008 was tagged at Bench Island in March 2009. This shark's tag (88235) released over Kelso Bank on December 20, 2009, and the shark was resighted at Bench Island on March 25, 2010.

Five sharks tagged at the Chatham Islands moved north or northeast of New Zealand to the Louisville Ridge, north Kermadec Trench, and Tonga (Figure 21.3). One shark, a 450 cm TL male (78470) left the Chatham Islands 9 d after tagging and spent 105 d roving over the Southwest Pacific Basin east of the Louisville Ridge. On August 6, 2008, it was about 560 km southwest of the Cook Islands (22°S). From there it moved to the Tonga Ridge north of Tongatapu, traversing a distance of more than 1000 km in about 9 d.

Sharks 78468 and 78469 remained at the Chatham Islands for about 5 months before moving to Tonga. Shark 78468 appears to have crossed the north Kermadec Ridge and then turned to follow the trend of the Kermadec and Tonga Ridges to Tongatapu, whereas shark 78469 moved almost directly to an area of shallow seamounts located about 50 km south of Ata Island. Examination of the archival record for shark 78468 shows that the day before it arrived in Tonga, it began making frequent dives, possibly searching for the bottom (Chapter 22, this book). In contrast, the archival record does not show any change in diving behavior over the Kermadec Ridge, suggesting the real track may have been located over deep water. Shark 64037, a 300 cm TL female, remained at the Chatham Islands for about 5 months before moving almost directly north to the Kermadec Trench. It then meandered over a large area southwest of the Kermadec Islands for 123 d before its tag released.

The remaining four sharks tagged at the Chatham Islands all moved northwest of New Zealand to Norfolk Island, New Caledonia, Vanuatu, and Queensland, Australia (Figure 21.3). All of these sharks appear to have passed East Cape, North Island, although shark 64032's track is uncertain given the small number of estimated daily positions. At 250–280 cm TL, this shark was the smallest tagged (Table 21.1). It completed a return migration of more than 7,100 km travelling to the South Fiji Basin and Queensland Plateau before returning to New Zealand (South Taranaki Bight) via southern Queensland and northern New South Wales. This shark's departure date from the Chatham Islands is unknown; however, the journey from the South Fiji Basin back to New Zealand (~5,000 km) took 160 d. A 350-cm female (64031) made a return migration to the Norfolk Ridge, its tag releasing in Pitt Strait about 15 km from the tagging location 271 d after deployment (Table 21.1). This shark departed Chatham Islands on April 19, four days after tagging, and took 20–23 d to reach the Norfolk Ridge. It remained there until November 12, probably spending most of its time on the Norfolk Island shelf (Chapter 22, this book), before almost retracing the outward leg of its track to the Chatham Islands.

Table 21.2 Minimum Sustained Swimming Speeds of White Sharks Moving across Ocean Basins to and from New Zealand Waters

Shark	TL (cm)	Distance (km)	km·day^{-1}	km·h^{-1}	m·s^{-1}	TL·s^{-1}
57033	420	1,035	148	6.18	1.71	0.41
57036	320	2,883	107	4.45	1.24	0.39
57035	400	2,841	92	3.82	1.06	0.27
64036	440	3,053	95	3.98	1.10	0.25
78468	400	2,562	102	4.27	1.19	0.30
64031	350	2,197	110	4.58	1.27	0.36
64031	350	2,197	88	3.66	1.02	0.29
78469	450	2,372	132	5.49	1.53	0.34
88234	480	2,372	148	6.18	1.72	0.36
78475	330	3,289	89	3.70	1.03	0.31
78472	400	2,949	105	4.39	1.22	0.30

Ocean swimming speeds were estimated for eleven ocean transits made by ten sharks (Table 21.2). Point-to-point displacement distances ranged from 1,035 to 3,289 km. Travelling times ranged from 16 to 37 d (average 25.9 ± 2.08 S.E. d). Average speed was 110 ± 6.7 S.E. km per day or 0.32 ± 0.02 S.E. total length per second (TL·s^{-1}). The maximum speed of 148 km per day was recorded by a 420 cm TL female (57033) that swam from the Chatham Islands to the south Louisville Ridge in 7 d, and a 480 cm TL male (88234) that swam from the southwest South Island to the Bellona Plateau, Coral Sea, in 16 d. TL·s^{-1} was independent of estimated TL, whereas km·h^{-1} was positively correlated with TL and negatively correlated with displacement distance (Figure 21.4).

DISCUSSION

Our data confirm that most White Sharks aggregating at New Zealand pinniped colonies migrate annually to subtropical and tropical regions of the Southwest Pacific and Coral Sea and that at least some return to the tagging location. All except one shark exhibited a period of temporary residency at or near the tag site, followed by an oceanic phase associated with movement to subtropical and tropical habitats. This pattern of movement applied across a wide size range (250–480 cm TL) and to mature males and immature individuals of both sexes. Shark 78471 was the only shark that did not undergo an oceanic phase. It is not known whether this was due to its injuries or if it migrated after its tag released.

The migratory behavior of New Zealand White Sharks is most similar to that of northeast Pacific White Sharks, most of which migrate annually to a large, relatively well defined mid-ocean foraging area (Boustany et al., 2002; Weng et al., 2007a; Domeier and Nasby-Lucas, 2008; Jorgensen et al., 2009; Nasby-Lucas et al., 2009; Chapters 11, 12, 13, and 16, this book). Oceanic swimming speeds were comparable to those reported previously for White Sharks (Boustany et al., 2002; Bonfil et al., 2005, 2010; Weng et al., 2007a; Domeier and Nasby-Lucas, 2008; Jorgensen et al., 2009), and the distances traveled and days spent migrating were similar to those reported for White Sharks in the northeast Pacific (Weng et al., 2007a; Domeier and Nasby-Lucas, 2008; Jorgensen et al., 2009). Unlike most White Sharks in the northeast Pacific, the sharks we tagged did not migrate to an oceanic focal area (see Chapter 12, this book). Instead they dispersed throughout the Southwest Pacific, occupying a broad band between 17.7°S and about 29°S, extending from eastern Australia to at least 156°32.4'W. Most of the sharks we tagged also moved to shelf, upper slope, seamount, and insular habitats and spent substantial amounts of time in deep water, possibly foraging close to the bottom

(Chapter 22, this book). Only two sharks spent all or most of the tag deployment over deep ocean basins (Figure 21.3). The frequent use of oceanic ridge, seamount, and insular habitats by White Sharks in the Southwest Pacific is probably a reflection of the region's complex bathymetry and the influence this has on the distribution of prey (Morato et al., 2010). Northeast Pacific White Sharks occasionally disperse as far as Hawaii, where their diving behavior implies foraging on pelagic prey aggregating near the islands (Weng et al., 2007a; Domeier and Nasby-Lucas, 2008; Jorgensen et al., 2009; Chapter 22, this book).

Although the purpose of oceanic migration in White Sharks remains speculative, data on diving behavior in the northeast Pacific focal area, Hawaii, and the southwest Pacific suggest they spend substantial amounts of time foraging (Weng et al., 2007a; Domeier and Nasby-Lucas, 2008; Jorgensen et al., 2009; Nasby-Lucas et al., 2009; Bonfil et al., 2010; Chapters 12 and 16, this book). White Sharks that we tracked to the tropics spent the most time (mean, 66.6%) in the upper 75 m but made frequent deep dives, with some spending extended periods apparently following the bottom (Chapter 22, this book). Similar behavior by White Sharks in the northeast Pacific has been interpreted as foraging at a target depth (Weng et al., 2007a; Domeier and Nasby-Lucas, 2008; Nasby-Lucas et al., 2009). Foraging at depths below visual contact with the surface implies feeding on fishes and possibly squids (Weng et al., 2007a; Chapters 4, 12, and 22, this book). Morato et al. (2010) found evidence that at least 12 species of pelagic elasmobranchs and teleosts aggregate near seamounts in the Southwest Pacific, with high pelagic species diversity predicted near seamounts on the Kermadec Ridge and along the Lord Howe Island Seamount Chain, in Tonga, Fiji, and southern New Caledonia and at Norfolk Island. With the exception of Fiji, these areas correspond to those that White Sharks migrating from New Zealand either passed through or were temporarily resident in. White Shark movements to eastern Australia and the Coral Sea also coincide with the seasonal distribution of Striped Marlin (*Tetrapturus audax*) and Swordfish (*Xiphias gladius*) (Molony, 2005; Domeier, 2006). Large deepwater fishes may also be abundant in these habitats. Taxa most frequently taken in seamount fisheries in the region include spiny dogfishes (*Squalus* spp., *Centrophorus* spp.), groupers (*Epinephelus septemfasciatus*, *E. octofasciatus*), deepwater snapper (*Etelis* spp., *Pristipomoides* spp., *Aphareus rutilans*), Alfonsino (*Beryx* spp.), gemfishes (*Rexea prometheoides*, *Tongaichthys robustus*), Escolar (*Lepidocybium flavobrunneum*), and Oil Fish (*Ruvettus pretiosus*) (Japan International Cooperation Agency, 1987; Laboute and Grandperrin, 2000).

White Sharks also predate small cetaceans and readily scavenge cetacean carcasses (Carey et al., 1982; Bruce, 1992; Long and Jones, 1996; Dudley et al., 2000; Chapter 16, this book). Their spatiotemporal distribution in the southwest Pacific overlaps that of a number cetaceans, including the tropical calving grounds of Humpback Whales (*Megapatera novaeangliae*) (Dawbin, 1966; Gaskin, 1976; Garrigue et al., 2002, 2010; Borsa, 2006; Bonfil et al., 2010; Tirard et al., 2010; Clua and Séret, 2010, Chapter 23, this book). Corkeron et al. (1987) reported bites attributed to White Sharks on Bottlenose Dolphins (*Tursiops truncatus*) in Queensland, and Randhawa (2011) linked adult cestodes collected from a juvenile female White Shark caught in northern New Zealand to larvae collected from two species of tropical dolphin. White Sharks have also been observed scavenging on a Sperm Whale (*Physeter macrocephalus*) carcass in New Caledonia (Tirard et al., 2010; Chapter 23, this book). Peak White Shark abundance in New Caledonian waters coincides with peak abundance and calving of Humpback Whales, and scavenging on Humpback Whales undoubtedly occurs (Tirard et al., 2010; Chapter 23, this book). Because large White Sharks have been observed to attack incapacitated juvenile Southern Right Whales (*Eubalaena australis*) aggregating at Head of Bight, South Australia (S. Burnell, Macquarie University, personal communication), the possibility that unattended Humpback calves or incapacitated yearlings may occasionally be predated by White Sharks cannot be discounted. Paterson and Van Dyck (1991) and Paterson et al. (1993) reported shark attack as the cause of death of two recently born Humpback Whale calves stranded in Queensland but did not identify the species of shark involved. Although

Humpback Whales migrate along the east and west coasts of New Zealand (Dawbin, 1966), White Shark movements were not synchronized with those of Humpback Whales. Three sharks (16%) began oceanic migration in mid- to late April, about 1 month prior to the peak abundance of northbound Humpback Whales in southern New Zealand; five (26%) departed between mid-June and mid-July when Humpback Whale abundance at 46°S would have been declining; and four (21%) departed between mid-August and mid-September when few, if any, northbound whales would have been present (Figure 21.4) (Dawbin, 1966). One shark departed in late February, at least 2 months prior to the arrival of the first Humpback Whales in New Zealand, and one did not migrate at all (Figure 21.4). The arrival of several sharks in the tropics in late September/early October coincides with the departure of most Humpback Whales from their breeding grounds, and at least three sharks remained in the tropics well after the last whales would have left for Antarctica (Figure 21.4) (Dawbin, 1966).

Although it seems unlikely that overall prey availability is higher in the oligotrophic waters north of New Zealand, the movement of White Sharks away from areas of high biological productivity such as the Subtropical Front implies that migration to warmer waters confers a considerable energetic benefit. White Sharks are endothermic and defend body core temperatures up to 14.3°C above ambient water temperatures (Carey et al., 1982; Goldman, 1997). Lower cardiac output and increased retial heat-exchanger efficiency enables lamnids to reduce heat loss at low ambient temperatures (Goldman, 1997; Bernal et al., 2001); however, prolonged defense of body core temperatures well above the surrounding water temperature is likely to incur a metabolic cost (Blank et al., 2007).

In shallow, well-mixed temperate waters, the ability of White Sharks to select water temperatures that allow them to minimize the cost of thermoregulation is probably limited. Movement to warmer water may allow White Sharks to forage more efficiently. Behavioral indicators of this might include lower swimming speeds and active depth-temperature selection (Bernal et al., 2001; Blank et al., 2007). Although our ocean swimming speed data are suggestive of slower swimming speeds in the tropics (Figure 21.5), changes in behavior between oceanic and shelf phases and the imprecise nature of light based geolocation mean that data on point-to-point swimming speeds of White Sharks in a variety of tropical habitats are required to confirm this (Bruce et al., 2006; Weng et al., 2007a, 2007b; Domeier and Nasby-Lucas, 2008; Johnson et al., 2009; Chapter 22, this book). During the temperate shelf phase, the sharks we tagged spent almost all (98.5%) of their time between 10 and 16°C, whereas in the tropics they spent most of their time between 18 and 26°C but made excursions of variable duration below the mixed layer (Bonfil et al., 2010; Chapter 22, this book). White Sharks in the Northeast Pacific exhibit similar behavior in tropical offshore habitats, spending most of their time between 17.5 and 25°C (Weng et al., 2007a; Domeier and Nasby-Lucas, 2008). This apparent temperature preference in tropical waters overlaps the range of body core temperatures, 23.4–27.6°C (mean, 26.6°C), reported for three free-swimming White Sharks at the South Farallon Islands, California, by Goldman (1997).

No data are available on metabolic rate or thermoregulation by White Sharks in subtropical and tropical habitats (Carey et al., 1982; Goldman, 1997). Bruce (2008) observed that correlations between White Shark presence or activity and temperature are generally poor and commented that reported responses to temperature change may reflect the movements of primary prey species. However, this is based upon the results of research conducted at temperate aggregation sites and not comparisons made between these and tropical offshore habitats. Seasonal movements of adult White Sharks may also be related to mating and parturition (Chapter 16, this book).

Although more than twice as many females were tagged at the Chatham Islands as males, White Shark populations at both tagging sites do not appear to be segregated by size or sex (C. Duffy, unpublished data). With the exception of mature females and young of the year, sharks of all sizes and sexes were present. However, smaller sharks were usually competitively excluded from the tagging boat by larger sharks. Because of the small sample size, further research is required to

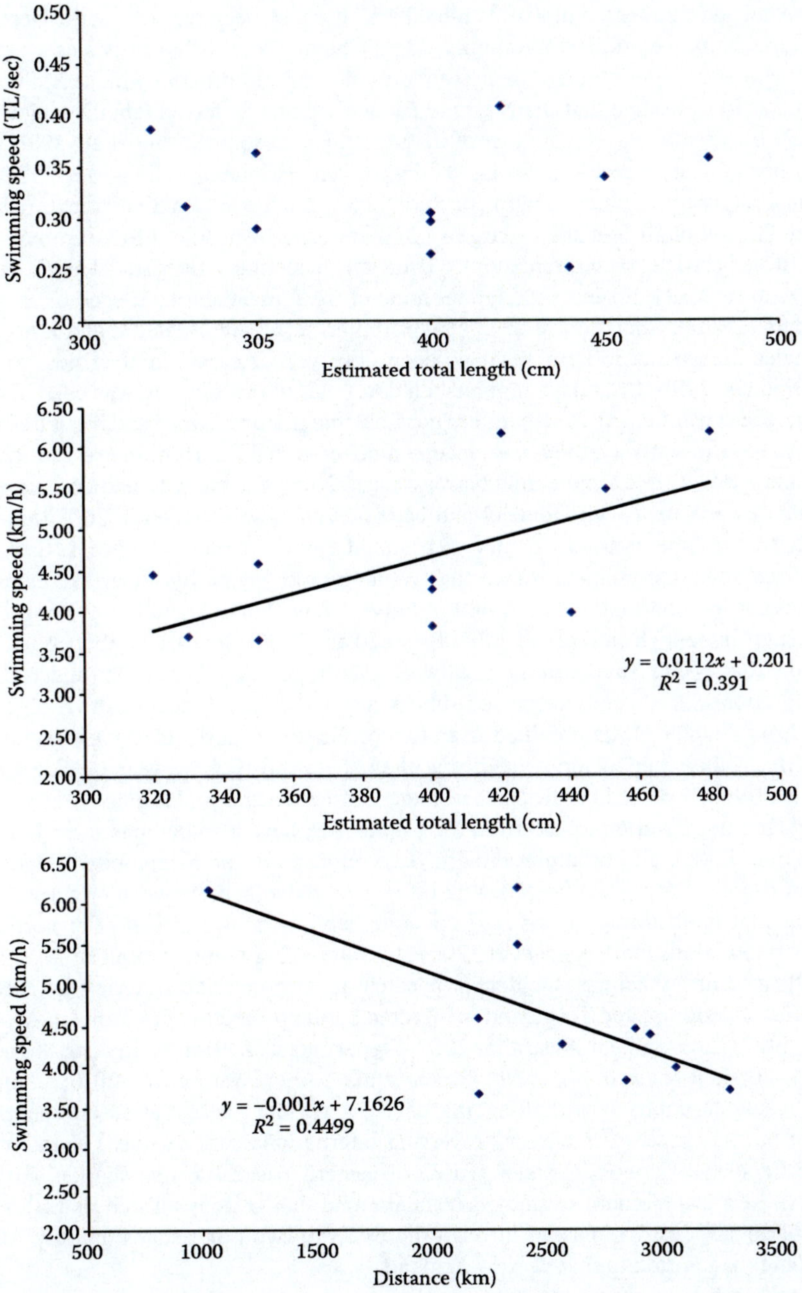

Figure 21.5 Relationships among total length, distance traveled, and minimum sustained swimming speed of White Sharks migrating to and from New Zealand waters.

determine whether males actually depart from Stewart Island earlier than females. Weng et al. (2007a) observed that tagged females also tended to leave the Farallon Islands later than males, but this difference was not statistically significant. Similarly Domeier and Nasby-Lucas (2007, 2008; Chapter 25, this book) found that mean departure dates for males and females were not significantly different, even though males arrive at Guadalupe Island up to 2 months earlier than females.

Temporal sexual and size segregation of White Sharks has also been reported at the Neptune Islands, South Australia (Strong et al., 1996; Robbins, 2007; Chapter 20, this book). Males are present there year-round, whereas females tend to be present only during autumn and winter, and then only during some years. It is possible that the Neptune Islands are not preferred female habitat or that they actively avoid the area due to the number of mature males present (Strong et al., 1996; Chapters 11 and 16, this book). The largest females tagged by us were estimated to be 440 cm TL (64036) and 450–480 cm TL (78469). Francis (1996) concluded most female White Sharks mature between 450 and 500 cm TL, although females as large as 520 cm have been found to be immature (Francis, 1996; B. Bruce, personal communication). Given the inaccuracy of visual length estimates and the large size at maturity, it seems likely that none of the female sharks tagged by us was mature. Pregnant female White Sharks have been caught in northern New Zealand, and large presumably mature females exceeding 500 cm TL have been observed or captured at or near all of the tagging sites (Francis, 1996; C. Duffy, unpublished data). All of our tagging was carried out between February and the start of April. It is therefore possible that mature females arrive at the tagging sites later in the year. Alternatively, they may occupy a different habitat from the rest of the population (Anderson and Pyle, 2003; Domeier and Nasby-Lucas, 2007; Chapters 11 and 16, this book).

Our data are consistent with Domeier and Nasby-Lucas's observation in 2008 that white shark migratory behavior appears to be adapted to regional environmental conditions. Satellite tagging in South Africa and Australia has shown that White Sharks generally undertake extensive return migrations along the continental shelf, with relatively few sharks spending prolonged periods in the pelagic environment (Bonfil et al., 2005; Bruce et al., 2006; Chapter 17, this book). In contrast, our data show that White Sharks in the Southwest Pacific are largely pelagic, undertaking regular seasonal migrations from widely dispersed subtropical and tropical locations to forage at pinniped colonies in New Zealand. Data obtained from two of the three sharks that made return migrations within the life of their tags demonstrated that they spent most of the year outside New Zealand waters. Both exhibited periods of tropical residency of between 5 and almost 7 months, with over a month engaged in migration to and from the tropics. We have also demonstrated interannual site fidelity by three sharks, all immature females. Two returned to the Muttonbird Islands and one to the Chatham Islands. Strong site fidelity with little or no movement between preferred coastal sites but with potential intermixing occurring in pelagic habitats is also a feature of northeast Pacific White Shark populations (Jorgensen et al., 2009; Chapters 11, 13, and 16, this book).

White Sharks are taken as incidental bycatch in commercial fisheries in New Zealand, Australia, New Caledonia, and Tonga and in directed bather protection programs in Australia (Reid and Krogh, 1992; Environment Australia, 2002; Department of Primary Industries and Fisheries, 2006; Duffy, 2004; Juncker et al., 2006; Green et al., 2009; Tirard et al., 2010; Chapter 23, this book). Our research emphasizes the ongoing need for regional cooperation on conservation and management of this species. As well as traversing international waters, these sharks moved to or through the Exclusive Economic Zones (EEZ) of eastern Australia (five sharks), Norfolk Island (three sharks), New Caledonia (five sharks), Vanuatu (one shark), Tonga (three sharks), and possibly Niue (one shark). At least five passed through the EEZs of two other states (mainly Australia and New Caledonia) in addition to that of New Zealand.

Further research is required to determine the movements and habitat requirements of young-of-the-year and mature female White Sharks in New Zealand waters or whether sharks tagged later in the year exhibit similar migratory behavior to those reported here. We have used the term *aggregation site* simply to describe areas where we were able to locate and tag multiple sharks. Further work is required to define the spatiotemporal dynamics of these aggregations and levels of connectivity between them and other parts of the New Zealand EEZ. Only one of the sharks we tagged moved south into subantarctic waters, yet they are reported relatively frequently as far south as Campbell Island (C. Duffy, unpublished data; Chapter 22, this book). White Sharks are also relatively common in shallow shelf waters of the northwest North Island, including the Manukau and Kaipara Harbours. Bruce et al. (2006) reported the movement of a subadult female between South Australia and the northwest North Island, but none of the sharks we have tagged has moved to either location. Bruce et al. (2006) postulated the occurrence of a migratory corridor along the east Australian coast. Our data point to the possibility of migratory corridors extending from southwest Fiordland along the southern margin of Lord Howe Rise to northern New South Wales and southern Queensland and one extending from the Chatham Islands along northeast North Island to the Norfolk Ridge and New Caledonia.

ACKNOWLEDGMENTS

This project was funded by the New Zealand Department of Conservation, National Institute of Water and Atmospheric Research Ltd., the Foundation for Research Science and Technology, Wildlife Conservation Society NY, National Geographic Society, and NABU/SharkTracker. We particularly wish to thank Kina Scollay and Tim Gregory-Hunt for their invaluable assistance with locating and tagging the sharks and Shannon O'Brien and Barbara Mangold for their technical assistance, good humor, and patience in the field. Tim Sippel, Tim Lam, and Murray Smith provided advice and assisted with analysis of the tracks. We also gratefully acknowledge the assistance of Department of Conservation staff from the Southern and Chatham Islands Areas, particularly Clint "Barney" Brown and Phred Dhobbins; Stu Cave; Peter Scott, Shark Dive NZ; Boyd McGregor, Surfit Charters; members of the Nukualofa Sport Fishing Club; Westfleet Ltd.; Aratuna Freighters and Hall's Transport; Southland Coolstores and Southern Seafoods Ltd.

REFERENCES

Anderson, S. D.; Pyle, P. 2003. A temporal, sex-specific occurrence pattern among white sharks (*Carcharodon carcharias*) at the South Farallon Islands, California. *California Fish and Game* 89: 96–101.

Baum, J. K.; Myers, R. A.; Kehler, D. G.; Worm, B.; Harley, S. J.; Doherty, P. A. 2003. Collapse and conservation of shark populations in the Northwest Atlantic. *Science* 299: 389–392.

Bernal, D.; Sepulveda, C.; Graham, J. B. 2001. Water-tunnel studies of heat balance in swimming mako sharks. *Journal of Experimental Biology* 204: 4043–4054.

Blank, J. M.; Morrissette, J. M.; Farwell, C. J.; Price, M.; Schallert, R. J.; Block, B. A. 2007. Temperature effects on metabolic rate of juvenile Pacific bluefin tuna *Thunnus orientalis*. *Journal of Experimental Biology* 210: 4254–4261.

Bonfil, R.; Meÿer, M.; Scholl, M. C.; Johnson, R.; O'Brien, S.; Oosthuizen, H.; Swanson, S.; Kotze, D.; Paterson, M. 2005. Transoceanic migration, spatial dynamics and population linkages of white sharks. *Science* 310: 100–103.

Bonfil, R.; Francis, M. P.; Duffy, C.; Manning, M. J.; O'Brien, S. 2010. Large-scale tropical movements and diving behavior of white sharks *Carcharodon carcharias* tagged off New Zealand. *Aquatic Biology* 8: 115–123.

Borsa, P. 2006. Marine mammal strandings in the New Caledonia region, Southwest Pacific. *Comptes Rendus Biologies* 329: 277–288.

Boustany, A. M.; Davis, S. F.; Pyle, P.; Anderson, S. D.; Le Boeuf, B. J.; Block, B. A. 2002. Expanded niche for white sharks. *Nature* 415: 35–36.

Bruce, B. D. 1992. Preliminary observations on the biology of the white shark, *Carcharodon carcharias*, in South Australian waters. *Australian Journal of Marine and Freshwater Research* 43: 1–11.

Bruce, B. D. 2008. The biology and ecology of the white shark, *Carcharodon carcharias*. In *Sharks of the Open Ocean*, edited by M. D. Camhi, E. D. Pikitch, and E. A. Babcock, pp. 69–81. Oxford, UK: Blackwell Publishing Ltd.

Bruce, B. D.; Stevens, J. D.; Malcolm, H. 2006. Movements and swimming behaviour of white sharks (*Carcharodon carcharias*) in Australian waters. *Marine Biology* 150: 161–172

Carey, F. G.; Kanwisher, J. W.; Brazier, O.; Gabrielson, G.; Casey, J. G.; Pratt, H. L. Jr. 1982. Temperature and activities of a white shark, *Carcharodon carcharias*. *Copeia* 1982: 254–260.

Chiswell, S. M. 1994. Variability in sea surface temperature around New Zealand from AVHRR images. *New Zealand Journal of Marine and Freshwater Research* 28: 179–192.

Clua, E.; Séret, B. 2010. Unprovoked fatal shark attack in Lifou Island (Loyalty Islands, New Caledonia, South Pacific) by a great white shark, *Carcharodon carcharias*. *The American Journal of Forensic Medicine and Pathology* 31: 281–286.

Compagno, L. J. V. 2001. Sharks of the world: An illustrated and annotated catalogue of shark species known to date: Vol. 2. Bullhead, mackerel and carpet sharks (Heterodontiformes, Lamniformes and Orectolobiformes). *FAO Species Catalogue for Fishery Purposes 1*. Rome, Italy: FAO.

Corkeron, P. J.; Morris, R. J.; Bryden, M. M. 1987. Interactions between bottlenose dolphins and sharks in Moreton Bay, Queensland. *Aquatic Mammals* 13: 109–113.

Dawbin, W. H. 1966. The seasonal migratory cycle of humpback whales. In *Whales, Dolphins, and Porpoises*, edited by K. S. Norris, pp. 145–170. Los Angeles, CA: University of California Press.

Department of Primary Industries and Fisheries. 2006. *A Report on the Queensland Shark Safety Program*. Brisbane, Australia: Department of Primary Industries and Fisheries.

Domeier, M. L. 2006. An analysis of Pacific striped marlin (*Tetrapturus audax*) horizontal movement patterns using pop-up satellite archival tags. *Bulletin of Marine Science* 79: 811–825.

Domeier, M. L.; Kiefer, D.; Nasby-Lucas, N.; Wagschal, A.; O'Brien, F. 2005. Tracking Pacific bluefin tuna (*Thunnus thynnus orientalis*) in the northeastern Pacific with an automated algorithm that estimates latitude by matching sea-surface-temperature data from satellites with temperature data from tags on fish. *Fishery Bulletin* 103: 292–306.

Domeier, M. L.; Nasby-Lucas, N. 2007. Annual re-sightings of photographically identified white sharks (*Carcharodon carcharias*) at an eastern Pacific aggregation site (Guadalupe Island, Mexico). *Marine Biology* 150: 977–984.

Domeier, M. L.; Nasby-Lucas, N. 2008. Migration patterns of white sharks *Carcharodon carcharias* tagged at Guadalupe Island, Mexico, and identification of an eastern Pacific shared offshore foraging area. *Marine Ecology Progress Series* 370: 221–237.

Dudley, S. F. J.; Anderson-Reade, M. D.; Thompson, G. S.; McMullen, P. B. 2000. Concurrent scavenging off a whale carcass by great white sharks, *Carcharodon carcharias*, and tiger sharks, *Galeocerdo cuvier*. *Fishery Bulletin* 98: 646–649.

Dudley, S. F. J.; Simpfendorfer, C. A. 2006. Population status of 14 shark species caught in the protective gillnets off KwaZulu-Natal beaches, South Africa, 1978–2003. *Marine and Freshwater Research* 57: 225–240.

Duffy, C. A. J. 2004. International trade in white shark *Carcharodon carcharias* products from New Zealand. *Shark News* 16: 1–3.

Environment Australia. 2002. *White Shark (Carcharodon carcharias) Recovery Plan*. Canberra, Australia: Environment Australia.

Francis, M. P. 1996. Observations on a pregnant white shark with a review of reproductive biology. In *Great White Sharks: The Biology of Carcharodon carcharias*, edited by A. P. Klimley, and D. G. Ainley, pp. 157–172. San Diego, CA: Academic Press.

Garrigue, C.; Aguayo, A.; Amante-Helweg, V. L. U.; Baker, C. S.; Caballero, S.; Clapham, P.; Constantine, R.; Denkinger, J.; Donoghue, M.; Flórez-González, L.; Greaves, J.; Hauser, N.; Olavarría, C.; Pairoa, C.; Peckham, H.; Poole, M. 2002. Movements of humpback whales in Oceania, South Pacific. *Journal of Cetacean Research and Management* 4: 255–260.

Garrigue, C.; Zerbini, A. N.; Geyer, Y.; Heide-Jørgensen, M.-P.; Hanaoka, W.; Clapham, P. 2010. Movements of satellite-monitored humpback whales from New Caledonia. *Journal of Mammalogy* 91:109–115.

Gaskin, D. E. 1976. The evolution, zoogeography and ecology of cetacea. *Oceanography and Marine Biology Annual Reviews* 14: 247–346.

Goldman, K. J. 1997. Regulation of body temperature in the white shark, *Carcharodon carcharias*. *Journal of Comparative Physiology B* 167: 423–429.

Green, M.; Ganasin, C.; Reid, D. D. 2009. *Report into the NSW Shark Meshing (Bather Protection) Program*. Sydney, Australia: New South Wales Department of Primary Industries.

Japan International Cooperation Agency. 1987. The fisheries resources survey in Fiji and Tuvalu. *Japan International Cooperation Agency Reports No. 87-016-33*. Tokyo, Japan: Japan International Cooperation Agency.

Johnson, R.; Bester, M. N.; Dudley, S. F. J.; Oosthuizen, W. H.; Meÿer, M.; Hancke, L.; Gennari, E. 2009. Coastal swimming patterns of white sharks (*Carcharodon carcharias*) at Mossel Bay, South Africa. *Environmental Biology of Fishes* 85: 189–200.

Jorgensen, S. J.; Reeb, C. A.; Chapple, T. K.; Anderson, S.; Perle, C.; Van Sommeran, S. R.; Fritz-Cope, C.; Brown, A. C.; Klimley, A. P.; Block, B. A. 2009: Philopatry and migration of Pacific white sharks. *Proceedings of the Royal Society B Biological Sciences* 277: 679–688.

Juncker, M.; Robert, M.; Clua, E. 2006. *Coastal Shark Fisheries in the Pacific: A Brief Overview of Current Knowledge.* Noumea, New Caledonia: CRISP Coordination Unit.

Laboute, P.; Grandperrin, R. 2000. *Poissons de Nouvelle-Calédonie*. Noumea, New Caledonia: Editions Catherine Ledru, IRD.

Lam, C. H.; Nielsen, A.; Sibert, J. R. 2008. Improving light and temperature based geolocation by unscented Kalman filtering. *Fisheries Research* 91: 15–25.

Long, D. L.; Jones, R. E. 1996. White shark predation and scavenging on cetaceans in the eastern North Pacific Ocean. In *Great White Sharks: The Biology of Carcharodon carcharias*, edited by A. P. Klimley and D. G. Ainley, pp. 293–307. San Diego, CA: Academic Press.

McPherson, J. M.; Myers, R. A. 2009. How to infer population trends in sparse data: Examples with opportunistic sighting records for great white sharks. *Diversity and Distributions* 15: 880–890.

Molony, B. 2005. Summary of the biology, ecology and stock status of billfishes in the WCPFC, with a review of major variables influencing longline fishery performance. *Report to the 1st Meeting of the Scientific Committee of the Western and Central Pacific Fisheries Commission WCPFC–SC1*. EB WP–2. Noumea, New Caledonia.

Morato, T.; Hoyle, S. D.; Allain, V.; Nicol, S. J. 2010. Seamounts are hotspots of pelagic biodiversity in the open ocean. *PNAS* 107: 9707–9711.

Nasby-Lucas, N.; Dewar, H.; Lam, C. H.; Goldman, K. J.; Domeier, M. L. 2009. White shark offshore habitat: A behavioral and environmental characterization of the eastern Pacific shared offshore foraging area. *PLoS ONE* 4: e8163. doi:10.1371/journal.pone.0008163.

Natanson, L. J. 2002. Preliminary investigations into the age and growth of the shortfin mako, *Isurus oxyrinchus*, white shark, *Carcharodon carcharias*, and thresher shark, *Alopias vulpinus*, in the western North Atlantic Ocean. SCRS/2001/066. *Collective Volume of Scientific Papers ICCAT* 54: 1280–1293.

Nielsen, A.; Bigelow, K. A.; Musyl, M. K.; Sibert, J. R. 2006. Improving light-based geolocation by including sea surface temperature. *Fisheries Oceanography* 15: 314–325.

Pardini, A. T.; Jones, C. S.; Noble, L. R.; Krieser, B.; Malcolm, H.; Bruce, B. D.; Stevens, J. D.; Cliff, G.; Scholl, M. C.; Francis, M.; Duffy, C. A. J.; Martin, A. P. 2001. Sex-biased dispersal of great white sharks. *Nature* 412: 139–140.

Paterson, R. A.; Van Dyck, S. 1991. Studies of two humpback whales, *Megaptera novaeangliae*, stranded at Fraser Island, Queensland. *Memoirs of the Queensland Museum* 30: 343–350.

Paterson, R. A.; Quayle, C. J.; Van Dyck, S. M. 1993. A humpback whale calf and two subadult dense-beaked whales recently stranded in southern Queensland. *Memoirs of the Queensland Museum* 33: 291–297.

Pratt, H. L. Jr. 1996. Reproduction in the male white shark. In *Great White Sharks: The Biology of Carcharodon carcharias*, edited by A. P. Klimley and D. G. Ainley, pp. 131–138. San Diego, CA: Academic Press.

Reid, D. D.; Krogh, M. 1992. Assessment of catches from protective shark meshing off New South Wales beaches between 1950 and 1990. *Australian Journal of Marine and Freshwater Research* 43: 283–296.

Robbins, R. L. 2007. Environmental variables affecting the sexual segregation of great white sharks *Carcharodon carcharias* at the Neptune Islands South Australia. *Journal of Fish Biology* 70: 1350–1364.

Randhawa, H. S. 2011. Insights using a molecular approach into the lifecycle of a tapeworm infecting great white sharks. *Journal of Parasitology* 97: 275–280.

Smith, S. E.; Au, D. W.; Show, C. 1998. Intrinsic rebound potentials of 26 species of Pacific sharks. *Marine and Freshwater Research* 49: 663–678.

Strong, W. R. Jr.; Bruce, B. D.; Nelson, D. R.; Murphy, R. D. 1996. Population dynamics of white sharks in Spencer Gulf, South Australia. In *Great White Sharks: The Biology of Carcharodon carcharias*, edited by A. P. Klimley and D. G. Ainley, pp. 401–414. San Diego, CA: Academic Press.

Tirard, P.; Manning, M. J.; Jollit, I.; Duffy, C.; Borsa, P. 2010. Records of great white sharks (*Carcharodon carcharias*) in New Caledonian waters. *Pacific Science* 64: 567–576.

Walker, T. I. 1998. Can shark resources be harvested sustainably? A question revisited with a review of shark fisheries. *Marine and Freshwater Research* 49: 553–572.

Weng, K. C.; Boustany, A. M.; Pyle, P.; Anderson, S. D.; Brown, A.; Block, B. A. 2007a. Migration and habitat of white sharks (*Carcharodon carcharias*) in the eastern Pacific Ocean. *Marine Biology* 152: 877–894.
Weng, K. C.; O'Sullivan, J. B.; Lowe, C. G.; Winkler, C. E.; Dewar, H.; Block, B. A. 2007b. Movements, behavior and habitat preferences of juvenile white sharks in the eastern Pacific as revealed by electronic tags. *Marine Ecology Progress Series* 338: 211–224.

CHAPTER 22

The Third Dimension

Vertical Habitat Use by White Sharks, Carcharodon carcharias*, in New Zealand and in Oceanic and Tropical Waters of the Southwest Pacific Ocean*

Malcolm P. Francis*
National Institute of Water and Atmospheric Research

Clinton A. J. Duffy
Department of Conservation and University of Auckland

Ramon Bonfil
The Bronx High School of Science

Michael J. Manning†
National Institute of Water and Atmospheric Research

ABSTRACT

White Sharks from New Zealand typically make long-distance migrations across open ocean to tropical locations in the southwest Pacific Ocean, resulting in major changes in habitat type and temperature regime. This study explored how White Sharks use the vertical dimension of their habitat, and the thermal environment they encounter. Between 2005 and 2009, 25 sharks were tagged with pop-up satellite tags around New Zealand. Twenty of the tags reported useful data, including four that were recovered, thus providing high-resolution archival information on depth and temperature. We analyzed depth-related behavior patterns in the vicinity of New Zealand seal colonies (shelf phase) and while crossing open oceans (ocean phase) to tropical regions (tropics phase). The sharks exhibited major changes in behavior among phases. They mostly stayed in shallow water less than 50 m deep in temperatures of 10–16°C during the shelf phase but had a bimodal depth distribution in the ocean phase, switching between surface swimming and diving to 200–800 m.

Most sharks occasionally dived deeper than 800 m, with one reaching a maximum of 1,200 m, and temperatures ranged between 2.7 and 26.6°C. In the tropics phase, White Sharks spent most of their time in the upper 75 m but continued making deep dives. They mainly experienced temperatures of 18–26°C. In all phases, behavior varied among sharks and at different times for the same shark. Diving behavior often varied in a diel cycle, with more time being spent in deeper water during the day than at night. Explanations for the observed behavior remain speculative but most likely include navigation (perhaps using celestial and depth cues) and feeding.

INTRODUCTION

During the last decade, our understanding of the habitat of White Sharks (*Carcharodon carcharias*) has changed dramatically. Previously, these sharks were considered to inhabit mainly shallow coastal waters in temperate regions of the world (Compagno, 1984). However, it is now known that many White Sharks spend considerable time in subtropical and tropical waters (Boustany et al., 2002; Bonfil et al., 2005, 2010; Weng et al., 2007b; Domeier and Nasby-Lucas, 2008; Nasby-Lucas et al., 2009; Jorgensen et al., 2010). In fact, at least in some parts of their range, White Sharks probably spend more time in the open ocean and in tropical regions than they do in shallow temperate waters (Chapter 21, this book). Furthermore, white-shark behavior in oceanic and tropical habitats is completely different from that observed in temperate waters: instead of patrolling close to shore in depths usually shallower than 75 m (Goldman and Anderson, 1999; Boustany et al., 2002; Bonfil et al., 2010), they have been found to range extensively through the water column to depths as great as 980 m (Bonfil et al., 2005; Weng et al., 2007b; Nasby-Lucas et al., 2009).

A migratory link between temperate and tropical waters has been reported for most regions of White Shark abundance: California and Mexico (Boustany et al., 2002; Weng et al., 2007b; Domeier and Nasby-Lucas, 2008; Nasby-Lucas et al., 2009; Jorgensen et al., 2010; Chapters 11, 12, 13, and 16, this book), South Africa (Bonfil et al., 2005), Australia (Bruce et al., 2006), and New Zealand (Bonfil et al., 2010; Chapter 21, this book). However, significant regional differences exist. In South Africa, Australia, and New Zealand, the sharks migrate mainly in a latitudinal direction, whereas in California and Mexico they migrate longitudinally. In South Africa and Australia, the sharks mainly follow the continental shelf and rarely move into open ocean waters [albeit with some notable exceptions (Bonfil et al., 2005; Bruce et al., 2006; Chapter 17, this book)], whereas in New Zealand, California, and Mexico, the sharks typically cross deep ocean basins (Boustany et al., 2002; Weng et al., 2007b; Domeier and Nasby-Lucas 2008; Nasby-Lucas et al., 2009; Jorgensen et al., 2010; Chapters 11, 12, 13, 16, and 21, this book). Genetically distinct White Shark populations occur in South Africa, Australia–New Zealand, and the northeast Pacific (Pardini et al., 2001; Jorgensen et al., 2010; Chapter 24, this book), so it appears that differences in migratory behavior may have evolved among populations.

Previously, we revealed that White Sharks in the southwest Pacific Ocean make transoceanic migrations and dive to considerable depths en route (Bonfil et al., 2010). However, that study was based on only three sharks tagged in New Zealand in 2005, so the results require confirmation. Since then, further tagging in New Zealand has confirmed that most White Sharks make transoceanic migrations to tropical waters of the Southwest Pacific, and some subsequently return to New Zealand (Chapter 21, this book). We have also found that deep diving is a universal feature of transoceanic crossings in the region. This raises interesting questions about how and why White Sharks use the third dimension (depth) of their habitat. In this chapter, we explore depth-related behavior patterns of White Sharks, and the thermal environment they encounter, in the southwest Pacific Ocean.

MATERIALS AND METHODS

Twenty-five White Sharks were tagged with pop-up archival transmitting (PAT) tags at Chatham Islands (175°W, 44°S) and Stewart Island (168°E, 47°S), New Zealand, between 2005 and 2009 (Table 22.1). Our earlier tagging results (Bonfil et al., 2010) are included in the present study. Three tag types were used: PAT4 and MK10 tags from Wildlife Computers (Redmond, WA) and PTT100 tags from Microwave Telemetry (Columbia, MD). Free-swimming sharks were tagged with a pole from a boat after being attracted by chum and fish baits. Tags were programmed to record depth, temperature, and light for 3–9 months. Further details of the tagging procedures are provided by Bonfil et al. (2010) and in Chapter 21, this book. Sharks were sexed visually from underwater video, from diving cages, or from the surface, and their total lengths (TL) were estimated visually. The tagged sharks comprised 11 males (2.8–4.8 m TL) and 14 females (2.5–4.5 m TL). Based on the estimated lengths at maturity of White Sharks (males, 3.6 m; females, greater than 4.5 m; Francis, 1996; Pratt, 1996), five of the males and one of the females may have been mature.

PAT4 and MK10 tags record depth with a resolution of 0.5 m, and temperature (–40–60°C) with a resolution of 0.05°C. The maximum depths that can be recorded are 980 m (PAT4) and about 1,500 m (MK10) (Wildlife Computers, personal communication). Raw data, collected at 2–60-sec intervals, are archived in memory and can be downloaded if the tag is recovered following pop-up. For most sharks, only data aggregated on board the tags and transmitted via Argos satellites were available. In this study, the data were summarized across time bins of 3 or 6 h. Depth and temperature data were summarized into two different formats: PAT depth-temperature profiles (PDT data) and histograms (bin data). PDT data consist of eight absolute depth-temperature pairs ranging from the minimum depth to the maximum depth within each time bin. PDT depths were rounded to 4-m intervals (PAT4s) or 8-m intervals (MK10s), and temperatures were rounded to 0.2°C intervals for transmission. Bin data are frequency distributions of the proportion of time spent within user-specified depth and temperature bins. PAT4s have 12 bins and MK10s have 14 bins. The first 12 MK10 depth bins were essentially the same as the 12 PAT4 bins, with two extra bins appended to the deep end of the range. Consequently, we combined data from the two tag types for depth analyses (inspection of the data indicated that no sharks tagged with PAT4s exceeded the maximum depth limit of 980 m). Different temperature bin limits were used at the upper end of the range for PAT4 and MK10 tags, so only MK10 tags were used for detailed temperature analyses.

PTT100 tags do not summarize data on board before transmission; instead they transmit point measurements of depth and temperature recorded at 15-, 30-, or 60-min intervals (longer time intervals are required for longer tag deployments because of memory constraints). Depth and temperature data are rounded on board to 5.4-m and 0.17°C intervals, respectively. The maximum depth is about 1,250 m (Microwave Telemetry, personal communication).

Data Analysis

Points along the migratory routes were estimated from the light data collected by the tags, using the manufacturers' software. These points, which may have relatively large errors (especially around the equinoxes), were improved by fitting tracks to the full tag record using Kalman filter analysis with sea-surface temperature (SST) correction (Sibert et al., 2003; Nielsen and Sibert, 2007) as described by Bonfil et al. (2010) and in Chapter 21, this book. A detailed description and analysis of the tracks for the sharks included in the present study are given in Chapter 21, this book. For the purpose of analyzing White Shark diving behavior, we divided each track into three phases: shelf (the period spent near the tagging location, during which the sharks are believed to have remained on the continental shelf near seal colonies); ocean (the period spent travelling across open ocean); and tropics (the period spent at a subtropical or tropical location). One shark (64031) had two ocean phases, which we combined into one data set.

Table 22.1 Tagging Details for Twenty-Five White Sharks Tagged in New Zealand

Shark Tag	TL (m)	Sex	Tag Type	Tagging Region	Tagging Date	Start Ocean	End Ocean	End Tropics	Pop-up date	Ocean Days	Tropics Days	Total Days	Tagging Latitude	Tagging Longitude	Pop-up Latitude	Pop-up Longitude	Archive Record
57033	4.2	F	WC PAT4	Chatham	4/7/2005	6/27/2005	7/4/2005		7/5/2005	8		89	44.224	184.014	35.423	187.965	
57034	3.8	F	WC PAT4	Chatham	4/9/2005				4/14/2005			5	44.224	184.014	43.750	183.710	
57035	4.0	F	WC PAT4	Chatham	4/9/2005	7/8/2005	8/8/2005	9/4/2005	9/5/2005	32	27	149	44.224	184.014	22.755	167.155	Y
57036	3.2	M	WC PAT4	Chatham	4/8/2005	9/8/2005	10/4/2005		10/6/2005	27		181	44.224	184.014	21.186	169.645	
64029	4.0	M	WC PAT4	Stewart	3/5/2008				3/8/2008			4	46.835	168.225	45.569	171.281	Y
64030	3.0	F	WC PAT4	Stewart	3/13/2008				3/17/2008			4	46.830	168.213	46.816	168.627	
64031	3.5	F?	WC PAT4	Chatham	4/15/2008	4/19/2008	5/9/2008	11/16/2008	1/11/2009	47[a]	191	271	44.232	184.008	44.143	183.860	Y
64032	2.5	F	WC PAT4	Chatham	4/19/2008				10/24/2008			188	43.942	183.626	40.261	172.890	
64033	3.5	M	WC PAT4	Stewart	2/27/2008	5/11/2008	5/21/2008		8/25/2008	11		180	46.794	168.558	21.975	151.645	
64035	3.3	M	MT PTT100	Stewart	3/20/2007				12/20/2007?			274?	46.830	168.228	41.095	174.751	
64036	4.4	F	MT PTT100	Stewart	3/18/2007	8/31/2007	9/21/2007	12/16/2007	12/17/2007	22	86	274	46.908	168.255	22.562	152.488	
64037	3.0	F	MT PTT100	Chatham	4/12/2008	8/5/2008	1/10/2009		1/11/2009	159		274	43.943	183.626	31.694	183.076	
78467	2.8	M	WC MK10	Stewart	3/26/2009				7/20/2009			116	46.912	168.257	37.858	179.768	
78468	4.0	M	WC MK10	Chatham	4/14/2008	9/17/2008	10/12/2008	11/1/2008	11/2/2008	26	20	202	44.232	184.008	21.198	184.750	Y
78469	4.5	F	WC MK10	Chatham	4/19/2008	9/15/2008	10/03/2008	10/15/2008	10/16/2008	19	12	180	43.942	183.626	22.621	184.322	
78470	4.5	M	WC MK10	Chatham	4/15/2008	4/21/2008	8/26/2008		8/27/2008	128		134	44.232	184.008	20.863	184.367	
78471	4.0	F	WC MK10	Stewart	3/23/2009				10/25/2009			216	46.912	168.258	46.395	168.201	Y
78472	4.0	M	WC MK10	Stewart	3/31/2009	5/25/2009	6/22/2009	8/23/2009	8/24/2009	29	62	146	46.833	168.228	21.346	159.204	
78473	2.8	M	WC MK10	Stewart	3/26/2009	6/04/2009	7/1/2009	12/18/2009	12/19/2009	28	170	268	46.912	168.257	31.448	164.422	
78474	3.8	F	WC MK10	Stewart	3/26/2009	7/4/2009	7/10/2009		7/11/2009	7		107	46.912	168.257	36.957	180.450	
78475	3.3	F	WC MK10	Stewart	3/27/2009	8/18/2009	9/24/2009	10/7/2009	10/8/2009	38	13	195	46.912	168.258	17.706	162.584	
78476	2.9	M	WC MK10	Stewart	3/30/2009	5/21/2009	6/10/2009	8/6/2009	8/7/2009	21	57	130	46.911	168.257	22.109	167.198	
78477	3.0	F	WC MK10	Gisborne	9/9/2009				9/25/2009			17	38.726	178.298	39.025	177.959	Y
88234	4.8	M	WC MK10	Stewart	3/24/2009	6/21/2009	7/7/2009		7/11/2009	17		109	46.830	168.229	28.596	154.061	
88235	4.0	F	WC MK10	Stewart	3/24/2009	7/08/2009	10/9/2009	12/19/2009	12/20/2009	57[b]	71	271	46.912	168.258	24.121	159.327	

The temporal limits and durations of the ocean and tropics phases are shown. Archive record indicates high resolution data available from recovered tags. TL, total length; WC, Wildlife Computers; MT, Microwave Telemetry. Blanks indicate not applicable or insufficient data for estimation. Latitudes are in degrees south, and longitudes are in degrees east.

[a] Includes 26 days (November 17 to December 12, 2008) on the return leg to Chatham Islands.

[b] Excludes 37 days on the Australian coast.

The dates marking the transition between each phase were estimated by simultaneous examination of temporal profiles of longitude, latitude, depth, and temperature. Missing data often prevented exact determination of the departure dates, but errors were probably minimal (about ±2 d). Post-tagging observation of sharks and examination of high resolution archive data indicated that tagging had little effect on their behavior. Nevertheless, we omitted all data collected on the day of and the day following tagging and the day of tag pop-up. All dates and times are reported in universal time (UT) unless specified as local time, which was calculated from UT and the degrees of longitude. Hours of darkness were defined as the period between the end of civil twilight at dusk and the start of civil twilight at dawn. Incomplete data transmission to the satellite meant that PDT data were not always available for determining the maximum depth and minimum and maximum temperatures attained by each shark. Occasionally, bin data indicated more extreme values than PDT data. In those cases, maximum depth was estimated as the lower (shallower) bound of the deepest depth bin, maximum temperature was estimated as the lower (colder) bound of the warmest temperature bin, and minimum temperature was estimated as the upper (warmer) bound of the coolest temperature bin.

High-resolution archive data from recovered tags were analyzed using Wildlife Computers software (Instrument Helper) to determine shark diving patterns. A dive was defined as a descent to over 200-m depth that started and finished in less than 5 m of water. Differences among distributions of dive parameters were tested for statistical significance using Kolmogorov-Smirnoff two-sample tests (Sokal and Rohlf, 1981). The time of day at which a dive started (truncated to the whole hour) was tested against the null hypothesis that start time was uniformly distributed across a 24-h period using *G* tests of goodness of fit (Sokal and Rohlf, 1981).

RESULTS

Twenty-five White Sharks were fitted with tags programmed for 3–9-month deployments, but some tags came off early; actual deployments were 4–274 d (Table 22.1). Four tags having deployments shorter than 3 weeks and one tag that transmitted negligible data were omitted from this study, leaving twenty sharks with deployments of 89–274 d. One shark (78471) remained in the vicinity of its Stewart Island tagging location for the entire 216-d deployment and therefore did not experience ocean and tropics phases. The remaining 19 sharks all showed evidence (track locations and/or high temperatures) indicating that they migrated to subtropical or tropical waters north of New Zealand. However, not all of these tags transmitted sufficient data to define ocean and tropics phases adequately, so our results were sometimes based on reduced sample sizes (Table 22.1). The tracks of sharks analyzed in this study are illustrated in Chapter 21, this book, and the start and finish dates of ocean and tropics phases are shown in Table 22.1.

Transmitted Data

While in the vicinity of the tagging site (shelf phase), all sharks remained in shallow water, with 40.0% of their time (on average) being spent at 11–25 m depth and 95.6% of their time being spent at 0–50 m (Figure 22.1). They rarely ventured beyond 100 m. During the ocean phase, sharks exhibited a completely different bimodal depth pattern (Figure 22.1): most time (mean 63.1%) was spent at the surface (0–1 m), with a second mode at 201–800 m (mean 19.3%). Deep diving was a consistent feature of the ocean phase: 16 out of 20 sharks dived to maximum depths of 601–1,200 m (Table 22.2); the remaining four sharks either transmitted few data or did not experience an ocean phase. The proportion of time spent in deep water was small: mean 6.4% greater than 600 m (range, 0.2–17.7%; $n = 14$) and mean 1.0% greater than 800 m (range, 0.0–5.2%; $n = 14$). Four sharks exceeded depths of 1,000 m (Table 22.2).

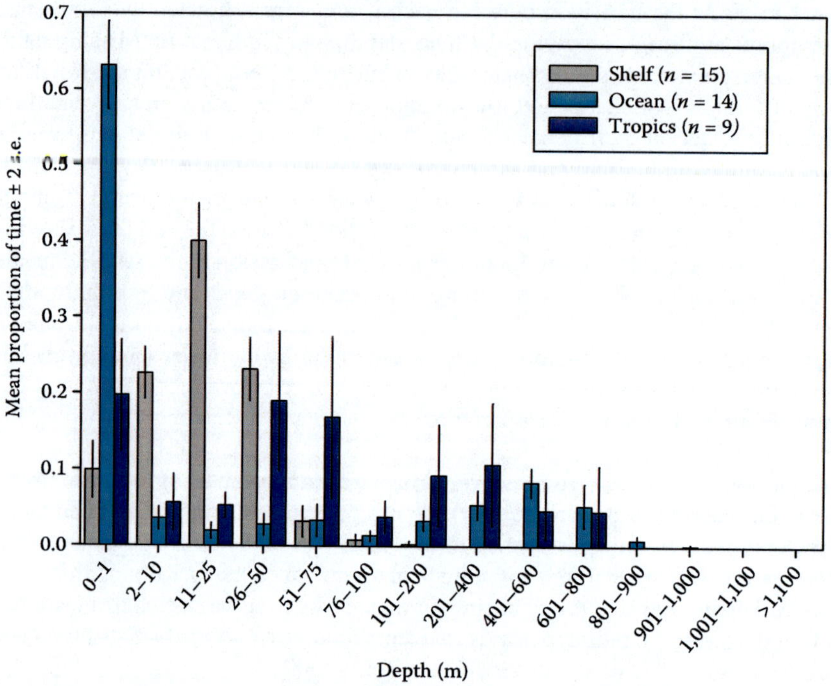

Figure 22.1 Mean proportion (and 95% confidence limits) of time spent by White Sharks in each depth range during the shelf, ocean, and tropics phases. n = number of sharks in each phase.

The 1,200 m PDT depth for shark 88234 was corroborated by a bin depth of greater than 1,100 m, and the calibration curve for the tag indicated that the error in the PDT depth was less than ± 15 m (Wildlife Computers, personal communication). Furthermore, ambient temperatures recorded by the tag support the depth measurement: on June 22, 2009, shark 88234 recorded depth-temperature pairs (PDT data) of 984 m and 5.4°C; 1,000 m and 5.8°C; 1,056 m and 5.4°C; and 1,200 m and 4.2°C. Thus, the greatest depth for this shark coincided with its lowest temperature (Figure 22.2). We therefore believe that the 1,200-m depth record is reliable and moderately accurate.

After arrival in subtropical or tropical regions, individual sharks displayed different depth preferences, leading to variable mean proportions at depth with large confidence limits (Figure 22.1). Four different depth patterns were identifiable: shallow water, mostly less than 100 m (two sharks at Norfolk Island and two at New Caledonia); broad distribution between the surface and 400–800 m (one shark at New Caledonia and two in the Coral Sea); bimodal distribution (one shark at Tonga); and deep water, mostly 100–400 m (one shark at Tonga).

Diel variation in depth distribution was explored for eleven White Sharks whose tags summarized data into 3-h time bins. Data were analyzed for shark-phase combinations having samples sizes greater than 4 d. During the shelf phase, most sharks (eight out of nine) showed an apparent diel depth cycle, with more time than average being spent deeper than 25 m during dawn–dusk (0600–1800 h local time) and more time than average shallower than 25 m at night (1800–0600 h local time) (Figure 22.3). The one exception (shark 78469) showed a slight inverse pattern. Some sharks (four out of six) in the ocean phase showed similar, although weaker, diel patterns, being deeper on average during dawn–dusk than at night. In the tropics phase, diel patterns ranged from very strong (deeper during the day) to negligible.

As expected from their large latitudinal and depth ranges, White Sharks experienced a wide range of ambient temperatures. In the shelf phase, the modal temperature was 12–14°C (mean

Table 22.2 **Maximum Depth and Minimum and Maximum Temperatures Recorded for 20 White Sharks**

Shark Tag	Maximum Depth (m)	Minimum Temperature (°C)	Maximum Temperature (°C)
57033	901	6.4	16.1
57035	820	7.2	23.8
57036	901	6.6	24.0
64031	901	6.8	22.6
64032[a]	384	11.4	25.0
64033[a]	601	8.0	25.4
64035[a]	378	9.6	23.6
64036	1001	2.7	26.6
64037[a]	113	7.6	23.9
78468	1003	5.4	26.0
78469	1048	5.6	23.6
78470	832	6.0	24.8
78471[b]	128	9.0	14.8
78472	904	6.6	25.4
78473	840	7.2	22.8
78474[b]	801	6.6	13.8
78475	696	7.4	25.8
78476	901	6.6	25.2
88234	1200	4.2	21.2
88235	872	6.6	26.1

[a] Minimal data transmitted by tag.
[b] Shark did not leave New Zealand waters.

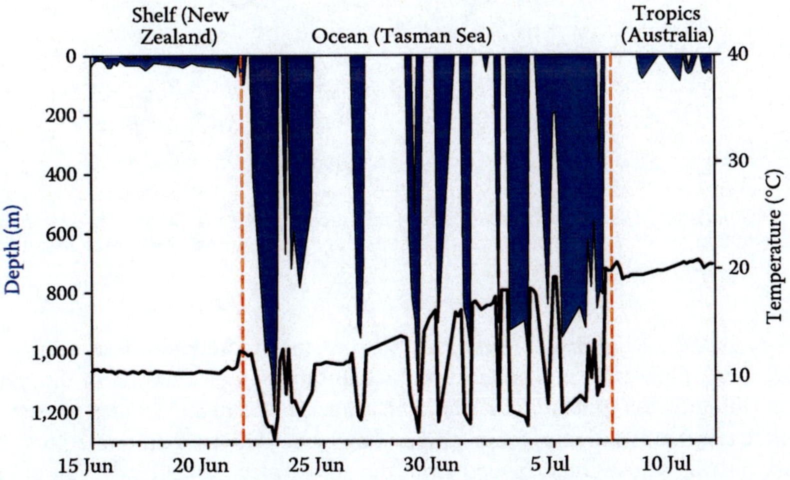

Figure 22.2 Depth and temperature profiles during 1 month of the 109-d track of a 4.8-m male White Shark (88234) between Stewart Island and Australia, June through July 2009. Data resolution = 3 h. Red dashed lines indicate approximate transitions between track phases.

58.3% of time), with nearly all of their time (98.5%) spent at 10–16°C (Figure 22.4). During the ocean phase, the sharks encountered a broad range of temperatures, mostly between 6 and 24°C, as they travelled from temperate New Zealand to warm subtropical and tropical waters, and dived to great depths. In the tropics phase, temperatures were mainly 18–26°C, but with a long tail of lower

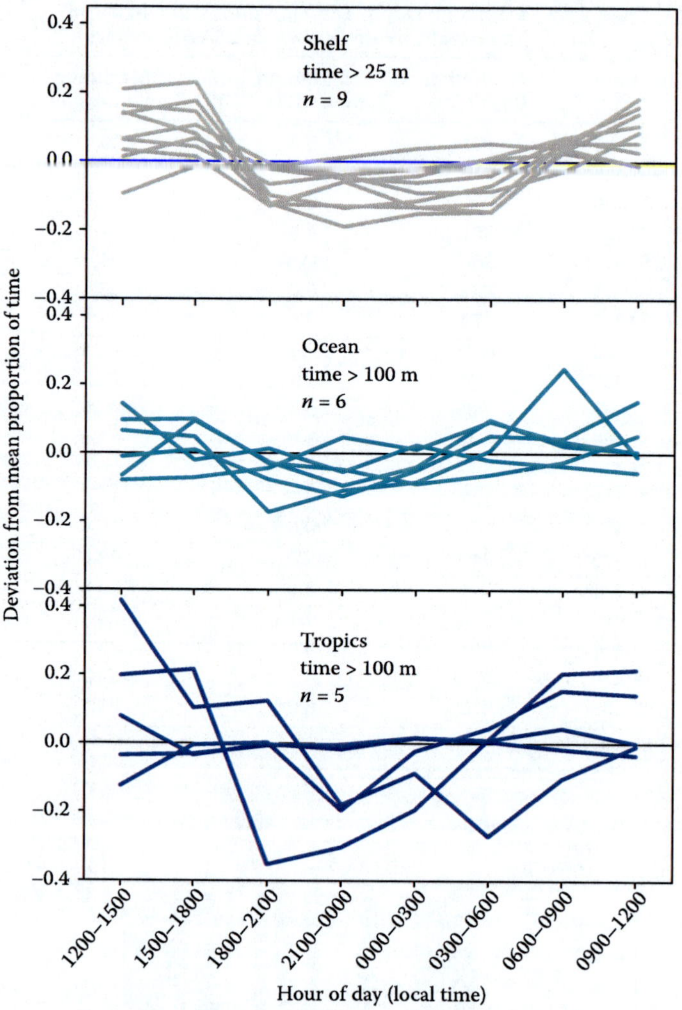

Figure 22.3 Diel variation in the mean proportion of time spent by White Sharks deeper than 25 m (shelf) or 100 m (ocean and tropics), expressed as the deviation from the overall mean for each shark. *n* = number of sharks.

temperatures recorded during dives. Most sharks experienced temperature minima of 4–8°C and maxima of 22–26°C (Table 22.2). The few sharks with minima or maxima in the range 10–22°C either had very limited data transmission or did not undergo ocean and tropics phases.

The absolute minimum and maximum temperatures recorded were 2.7 and 26.6°C, respectively, both for shark 64036, whose track ranged from the subantarctic Auckland Islands south of New Zealand to the Great Barrier Reef off Australia. The 2.7°C temperature was recorded within a 4-d period during which temperatures oscillated mainly between 7–8 and 3–5°C. On August 13, 2007, this shark spent 10 h at temperatures of 3.2°C or less, nearly all of it at the surface. In mid-August, satellite-sensed SSTs below 4°C occurred south of the Antarctic Convergence, which was about 450 km south of the shark's southernmost estimated position. However, tongues of Antarctic water penetrated north from the Convergence to within 200 km of the shark's position. Furthermore, the shark was apparently near or over the steep-sided Solander Trough, which may generate upwelling of cold subsurface water.

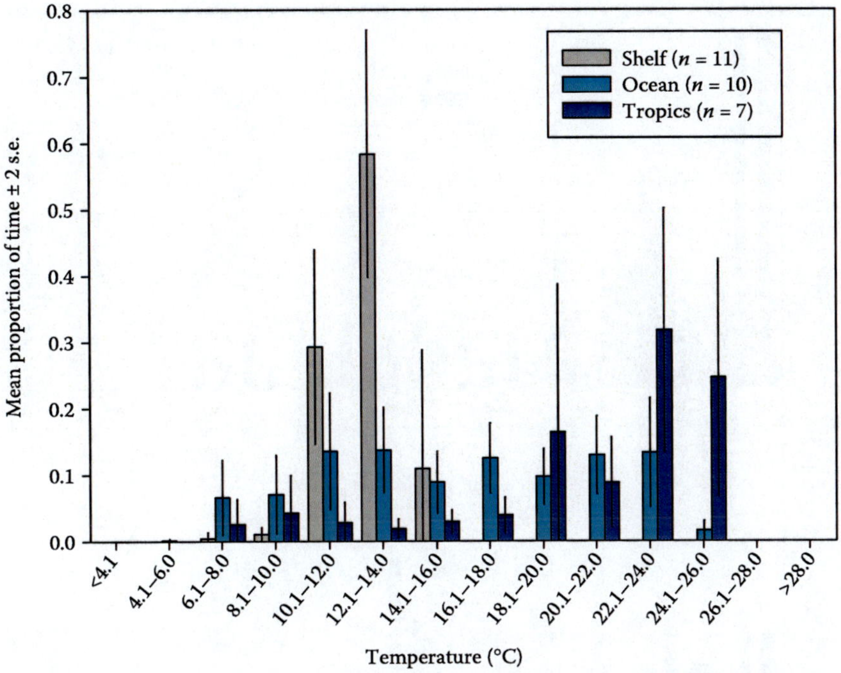

Figure 22.4 Mean proportion (and 95% confidence limits) of time spent by White Sharks in each temperature range during the shelf, ocean, and tropics phases (MK10 tag data only). n = number of sharks in each phase.

Archival Data from Recovered Tags

Archival records were obtained from six recovered tags (Table 22.1). However, two sharks had short (4- and 17-d) records, and one of the sharks (78471) did not make an offshore migration. Dives deeper than 200 m were not made by any shark during the shelf phase, so our dive analysis was limited to the three remaining sharks with 5–9-month archive records (57035, 64031, and 78468) and to the ocean and tropics phases.

Diving patterns varied among the three sharks and between ocean and tropics phases. In the ocean phase, the maximum depth of each dive varied widely from 200 m (the defined minimum) to over 800 m in all three sharks, with no clear modal pattern (Figure 22.5). In the tropics phase, most dives were to 200–400 m for one of the sharks and 200–600 m for the other two sharks. Dives were significantly shallower in the tropics phase (median maximum depth, 257–361 m) than in the ocean phase (432–584 m) (Table 22.3).

Dive duration was mainly 1–6 h for shark 57035, 0–4 h for shark 64031, and 0–3 h for shark 78468, but the last two sharks had long tails of dives longer than 12 h, including a few of 20–25 h (Figure 22.6). There were no significant differences in dive duration between the ocean and tropics phases (Table 22.3), although shark 78468 made frequent dives longer than 10 h during the tropics phase but none in the ocean phase.

Dives deeper than 200 m were initiated at all times of the day and night (Figure 22.7). However, the distributions of start time were significantly different from a uniform distribution for all three sharks in both ocean and tropics phases (G test, $p < 0.05$), indicating the presence of diel patterns. Shark 57035 had strong modes at 0400–0700 h (dawn) and 1400–1700 h (afternoon) in both ocean and tropics phases. Shark 78468 had a very strong mode at 0500–0600 h (dawn) and a secondary mode at 1600–1800 h (evening) during the ocean phase but dived mainly at night during the tropics

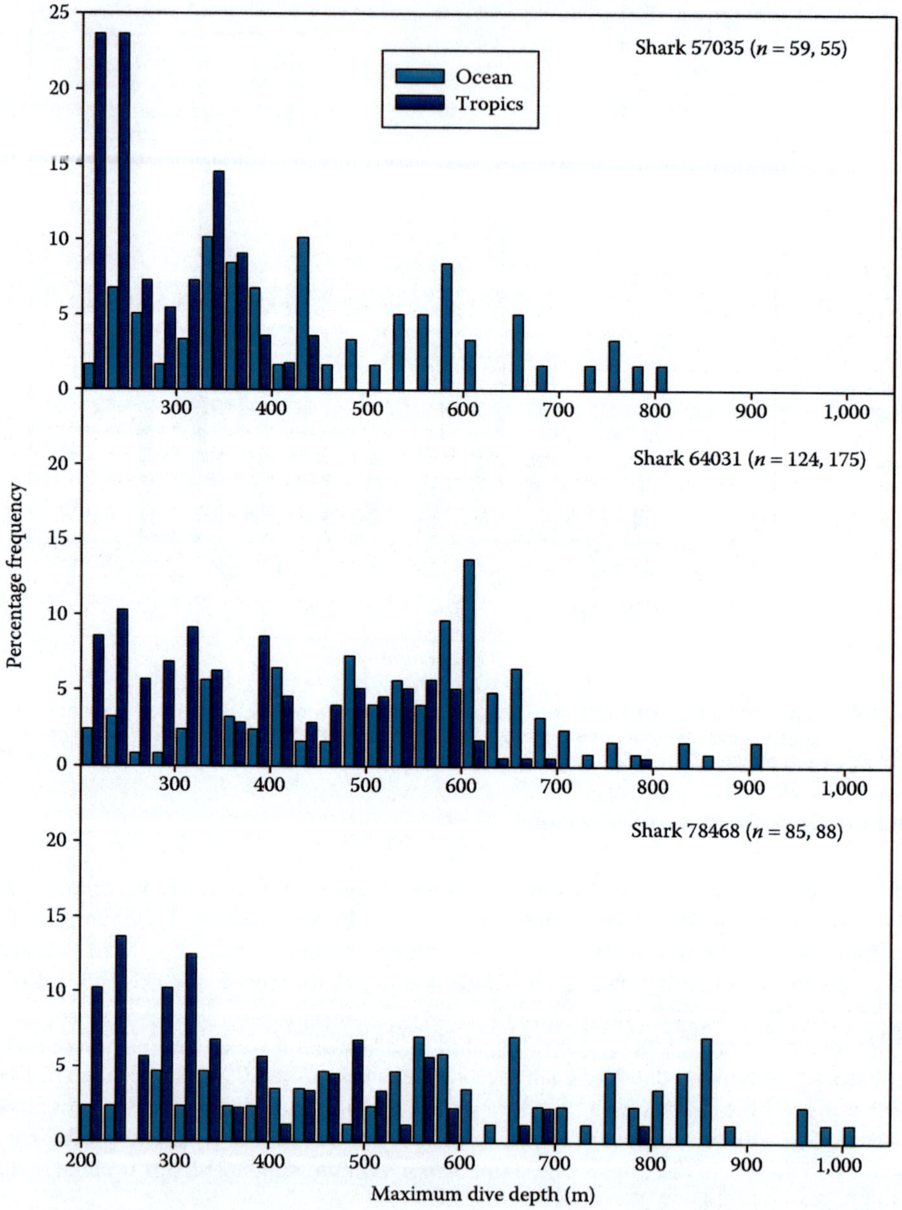

Figure 22.5 Maximum depths of dives of three sharks (57035, 64031, and 78468) during ocean and tropics phases. n = number of dives for ocean and tropics phases, respectively.

phase. Shark 64031 showed no strong diel patterns, although there may have been reduced diving activity during the morning (0700–1200 h) in both the ocean and tropics phases.

For most shark-phase combinations, there were no significant differences between descent and ascent speeds; the exception was for shark 64031, which made faster descents (median, 0.6 $m \cdot s^{-1}$) than ascents (0.4 $m \cdot s^{-1}$) in the ocean phase (Table 22.3 and Figure 22.8). By contrast, dive speeds (both descents and ascents) were greater during the ocean phase than the tropics phase for all three sharks. Speeds exceeding 0.8 $m \cdot sec^{-1}$ were achieved much more often during descents than during ascents (Figure 22.8). The distributions of dive speeds were skewed toward low speeds in the tropics

Table 22.3 Medians of Dive Parameters for Three Sharks in Ocean and Tropics Phases

			Medians of Dive Parameters				
Shark	TL (m) and Sex	Phase	Max Dive Depth (m)	Dive Duration (h)	Descent Speed ($m \cdot s^{-1}$)	Ascent Speed ($m \cdot s^{-1}$)	Sig. of Descent and Ascent Speeds
57035	4.0 F	Ocean	432.0	2.0	0.2	0.2	NS
		Tropics	257.0	3.1	0.1	0.1	NS
		Sig.	[a]	NS	[a]	[a]	
64031	3.5 F?	Ocean	562.0	2.3	0.6	0.4	[a]
		Tropics	361.0	2.6	0.2	0.2	NS
		Sig.	[a]	NS	[a]	[a]	
78468	4.0 M	Ocean	584.0	1.1	0.5	0.5	NS
		Tropics	321.0	1.3	0.3	0.2	NS
		Sig.	[a]	NS	[a]	[a]	

Sig, statistical significance of Kolmogorov-Smirnov distribution tests; NS, not significant.
[a] $p < 0.05$.

phase but were broad and unimodal for two of the three sharks in the ocean phase. The number of dives per day exceeding 200 m varied greatly, from 0 to 11 dives per day in the ocean phase and from 0 to 17 dives per day in the tropics phase (Figure 22.9).

Shark 57035 spent almost 3 months near the tagging site (shelf phase), during which she rarely dived beyond 100 m (Figure 22.10). She then headed northwest toward New Caledonia (ocean phase) and immediately began diving deeper than 100 m, building up to depths exceeding 700 m within a few days, and then made a number of dives to 700–800 m over the next week (Figures 22.10 and 22.11). Maximum dive depth declined gradually to around 400 m when she arrived at New Caledonia. During the ocean phase, the shark's behavior ranged from occasional, short (1–2 h) dives to dives that lasted many hours. On long dives, she sometimes remained at a similar depth throughout the dive, and sometimes engaged in regular vertical oscillations of 200–300 m (Figure 22.11). After arrival at New Caledonia (tropics phase), the shark spent much less time at the surface and more time diving, although maximum dive depths were reduced to 200–400 m (Figures 22.10 and 22.12). The mean number of dives in the ocean phase was 1.8 per day, and the mean number in the tropics phase was 2.0 per day (Figure 22.9).

Shark 64031 departed from the Chatham Islands 4 d after tagging. She spent three weeks in the ocean phase en route to Norfolk Island and on most days made 2–4 dives per day (mean, 2.5) deeper than 200 m (Figures 22.9 and 22.10). Many of the dives were to depths of 600–900 m. After arrival at Norfolk Island, her behavior changed abruptly, with large amounts of time spent at 50–70 m, punctuated by occasional deeper dives of 400–550 m. This pattern may result from most of her time being spent over the shallow plateau surrounding Norfolk Island, interspersed with periodic forays over the edge of the continental shelf. After about 5 months in the vicinity of Norfolk Island, the frequency and depth of dives increased, with deep dives on most days and considerably more time being spent in depths of 400–600 m. On the day before her departure from Norfolk Island, the shark made 14 deep dives. The overall mean number of dives per day in the tropics phase was 0.9, but it was only 0.5 in the first 160 d, increasing to 3.2 for the last 31 d. The return ocean crossing to Chatham Islands (second ocean phase) was similar to the first, with several dives on most days (mean, 2.8) to depths of 600–700 m. After arrival back at Chatham Islands, the depth profile changed abruptly again to a shallow water shelf pattern.

Shark 78468 had a diving pattern similar to that of shark 57035. He spent 5 months near the tagging site (shelf phase), during which he rarely dived beyond 120 m. He then headed north to Tonga (ocean phase) and immediately began diving deeper than 500 m, with many dives of 600–1,000 m. Diving frequency increased toward the end of the phase (Figure 22.9), and maximum dive depth

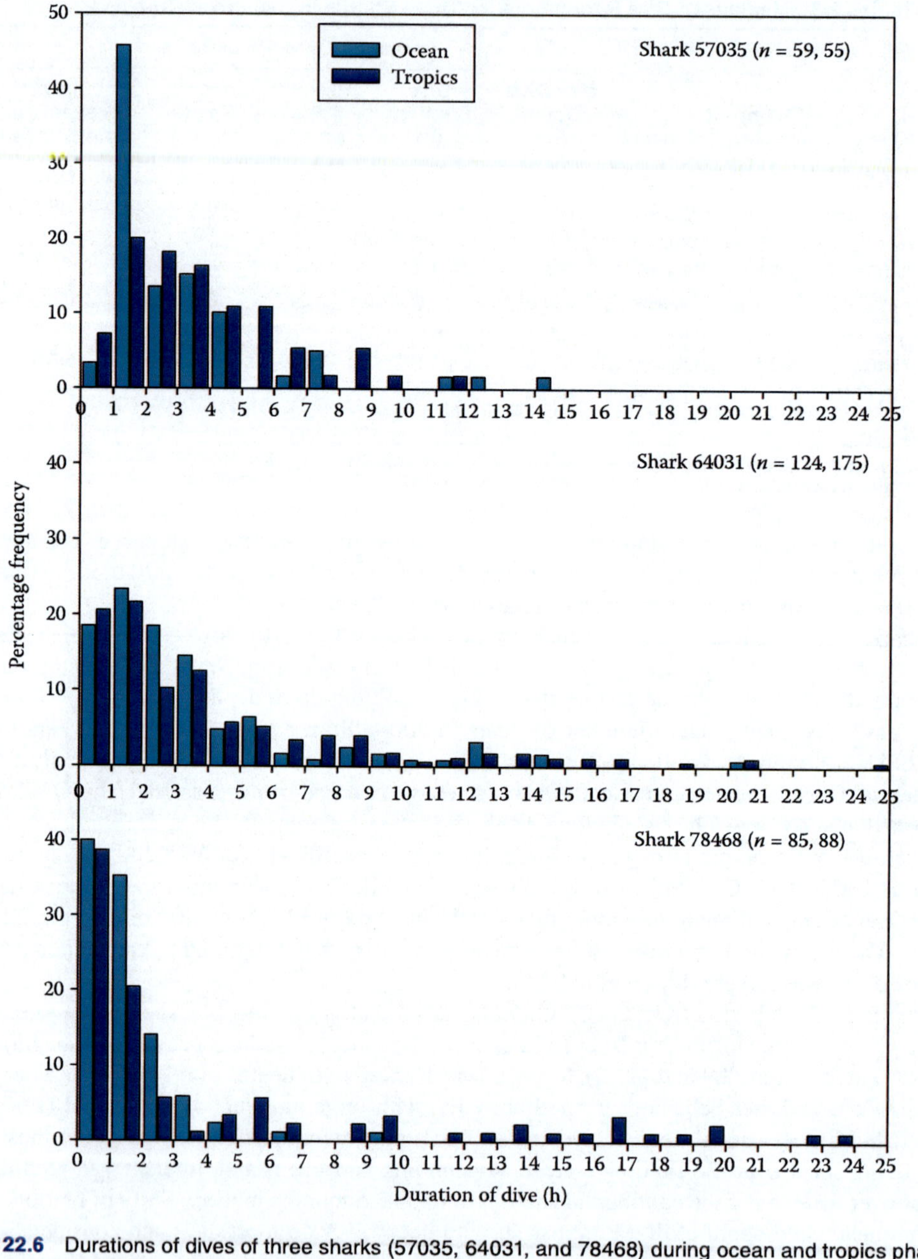

Figure 22.6 Durations of dives of three sharks (57035, 64031, and 78468) during ocean and tropics phases. *n* = number of dives for ocean and tropics phases, respectively.

declined gradually to around 500–600 m. The mean number of dives in the ocean phase was 3.2 per day. After arrival at Tonga (tropics phase), the shark spent much less time at the surface and more time diving, although maximum dive depths were around 300–500 m. The mean number of dives in the tropics phase was 4.2 per day, including 1 d with 17 dives (Figure 22.9). Close inspection of the high resolution depth data reveals the periods when shark 78468 departed from Chatham Islands and when he arrived at Tonga. On departure day (September 17), he made two short dives to over 100 m, followed by a deeper dive to about 300 m, at which point he appears to have encountered the seabed and followed it down the slope for over 9 h to almost 600 m before returning to

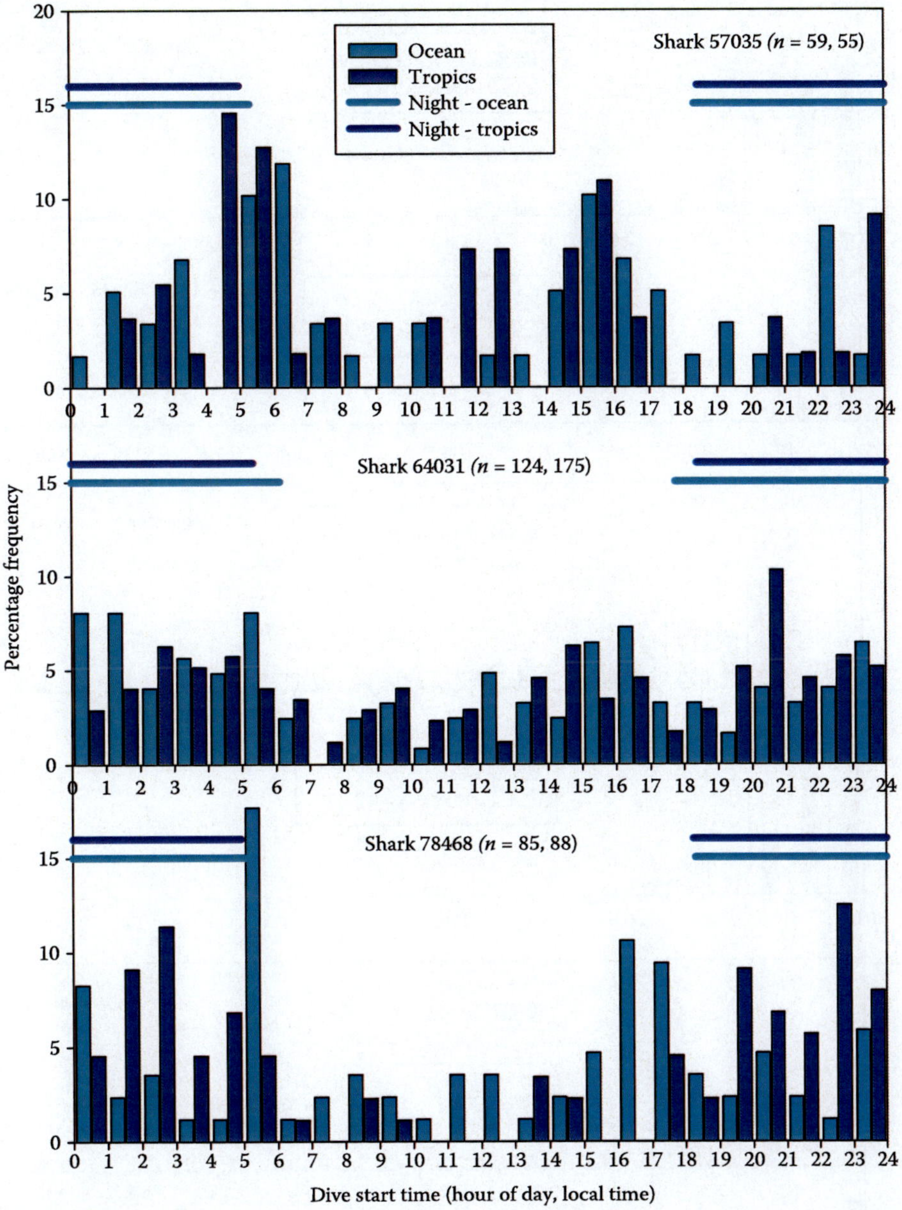

Figure 22.7 Approximate local start time of dives of three sharks (57035, 64031, and 78468) during ocean and tropics phases. Estimated hours of darkness are indicated by the horizontal bars. n = number of dives for ocean and tropics phases, respectively.

the surface (Figure 22.13). In another dive soon after, he may have briefly encountered the seabed again at about 650 m. Subsequent dives did not appear to reach the seabed, judging by the shape of the dive profiles. Before arrival in Tonga, the shark was making repeated dives to 300–500 m (Figure 22.13). He probably encountered the seabed at about 580 m on October 12 and, after two brief trips to the surface, followed the seabed up to about 200 m. He may then have remained near the seabed at around 200 m for 2 d, apart from two brief excursions to the surface.

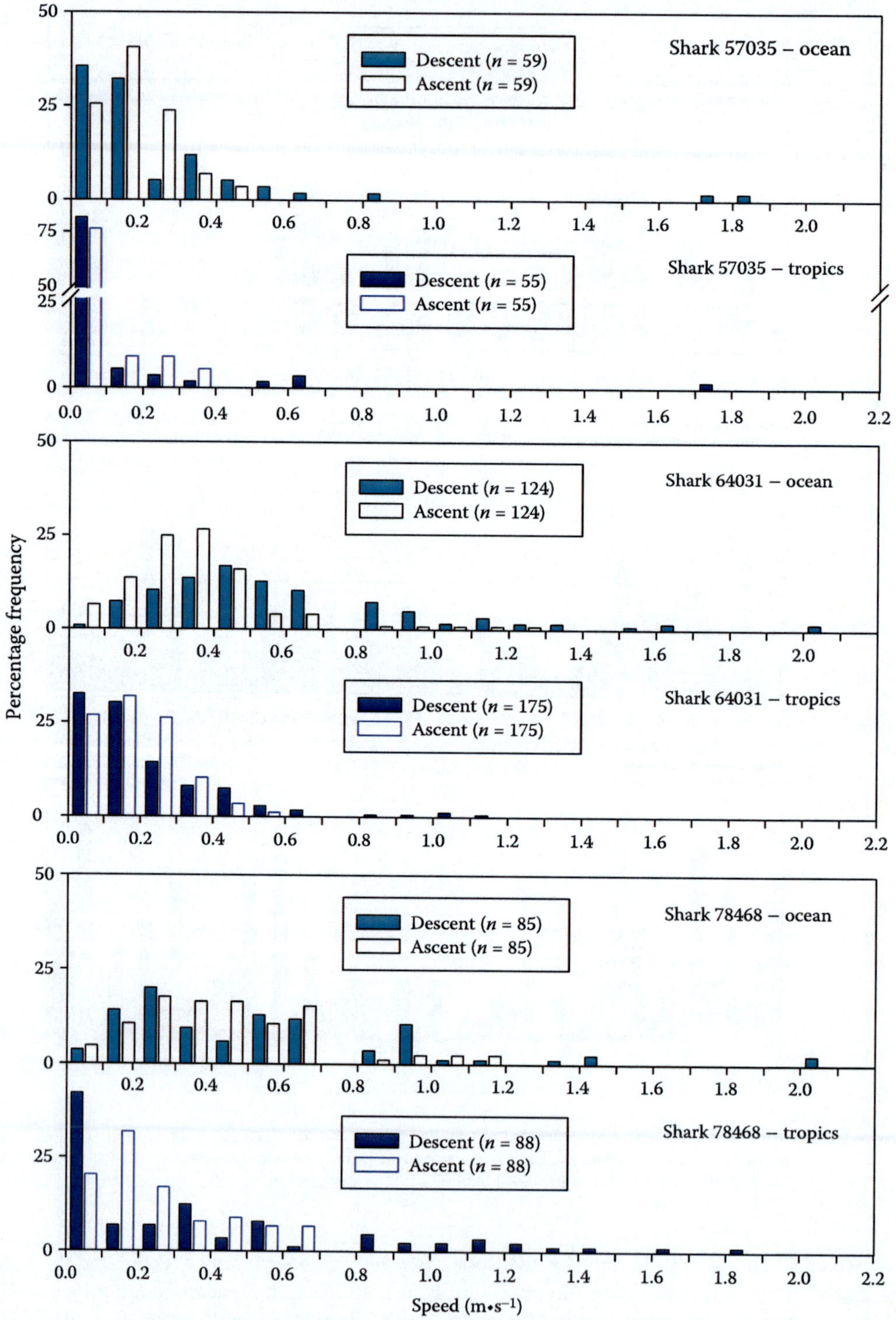

Figure 22.8 Descent and ascent speeds during dives by three sharks (57035, 64031, and 78468) during ocean and tropics phases. Note different *y* axis scale for shark 57035 tropics phase. *n* = number of dives.

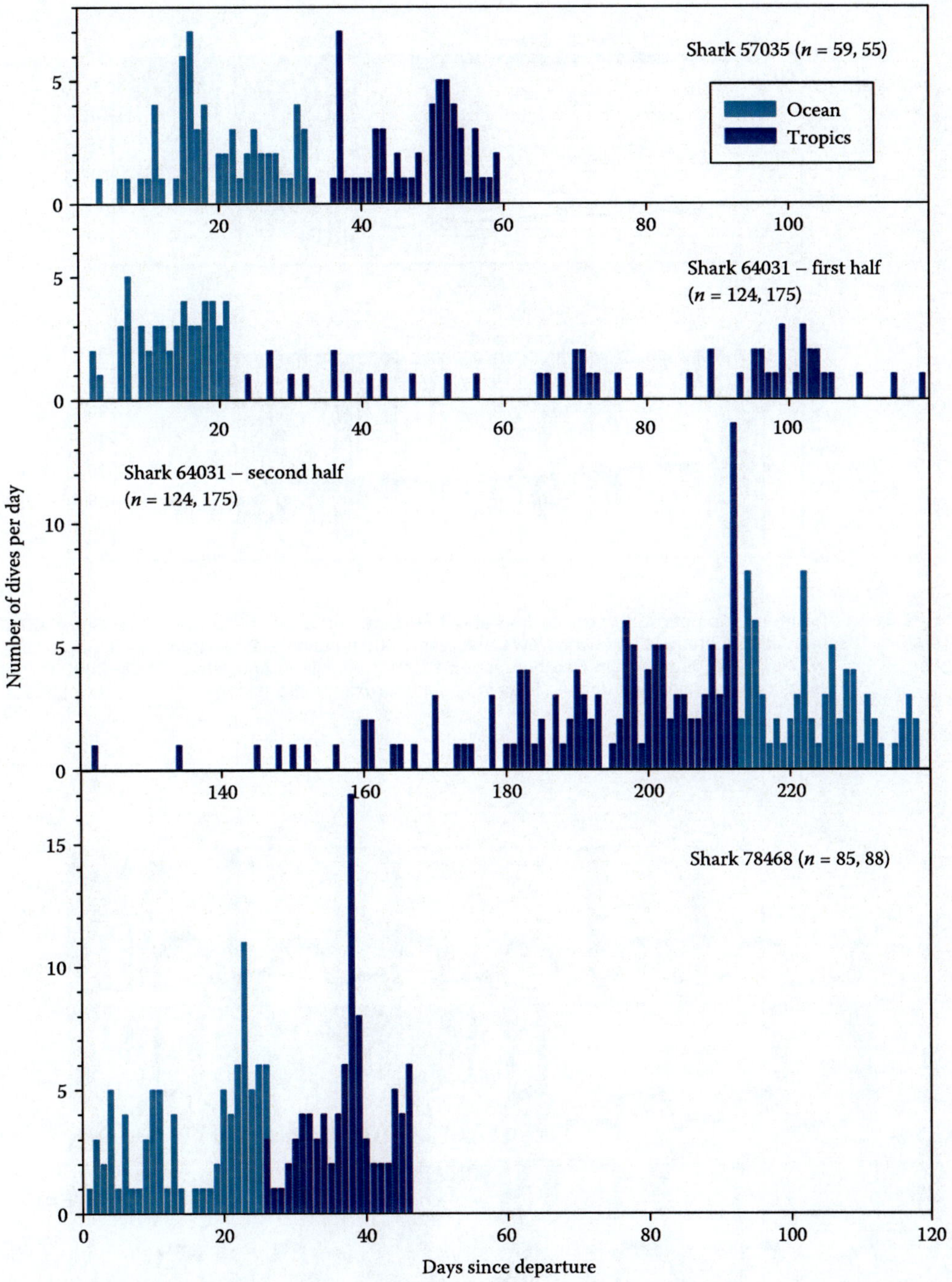

Figure 22.9 Number of dives per day of three sharks (57035, 64031, and 78468) during ocean and tropics phases. n = number of dives for ocean and tropics phases, respectively.

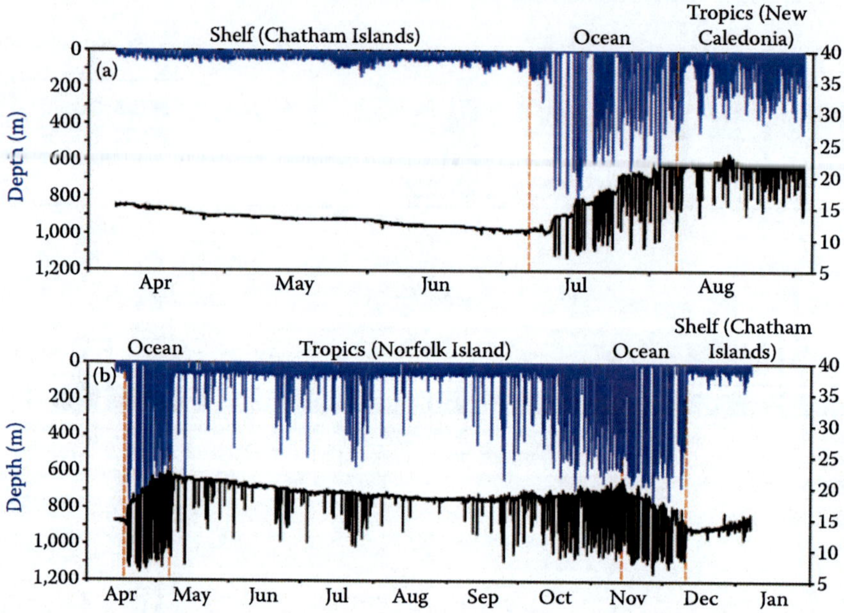

Figure 22.10 Depth and temperature profiles from the 149-d track of a 4.0-m female White Shark (57035) between Chatham Islands and New Caledonia, 2005 (a) and the 271-d track of a 3.5-m female (?) White Shark (64031) from Chatham Islands to Norfolk Island and return, 2008–2009 (b). Data resolution = 1 min. Red dashed lines indicate approximate transitions between track phases.

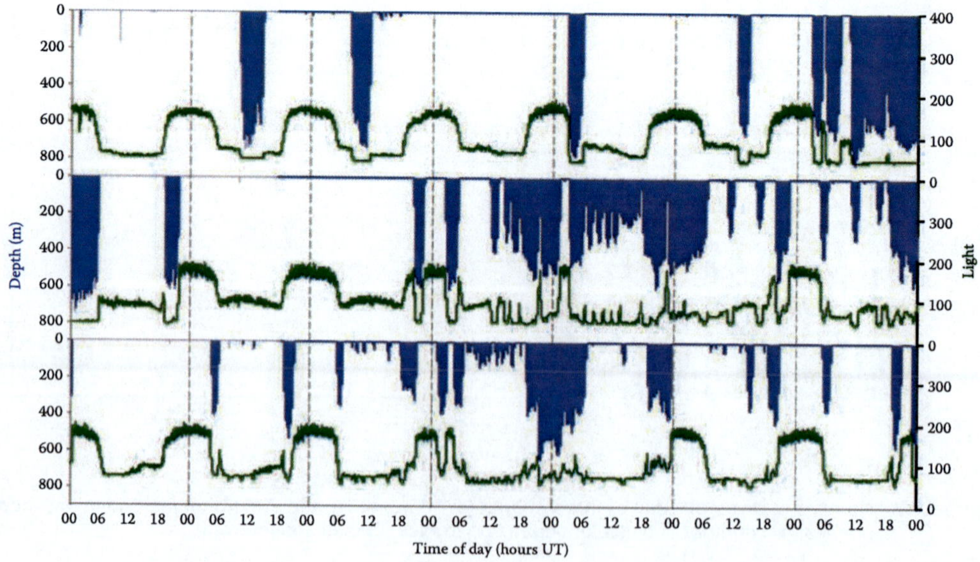

Figure 22.11 Depth and light profiles from three consecutive 1-week periods in the track of a 4.0-m female White Shark (57035) between Chatham Islands and New Caledonia, July through August 2005. Data resolution = 1 min. Dashed vertical lines indicate midnight UT, which is approximately midday local time.

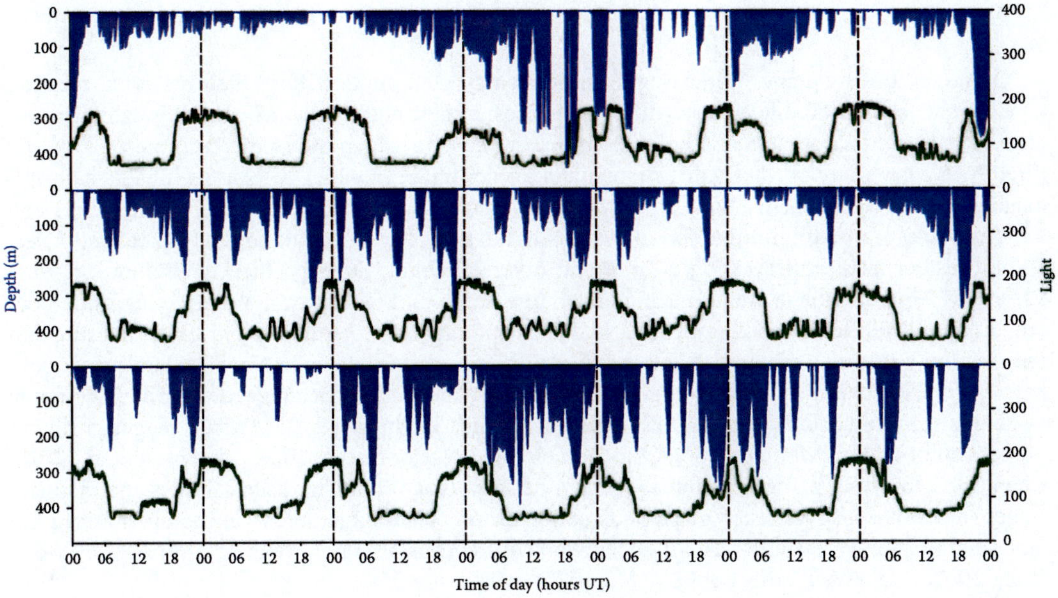

Figure 22.12 Depth and light profiles from three consecutive 1-week periods in the track of a 4.0-m female White Shark (57035) near New Caledonia, August 2005. Data resolution = 1 min. Dashed vertical lines indicate midnight UT, which is approximately midday local time.

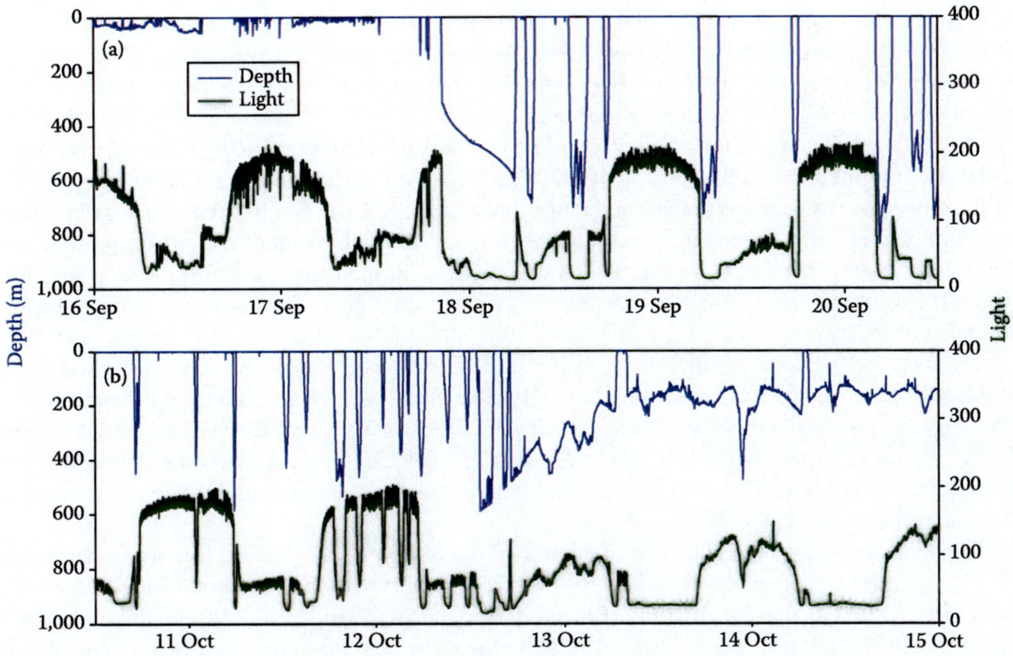

Figure 22.13 Depth and light profiles during the track of a 4.0-m male White Shark (78468) around the day of departure from the Chatham Islands on an ocean crossing to Tonga, September 2008 (a), and the day of arrival at Tonga, October 2008 (b). Data resolution = 6 sec.

DISCUSSION

In the southwest Pacific, White Sharks undergo major temporal shifts in their behavior, creating a niche that consists of at least three different phases: a shelf phase in which sharks inhabit temperate, coastal waters around New Zealand; an ocean phase in which the sharks migrate rapidly and directly in a generally northerly direction; and a tropics phase in which the sharks inhabit subtropical and tropical locations in a wide arc from Australia to Tonga (Bonfil et al., 2010; Chapters 21 and 23, this book). Latitudinal migration of White Sharks from New Zealand to the tropical southwest Pacific results in them traversing a large and diverse habitat, during which they exhibit a wide variety of vertical habitat use and experience large depth and temperature ranges. Our study covered mainly juvenile, subadult, and adult males (about half were mature) and juvenile and subadult females (only one was mature). Adult female White Sharks do occur in New Zealand waters but have only rarely been seen at our tagging sites. A study in South Africa tagged a similar size range of sharks to ours (Bonfil et al., 2005), whereas studies in the northeast Pacific tagged a higher proportion of adults (Weng et al., 2007b; Domeier and Nasby-Lucas, 2008; Jorgensen et al., 2010; Chapter 12, this book). Tagging studies worldwide show that White Sharks of all sizes and maturity stages undertake long-distance migrations and dive to considerable depths in the open ocean but that young-of-the-year sharks usually occur in shallow coastal waters (Dewar et al., 2004; Weng et al., 2007a; Chapter 17, this book).

Shelf Phase

In the shelf phase, sharks in our study were presumed to have remained in coastal waters near the tagging sites, which were adjacent to colonies of New Zealand fur seals *Arctocephalus forsteri* (precise locations were not available because of the errors inherent in light-based geolocation). At these times, the sharks showed a modal depth range of 11–25 m, rarely dived deeper than 50 m (mean 4.4% of the time), and never dived deeper than 200 m, despite the availability of water deeper than 100 m within 10–30 km of the tagging sites. Preferences for depths shallower than 50 m are also displayed by White Sharks near pinniped colonies elsewhere (Strong et al., 1992; Goldman and Anderson, 1999; Boustany et al., 2002; Bruce et al., 2006; Hammerschlag et al., 2006; Weng et al., 2007a). Goldman and Anderson (1999) found that White Sharks near a pinniped colony in California remained near the seabed in depths down to about 30 m, but in deeper water they swam above the seabed. Hammerschlag et al. (2006) reported that white-shark attacks on pinnipeds at Seal Island, South Africa, peaked in depths of 26–30 m, despite the availability of deeper water. These studies suggest that 20–30 m may represent an optimum depth from which to stalk and attack surface-swimming prey.

We observed an apparent diel depth cycle in the shelf phase for all but one shark. More time was spent deeper than 25 m between dawn and dusk, possibly because higher light levels made it advantageous for sharks to hunt surface prey from deeper water where they would not be noticed. Conversely, sharks spent more time shallower than 25 m at night, perhaps because they could hunt under cover of darkness. However, inverse diel depth patterns (deeper at night) have been reported elsewhere (Strong et al., 1992).

Temperatures experienced by New Zealand White Sharks during the shelf phase were mainly (98.5% of time) in a narrow range of 10–16°C. Similar ranges have been reported for White Sharks in coastal Northern California (Boustany et al., 2002; Weng et al., 2007b), but higher temperatures (14–22°C) have been reported in New South Wales, Southern California, and Mexico (Dewar et al., 2004; Domeier and Nasby-Lucas, 2008; Chapter 17, this book). The wide temperature range experienced by White Sharks in different regions reflects the environment inhabited by their prey.

Ocean Phase

While crossing open oceans, White Sharks exhibited a bimodal depth distribution that consisted of a high proportion of time spent at the surface punctuated by deep dives beyond the photic zone. Sharks spent 63% of their time on average in the top 1 m of water and 19% of their time between 201 and 800 m during the ocean phase. Similar bimodal patterns have been reported for both sexes and a wide range of sizes and from various parts of the globe (Boustany et al., 2002; Bonfil et al., 2005; Weng et al., 2007b; Domeier and Nasby-Lucas, 2008; Nasby-Lucas et al., 2009). Some sharks showed a weak diel pattern, being deeper on average during the day than at night, whereas others showed no diel pattern.

Most sharks in our study exceeded depths of 800 m. Dives to 800–1,000 m have also been reported for White Sharks in the Indian Ocean, northeast Pacific, and off Australia (Bonfil et al., 2005; Weng et al., 2007b; Nasby-Lucas et al., 2009; Chapter 17, this book). However, little time is spent by sharks at these depths. In the southwest Pacific, the mean times spent deeper than the indicated depths were: 400 m, 14.8%; 600 m, 6.4%; and 800 m, 1.0%. In the northeast Pacific off California, 12 White Sharks spent an average of 1.1% of their time deeper than 500 m while "traveling," and 10 of the same sharks spent no time below 500 m while in the "offshore focal area" (K. Weng, personal communication). Off Mexico, Nasby-Lucas et al. (2009) found that four sharks in the open ocean spent only 0.3% of their time below 500 m, and they suggested that white-shark vertical habitat may be constrained by low oxygen levels in deeper water. Lack of oxygen could affect the sharks directly (by limiting their aerobic respiration) or indirectly (by restricting their prey to shallower waters). The depth and concentration of the oxygen minimum layer are much greater in the southwest Pacific (about 2,000 m and 3–4 $ml \cdot L^{-1}$ respectively) than in the northeast Pacific (about 500–1,000 m and 0.1–0.5 $ml \cdot L^{-1}$) (Weyl, 1965). Nasby-Lucas et al. (2009) suggested that white-shark activity is largely limited to oxygen concentrations above 1.5–2.0 $ml \cdot L^{-1}$, so oxygen is unlikely to limit the diving of sharks in the southwest Pacific. In the Indian Ocean, where oxygen levels are higher than in the southwest Pacific, one shark spent 22.3% of its time deeper than 500 m (Bonfil et al., 2005). The oxygen-limitation hypothesis may explain why White Sharks spend longer at depths greater than 500 m in the southwest Pacific and Indian oceans than they do in the northeast Pacific.

Archival records gave detailed insights into the vertical behavior of three sharks. The maximum depth of individual dives varied greatly between 200 m and over 800 m, with no strong modal peaks. Dive duration also varied considerably, being mainly 0–6 h long, but could last as long as 25 h. Dives were made at any time of the day or night but were not uniformly distributed throughout a 24-h period. Two sharks had dawn and afternoon–evening peaks of diving, whereas a third shark showed no clear diel pattern. Descent and ascent speeds also varied widely and among sharks, but in all three sharks, these speeds were significantly greater in the ocean phase than in the tropics phase. The wide range of variation in all of the dive parameters examined suggests that diving behavior while in the ocean phase is quite flexible and not consistent through time. Consistent diving strategies may be employed for days or even weeks at a time but may not have been revealed by our analyses because of aggregation of results over the entire ocean phase.

White Sharks experienced a much wider range of temperatures during the ocean phase than the shelf phase: most sharks recorded temperature minima of 4–8°C and maxima of 22–26°C, and 98.3% of time was spent between 6 and 24°C. This wide range reflects the latitudinal and seasonal transition from cool temperate waters in autumn–spring to tropical waters in spring–summer and the behavioral shift from being confined to the shallow mixed layer to diving to depths of hundreds of meters. White Sharks sometimes visit subantarctic regions south of New Zealand. White Shark records (sightings, attacks on sea lions, and an attack on a human) peaked in April–August at Campbell Island (52.6°S, 169.2°E), and mean monthly SSTs during these months were 6.3–8.2°C

(M. Fraser, unpublished data). One of our tagged sharks (64036) experienced an average temperature of 7.1°C (range, 2.7–8.4°C) during a 21-d period west of the Auckland Islands (50.7°S, 166.0°E). Thus, the low temperatures recorded for deep-diving White Sharks may not necessarily represent physiological extremes that they can tolerate for only short periods.

Tropics Phase

In the tropics phase, White Sharks preferred the upper 75 m of the water column (66.6% of time on average) but continued making deep dives. Four distinct depth distribution patterns were identified: shallow, broad, bimodal, and deep. Although our data suggest that there may be regional variation in tropical diving patterns (i.e., shallow or broad in the Coral Sea, New Caledonia, and Norfolk Island and bimodal or deep in Tonga), the sample size was small ($n = 9$), and this result requires confirmation.

Archival tag records revealed differences in the patterns of maximum dive depths, dive durations, and dive start times among three sharks. One shark had dawn and afternoon diving peaks, one dived mainly at night, and the third showed no diel pattern. These differences are not surprising, because the sharks were sometimes near shore and may have been associated with shallow coral reefs or were offshore and perhaps associated with seamounts or ridges. Different foraging strategies would be expected for these different habitats. However, our light-based estimates of shark positions are too imprecise to enable us to interpret their behavior in relation to specific habitat features. Descent and ascent speeds were consistently lower in the tropics phase than in the ocean phase. Temperatures experienced in the tropics phase were mostly 18–26°C but with a long tail of low temperature resulting from deep dives.

Depth and Temperature Extremes

Previously, the maximum recorded depths for tagged White Sharks were 984 m for a juvenile off Australia (Chapter 17, this book), and 980 m for four sharks traversing the Indian Ocean and the northeast Pacific Ocean (Bonfil et al., 2005; Weng et al., 2007b; Nasby-Lucas et al., 2009). However, 980 m is the depth limit for the Wildlife Computers PAT4 tags used for these last four sharks, so they probably all dived deeper than that. The only literature report of a greater depth of which we are aware comes from a large Cuban shark, said by Bigelow and Schroeder (1948) to have been caught on a line at 700 fathoms (1,280 m). However, the actual seabed depth was about 300 fathoms (550 m), and the length of line used was 400 fathoms (730 m) (Guitart and Fernandez, 1974), so the capture depth was much shallower than 1,280 m. At least four of our tagged animals dived to 1,000 m or deeper, and our deepest record was 1,200 m. The latter appears to be the greatest depth reliably recorded for a White Shark. However, this will undoubtedly be exceeded as more electronic tags capable of recording greater depths are deployed. Even juveniles may dive to remarkable depths: in our study, sharks of 2.8- and 2.9-m TL dived to 840 and 896 m, respectively, and in eastern Australia, a 1.9-m juvenile dived to 984 m (Chapter 17, this book).

The extreme temperatures recorded in our study (by the same shark, 64036) spanned almost 24°C (2.7–26.6°C). Elsewhere, temperature minima of 3.4–4.5°C have been reported (Bonfil et al., 2005; Weng et al., 2007b; Domeier and Nasby-Lucas, 2008; Nasby-Lucas et al., 2009). Very low temperatures experienced by White Sharks are associated with both cold SSTs and deep diving. Eight of our sharks recorded maxima of 25°C or higher, and three had maxima of 26.0°C or higher. Previous reported maxima have been around 25°C (Domeier and Nasby-Lucas, 2008; Nasby-Lucas et al., 2009), 26.2°C (Boustany et al., 2002), and 26.4°C (Bonfil et al., 2005). Thus, our extreme temperatures of 2.7 and 26.6°C both appear to be records for White Sharks. The upper limit at least is likely to be extended with further study of White Sharks in tropical waters. The thermal tolerance of White Sharks is remarkable and is likely achieved through its ability to thermoregulate (Goldman, 1997).

Diel Behavior

We found that some sharks had strong diel patterns in their use of the water column, whereas others had no clear pattern. Individual sharks showed variation in their diel pattern among phases and within phases. The main pattern we observed was for sharks to spend longer in deep water during the day than at night, which is consistent with visual foraging in dim light in deep water. Studies elsewhere have also observed similar diel patterns in White Sharks (Weng et al., 2007a; Domeier and Nasby-Lucas, 2008; Nasby-Lucas et al., 2009; Jorgensen et al., 2010; Chapters 10 and 18, this book), including a 1.5 m TL young-of-the-year (Dewar et al., 2004). Interestingly, deeper diving during daylight hours is often accompanied by longer periods at the surface, thus generating a stronger bimodal distribution by day than at night. Observations that White Sharks dive deeper on the full moon than the new moon, albeit with relatively small differences in mean depth (Weng et al., 2007a; Nasby-Lucas et al., 2009), support the notion that White Sharks can hunt visually in very dim light while on deep dives. White Sharks in the vicinity of coastal pinniped colonies are active by both day and night (Klimley et al., 2001; Hammerschlag et al., 2006). Changes in light intensity with depth produce a diel vertical migration of the deep scattering layer, a moving aggregation of planktonic animals and their progressively larger predators. Diel variation in depth distribution has been observed in a variety of tunas, squid, swordfish, and other sharks (Carey and Robison, 1981; West and Stevens, 2001; Musyl et al., 2003; Gilly et al., 2006), and these animals may be prey for foraging White Sharks.

Despite the high frequency of diel behavior in most White Shark studies, it is not universal. One of our sharks in the shelf phase showed a weak inverse diel pattern, with more time spent in deep water during the night, and others showed no clear diel pattern. There have also been reports of no or inverse diel patterns elsewhere (Weng et al., 2007a; Chapters 17 and 18, this book), and sharks can switch between diel and nondiel behavior: Jorgensen et al. (2010) observed diel behavior on 51 out of 93 d for a shark near Hawaii, Weng et al. (2007a) reported greater daytime depth in summer and autumn but no diel pattern in winter, and Chapter 18, this book reported diel behavior in only a portion of the data from six juvenile White Sharks. Intensive study of the behavior of White Sharks in coastal Australia has revealed rapid switching among a range of different foraging strategies and between diel and nondiel behavior (Bruce et al., 2006; Chapter 17, this book). Aggregating depth-time behavior over lengthy periods is likely to obscure any short-term behavior patterns (Chapter 17, this book), and we plan to conduct detailed analyses of shorter time segments using fast Fourier transforms to identify the periodicity of temporal cycles (Shepard et al., 2006).

Explanations for Open-Ocean Vertical Behavior

Surface swimming punctuated by deep diving appears to be universal for White Sharks while crossing open oceans, suggesting that these two behavior modes have important functions. Many hypotheses have been proposed to explain this behavior, but most are difficult to test, and they remain speculative. Surface swimming could serve a variety of functions, including geomagnetic navigation using the Earth's dipole field (Klimley et al., 2002), navigation using celestial cues (Klimley et al., 2002; Bonfil et al., 2005; Weng et al., 2007b), thermoregulation (Carey and Scharold, 1990), oxygen recovery following deep dives (Jorgensen et al., 2010), and maximizing opportunities for feeding on floating carcasses, seabirds and schooling pelagic fishes. Weng et al. (2007b) found no correlation between White Shark tracks and maps of magnetic and gravitational anomalies in the northeast Pacific, and there is no evidence that White Sharks use celestial cues for navigation. However, White Sharks navigate directly and accurately, as evidenced by their often straight tracks and ability to return repeatedly to the same locations after journeys of thousands of kilometers.

With about two-thirds of their time being spent at the surface, it seems likely that White Sharks are utilizing environmental cues for navigation and maximizing their chances of encountering an odor trail from a floating cetacean carcass.

Deep diving may also serve many purposes, including searching for prey (Carey and Scharold, 1990; Takahashi et al., 2003), traversing multiple horizontally stratified ocean layers to increase the chances of encountering odor trails entrained within them (Carey and Scharold, 1990; Klimley et al., 2002), energy conservation (Weihs, 1973), using small-scale anomalies in the Earth's magnetic field as navigational beacons or waypoints (Klimley et al., 2002), and seabed tracking (Bruce et al., 2006; Chapter 17, this book). The energy-conservation swim-glide hypothesis holds that negatively buoyant fishes "glide" slowly downwards (and forwards) under the influence of gravity and then swim actively upwards, before repeating the process (Weihs, 1973). We can reject this hypothesis because descent and ascent speeds were not usually significantly different (Dewar et al., 2004; Nasby-Lucas et al., 2009).

Deep dives may also enable bioacoustic navigation. Humpback Whales and other cetaceans vocalize at low frequencies (Mercado and Frazer, 1999) while migrating northwards past New Zealand during May–August (McDonald, 2006). The frequencies heard by White Sharks are unknown, but a variety of other shark species can hear low frequencies (Myrberg, 1978). Variations in water density with depth create a sound fixing and ranging (SOFAR) channel that focuses sound waves by refraction. Between New Zealand and New Caledonia/Vanuatu, the SOFAR channel is about 600–1,900 m deep (S. Chiswell, personal communication). White Sharks diving deeper than 600 m may use whale vocalizations as a timing and directional cue for following whales to their tropical calving areas. However, the departure of White Sharks from New Zealand on their ocean migration is not synchronized, being spread over about 6 months (Chapter 21, this book), so this hypothesis would not explain the behavior of all sharks. Furthermore, some sharks spent little time near the surface in tropical waters, which is inconsistent with a foraging strategy directed mainly at cetaceans.

Feeding in deep oceanic waters is probably a major driver of deep dives. Many other open-ocean species, including tunas, swordfish, and cetaceans, carry out similar deep dives to feed on mesopelagic organisms, often in association with the deep scattering layer. In the southwest Pacific, Swordfish (*Xiphias gladius*) spend most daylight hours in depths of 600–900 m (Holdsworth et al., 2010), and they are a known prey of White Sharks in the region. Mesopelagic and oceanic squids have been found in the stomachs of White Sharks caught off South Africa (Chapter 4, this book), and spawning aggregations of several squid species occur in the offshore habitat of White Sharks in the northeastern Pacific Ocean (Chapters 12 and 16, this book). Depleted ^{14}C in the vertebrae of White Sharks from the eastern North Pacific can be attributed to inclusion of deepwater organisms in the diet (Kerr et al., 2006). Observations that White Sharks vary their diving depth with time of day and moon phase (Weng et al., 2007a) also provide circumstantial evidence for visual feeding while at depth.

The authors of Chapter 17, this book observed that juvenile White Sharks off New South Wales frequently travelled in a depth band (60–120 m) that coincided with a change in slope of the continental shelf and made regular dives to the seabed. They suggested that White Sharks may use this feature as a navigational cue. In the present study, a White Shark appeared to follow the seabed downwards at the time of departure on an ocean crossing and to follow the seabed upwards on arrival at the tropical destination. Diving frequency also changed around the transition between phases for two of our three sharks with archival records. White Sharks may dive to locate the seabed and use it to assist with navigation at the start and finish of an ocean crossing.

The reasons why White Sharks exhibit surface swimming and deep diving in the open ocean remain speculative until further data can be obtained on their behavior during their ocean-crossing phase and on their sensory and navigational abilities. However, we believe the most likely explanations involve feeding and navigation.

ACKNOWLEDGMENTS

This project was made possible by funding from the National Institute of Water and Atmospheric Research Ltd, the Foundation for Research Science and Technology, New Zealand Department of Conservation, Wildlife Conservation Society NY, National Geographic Society, and NABU/SharkTracker, for which we are grateful. We thank Kina Scollay and Tim Gregory-Hunt for their enthusiasm, expertise, and support while shark tagging; Warrick Lyon, Shannon O'Brien, and Barbara Mangold for their technical assistance, good humor, and patience in the field; and Tim Sippel, Tim Lam, and Murray Smith for advice on the analysis of tracks. We also gratefully acknowledge the assistance of Department of Conservation staff from the Southern and Chatham Islands Areas, particularly Clint "Barney" Brown and Phred Dhobbins; Stu Cave; Peter Scott, Shark Dive NZ; Boyd McGregor, Surfit Charters; members of the Nuku'alofa Sport Fishing Club; Westfleet Ltd; Aratuna Freighters and Hall's Transport; Southland Coolstores; and Southern Seafoods Ltd. We thank Andre Boustany and an anonymous reviewer for helpful comments on the manuscript.

REFERENCES

Bigelow, H. B.; Schroeder, W. C. 1948. Fishes of the Western North Atlantic: Part 1. Lancelets, cyclostomes, sharks. *Memoir of the Sears Foundation for Marine Research* 1: 59–546.

Bonfil, R.; Meÿer, M.; Scholl, M. C.; et al. 2005. Transoceanic migration, spatial dynamics, and population linkages of white sharks. *Science* 310: 100–103.

Bonfil, R.; Francis, M. P.; Duffy, C.; et al. 2010. Large-scale tropical movements and diving behavior of white sharks *Carcharodon carcharias* tagged off New Zealand. *Aquatic Biology* 8: 115–123.

Boustany, A. M.; Davis, S. F.; Pyle, P.; et al. 2002. Expanded niche for white sharks. *Nature* 415: 35–36.

Bruce, B. D.; Stevens, J. D.; Malcolm, H. 2006. Movements and swimming behaviour of white sharks (*Carcharodon carcharias*) in Australian waters. *Marine Biology* 150: 161–172.

Carey, F. G.; Robison, B. H. 1981. Daily patterns in the activities of swordfish, *Xiphias gladius*, observed by acoustic telemetry. *Fishery Bulletin* 79: 277–292.

Carey, F. G.; Scharold, J. V. 1990. Movements of blue sharks (*Prionace glauca*) in depth and course. *Marine Biology* 106: 329–342.

Compagno, L. J. V. 1984. Sharks of the world. An annotated and illustrated catalogue of shark species known to date, Vol. 4. *FAO Fisheries Synopsis*. Rome, Italy: FAO.

Dewar, H.; Domeier, M.; Nasby-Lucas, N. 2004. Insights into young of the year white shark, *Carcharodon carcharias*, behavior in the Southern California Bight. *Environmental Biology of Fishes* 70: 133–143.

Domeier, M. L.; Nasby-Lucas, N. 2008. Migration patterns of white sharks *Carcharodon carcharias* tagged at Guadalupe Island, Mexico, and identification of an eastern Pacific shared offshore foraging area. *Marine Ecology Progress Series* 370: 221–237.

Francis, M. P. 1996. Observations on a pregnant white shark with a review of reproductive biology. In *Great white sharks: The biology of Carcharodon carcharias*, edited by Klimley, A. P., and Ainley, D. G., pp. 157–172. San Diego, CA: Academic Press.

Gilly, W. F.; Markaida, U.; Baxter, C. H.; et al. 2006. Vertical and horizontal migrations by the jumbo squid *Dosidicus gigas* revealed by electronic tagging. *Marine Ecology Progress Series* 324: 1–17.

Goldman, K. J. 1997. Regulation of body temperature in the white shark, *Carcharodon carcharias*. *Journal of Comparative Physiology B* 167: 423–429.

Goldman, K. J.; Anderson, S. D. 1999. Space utilization and swimming depth of white sharks, *Carcharodon carcharias*, at the South Farallon Islands, central California. *Environmental Biology of Fishes* 56: 351–364.

Guitart, D.; Fernandez, J. 1974. El monstruo marino de Cojímar. *Mar y Pesca* 104: 10–11.

Hammerschlag, N.; Martin, R. A.; Fallows, C. 2006. Effects of environmental conditions on predator-prey interactions between white sharks (*Carcharodon carcharias*) and Cape fur seals (*Arctocephalus pusillus pusillus*) at Seal Island, South Africa. *Environmental Biology of Fishes* 76: 341–350.

Holdsworth, J. C.; Sippel, T. J.; Saul, P. J. 2010. Movement of broadbill swordfish from New Zealand tagged with pop-up satellite archival tags. *New Zealand Fisheries Assessment Report 2010/4*. Wellington, New Zealand: Ministry of Fisheries.

Jorgensen, S. J.; Reeb, C. A.; Chapple, T. K.; et al. 2010. Philopatry and migration of Pacific white sharks. *Proceedings of the Royal Society B* 277: 679–688.

Kerr, L. A.; Andrews, A. H.; Cailliet, G. M.; et al. 2006. Investigations of $\Delta^{14}C$, $\delta^{13}C$, and $\delta^{15}N$ in vertebrae of white shark (*Carcharodon carcharias*) from the eastern North Pacific Ocean. *Environmental Biology of Fishes* 77: 337–353.

Klimley, A. P.; Le Boeuf, B. J.; Cantara, K. M., et al. 2001. The hunting strategy of white sharks (*Carcharodon carcharias*) near a seal colony. *Marine Biology* 138: 617–636.

Klimley, A. P.; Beavers, S. C.; Curtis, T. H.; et al. 2002. Movements and swimming behavior of three species of sharks in La Jolla Canyon, California. *Environmental Biology of Fishes* 63: 117–135.

McDonald, M. A. 2006. An acoustic survey of baleen whales off Great Barrier Island, New Zealand. *New Zealand Journal of Marine and Freshwater Research* 40: 519–529.

Mercado, E.; Frazer, N. 1999. Environmental constraints on sound transmission by humpback whales. *Journal of the Acoustical Society of America* 106: 3004–3016.

Musyl, M. K.; Brill, R. W.; Boggs, C. H.; et al. 2003. Vertical movements of bigeye tuna (*Thunnus obesus*) associated with islands, buoys, and seamounts near the main Hawaiian Islands from archival tagging data. *Fisheries Oceanography* 12: 152–169.

Myrberg, A. A. 1978. Underwater sound: Its effect on the behavior of sharks. In *Sensory biology of sharks, skates and rays*, edited by Hodgson, E. S., and Mathewson, R. F., pp. 391–417. Arlington, VA: Office of Naval Research, Department of the Navy.

Nasby-Lucas, N.; Dewar, H.; Lam, C. H.; et al. 2009. White Shark offshore habitat: A behavioral and environmental characterization of the eastern Pacific shared offshore foraging area. *PloS ONE* 4: 1–14.

Nielsen, A.; Sibert, J. R. 2007. State-space model for light-based tracking of marine animals. *Canadian Journal of Fisheries and Aquatic Sciences* 64: 1055–1068.

Pardini, A. T.; Jones, C. S.; Noble, L. R.; et al. 2001. Sex-biased dispersal of great white sharks. *Nature* 412: 139–140.

Pratt, H. L. 1996. Reproduction in the male white shark. In *Great white sharks: The biology of Carcharodon carcharias*, edited by Klimley, A. P., and Ainley, D. G., pp. 131–138. San Diego, CA: Academic Press.

Shepard, E. L. C.; Ahmed, M. Z.; Southall, E. J.; et al. 2006. Diel and tidal rhythms in diving behaviour of pelagic sharks identified by signal processing of archival tagging data. *Marine Ecology Progress Series* 328: 205–213.

Sibert, J. R.; Musyl, M. K.; Brill, R. W. 2003. Horizontal movements of bigeye tuna (*Thunnus obesus*) near Hawaii determined by Kalman filter analysis of archival tagging data. *Fisheries Oceanography* 12: 141–151.

Sokal, R. R.; Rohlf, F. J. 1981. *Biometry: The principles and practice of statistics in biological research*, 2nd edition. New York: Freeman.

Strong, W. R.; Murphy, R. C.; Bruce, B. D.; et al. 1992. Movements and associated observations of bait-attracted white sharks, *Carcharodon carcharias*: A preliminary report. *Australian Journal of Marine and Freshwater Research* 43: 13–20.

Takahashi, M.; Okamura, H.; Yokawa, K.; et al. 2003. Swimming behaviour and migration of a swordfish recorded by an archival tag. *Marine and Freshwater Research* 54: 527–534.

Weihs, D. 1973. Mechanically efficient swimming techniques for fish with negative buoyancy. *Journal of Marine Research* 31: 194–209.

Weng, K. C.; O'Sullivan, J. B.; Lowe, C. G.; et al. 2007a. Movements, behavior and habitat preferences of juvenile white sharks *Carcharodon carcharias* in the eastern Pacific. *Marine Ecology Progress Series* 338: 211–224.

Weng, K. C.; Boustany, A. M.; Pyle, P.; et al. 2007b. Migration and habitat of white sharks (*Carcharodon carcharias*) in the eastern Pacific Ocean. *Marine Biology* 152: 877–894.

West, G. J.; Stevens, J. D. 2001. Archival tagging of school shark, *Galeorhinus galeus*, in Australia: Initial results. *Environmental Biology of Fishes* 60: 283–298.

Weyl, P. K. 1965. On the oxygen supply of the deep Pacific Ocean. *Limnology and Oceanography* 10: 215–219.

CHAPTER 23

New Caledonia (South Pacific) as a Potential Tropical Wintering Ground for the White Shark, *Carcharodon carcharias*

Eric Clua*
Secretariat of the Pacific Community

Bernard Séret
National Museum of Natural History, Paris

ABSTRACT

Several occurrences of White Sharks in tropical waters have been reported, but in the context of gaps in the knowledge of the ecology of that top predator, they appear as unusual and unexpected destinations. The New Caledonian exclusive economic zone covers 1,400,000 km^2, with sea-surface temperatures oscillating between 20 and 21°C in winter and 28 and 32°C in summer. We have validated 53 sightings of White Sharks between 1943 and 2010, mainly based on commercial fishing bycatches and observations at sea. We obtained information on season in 39 cases, shark size in 47 cases, and gender in 12 cases. To this, we added information for six different animals that were tagged with pop-up archival tags in New Zealand in April 2005 (two sharks) and March 2009 (four sharks), before releasing their tags in New Caledonian waters a few months later. Fifty-two percent of the animals were spotted in winter (July to September), 26% in spring (October to December), 23% in summer (January to March), and none in fall (April to June), which corresponds with the best season for tagging them in New Zealand. Thirty-two percent of the sharks were shorter than 3.8 m total length (TL), 45% between 3.8 and 4.4 m TL, 15% between 4.5 and 5 m TL, and 8% above 5 m TL. The smallest animal was 2.6 m TL and the largest was 5.5 m TL. The sex ratio was unbalanced, with 12 males and six females. As in other tropical islands, the peak presence of White Sharks in winter coincided with that of Humpback Whales that come to mate and calve in New Caledonian waters in winter before leaving in October. The extended presence of White Sharks until the end of summer could be linked to other sources of food such as spawning aggregations of reef fishes and deep-sea snappers, in addition to other usual prey. New Caledonian waters appear to be a wintering ground for South Pacific White Sharks, and the study of their ecology should be addressed at a regional level for providing critical information aimed at ensuring their sustainable management in this part of the world.

INTRODUCTION

The largest of the carnivorous fishes, the White Shark is considered to be mainly distributed in temperate waters (Last and Stevens, 1994; Compagno et al., 1997; Compagno, 2001), but with a wide thermal tolerance that allows it, like other lamnids, to reach boreal latitudes (Weng et al., 2005; Chapter 22, this book). Satellite tracking technology has enabled several authors to show that White Sharks can travel thousands of kilometers (Bonfil et al., 2005) and on occasion penetrate tropical waters and swim around coral reefs (Figure 23.1). Boustany et al. (2002) and Domeier and Nasby-Lucas (2008), for example, showed that several large White Sharks regularly migrate from the west coast of North America to Hawaii, where an adult individual stayed up to 122 d, and another was able to cope with surface temperatures up to 27.2°C (Weng et al., 2007a). An affinity for warm water seems to be more pronounced for young animals, which live in cold waters down to 10°C but seem to prefer higher temperatures up to 23°C (Dewar et al., 2004; Weng et al., 2007b). Young-of-the-year animals from the west coast of North America use tropical environments for their development, migrating South from the Southern California Bight (at 34°S) toward Baja California, Mexico (29°S) (Weng et al., 2007b; Chapters 16 and 28, this book). All of the information scientifically collected so far about White Sharks spending time in warm waters was obtained through studies that followed the animals from their temperate, and presumed usual, habitats toward unexpected tropical destinations. Information gathered from the perspective of potentially important tropical habitats, however, contributes toward a better understanding of the ecology of White Sharks.

New Caledonia is an archipelago composed of a main island (16,800 km^2) surrounded by smaller islands of tens to hundreds of km^2, at the center of an exclusive economic zone (EEZ) ranging from 15 to 25°S and 155 to 175°E. This represents an area of 1,400,000 km^2 of tropical waters with a temperature oscillating between 20 and 22°C in winter and 28 and 32°C in summer (Figure 23.2). The main island is 3,000 km east of Queensland, Australia, and around 2,500 km northeast of the North Island of New Zealand. The first official identification of a White Shark in the tropical waters

Figure 23.1 Male White Shark [4.0 m total length (TL)] observed September 2007 over a coral reef in the Dumbea Pass, southern lagoon of New Caledonia. This observation was made by a group of five scuba divers at an average depth of 25 m. (Courtesy of Luc Bourdil.)

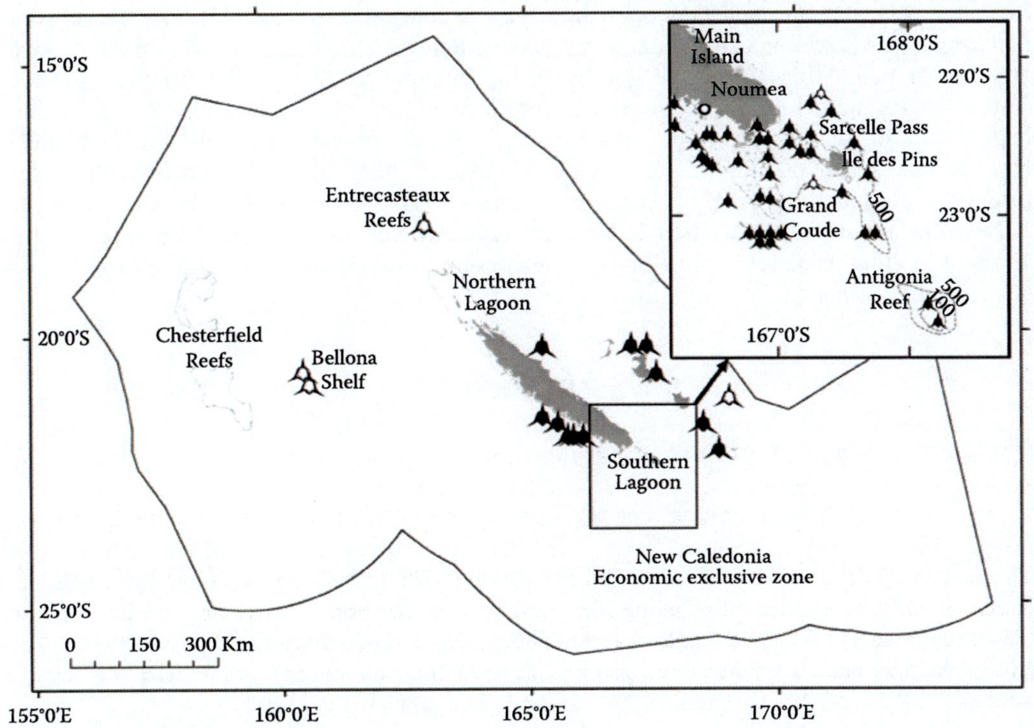

Figure 23.2 Location of confirmed White Shark records within the New Caledonian EEZ. Black icons represent shark records Table 23.1, and white icons represent the pop-up locations of tags from sharks in Table 23.2.

of New Caledonia was recorded in 1943 (Anonymous, 2005), and since then, despite its presence being well known in New Caledonian waters (Fourmanoir and Laboute, 1976; Séret, 1994), no specific investigation has been conducted to better understand the ecology of this top predator in this area. Furthermore, the role that this part of the south Pacific Ocean may play at a regional level in the life cycle and ecology of this species remains unknown. In this paper, we analyzed all of the available data on White Shark occurrences in New Caledonia to test the hypothesis that they use New Caledonian waters as a long-duration wintering ground.

MATERIAL AND METHODS

Observations of White Sharks in New Caledonian waters were compiled through several channels, including interviews with recreational and professional fishermen, spear fishermen, and scuba divers and other scientists and naturalists working at sea. Records provided by professional fishermen are mainly based on bycatch of White Sharks, which are hooked and tangled on longlines. Other records are mainly based on sightings of the animals either on the surface or underwater by scuba and free divers. When a photo was provided, it allowed confirmation of the identification and sometimes an assessment of the size and gender. Given the perceived status of this animal as a threat to humans, the occurrence of a White Shark is usually highlighted by the local newspaper *Les Nouvelles Calédoniennes*, which was a useful source of information through original copies stored at the Bernheim Library in Noumea. Even more sparse is information found in naturalist publications such as fish identification books (Fourmanoir and

Laboute, 1976; Laboute and Grandperrin, 2000). A nonexhaustive list of White Shark occurrences in New Caledonia can be found in Clua and Séret (2010), which deals with a fatal shark attack attributed to this species at Lifou Island in September 2007, as well as another fatal accident with a young surfer along the west coast of the main island in March 2009 (E. Clua, unpublished data). In addition, some scientific data were recently provided from studies conducted in New Zealand since 2005, aiming to provide information about the movements of White Sharks through pop-up archival tagging. These studies showed that six different White Sharks tagged in New Zealand released their tags in New Caledonian waters, two in 2005 (Bonfil et al., 2010) and four in 2009 (Chapter 21, this book). These studies provided accurate data on the size and gender of the animals.

RESULTS

Number and Location of the Sightings

We compiled 53 White Shark records from the New Caledonian EEZ (Figure 23.2) dating between 1943 and 2010, four of which were identified as resightings of the same animal (see below). The remaining 49 occurrences comprised professional fishing bycatch, sightings by recreational boats, scientific assessments (including forensics), and observations of scavenging events on large marine mammals (Table 23.1). Eighty percent of the records were concentrated around the southern part of the mainland. This area (enlarged in Figure 23.2) encompasses the Southern-West lagoon, the Iles des Pins and the surrounding passage, and sea mounts to the south.

Size and Gender of the Animals

An assessment of shark size was possible in 44 cases (Figure 23.3). Sharks smaller than 3.8 m TL were assumed to be immature because this is the maturity threshold for males (Pratt, 1996). Sharks above 4.5 m TL were potentially mature females because this is the threshold for female maturity (Francis, 1996). About a third (32%) were immature, but only two animals were shorter than 3 m TL. The smallest animal was 2.6 m TL, and the most reliable record of the largest one was around 5.5 m TL. This last observation was made in February 1992 by comparison with the length of the hull of the boat along which the shark passed several times (P. Simoni, personal communication). Most of the animals (45%) had a total length between 3.8 and 4.4 m, a few (15%) were between 4.5 and 5 m, and only 8% were longer than 5 m TL. The gender of the animals was available in 18 records, combining the visual observations (Table 23.1) and the tagging programs (Table 23.2). There was an unbalanced sex ratio of 12 males to 6 females.

Table 23.1 Observations of White Shark (*Carcharodon carcharias*) in New Caledonian Waters, 1943–2010

Years	Fishing Bycatch	Recreational	Scientific and Forensics	Scavenging Event	Total
1943–1970	5	0	0	2	7
1971–1990	4	0	0	0	4
1991–2000	12	2	0	2	16
2001–2010	5	10	5	2	22
Total	26	12	5	6	49

The table shows observations of White Shark (*Carcharodon carcharias*) in New Caledonian waters, 1943–2010, classified by time period and source. All observations were verified by a shark researcher and had accurate locations.

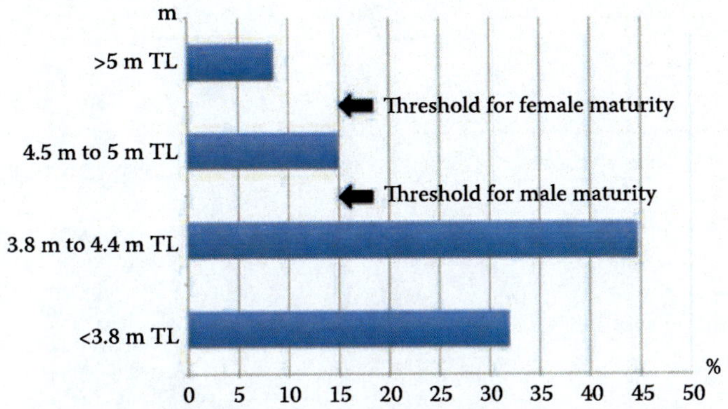

Figure 23.3 Number of sharks per length intervals (n = 47). The length of 3.8 m TL was chosen as threshold for mature males (Pratt, 1996). The interval 3.8–4.4 m TL represents the animals that are mature as male but still immature as female. The length of 4.5 m TL was chosen as threshold for mature females (Francis, 1996). The length of 5 m was chosen as a threshold for very large animals.

Table 23.2 Location in the New Caledonian EEZ and Date of Release of Pop-Up Archival Tags

Tagging Date	Tagging Location	Pop-up Date	Pop-up Location	Area	Gender	Size (m)	Reference
4/8/2005	Chatham Islands	10/5/2005	21°11′S, 169°38′E	East of Mare Island	M	3.4	Bonfil et al. (2010)
4/9/2005		9/5/2005	22°45′S, 167°09E′	Southern Shelf of Isle of Pine	F	4.0	
3/24/2009	Stewart Islands	12/20/2009	21°33′S, 159°16′E	Bellona Shelf	F	4.0	Chapter 21, this book
3/27/2009		10/8/2009	17°44′S, 162°45′E	Entrecasteaux Archipelago	F	3.3	
3/30/2009		8/12/2009	22°12′S, 167°12′E	North of Yate Pass	M	2.9	
3/31/2009		8/24/2009	21°30′S, 159°14′E	Bellona Shelf	M	4.0	

The table shows the locations in the New Caledonian EEZ and dates of release of pop-up archival tags set on six White Sharks in New Zealand in April 2005 (Chatham Islands) and March 2009 (Stewart Islands).

Seasonality and Local Occurrence

The month of White Shark occurrences in New Caledonian waters was recorded for 39 cases. Around a half (52%) were observed in winter (July to September), and a quarter were observed either in spring (October to December) (26%) or in summer (January to March) (23%) (Figure 23.4). There was not a single White Shark record in New Caledonian waters during fall (April to June).

There is little information about resightings of animals that could permit reliable descriptions of specific movements within the New Caledonian EEZ. In two cases, probably the same animal was spotted twice on the same or following day. On August 23, 2008, a 4 m TL female was spotted by a helicopter pilot in the southern lagoon (B. Civet, personal communication) and few hours later in a neighboring pass (less than five nautical miles away) in the barrier reef by a group of spear fishermen (A. Leconte, personal communication). On March 6, 2009, the White Shark responsible for the fatal attack on a young surfer at 11 a.m., was spotted 10 min. before by researchers conducting an aerial survey of Dugongs, *Dugong dugong*, several miles south of the surfing spot (C. Garrigue, personal communication). Another record some 10 nautical miles from the attack was made the day after by an ultra-light aircraft pilot (S. Timboni, personal communication), unfortunately without any accurate assessment of the size that could confirm that it was the same shark.

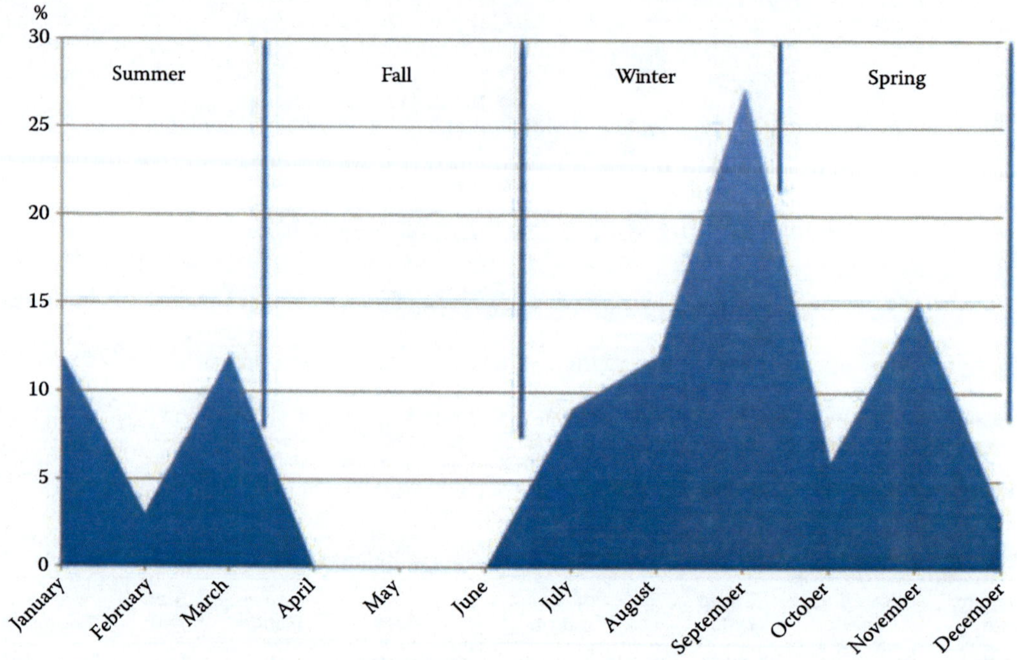

Figure 23.4 Seasonal occurrences ($n = 39$) of White Sharks in New Caledonia with regard to the season: 52% in winter (July to September), 23% in spring (October to December), 25% in summer (January to March), and none in fall (April to June).

DISCUSSION

The record of 49 occurrences in more than 50 yrs. cannot be considered an accurate assessment of the density of White Sharks in New Caledonia. In fact there were only seven reliable records in almost 30 yrs. between 1943 and 1970 (Table 23.1). This lack of information is due to the low interest given to such observations, exacerbated by the poor capability for recognizing this specific species. The information became more reliable by the 1970s with the development of local longline fisheries. An average of four boats per year reported bycatches of White Shark until the late 1990s, when this type of fishing was replaced by deep-sea snapper lining with electric reels. Large animals break the snapper lines when they predate a hooked fish, reducing the captures of White Sharks. These bycatch records, however, represent most of the 26 records for the three decades between 1970 and 2000, showing an average of almost one sighting of White Sharks per year, based on a similar annual fishing effort. The 10-yr. period between 2000 and 2010 showed 22 occurrences, that is, an average of 2.2 sightings per year (Table 23.1). This higher figure can be explained by the increasing number of people, mainly recreational users (including divers), traveling long distances thanks to GPS technology, increasing the probability of shark encounters. Despite the increased number of potential observers, the chance of seeing a White Shark among reef ecosystems remains very low. White Sharks spend less than 8% of their time in shallow, warm waters of 0–5 m, whereas this surface time is around 22% in cooler waters during other periods of the year (Weng et al., 2007a). Francis et al. (Chapter 22, this book and personal communication) showed that for tagged White Sharks in tropical waters ($n = 9$), the mean time in the 0–1-m depth range was 19.7%, but the amount varied considerably among sharks. The appearance of White Sharks at the surface is significantly increased in the presence of a whale carcass on which they regularly feed (Cliff et al., 1996; Long and Jones, 1996; Curtis et al., 2006; Dicken, 2008). Such a situation occurred at least twice in New Caledonia (1997 and 2006) and produced sightings of two White Sharks each time (Table 23.1).

The unbalanced location of sightings between the southwest area of New Caledonia (80%) and the rest of the EEZ (20%) is explained by the proximity of the capital Noumea, which contains 60% (around 120,000 people) of the overall population and the highest density of recreational sea users in New Caledonia. Noumea harbor is also the base for most of the professional fishing boats that, for economic reasons, use the closest fishing area between 100- and 600-m depth, off the Grand Coude, south to the Ile des Pins, and the Antigonia seamount (Figure 23.2). However, this area might also constitute important habitat for the White Shark given its geomorphology: a wide passage with strong currents and local upwellings separates the main island from the Ile des Pins, and the lagoon offers an open barrier reef, several islets, and intermediate reefs in deep waters (up to 70 m). This "intermediate" area between the reef and open sea is the preferred mating and nursing ground for Humpback Whales, *Megaptera novaeangliae*, in New Caledonia. Humpback Whales utilize the shelter and easy access to the open ocean offered by this area. Of the remaining 20% of sightings of White Sharks outside the southern area, half were within a radius of 150 km on the western coast of the southern province of New Caledonia, still in the most inhabited area of the territory. The human-density factor is a driving factor for White Shark records; this assumption is supported by the fact that only two pop-up tags were recorded in the southern area (around the Ile des Pins), with a more widespread distribution of the other four animals within the New Caledonian EEZ, in the north (around Entrecasteaux Archipelago), the west (around the Bellona Shelf), and the northeast (off Mare Island) (Figure 23.2). This suggests that the density of White Sharks in the Caledonian EEZ might be higher than shown by the average yearly records.

Given the demonstrated philopatry of female White Sharks (Pardini, 2001; Domeier and Nasby-Lucas, 2007; Chapters 11, 16, and 25, this book), the unbalanced sex ratio in favor of males (2:1), and the apparently low presence in New Caledonian waters of potentially mature females over 4.5 m TL (Francis 1996), this region does not appear to play a reproductive role. Considering that the sharks were elsewhere before arriving in New Caledonia, such as New Zealand as determined by the tagging programs, their arrival coincides with the calving period of Humpback Whales. White Shark presence in New Caledonian waters peaks in July and August and then decreases until November when most of the cetaceans have left the area (Garrigue et al., 2002). This pattern is consistent with similar observations in the Hawaiian Islands (November to May; Taylor, 1985; Boustany et al., 2002; Weng et al., 2007a; Domeier and Nasby-Lucas, 2008) or for Western Australia and at similar latitudes to New Caledonia in subtropical Queensland, Australia (Paterson and Paterson, 1989; Bruce et al., 2006). White Sharks may follow large cetaceans, expecting feeding opportunities based on vulnerable newborn animals and males exhausted after mating fights or just scavenging on the huge placentas of Humpback Whales (Bonfil et al., 2010).

However, some White Sharks remain in New Caledonia until the end of summer, long after the Humpback Whales have left the area in October, requiring a complementary hypothesis to explain their extended presence. It has been suggested that White Sharks are able to spend extended periods in a given area when food resources are available but then move rapidly in search of other prey sources when it disappears (Long et al., 1996; Bruce et al., 2006). In spite of the departure of Humpback Whales from New Caledonian waters before summer, White Sharks remain present in densities consistent with spring (Figure 23.4). Werry et al. (Chapter 19, this book) noted a similar trend with the catch records of White Sharks (1993–2010) in the Queensland Shark Control Program. This trend might be explained by another source of food, starting to increase by November and lasting until March. A potential food source is the spawning aggregations of reef fishes. White Sharks forage on schools of fishes elsewhere, such as on Sparids, *Pagrus auratus*, in southern Australia (Malcolm et al., 2001). An observation of a White Shark feeding on Orange Spot Groupers, *Epinephelus coioides*, hooked by a handline fisher on a spawning aggregation in Prony Bay, southern lagoon of New Caledonia, in November 2008 (L. Pascal, personal communication) supports this hypothesis. As bycatch on longlines set at over 300-m depth revealed in the 1980s and 1990s, White Sharks also feed on deep-sea snappers, in particular Queen and Ruby Snappers, *Etelis* spp. In the South Pacific, spawning aggregations for these two species start in November and last until May (Moffitt, 1993). In New Caledonia, above-average catch rates of these deep-sea snappers are experienced by fishermen from November until February through March (P. Boblin, personal communication) and also for Crimson Snappers, *Pristipomoides* spp., which peak in January (A. Desurmont, unpublished data). These aggregations may therefore constitute a reliable source of food for White Sharks before they start migrating away from New Caledonian waters in the fall. Other potential prey of Great White Sharks are marine mammals such as Pygmy Sperm Whales, *Kogia breviceps* (Long, 1991), sea turtles (Fergusson et al., 2000), and large fishes such as tunas, which all occur year-round in New Caledonian waters. The presence of these preys could help to explain the extended residence of White Sharks.

In the southwest Pacific, a seasonal pattern of migration between temperate and tropical waters has been shown through several tagging programs by Bruce et al. (2006). Australian White Sharks occupy the Great Australian Bight during the southern spring and summer, make northward movements during autumn and winter through New South Wales (NSW) into tropical waters of Queensland, and return to the southern region by early summer. This seasonal pattern of latitudinal migration between temperate and tropical waters is supported by the capture of White Sharks by shark control programs in NSW and Queensland and even attacks on humans attributed to White Sharks that peak in late summer/autumn in NSW and in spring in Queensland (West, 1996). Bruce et al. (2006) also showed, through the example of a single shark that returned to Neptune Island in Southern Australia 286 d after it was tagged there, that White Sharks undertaking such extensive movements remain part of a southern Australian population and may show patterns of site fidelity similar to that reported for White Sharks in Californian, Mexican, and South African waters (Klimley, 1985; Long et al., 1996; Bonfil et al., 2005; Domeier and Nasby-Lucas, 2007; Chapter 25, this book). For such movements, these sharks spend the majority of time near the Australian coast, but Bruce et al. (2006) also reported a single pelagic movement from Australia to New Zealand, showing connectivity between these two areas. That same connectivity is supported by the example of a 4.0 m TL male that was tagged in Stewart Island in March 2009 and had a pop-up release in the Bellona Shelf, New Caledonia, in August 2009 (Table 23.2), after passing through the Coral Sea off the Queensland coast (Chapter 21, this book). Results from the tagging program from New Zealand support the hypothesis of an interconnected population of White Sharks in the southwest Pacific, encompassing southern and western Australia, New Zealand, and New Caledonia and eastward to Fiji and Tonga (Bonfil et al., 2010; Chapter 21, this book).

CONCLUSION

Evidence for interconnectivity among Australian, New Zealand, and New Caledonian White Sharks is sparse but convincing. In the absence of a specific monitoring program, White Sharks are difficult to observe in tropical waters. The occurrences in New Caledonia indicate that their density is quite low but that their residency time is significant (up to 9 months per year), with a strong peak in winter. Therefore, New Caledonian waters appear to play an important role as a tropical winter residency ground for South Pacific White Sharks, including young animals. These consistent and long-lasting residencies result in Great White Sharks mixing with other large predators such as Tiger Sharks, *Galeocerdo cuvier*, particularly during scavenging events on cetacean carcasses (Dudley et al., 2000). The presence of two White Sharks, a 3.5 m TL male and a 4.0 m TL female, around the carcass of a Dwarf Sperm Whale, *Kogia breviceps*, carcass stranded on the fringing reef of Ugo Islet in September 2006 scared away several Tiger Sharks that were already in the vicinity (E. Clua, unpublished data). In December 1998, two White Sharks were observed in the La Sarcelle

Figure 23.5 White Shark and Tiger Shark scavenging together on a Sperm Whale carcass in the pass of La Sarcelle, December 1997. Both sharks were around 4.0 m TL. (Courtesy of Bruno Civet.)

pass (between the mainland and the Ile des Pins), scavenging on the drifting carcass of a Sperm Whale, *Physeter macrodon*, surrounded by tens of Tiger Sharks. No particular hostility was witnessed between these two top predators, and they were sometimes feeding on the carcass together (Figure 23.5) (B. Civet, personal communication).

The regular and extended presence of White Sharks, including young animals, in New Caledonian waters suggests there is potential for studying their interactions with large cetaceans and other top predators. The development of regional programs for tagging White Sharks in the New Caledonian EEZ appears to be a critical step toward a better understanding of the ecology of this endangered species in this part of the world. These data will be paramount for supporting an effective management plan at a regional scale and the implementation of specific regulations for protecting this top pelagic predator while utilizing tropical waters in New Caledonia.

ACKNOWLEDGMENTS

We are grateful to all the people from New Caledonia who shared their data on White Sharks, especially Pierre Larue. We also thank Bruno Civet and Luc Bourdil for allowing us to publish their photos of White Sharks. The information provided by our colleagues from New Zealand, Clinton Duffy and Malcolm Francis, was critical, and we warmly thank them for their confidence. Finally, we gratefully acknowledge comments by our Australian colleagues, Jonathan Werry from Griffith University and Clive Wilkinson from the Reef Rainforest Research Center in Townsville, on an earlier draft of this paper. This study benefitted from the financial support of the CRISP (Coral Reef InitiativeS for the Pacific) Programme Coordinating Unit at the Secretariat of the Pacific Community in Noumea, New Caledonia.

REFERENCES

Anonymous. 2005. Requins du lagon: Vraies et fausses raisons de s'alarmer. *Les Nouvelles Calédoniennes* du 30/11/2005: p. 3.

Bonfil, R., Meÿer, M., Scholl, M. C., Johnson, R., O'Brien, S., Oosthuizen, H., et al. 2005. Transoceanic migration, spatial dynamics and population linkages of white sharks. *Science* 310: 100–103.

Bonfil, R., Francis, M. P., Duffy, C., Manning, M., and O'Brien, S. 2010. Large-scale tropical movements and diving behavior of white sharks *Carcharodon carcharias* tagged off New Zealand. *Aquatic Biology* 8: 115–123.

Boustany, A. M., Davis, S. F., Pyle, P., Anderson, S. D., Le Boeuf, B. J., and Block, B. A. 2002. Expanded niche for white sharks. *Nature* 415: 35–36.

Bruce, B. D. 2008. The biology and ecology of the white shark (*Carcharodon carcharias*). In *Sharks of the open ocean*, edited by M. Camhi and E. K. Pikitch, pp. 69–81. Oxford, UK: Blackwell Scientific.

Bruce, B. D., Stevens, J. D., and Malcolm, H. 2006. Movements and swimming behaviour of white sharks (*Carcharodon carcharias*) in Australian waters. *Mar. Biol.* 150: 161–172.

Clua, E., and Séret, B. 2010. Unprovoked fatal shark attack in Lifou Island (Loyalty Islands, New Caledonia, South Pacific) by a great white shark, *Carcharodon carcharias. The American Journal of Forensic Medicine and Pathology* 31: 281–286.

Cliff, G., Van der Elst, R. P., Govender, A., Witthuhn, T. K., and Bullen. E. M. 1996. First estimates of mortality and population size of white sharks on the South African coast. In *Great white sharks: The biology of Carcharodon carcharias*, edited by A. P. Klimley and D. G. Ainley, pp. 393–400. San Diego, CA: Academic Press.

Compagno, L. J. V. 2001. Sharks of the world: An annotated and illustrated catalogue of shark species known to date: Volume 2. Bullhead, mackerel and carpet sharks (Heterodontiiformes, Lamniformes and Orectolobiformes). *FAO Species Catalogue for Fisheries Purposes*. Rome, Italy: FAO.

Compagno, L. J. V., Marks, M. A., and Fergusson, I. K. 1997. Threatened fishes of the world: *Carcharodon carcharias* (Linnaeus, 1758) (Lamnidae). *Environ. Biol. Fishes* 50: 61–62.

Curtis, T. H., Kelly, J. T., Menard, K. L., Laroche, R. K., Jones, R. E., and Klimley, A. P. 2006. Observations on the behavior of white sharks scavenging from a whale carcass at Point Reyes, California. *Cal. Fish Game* 92: 113–124.

Dewar, H., Domeier, M. L., and Nasby-Lucas, N. 2004. Insights into young of the year white shark, *Carcharodon carcharias*, behavior in the Southern California Bight. *Environ. Biol. Fishes* 70: 133–143.

Dicken, M. L. 2008. First observations of young of the year and juvenile great white sharks (*Carcharodon carcharias*) scavenging from a whale carcass. *Mar. Freshw. Res.* 59: 596–602.

Domeier, M. L., and Nasby-Lucas, N. 2007. Annual re-sightings of photographically identified white sharks (*Carcharodon carcharias*) at an eastern Pacific aggregation site (Guadalupe Island, Mexico). *Mar. Biol.* 150: 977–984.

Domeier, M. L., and Nasby-Lucas, N. 2008. Migration patterns of white sharks *Carcharodon carcharias* tagged at Guadalupe Island, Mexico, and identification of an eastern Pacific shared offshore foraging area. *Mar. Ecol. Prog. Ser.* 370: 221–237.

Dudley, S. F. J., Anderson-Reade, M. D., Thompson, G. S., and McMullen, P. B. 2000. Concurrent scavenging off a whale carcass by great white sharks, *Carcharodon carcharias*, and tiger sharks, *Galeocerdo cuvier*. *Fish. Bull.* 98: 646–649.

Fergusson, I. K., Compagno, L. J. V., and Marks, M. A. 2000. Predation by white sharks *Carcharodon carcharias* (Chondrichthyes: Lamnidae) upon chelonians, with new records from the Mediterranean Sea and a first record of the ocean sunfish *Mola mola* (Osteichthyes: Molidae) as stomach contents. *Env. Biol. Fishes* 58: 447–453.

Fourmanoir, P., and Laboute, P. 1976. *Poissons de Nouvelle-Calédonie et des Nouvelles-Hébrides*. Singapore: Editions du Pacifique.

Francis, M. P. 1996. Observations on a pregnant white shark with a review of reproductive biology. In *Great white sharks: The biology of Carcharodon carcharias*, edited by A. P. Klimley and D. G. Ainley, pp. 157–172. San Diego, CA: Academic Press.

Garrigue, C., Aguayo, A., Amante-Helweg, V., Baker, C. S., Caballero, S., Claphma, P., et al. 2002. Movements of humpback whales in Oceania, South Pacific. *J. Cetacean Res. Manage.* 4: 255–260.

Klimley, A. P. 1985. The areal distribution and autoecology of the white shark, *Carcharodon carcharias*, on the west coast of North America. *Sthn. Calif. Acad. Sci. Mem.* 9: 15–40.

Laboute, P., and Grandperrin, R. 2000. Requins et raies. In *Poissons de Nouvelle-Calédonie*, edited by C. Ledru, p. 85. Nouméa, New Caledonia: Editions Catherine Ledru.

Last, P., and Stevens, J. 1994. *Sharks and rays of Australia*, Vol 1. Melbourne, Australia: Commonwealth Scientific and Industrial Research Organisation.

Long, D. J. 1991. Apparent predation by a white shark *Carcharodon carcharias* on a pygmy sperm whale *Kogia breviceps*. *Fishery Bull. U.S.* 89: 538–540.

Long, D. J., and Jones, R. E. 1996. White shark predation and scavenging on cetaceans in the eastern North Pacific Ocean. In *Great white sharks: The biology of Carcharodon carcharias*, edited by A. P. Klimley and D. G. Ainley, pp. 293–307. San Diego, CA: Academic Press.

Long, D. J., Hanni K. D., Pyle P., Roletto J., Jones R. E., and Bandar, R. 1996. White shark predation on four pinniped species in central California waters: Geographic and temporal patterns inferred from wounded carcasses. In *Great white sharks: The biology of Carcharodon carcharias*, edited by A. P. Klimley and D. G. Ainley, pp. 263–274. San Diego, CA: Academic Press.

Malcolm, H., Bruce, B. D., and Stevens, J. D. 2001. *A review of the biology and status of white sharks in Australian waters. I: Report to Environment Australia.* Hobart, Australia: Marine Species Protection Programme, CSIRO Marine Research.

Moffitt, R. B. 1993. Deepwater demersal fish. In *Nearshore marine resources of the South Pacific: Information for fisheries development and management*, edited by A. Wright and L. Hill, pp. 73–95. Institute of Pacific Studies, Suva, Fiji. Forum Fisheries Agency, Honiara, Solomon Islands. International Centre for Ocean Development, Canada.

Pardini, A. T., Jones, C. S., Noble, L. R., Kreiser, B., Malcolm, H., Bruce, B. D., et al. 2001. Sex-biased dispersal of great white sharks. *Nature* 412: 139–140.

Paterson, R., and Paterson, P. 1989. The status of the recovering stock of humpback whales *Megaptera novaeangliae* in east Australian waters. *Biol. Conserv.* 47: 33–48.

Pratt, H. L. 1996. Reproduction in the male white shark. In *Great white sharks: The biology of Carcharodon carcharias*, edited by A. P. Klimley and D. G. Ainley, pp. 131–138. San Diego, CA: Academic Press.

Séret, B. 1994. Chondrichthyan fishes of New Caledonia. *Chondros* 5: 6–9.

Taylor, L. 1985. White sharks in Hawai'i: Historical and contemporary records. *Mem. South. Calif. Acad. Sci.* 9: 41–48.

Weng, K. C., Castilho, P. C., Morrissette, J. M., Landeira, A., Schallert, R. J., Holts, D. B., et al. 2005. Satellite tagging and cardiac physiology reveal niche expansion in salmon sharks. *Science* 310: 104–106.

Weng, K. C., Boustany, A. M., Pyle, P., Anderson, S. D., Brown A., and Block, B. A. 2007a. Migration and habitat of white sharks (*Carcharodon carcharias*) in the eastern Pacific Ocean. *Mar. Biol.* 152: 877–894.

Weng, K. C., O'Sullivan, J. B., Lowe, C. G., Winkler, C. E., Dewar, H., and Block, B. A. 2007b. Movements, behavior and habitat preferences of juvenile white sharks in the eastern Pacific as revealed by electronic tags. *Mar. Ecol. Prog. Ser.* 338: 211–224.

West, J. 1996. White shark attacks in Australian waters. In *Great white sharks: The biology of Carcharodon carcharias*, edited by A. P. Klimley and D. G. Ainley, pp. 449–455. San Diego, CA: Academic Press.

SECTION III

Population Monitoring, Policy, and Review

CHAPTER 24

Application of Molecular Genetics for Conservation of the White Shark, *Carcharodon carcharias*, L. 1758

Chrysoula Gubili*
Aberdeen University and Marine Biological Association of the United Kingdom

Clinton A. J. Duffy
Department of Conservation and University of Auckland

Geremy Cliff
KwaZulu-Natal Sharks Board and University of KwaZulu-Natal

Sabine P. Wintner
KwaZulu-Natal Sharks Board and University of KwaZulu-Natal

Mahmood Shivji
Guy Harvey Research Institute and Nova Southeastern University

Demian Chapman
Guy Harvey Research Institute and Nova Southeastern University

Barry D. Bruce
CSIRO Marine and Atmospheric Research

Andrew P. Martin
University of Colorado

David W. Sims
Marine Biological Association of the United Kingdom and the University of Plymouth

Catherine S. Jones
Aberdeen University

Leslie R. Noble*
Aberdeen University

ABSTRACT

Although a species capable of long-distance dispersal, the Great White Shark, *Carcharodon carcharias*, exhibits philopatric behavior, with females known to return to natal sites in at least part of its range. However, intractability and rarity mean knowledge of its population dynamics remains limited. Direct observation, by tagging or photographic tracking, often lack, respectively, repeatability and objectivity, however, integrating these with advances in molecular genetics provides an indirect assessment of population connectivity, substructuring, and dispersal. Here we review the utility of population genetic structure inferred from molecular markers and discuss its application in the development and implementation of effective conservation strategies. Application of nuclear markers and population genetic analysis combined with published mitochondrial sequences afford the most highly resolved view of White Shark population structure to date, allowing further insight into the extent of sex-biased dispersal and the role of mating behavior in maintaining the observed population structure. The distribution of matrilineal clades across oceans gives cause for concern, emphasizing the existence of endangered populations in enclosed areas. Current genetic profiling and species diagnostic techniques are reviewed, and their application in law enforcement is discussed. Although concordance of direct (photographic identification) and indirect (molecular tools) methods of individual identification has been used to validate proposed White Shark local movements, extensive sampling, new markers, and alternative DNA sources should also be considered. New methods and next generation DNA sequencing technologies may be developed to afford new insights into elasmobranch biology. Findings from these approaches are essential for formulation of White Shark population management plans in a species that is classified by the International Union for Conservation of Nature as "vulnerable."

INTRODUCTION

Genetic Tools for Determining Diversity and Relatedness in Sharks

Analysis of genomic variation has proven a useful tool in establishing the population genetic structure of many shark species over the last three decades (Smithy, 1986; Lavery and Shaklee, 1989; Heist et al., 1995; Pardini et al., 2000, 2001; Feldheim et al., 2001; Saville et al., 2002; Schrey and Heist, 2003; Portnoy et al., 2010). Following the application of allozymes during the early years of genetic analysis of elasmobranchs, contemporary studies focus on mitochondrial and nuclear genomic DNA, utilizing markers such as microsatellite loci, some of which have been conserved in sharks for hundreds of millions of years (Martin et al., 2002). Allozymes exhibit less variability and statistical power in elasmobranchs because of the limited number of polymorphic loci and presumably a lower mutation rate (Heist, 1999). Recent studies of sharks have kept pace with genetic advances, utilizing either one of these markers such as in the Lemon Shark, *Negaprion brevirostris* (Feldheim et al., 2001); Sandbar, *Carcharhinus plumbeus* (Heist et al., 1995; Heist and Gold, 1999); Mako, *Isurus oxyrinchus* (Schrey and Heist, 2002, 2003); Nurse Shark, *Ginglymostoma cirratum* (Heist et al., 2003); and Blacktip Shark, *C. limbatus* (Keeney and Heist, 2003, 2006; Keeney et al.,

2003), or both markers, such as in *C. limbatus* (Keeney et al., 2005); White Shark, *Carcharodon carcharias* (Pardini et al., 2001); and Sandbar Shark, *C. plumbeus* (Portnoy et al., 2010).

Insights from genetic analyses can be useful for addressing population genetic issues, revealing whether two disjunct populations share a common gene pool (Heist et al., 1995) or clarifying taxonomic relationships between species (Gardner and Ward, 2002). Genetic tools can also validate phylogenetic relationships among shark species (Lavery, 1992; Martin et al., 2002), particularly where morphological differences have yet to be established (Moura et al., 2008), identifying cryptic species (Quattro et al., 2006; Corrigan et al., 2008; Griffiths et al., 2010). Markers such as microsatellites can be extremely valuable for inferring population demography due to their high polymorphism yet are commonly used for paternity tests and individual assignment (Schlötterer, 2000). Mitochondrial sequences also yield useful polymorphisms applicable for population identification, fingerprinting, and phylogeographic studies. However, mitochondrial nucleotide diversity seems to be lower in elasmobranchs than in teleosts (Heist, 1999). Similarly, a recent survey of microsatellite analyses in sharks and marine teleosts showed that shark studies use fewer loci and detect lower heterozygosity but not fewer alleles than teleosts (Hauser, 2009; Table 24.1), suggesting that microsatellites are rarer and more difficult to isolate in sharks than teleosts (Heist, 1999; Heist et al., 2003; Hauser, 2009). However, cross-species transferability of microsatellite markers is very common among shark studies as the success within and between genera or families can reach 25% in fishes (Barbará et al., 2007). This relatively high rate (higher than in birds and plants, but lower than in mammals; Barbará et al., 2007) is due to the presence of highly conserved flanking regions among shark species (Martin et al., 2002). Comprehensive reviews on the applicability of molecular markers for shark studies are given by Heist (2005) and Hauser (2009).

Molecular genetics have become increasingly important for addressing conservation issues (Alves et al., 2001). For example, monitoring the trade of shark-derived products has been problematic because of the difficulties in identifying the species involved (Chapman et al., 2003; Abercrombie et al., 2005), preventing quantitative stock assessments (Clarke et al., 2006a). However, mixtures of genomic DNAs can be successfully separated, discriminating protected species from those that are legitimate commercial taxa (Pank et al., 2001; Shivji et al., 2002; Chapman et al., 2003). Such work is instrumental in enforcing legislation covering the most widely protected elasmobranch in the world, the Great White Shark.

The application of modern genetic methods combined with previous biological (Bruce, 1992; Cliff et al., 1996; Compagno et al., 1997) and dietary findings (Dudley et al., 2000) has delineated a more realistic and holistic picture of the White Shark's current biology. Analysis of genomic DNA in lamniform sharks is an approach developed recently (Pardini et al., 2000, 2001; Martin et al., 2002; Schrey and Heist, 2002, 2003; Jorgensen et al., 2009). Only 25 yrs. ago little was known of the usefulness of genetic information and its application for determining relationships among shark populations (Smith, 1986; Lavery and Shaklee, 1989), and care is still required when interpreting data from understudied species, particularly those exhibiting small sample sizes. Nonetheless, over the last two decades, there has been an upsurge of interest in investigating elasmobranch population biology and ecology, making it timely to select highly polymorphic molecular markers with which to address questions regarding the status, population structure, and origin of the Great White, a threatened shark species with a complex behavior.

Dispersal ability and associated traits such as habitat preference and size are regarded as major factors influencing the population genetic structure of elasmobranchs (Veríssimo et al., 2010a; Karl et al., 2011). Significant intraspecific genetic structuring has been reported for small, regionally restricted species (*Raja clavata*, Chevolot et al., 2006; *Squalus acanthias*, Veríssimo et al., 2010a), whereas more subtle population differentiation is evident over at least part of the ranges of larger, highly vagile, oceanic fish with global distributions (*I. oxyrinchus*, Schrey and Heist, 2003; *C. maximus*, Hoelzel et al., 2006; *R. typus*, Castro et al., 2007; Schmidt et al., 2009). However, there are exceptions to this trend where environmental factors directly affect the life history characteristics

Table 24.1 Summary Information on Mitochondrial and Microsatellite Data for Elasmobranchs on a Global Scale

Species	Number of Samples (mtDNA)	Number of Haplotypes	mtDNA Gene Region	Nucleotide Diversity (π)	Haplotypic Diversity (h)	Number of Samples (msats)	L	A	*HE*	Reference
Carcharhinus leucas (EA)	102	14	CR	0.0028 ± 0.0017	0.76 ± 0.03	136	5[a]	30.4	0.84	Karl et al. (2011)
Carcharhinus limbatus	364	37	CR	0.00413 ± 0.00226	0.843 ± 0.015	na	na	na	na	Keeney & Heist (2006)
Carcharhinus limbatus (NWA)	na	na	na	na	na	418	8	14	0.500	Keeney et al. (2003)
Carcharhinus obscurus (IWP)	28	7	CR	0.0050 ± 0.0048	0.60 ± 0.29	25	4[a]	7.9	0.732	Ovenden et al. (2009)
Carcharhinus plumbeus	329	67	CR	0.00475	0.959	329	8	na	na	Portnoy et al. (2010)
Carcharhinus plumbeus	na	na	na	na	na	51	5	22.6	0.852	Portnoy et al. (2006)
Carcharhinus sorrah (IWP)	49	12	CR	0.0030 ± 0.0019	0.60 ± 0.24	43	2+3[a]	12.6	0.55	Ovenden et al. (2009)
Carcharhinus sorrah	na	na	na	na	na	42	1+4[a]	9.8	0.54	Ovenden et al. (2006)
Carcharhinus tilstoni	na	na	na	na	na	59	1+4[a]	10.8	0.64	Ovenden et al. (2006)
Carcharias taurus	193	9	CR	0.004 ± 0.00003	0.725 ± 0.002	186	5+1[a]	10.5	0.737	Ahonen et al. (2009)
Carcharias taurus (EAu)	19	1	CR	0	0	na	na	na	na	Stow et al. (2006)
Carcharias taurus (WAu)	16	2	CR	0.0031 ± 0.0001	0.458 ± 0.024	na	na	na	na	Stow et al. (2006)
Carcharias taurus (SA)	26	4	CR	0.0030 ± 0.0001	0.717 ± 0.010	na	na	na	na	Stow et al. (2006)
Carcharias taurus (SA)	na	na	na	na	na	27	8+1[a]	8.3	0.731	Feldheim et al. (2007)
Carcharodon carcharias (IWP)	88	na	CR	0.0203	na	na	na	na	na	Pardini et al. (2001)
Carcharodon carcharias	na	na	na	na	na	20	5	5.4	0.658	Pardini et al. (2000)

Carcharodon carcharias	54	45	CR	0.0223 ± 0.0110	0.9888 ± 0.0075	na	na	na	na	Gubili et al. (2011)
Carcharodon carcharias (SA)	na	na	na	na	na	110	6+1[a]	6.9	0.674	Gubili et al. (2009)
Carcharodon carcharias (NEP)	59	20	CR	0.0013 ± 0.0009	0.79	na	na	na	na	Jorgensen et al. (2010)
Cetorhinus maximus	62	6	CR	0.0013 ± 0.0009	0.720 ± 0.028	na	na	na	na	Hoelzel et al. (2006)
Chiloscyllium plagiosum	na	na	na	na	na	34	12	7	0.693	Ding et al. (2009)
Galeorhinus galeus	116	38	CR	0.0071 ± 0.0037	0.92 ± 0.01	na	na	na	na	Chabot & Allen (2009)
Ginglymostoma cirratum	na	na	na	na	na	28	9	5	0.54	Heist et al. (2003)
Hexanchus griseus	na	na	na	na	na	143	14+1[a]	45.4	0.888	Larson et al. (2009)
Isurus oxyrinchus	na	na	na	na	na	426	4	25.5	0.873	Schrey & Heist (2003)
Mustelus schmitti	99	11	cyt b	0.0015	0.226	na	na	na	na	Pereyra et al. (2010)
Negaprion acutidens	58	4	CR	0.0006	0.280	58	6[a]	10.5	0.67	Schultz et al. (2008)
Negaprion brevirostris	80	11	CR	0.0059	0.780	80	9	19.8	0.81	Schultz et al. (2008)
Negaprion brevirostris	na	na	na	na	na	1142	9	25.6	0.79	Feldheim et al. (2002)
Notorynchus cepedianus	na	na	na	na	na	97	15[a]	24.5	0.794	Larson et al. (2009)
Prionace glauca (IWP)	60	16	CR	0.0054 ± 0.0011	0.92 ± 0.005	na	5[a]	6.6	0.58	Ovenden et al. (2009)
Orectolobus halei	12	5	CR	na	na	10	2	1	0	Corrigan et al. (2008)
Orectolobus maculatus	8	1	CR	na	na	7	2	1	0	Corrigan et al. (2008)
Orectolobus ornatus	6	2	CR	na	na	10	2	1	0	Corrigan et al. (2008)

continued

Table 24.1 (Continued) Summary Information on Mitochondrial and Microsatellite Data for Elasmobranchs on a Global Scale

Species	Number of Samples (mtDNA)	Number of Haplotypes	mtDNA Gene Region	Nucleotide Diversity (π)	Haplotypic Diversity (h)	Number of Samples (msats)	L	A	*HE*	Reference
Rhincodon typus	70	44	CR	0.011 ± 0.006	0.974 ± 0.008	na	na	na	na	Castro et al. (2007)
Rhincodon typus	na	na	na	na	na	>68	8	9	0.687	Schmidt et al. (2009)
Rhincodon typus (GM)	36	14	CR	0.005	0.9	na	na	na	na	Ramirez-Macias et al. (2007)
Rhincodon typus (GM)	na	na	na	na	na	50	9	6.2	0.553	Ramirez-Macias et al. (2009)
Rhizoprionodon lalandii (SWA)	94	16	CR	0.0048 ± 0.0029	0.8239	na	na	na	na	Mendonça et al. (2009)
Somniosus microcephalus (NA)	16	7	cyt b	0.0022 ± 0.0015	0.7750 ± 0.0876	na	na	na	na	Murray et al. (2008)
Somniosus pacificus (TW)	28	8	cyt b	0.0031 ± 0.0020	0.7937 ± 0.0424	na	na	na	na	Murray et al. (2008)
Somniosus pacificus (AL)	15	7	cyt b	0.0043 ± 0.0026	0.8381 ± 0.0680	na	na	na	na	Murray et al. (2008)
Somniosus antarcticus (SO)	16	5	cyt b	0.0023 ± 0.0015	0.6667 ± 0.1130	na	na	na	na	Murray et al. (2008)
Squalus acanthias	412	103	ND2	0.0086	0.839	467	4+4[a]	13.5	0.61	Verissimo et al. (2010)
Squalus acanthias	na	na	na	na	na	20	8	5.5	0.70	McCauley et al. (2004)
Squalus mitsukurii (HA)	112	6	CR	0.0010 ± 0.0008	0.541 ± 0.0015	96	6+2[a]	10.4	0.567	Daly-Engel et al. (2010)
Squatina californica	na	na	na	na	na	3	11[a]	2.4	0.59	Larson et al. (2009)
Sphyrna lewini	271	24	CR	0.0130 ± 0.0068	0.805 ± 0.018	na	na	na	na	Duncan et al. (2006)
Sphyrna lewini (EP)	na	na	na	na	na	80	15	14	0.76	Nance et al. (2009)
Sphyrna lewini (IWP)	47	9	CR	0.0098 ± 0.0072	0.61 ± 0.12	53	3[a]	6.4	0.54	Ovenden et al. (2009)

Sphyrna tiburo (NWA)	na	na	na	na	na	119	4	13.5	0.686	Chapman et al. (2004)
Stegostoma fasciatum (IWP)	75	8	ND4	0.14 ± 0.14	0.72 ± 0.35	176	13	5.6	0.725	Dudgeon et al. (2009)
Triakis semifasciata	na	na	na	na	na	15	13[a]	7.2	na	Larson et al. (2009)
Aerobatus narinari (IWP)	129	17	cyt b	0.0126	0.8	na	na	na	na	Schluessel et al. (2010)
Aerobatus narinari (IWP)	66	16	ND4	0.0085	0.81	na	na	na	na	Schluessel et al. (2010)
Dipturus batis (north clade)	38	2	CR	0.00027	0.193	33	6[a]	3.4	0.366	Griffins et al. (2010)
Dipturus batis (south clade)	85	13	CR	0.00159	0.717	83	6[a]	4.6	0.349	Griffins et al. (2010)
Dipturus batis (NEA)	na	na	na	na	na	21	11[a]	4.7	0.388	El Nagar et al. (2010)
Leucoraja naevus (NEA)	na	na	na	na	na	17	10[a]	5.2	0.516	El Nagar et al. (2010)
Raja asterius (MED)	18	2	CR	0.0092	0.29	na	na	na	na	Valsecchi et al. (2005)
Raja clavata (NA)	337	34	cyt b	0.0056	0.796	na	na	na	na	Chevolot et al. (2007)
Raja clavata (NEA)	na	na	na	na	na	483	5	10.9	0.693	Chevolot et al. (2006)
Raja clavata (MED)	26	6	CR	0.0072	0.61	na	na	na	na	Valsecchi et al. (2005)
Raja clavata (NEA)	na	na	na	na	na	22	11[a]	4.5	0.486	El Nagar et al. (2010)
Raja miraletus (MED)	18	2	CR	0.0031	0.17	na	na	na	na	Valsecchi et al. (2005)
Raja montagui (NEA)	na	na	na	na	na	23	10[a]	5.9	0.661	El Nagar et al. (2010)

When no worldwide sampling is performed for one species, the sampling location is included in parentheses. The numbers of samples and haplotypes, mtDNA gene region (CR, control region; cyt b, cytochrome b), nucleotide diversity and standard deviation (π), and haplotype diversity and standard deviation (h) are cited from studies using mitochondrial data. Additionally, the numbers of samples, expected heterozygosity (*H*E), number of loci (L), and number of alleles (A) from microsatellite studies (msats) of sharks are given. AL, Alaska; EA, east Atlantic; EAu, east Australia; EP, east Pacific; GM, Gulf of Mexico; HA, Hawaii; IWP, Indo-Australian archipelago; MED, Mediterranean; NA, North Atlantic; NEA, northeast Atlantic; NEP, northeast Pacific; NWA, northwest Atlantic; SA, South Africa; SO, Southern Ocean; SWA, southwest Atlantic; TW, Taiwan; WAu, west Australia.

[a] Cross-species amplification.

of fishes (Hauser and Ward, 1998). Female site fidelity to specific habitats or nursery areas encourages stronger population structure, with discrete populations developing as seen in the Blacktip *C. limbatus* (Schrey and Heist, 2003) and Lemon Sharks *N. brevirostris* (Feldheim et al., 2002, 2004). Even highly vagile species may show limited dispersal through unsuitable environments, exemplified by tropical waters acting as a dispersal barrier for the temperate Tope (*Galeorhinus galeus*; Chabot and Allen, 2009). Habitat discontinuities such as oceanic waters may similarly restrict gene flow, as is the case with the Scalloped Hammerhead, *Sphyrna lewini* (Duncan et al., 2006). In general, climate change throughout geological time has presented high latitude species with more environmental variation than experienced by species in more stable tropical climes, where uninterrupted gene flow has curtailed the potential for genetic divergence (Zachos et al., 2001). The globally distributed Great White Shark displays population structuring despite its ability to undertake transoceanic migrations, proving to be another exception to the low genetic differentiation of highly vagile elasmobranchs (Veríssimo et al., 2010a). Indeed, the genetic consequences of an environmental barrier are shown dramatically in this species (Gubili et al., 2011). Over the last 700,000 yrs., the frequent rise of the Sunda and Sahul shelves during glacial events caused long-term separation, often for up to 100,000 yrs., of Atlantic and Pacific stocks (Alvarado Bremer et al., 2005), contributing to a 4% sequence divergence in the mitochondrial control region of Great Whites (Pardini et al., 2001).

Depending on the level of site fidelity or philopatry (the tendency of an individual to return to or stay in its home area or birth place) and to a lesser extent site segregation, such population structuring may be reinforced or even promoted by activities, such as parturition or mating, associated with discrete locations. Genetic evidence for philopatry in sharks is steadily growing (Table 24.2) and, with certain caveats, comparing the level of population genetic disconcordance revealed by nuclear and mitochondrial markers may help establish which sex is philopatric.

Spatial Ecology and Biology of the White Shark

Knowledge of the species' global distribution has changed significantly as a result of satellite tagging studies conducted over the last decade, whereas their population structure is still poorly known. Their ability to travel long offshore distances has been documented with deployment of satellite transmitters. However, all tracking has been limited to the Indo-Pacific basin (Boustany et al., 2002; Bonfil et al., 2005, 2010; Bruce et al., 2006; Weng et al., 2007a; Jorgensen et al., 2009), leaving a huge gap in knowledge regarding possible migration routes in the Atlantic and the Mediterranean. Moreover, return to the tagging point of origin has been documented where sufficient research

Table 24.2 Genetic Evidence for Philopatric Behavior in Elasmobranchs

Species	Common Name	Reference
Carcharodon carcharias (Indian Ocean)	White Shark	Pardini et al. (2001)
Carcharodon carcharias (northeast Pacific)	White Shark	Jorgensen et al. (2009)
Isurus oxyrinchus	Shortfin Mako	Schrey and Heist (2003)
Sphyrna lewini	Scalloped Hammerhead	Duncan et al. (2006)
Negaprion brevirostris (northwest Atlantic)	Lemon Shark	Feldheim et al. (2002)
Negaprion brevirostris (northwest Atlantic)	Lemon Shark	DiBattista et al. (2008)
Genus *Negaprion*		Schultz et al. (2008)
Carcharhinus leucas (east Atlantic)	Bull Shark	Karl et al. (2011)
Carcharhinus plumbeus (world)	Sandbar Shark	Portnoy et al. (2010)

has been conducted, for example in Australia (Bruce et al., 2006) and the northeast Pacific (Weng et al., 2007a; Domeier and Lucas-Nasby, 2007). Bonfil et al. (2005) describe the tagging of several animals from South African waters, with one returning after a very-long-distance migration (crossing the Indian Ocean basin to northwest Australia and returning). This latter event supports the suggestions of Pardini et al. (2001) regarding female philopatric behavior, the first report of elasmobranchs of staying or returning to a specific location. Observation of significant genetic differentiation between the southwest Pacific and western Indian Ocean, inferred from maternally and biparentally inherited markers, indicated rare female-mediated gene flow versus frequent male gene flow. Emigration and philopatric patterns of South Pacific animals (Jorgensen et al., 2009) also corroborate this hypothesis. Additionally, Jorgensen et al. (2009) and Gubili et al. (2011) suggest that *C. carcharias* is capable of longer migrations both transequatorial and transoceanic when conditions are appropriate, extending the species distribution from the southwest Pacific and establishing new populations (northeast Pacific and Mediterranean, respectively).

Present Review

It is widely accepted that habitat loss, fragmentation, fisheries overexploitation, and climate change have a serious effect on population declines and migration pathways (Wilcove and Wikelski, 2008), consequently shaping the genetic structure of species by restricting animal movement and gene flow (Keller et al., 2004; Van Rossum et al., 2004; Vandergast et al., 2007). The effects of such "schism" can only be evaluated when compared with other existing populations or where historical samples of the examined population exist. Unfortunately, acquiring specimens becomes more difficult because of the scarcity of encounters between animals and researchers. Therefore, biodiversity of an endangered or threatened species cannot be easily defined, particularly of those exhibiting long migrations, inhibiting conservation strategies and management regimes.

In this review we will discuss the current genetic approaches used to characterize *C. carcharias*, each attempting to address an issue pertinent to White Shark conservation. Firstly, we refer briefly to the controversial evolutionary origin and relationships of the species. Secondly, the White Shark global population structure is examined from sequence data from the control region of mitochondrial DNA and allelic scoring at several microsatellite loci, from individuals sampled from across ocean basins (north Pacific Ocean, Indian Ocean, northwest Atlantic Ocean, and Mediterranean Sea) (Pardini et al., 2000, 2001; Martin et al., 2002; Jorgensen et al., 2009; Gubili et al., 2011). Thirdly, the application of molecular approaches for the genetic profiling and species identification will be described, methods that have been applied successfully contributing to the global conservation and trade monitoring of the White Shark (Chapman et al., 2003; Shivji et al., 2005). We then consider the accuracy of photographic identification of individual *C. carcharodon*, by comparing the concordance between photographic and microsatellite-based, genetic identification (Gubili et al., 2009). Finally, we discuss what requirements are needed for future fruitful white-shark research and consider what progress is possible in the field of elasmobranch genetics because of significant advances in genomic technology.

MOLECULAR GENETIC APPROACHES TOWARD THE CONSERVATION GENETICS OF GREAT WHITE SHARKS

Up-to-Date Work

The Great White Shark is found throughout coastal and oceanic waters, at a variety of temperatures and depths (Compagno, 1984; Boustany et al., 2002; Bonfil et al., 2005, 2010; Bruce et al., 2006; Weng et al., 2007a). However, despite information on the biology of this species (Compagno,

2001; Bruce, 2008), little is known about its population structure or how it uses the oceans. A single realistic population estimate has been made, restricted to the South African coast, suggesting the presence of about 800–1600 animals (Cliff et al., 1996). Such estimates are based only on White Sharks inhabiting coastal areas, knowledge of this species' pelagic behavior being restricted to tagged individuals (Boustany et al., 2002; Bonfil et al., 2005; Bruce et al., 2006). Therefore, genetics can become a complementary tool toward elucidating behavioral (migration) and biological (population structure, mating strategies, and stock health) patterns of the animal.

Evolutionary Origins

Despite the existence of an earlier ancestor of *C. carcharias* (Late Cretaceous), *Cretoxyrhina mantelli* (based on body form and length similarities; Shimada, 2008; Shimada et al., 2006), there are only speculations concerning the evolutionary origin of the species. Indeed, much debate and uncertainty surround the evolution of the family Lamnidae. With the divergence of *Carcharodon* estimated to be around 60 Ma (Paleocene or early Eocene) by both mitochondrial DNA (mtDNA) (cytochrome *b*, cyt *b*; Martin, 1996) and nuclear data (conserved microsatellite locus Loc6; Martin et al., 2002), the ancestry of *C. carcharias* is described as either from the extinct Megatoothed Sharks, such as *C. megalodon*, or from another extinct Mako Shark species, such as *Isurus hastalis* (Nyberg et al., 2006), both from the late Miocene. Nyberg et al. (2006) found support for the *hastalis* hypothesis based on similarities in dentition (morphometrics and structure of teeth).

Conversely, Applegate and Espinosa-Arrubarrena (1996) and Purdy (1996) favor the *megalodon* hypothesis on almost the same criteria, insinuating that interpretation of dentition data is not an objective approach, stressing the need for a more robust and reliable methodology. However, none of the above suggestions should be disregarded because interpretation could be difficult; identical tooth development could be attributed to foraging for the same prey. Despite controversies in the evolutionary history of the species, Ehret et al. (2009) were able to place an early Pliocene specimen (4 Ma) found in southwestern Peru in an intermediate position between *Isurus* and *Carcharodon* based once again on tooth morphology, dismissing the *megalodon* scenario. Nevertheless, the lack of data and appropriate sample size prevents validation of these findings, meaning the interpretation of fossilized teeth will remain a contentious issue.

Population Structure of the Great White Shark

A global phylogeography and definition of population dynamics are fundamental to formulation of management and conservation measures. To understand whether individual White Sharks display site fidelity, partitioned by sex, or travel and mix with other populations, a series of analyses have been performed to investigate genetic differences between disjunct populations using both mitochondrial DNA sequences (Pardini et al., 2001; Jorgensen et al., 2009; Gubili et al., 2011) and results from microsatellite data (Pardini et al., 2001). A complex pattern of White Shark population structure is evident.

The matrilineal population structure of White Sharks was examined initially by Pardini et al. (2001) and subsequently by Jorgensen et al. (2009) and Gubili et al. (2011) using mtDNA control region sequences from animals sampled in five main localities: the northeast Pacific (NEP), the southwest Pacific [Australia (AU) and New Zealand (NZ)], West Indian Ocean [South Africa (SA)], northwest Atlantic (NWA), and the Mediterranean (MED). *C. carcharias* is a species with high mtDNA haplotype and nucleotide diversities ($h = 0.9888 \pm 0.0075$; $\pi = 0.02231 \pm 0.01104$; Gubili et al., 2011), with all phylogenetic analyses showing strong evidence of two main lineages (Pardini et al., 2001; Jorgensen et al., 2009; Gubili et al., 2011). It is evident from a haplotype network that the SA/NWA populations form a distinct group from the Indo-Pacific/MED cluster, separated by 43 mutational steps (Gubili et al., 2011), consistent with phylogenetic analyses. Additionally, a second smaller clade was created within the South African group, with the northwest Atlantic lineage

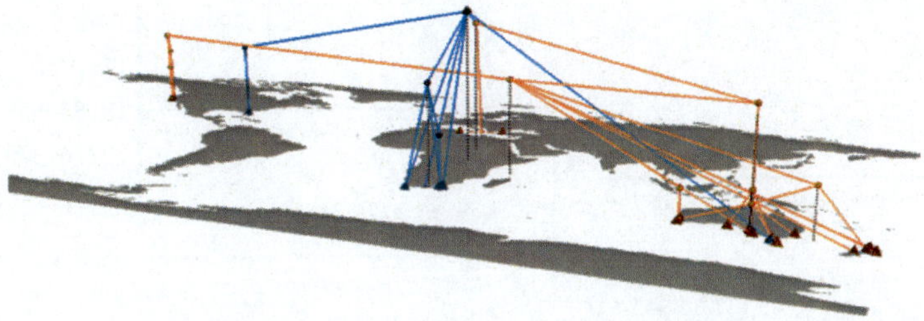

Figure 24.1 A visual "geophylogeny" revealing the presence of two divergent genetic lineages within *C. carcharias* and their relationship to collecting localities (Gubili et al., 2011). Following tree construction, the Bayesian tree and the associated geographical data (Gubili et al., 2011) were applied in GeoPhyloBuilder v1.1 (Kidd and Liu, 2008) to construct a different geophylogenetic model with two main lineages present. The SA/NWA clade is represented with the blue branches (including one displaced animal, a 3.5-m male captured in Tasmania (AU) with a SA haplotype; Pardini et al., 2001), whereas the north-southwest Pacific/Mediterranean clade is shown in red. The dotted vertical lines are droplines. The top gray diamond was used for the root node, spheres indicate internal nodes, cubes indicate tips, and tetrahedra positioned on top of the map are for observations.

separated by seven mutational steps from the South African haplotypes. Conversely, there were very few mutational steps differentiating the southwest Pacific (AU and NZ) and NEP. Nevertheless, estimates of gene flow, as indicated by F_{ST} analysis, suggest that these two populations (southwest Pacific and NEP) are largely reproductively isolated (Jorgensen et al., 2009). Surprisingly, recent analyses from Gubili et al. (2011) suggest a close genetic similarity of the Mediterranean with the western Indo-Pacific haplotypes (Figure 24.1), an event attributed to historical migration of "founder" females during Late Pleistocene, an environmentally extremely dynamic period that may have also played a role in the founding of the NEP stock (Jorgensen et al., 2009). Overall, at least three White Shark matrilineal populations (South Africa/northwest Atlantic, northeast Pacific, and southwest Pacific) are indicated by analysis of DNA sequences from the mitochondrial control region locus (Jorgensen et al., 2009; Gubili et al., 2011).

Our phylogenetic analysis of another mtDNA gene region, cytochrome oxidase subunit I gene (COI), involving 11 partial sequences of 652 base pairs (accession numbers DQ108328, DQ884985, EU398646, FJ518939-44, GU440260, and HQ167639) (Figure 24.2) also identified two main lineages. The north Pacific samples cluster with several from southwest Pacific (Australia), yet within this cluster they form a distinct but poorly supported monophyletic clade (67% bootstrap support). Hence these animals are more similar to and a clear descendant of the southwest Pacific group (Australian), consistent with findings from Jorgensen et al. (2009). However, a second main lineage (with a 91% bootstrap support) consists of samples from southwest Pacific and Indian Ocean (AU, NZ, and SA), one from the Mediterranean (Turkey), and one from the southwest Atlantic (Brazil), inconsistent with previous findings (Pardini et al., 2001; Jorgensen et al., 2009; Gubili et al., 2011). Thus, despite the presence of two clusters and the proximity of the northeast with the southwest Pacific samples, the gene's slow evolution rate does not always render COI an appropriate locus for intraspecific phylogenetic analysis, although inclusion of more samples may resolve some branch points.

Although mtDNA gives an indication of matrilineal contributions to phylogeography and population structure, a better understanding of the relative contributions of each sex may be derived from comparison of maternally and biparentally inherited markers. Development of five nuclear microsatellite dinucleotide markers (Pardini et al., 2000) allowed genotyping of 95 White Sharks from the Indian Ocean and southwest Pacific (SA, AU, and NZ) (Pardini et al., 2001). Whereas mtDNA analyses define a strong population structure between sampling localities (SA, AU, NZ, NEP, NWA, and MED), the five microsatellite loci instead show genetic homogeneity across the Indian Ocean, suggesting that population differences could be a consequence of female philopatry (Pardini et al.,

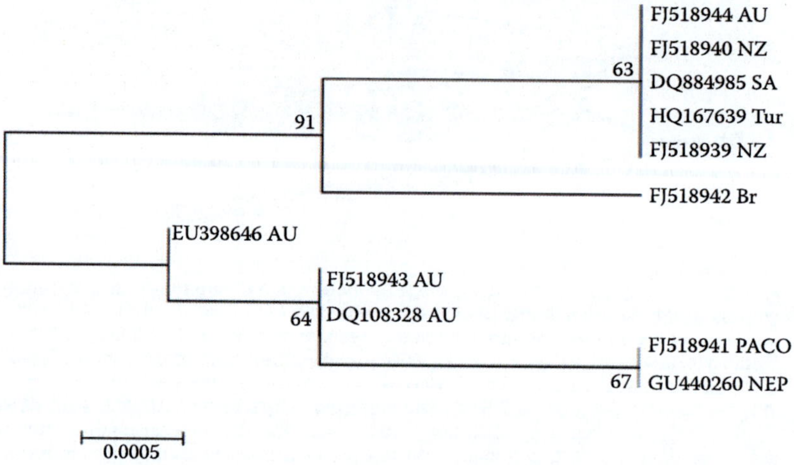

Figure 24.2 Genetic divergences between samples from cytochrome oxidase subunit I gene (COI) sequences of *C. carcharias* were calculated using Kimura 2-parameter distance (K2P; Kimura, 1980) and clustered by neighbor-joining (NJ). Trees were generated on MEGA v.4 (Tamura et al., 2007) and tested with 1,000 bootstrap replicates. The numbers at nodes are bootstrap values derived from 1,000 replicates. Bootstrap values under 60% are not shown. Sampling sites: AU, Australia; Br, Brazil; NEP, northeast Pacific; NZ, New Zealand; PACO, Pacific Ocean; SA, South Africa; Tur, Turkey.

2001). This discordant genetic differentiation of the two markers suggests frequent male-mediated gene flow between the West Indian Ocean (SA) and southwest Pacific (AU/NZ) populations (nuclear F_{ST} N.S., Pardini et al., 2001; mtDNA $F_{ST} = 0.93302$, $p < 0.0001$, Jorgensen et al., 2009). Telemetry and direct observation are concordant with the genetic interpretation, implying that both male and female White Sharks undertake long transoceanic return migrations (Bonfil, et al., 2005, 2010; Weng et al., 2007a, 2007b; Jorgensen et al., 2009). The strong phylogenetic mitochondrial signal implies that pregnant females return to natal sites frequently.

Genetic Profiling and Species Identification: A Tool for Conservation and Trade Monitoring

The restriction of uncontrolled exploitation (legal or illegal) of threatened and vulnerable species has been an important issue for conservation managers. Despite all of the efforts of animal protection over the last 30 yrs., the species listed on the protective catalogue is increasing. Elasmobranchs fall into the category of highly unregulated and exploited species, with worldwide declines in catches reaching over 80% (Baum et al., 2003; Dulvy et al., 2008; Ferretti et al., 2008). However, a proposal to list an additional eight species (Porbeagle, Oceanic Whitetip, Scalloped Hammerhead, Great, Smooth, and Dusky hammerhead, Sandbar, and Spiny Dogfish Sharks) in the 2010 Convention on International Trade in Endangered Species (CITES) to increase the limited listing beyond the original three species (Whale, Basking, and White Sharks) was not supported.

The need for shark body-part identification, mainly dried fins and processed carcasses, is essential for better development and practice of management programmes and law enforcement, especially for species that show morphological similarities. Molecular methods have been developed recently to contribute toward management of trade and shark catches utilizing mtDNA (restriction fragment length polymorphisms: Heist and Gold, 1999; Sequencing: 12–16S ribosomal DNA, Greig et al., 2005; Rodrigues-Filho et al., 2009; COI, Ward et al., 2005, 2008; Holmes et al., 2009; Wong et al., 2009; cyt b, Blanco et al., 2008; cyt b and NADH dehydrogenase subunit 2, Hoelzel, 2001; cyt b and COI, Filonzi et al., 2010). Additionally nuclear DNA (ribosomal internal transcribed spacer 2, ITS2, Pank et al., 2001; Shivji et al., 2002; Abercrombie et al., 2005; Clarke et al., 2006b; Magnussen et al., 2007; 5S ribosomal DNA, Pinhal et al., 2008) or both markers (Chapman et al., 2003; Iglésias et al., 2005) have also been used based either on species-specific primers or sequence data for forensic analysis.

A technique further aiding trade monitoring of White Shark parts and facilitating species identification was developed by Chapman et al. (2003). This sensitive molecular diagnostic assay allows the application of two species-specific PCR primers targeting both mtDNA (cyt b) and nuclear DNA (ITS2). The method's reliability was tested by Shivji et al. (2005) in a case of illegal White Shark fin trade in late 2003. Such cases show clearly that harvesting of White Shark parts is a contemporary incident, despite the species' protection along the U.S. Atlantic coast since 1993. Additionally, cases of fraud in mislabeling commercially exploited elasmobranchs have been reported in Italy, with the most alarming case of smooth-hound samples being identified as White Shark (Filonzi et al., 2010). All samples were taken in 2008, 4 yrs. after the listing of *C. carcharias* on CITES. Furthermore, Wong et al. (2009) developed a nucleotide-diagnostic–sensitive approach for barcode identification, focussing mainly on Carcharhiniformes and Lamniformes. Direct sequence screening of mtDNA COI gene from various species would readily allow the identification of *C. carcharias* by using a diagnostic amino acid, isoleucine (nucleotide diagnostic 197-A position; COI gene). An additional marker, the 12–16S mtDNA region was verified as suitable for species identification containing limited intraspecific variability combined with sufficient interspecific divergence (Greig et al., 2005). This rapid, cost-effective, and sensitive method can allow the discrimination of species within the Lamnidae family among 35 shark species in the Atlantic fisheries, facilitating reliable forensic cases, species-specific landing, and survey data commonly mislabeled because of species similarities. All of these methods have therefore proven that species identification is feasible; however, despite the means (techniques) and protective legislation, products are still freely available to purchase.

Concordance of Genetic and Fin Photo Identification

Photo identification is a popular technique much used over the last 40 yrs. for both terrestrial and marine animals that bear distinctive features, mainly natural marks. It is assumed to be one of the most stress-free and least invasive study techniques implemented, particularly with the advance of imaging technology. However, photographic inconsistencies have been attributed to poor quality photographs because of bad weather/surface conditions, incorrect shooting angle, and matches with the wrong side of the fin (Van Tienhoven et al., 2007). Multilocus genetic profiling of individuals provides an objective, independent arbiter. Nevertheless, to date only one study (Gubili et al., 2009) has tested the reliability of photographic identification by molecular techniques in elasmobranchs. Indeed, thus far the reliability of the method has only been evaluated twice in marine mammals (Gowans et al., 2000; Stevick et al., 2001) using genetic markers as the objective authority.

Over a 5-yr. period, 110 biopsies were collected for genetic assessment of 314 shark photographs (Gubili et al., 2009). The animals were identified from dorsal fin features and marks. Six species-specific (Pardini et al., 2000; Gubili et al., 2009) and one heterologous microsatellite marker (Schrey and Heist, 2002) were used to genotype each biopsied individual. Analyses were conducted as double-blind tests to minimize potential bias, as described in Gubili et al. (2009). The results were similar to those found by Stevick et al. (2001) in Humpback Whales. Six pairwise photographic discrepancies out of 37 cases (16.22%) were noted when identification was compared, whereas association of genetic and photographic identification of 15 possible individuals showed four false positives, which translates into 26.67% unsuccessful matches (Gubili et al., 2009). Stevick et al. (2001) reported 88 discrepancies (21.26%) between photographic and genetic identifications in over 414 cases; hence, the same difficulties and restrictions found in marine mammal photo identification are also inherent in White Shark fin photo identification and perhaps potentially in other sharks. Conversely, both methods had an almost seven times higher mismatch rate than the one reported by Gowans et al. (2000) in Northern Bottlenose Whales, *Hyperoodon ampullatus*. However, the latter analysis was based on a physiological trait of sexual maturity that is easy to score and remains unchanged after maturity, whereas identification based on an individual's natural marks increases potential mismatches caused by a higher rate of change over time. Using genetic markers that restrict the probability of two individuals sharing the same genotypes to less than 1% means that such mismatches could probably be attributed to human error and accidental mislabeling of photographs or sample tubes.

The initial aim of such work was to evaluate the accuracy and utility of photographic identification of individual White Sharks using naturally acquired marks and not on distinctive patterns and also to eliminate possible duplicate entries of the same individual in the analyzed data set (Gubili et al., 2009). These findings emphasize that the use of fin photographs is a valid and accurate identification method but that care is needed when analyzing large data sets of species and that results must be interpreted with caution.

Future Work

Preliminary studies have documented the complexity of White Shark population genetic structure, suggesting male dispersal and female philopatry (Pardini et al., 2001; Jorgensen et al., 2009). However, these attempts were restricted to few animals sampled in the Indo-Pacific basin using different markers; five microsatellite loci, mtDNA control region sequences, and restriction fragment length polymorphism analyses. The first steps toward a better understanding of the population genetic structure and trade monitoring of the Great White Shark have proven very promising; nevertheless, much work is still required to protect and ensure the future of *C. carcharias*.

Numbers: Samples and Molecular Markers

Why does the North East Pacific stock(s) seem to be isolated from the southwest Pacific? Is there any gene flow among the Atlantic and the Mediterranean populations? What is the structure in the South Atlantic? A larger sample size including individuals from existing and new sites, such as northwest Pacific, the Mediterranean, and the whole of the Atlantic would establish and better characterize the Great White's population genetic structure over all oceans. A more detailed data set than the one currently gathered, including historical samples and further biological information such as sex and length, would support or refute earlier findings and elucidate the stability of population structure over a time frame of several generations, illustrating the utility of genetic analyses in assessing the temporal stability or diversity of stocks.

Additionally, more genetic markers, both nuclear and mitochondrial, should be examined (Chenuil, 2006). New polymorphic loci may establish subdivisions within and between populations and indicate separation of reproductive stocks, tasks that were not demonstrated clearly using the available seven microsatellites, the low expected heterozygosity (0.673; Gubili et al., 2009) of which restricts power to detect independent stocks. Hence with significantly fewer loci, with lower heterozygosity in sharks than teleosts (Hauser, 2009) because of low mutation rates (Martin, 1999), there is an urgent need for more markers to increase variability and power of analysis. Similarly, more mtDNA genes should be examined because they also exhibit lower variability in sharks than teleosts (Heist, 1999); indications of the species' origins and more accurate time since divergence could be elucidated with the use of additional gene sequences such as cytochrome b (cyt b) or cytochrome oxidase I (COI). However, it is important to note that mtDNA is inherited as a single locus limiting the usefulness of this marker. Future studies employing more than one class of markers are especially important for species with an unknown population size and potential sex-biased dispersal. Zenger et al. (2006) report the successful application of amplified fragment length polymorphisms to examine genetic diversity in seven White Sharks, perhaps indicating a potentially useful approach to elucidating White Shark population structure.

Genomic approaches significantly enhance the information available from each individual. More affordable commercial sequencing technology, driven by the Human Genome Project, has become available to other species in the last few years. Because elasmobranchs are not considered model species, the shift from lower variability markers to higher volumes of genetic information, although technically feasible, is of lower priority than in intensively exploited commercial fish. However, the Little Skate (*Leucoraja erinacea*) and Spiny Dogfish (*S. acanthias*) have been targeted for full-genome sequencing (http://www.genome.gov/Pages/Research/Sequencing/SeqProposal/SharksSkateSeq.pdf).

Meanwhile, next generation sequencing and expressed sequence tags (ESTs) are the most common short DNA fragments publicly available (GenBank, EMBL), which may uncover new genes and molecular markers that could be useful in revealing gene diversity. They could also be used to produce full-length cDNAs to provide probes for cDNA expression arrays, both helping in the development of molecular markers from full sequences, as well as identifying the function of these transcripts (Parton et al., 2010). Elasmobranchs are primitive, jawed, cartilaginous fish and as such could be ideal for biomedical research because they have many highly conserved sequences (Martin et al., 2002). Currently, EST libraries are available for species such as the Spiny Dogfish, *S. acanthias,* and Little Skate. *Leucoraja erinacea* (Parton et al., 2010), as well as the Pacific Electric Ray, *Torpedo californica* (Nazarian et al., 2007), revealing a significant number of ESTs with no homologue in other organisms (Parton et al., 2010). We can only imagine what other species can reveal. These sequences have the potential to provide hundreds of polymorphic sites (single nucleotide polymorphisms), as well being useful for microsatellite characterization (Spiny Dogfish; Veríssimo et al., 2010a). However, because of the costs involved, cartilaginous fish ecologists will probably have to wait few more years to expand the list of studied species.

DNA Extraction from Alternative Sources

Direct monitoring of elusive species can be expensive, time consuming, and difficult. With molecular tools, hypotheses concerning species dispersal or population fragmentation can be examined more effectively indirectly as seen from the increase in popularity of genetic studies on marine species over the last 30 yrs. The success of such studies is dependent upon quality of the extracted DNA. Although soft tissue and blood are common DNA sources, these tissue types are often difficult to obtain from wild endangered species; in such cases it is imperative that alternative biological material is utilized. Sources such as hair, feathers, feces, teeth, and urine are of increasing interest because they do not cause stress when collected because there is no immediate interaction with the individual (Merriwether et al., 1994; Rosenbaum et al., 1997; Taberlet and Luikart, 1999; Rohland et al., 2004; Hausknecht et al., 2007).

These new DNA sources should be considered as serious alternatives when dealing with endangered species, often characterized by limited encounters. This may help determine sample provenance where their mtDNA haplotypes characterize discrete locations (Chapman et al., 2009). Teeth still remain the most accessible source of DNA for sharks for many reasons; they are numerous, are easily and often naturally detached from the animal, and can be found on beaches, the sea bed, in museums, the tooth trade, and occasionally even in shark-attack victims. These as well as other preserved samples can be used to infer current and historical population structure, investigate potential loss of genetic diversity (Pichler and Baker, 2000), monitor tooth trade, and identify an attacking species from fragments.

Ahonen and Stow (2008) report DNA extraction from teeth and jaws of four taxa: Grey Nurse (*Carcharias taurus*), whaler (Carcharhinidae family), Tiger (*Galeocerdo cuvier*), and School Shark (*Galeorhinus galeus*). They amplified a 608-base pair fragment of mtDNA control region from

C. taurus. DNA extraction from teeth can be valuable in many aspects of research, particularly when important historical/museum samples can be used, especially for species lacking pulp cavity, such as the White Shark (Vennemann et al., 2001). The first attempt to extract and assess the size of the largest possible mitochondrial and nuclear DNA fragments recovered from White Shark teeth gave promising results, facilitating haplotype identification by reconstituting larger mtDNA control region sequences from several smaller amplicons (about 280 base pairs) (C. Gubili, personal communication; 14th European Elasmobranch Association). However, microsatellite genotyping error and amplification failure of loci with larger fragment sizes from the tooth DNA were common. Such problems are not restricted to White Sharks because false alleles, high allelic dropout rates, and artefacts have been reported previously when using low quality and minimum quantity of DNA (Ramos et al., 1995; Sefc et al., 2003; Pompanon et al., 2005).

Despite these difficulties, redesigning primers to target smaller amplicons allows DNA extracted from teeth and jaws to be useful in establishing species and individual identifications and for assessing genetic structure and variation. Therefore, the potential utility of samples held in contemporary and historical collections could be of great assistance, particularly in the case of White Sharks, because they are characterized by small population sizes, with scarce and opportunistic sampling only.

Historical Population Demography

Some key concerns regarding species conservation include calculation of population size, migration, and divergence rates using genetic data (Wang and Whitlock, 2003). Unfortunately, such estimates would require a large sample size, temporal genetic data composed by contemporary and historical samples, assayed by more than one gene (Wang and Whitlock, 2003; Waples, 2006). However, a first attempt to calculate the evolutionary rates of White Shark mtDNA control region by Gubili et al. (2011) gave values ranging from 0.74 to 1.19% divergence between lineages per million years based on different calibration points; the lower rate was similar to that calculated for other large elasmobranchs (Schultz et al., 2008; Duncan et al., 2006). However, using such estimated evolutionary rates to calculate divergence times with a single locus must be treated with caution because such values may be locus-specific and not a true reflection of the species/population divergence time, particularly for the mtDNA control region, which is not well conserved. Moreover, there is a poor fossil calibration for the White Shark.

Currently, only the haplotype and nucleotide diversities give an insight to the population genetic status of *C. carcharias*. A large value for both diversities globally (Gubili et al., 2011) suggests the presence of a stable population with long evolutionary history or secondary contact between lineages (Grant and Bowen, 1998). However, values for regional populations were high for haplotype diversities (h) but low for nucleotide diversities (π) (South Africa $h = 0.8590$ and $\pi = 0.00382$: Australia $h = 1.0000$ and $\pi = 0.00340$: New Zealand $h = 1.0000$ and $\pi = 0.00373$; northeast Pacific $h = 0.9895$ and $\pi = 0.00222$; Gubili et al., 2011), perhaps indicating population bottlenecking followed by rapid population growth and mutation accumulation. In the meantime, other techniques such as mark-recapture data collected longitudinally (by tagging and photo identification) should be considered when calculating population parameters. However, although a useful approach, such an estimate is available only for White Sharks in South African waters (Cliff et al., 1996).

Multiple Paternity in Elasmobranchs

The mating behavior and reproductive biology of elasmobranchs have attracted much attention, not only because of the difficulty of tracking some species but also as necessary information for formulation of management and conservation strategies, particularly in endangered species still heavily harvested. Despite the restricted number of mating observations, it is known that sharks have internal fertilization, in contrast to most teleost fish (Pratt, 1993; Pratt and Carrier, 2001; Whitney et al., 2004). Mating behavior could explain the social interactions of the species studied and also help elucidate attempts of individuals to maximize their direct fitness.

Evidence for multiple mating in elasmobranchs is accumulating, with polyandry (a mating pattern where females mate with more than one male per breeding season) apparently a common female mating strategy in sharks (Saville et al., 2002; Chapman et al., 2004; Feldheim et al., 2004; Daly-Engel et al., 2006, 2007, 2010; Portnoy et al., 2007; Lage et al., 2008; Veríssimo et al., 2010b), rays and oviparous species (Chevolot et al., 2007). Multiple mating has been detected in all but two elasmobranchs studied to date: the Galapagos Shark, *Carcharhinus galapagensis* (Daly-Engel et al., 2006), and the Whale Shark, *Rhincodon typus* (Schmidt et al., 2010); however, these observations are based on single litters. The percentage of multiple male contributions to natural litters ranges from low in Squalid Sharks 11–17% of litters (*Squalus mitsukurii*, Daly-Engel et al., 2010; *S. acanthias*, Veríssimo et al., 2010b) up to 100% in the Thornback Ray, *Raja clavata* (Chevolot et al., 2007).

Although the suggested benefits of multiple partners are well documented for both sexes in the animal kingdom (Arnqvist and Nilsson, 2000; Muller et al., 2006; Fokidis et al., 2007), how mating encounters with multiple males would directly advantage female sharks remains unclear (Chapman et al., 2004). To date there are no direct correlations between multiple paternity and genetic diversity in shark offspring, although sample sizes remain very small (Daly-Engel et al., 2010).

Preliminary analysis from one *C. carcharias* (eight pups) and four *I. oxyrinchus* litters (16, 13, 10, and 11 pups, respectively) indicated multiple paternity in all cases (Gubili, 2008). Despite this initial evidence of promiscuity in female lamnids, the reasons for such a strategy remain unclear, because current sample sizes are insufficient to determine any benefits. Ideally, larger litters and parent samples representing a wider variety of lamnids are required before the function of mating strategies can be placed in a broader context across the family. Obtaining a larger sample of litters presents obvious difficulties for *C. carcharias* because of the scarcity of pregnant females, although such a goal is more feasible in other lamnids because some remain targeted species, especially by game fisheries.

Summary

Population Status and Conservation

Despite the identification of gene flow among the Indo-South Pacific oceans by microsatellite analyses, the very marked population differentiation revealed by the mtDNA analyses indicate the existence of a more fine-scale matrilineal structure, including populations from South Africa, New Zealand/Australia, northeast Pacific, northwest Atlantic, and the Mediterranean (Pardini et al., 2001; Gubili, 2008; Jorgensen et al., 2009; Gubili et al., 2011). Nevertheless, more information on how White Sharks use the oceans is necessary before defining management regimes using contrasting neutral markers. By combining data from mtDNA, nuclear genomic DNA, and telemetry, population structure can be defined between different groups, highlighting differences in behavior between the sexes. Despite the genetic subdivisions detected, tagging data has shown that White Sharks are capable of long-distance migrations, particularly in the Indo-Pacific basin stressing the need for a more global management perspective, whereas local measurements should not be disregarded particularly for small confined populations such as the Mediterranean (Gubili et al., 2011).

Classification of the *C. carcharias* by International Union for Conservation of Nature (2007) as "vulnerable" and its placement on Appendix II of CITES in 2004 testifies to increased concerns about dramatically reduced population sizes (Baum et al., 2003). Despite the belief that populations are declining generally, White Shark protection is not fully implemented. Both Shivji et al. (2005) and Filonzi et al. (2010) clearly demonstrate the presence of an illegal international fin trade, particularly of young individuals and populations of unknown status, even in areas where White Sharks are protected, stressing that monitoring should be international.

It is increasingly recognized that management policies are formulated by cooperation between scientists and decision makers, who must have up-to-date relevant information on which to base legislation. Here, we discuss how aspects of White Shark biology, particularly population structure and dispersal patterns and reliable species and individual identification, may be elucidated using existing molecular tools and reflect on how new approaches have the potential to offer a better understanding of the species' biology and contribute to improved *C. carcharias* conservation.

REFERENCES

Abercrombie, D. L., Clarke, S. C., and Shivji, M. S. 2005. Global-scale identification of hammerhead sharks: Application to assessment of the international fin trade and law enforcement. *Conserv Genet* 6: 775–788.

Ahonen, H., and Stow, A. J. 2008. Shark jaws and teeth: An unexploited resource for population genetic studies. *J Fish Biol* 73: 450–455.

Alvarado Bremer, J., Vinas, J., Mejuto, J., Ely, B., and Pla, C. 2005. Comparative phylogeography of Atlantic bluefin tuna and swordfish: The combined effects of vicariance, secondary contact, introgression, and population expansion on the regional phylogenies of two highly migratory pelagic fishes. *Mol Phylogenet Evol* 36: 169–187.

Alves, M. J., Coelho, H., Collares-Pereira, M. J., and Coelho, M. M. 2001. Mitochondrial DNA variation in the highly endangered cyprinid fish *Anaecypris hispanica*: Importance for conservation. *Heredity* 87: 463–473.

Applegate, S. P., and Espinosa-Arrubarrena, L. 1996. The fossil history of *Carcharodon* and its possible ancestor, *Cretolamna*: A study in tooth identification. In *Great White Sharks: The Biology of Carcharodon carcharias*, edited by A. P. Klimley and D. G. Ainley, pp. 19–36. San Diego, CA: Academic Press.

Arnqvist, G., and Nilsson, T. 2000. The evolution of polyandry: Multiple mating and fitness in insects. *Anim Behav* 60: 145–164.

Barbará, T., Palma-Silva, C., Paggi, G. M., Bered, F., Fay, M. F., and Lexer, C. 2007. Cross-species transer of nuclear microsatellite markers: Potential and limitations. *Mol Ecol* 16: 3759–3767.

Baum, J. K., Myers, R. A., Kehler, D. G., Worm, B., Harley. S. J., and Doherty, P. A. 2003. Collapse and conservation of shark populations in the northwest Atlantic. *Science* 299: 389–392.

Blanco, M., Pérez-Matrín, R. I., and Sotelo, C. G. 2008. Identification of shark species in seafood products by Forensically Informative Nucleotide Sequencing (FINS). *J Agric Food Chem* 56: 9868–9874.

Bonfil, R., Meÿer, M., Scholl, M. C., Johnson, R., O'Brien, S., Oosthuizen, H., Swanson, S., Kotze, D., and Paterson, M. 2005. Transoceanic migration, spatial dynamics, and population linkages of white sharks. *Science* 310: 100–103.

Bonfil, R., Francis, M. P., Duffy, C., Manning, M. J., and O'Brien, S. 2010. Large-scale tropical movements and diving behavior of white sharks *Carcharodon carcharias* tagged off New Zealand. *Aquat Biol* 8: 115–123.

Boustany, A. M., Davis, S. F, Pyle, P., Anderson, S. D., Le Boeuf, B. J., and Block, B. A. 2002. Expanded niche for white sharks. *Nature* 415: 35–36.

Bruce, B. D. 1992. Preliminary observations on the biology of the white shark, *Carcharodon carcharias*, in South Australian Waters. *Aust J Mar Fresh Res* 43: 1–11.

Bruce, B. D. 2008. The biology and ecology of the white shark, *Carcharodon carcharias*. In *Sharks of the Open Ocean: Biology, Fisheries and Conservation*, edited by M. D. Camhi, E. K. Pikitch, and E. A. Babcock, pp. 69–81. Fish and Aquatic Resources Series 13. Oxford, UK: Blackwell Publishing.

Bruce, B. D., Stevens, J. D., and Malcolm, H. 2006. Movements and swimming behaviour of white sharks (*Carcharodon carcharias*) in Australian waters. *Mar Biol* 150: 161–172.

Castro, A. L. F., Stewart, B. S., Wilson, S. G., Heuter, R. E., Meekan, M. G., Motta, P. J., Bowen, B. W., and Karl, S. A. 2007. Population genetic structure of Earth's largest fish, the whale shark (*Rhincodon typus*). *Mol Ecol* 16: 5183–5192.

Chabot, C. L., and Allen, L. G. 2009. Global population structure of the tope (*Caleorhinus galeus*) inferred by mitochondrial control region sequence data. *Mol Ecol* 18: 545–552.

Chapman, D. D., Abercrombie, D. L., Douady, C. J., Pikitch, E. K., Stanhope, M. J., and Shivji, M. S. 2003. A streamlined, bi-organelle, multiplex PCR approach to species identification: Application to global conservation and trade monitoring of the great white shark, *Carcharodon carcharias. Conserv Genet* 4: 415–425.

Chapman, D. D., Prodöhl, P. A., Gelsleichter, J., Manire, C. A., and Shiviji, M. S. 2004. Predominance of genetic monogamy by females in a hammerhead shark, *Sphyrna tiburo*: Implications for shark conservation. *Mol Ecol* 13: 1965–1974.

Chapman, D. D., Pinhal, D., and Shivji, M. S. 2009. Tracking the fin trade: Genetic stock identification in western Atlantic scalloped hammerhead sharks, *Sphyrna lewini. Endang Species Res* 9: 221–228.

Chapuis, E., Trouve, S., Facon, B., Degen, L., and Goudet, J. 2007. High quantitative and no molecular differentiation of the freshwater snail (*Galba truncatula*) between temporary and permanet water habitats. *Mol Ecol* 16: 3484–3496.

Chenuil, A. 2006. Choosing the right molecular genetic markers for studying biodiversity: From molecular evolution to practical aspects. *Genetica* 127: 101–120.

Chevolot, M., Hoarau, G., Rijnsdorp, A. D., Stam, W. T., and Olsen, J. L. 2006. Phylogeography and population structure of thornback rays (*Raja clavata* L., Rajidae). *Mol Ecol* 15: 3693–3705.

Chevolot, M., Ellis J. R., Rijnsdorp A. D., Stam W. T., and Olsen, J. L. 2007. Multiple paternity analysis in the thornback ray, *Raja clavata* L. *J Hered* 98: 712–715.

Clarke, S. C., McAllister, M. K., Milner-Gulland, E. J., Kirkwood, G. P., Michielsens, C. G. J., Agnew, D. J., Pikitch, E. K., Nakano, H., and Shivji, M. S. 2006a. Global estimates of shark catches using trade records from commercial markets. *Ecol Let* 9: 115–1126.

Clarke, S. C., Magnussen, J. E., Abercrombie, D. L., McAllister, M. K., and Shivji, M. S. 2006b. Identification of shark species composition and proportion in the Hong Kong shark fin market based on molecular genetics and trade records. *Conserv Biol* 20: 201–211.

Cliff, G., van der Elst, R. P., Govender, A., Witthuhn, T. K., and Bullen, E. M. 1996. First estimates of mortality and population size of white sharks on the on the South African coast. In *Great White Sharks: The Biology of Carcharodon carcharias*, edited by A. P. Klimley, and D. G. Ainley, pp. 393–400. San Diego, CA: Academic Press.

Compagno, L. J. V. 1984. Sharks of the world: An annotated and illustrated catalogue of shark species known to date. Part 1. Hexanchiformes to Lamniformes. *Food and Agriculture Organization (FAO) of the United Nations Species Catalogue,* Vol. 5. Fisheries Synopsis 125, Vol. 4. Rome, Italy: FAO.

Compagno, L. J. V. 2001. Sharks of the world: An annotated and illustrated catalogue of shark species known to date. Vol. 2. Bullhead, mackerel and carpet sharks (Heterodontiformes, Lamniformes and Orectolobiformes). *FAO Species Catalogue for Fishery Purposes 1*. Rome, Italy: FAO.

Compagno, L. J. V., Marks, M. A., and Fergusson, I. K. 1997. Threatened fishes of the world: *Carcharodon carcharias* (Linnaeus, 1758) (Lamnidae). *Environ Biol Fish* 50: 61–62.

Corrigan, S., Huveneers, C., Schwartz, T. S., Harcourt, R. G., and Beheregaray, L. B. 2008. Genetic and reproductive evidence for two species of ornate wobbegong shark *Orectolobus* spp. on the Australian east coast. *J Fish Biol* 73: 1662–1675.

Daly-Engel, T. S., Grubbs, R. D., Holland, K N., Toonen, R. J., and Bowen, B. W. 2006. Assessment of multiple paternity in single litters from three species of carcharhinid sharks in Hawaii. *Environ Biol Fish* 76: 419–424.

Daly-Engel, T. S., Grubbs, R. D., Bower, B. W., and Toonen, R. J. 2007. Frequency of multiple paternity in an unexploited tropical population of sandbar sharks (*Carcharhinus plumbeus*). *Can J Fish Aquat Sci* 64: 198–204.

Daly-Engel, T. S., Grubbs, R. D., Feldheim, K. A., Bower, B. W., and Toonen, R. J. 2010. Is multiple mating beneficial or unavoidable? Low multiple paternity and genetic diversity in the shortspine spurdog *Squalus mitsukurii. Mar Ecol Prog Ser* 403: 255–267.

DiBattista, J. D., Feldheim, K. A., Thibert-Plante, X., Gruber, S. H., and Hendry, A. P. 2008. A genetic assessment of polyandry and breeding site fidelity in lemon sharks. *Mol Ecol* 17: 3337–3351.

Ding, S. X., Pan, Y., Zeng, H S., and Wang, J. 2009. Characterization of 12 polymorphic microsatellite loci for the whitespotted bamboo shark (*Chiloscyllium plagiosum* Bennett). *Mol Ecol Resour* 9: 1398–1400.

Domeier, M. L., and Nasby-Lucas, N. 2007. Annual re-sightings of photographically identified white sharks (*Carcharodon carcharias*) at an eastern Pacific aggregation site (Guadalupe Island, Mexico). *Mar Biol* 150: 977–984.

Dudgeon, C. L, Broderick, D., and Ovenden, J. R. 2009. IUCN classification zones concord with, but underestimate, the population genetic structure of the zebra shark *Stegostoma fasciatum* in the Indo-West Pacific. *Mol Ecol* 18: 248–261.

Dudley, S. F. J., Anderson-Reade, M. D., Thompson, G. S., and McMullen, P. B. 2000. Concurrent scavenging off a whale carcass by great white sharks, *Carcharodon carcharias*, and tiger shark, *Galeocerdo cuvier. Fish Bull* 98: 646–649.

Duncan, K. M., Martin, A. P., Bowen, B. W., and De Couet, H. G. 2006. Global phylogeography of the scalloped hammerhead shark (*Sphyrna lewini*). *Mol Ecol* 15: 2239–2251.

Dulvy, N. K., Baum, J. K., Clarke, S., Compagno, L. J. V., Cortés, E., Domingo, A., Fordham, S., Fowler, S., Francis, M. P., Gibson, C., Martinez, J., Musick, J. A., Soldo, A., Stevens, J. D., and Valenti, S. 2008. You can swim but you can't hide: The global status and conservation of oceanic pelagic sharks and rays. *Aquat Conserv* 18: 459–482.

Ehret, D. J., Hubbell, G., and MacFadden, B. J. 2009. Exceptional preservation of the white shark *Carcharodon* (Lamniformes, Lamnidae) from the early Pliocene of Peru. *J Vertebr Paleont* 29: 1–13.

El Nagar, A., McHugh, M., Rapp, T., Sims, D. W., and Genner, M. J. 2010. Characterisation of polymorphic microsatellite markers for skates (Elasmobranchii: Rajidae) from expressed sequence tags. *Conserv Genet* 11: 1203–1206.

Feldheim, K. A., Gruber, S. H., and Ashley, M. V. 2001. Population genetic structure of the lemon shark (*Negaprion brevirostris*) in the western Atlantic: DNA microsatellite variation. *Mol Ecol* 10: 295–303.

Feldheim, K. A., Gruber, S. H., and Ashley, M. V. 2002. The breeding biology of lemon sharks at a tropical nursery lagoon. *Proc R Soc B* 269: 1655–1661.

Feldheim, K. A., Gruber, S. H., and Ashley, M. V. 2004. Reconstruction of parental microsatellite genotypes reveals female polyandry and philopatry in the lemon shark, *Negaprion brevirostris. Evolution* 10: 2332–2342.

Feldheim, K. A., Stow, A. J., Ahonen, H., Chapman, D. D., Shiviji, M., Peddemors, V., and Wintner, S. 2007. Polymorphic microsatellite markers for studies of the conservation and reproductive genetics of imperiled sand tiger sharks (*Carcharias taurus*). *Mol Ecol Notes* 7: 1366–1368.

Ferretti, F., Myers, R. A., Serena, F., and Lotze, H. K. 2008. Loss of large predatory sharks from the Mediterranean Sea. *Conserv Biol* 22: 952–964.

Filonzi, L., Chiesa, S., Vaghi, M., and Marzono, F. N. 2010. Molecular barcoding reveals mislabelling of commercial fish products in Italy. *Food Res Int* 43: 1383–1388.

Fokidis, H. B., Risch, T. S., and Glenn, T. C. 2007. Reproductive and resource benefits to large female body size in a mammal with female-biased sexual size dimorphism. *Anim Behav* 73: 479–488.

Gardner, M. G., and Ward, R. D. 2002. Taxonomic affinities within Australian and New Zealand *Mustelus* sharks (Chondrichthyes: Triaakidae) inferred from allozymes, mitochondrial DNA and precaudal vertebrae counts. *Copeia* 2: 356–363.

Gowans, S., Dalebout, M. L., Hooker, S. K., and Whitehead, H. 2000. Reliability of photographic and molecular techniques for sexing northern bottlenose whales (*Hyperoodon ampullatus*). *Can J Zoolog* 78: 1224–1229.

Grant W. S., and Bowen, B. W. 1998. Shallow population histories in deep evolutionary lineages of marine fishes: Insights from sardines and anchovies and lessons for conservation. *J Hered* 89: 415–426.

Greig, T. W., Moore, M. K., Woodley, C. M., and Quattro, J. M. 2005. Mitochondrial gene sequences useful for species identification of western North Atlantic Ocean sharks. *Fish Bull* 103: 516–523.

Griffiths, A. M., Sims, D. W., Cotterell, S. P., El Nagar, A., Ellis, J. R., Lynghammar, A., McHugh, M., Neat, F. C., Pade, N. G., Queiroz, N., Serra-Pereira B., Rapp, T., Wearmouth, V. J., and Genner, M. J. 2010. Molecular markers reveal spatially segregated cryptic species in a critically endangered fish, the common skate (*Dipturus batis*). *Proc R Soc B* 227: 1497–1503.

Gubili, C. 2008. Application of Molecular Genetics for Conservation of the Great White Shark, *Carcharodon carcharias*, L. 1758. Ph.D. dissertation, Aberdeen University.

Gubili, C., Bilgin, R., Kalkan, E., Karhan, S. Ü., Jones, C. S., Sims, D. W., Kabasakal, H., Martin, A. P., and Noble, L. R. 2011. Antipodean White Sharks on a Mediterranean walkabout? Historical dispersal leads to genetic discontinuity and an endangered anomalous population. *Proc R Soc B*. 278: 1679–1686.

Gubili, C., Johnson, R., Gennari, E., Oosthouizen, W. H., Kotze, D., Meÿer, M., Sims, D. W., Jones, C. S., and Noble, L. R. 2009. Concordance of genetic and fin photo identification in the great white shark, *Carcharodon carcharias*, off Mossel Bay, South Africa. *Mar Biol* 156: 2199–2207.

Hauser, L. 2009. The molecular ecology of dogfish sharks. In *Biology and Management of Dogfish Sharks*, edited by V. F. Gallucci, G. A. McFarlana, and G. G. Bargmann, pp. 229–252. Bethesda, MD: American Fisheries Society.

Hauser, L., and Ward, R. D. 1998. Population identification in pelagic fish: The limits of molecular markers. In *Advances in Molecular Ecology*, edited by G. R. Carvalho. Fairfax, VA: IOS Press.

Hausknecht, R., Gula, R., Pirga B., and Kuehn, R. 2007. Urine: A source for non-invasive genetic monitoring in wildlife. *Mol Ecol Notes* 7: 208–212.

Heist, E. J. 1999. A review of population genetics in sharks. In *Life in the slow lane: Ecology and conservation of long-lived marine animals*, edited by J. A. Musick, pp. 161–168, Symposium 23. Bethesda, MD: American Fisheries Society.

Heist, E. J. 2005. Population and reproductive genetics in Chondrichthyes. In *Reproductive biology and phylogeny of Chondrichthyes: Sharks, skates, stingrays, and chimaeras*, edited by W. C. Hamlett. Plymouth, MA: Science Publishers.

Heist, E. J., and Gold, J. R. 1999. Microsatellite DNA variation in sandbar sharks (*Carcharhinus plumbeus*) from the Gulf of Mexico and Mid-Atlantic Bight. *Copeia* 1999: 182–186.

Heist, E. J., Graves, J. E., and Musick, J. A. 1995. Population genetics of the sandbar shark (*Carcharinus plumbeus*) in the Gulf of Mexico and Mid-Atlantic Bight. *Copeia* 1995: 555–562.

Heist, E. J., Jenkot, J. L., Keeney, D. B., Lane, R. L., Moyer, G. R., Reading, B. J., and Smith, N. L. 2003. Isolation and characterization of polymorphic microsatellite loci in nurse shark (*Ginglymostoma cirratum*). *Mol Ecol Notes* 3: 59–61.

Hoelzel, A. R. 2001. Shark fishing in fin soup. *Conserv Genet* 2: 69–72.

Hoelzel, A. R., Shivji, M. S., Magnussen, J., and Francis, P. F. 2006. Low worldwide genetic diversity in the basking shark (*Cetorhinus maximus*). *Biol Lett* 2: 639–642.

Holmes, B. H., Steinke, D., and Ward, R. D. 2009. Identification of shark and ray fins using DNA barcoding. *Fish Res* 95: 280–288.

Iglésias, S. P., Lecointre, G., and Sellos, D. Y. 2005. Extensive paraphylies within sharks of the order Carcharhiniformes inferred from nuclear and mitochondrial genes. *Mol Phylogenet Evol* 34: 569–583.

International Union for Conservation of Nature. 2007. 2007 IUCN red list of threatened species. Gland, Switzerland: IUCN. http://www.iucnredlist.org.

Jorgensen, S. J., Reeb, C. A., Chapple, T. K., Anderson, S., Perle, C., Van Sommeran, S. R., Fritz-Cope, C., Crown, A. C., Klimley, A. P., and Block, B. A. 2009. Philopatry and migration of Pacific white sharks. *Proc R Soc B* 277: 679–688.

Karl, S. A., Castro, A. L. F., Lopez, J. A., Charvet, P., and Burgess, G. H. 2011. Phylogeography and conservation of the bull shark (*Carcharhinus leucas*) inferred from mitochondrial and microsatellite DNA. *Conserv Genet* 12: 371–382.

Keeney, D. B., and Heist, E. J. 2003. Characterization of microsatellite loci isolated from the blacktip shark and their utility in requiem and hammerhead sharks. *Mol Ecol Notes* 3: 501–504.

Keeney, D. B., and Heist, E. J. 2006. Worldwide phylogegraphy of the blacktip shark (*Carcharhinus limbatus*) inferred from mitochondrial DNA reveals isolation of western Atlantic populations coupled with recent Pacific dispersal. *Mol Ecol* 15: 3669–3679.

Keeney, D. B., Heupel, M., Hueter, R. E., and Heist, E. J. 2003. Genetic heterogeneity among blacktip shark, *Carcharinus limbatus*, continental nurseries along the U.S Atlantic and Gulf of Mexico. *Mar Biol* 143: 1039–1046.

Keeney, D. B., Heupel, M. R., Hueter, R. E., and Heist, E. J. 2005. Microsatellite and mitochondrial DNA analyses of the genetic structure of blacktip shark (*Carcharhinus limbatus*) nurseries in the northwestern Atlantic, Gulf of Mexico, and Caribbean Sea. *Mol Ecol* 14: 1911–1923.

Keller, I., Nentwig, W., and Largiadèr, R. C. 2004. Recent habitat fragmentation due to roads can lead to significant genetic differentiation in an abundant flightless ground beetle. *Mol Ecol* 13: 2983–2994.

Kidd, D. M., and Liu, X. 2008. GEOPHYLOBUILDER 1.0: An ARCGIS extension for creating "geophylogenies." *Mol Ecol Resourc* 8: 88–91

Kimura, M. 1980. A simple method of estimating evolutionary rate of base substitutions through comparative studies of nucleotide sequences. *J Mol Evol* 16: 111–120.

Larson, S., Tinnemore, D., and Amemiya, C. 2009. Microsatellite loci within sixgill sharks, *Hexanchus griseus*. *Mol Ecol Resour* 9: 978–981.

Lavery, S. 1992. Electrophoretic analysis of phylogenetic relationships among Australian Carcharhinid sharks. *Aust J Mar Fresh Res* 43: 97–108.

Lavery, S., and Shaklee, J. B. 1989. Population genetics of two tropical sharks, *Carcharhinus tilstoni* and *C. sorrah*, in Northern Australia. *Aust J Mar Fresh Res* 40: 541–557.

Magnussen, J. E., Pikitch, E. K., Clarke, S. C., Nicholson, C., Hoelzel, A. R., and Shivji, M. S. 2007. Genetic tracking of basking shark products in international trade. *Anim Conserv* 10: 199–207.

Martin, A. F. 1996. Systematics of the Lamnidae and origination time of *Carcharodon carcharias* inferred from the comparative analysis of mitochondrial DNA sequences. In *Great White Sharks: The Biology of Carcharodon carcharias*, edited by A. Kimley, and D. Ainley, pp. 49–53. San Diego, CA: Academic Press.

Martin, A. P. 1999. Substitution rates of organelle and nuclear genes in sharks: Implicating metabolic rate (again). *Mol Biol Evol* 16: 996–1002.

Martin, A. P., Pardini, A. T., Noble, L. R., and Jones, C. S. 2002. Conservation of a dinucleotide simple sequence repeat locus in sharks. *Mol Phylogenet Evol* 23: 205–213.

McCauley, L., Goecker, C., Parker, P., Rudolph, T., Goetz, F., and Gerlach, G. 2004. Characterization and isolation of DNA microsatellite primers in the spiny dogfish (*Squalus acanthias*). *Mol Ecol Notes* 4: 494–496.

Merriwether, D. A., Rothhammer, F., and Ferrell, R. E. 1994. Genetic variation in the New World: Ancient teeth, bone, and tissue as sources of DNA. *Experimentia* 50: 592–601.

Moura, T., Silva, M. C., Figueiredo, I., Neves, A., Muñoz, P. D., Coelho, M. M., and Gordo, L. S. 2008. Molecular barcoding of north-east Atlantic deep-water sharks: Species identification and application to fisheries management and conservation. *Mar Freshwater Res* 59: 214–223.

Muller, M. N., Kahlenberg, S. M., Emery Thompson, M., and Wrangham, R. W. 2006. Male coercion and the costs of promiscuous mating for female chimpanzees. *Pro R Soc B* 274: 1009–1014.

Murray, B. W., Wang, J. Y., Yang, S.-C., Stevens, J. D., Fisj, A., and Svavarsson, J. 2008. Mitochondrial cytochrome b variation in sleeper sharks (Squaliformes: Somniosidae). *Mar Biol* 153: 1015–1022.

Nance, H. A., Daly-Engel, T. S., and Marko, T. B. 2009. New microsatellite loci for the endangered scalloped hammerhead shark, *Sphyrna lewini*. *Mol Ecol Resour* 3: 959–957.

Nazarian, J., Hathout, Y., Vertes, A., and Hoffman, E. P. 2007. The proteome survey of an electricity-generating organ (*Torpedo californica* electric organ). *Proteomics* 7: 617–627.

Nyberg, K. G., Ciampaglio, C. N., and Wray, G. A. 2006. Tracing the ancestry of the great white shark, *Carcharodon carcharias*, using morphometric analyses of fossil teeth. *J Vertebr Paleontol* 26: 806–814.

Ovenden, J. R., Street, R., and Broderick, D. 2006. New microsatellite loci for Carcharhinid sharks (*Carcharhinus tilstoni* and *C. sorrah*) and their cross-amplification in other shark species. *Mol Ecol Notes* 6: 415–418.

Ovenden, J. R., Kashiwagi, T., Broderick, D., Giles, J., and Salini, J. 2009. The extent of population genetic subdivision differs among four co-distributed shark species in the Indo-Australian archipelago. *BMC Evol Biol* 9: 40.

Pank, M., Stanhope, M., Natanson, L., Kohler, N., and Shivji, M. 2001. Rapid and simultaneous identification of body parts from the morphologically similar sharks *Carcharhinus obscurus* and *Carcharhinus plumbeus* (Carcharhinidae) using multiplex PCR. *Mar Biotechnol* 3: 231–240.

Pardini, T. A., Jones, C. S., Scholl, M. C., and Noble, L. R. 2000. Isolation and characterization of dinucleotide microsatellite loci in the great white shark, *Carcharodon carcharias*. *Mol Ecol* 9: 1176–1178.

Pardini, A. T., Jones, C. S., Noble, L. R., Kreiser, B., Malcolm, H., Bruce, B. D., Stevens, J. D., Cliff, G., Scholl, M. C., Francis, M., Duffy, C. A. J., and Martin, A. P. 2001. Sex-biased dispersal of great white sharks. *Nature* 412: 139–140.

Parton, A., Bayne, C. J., and Barnes, D. W. 2010. Analysis and functional annotation of expressed sequence tags from *in vitro* cell lines of elasmobranchs: Spiny dogfish shark (*Squalus acanthias*) and little skate (*Lucoraja erinacea*). *Comp Biochem Physiol Part D* 5: 199–206.

Pereyra, S., García, G., Miller, P., Oviedo, S., and Domingo, A. 2010. Low genetic diversity and population structure of the narrownose shark (*Mustelus schmitti*). *Fish Res* 106: 468–473.

Pichler, F. B., and Baker, C. S. 2000. Loss of genetic diversity in the endemic Hector's dolphin due to fisheries-related mortality. *Proc R Soc B* 267: 97–102.

Pinhal, D., Gadig, O. B. F., Wasko, A. P., Oliveira, C., Ron, E., Foresti, F., and Martins, C. 2008. Discrimination of shark species by simple PCR of 5S rDNA repeats. *Genet Mol Biol* 31 (Suppl. 1): 361–365.

Pompanon, F., Bonin, A., Bellemain, E., and Taberlet, P. 2005. Genotyping errors: Causes, consequences and solutions. *Nat Rev Genet* 6: 847–859.

Portnoy, D. S., Piercy, A. N., Musick, J. A., Burgess, G. H., and Graves, J. E. 2007. Genetic polyandry and sexual conflict in the sandbar shark, *Carcharhinus plumbeus*, in the western North Atlantic and Gulf of Mexico. *Mol Ecol Notes* 16: 187–197.

Portnoy, D. S., McDowell, J. R., Heist, E. J., Musick, J. A., and Graves, J. E. 2010. World phylogeography and male-mediated gene flow in the sandbar shark, *Carcharhinus plumbeus*. *Mol Ecol* 19: 1994–2010.

Pratt, H. L. Jr. 1993. The storage of spermatozoa in the oviducal glands of western North Atlantic sharks. *Environ Biol Fish* 38: 139–149.

Pratt H. L. Jr., and Carrier, J. C. 2001. A review of elasmobranch reproductive behaviour with a case study on the nurse shark, *Ginglymostoma cirratum*. *Environ Biol Fish* 60: 157–188.

Purdy, R. W. 1996. Paleoecology of fossil white sharks. In *Great White Sharks: The Biology of Carcharodon carcharias*, edited by A. P. Klimley and D. G. Ainley, pp. 67–78. San Diego, CA: Academic Press.

Quattro, J. M., Stoner, D. S., Driggers, W. B., Anderson, C. A., Priede, K. A., Hoppmann, E. C., Campbell, N. H., Duncan, K. M., and Grady, J. M. 2006. Genetic evidence of cryptic speciation within hammerhead sharks (genus *Sphyrna*). *Mar Biol* 148: 1143–1155.

Ramos, M. D., Lalueza, C., Girbau, E., Pérez-Pérez, A., Quevedo, S., Turbón, D., and Estivill, X. 1995. Amplifying dinucleotide microsatellite loci from bone and tooth simples of up to 5000 years of age: More inconsistency than usefulness. *Hum Genet* 96: 205–212.

Ramírez-Macías, D., Vázquez-Juárez, R., Galván-Magaña, F., and Munguía-Vega, A. 2007. Variations of the mitochondrial control region sequence in whale sharks (*Rhincodon typus*) from the Gulf of California, Mexico. *Fish Res* 84: 87–95.

Ramírez-Macías, D., Shaw, K., Ward, R., Galván-Magaña, F., and Vázquez-Juárez, R. 2009. Isolation and characterization of microsatellite loci in the whale shark (*Rhincodon typus*). *Mol Ecol Resour* 9: 798–800.

Rodrigues, L. F. S., da Rocha, T. C., do Rêgo P. S., Schneider, H., Sampaio, I., and Vallinoto, M. 2009. Identification and phylogenetic inferences on stocks of sharks affected by the fishing industry off the Northern coast of Brazil. *Genet Mol Biol* 32: 405–413.

Rohland, N., Siedle, H., and Hofreiter, M. 2004. Nondestructive DNA extraction method for mitochondrial DNA analyses of museum specimens. *Biotechniques* 36: 814–821.

Rosenbaum, H. C., Egan, M. G., Clapham, P. J., Brownell, R. L., and Desalle, R. 1997. An effective method for isolating DNA from historical specimens of baleen. *Mol Ecol* 6: 677–681.

Saville, K. J., Lindley, A. M., Maries, E. G., Carrier, J. C., and Pratt, H. L., Jr. 2002. Multiple paternity in the nurse shark, *Ginglymostoma cirratum*. *Environ Biol Fish* 63: 347–351.

Schlötterer, C. 2000. Evolutionary dynamics of microsatellite DNA. *Chromosoma* 109: 365–371.

Schluessel, V., Broderick, D., Collin, S. P., and Ovenden, J. R. 2010. Evidence for extensive population structure in the white-spotted eagle ray within the Indo-Pacific inferred from mitochondrial gene sequences. *J Zool* 281: 46–55.

Schmidt, J. V., Schmidt, C. L., Ozer, F., Ernst, R. E., Feldheim, K. A., Ashley, M. V., and Levine, M. M. 2009. Low genetic differentiation across three major ocean populations of the whale shark, *Rhincodon typus*. *PLoS ONE* 4 e4988. doi:10.1371/journal.pone.0004988.

Schmidt, J. V., Chen, C.-C, Sheikh, S. I., Meekan, M. G., Norman, B. M., and Joung, S.-J. 2010. Paternity analysis in a litter of whale shark embryos. *Endang Species Res* 12: 117–124.

Schrey, A. W., and Heist, E. J. 2002. Microsatellite markers for the shortfin mako and cross-species amplification in lamniformes. *Conserv Genet* 3:459–461.

Schrey, A. W., and Heist, E. J. 2003. Microsatellite analysis of the population structure in the shortfin mako (*Isurus oxyrinchus*). *Can J Fish Aquat Sci* 60: 670–675.

Schultz, J. K., Feldheim, K. A., Gruber, S. H., Ashley, M. V., McGovern, T. M., and Bowen, B. W. 2008. Global phylogeography and seascape genetics of the lemon shark (genus *Negaprion*). *Mol Ecol* 17: 5336–5348.

Sefc, K. M., Payne, R. B., and Sorencon, M. D. 2003. Microsatellite amplification from museum feather samples: Effects of fragment size and template concentration on genotyping errors. *The Auk* 120: 982–989.

Shimada, K. 2008. Ontogenetic parameters and life history strategies of the late cretaceous lamniform shark, *Cretoxyrhina mantelli*, based on vertebral growth increments. *J Vertebr Paleontol* 28: 21–33.

Shimada, K., Cumbaa, S. L., and Van Rooyen, D. 2006. Caudal fin skeleton of the late cretaceous lamniform shark, *Cretoxyrhina mantelli*, from the niobrara chalk of Kansas. *New Mexico Museum of Natural History and Science Bulletin* 35: 185–192.

Shivji, M., Clarke, S., Pank, M., Natanson, L., Kohler, N., and Stanhopes, M. 2002. Genetic identification of pelagic shark body parts for the conservation and trade monitoring. *Conserv Biol* 16: 1036–1047.

Shivji, M. S., Chapman, D. D., Pikitch, E. K., and Raymond, P. W. 2005. Genetic profiling reveals illegal international trade in fins of the great white shark, *Carcharodon carcharias*. *Conserv Genet* 6: 1035–1039.

Smith, P. J. 1986. Low genetic variation in sharks (Chondrichthyes). *Copeia* 1986: 202–207.

Stevick, P. T., Palsbøll, P. J., Smith, T. D., Bravington, M. V., and Hammond, P. S. 2001. Errors in identification using natural markings: rates, sources, and effects on capture-recapture estimates of abundance. *Can J Fish Aquat Sci* 58: 1861–1870.

Stow, A., Zenger, K., Briscoe, D., Gillings, M., Peddemors, V., Otway, N., and Harcourt, R. 2006. Isolation and genetic diversity of endangered grey nurse shark (*Carcharias taurus*) populations. *Biol Lett* 2: 308–311.

Taberlet, P., and Luikart, G. 1999. Non-invasive genetic sampling and individual identification. *Biol J Linn Soc* 68: 41–55.

Tamura, K., Dudley, J., Nei, M., and Kumar, S. 2007. MEGA4: Molecular evolutionary genetics analysis (mega) software version 4.0. *Mol Biol Evol* 24: 1596–1599.

Valsecchi, E., Vacchi, M., and Di Sciara, D. N. 2005. Characterization of a new molecular marker for investigating skate population genetics: Analysis of three Mediterranean skate species (genus *Raja*) of commercial interest as a test case. *J Northwest Atl Fish Sci* 35: 225–231.

Vandergast, A. G., Bohonak, A. J., Weissman, D. B., and Fisher, R. N. 2007. Understanding the genetic effects of recent habitat fragmentation in the context of evolutionary history: Phylogeography and landscape genetics of a southern California endemic Jerusalem cricket (Orthoptera: Stehopelmatidae: *Stenopelmatus*). *Mol Ecol* 16: 977–992.

Van Rossum, F., Campos De Sousa, S., and Triest, L. 2004. Genetic consequences of habitat fragmentation in an agricultural landscape on the common *Primula veris*, and comparison with its rare congener, *P. vulgaris*. *Conserv Genet* 5: 231–245.

Van Tienhoven, A. M., Den Hartog, J. E., Reijns, R. A., and Peddemors, V. M. 2007. A computer-aided program for pattern-matching of natural marks on the spotted raggedtooth shark *Carcharias taurus*. *J Appl Ecol* 44: 273–280.

Vennemann, T. W., Hegner, E., Cliff, G., and Benz, G. W. 2001. Isotopic composition of recent shark teeth as a proxy for environmental conditions. *Geochim Cosmochim Acta* 65: 1583–1599.

Veríssimo, A., McDowell, J. R., and Graves, J. E. 2010a. Global population structure of the spiny dogfish *Squalus acanthias*, a temperate shark with an antitropical distribution. *Mol Ecol* 10: 1651–1662.

Veríssimo, A., Grubbs, D., McDowell, J. R., Musick, J., and Portnoy, D. 2010b. Frequency of multiple paternity in the spiny dogfish *Squalus acanthias* in the Western North Atlantic. *J Hered* 102: 88–93.

Wang, J. L., and Whitlock, M. C. 2003. Estimating effective population size and migration rates from genetic samples over space and time. *Genetics* 163: 429–446.

Waples, R. S. 2006. A bias correction for estimates of effective population size based on linkage disequilibrium at unlinked gene loci. *Conserv Genet* 7: 167–184.

Ward, R. D., Zemlak, T. S., Innes, B. H., and Last, P. R. 2005. DNA barcoding Australia's fish species. *Phil Trans R Soc B* 360: 1847–1857.

Ward, R. D., Costa, F. O., Holmes, B. H., and Steinke, D. 2008. DNA barcoding of shared fish species from the North Atlantic and Australasia: Minimal divergence for most taxa, but *Zeus faber* and *Lepidopus caudatus* each probably constitute two species. *Aquat Biol* 3: 71–78.

Weng, K. C., Boustany, A. M., Pyle, P., Anderson, S. D., Brown, A., and Block, B. A. 2007a. Migration and habitat of white sharks (*Carcharodon carcharias*) in the eastern Pacific Ocean. *Mar Biol* 152: 877–894.

Weng, K. C., O'Sullivan, J. B., Lowe, C. G., Winkler, C. E., Dewar, H., and Block, B. A. 2007b. Movements, behaviour and habitat preferences of juvenile white sharks *Carcharodon carcharias* in the eastern Pacific. *Mar Ecol Prog Ser* 338: 211–224.

Whitney, N. M., Pratt, H. L. Jr., and Carrier, J. C. 2004. Group courtship, mating behaviour and siphon sac function in the whitetip reef shark, *Triaenodon obesus*. *Anim Behav* 68: 1435–1442.

Wilcove, D. S., and Wikelski, M. 2008. Going, going, gone: Is animal migration disappearing? *Plos Biol* 6: 1361–1364.

Wong, E. H.-K., Shivji, M., and Hanner, R. H. 2009. Identifying sharks with DNA barcodes: Assessing the utility of a nucleotide diagnostic approach. *Mol Ecol Resourc* 9 (Suppl. 1): 243–256.

Zachos, J., Pagani, M., Sloan, L., Thomas, E., and Billups, K. 2001. Trends, rhythms, and aberrations in global climate 65 Ma to present. *Science* 292: 686–693.

Zenger, K. R., Stow, A. J., Peddemors, V., Briscoe, D. A., and Harcourt, R. G. 2006. Widespread utility of highly informative AFLP molecular markers across divergent shark species. *J Hered* 97: 607–611.

CHAPTER 25

Use of Photo Identification to Describe a White Shark Aggregation at Guadalupe Island, Mexico

Nicole Nasby-Lucas*
Marine Conservation Science Institute

Michael L. Domeier
Marine Conservation Science Institute

ABSTRACT

Photo-identification (photo-ID) techniques have been used to monitor and track northeastern Pacific White Sharks (*Carcharodon carcharias*) at a seasonal adult aggregation site at Guadalupe Island, Mexico. Between 2001 and 2009, 142 sharks have been photographed, 113 of which have been entered into a catalogue that requires both the left and right sides to be photographed. This catalog consists of 67 males (59%) and 46 females (41%). The photo-ID data demonstrated White Sharks to be present at Guadalupe Island between late July and December, although effort past December was minimal. Sharks exhibited seasonal site fidelity, with 83% of the sharks in the catalog sighted in at least 2 yrs. and 28% sighted in at least 5 yrs. of the study. Males were found to return to the island earlier than females, with the sex ratio reaching parity in November. Males were more likely to be sighted in consecutive years, whereas females were more likely to exhibit a non-consecutive year visitation pattern. However, only a small percentage of sexually mature females exhibited a consistent every-other-year pattern, which has been proposed to be associated with an 18-month gestation cycle. The continuation of this photo-ID program will provide important data on life history, migratory patterns, and abundance, which will aid in the management and conservation of this vulnerable species.

INTRODUCTION

Over the past decade, worldwide numbers of large pelagic sharks have been declining because of targeted fishing, trophy hunting, and fishery bycatch (Baum et al., 2003; Myers and Worm, 2003). Included in these statistics are White Sharks (*Carcharodon carcharias*), which are particularly vulnerable because of the high market value of their jaws and fins (Shivji et al., 2005; Dulvy et al., 2008). These threats, along with biological characteristics such as low intrinsic rate of population increase (Cailliet et al., 1985; Francis, 1996; Pratt, 1996; Smith et al., 1998; Compagno, 2001), late maturation, and natural rarity, make this species susceptible to overexploitation (Smith et al., 1998; Myers and Worm, 2003). White Sharks are now listed in Appendix II of the Convention on International Trade in Endangered Species of Wild Flora and Fauna and are also listed as vulnerable by the World Conservation Union (IUCN, IUCN Global Red List status VU A1cdþ+2cd) (Hilton-Taylor, 2000).

White Sharks are a global apex predator found in relatively low numbers. In the northeastern Pacific Ocean, adult White Sharks are known to seasonally aggregate at two sites off the coast of North America: Guadalupe Island (GI), Mexico, and Central California (CC), United States. Previous studies have shown that sharks from both of these regions make a seasonal winter/spring migration to an area termed the Shared Offshore Foraging Area (SOFA), centered midway between the Hawaiian Islands and Baja California (Boustany et al., 2002; Weng et al., 2007; Domeier and Nasby-Lucas, 2008). Pop-up satellite tagging and photo-identification (photo-ID) data have shown philopatry in eastern Pacific White Sharks, with individual sharks returning to their specific aggregation site in late summer/early fall (Boustany et al., 2002; Domeier and Nasby-Lucas, 2007, 2008; Weng et al., 2007). The act of seasonally aggregating at a known time and place creates a period of local vulnerability when directed or incidental harvest could lead to local, if not regional, sharp population decline. This presents a serious threat to White Sharks, and the monitoring and management of these aggregation sites are critical for the conservation of eastern Pacific White Sharks.

Photo-ID of individual White Sharks is an important long-term tracking tool that can aid in the monitoring of these aggregation sites. Previous photo-ID studies have provided information on residency patterns (Klimley and Anderson, 1996; Domeier and Nasby-Lucas, 2007) and spatiotemporal differences in sex ratio (Strong et al., 1992; Domeier and Nasby-Lucas, 2007), as well as short- and long-distance movements (Anderson and Goldman, 1996; Bonfil et al., 2005). Recently, two photo-ID databases in the northeastern Pacific were used to estimate the abundance of White Sharks at the GI aggregation site (Chapter 26, this book) and the CC site (Chapple et al., 2011). The long-term use of this information will be an important tool in tracking trends in abundance of White Sharks over time and will be essential for the establishment of management policies.

The GI, Mexico White Shark population has been monitored via photo-ID since 2001 (Domeier and Nasby-Lucas, 2007). GI is 407 km south-southwest of San Diego, California and 260 km offshore of Baja California, Mexico and is both a Mexican nature preserve (dedicated in 1925) and a pinniped sanctuary (1975). The island serves as a haulout and pupping site for the Northern Elephant Seal (*Mirounga angustirostris*), the Guadalupe Fur Seal (*Arctocephalus townsendi*), and the California Sea Lion (*Zalophus californianus*). The photo-ID project at GI uses pigment patterns to identify individual White Sharks, and as of 2005, Domeier and Nasby-Lucas (2007) had collected 5 yrs. of data resulting in a catalog of 73 individual White Sharks. These data showed repeat patterns in seasonal visitation, with 78% of the sharks being resighted at the island in multiple years. Through both photo-ID tracking and previous satellite tagging studies (Domeier and Nasby-Lucas, 2008), it has been shown that male White Sharks return to GI every year in late summer, whereas sexually mature females may return in alternate years (Domeier and Nasby-Lucas, 2007; Chapter 16, this book). The 2-yr. migration pattern for females is proposed to be associated with a presumed 18-month gestation cycle (Mollet et al., 2000).

This paper provides a valuable update of the photo-ID database from GI. The goals for the continuation of the GI photo-ID database include the following:

1. Continue to monitor the visitation patterns of previously identified White Sharks at GI, as well as add new sharks to the database.
2. Monitor the visitation pattern of sexually mature females to examine the hypothesized every-other-year visitation pattern.

3. Examine the observed sex ratio for GI White Sharks.
4. Provide mark-recapture type data that can be used for annual population estimates (Chapter 26, this book).

MATERIALS AND METHODS

Between 2001 and 2009, photographic images of White Sharks at GI, Mexico, were obtained from video and still cameras operated by caged divers or through the use of a camera mounted on a hand-held pole. Images were taken by the authors and divers who volunteered images after visiting the site on commercial cage-dive operations that began in 2002. Photographer, date, and location were logged for each photograph. Pigment patterns from three regions of the body were used to identify sharks: the gill flaps (extension of the interbrachial septum), pelvic fin region, and caudal fin region. Two nonpigment characters were also recorded to aid reidentification: the sex of the fish and permanent trauma-induced markings (i.e., major scars, damage and notches on the trailing edge of the dorsal fin, and any mutilations) (Figure 25.1) (see Domeier and Nasby-Lucas (2007) for a detailed description of pigment pattern identification methods).

A catalog of identified sharks was maintained, and sharks were assigned numbers in the order they were identified. Only sharks with photos from both the left and right side were added to the catalog, because pigment patterns are not bilaterally symmetrical. Length data were added when available either using actual measurements from captured sharks (Chapter 11, this book), calculated using a paired-laser system, or estimated while the shark was next to a boat or shark cage of known length. Length values are reported as total length. Sexual maturity was based on total length and was designated for males of at least 3.6 m and females of at least 4.5 m in length (Francis, 1996; Pratt, 1996). A summary sighting log was kept to track the years and months the individual sharks were sighted at the island, and the complete database was compiled indicating each date that each individual shark was sighted at the island.

Sighting data were analyzed to determine number of sharks per week, month, and year as well as to calculate the sex ratio and examine repeat-visitation patterns of individual sharks. The number of sharks sighted per week was adjusted for effort by dividing the number of sharks by the effort for that particular week. Effort was dependent upon the number of days that dive boats were present at the island and was determined by counting the number of individual days in which photos were taken. The year 2001 was omitted from the analysis of sightings by week because of very little effort. Sex ratio was calculated from the total number of sharks in the database and by sharks sighted each year. Sex ratio was also calculated by month from the average number of male and female sightings from 2002 to 2009. Repeat-visitation patterns were examined to determine whether sharks were sighted in consecutive years, nonconsecutive years, or only once. Any gap between years of sighting was considered nonconsecutive.

Figure 25.1 Characters used in identification of White Sharks include pigment pattern at the gill flaps (a), pelvic fin (b), caudal fin (c), the sex of the shark (d), and dorsal fin notches or any additional mutilations (e). (Courtesy of Stephen McCulloch, Harbor Branch Oceanographic Institute.)

RESULTS

The photographic and video records collected from 2001 to 2009 led to the identification of 113 individual White Sharks from GI consisting of 67 males (59%) and 46 females (41%) (Table 25.1). Ninety-four (83%) individuals were sighted in more than 1 yr. (females $n = 38$, males $n = 56$), and of these, 12 were identified in at least 7 yrs. (all male), 7 in 6 of the 9 yrs. (4 males, 3 female), 13 in 5 of the 9 yrs. (8 males, 5 females), 18 in 4 of the 9 yrs. (12 males, 6 females), 26 in 3 of the 9 yrs. (15 males, 11 females), and 18 in 2 of the 9 yrs. (5 males, 13 females) (Figure 25.2). Effort varied by year, but the maximum number of sharks sighted in a single year was 62 in 2005 (39 males, 23 females). The maximum number of individual sharks sighted in one month was 40 in September 2005 (29 males, 11 females) (Table 25.2), and the maximum effort was also in September 2005, with 23 d (Table 25.3). Twenty-one sharks have not been sighted since 2005 or earlier. This includes 12 females and 9 males (Table 25.1). The longest gap between sightings was 6 yrs. for a female (#6) and 4 yrs. for males (#17, #71, #73, and #96; Table 25.1).

Between 2006 and 2009, 40 new sharks were added to the database. Eight of the new sharks included a corresponding length resulting in the determination of two sexually immature males, two sexually mature females, and four sexually mature males added to the database. Ten of the new sharks were found to previously exist in our records as orphans. Orphans are defined as sharks for which there are photographs of only one side and therefore cannot be entered into the catalogue (Domeier and Nasby-Lucas, 2007). Once an orphan is matched with images of the opposing side, years worth of data can be recovered for that individual. In this case, all 10 recovered orphans allowed for data for these new sharks to be entered from 2005 or earlier (Table 25.1). To date, we have 27 left-side and 29 right-side orphans remaining, representing a minimum of 29 additional individual sharks. Adding the 29 orphans to the 113 fully identified individual sharks results in a minimum of 142 sharks photographed at GI since 2001.

Male White Sharks seasonally arrived at GI earlier than females. The earliest male sighting was July 14, whereas the earliest female sighting was August 20. August sightings of females were quite rare (maximum of two in any given year), but males were commonly observed in August (maximum of 25 in any given year) (Table 25.2). The rate (number of individuals sighted per day of effort) of female sightings increased as the season progressed, peaking in November. Male sightings peaked in late August to early September and decreased thereafter, but peaked again in December (note that there were very few days of effort in December, and most photos taken in December were by the authors on scientific expeditions) (Figure 25.3).

The overall sex ratio for the GI White Shark photo-ID catalog was 1.5:1 (male:female) and was significantly different than 1:1 (chi-squared $P = 0.05$). The sex ratio by year from 2002 to 2009 ranged from 1.4:1 to 2.9:1 (Figure 25.4). The ratio of males to females by month for all years combined (Table 25.2) was significantly different in August and September (ratios 18.8:1 and 3.1:1, t-test $P = 0.0001$, 0.0008). The overall sex ratio was more uniform in October, November, and December (ratios 1.6:1, 0.9:1, and 2.7:1, respectively; t-test $P = 0.08$, 0.88, and 0.16, respectively).

The return visitation patterns exhibited in the GI photo-ID database were examined for all sharks. A similar percentage of males and females were sighted in only 1 yr. (males 16.4%, females 17.4%), but they differed for sharks that were sighted in consecutive years (males 53.7%, females 26.1%) with females having a higher percentage of sightings in nonconsecutive years (males 29.9%, females 56.5%) (Figure 25.5). Six females with nonconsecutive visitation patterns exhibited a consistent every-other-year pattern. Two of the females exhibited a pattern of every other year for 5 yrs. and then returned in the following consecutive year (both of these sharks were over 5.1 m and presumed sexually mature), seven females were present in consecutive years and then began an every-other-year pattern (length was unknown), three females were present every third year (one

Table 25.1 Sighting Log of All Sharks by Year

		2001	2002	2003	2004	2005	2006	2007	2008	2009
1	F			■	■	■	■	■		■
2	F		■	■	■	■				
3	F			■	■			■		
4	F		■	■	■		■			■
5	F			■	■	■				
6	F			■						■
7	M		■	■	■	■	■	■	■	■
8	M			■	■	■	■	■		
9	M	■	■	■	■	■	■	■	■	
10	M			■	■	■				
11	M			■	■	■	■	■	■	■
12	M			■	■	■				
13	M	■	■	■	■	■	■	■	■	■
14	M		■	■	■	■	■	■	■	■
15	M	■	■	■	■	■	■	■		■
16	M	■	■	■		■				
17	M	■	■			■				■
18	M			■	■	■		■		■
19	M	■	■	■	■	■	■	■	■	■
20	F			■	■	■	■		■	
21	M				■	■	■	■	■	■
22	M		■		■	■	■			■
23	M	■	■	■	■	■	■	■	■	■
24	M				■	■	■	■		
25	M			■	■	■	■	■		
26	M			■	■	■	■			
27	M				■		■	■	■	■
28	F			■	■			■	■	■
29	F				■	■				
30	F				■	■	■	■		■
31	F				■	■			■	
32	M			■	■	■	■	■	■	■
33	M			■	■	■			■	
34	M				■	■	■	■	■	
35	F			■	■	■	■	■	■	
36	F		■	■	■		■		■	
37	F		■		■		■			
38	F				■			■		
39	F				■		■		■	■
40	F		■		■		■			
41	F				■	■				
42	F		■		■		■		■	
43	F	■	■		■	■				
44	F				■	■				
45	M		■	■	■			■		■
46	M	■		■	■	■	■	■	■	■
47	F				■	■	■	■	■	■

continued

Table 25.1 (Continued) Sighting Log of All Sharks by Year

		2001	2002	2003	2004	2005	2006	2007	2008	2009
48	M			■	■	■	■	■	■	■
49	M	■		■	■			■	■	■
50	M			■						
51	F		■		■					
52	M		■		■	■	■			
53	F		■							
54	F		■	■		■		■		
55	M		■							
56	M	■		■		■				
57	F			■						
58	F					■		■	■	
59	M		■		■	■	■	■	■	
60	F					■				
61	M				■	■		■		■
62	F	■				■				
63	F		■			■			■	
64	F					■				
65	M					■		■	■	■
66	M				■	■	■	■		
67	F			■	■	■		■		
68	M	■		■	■					
69	M					■				
70	M				■	■				
71	F					■				■
72	M		■	■		■				
73	M		■	■				■		
74	M				■	■	■	■	■	■
75	F					■				
76	M						■	■	■	■
77	F				■		■		■	■
78	M					■	■	■	■	
79	M							■	■	■
80	M							■		■
81	M							■	■	■
82	M					■		■		
83	M							■	■	
84	F							■		■
85	M							■		
86	M								■	
87	F								■	■
88	M					■	■	■	■	■
89	M							■	■	■
90	M								■	
91	F							■	■	■
92	M						■	■	■	■
93	M							■	■	
94	M							■	■	■
95	M			■	■	■	■	■	■	■

Table 25.1 (Continued) Sighting Log of All Sharks by Year

		2001	2002	2003	2004	2005	2006	2007	2008	2009
96	M		■			■				■
97	M							■	■	■
98	F								■	
99	M							■	■	■
100	F									■
101	M									■
102	M									■
103	M									■
104	F									■
105	M							■	■	■
106	F								■	■
107	M							■	■	■
108	F					■		■		■
109	F								■	■
110	M								■	■
111	M									■
112	F					■		■		■
113	M									■

This table lists all sharks by year and indicates the number of the shark and whether it is a male or a female. Gray shading indicates that the shark was sighted at Guadalupe Island in that year.

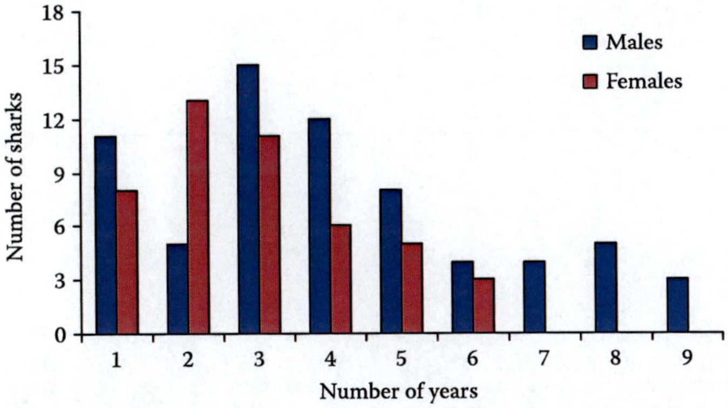

Figure 25.2 Number of annual visitations documented for individual male and female White Sharks at Guadalupe Island (n = 113).

shark was 4.8 m and two were of unknown length), one female was sighted every fourth year (length unknown), and one was sighted every sixth year (4.6 m).

DISCUSSION

The long-term photo-ID monitoring of White Sharks at GI has resulted in a 9-yr. data set that tracks the seasonal presence of individual White Sharks. Forty sharks have been added to the database since 2007 (Domeier and Nasby-Lucas, 2007), bringing the total to 113 identified sharks. An additional 29 discrete orphan records exist for a minimum total of 142 photographed individuals. Eighty-three percent of the sharks in the database were sighted in more than 1 yr., with 11% of

Table 25.2 Sighting Log of All Sharks by Month

	July	August	September	October	November	December	Total
2001	0/0	0/0	4/0	6/1	1/1	0/0	11/2
2002	2/0	4/0	6/3	5/2	8/8	0/1	17/11
2003	0/0	13/1	11/1	9/4	15/11	10/3	27/13
2004	0/0	11/0	20/8	19/11	11/16	12/6	33/24
2005	0/0	10/1	29/11	16/10	10/11	11/3	39/23
2006	0/0	15/1	17/5	9/11	3/8	0/0	28/12
2007	0/0	25/1	29/4	25/10	7/7	3/1	43/15
2008	0/0	18/2	20/7	14/11	1/1	7/5	34/17
2009	2/0	17/0	22/10	15/14	9/9	0/0	40/18

The total number of individual sharks (male/female) sighted for each month and year of sampling. Annual totals reflect the fact that many individuals were sighted in more than one month.

Table 25.3 Days of Effort by Month and Year

	July	August	September	October	November	December	Total
2001	0	0	1	1	1	0	3
2002	2	2	6	6	5	1	22
2003	0	9	5	6	15	2	37
2004	0	3	10	13	19	5	50
2005	0	6	23	12	13	3	57
2006	0	11	16	12	11	0	50
2007	0	6	18	21	6	2	53
2008	0	13	17	12	2	5	49
2009	2	10	20	20	12	0	64

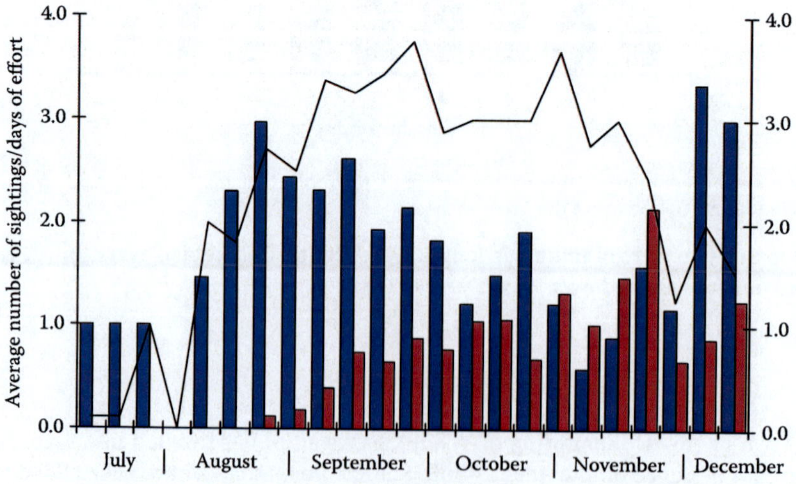

Figure 25.3 Average number of sharks sighted by week adjusted by weekly sighting effort. Blue indicates male, pink indicates female, and the black line shows average days of effort for that week over all years of data.

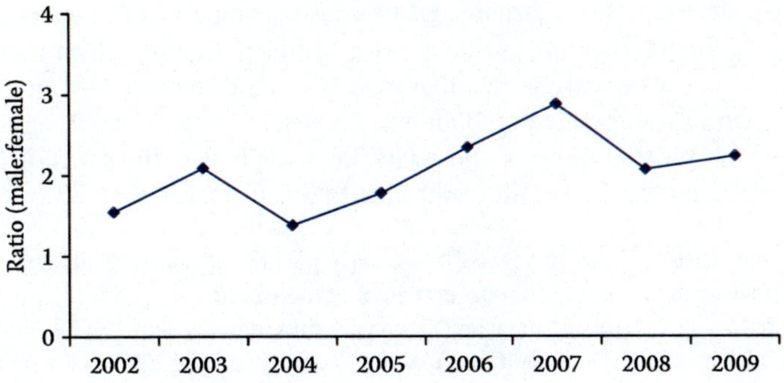

Figure 25.4 Sex ratio of sharks sighted by year.

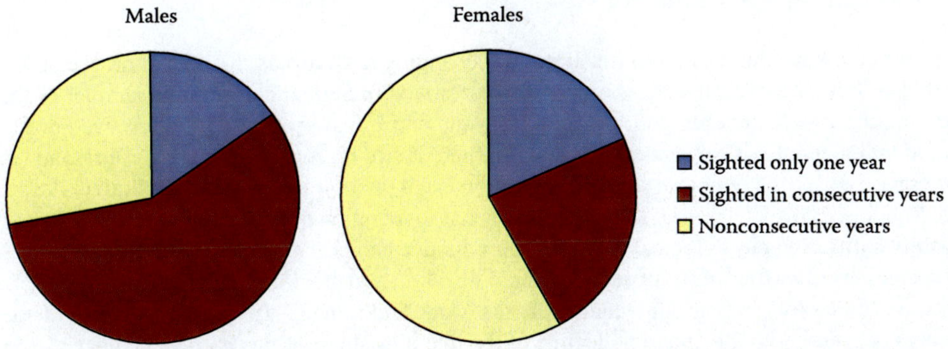

Figure 25.5 Repeat sighting pattern for all sharks in the catalog; males were more frequently sighted in consecutive years than females.

the identified sharks recorded in 7 yrs. or more, providing clear evidence of long-term, seasonal philopatry of White Sharks around GI.

The aggregation of White Sharks at GI consists of adults and subadults; small juveniles have not been recorded by our program. The majority of the sharks added to our database since 2005 have no length estimate, but two were known to be subadult males, perhaps an indication of new recruitment to GI. Young-of-the-year, juvenile, and subadult White Sharks are believed to inhabit coastal regions year round (Klimley, 1985; Chapter 16, this book), but it is not known how or when subadults recruit to the adult aggregations found at GI and CC.

Six of the newly catalogued individuals were large enough to be considered sexually mature. One of these mature sharks was a 5.4-m male, possibly the largest male recorded anywhere in the world (Chapter 16, this book); it is unlikely that this shark could have been present at GI for each of the previous 9 yrs. without being sighted, leading to the possibility that some sharks may recruit to GI from elsewhere, as fully mature adults. It is also possible that the absence of this large male in previous years is an artifact of different behavioral responses between individual sharks; however, larger sharks are typically dominant in the presence of a bait source, which would increase the probability of sighting a large individual (Laroche et al., 2007).

Sex-Specific Differences

The overall sex ratio of White Sharks at GI was biased toward males (1.5:1). This may be a genuine male bias but is likely to be an artifact of sampling. Females arrive later in the season when photographic effort declines. Furthermore, effort ceases prior to the seasonal dissolution of the aggregation, so the entire aggregation period is never sampled. For any given year, we would expect a male-biased sex ratio because of the fact that mature females have a 2-yr. migratory pattern that causes a large portion of the female population to be absent from the adult aggregation (Chapters 11 and 16, this book). Over the course of this study, the male bias in the sex ratio has increased over time. There are several factors that could contribute to this observation:

1. Beginning around 2005, effort increased in the early part of the season and decreased in the latter part, which is when the females are most prevalent at the island.
2. Mature females are not present at the island each year making it easier to "miss" individual females.
3. Males travel back and forth from the middle of the Pacific Ocean (SOFA) to GI, where there are few interactions with fisheries, but it is assumed that mature females travel to areas along the coast of Southern California and Baja California, Mexico to give birth (Chapter 16, this book). Females are then more likely to have an incidental fatal encounter with fishing activities that are concentrated along the coast of North America.

Analysis of sex ratio by month illustrates the changing sex ratio as the season progresses. Males arrived late July or early August, whereas females arrived in September with the number of females increasing through September and October. Over the entire history of the program, we have a 0.9:1 sex ratio for the month of November, a time when there is still reasonable sampling effort, and females are present. Spatial and temporal sexual segregation is not uncommon in sharks (Klimley, 1987; Sims, 2005; Robbins, 2007; Chapter 11, this book). Seasonal segregation of White Sharks may be due to prey availability, differing physiological needs, or reproductive habits. The diminishing rate of male White Shark sightings after the return of females could also be an artifact of intraspecific behavior. Larger sharks are often observed chasing smaller sharks away, and females are typically larger than males, possibly contributing to the apparent decline of the rate of male sightings until December.

Differences in visitation patterns between the sexes were noted, with the females more likely to be sighted in nonconsecutive years. This pattern has been observed at the White Shark aggregation in CC (Anderson and Pyle, 2003), as well as in other species of sharks. Pratt and Carrier (2001) found that in Nurse Sharks (*Ginglymostoma cirratum*), the adult males visit every year whereas adult females visit the study area to mate in alternate years. This corresponds with patterns observed in eastern Pacific White Sharks using satellite telemetry data (Chapter 11, this book), and in light of newly emerging reproductive data (Chapters 16 and 30, this book), the eastern Pacific adult White Shark aggregation sites may be mating sites.

Although prior studies have suggested that mature females have a strong every-other-year pattern of visitation to the adult aggregation sites (Anderson and Pyle, 2003; Domeier and Nasby-Lucas, 2007), the longer time series presented in this study revealed that established individual female visitation patterns are malleable. While females were more likely to exhibit a nonconsecutive year visitation pattern, only a small percentage exhibited a consistent every-other-year pattern. Two sexually mature females appeared in consecutive years after a history of appearing every-other-year, whereas another mature female was sighted every third year. This variation in mature female visitation patterns could indicate that mating is not always successful or that females may sometimes skip a reproductive cycle. There is much to learn regarding reproduction in White Sharks. Continued monitoring via photo-ID combined with multiyear tracking via electronic tagging will help to answer these important life history questions.

CONCLUSIONS

The GI White Shark photo-ID project is an important monitoring tool for the GI White Shark aggregation. This database has allowed us to produce the first ever index of abundance for the GI

White Shark aggregation (Chapter 16, this book). Although we do not know the absolute number of adult and subadult White Sharks that visit GI, clearly it is a relatively low number of individuals, which indicates the importance of managing and protecting this population. It is currently prohibited to kill White Sharks in U.S. and Mexican waters, but there is evidence of incidental catch on juveniles along both coasts that will have an effect on the adult population (Galván-Magaña et al., 2010; Chapters 14 and 15, this book). This study is important for monitoring the size and stability of the GI White Shark population and is critical for long-term management. Learning more about the multiyear migration patterns and where female White Sharks go to give birth is important in protecting this vulnerable shark, which travels great distances and is known to cross international boundaries.

ACKNOWLEDGMENTS

We would like to thank T. Pfleger, J. Harper, M. Graf, P. Douglas, L. Groth, and S. Drogin for their assistance in taking and collecting photographs. We also thank all cage divers who allowed us to view their photos. This project was funded through the George T. Pfleger Foundation, the Offield Family Foundation, and the Guadalupe Island Conservation Fund. We thank F. Galván-Magaña, M. Hoyos-Padilla, and O. Sosa-Nishizaki for Mexican permitting support. Research was conducted in accordance with permits through the Comisión Nacional de Áreas Naturales Protegidas.

REFERENCES

Anderson, S.D., and K. J. Goldman. 1996. Photographic evidence of white shark movements in California water. *California Fish and Game* 82: 182–186.

Anderson, S. D., and P. Pyle. 2003. A temporal, sex-specific occurrence pattern among white sharks (*Carcharadon carcharias*) at the South Farallon Islands, California. *California Fish and Game* 89: 96–101.

Baum, J. K., R. A. Myers, D. G. Kehler, B. Worm, S. J. Harley, and P. A. Doherty. 2003. Collapse and conservation of shark populations in the Northwest Atlantic. *Science* 299: 389–392.

Bonfil, R., M. Meyer, M. C. Scholl, R. Johnson, S. O'Brien, H. Oosthuizen, et al. 2005. Transoceanic migration, spatial dynamics, and population linkages of white sharks. *Science* 310: 100–103.

Boustany, A. M., S. F. Davis, P. Pyle, S. D. Anderson, B. J. Le Boeuf, and B. A. Block. 2002. Expanded niche for white sharks. *Nature* 415: 35–36.

Cailliet, G. M., L. J. Natanson, B. A. Weldon, and D. A. Ebert. 1985. Preliminary studies on the age and growth of the white shark *Carcharodon carcharias*, using vertebral bands. *Memoirs of the Southern California Academy of Sciences* 9: 49–60.

Chapple, T. K., S. J. Jorgensen, S. D. Anderson, P. E. Kanive, A. P. Klimley, L. W. Botsford, et al. 2011. A first estimate of white shark, *Carcharodon carcharias*, abundance off Central California. *Biology Letters* 7(2). doi: 10.1098/rsbl.2011.0124.

Compagno, L. J. V. 2001. Sharks of the world. An annotated and illustrated catalogue of the shark species known to date: Vol 2. Bullhead, mackerel and carpet sharks (Heterodontiformes, Lamniformes and Orectolobiformes). *FAO Species Catalogue for Fisheries Purposes, No. 1,* Vol. 2. Rome, Italy: FAO.

Domeier, M. L., and N. Nasby-Lucas. 2007. Annual re-sightings of photographically identified white sharks (*Carcharodon carcharias*) at an eastern Pacific aggregation site (Guadalupe Island, Mexico). *Marine Biology* 150: 970–984.

Domeier, M. L., and N. Nasby-Lucas. 2008. Migration patterns of white sharks *Carcharodon carcharias* tagged at Guadalupe Island, Mexico, and identification of an eastern Pacific shared offshore foraging area. *Marine Ecology Progress Series* 370: 221–237.

Dulvy, N. K., J. K. Baum, S. Clarke, L. J. V. Compagno, E. Cortés, A. Domingo, et al. 2008. You can swim but you can't hide: The global status and conservation of oceanic pelagic sharks and rays. *Aquatic Conservation: Marine and Freshwater Ecosystems* 18: 459–482.

Francis, M. P. 1996. Observations on a pregnant white shark with a review of reproductive biology. In *Great white sharks: The biology of Carcharodon carcharias*, edited by A. P. Klimley and D. G. Ainley, pp. 157–172. San Diego, CA: Academic Press.

Galván-Magaña, F., E. M. Hoyos-Padilla, C. J. Navarro-Serment, and F. Márquez-Farías. 2010. Records of white shark, *Carcharodon carcharias*, in the Gulf of California, Mexico. *Marine Biodiversity Records* 3: e111. doi:10.1017/S1755267210000977.

Hilton-Taylor, C. 2000. *2000 IUCN Red List of threatened species*. Gland, Switzerland: IUCN.

Klimley, A. P. 1985. The aerial distribution and autoecology of the white shark (*Carcharodon carcharias*), off the west coast of North America. *Memoirs of the Southern California Academy of Sciences* 9: 15–40.

Klimley, A. 1987. The determinants of sexual segregation in the scalloped hammerhead shark, *Sphyrna lewini*. *Environmental Biology of Fishes* 18: 27–40.

Klimley, A. P., and S. D. Anderson. 1996. Residency patterns of white sharks at the South Farallon Islands, California. In *Great white sharks: The biology of Carcharodon carcharias*, edited by A. P. Klimley and D. G. Ainley, pp. 365–373. San Diego, CA: Academic Press.

Laroche, R. K., A. Kock, L. M. Dill, and W. H. Oosthuizen. 2007. Effects of provisioning ecotourism activity on the behaviour of white sharks *Carcharodon carcharias*. *Marine Ecology Progress Series* 338: 199–209.

Mollet, H. F., G. Cliff, H. L. Pratt, Jr., and J. D. Stevens. 2000. Reproductive biology of the female shortfin mako, *Isurus oxyrinchus* Raninesque, 1810, with comments on the embryonic development of lamnoids. *Fisheries Bulletin* 98: 299–318.

Myers, R., and B. Worm. 2003. Rapid worldwide depletion of predatory fish communities. *Nature* 423: 280–283.

Pratt, H. L., Jr. 1996. Reproduction in the male white shark. In *Great white sharks: The biology of Carcharodon carcharias*, edited by A. P. Klimley and D. G. Ainley, pp. 131–138. San Diego, CA: Academic Press.

Pratt, H. L., Jr., and J. C. Carrier. 2001. A review of elasmobranch reproductive behavior with a case study on the nurse shark, *Ginglymostoma cirratum*. *Environmental Biology of Fish*es 60: 157–188.

Robbins, R. 2007. Environmental variables affecting the sexual segregation of great white sharks *Carcharodon carcharias* at the Neptune Islands South Australia. *Journal of Fish Biology* 70: 1350–1364.

Shivji, M. S., D. D. Chapman, E. K. Pikitch, and P. W. Raymond. 2005. Genetic profiling reveals illegal international trade in fins of the great white shark, *Carcharodon carcharias*. *Conservation Genetics* 6: 1035–1039.

Sims, D. W. 2005. Differences in habitat selection and reproductive strategies of male and female sharks. In *Sexual Segregation in Vertebrates: Ecology of the Two Sexes*, edited by K. E. Ruckstuhl and P. Neuhaus, pp. 127–147. Cambridge, UK: Cambridge University Press.

Smith, S. E., D. W. Au, and C. Show. 1998. Intrinsic rebound potentials of 26 species of Pacific sharks. *Marine and Freshwater Research* 49: 663–678.

Strong, W. R., Jr., R. C. Murphy, B. D Bruce, and D. R. Nelson. 1992. Movements and associated observations of bait-attracted white sharks, *Carcharodon carcharias*: A preliminary report. *Australian Journal of Marine and Freshwater Research* 43:13–20.

Weng, K. C., A. M. Boustany, P. Pyle, S. D. Anderson, A. Brown, and B. A. Block. 2007. Migration and habitat of white sharks (*Carcharodon carcharias*) in the eastern Pacific Ocean. *Marine Biology* 152: 877–894.

CHAPTER 26

Problems with Photo Identification as a Method of Estimating Abundance of White Sharks, *Carcharodon carcharias*

An Example from Guadalupe Island, Mexico

Oscar Sosa-Nishizaki*
Centro de Investigación Científica y de Educación Superior de Ensenada

Enrique Morales-Bojórquez
Centro de Investigaciones Biológicas del Noroeste

Nicole Nasby-Lucas
Marine Conservation Science Institute

Erick C. Oñate-González
Centro de Investigación Científica y de Educación Superior de Ensenada

Michael L. Domeier
Marine Conservation Science Institute

ABSTRACT

White Shark aggregation sites offer some of the best opportunities to gather data on species biology and ecology. Mark-recapture methods are useful for estimating abundance, as long as their underlying assumptions are met. In this chapter, we employed the encounter histories of 113 (67 males and 46 females) photo-identified White Sharks as the base information for a mark-recapture analysis for the Guadalupe Island aggregation site. We estimated apparent survival, recapture probabilities, permanent entry into the population, and the superpopulation size by using the Cormack-Jolly-Seber (CJS) model and a generalization of the Jolly-Seber (JS) model for the open-population mark-recapture experiments. We fitted several model structures for each of the CJS and JS types and selected the best-fit model structure using the corrected Akaike's Information Criterion. Our results indicated that although apparent survival and permanent entry into the population varied with time, the recapture probability varied with sex ($p_{female} = 0.51 \pm 0.044$; $p_{male} = 0.79 \pm 0.029$). The superpopulation size by sex were 51 females (SE = 2.7) and 69 males (SE = 1.6). However, the assumption of equal recapture probability was violated by our encounter histories of males, producing an underestimation of the population size by the model. We discuss all the possible explanations for the violation of the assumption, based on this and another recent study. We recommend the use of this type of analysis as an index of abundance only, which can provide a long-term population trend but not an absolute abundance until methods can be improved.

INTRODUCTION

Elasmobranch aggregation sites offer some of the best opportunities to gather data on species whose population sizes or distributions are otherwise unclear or unknown (Marshall et al., 2011). For the White Shark (*Carcharodon carcharias*), such aggregations have been identified off of Central California (CC), California (Klimley et al., 1992; Klimley and Anderson, 1996); the southern coast of South Africa (Cliff et al., 1996; Ferreira and Ferreira, 1996); Spencer Gulf, South Australia (Strong et al., 1992, 1996); and Guadalupe Island (GI), Mexico (Domeier and Nasby-Lucas, 2007; Chapter 25, this book).

Satellite tagging studies in the northeastern Pacific have shown annual offshore migrations of subadult and adult White Sharks tagged at the aggregation sites in CC and GI (Boustany et al., 2002; Weng et al., 2007; Domeier and Nasby-Lucas, 2008; Jorgensen et al., 2010). The return of tagged White Sharks to their original tagging location, after an offshore migration that lasted many months, documented seasonal philopatry in White Sharks (Weng et al., 2007; Jorgensen et al., 2010; Chapters 11 and 16, this book). This site fidelity has allowed for long-term monitoring of individual White Shark visitation patterns, via photo identification (photo-ID) methods, at both the CC (Anderson and Goldman, 1996; Anderson et al., 1996, 2011) and GI (Domeier and Nasby-Lucas, 2007; Chapter 25, this book) aggregation sites.

Photo ID, which uses natural marks to recognize and catalogue individuals, has been widely applied as a population biology tool, particularly for the study of cetaceans (Stevick et al., 2001). Photo-ID studies of elasmobranchs are relatively recent (Marshall et al., 2011) and have examined a suite of species using natural pigment patterns, dorsal fin shape, and scars as natural tags (e.g., Anderson and Goldman, 1996; Sims et al., 2000, Castro and Rosa, 2005; Porcher, 2005; Domeier and Nasby-Lucas, 2007; Dudgeon et al., 2008; Holmberg et al., 2009; Anderson et al., 2011; Marshall et al., 2011; Chapter 25, this book). Photo-ID methods have allowed the description of individual growth rates, reproductive cycles, sex difference in aggregations, demography analysis, and estimations of population structure and size.

The ability to assess the number of individuals in a given population is one of the most important process to effectively managing marine fish species, and mark-recapture techniques are an established means of conducting such studies (Castro and Rosa, 2005). After reviewing several approaches for conducting estimates of White Shark populations, Cailliet (1996) suggested that mark-recapture methods may be the most viable. Conventional tags were used to estimate the White Shark populations off the coast of South Africa (Cliff et al., 1996) and off Spencer Gulf, South Australia (Strong et al., 1996). More recently a photo-ID study used unique marks on the trailing edge of the dorsal fin to estimate the population of White Sharks off CC (Chapple et al., 2011).

Mark-recapture methods are powerful for estimating abundance, as long as their underlying assumptions are met (Letting and Armstrong, 2003). These methods consider different assumptions depending on whether the population is closed or open. Open populations are those subject to births, death, immigration, and emigration during the time of the study and require four main assumptions to be met: individuals do not lose their mark and are not overlooked; the duration of each capture occasion is instantaneous in relation to the interval between occasions; all animals have the same survival probability; and all animals (tagged or not) have the same capture probability (Krebs, 1990). The heterogeneity of survival and capture probabilities is likely to exist, in some point, in all mark-recapture studies (Pollock et al., 1990). However, if assumptions are violated in some degree, you might still use the mark-recapture methods and recognize that the resulting estimates are biased, but you should be careful in using them only as indices of population size (Krebs, 1990).

In this chapter, we employed encounter histories of photo-identified White Sharks as mark-recapture information (Chapter 25, this book) to estimate their apparent survival and resighting (hereafter termed recapture for the purpose of this study) probabilities as well as annual and superpopulation sizes at GI. We applied the Cormack-Jolly-Seber (CJS) model and a generalization of the Jolly-Seber (JS) model for the open-population mark-recapture experiments (Schwarz and Arnason, 1996). Our results are discussed based on the methodology assumptions. In terms of biodiversity, GI is one of the most important Mexican islands in the Pacific Ocean, earning Biosphere Reserve status in 2003 (Ezcurra et al., 2005). White Sharks are protected under Mexican law (Chapter 15, this book), and the conservation of this species is one of the main objectives of the Biosphere Reserve management plan (Nadia Olivares, personal communication).

MATERIALS AND METHODS

The encounter histories for the 113 White Sharks described by Nasby-Lucas and Domeier (Chapter 25, this book) were analyzed by a mark-recapture approach to obtain population estimates of the White Sharks that aggregate at GI. Using pigment patterns and external reproduction organs to identify individual sharks, they photographically identified 67 males (59%) and 46 females (41%) between 2001 and 2009. Each White Shark was considered "tagged" at the time of first identification, and subsequent resightings were considered "recaptures." Because of the low number of tagged sharks during 2001, sharks of years 2001 and 2002 were pooled for all analyses (see Table 25.1 of Chapter 25, this book for encounter histories).

Closed-Population Test

The assumptions for a closed population (i.e., emigration, immigration, mortality, or recruitment do not occur during the study period) were tested using the Stanley and Burnham (1999) and Otis et al. (1978) developments implemented in the CloseTest program (Stanley and Richards, 2005). Test results indicated that closed-population assumptions were violated by the encounter histories of the White Sharks that aggregate in Isla Guadalupe ($\chi^2 = 145.98$, $p < 0.0001$; z value = -8.21, $p < 0.0001$), so open-population models were implemented.

Open-Population Models

CJS models were used to estimate apparent survival (φ) between annual shark-sighting seasons, and recapture probabilities (p) for each shark during a sighting season. The JS model provided estimates by sex for apparent survival (φ), initial capture (tag), and recapture probability (p), permanent entry into the population (β) and initial population size (N). Therefore, allowing us to estimate annual population size and superpopulation size, defined as the number of individuals visiting the study area each sighting season and the total number of individuals visiting the study area over the duration of the study, respectively (Marshall et al., 2011). Data analysis and model selection were carried out using the CJS and POPAN (for JS) routines in program MARK version 6.1 (White and

Burnham, 1999). To select the best-fit model, the corrected Akaike's Information Criterion (AICc; a second order AIC necessary for small samples) was used (Burnham and Anderson, 2002).

Test of Assumptions

The use of open-population models to estimate demographic parameters is based principally on the following main assumptions (Krebs, 1990):

1. Individuals do not lose their marks, and marks are not overlooked at capture.
2. Sampling time is negligible (instantaneous sampling) in relation to intervals between samples.
3. Every individual has the same probability (p_t) of being caught in the tth sample, regardless of whether it is marked or unmarked (JS model).
4. Every marked individual has the same probability of surviving (φ_t) from the tth to the (t + 1)th sample.

For the CJS model, the third assumption refers to only marked individuals (Marshall et al., 2011). The achievement of the first and second assumption is discussed in the Discussion section. To test the third and fourth assumption for the CJS model, Tests 2 and 3 were applied to the encounter histories, using the program RELESE, implemented through MARK (White and Burnham, 1999). Additionally, to assess overdispersion (lack of independence among individuals) in the data for the CJS model, the variance inflation factor c-hat was calculated by dividing the combined Test 2 and Test 3 χ^2 statistic by the degrees of freedom (*df*). When c-hat > 1.0, there is indication of overdispersion, and CJS and JS models have to be adjusted using this estimated value (Burnham et al., 1987).

RESULTS

Test of Assumptions

Application of Test 2 to the overall encounter histories showed a nonsignificant heterogeneity in capture probabilities ($\chi^2 = 17.61$, $p = 0.062$) considering $\alpha = 0.05$. When analyzing by sex, the female group gave nonsignificant heterogeneity ($\chi^2 = 3.59$, $p = 0.601$), and the males showed significant heterogeneity ($\chi^2 = 14.02$, p = 0.0155); however, the test also indicated that the data were insufficient, so open models can be applied, but this result must be considered during the data interpretation. Test 3, in its two components (Test 3 Sm and Test 3 Sr), showed nonsignificant heterogeneity in survival probabilities in the overall data and by sex, yielding p values > 0.22. These results support the underlying assumption for the open model. The combined goodness-of-fit testing (Test 2 + Test 3) for the full-time dependent CJS model showed a good fit to the data ($\chi^2 = 28.13$, $df = 30$, $p = 0.564$) indicating no overdispersion (c-hat = 0.938), so no further adjustment to the CJS and JS models were need.

CJS Models

CJS models were constructed considering the apparent survival (φ) and the recapture (p) probabilities as constant (.) or varied with sampling interval (t) or by sex (*sex*) or their combination (t**sex*). From 16 possible model variants, seven were selected based on their AICc weight and model likelihood values (Table 26.1). Burnham and Anderson (2002) state that values of ΔAICc < 2 indicate approximately equal support of candidate models and parameters of the models should be averaged. The difference between our model with the minimum AICc and the following one was 6.04 (Table 26.1), so there was no need for averaging. We found the best-fit model of $\varphi(t)$ $p(sex)$ with an AICc weight of 0.915 compared with other model variants. These results indicate that apparent survival varied between years independent of the sex and without a clear tendency through time ($\varphi_1 = 0.97$

Table 26.1 Model Selection for the Cormack-Jolly-Seber Models of Survival and Capture Probability

Model	AICc	ΔAICc	AICc Weights	Model Likelihood	Number of Parameters
φ(*t*) *p*(*sex*)	643.17	0	0.915	1	8
φ(.) *p*(*sex*)	649.21	6.04	0.045	0.049	3φ(*t***sex*)
	650.82	7.65	0.020	0.022	12
φ(*sex*) *p*(*sex*)	651.25	8.08	0.016	0.018	4
φ(t) *p*(*t***sex*)	654.82	11.64	0.003	0.003	19
φ(.) *p*(*t***sex*)	657.11	13.94	0.001	0.001	15
φ(*sex*) *p*(*t***sex*)	659.06	15.89	0.000	0.000	16

The table shows model selection for the CJS models of survival (φ), capture probability (*p*), considered as constant (.) or varied with time (*t*) or with sex (*sex*) or a combination. AICc, corrected Akaike information criterion; ΔAICc, difference in the AICc of a model from the minimum AICc model; AICc weight, Akaiki weight.

± 0.028 SE, $\varphi_2 = 0.98 \pm 0.300$, $\varphi_3 = 0.99 \pm 0.000$, $\varphi_4 = 0.80 \pm 0.052$, $\varphi_5 = 0.97 \pm 0.042$, $\varphi_6 = 0.88 \pm 0.056$, and $\varphi_7 = 0.98 \pm 0.082$). Recapture probabilities varied depending on sex, where females recapture probability ($p_{female} = 0.51 \pm 0.044$) was smaller than males ($p_{male} = 0.79 \pm 0.029$). The second most parsimonious model φ(.) *p*(*sex*), with only 0.04 of AICc weight (Table 26.1), supported a constant survival over time ($\varphi = 0.93 \pm 0.014$ SE) with recapture probabilities that varied depending on sex ($p_{female} = 0.50 \pm 0.044$; $p_{male} = 0.79 \pm 0.029$) and were equal with the most parsimonious model considering the SE overlaps.

JS Models

In total, seven JS models were constructed considering apparent survival (φ) and recapture (*p*) probabilities and the permanent entry into the population (β) as varied with sampling interval (*t*) or by sex (*sex*) or their combination (*t*sex*). The models were fitted using the logit link function for φ and *p*, the log link function for *N*, and the multinomial logit link function for β to constrain the set of parameters to $\beta < 1$ (White and Burnham, 1999). The best-fit model was φ(*t*) *p*(*sex*) β(*t*)*N* with an AICc weight of 0.932 compared to other model variants (Table 26.2). Estimations of the apparent survival and recapture probabilities from this model were similar to those of the CJS model considering overlapping SE margins (Table 26.3). There were pulses of permanent entry into the populations (β, recruitment + immigration) (Table 26.3), likely an artifact of sampling effort. The examination of the parameter estimates from the first, less parsimonious model φ(*t*) *p*(*sex*) β(*t*sex*)*N* with a AICc weight = 0.029 (Table 26.2) allowed an understanding of the variations of β deeper, as values varied with time and sex in the model. The tendency of the estimated values showed an every-other-year cycle and was mainly related with the permanent entry of females ($\beta_{2005\text{–}2006\ female} = 0.49 \times 10^{-7} \pm 0.000$ contrasted with $\beta_{2006\text{–}2007\ female} = 0.10 \pm 0.052$).

White Shark Population Estimates

White Shark annual population estimates by sex using the best-fitted model showed a general increasing tendency with values varying between 21 and 36 females and 28 to 49 males, with a decline in 2006, associated with the decline of β between 2005 and 2006 (Table 26.4). The numbers of White Sharks visiting the study area over the duration of the study (superpopulation) (Schwarz and Arnason, 1996) by sex were 51 females (SE = 2.7) and 69 males (SE = 1.6) (Table 26.4).

Table 26.2 Model Selection for the Jolly-Seber Models of Survival, Capture Probability, and Permanent Entry into the Population

Model	AICc	ΔAICc	AICc Weights	Model Likelihood	Number of Parameters
φ(*t*) *p*(*sex*) β(*t*)}	720.00	0	0.932	1	17
φ(*t*) *p*(*sex*) β(*t*sex*)	726.97	6.97	0.029	0.031	22
φ(*t*sex*) *p*(*sex*) β(*t*)	727.49	7.49	0.022	0.024	21
φ(*sex*), *p*(*sex*) β(*t*)	728.85	8.85	0.011	0.012	13
φ(*t*) *p*(*t*sex*) β(*t*)	732.18	12.17	0.002	0.002	29
φ(*t*) *p*(*t*sex*) β(*t*sex*)	732.80	12.80	0.002	0.002	31
φ(*t*sex*) *p*(*t*sex*) β(*t*)	743.29	23.29	0.000	0.000	34

The table shows model selection for the JS models of survival (φ), capture probability (*p*), and permanent entry into the population (β) considered as varied with time (*t*) or with sex (*sex*) or a combination. AICc, corrected Akaike information criterion; ΔAICc, difference in the AICc of a model from the minimum AICc model; AICc weight, Akaiki weight.

Table 26.3 Parameter Estimates from the Best-Fitted Jolly-Seber Model φ(*t*) *p*(*sex*) β(*t*)

Parameter	Year/Sex	Estimate	SE
ϕ	01&02-03	0.97	0.026
	03-04	0.98	0.029
	04-05	0.99	2.8×10^{-5}
	05-06	0.79	0.052
	06-07	0.97	0.042
	07-08	0.88	0.055
	08-09	0.98	0.082
p	Female	0.53	0.042
	Male	0.80	0.029
β	01&02-03	0.15	0.058
	03-04	0.12	0.050
	04-05	0.06	0.039
	05-06	0.01	0.018
	06-07	0.16	0.043
	07-08	0.03	0.032
	08-09	0.06	0.030

Table 26.4 Annual Population Size Estimations and Superpopulation Size Estimates by Sex

Year	Female (SE)	Male (SE)
2001 and 2002	21 (3.2)	28 (4.2)
2003	27 (3.1)	37 (4.0)
2004	33 (2.9)	44 (3.7)
2005	36 (2.7)	48 (3.3)
2006	29 (2.8)	39 (3.5)
2007	36 (3.1)	49 (3.7)
2008	33 (3.0)	45 (3.7)
2009	36 (3.8)	49 (4.7)
Superpopulation	51 (2.7)	69 (1.6)

DISCUSSION

Using 9 yrs. of encounter histories of photographically identified White Sharks allowed us to estimate a superpopulation size of 120 subadult and adult sharks with $N = 51$ females (95% confidence interval = 47, 59), and $N = 69$ males (95% confidence interval = 67, 75) that aggregate at GI. Chapple et al. (2011) conducted a population estimate for White Sharks that aggregate off CC, and using a Bayesian closed-population model they arrived at a number of 219 (range, 130–275). Together these population estimates suggest a total of 349 subadult and adult White Sharks visiting these two aggregation sites. Because this likely constitutes the majority of White Sharks in the northeastern Pacific (within these age classes), the number is alarming low. However, before we accept this as an absolute population size, we have to evaluate how well the assumptions of the models were met.

Closed versus Open Population

A closed population is one that does not change in size during the study period; that is, the effect of births, deaths, and movements are negligible. Thus the populations are typically closed over only a short period of time. An open population is the more usual case, a population that changes in size and composition from birth, death, and movements (Krebs, 1990). Our estimate assumed an open population, while that of Chapple et al. (2011) assumed a closed population over a 3-yr. sampling period. Our long term photo-ID monitoring at Guadalupe Island (Domeier and Nasby-Lucas, 2007; Chapter 25, this book), as well as our satellite-linked radio transmitting tags, have shown individuals to disappear (Chapter 16, this book); these sharks either died or migrated elsewhere. We have also seen the sudden appearance of new sharks that are so big it is unlikely they have been present for many years and never before observed (Chapter 16, this book). Furthermore, Chapter 13, this book documented the migration of a White Shark from CC to GI; this shark did not return to CC. Clearly these populations are not closed, and our data did not pass the closed-model test, but how much of a difference will this ultimately make in the resulting population estimate?

Initial testing of our data under the assumption that the population is closed (Petersen maximum likelihood version) resulted in a population estimate of one hundred thirty-two, only 9% greater than the open model method presented here. The data from Chapple et al. (2011) passed the closed-population test, but this could be a result of their use of only 3 yrs. of data, compared with our 9 yrs. of data, or perhaps the population is so marginally open that the data can be modeled either way. Modeling an open population as closed would force the population size to be constant during the time of the study, restricting the analysis year by year without the possibility of estimating the superpopulation size.

Assumptions

The use of individual identification for tag-recapture estimates of abundance is based on four assumptions (Krebs, 1990). The first assumption is that no animal will lose its tag during the study or be overlooked during efforts to recapture tagged individuals. The violation of this assumption is likely to result in the underestimation of recapture probabilities and an inability to calculate survival/mortality and will result in an overestimation of abundance estimates (Pollock et al., 1990; Stevick et al., 2001; Marshall et al., 2011). Both photo-ID methods used on northeastern Pacific White Sharks have proven to be very effective over long time periods (Domeier and Nasby-Lucas, 2007; Anderson et al., 2011; Chapter 25, this book), validating this first assumption.

The second assumption requires instantaneous sampling that allows sufficient time between sampling intervals for mixing of the tagged and untagged sharks; this ensures the homogeneity of recapture probability (Pollock et al., 1990; Strong et al., 1996). Photographically identified White Sharks were recaptured during the annual aggregation season at GI (August to January) with most recaptures occurring between August and November (Figure 26.1). The 8-month period between sampling was sufficient for population mixing and allowed robust survival estimates (see below). Mixing was further insured by the fact that the sharks leave the GI aggregation for an extended migration to offshore environments (Chapters 11, 12, and 16, this book) before sampling begins again upon their return. This is very similar to the situation off CC, and therefore both our study and that of Chapple et al. (2011) likely meet this assumption. This assumption is important for establishing conditions that allow for an equal probability of sighting all individuals during a sampling period (the third assumption).

The third assumption is that there is equal probability of recapture for all individuals during sampling, regardless of whether it is tagged or not. Although our entire encounter history resulted in a homogenous recapture probability, the result was very close to failing this test ($p = 0.062$). Furthermore, when the sexes were tested independently, the males failed the test of homogeneous recapture probability. To some degree, heterogeneity of capture probabilities is likely to exist in all tag-recapture studies (Pollock et al., 1990) and is generally the result of three factors (Krebs, 1990):

1. Learning by captured animals such that they avoid recapture in the future (called trap-happy or trap-shy individuals).
2. The behavior of individuals in the vicinity of the trap influences the recapture rate.
3. The position of the trap causes unequal opportunity for capture (heterogeneous sampling intensity).

Violating this assumption will result in an underestimate of the true population; because we empirically know our estimate is biased (because the modeled population is smaller than the number of known sharks; Chapter 25, this book), we must closely look at each of these causes of unequal catchability and assess whether any or all of them may be influencing our results.

Trap-happy responses are often associated with the use of baited traps (Pollock et al., 1990); although cage dive operators at GI and researchers off CC use bait to attract White Sharks within photographic range (Chapple et al., 2011; Chapter 25, this book), the fish are not actually captured, and therefore it is unlikely that this would cause individuals to avoid the vicinity of the bait in the future. Furthermore, it is not logical that cage-dive operations would produce a trap-happy response for males but not females. To the contrary, some individuals may become habituated to the bait and therefore be more likely to be attracted to the baiting site than nonhabituated individuals.

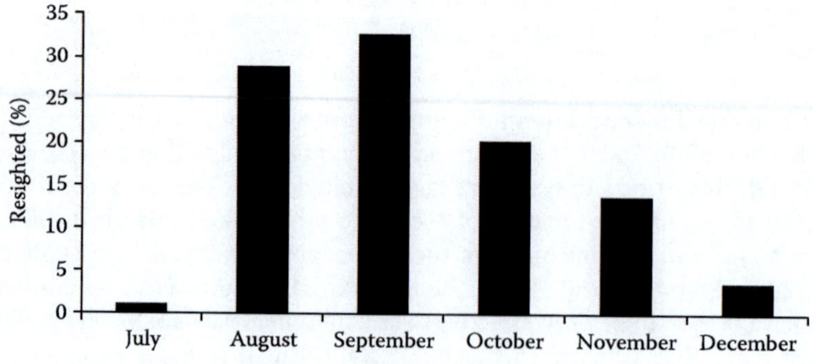

Figure 26.1 Monthly subadult and adult White Shark resighting percentage at Guadalupe Island, Mexico, during the 2001–2009 aggregation seasons.

Individual shark behavior may be influencing our results. Strong et al. (1996) suggested that social dominance influenced the probability of sighting individual White Sharks once they arrived at the bait source, based upon their observation of a size-based feeding hierarchy. They observed smaller White Sharks defer to larger conspecifics during direct interactions; however, subordinates were not completely excluded from the area. The dominance of larger animals over smaller individuals has often been observed at GI (M. Domeier, personal observation), and the male encounter history revealed a tendency for certain individuals to consistently have a higher recapture probability. Sex-specific migratory behaviors (Chapter 11, this book) also demonstrated a violation of equal recapture probability. For example, adult females have a 2-yr. migration cycle that may not bring them to the sampling sites (adult aggregation sites at GI and CC) every year (Anderson and Pyle, 2003; Domeier and Nasby-Lucas, 2007; Chapters 11, 16, and 25, this book). This would imply that up to 50% of the adult female population has a 0% chance of being resighted during every sampling period. In light of this, it is interesting that the female-recapture probability did not show recapture heterogeneity when tested, whereas the males failed the test, the opposite of what we would expect and perhaps an indication of inadequate sampling despite 9 yrs. of effort. However, this difference between the sexes could also be explained by the pattern of cage-dive operations that tend to be more intensive during the early portion of the aggregation season, when females are largely absent (Domeier and Nasby-Lucas, 2007; Chapters 16 and 25, this book). Although sampling occurs during the months females are present, the effort is less, and sampling does not occur through the end of the aggregation season.

The third potential cause of biased recapture probability involves the position of the trap or in this case the boat that is dispersing an attractant to lure White Sharks within photographic range. The boat and bait are at the surface, and sharks must be attracted close to the surface to be photographed from cages moored behind the vessel. Because White Sharks that visit the GI aggregation site tend to remain at the island for the duration of the aggregation season (Chapter 16, this book), and the cage-dive vessels operate at the same site year after year (northeastern corner of GI), this potential cause of heterogeneity of recaptures is not likely a factor at GI. For the Chapple et al. (2011) study, the position of the trap might have influenced the heterogeneity of the recapture probability. Jorgensen et al. (2010) described four sites of high White Shark density within the CC area and indicated that individual White Sharks showed some preference for one or more of these sites, and yet sampling for the CC population estimate only occurred at two of the four known sites of high White Shark abundance (Southeast Farallon Island and Tomales Point) (Chapple et al., 2011). For example, a shark with a preference for Tomales Point would not have the same probability of being reidentified at Southeast Farallon Island when compared with a shark with a known preference for Southeast Farallon Island. Furthermore, a third important CC site, Año Nuevo, was not sampled for the CC population estimate.

The fourth and final assumption used for estimating the population of White Sharks at adult aggregation sites requires equal apparent survival among tagged individuals, for the entire population as well as each sex separately. It is important to emphasize that the JS analysis cannot discriminate between survival and emigration, so the estimated apparent survival includes both permanent and temporary emigration as well as low-level mortality (Schwarz and Arnason, 1996). Our test result showed a homogeneous apparent survival between individuals, but the best-fit model indicated that permanent entry to the GI White Shark population varied over time. However, these pulses of recruitment match the pattern of newly identified individuals and are therefore explained by the variation in sampling effort between years (Chapter 25, this book). Given the different migratory patterns of males and females, we might have expected differing survival rates between the sexes. For example, females must migrate to coastal nursery areas where they may be more susceptible to incidental take by longline or gillnet gear (Chapter 16, this book), and females also spend more time in the open ocean where there are no protective measures in place that would restrict the take of White Sharks. However, the best-fit model indicated that there were no significant differences between the sexes.

Population Estimate versus Index of Abundance

The White Shark is a highly migratory, widely distributed apex predator with naturally low abundance, low natural mortality, and low reproductive potential (Bruce, 2008). These life-history traits, along with the fact that they form seasonally predictable adult aggregations, make them vulnerable to any anthropogenic mortality. Management and monitoring of aggregation sites is essential for the long-term survival of these populations, and modeling the populations should be an integral monitoring tool. However, both the analysis presented here and the analysis of Chapple et al. (2011) violate the assumption of equal probability of recapture and thus result in underestimates of the actual population; in our case, this fact has been empirically established by the raw photo-ID data from GI that provide a number of known individuals that is 8.5% higher than the modeled population (Chapter 25, this book). In our case we statistically failed this assumption while the sampling design of Chapple et al. (2011) creates a violation of this assumption. Chapple et al. (2011) used eight different models with similar results while stating that the presences of heterogeneity in recapture was limited; however, by excluding Año Nuevo and not considering the differing recapture probabilities relative to their study sites, their analysis may not be valid for the entire region. Furthermore, neither of the attempts to estimate adult and subadult populations attempted to develop a life history table that would provide population estimates for each age class of the overall White Shark population. Not all subadults visit adult aggregation sites, and virtually no juveniles visit these sites (Chapters 14, 15, and 16, this book), so the vast majority of the overall White Shark population is not included in either of these models.

Although efforts to estimate the number of individuals that visit adult aggregation sites are admittedly biased, they are valuable in establishing a baseline index of abundance for each of the sites. The White Shark is a protected species, highly mobile, and random sampling is very difficult to achieve (Cailliet, 1996). Until other methods that overcome these problems are developed, mark-recapture methods might be the only sources of estimation. However, considering the White Shark's multiyear migration patterns, and having access to long time series, such as the photo-ID data available from adult White Shark aggregations, the use of closed models is not justified. Furthermore, tagging studies have demonstrated that the northeastern Pacific White Shark adult aggregation sites are open populations (Chapter 13, this book); therefore we suggest standardizing population estimates at both sites by using a JS model that consider site and shark size as covariate (Pollock et al., 1990; Holmberg et al., 2009). Because the assumption of equal probability of recapture has been violated in several ways (resulting in underestimates of the true population), future modeling and/or sampling methods must be sought to increase the accuracy these population estimates. Although we identified bias in the application of photo-ID data to current efforts to model absolute population levels, our results, and that of Chapple et al. (2011), indicate that adult White Shark populations in the northeastern Pacific are small, highlighting the need for continued monitoring and precautionary management.

ACKNOWLEDGMENTS

We gratefully acknowledge the financial support of the Comisión Nacional de Áreas Naturales Protegidas and Centro de Investigación Científica y de Educación Superior de Ensenada.

REFERENCES

Anderson, C. D., and K. J. Goldman. 1996. Photographic evidence of white shark movements in California waters. *California Fish and Games* 84: 182–186.

Anderson, S. D., R. P. Henderson, P. Pyle, D. G. Ainley, and A. P. Klimley. 1996. White shark reactions to unbaited decoys. In *Great white sharks: The biology of Carcharodon carcharias*, edited by A. P. Klimley and D. G. Ainley, pp. 223–228. San Diego, CA: Academic Press.

Anderson, S. D., T. K. Chapple, S. J. Jorgensen, A. P. Klimley, and B. A. Block. 2011. Long-term individual identification and site fidelity of white sharks, *Carcharodon carcharias*, off California using dorsal fins. *Marine Biology*. doi:10.1007/s00227-011-1643-5.

Anderson, S. D., and P. Pyle. 2003. A temporal, sex-specific occurrence pattern among white sharks (*Carcharadon carcharias*) at the South Farallon Islands, California. *California Fish and Game* 89: 96–101.

Boustany, A. M., S. F. Davis, P. Pyle, S. D. Anderson, B. J. Le Boeuf, and B. A. Block. 2002. Satellite tagging: Expanded niche for white sharks. *Nature* 415: 35–36.

Bruce, B. D. 2008. The biology and ecology of the white shark, *Carcharodon carcharias*. In *Sharks of the open ocean: Biology, fisheries and conservation*, edited by M. D. Camhi, E. K. Pikitch, and E. A. Babcock, pp. 69–81. Oxford, UK: Blackwell Publishing.

Burnham, K. P., and D. R. Anderson. 2002. *Model selection and multimodel inference: A practical information-theoretic approach*. New York, NY: Springer-Verlag.

Burnham, K. P., D. R. Anderson, G. C. White, C. Brownie, and K. H. Pollock. 1987. *Design and analysis methods for fish survival experiments based on release-recapture, Monograph* 5. Bethesda, MD: American Fisheries Society.

Cailliet, G. M. 1996. An evaluation of methodologies to study the population biology of white sharks. In *Great white sharks: The biology of Carcharodon carcharias*, edited by A. P. Klimley and D. G. Ainley, pp. 415–416. San Diego, CA: Academic Press.

Castro, A. L. F., and R. S. Rosa. 2005. Use of natural marks on population estimates of the nurse shark, *Ginglymostoma cirratum*, at Atol da Rocas Biological Reserve, Brazil. *Environmental Biology of Fishes* 72: 213–221.

Chapple, T. K., S. J. Jorgensen, S. D. Anderson, P. E. Kanive, J. P. Klimley, L. W. Botsford, et al. 2011. A first estimate of white shark, *Carcharodon carcharias*, abundance off Central California. *Biology Letters*. doi:10.1098/rsbl.2011.0124.

Cliff, G., R. P. Van der Elst, A. Govender, T. K. Witthumh, and E. M. Bullen. 1996. First estimates of mortality and population size of white sharks on the South African coast. In *Great white sharks: The biology of Carcharodon carcharias*, edited by A. P. Klimley and D. G. Ainley, pp. 393–400. San Diego, CA: Academic Press.

Domeier, M. L., and N. Nasby-Lucas. 2007. Annual re-sightings of photographically identified white sharks (*Carcharodon carcharias*) at an eastern Pacific aggregation site (Guadalupe Island, Mexico). *Marine Biology* 150: 997–984.

Domeier, M. L., and N. Nasby-Lucas. 2008. Migration patterns of the white shark *Carcharodon carcharias* tagged at Guadalupe Island, Mexico, and identification of an eastern Pacific shared offshore foraging area. *Marine Ecology Progress Series* 370: 221–237.

Dudgeon, C. L., M. J. Noad, and J. M. Lanyon. 2008. Abundance and demography of a seasonal aggregation of zebra sharks *Stegostoma fasciatum*. *Marine Ecology Progress Series* 368: 269–281.

Ezcurra, E., A. Aguirre Muñoz, L. M. Salas Flores, C. García, L. Luna, K. Santo del Prado, et al. 2005. Plan de erradicación de especies introducidas en Isla Guadalupe. In *Isla Guadalupe, restauración y conservación*, edited by K. Santos del Prado and E. Peters, pp. 263–277. Distrito Federal: INE-SEMARNAT.

Ferreira, C. A., and T. P. Ferreira. 1996. Populations dynamics of white sharks in South Africa. In *Great white sharks: The biology of Carcharodon carcharias*, edited by A. P. Klimley and D. G. Ainley, pp. 381–391. San Diego, CA: Academic Press.

Holmberg, J., B. Norman, and Z. Arzoumanian. 2009. Estimating population size, structure, and residency time for whale shark *Rhincodon typus* through collaborative photo-identification. *Endangered Species Research* 7: 39–53.

Jorgensen, S. L., C. A. Reeb, T. K. Chapple, S. Anderson, C. Perle, S. R. Van Sommeran, et al. 2010. Philopatry and migration of Pacific white sharks. *Proceedings of the Royal Society B* 277: 679–688.

Klimley, A. P., and S. D. Anderson. 1996. Residency patterns of white sharks at the South Farallon Islands, California. In *Great white sharks: The biology of Carcharodon carcharias*, edited by A. P. Klimley and D. G. Ainley, pp. 365–373. San Diego, CA: Academic Press.

Klimley, A. P., S. D. Anderson, P. Pyle, and R. P. Henderson. 1992. Spatio temporal patters of the white shark (*Carcharodon carcharias*) predation at the South Farallon Islands, California. *Copeia* 1992: 680–690.

Krebs, C. J. 1990. *Ecological methodology*, 2nd edition. Menlo Park, CA: Addison Wesley Longman Inc.

Letting, M., and D. P. Armstrong. 2003. An introduction to using mark-recapture analysis for monitoring threatened species. *Department of Conservation Technical Series. New Zealand* 28A: 5–32.

Marshall, A. D., C. L. Dudgeon, and M. B. Bennett. 2011. Size and structure of photographically identified population of manta rays *Manta alfredi* in southern Mozambique. *Marine Biology* 158: 1111–1124.

Otis, D. L., K. P. Burnham, G. C. White, and D. R. Anderson. 1978. Statistical inference from capture data on closed animal populations. *Wildlife Monographs* 62: 1–135.

Pollock, K. H., J. D Nichols, C. Brownie, and J. E. Hines. 1990. Statistical inference for capture-recapture experiments. *Wildlife Monographs* 107: 1–97.

Porcher, I. F. 2005. On the gestation period of the black fin reef shark, *Carcharhinus melanopterus*, in waters off Moorea, French Polynesia. *Marine Biology* 146: 1207–1211.

Schwarz, C. J., and Arnason, A. N. 1996. A general methodology for the analysis of capture-recapture experiments in open populations. *Biometrics* 52: 860–873.

Sims, D. W., C. D. Speedie, and A. M. Fox. 2000. Movements and growth of a female basking shark re-sighted after a three year period. *Journal of the Marine Biological Association of the UK* 80: 1141–1142.

Stanley, T. R., and K. P. Burnham. 1999. A closure test for time-specific capture-recapture data. *Environmental and Ecological Statistics* 6: 197–209.

Stanley, T. R., and J. D. Richards. 2005. Software review: A program for testing capture-recapture data closure. *Wildlife Society Bulletin* 33: 782–785.

Stevick, P. T., P. J. Palsboll, T. D. Smith, M. V. Bravington, and P. S. Hammond. 2001. Errors in identification using natural markings: Rate, sources, and effect on capture-recapture estimates and abundance. *Canadian Journal of Fisheries and Aquatic Sciences* 58: 1861–1870.

Strong, W. R., Jr., R. C. Murphy, B. D. Bruce, and D. R. Nelson. 1992. Movements and associated observations of bait-attracted white sharks, *Carcharodon carcharias*: A preliminary report. *Australian Journal of Marine and Freshwater Research* 43: 13–20.

Strong, W. R., Jr., B. D. Bruce, D. R. Nelson, and R. D. Murphy. 1996. Population dynamics of the white shark in Spencer Gulf, South Australia. In *Great white sharks: The biology of Carcharodon carcharias*, edited by A. P. Klimley and D. G. Ainley, pp. 401–414. San Diego, CA: Academic Press.

Weng, K. C., A. M. Boustany, P. Pyle, S. D. Anderson, A. Brown, and B. A. Block. 2007. Migration and habitat of white sharks (*Carcharodon carcharias*) in the eastern Pacific Ocean. *Marine Biology* 152: 877–894.

White, G. C., and K. P. Burnham. 1999. Program Mark: Survival estimation from populations of marked animals. *Bird Study* 46(supplement): S120–S139.

CHAPTER 27

Implications of Increasing Pinniped Populations on the Diet and Abundance of White Sharks off the Coast of Massachusetts

Gregory B. Skomal*
Massachusetts Division of Marine Fisheries

John Chisholm
Massachusetts Division of Marine Fisheries

Steven J. Correia
Massachusetts Division of Marine Fisheries

ABSTRACT

Although the occurrence of the White Shark, *Carcharodon carcharias*, is well documented in the North Atlantic, the species is relatively rare, and much of what is known about its distribution and movements is based on historical sightings data. The advent of new tagging technology coupled with the existence of White Shark "hot spots" near pinniped colonies have allowed researchers to investigate the ecology of this species in the Pacific and Indian Oceans, but its elusive nature in the Atlantic has hampered such studies in this region. However, the numbers of White Shark sightings and White Shark pinniped predation events have been rising off the coast of Massachusetts in recent years and, in particular, near Monomoy Island on Cape Cod, which hosts a large growing population of Gray Seals (*Halichoerus grypus*). Although the perceived increase in shark predation on Gray Seals can be attributed to several factors, it is feasible that White Sharks, which were thought to primarily scavenge cetaceans in the Atlantic, are expanding their diet in response to regional changes in seal abundance. Based on documented changes in White Shark populations exhibited in other parts of the world, we anticipate that the number of White Shark sightings and seal interactions will continue to rise off the coast of Massachusetts.

INTRODUCTION

In the western North Atlantic, the White Shark (*Carcharodon carcharias*) is well documented from Newfoundland to the Gulf of Mexico, including the Bahamas and parts of the Caribbean (Bigelow and Schroeder, 1948; Templeman, 1963; Casey and Pratt, 1985; Compagno, 1984). Despite its well-established presence in the North Atlantic, the White Shark is not considered an abundant species, and efforts to study its life history and ecology have been hampered by the inability of researchers to predictably encounter these sharks. Indeed, much of what is known of this species in the North Atlantic comes from the analysis of distribution records (Templeman, 1963; Casey and Pratt, 1985), rare behavioral observations (Pratt et al., 1982; Carey et al., 1982), and the opportunistic examination of dead specimens (Pratt, 1996).

In their review of White Shark distribution in the western North Atlantic, Casey and Pratt (1985) compiled observations from numerous sources, including fisheries interactions, confirmed sightings, and published accounts dating back to 1874. The rarity of this species in this region is exemplified by their compilation of only 380 records and their observation that White Sharks represented only 0.04% of the sharks taken by over 2.1 million hooks of pelagic longline effort from the Grand Banks to the Gulf of Mexico (1963–1983). Nonetheless, Casey and Pratt (1985) concluded that White Sharks in the western North Atlantic were most abundant on the continental shelf in the Mid-Atlantic Bight (from Cape Hatteras, North Carolina to Cape Cod, Massachusetts). Moreover, the species is thought to exhibit seasonal movements, mediated by water temperature, into northern latitudes.

In the Pacific and Indian Oceans, the White Shark is known to feed on pinnipeds, and its ecology as it relates to large pinniped colonies is well studied (Klimley and Ainley, 1996; Chapter 9, this book). The high seasonal abundance of White Sharks near seal and sea lion colonies has also allowed researchers in these regions to study White Shark movements over broad spatial and temporal scales (Domeier and Nasby-Lucas, 2008; Jorgenson et al., 2010; Chapters 10, 11, 12, 13, 14, 16, 21, and 22, this book). The only behavioral observations of White Sharks in the North Atlantic come from a single acoustic tracking study conducted by Carey et al. (1982). These researchers acoustically tagged a 4.6 m total length White Shark that was scavenging a Fin Whale carcass 39 km southwest of Montauk Point, New York and tracked the shark for 83 h as it moved 190 km southwest along the 25-fathom bathymetric curve. The vertical movements of the shark were of particular interest in that it remained largely associated with the thermocline at approximately 10–20 m but made periodic excursions to the bottom. At the time, Carey at al. (1982) noted that "the seals, sea lions, and elephant seals, which are common items in the diet of White Sharks in other regions, are not available" in the western North Atlantic. Hence, they concluded that the observed diving behavior may be associated with searching for dead whales, which are an important food resource for large White Sharks (>3 m) on the continental shelf. Indeed, the presence of scavenging White Sharks on whale carcasses is well documented in this region (Pratt et al., 1982).

It can be argued that the elusive nature of the White Shark in the North Atlantic may be a result of the lack of strong spatial overlap between White Sharks and large pinniped colonies. This may not have always been the case because at least one large pinniped, the Gray Seal (*Halichoerus grypus*), was once quite abundant in southern New England waters, an area identified by Casey and Pratt (1985) as having the highest White Shark abundance. If this were indeed the case, White Sharks may have exhibited a dietary shift to other prey, including dead cetaceans, as a result of the demise of the Gray Seal population in the seventeenth century (Wood LaFond, 2009). With the protection of marine mammals over the last 40 years, the western North Atlantic Gray Seal population has rebounded (National Marine Fisheries Service, 2009a; Wood LaFond, 2009). Hence, it is entirely plausible that White Sharks are expanding their foraging strategies to once again include active predation on pinnipeds, which may be becoming a viable food resource.

The purpose of this paper is to explore the dynamic predatory relationship of White Sharks and pinnipeds in the western North Atlantic. Based on documented changes in White Shark populations exhibited in other parts of the world, we have compiled the best available information on temporal and spatial changes in Gray Seal and White Shark abundance as well as shark-seal interactions to define the current and future status of this relationship.

MATERIALS AND METHODS

White Shark Sightings

Since the establishment of the Massachusetts Shark Research Project in 1987, we have been tabulating and investigating reports of White Shark sightings in New England waters. In most cases, these reports comprised fisheries gear interactions or observations by fisheries observers, spotter pilots working with commercial fishermen, whale-watch vessels, boaters, beach users, and recreational and commercial fishermen (Figure 27.1). To confirm species identification, we have taken into consideration physical evidence (i.e., a dead specimen), photographic/video evidence, eyewitness accounts, and observer experience. In most cases, those species typically confused for White Sharks included Basking Sharks (*Cetorhinus maximus*), Ocean Sunfish (*Mola mola*), and a variety of marine mammals (dolphins, porpoises). Sightings that were clearly not White Sharks based on descriptions and/or photographic evidence were discarded and not classified. All other sightings reports were categorized as follows:

1. Class A: Positive identification based on valid description, dead specimen, and/or photo/video support
2. Class B: Accurate description, but no photographic evidence
3. Class C: Suspect description, no corroborative support
4. Class F: False report or misidentification (witness acknowledges after being shown photos)

Seal Interactions

In the United States, marine mammal strandings data are collected and compiled by regional strandings networks with oversight from the National Marine Fisheries Service. In the northeast region and, more specifically, Massachusetts, marine mammal strandings are primarily investigated by the New England Aquarium and the International Fund for Animal Welfare (IFAW)/Cape Cod

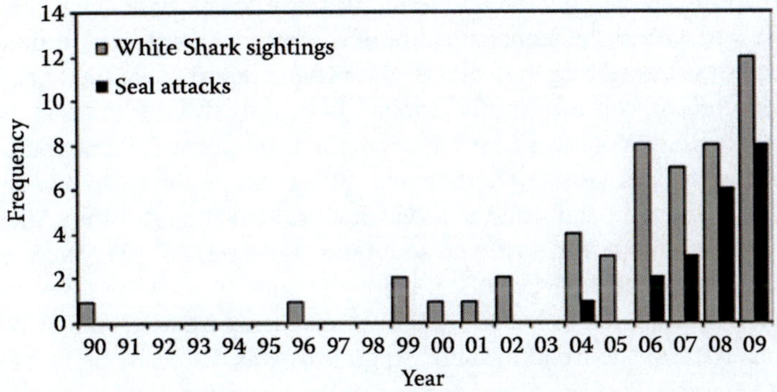

Figure 27.1 The annual number of White Shark sightings and attacks on seals based on information reported to the Massachusetts Shark Research Project, 1990–2009.

Stranding Network. The responsibilities of each program are divided geographically. The New England Aquarium's Marine Animal Rescue Program is responsible for much of the eastern shoreline of Massachusetts and New Hampshire but also includes Martha's Vineyard and Nantucket Islands. The IFAW/Cape Cod Stranding Network (CCSN) covers the Cape Cod region, extending west to the Rhode Island border.

From the mid 1980s to 1998, marine mammal strandings on Cape Cod were compiled by several animal-welfare and conservation organizations under the authorization of the New England Aquarium. In 1998, the nonprofit CCSN was formed with dedicated staff and over 300 volunteers trained to respond to marine mammal and turtle strandings on Cape Cod. More recently, the CCSN has become a project of the IFAW. These organizational changes did not result in greater observational effort because virtually all pinniped strandings have been and continue to be reported by the general public (e.g., beach users, lifeguards, etc.).

Among the goals of these programs is to respond to live and dead stranded marine mammals, including seals, and, when possible, perform postmortem necropsies to determine the cause of death. Over the last decade, we have been working closely with these programs to identify seal injuries and mortalities that may be indicative of interactions with sharks. If it was suspected that injuries sustained by a seal may have been the result of a shark attack, the stranding network contacted our program to examine the seal. Given the prey item (seal), the temperate location (New England), the size and nature of the lacerations [based on examples given by Long and Jones (1995)], and the limited number of shark species that inhabit this region (Bigelow and Schroeder, 1948), the White Shark was the most likely species implicated in these attacks.

Data Analysis

Annual White Shark counts were modeled with a Poisson regression using year as a covariate. To take into consideration uncertainty over effort, we deployed two methods. First, we refit the generalized linear model (GLM) with a 3% annual increase in effort as an offset. Second, we used an approach recently developed by McPherson and Myers (2009) to examine population trends from observational data. In short, this method fits a GLM to extract the abundance trend in relative terms but also tests the sensitivity of trend estimates to changes in observer effort (McPherson and Myers, 2009). We ran the model for the time period of 1990–2009 and tested the sensitivity of the results to annual observational effort changes of +50%, 0%, and –50%.

RESULTS

From 1990 to 2009, we recorded 63 White Shark sightings classified as A ($n = 32$), B ($n = 18$), C ($n = 8$), and F ($n = 5$). Given the tenuous nature of the latter two categories, they were eliminated from subsequent analyses, resulting in a total of 50 credible reports (Figure 27.1). The bulk of these fish (70%) was reported by commercial fisherman (44%: gillnet, bottom trawl, trap, longline, and tuna rod and reel), spotter pilots working with Bluefin Tuna purse seiners and harpooners (16%), and chartered fishing vessels (10%). The balance (30%) was reported by beach users, kayakers, paddleboarders, recreational boaters, and seal-watch vessels. Although White Sharks were reported over a broad geographic area north, east, and south of Cape Cod, 26 (52%) were in close proximity to Monomoy Island, an established Gray Seal colony (Figure 27.2). Using Poisson regression with year as a covariate, we found a significant ($p < 0.001$) increasing trend in annual White Shark counts (Figure 27.3). Sensitivity of the trend to increasing effort was tested by using an effort series that increased 3% per year as an offset. A comparison of the predicted counts from the model with no effort changes to the sensitivity model suggests that the direction of the trend is not sensitive to the 3% increase in effort (Figure 27.3). The approach developed by McPherson and Myers (2009) also

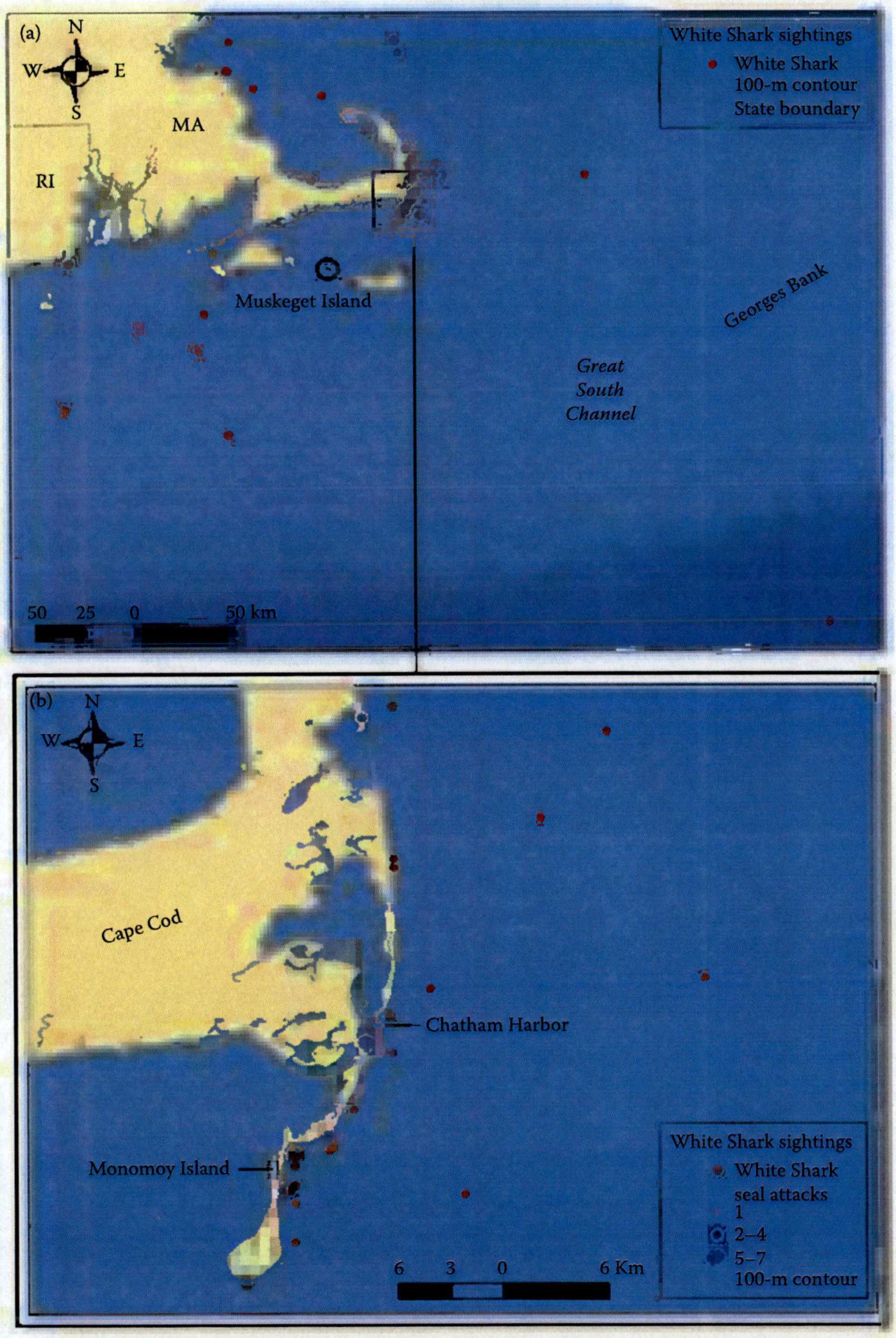

Figure 27.2 (a) Locations of White Shark sightings off the coast of Massachusetts, 1990–2009 (n = 50). (b) Locations of White Shark sightings and attacks on seals adjacent to the large Gray Seal colony on Monomoy Island, Massachusetts.

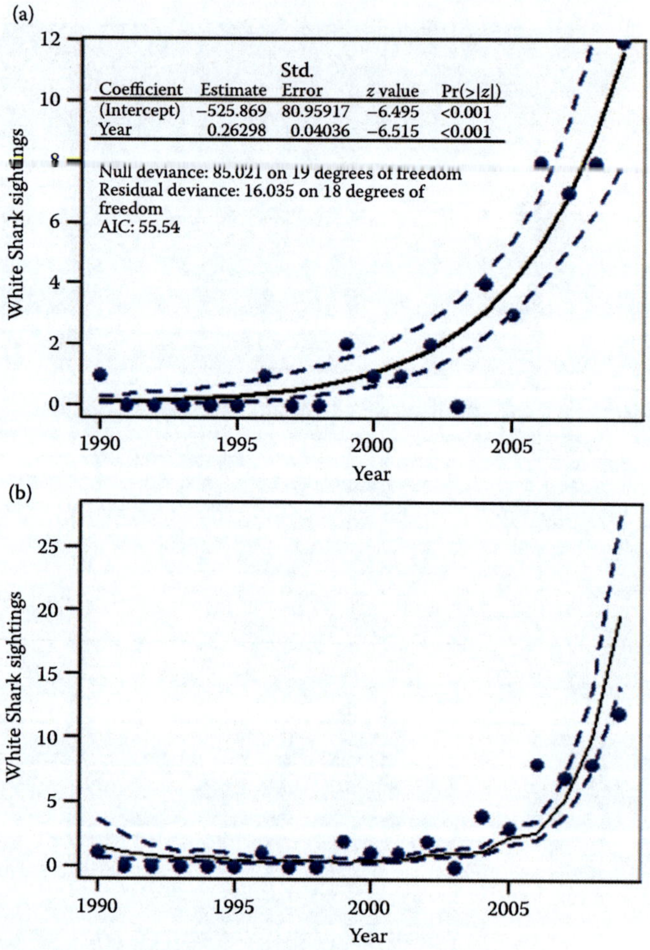

Figure 27.3 (a) Predicted counts of White Sharks modeled with a GLM using year as a covariate. The black line shows predicted counts, the dashed blue lines are approximate 95% confidence limits, and blue points are observed counts. (b) Predicted counts of White Sharks modeled with a GLM using year as a covariate with a 3% annual increase in effort as an offset. The black line shows predicted counts, the dashed blue lines are approximate 95% confidence limits, and blue points are observed counts.

produced a significant increasing trend (year effect, $p < 0.01$) in relative White Shark abundance off the coast of Massachusetts during the time period 1990–2009. The magnitude of change determined by the model varied with changes in observational effort, but the overall increasing trend remained (Figure 27.4).

In addition, 29 White Sharks were sighted in close proximity to this Gray Seal colony during the period of September 3–8, 2009 as a result of directed effort to locate (with a spotter plane) and to tag White Sharks in this area (Figure 27.5). Because of the intensified effort, these sighting were not included in the trend analyses (Figures 27.3 and 27.4). The number of sightings during this 5-d period does not necessarily reflect the actual number of individual White Sharks because many may have been counted more than once. Nonetheless, we believe that the group of White Sharks around Monomoy Island at that time represents a minimum of ten to twelve individual fish because of concurrent sightings by two spotter pilots and our tagging efforts during that period.

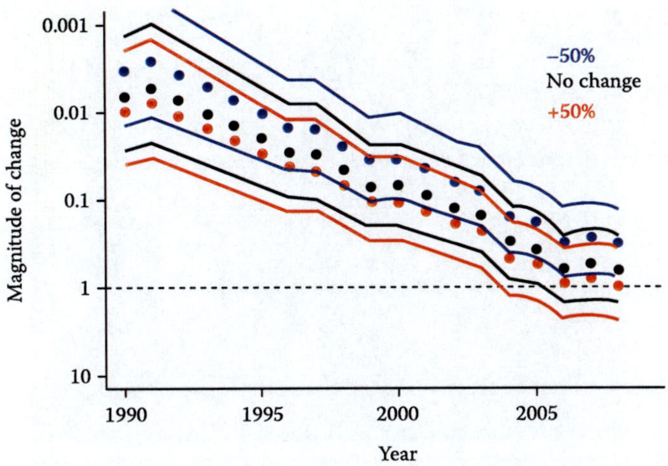

Figure 27.4 Ratio of predicted counts in reference year to predicted counts in 2009 using methodology of McPherson and Myers (2009). The black points and lines are with no changes in effort, blue points are sensitivity to 50% decline in effort from 1990 to 2009, red points are sensitivity to 50% increase in effort, and the dashed line = 1. Values less than or equal to 1 mean that the counts are increasing (<1) or stable (1).

Figure 27.5 Aerial photograph of White Shark swimming in close proximity to Gray Seals off Monomoy Island, Massachusetts, September 2009. (Courtesy of Dan McKiernan, Massachusetts Division of Marine Fisheries.)

The number of shark-bitten seals has been increasing off the coast of Massachusetts and, in particular, off Monomoy Island over the last decade (Figures 27.1 and 27.6). Working closely with the CCSN, we have compiled information on twenty attacks on seals. All of these interactions involved Gray Seals (Figure 27.6); the White Shark was the shark species implicated in the attacks. In at least five cases, fishermen or beach goers witnessed the attacks in close proximity to the large Gray Seal colony on Monomoy Island (Figure 27.2). These eyewitness accounts further bolster our contention that White Sharks are feeding on seals in this area.

DISCUSSION

Based on sightings data, there is evidence that the local seasonal population of White Sharks off Massachusetts and, particularly, off the east coast of Cape Cod has increased in recent years.

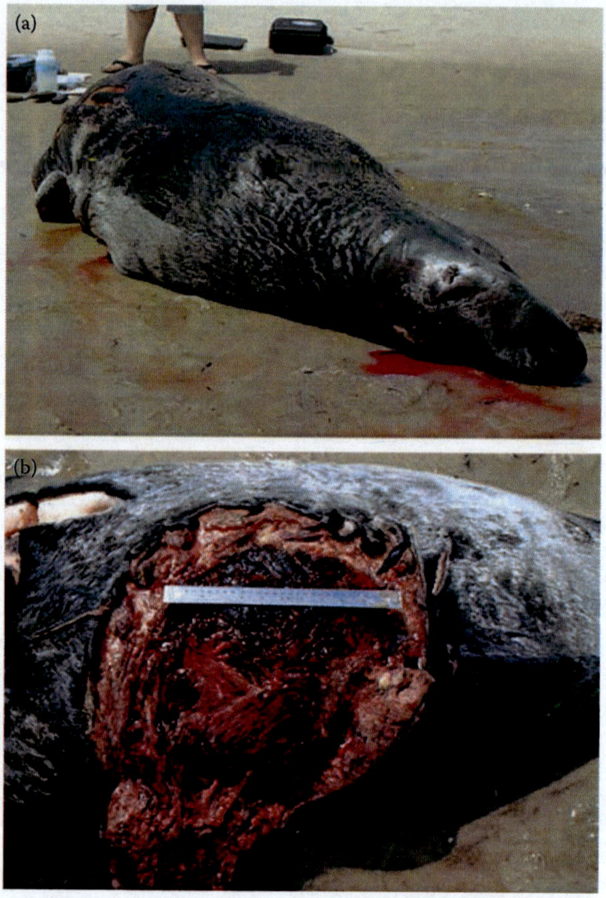

Figure 27.6 Stranded Gray Seal (a) on Monomoy Island, Massachusetts, with large wound (b) attributed to an attack by a White Shark, July 2005.

Although it is plausible that this trend may reflect an increase in effort, we do not believe this to be the case for the following reasons. The majority of sightings (70%) come from commercial fisherman targeting groundfish and Bluefin Tuna. It has been illegal for fishermen to capture and kill White Sharks since 1997 (National Marine Fisheries Service, 1997). Hence, there has been no incentive for commercial fishermen to report White Shark interactions with fishing gear because it may result in a fine. Moreover, fishing effort and landings by these commercial fishing groups over the last decade have been in steep decline because of struggling fish stocks, restrictive fisheries regulations, and changes in fish distribution.

Off the northeastern United States, regulatory efforts to reduce groundfish [a complex of demersal species, including Atlantic Cod (*Gadus morhua*), Haddock (*Melanogrammus aeglefinus*), and Yellowtail Flounder (*Limanda ferruginea*)] landings have resulted in a dramatic reduction in the number of days that fisheries participants are allowed to fish annually [termed "days at sea" (DAS)]. From 2001 to 2009, the number of DAS allocated to this fishery was reduced 72%, and the actual number of DAS used (i.e., fishing effort) declined by 52% (Figure 27.7; New England Fishery Management Council, 2009). Similarly, the commercial Bluefin Tuna (*Thunnus thynnus*) fishery off the east coast of the United States has experienced a decline in fishing effort and landings over the last decade. In short, this fishery comprises three gear-specific National Marine Fisheries Service-permitted categories: general (primarily rod and reel), purse seine, and harpoon. As evidenced by

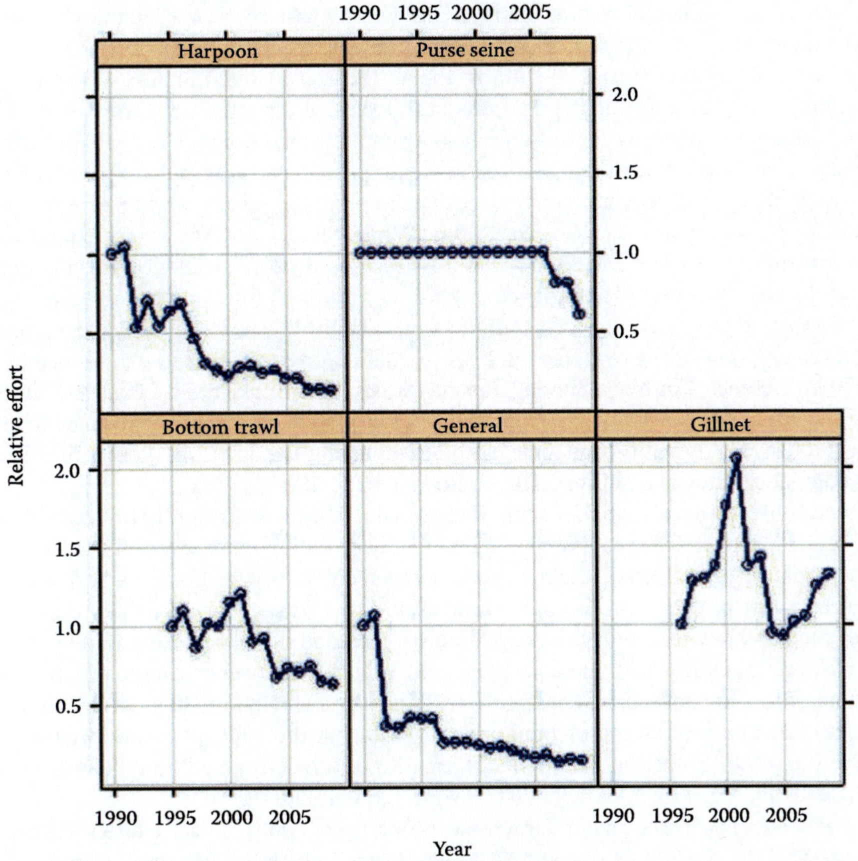

Figure 27.7 Trends in commercial groundfish and Bluefin Tuna fishing effort off the northeastern United States standardized as fraction of effort in first year; bottom trawl and gillnet in DAS; and general category tuna, tuna purse seine, and harpoon in number of permits.

the number of permits issued, fishing effort declined 45, 40, and 54% in the general, purse seine, and harpoon categories, respectively, from 1990 to 2009 (Figure 27.7; National Marine Fisheries Service, 2009b). This was also reflected in the Bluefin Tuna landings, which declined 49, 95, and 64%, respectively (National Marine Fisheries Service, 2009b). Although purse seine and harpoon fishermen do not catch White Sharks, they are the primary employers of spotter planes and, as such, are indicative of spotter-plane effort.

In summary, the amount of fishing effort by those fisheries that typically report White Sharks declined by roughly 50% during the period when White Shark sightings were dramatically increasing. Using GLM, we found a significant increasing trend in local White Shark abundance off the coast of Massachusetts (Figure 27.3). When we tested the sensitivity of this trend to changes in observational effort, the trend remained positive regardless of simulated increases in fishing effort (Figures 27.3 and 27.4). Given the strong evidence that fishing and, hence, observational effort declined by as much as 50% over this time period, the positive trend in White Shark sightings may be stronger than evidenced in our dataset. In contrast, when McPherson and Myers (2009) used this approach to model changes in White Shark abundance based on sightings data from eastern Adriatic and eastern Canada waters, they found a significant decreasing trend. We do not believe that the trend in our sightings data necessarily reflects actual population growth but rather a change in local abundance because of other factors, namely the growing Gray Seal population.

The increasing number of White Shark sightings in southern New England coupled with the concurrent increase in the number of White Shark attacks on Gray Seals may be indicative of a change in White Shark predatory behavior off the northeastern United States. White Sharks, which were thought to primarily scavenge cetaceans in this part of the Atlantic (Carey et al., 1982; Pratt et al., 1982), may be expanding their diet in response to regional changes in seal abundance. Shifts in predatory behavior have been documented in some species of sharks and are generally associated with ontogeny, changes in prey abundance, changes in predator density, and/or ease of prey capture (Heithaus, 2004; Wetherbee and Cortés, 2004). White Sharks are known to exhibit an ontogenetic dietary shift (Tricas and McCosker, 1984; Estrada et al., 2006) and have been documented to respond to changes in pinniped abundance (Pyle et al., 1996). White Shark predation on pinnipeds is well documented in the Pacific and Indian Oceans (Klimley and Ainley, 1996; Chapter 9, this book), but the only published evidence of shark predation on seals in the Atlantic occurs on Sable Island, 170 km southeast of Nova Scotia, Canada (Brodie and Beck, 1983; Lucas and Stobo, 2000). Although the species of shark involved remains unknown, such predation seems to be in response to a growing Gray Seal population and appears to be impacting the much smaller Harbor Seal (*Phoca vitulina*) population (Lucas and Stobo, 2000; Bowen et al., 2003).

Given our observations, we believe that this predatory behavior is tightly linked to the dramatic increase in Gray Seal abundance. Archeological evidence dating back several thousand years from Native American shell middens indicates that Gray Seal colonies were historically distributed along the northeast coast of the United States from Maine to Connecticut (Wood LaFond, 2009). By the end of the seventeenth century, however, subsistence and bounty hunting had extirpated these colonies (Wood LaFond, 2009). In subsequent years, Gray Seals were infrequently reported in the United States, and the population was largely sustained by a single breeding colony on Sable Island off Nova Scotia. The prohibition of bounty hunting during the 1960s followed by the passage of the Marine Mammal Protection Act in 1972 have allowed the Gray Seal to recolonize parts of the United States from Maine to Massachusetts (Wood LaFond, 2009).

There are currently three major Gray Seal breeding colonies in the United States: Seal and Green Islands off the coast of Maine and Muskegut Island, which lies just west of Nantucket Island, Massachusetts (Figure 27.2). The latter is the largest breeding colony, annually producing approximately 80% of the Gray Seal pups in the United States (Wood LaFond, 2009). Based on annual aerial-survey counts, pup production has steadily increased on Muskegut Island from six in 1991 to 2095 in 2008 (Figure 27.8). Much (~20% in 2008) of this pup production comes from female immigrants from Sable Island (Wood LaFond, 2009).

In addition to the large breeding colony on Muskegut Island, Gray Seals frequent other large haulouts along the coast of Massachusetts. The largest and most frequently used area is on

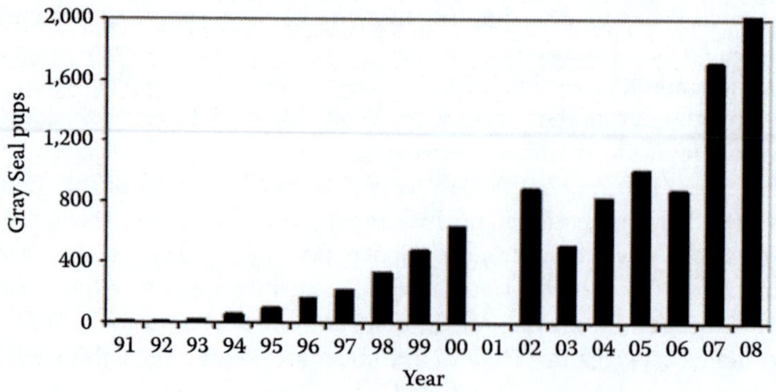

Figure 27.8 Annual Gray Seal pup counts on Muskegut Island, Massachusetts, the largest Gray Seal breeding colony in the United States, 1991–2008 [data are from Wood LaFond (2009)].

Figure 27.9 Thousands of Gray Seals fringe the intertidal zone of Monomoy Island, Massachusetts, 2010. (Courtesy of Charlotte Richardson.)

Monomoy Island, a large National Wildlife Refuge stretching 13 km south into the Atlantic from Chatham, Massachusetts (Figure 27.9). The refuge largely comprises the two barrier islands of North and South Monomoy. Gray Seals are most frequently seen on the east coast of the barrier islands and on adjacent shoals from Chatham Harbor to the southern tip. Although counts from formal seal surveys are currently lacking, the number of Gray Seals using these areas is typically in the thousands (Figure 27.9). Gray Seals have also periodically pupped on Monomoy Island (Wood LaFond, 2009).

With this expansion of the western Atlantic Gray Seal population, there has been an increase in White Shark sightings as well as seal interactions over the last decade. Although White Sharks will probably continue to scavenge whale carcasses, active predation on Gray Seals is likely a return to a preexisting trophic scenario, which changed with the demise of the Gray Seal population several hundred years ago. A similar scenario between White Sharks and their pinniped prey has been well-documented at the Farallon Islands on the west coast of the United States. Pyle et al. (1996) described two general phases that occurred in this area. As the Northern Elephant Seal (*Mirounga angustirostris*) population rebounded during the 1970s and early 1980s, increases in individual predation rates by a relatively static number of White Sharks caused the attack frequency to increase. It was hypothesized that there were probably six or less White Sharks in the area at that time, because the number of attacks declined markedly when four of these sharks were removed in 1982. As the seal population stabilized in the late 1980s, individual predation rates stabilized as well, but the number of White Sharks attacking and consuming this resource climbed (Pyle et al., 1996).

Although accurate Gray Seal population estimates are lacking, the western North Atlantic population is thought be in excess of 200,000 animals and increasing in the United States

(National Marine Fisheries Service, 2009a; Figure 27.9). Given this upward trajectory and the relatively low number of attacks on seals, the number of individual White Sharks utilizing this resource is likely low as well. If the western scenario is applicable to the east, we are in the early stages of this relationship but can anticipate additional White Sharks utilizing this food resource. However, the size and status of the White Shark population in the western North Atlantic remains unknown. Although the species has been prohibited from retention in this region for more than a decade, high levels of incidental mortality will result in population declines, which will also influence these trophic dynamics.

ACKNOWLEDGMENTS

The Massachusetts Shark Project is supported with funds from the Federal Aid in Sportfish Restoration Act. We thank those fishermen and members of the general public who reported White Shark sightings. We appreciate the cooperation and timely stranding reports from the IFAW/Cape Cod Stranding Network and the Town of Chatham harbormaster Stuart Smith and his crew. This is Massachusetts Division of Marine Fisheries Contribution No. 36.

REFERENCES

Bigelow, H.B., and Schroeder, W.C. 1948. *Fishes of the Western North Atlantic, Part 1*. New Haven, CT: Yale University Press.

Bowen, W.D., Ellis, S.L., Iverson, S.J., and Boness, D.J. 2003. Maternal and newborn life history traits during periods of contrasting population trends: Implications for explaining the decline of harbour seals (*Phoca vitulina*), on Sable Island. *J. Zool. Lond.* 261: 155–163.

Brodie, P., and Beck, B. 1983. Predation by sharks on grey seal (*Halichoerus grypus*) in eastern Canada. *Can. J. Fish. Aquat. Sci.* 40: 267–271.

Carey, F.G., Kanwisher, J.W., Brazier, O., Gabrielsen, G., Casey, J.G., and Pratt, H.L., Jr. 1982. Temperature and activities of a white shark, *Carcharodon carcharias*. *Copeia* 1982: 254–260.

Casey, J.G., and Pratt, H.L., Jr. 1985. Distribution of the white shark, *Carcharodon carcharias*, in the western North Atlantic. *South. Calif. Acad. Sci., Mem.* 9: 2–14.

Compagno, L.J.V. 1984. FAO Species Catalogue. Vol. 4. Sharks of the world. An annotated and illustrated catalogue of shark species known to date. Part 1: Hexanchiformes to Lamniformes. *FAO Fish. Synop.* 125: 1–249.

Domeier, M. L., and Nasby-Lucas, N. 2008. Migration patterns of white sharks *Carcharodon carcharias* tagged at Guadalupe Island, Mexico, and identification of an eastern Pacific shared offshore foraging area. *Mar. Ecol. Prog. Ser.* 370: 221–237.

Estrada, J.A., Rice, A.N., Natanson, L.J., and Skomal, G.B. 2006. The use of isotopic analysis of vertebrae in reconstructing ontogenetic feeding ecology in white sharks. *Ecology* 87: 829–834.

Heithaus, M.R. 2004. Predator-prey interactions. In *Biology of Sharks and Their Relatives*, edited by J.C. Carrier, J.A. Musick, and M.R. Heithaus, pp. 487–521. Boca Raton, FL: CRC Press.

Jorgenson S.J., Reeb, C.A., Chapple, T.K., Anderson, S., Perle, C., Van Sommeran, S.R., et al. 2009. Philopatry and migration of Pacific white sharks. *Proc. R. Soc. B.* 277: 679–688.

Klimley, P.A., and Ainley, D.G. 1996. *Great White Sharks: The Biology of Carcharodon carcharias*. San Diego, CA: Academic Press.

Long, D.J., and Jones, R.E. 1996. White shark predation and scavenging on cetaceans in the eastern North Pacific Ocean. In *Great White Sharks: The Biology of Carcharodon carcharias*, edited by A.P. Klimley and D.G. Ainley, pp. 293–307. San Diego, CA: Academic Press.

Lucas, Z., and Stobo, W.T. 2000. Shark-inflicted mortality on a population of harbour seals (*Phoca vitulina*) at Sable Island, Nova Scotia. *J. Zool. Lond.* 252: 405–414.

McPherson, J.M., and Myers, R.A. 2009. How to infer population trends in sparse data: Examples with opportunistic sightings records for great white sharks. *Diversity Distrib.* 15: 880–890.

New England Fishery Management Council. 2009. *Final Amendment 16 to the Northeast Multispecies Fishery Management Plan.* Newburyport, MA: New England Fishery Management Council.

National Marine Fisheries Service. 1997. *Fishery Management Plan for Sharks of the Atlantic Ocean.* Silver Spring, MD: Highly Migratory Species Management Division, Office of Sustainable Fisheries, NOAA Fisheries.

National Marine Fisheries Service. 2009a. *Gray Seal (Halichoerus grypus): Western North Atlantic Stock.* Silver Spring, MD: Marine Mammal Stock Assessment Report, Office of Protected Resources, NOAA Fisheries.

National Marine Fisheries Service. 2009b. *Stock Assessment and Fishery Evaluation (SAFE) Report for Atlantic Highly Migratory Species.* Silver Spring, MD: Highly Migratory Species Management Division, Office of Sustainable Fisheries, NOAA Fisheries.

Pratt, H.L., Jr. 1996. Reproduction in the male white shark. In *Great White Sharks: The Biology of Carcharodon carcharias*, edited by A.P. Klimley and D.G. Ainley, pp. 131–138. San Diego, CA: Academic Press.

Pratt, H.L., Casey, J.G., and Conklin, R.E. 1982. Observations on large white sharks, *Carcharodon carcharias*, off Long Island, New York. *Fish. Bull.* 80: 153–156.

Pyle, P., Anderson, S.D., and Ainley, D.G. 1996. Trends in white shark predation at the south Farallon Islands. In *Great White Sharks: The Biology of Carcharodon carcharias*, edited by A.P. Klimley and D.G. Ainley, pp. 375–379. San Diego, CA: Academic Press.

Templeman, W. 1963. Distribution of sharks in the Canadian Atlantic (with special reference to Newfoundland waters). *Bull. Fish. Res. Board Can.* 140: 1–77.

Tricas, T.C., and McCosker, J.E. 1984. Predatory behavior of the white shark (*Carcharodon carcharias*), with notes on its biology. *Proc. Calif. Acad. Sci.* 43: 221–238.

Wetherbee, B.M., and Cortés, E. 2004. Food consumption and feeding habits. In *Biology of Sharks and Their Relatives*, edited by J.C. Carrier, J.A. Musick, and M.R. Heithaus, pp. 225–246. Boca Raton, FL: CRC Press.

Wood LaFond, S.A. 2009. Dynamics of recolonization: A study of the gray seal (*Halichoerus grypus*) in the northeast U.S. Ph.D. dissertation, The University of Massachusetts, Boston, MA.

CHAPTER 28

Back to the Wild

Release of Juvenile White Sharks from the Monterey Bay Aquarium

Kevin C. Weng*
University of Hawaii at Manoa and Stanford University

John B. O'Sullivan
Monterey Bay Aquarium

Christopher G. Lowe
California State University, Long Beach

Chuck E. Winkler
Southern California Marine Institute

Mary E. Blasius
California State University, Long Beach

Kerri A. Loke-Smith
California State University, Long Beach

Timothy J. Sippel
University of Hawaii at Manoa

Juan M. Ezcurra
Monterey Bay Aquarium

Salvador J. Jorgensen
Monterey Bay Aquarium and Stanford University

Michael J. Murray
Monterey Bay Aquarium

ABSTRACT

The White Shark (*Carcharodon carcharias*) is an important apex predator and as an icon of the oceans generates public interest in marine science. There has been keen interest in bringing the species into captivity, but its biology, habitat, and size have prevented animals from being held successfully. From 2004 to 2010, the Monterey Bay Aquarium exhibited five young-of-the-year White Sharks for a cumulative 575 d that were collectively viewed by more than 2.5 million people. Upon release from the exhibit to the wild, four of the animals moved southward from California and into waters off Baja California, Mexico, and one remained in the Southern California Bight. They inhabited near-surface waters, and at times their behavior suggested foraging on benthic resources. Their geographic movements, depth, and thermal habitats were similar to those of wild juvenile White Sharks in the region. In both U.S. and Mexican waters, juvenile White Sharks are taken as bycatch in set and drift-gillnet fisheries. Because fishery interactions with larger White Sharks are rare, gillnet interactions with juveniles are the most serious management issue for the species in the eastern Pacific region. Current movement data suggest that management of the species must be coordinated between Mexico and the United States.

INTRODUCTION

Many shark species experience high mortality rates in fisheries (Stevens et al., 2000). White Sharks (*Carcharodon carcharias*) have late age at first reproduction, slow growth rate, and small litter size and are highly susceptible to overexploitation (Mollet and Cailliet, 2002). The species is listed as vulnerable by the International Union for Conservation of Nature (Fergusson et al., 2000) and is in Appendix II of the Convention on International Trade in Endangered Species (Inskipp and Gillett, 2005; Yalowitz, 2004). In 1997, the White Shark was given permanent protection in California's state waters (California Senate Bill 144, 1997). Conservation concerns and the charismatic public profile of the species have made it an attractive, yet challenging focus of aquariums and researchers around the world.

Because of its high profile and public appeal, the White Shark is an excellent ambassador for communicating ocean conservation messages (Falk et al., 2007). Over the past few decades, a number of aquariums have attempted to maintain White Sharks of various sizes in captivity but have been successful only over brief periods. Challenges have included insufficient captive facilities, incompatible animal size, and various husbandry objectives such as initiation of feeding.

Captures of adult White Sharks in the eastern Pacific in fisheries are rare, whereas captures of juveniles occur more commonly. In eastern Pacific waters off California, adult White Sharks have not been reported in longline fisheries (O'Brien and Sunada, 1994) and are rare in drift-gillnet fisheries (Hanan et al., 1993). In contrast, juvenile White Sharks are captured regularly as bycatch in set-gillnet fisheries of California (United States) and Baja California (Mexico) (Klimley, 1985; Weng et al., 2007b; Chapters 14, 15, and 16, this book).

Capture and handling can have negative impacts on fish, including mortality and various sublethal effects (Frick et al., 2010; Musyl et al., 2009; Skomal, 2007; Wood et al., 1983). Hooking and net entanglement can cause wounds, oxygen deprivation, and a variety of harmful stress responses. Gillnet capture of sharks appears to result in respiratory and metabolic acidosis and hypoglycemia and potentially cellular damage, with the severity of injury dependent on time in net, degree of struggling, occlusion of gills by netting, and water temperature (Manire et al., 2001).

Tagging is a very useful method for determining postrelease survivorship (Domeier et al., 2003) and can also help investigate potential changes in fitness (Hampton, 1986) and behavior after capture and handling by humans (Bestley et al., 2008; Sherrill-Mix and James, 2008), including captivity (Brennan et al., 2006). Exhaustive exercise, capture, handling, and injury may lead to delayed mortality. A variety of methods have been used to study this topic, including active acoustic tracking, pop-up archival transmitting (PAT) tags, and satellite-linked radio transmitting (SLRT) tags. SLRT tags alone are difficult to use to investigate survivorship because they only transmit when animals surface, and the signal can be lost as a result of changes in animal behavior, tag malfunction, or inclement weather, as well as mortality. Active acoustic tracking was the first method used to detect mortalities (Yuen et al., 1975). Early versions of PAT tags could not distinguish between mortality and tag failure, because negative data (i.e., no transmissions) could be interpreted as either (Graves et al., 2002; Kerstetter et al., 2003). Later versions of PAT tags were designed to detect prolonged periods at constant depth, detach, surface, and transmit. In these cases, transmitted data are obtained that record the mortality event (Domeier et al., 2003; Graves et al., 2002; Kerstetter et al., 2003; Campana et al., 2009; Musyl et al., 2009). In the event that no transmissions are received from the newer PAT tags, it is not possible to distinguish mortality from tag failure. In this study, we used PAT tags to determine postrelease status of juvenile White Sharks released from captivity in comparison with juveniles tagged in the wild. In addition, some of the animals were double-tagged using SLRT tags, providing an additional data stream for the evaluation of an animal's status.

This paper describes a program to bring White Sharks into captivity; documents what happened to each animal following its release back the wild; and compares the geographic movement, habitat use, and behavior of the five postcaptive sharks to similar aged individuals that were tagged in the wild. Our results are based on satellite telemetry using the same methods for postcaptive and wild animals and seek to shed light on the biology and conservation of the species.

MATERIALS AND METHODS

Field

In this chapter, we use life-stage definitions following Bruce and Bradford (Chapter 17, this book) as follows: young of the year, <1.75 m total length (TL); juvenile, 1.75–3 m TL; subadult male, 3–3.6 m TL; subadult female, 3–4.8 m TL; adult male, >3.6 m TL; and adult female, >4.8 m TL. Young-of-the-year White Sharks were captured in the Southern California Bight (Figure 28.1) using multiple methods, including targeted fishing by researchers and collaboration with commercial gillnet fishermen. The commercial set-gillnet fishery in the Southern California Bight targets White Seabass (*Atractoscion nobilis*) and California Halibut (*Paralichthys californicus*). These fishing nets are typically 460 m long and 2 m high and have 16–20-cm stretch mesh. Nets are required to be shorter than 1800 m when deployed north of Santa Barbara and shorter than 2700 m south of Santa Barbara. Set gillnets are generally retrieved at sunrise and immediately reset. Soak times are generally 24 h but 48 h sets are not uncommon. Weather conditions and equipment failures may cause longer soak times.

Targeted fishing was conducted using three methods in waters between Ventura and Los Angeles. Longlining was conducted using a chartered commercial longline vessel in 2002 using Pacific Mackerel (*Scomber japonicus*) as bait. After sighting White Sharks, rod and reel fishers offered Pacific Mackerel baits during 2002–2009. A small purse-seine vessel (seine-net dimensions were 370 m long and 40 m deep with 2-cm mesh) captured sharks sighted near the surface by spotter

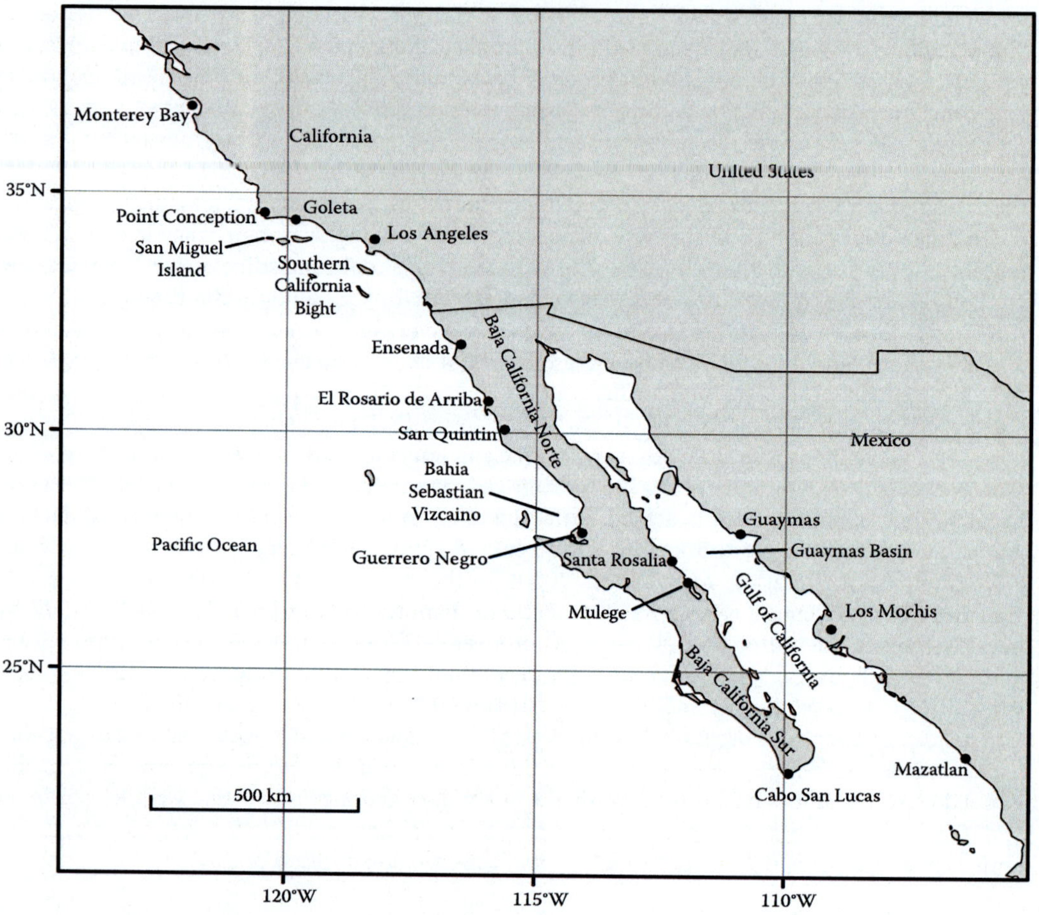

Figure 28.1 Map of the study area. Juvenile White Sharks were tagged and released from Monterey Bay and Goleta, California. Sharks occupied waters to the west of California (United States), Baja California (Mexico), and the Gulf of California (Mexico).

planes during 2007–2009. After capture, the sharks were transferred from the fishing vessel to the RV Lucille, a 7-m custom vessel outfitted with a shark-transportation system with oxygenated ram ventilation. The RV Lucille transported sharks to a holding pen off Malibu, California, to keep sharks in a semi-natural enclosure for evaluation prior to transfer to the Monterey Bay Aquarium (Chapter 1, this book). Transport to the Monterey Bay Aquarium was accomplished via a tank mounted on a tractor-trailer rig (Chapter 2, this book).

Captivity and Public Education

The aquarium maintained captive White Sharks in a large exhibit designed for the public display of pelagic species. The "Outer Bay Waters" exhibit has a volume of a 3,780,000 liters, is 10 m deep, and is maintained at 20°C (Farwell, 2001; Chapter 1, this book). A variety of other pelagic species were exhibited with the White Sharks, including scombrids and other sharks.

The aquarium used a combination of outreach and broadcast media along with visitor education programs to highlight White Shark biology and conservation. A 15-min. presentation was shown multiple times per day in the auditorium. The website added video clips, weekly updates, and links to research partners. A streaming web cam offered views of the exhibit, allowing the public to see the White Shark in real time from their computers.

The public-education value of exhibiting a White Shark was measured in a number of ways, including:

1. Increase in website hits
2. Increase in coverage of the aquarium by popular media
3. Visitor interviews and surveys gauging awareness of shark conservation issues

To estimate the number of people who viewed news media accounts of the exhibit White Sharks, the aquarium used print-media circulation numbers and Nielsen ratings numbers.

Release

Sharks were captured from the Monterey Bay Aquarium's exhibit using a large hoop net, placed in a water-filled transportation box, taken a short distance offshore in a 5-m skiff, and released into Monterey Bay. A pump circulated oxygenated water around the transport holding box, with the sharks faced into the current to facilitate ram ventilation. One shark (08-11) was driven south to Goleta, California, in a tractor-trailer transport tank and released (310-km straight distance southeast of Monterey). This animal was transported south to warmer waters because of concerns about its ability to adapt to the cold waters of Monterey Bay given its small size. Based on the size thresholds from Chapter 17, this book, sharks 04-01, 06-10, and 07-04 were 1 yr. old, whereas sharks 08-11 and 09-11 were young-of-the-year.

Satellite Telemetry

We used two types of satellite tags in this study. PAT tags archive data until a predetermined date, then release from the animal, float to the surface, and transmit data to the Argos system. While exposed to air when an animal surfaces, SLRT tags transmit nearly real-time data to Argos to calculate geopositions (Weng et al., 2005).

PAT tags (hardware versions PAT-2 and PAT-MK10; Wildlife Computers, Redmond, WA) were deployed on five juvenile White Sharks during 2002–2009 (Table 28.1). The tags were programmed to archive depth, ambient water temperature, and light intensity, at intervals between 5 and 60 s. These data were summarized on board the tag into 6-h bins for transmission to Argos. For each time interval, tags produced a depth-temperature profile and two histogram series, one of time-at depth and one of time-at temperature. In addition, a dawn and dusk light curve was transmitted for each day. Each PAT tag was tethered to a titanium dart (59 mm × 13 mm) with a 15-cm segment of 136-kg monofilament line (300-lb. test Extra-hard Hi-catch; Momoi Manufacturing, Kobe, Japan) covered with heat-shrink wrap. The dart was cleaned with povidone-iodine microbicide or isopropyl alcohol and inserted into the dorsal musculature at the base of the first dorsal fin using a stainless steel applicator, such that the PAT tag trailed behind the fin. To ease the insertion of the dart, a small slit in the skin was made using a surgical scalpel. When possible, attempts to recover PAT tags were made in order to obtain the high-frequency archive of ambient light, depth, and temperature data. From small boats, a radio direction finder (Doppler Systems, Inc.) tuned to the Argos frequency (~401.65 MHz) helped home in on the direction and range of the tag's signal to aid in recovery.

Two individuals were also tracked with SLRT tags (SPOT Hardware version 5; Wildlife Computers, Redmond, WA), which provide Argos geopositions with root-mean-squared errors ranging from 150 m to about 50 km (Argos, 2003; Hays et al., 2001; White and Sjoberg, 2002). SLRT tags were attached to the dorsal fin such that the antenna and conductivity switch of the tag would be exposed in air when the animal swam at the surface. Attachments were made using small

Table 28.1 White Sharks Held at the Monterey Bay Aquarium

Access Number	Date	Event	Location	Latitude	Longitude	Sex	Total Length (cm)	Mass (kg)	Tags
04-01	8/20/2004	Caught in gillnet	Huntington Flats	33.61	−118.02	F	141	28.0	
	3/31/2005	Released from aquarium	Monterey Bay	36.66	−121.92		184	73.4	PAT tag[a,b]
	4/30/2005	PAT tag transmits	Point Conception	34.59	−121.14				
06-10	8/17/2006	Caught by hook and line	Will Rogers State Beach	34.00	−118.57	M	164	47.0	
	1/16/2007	Released from aquarium	Monterey Bay	36.66	−121.92		187	77.6	PAT tag[a,c]
	4/15/2007	PAT tag transmits	Cabo San Lucas	22.89	−109.16				
07-04	8/04/2007	Captured in gillnet	Ventura	34.26	−119.43	M	143	30.6	
	2/05/2008	Released from aquarium	Monterey Bay	36.65	−121.95		176	63.6	PAT tag[a,c] SLRT tag
	6/16/2008	PAT tag transmits	Gulf of California, Santa Rosalia	27.43	−112.06				
08-11	8/16/2008	Caught in purse-seine net	Malibu	34.00	−118.71	F	137	25.2	
	9/07/2008	Released from aquarium	Goleta	34.40	−119.83		137	22.6	PAT tag[a,c]
	9/11/2008	Caught in gillnet and released alive, with PAT tag still attached	Ventura	34.18	−119.28				
	10/16/2008	PAT tag transmits	San Miguel Island	34.07	−120.37				
09-11	8/13/2009	Caught in purse seine	Malibu	34.02	−118.68	F	157	36.2	
	11/4/2009	Released from aquarium	Monterey Bay	36.65	−121.95		166	45.4	PAT tag[c] SLRT tag
	2/25/2010	Caught in gillnet	El Rosario, North of Punta Baja	30.06	−115.88				
	3/02/2010	PAT tag transmits, after drifting at surface	Punta Blanco	29.03	−115.13				
	3/07/2010	SLRT tag transmits from house, is recovered from fisherman	San Quintin						

[a] PAT tag recovered.
[b] Hardware version PAT2.
[c] Hardware version MK10-PAT.

stainless-steel bolts (3 mm × 40 mm) coated with plastic shrink wrap so that no metal was in contact with the tissues of the animal. The ventral surface of the SPOT tag and the bolts were cleaned with povidone-iodine or isopropyl alcohol prior to attachment.

Analysis

Geolocations from PAT tags were estimated in two steps. Initial positions were estimated by threshold light-level geolocation (Hill and Braun, 2001) implemented within Wildlife Computers' proprietary software WC-GPE (version 1.02.0005). Light-level geolocations were filtered to remove poor light-curves and outliers and sequential longitudes that were more than 1°/d apart. Positions that passed through these filters were used as inputs into an sea-surface temperature (SST)-referenced unscented Kalman filter (uKFSST; Lam et al., 2010) to estimate most probable tracks. Maximum daily water temperatures reported by tags while within the top 10 m of the water column were cross-referenced to blended SST data averaged over 5 d (0.1 degree spatial resolution) from National Oceanic and Atmospheric Administration Coastwatch (http://coastwatch.pfeg.noaa.gov). Travel speeds from SLRT tags were calculated as great circle distance per unit of time and standardized to 24-h intervals to assess daily movement rates. Model parameters common to all track analysis included activation of SST bias estimation parameters and advection constrained to 74 km/d (40 nautical miles/d) (this study and Chapter 17, this book).

Geolocation estimates resulting in movements greater than 1°/d and locations on land were filtered out of final tracks. The root-mean-squared error between uKFSST-derived geolocations and concurrently observed SLRT tag positions on double-tagged animals (within 24 h of each other) were computed as a measure of PAT-tag geolocation accuracy.

For all available locations, distance to the mainland shoreline (excluding islands) was calculated, and depths were extracted from a bathymetry grid at 2-min. resolution (Smith and Sandwell, 1997). Different position uncertainties from SLRT and PAT tags were taken into consideration when extracting bathymetric depths. Distributions of PAT tag longitude and latitude geolocation errors were used to calculate a spatial kernel of joint probability densities (Figure 28.2). The 95% confidence bounds of the spatial grid were fit over each PAT tag geolocation, and depths were sampled from positions within. A weighted mean of these depths was calculated, using the height of the kernel at each sampled grid position as the weighting factor. Bathymetric depths extracted from Argos positions were assumed to be the bathymetric grid pixel in which they were located.

Hovmoller diagrams were used to visualize the complete records of depth, temperature, and light from archival records (Hovmoller, 1949). Instead of using the traditional plot of time versus space, which is used to highlight the existence of waves, we plotted time of day versus date in order to highlight diel patterns.

Depth and temperature analyses characterized the vertical and thermal habitat use of the animals. Because the transmitted records from PAT tags utilize predetermined time bins, which do not exactly correspond with changes of the diel cycles through time, we utilized archival records only. The light record from the tag was used to determine dawn, day, dusk, and night periods, and the depth and temperature data were assigned to these diel periods accordingly, following the method of Weng et al. (2007b). Analyses of data were performed in R (R, 2008) and MATLAB® (The Mathworks, Natick, MA).

RESULTS

Capture of Juvenile White Sharks

Juvenile White Sharks captured and held in captivity as part of this program are detailed in Table 28.1. Two juvenile White Sharks were captured as bycatch in commercial set-gillnet fisheries,

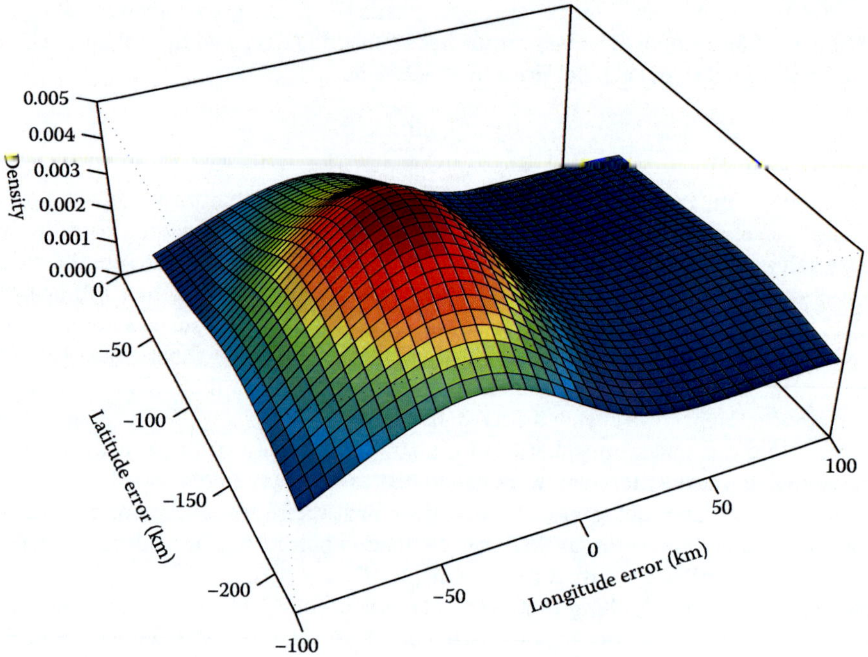

Figure 28.2 Distribution of geolocation errors for postcaptive juvenile White Sharks (n = 2) with respect to SLRT-tag Argos positions, PAT-tag pop-up positions (excluding LC = Z), and GPS starting positions. For all calculations, we filtered out data with speeds in excess of 1°/d. The peak in errors shows a southern bias.

one by targeted hook and line, and two by targeted purse-seine fishing. None were captured using longline gear.

Captivity

Juvenile White Sharks were held in the Monterey Bay Aquarium for durations ranging from 11 to 198 d and averaging 137 d (40–175 d) (median, interquartile range), compared to 4 d (1–10 d) for previous programs (Table 28.2). The news media covered the White Shark exhibit around the world. Detailed records of print and television news stories pertaining to the captive White Sharks were obtained for the period August 21, 2006 to October 30, 2006, and the audience was estimated at 23,961,200. Since the initiation of the White Shark exhibit, visits to the Aquarium's website have increased. In 2006, average page views to the web cam increased from 1,140/d to 5,514/d; furthermore, visits to the updates page increased from an average of 100/d to 2,500/d. Visitor surveys and interviews revealed that the White Shark was a factor in the decision to visit the Monterey Bay Aquarium. 45% of respondents were influenced by the first shark (shark 04-01), 32% by the second (shark 06-10), 18% by the third (shark 07-04), and 21% by the fifth (shark 09-11). The short time in captivity of the fourth shark (shark 08-11) meant that visitor interviews were not conducted.

Geolocation

Root-mean-squared errors for final tracks from uKFSST were 70.4 ± 61 km for longitude and 228 ± 39 km for latitude (Figure 28.2). Prior to application of the land mask and the speed filter, many position estimates were northwards of positions determined by SLRT tags, GPS release

Table 28.2 White Sharks Held in Captivity

Date	Aquarium	TL (m)	Sex	Days in captivity	Outcome	Reference
1955	Marine Land of the Pacific	1.4	NA	1	NA	Ellis and McCosker (1991)
March 1961	Waikiki Aquarium–Hawaii	4	NA	2	NA	Ellis and McCosker (1991)
December 1962	Marineland of Florida	2.4	Male	1.5	NA	Ellis and McCosker (1991)
1968	Durban Aquarium–South Africa	NA	Male	1	NA	Ellis and McCosker (1991)
May 1968	Manly Aquarium–Sydney	2.3	Male	10	NA	Malcolm et al. (2001)
1969	Sea World–San Diego	1.88	Male	8	NA	Powell (2001)
1974	Ocean World–Sydney	1.3	NA	10	Mortality	Malcolm et al. (2001)
1976	Sea World–Gold Coast Australia	3	NA	3	Mortality	Malcolm et al. (2001)
October 1978	Durban Aquarium–South Africa	2.1	Male	10	NA	Ellis and McCosker (1991)
8/12/1980	Steinhart Aquarium–San Francisco	2.3	Female	5	Released	Ellis and McCosker (1991)
Before 1980	Marineland of the Pacific	2	NA	NA	Released	R. Collier, personal communication
1981	Marineland of the Pacific	1.47	NA	7	NA	Ellis and McCosker (1991)
8/3/1981	Sea World–San Diego	1.6	NA	16	NA	Ellis and McCosker (1991)
1983	Steinhart Aquarium–San Francisco	NA	NA	NA	NA	Hewitt (1984)
1983	Ocean World–Sydney	2	Female	1	Mortality	Malcolm et al. (2001)
July 1984	Steinhart Aquarium–San Francisco	1.5	NA	4	NA	Hewitt (1984)
1984	Ocean World–Sydney	2	Female	1	Mortality	Malcolm et al. (2001)
9/10/1984	Monterey Bay Aquarium–Monterey	1.54	Male	11	NA	Ellis and McCosker (1991)
1994	Sea World–San Diego	1.55	Female	10	Released	Reidarson and McBain (1994)
1994	Ocean World–Sydney	2.1	Male	5	Released	ELASMO-L posting, I. Gordon
10/13/1997	National Museum of Marine Biology–Taiwan	0.5	NA	1	Mortality	V. Lin, personal communication
1998	Underwater World–Australia	1.6	Female	2	Released	Malcolm et al. (2001)
5/30/2000	Southern California Marine Institute, Los Angeles	1.4	Female	3.3	Released	C. Winkler, unpublished observation
	Median (interquartile range)	**1.9 (1.5–2.2)**		**4 (1–10)**		
9/14/2004	Monterey Bay Aquarium–Monterey	1.41	Female	198	Released	This chapter, shark 04-01

continued

Table 28.2 (Continued) White Sharks Held in Captivity

Date	Aquarium	TL (m)	Sex	Days in captivity	Outcome	Reference
8/31/2006	Monterey Bay Aquarium–Monterey	1.64	Male	138	Released	This chapter, shark 06-10
8/28/2007	Monterey Bay Aquarium–Monterey	1.43	Male	161	Released	This chapter, shark 07-04
8/27/2008	Monterey Bay Aquarium–Monterey	1.37	Female	11	Released	This chapter, shark 08-11
8/26/2009	Monterey Bay Aquarium–Monterey	1.57	Female	70	Released	This chapter, shark 09-11
	Median (interquartile range)	**1.4 (1.4–1.6)**		**137 (70–152)**		

In ten other captivity attempts by various aquariums, the sharks died in transport and are not listed in this table (J. O'Sullivan, unpublished observations).
Source: From Henry Mollet, http://homepage.mac.com/mollet/Cc/Cc_captive.html. With permission.

locations, and PAT-tag pop-up end points. After application of the land mask and speed filter, positions were biased south of SLRT-tag positions in double-tagged animals.

Geographic Movements

Satellite-tag data were collected from 2005 to 2010 (except 2006) and aggregately during all months (except July), ranging geographically from Monterey Bay, California, to Cabo San Lucas, Mexico, and into the Gulf of California (Figures 28.1 and 28.3a). Tracking durations for the five postcaptive sharks averaged 89 d (35–123 d) (median, interquartile range), as compared to 49 d (31–62 d) for the four wild young-of-the-year sharks tracked by Weng et al. (2007b). Although released during the months of January, February, March, August, and November, all of the postcaptive juvenile White Sharks moved south when liberated. They traveled 1.2 ± 4.1 km/d (mean ± SD; range, 0–32).

Shark 04-01, a 141 cm TL female, was captured in a commercial gillnet on August 20, 2004 and exhibited for 198 d. It was PAT-tagged on March 31, 2005 and released into Monterey Bay. It moved offshore to the southwest and beyond the 1,000-m contour before returning southeast to the coast. The PAT tag popped up on April 30, 2005 near Point Conception (Figure 28.3b) and was recovered.

Shark 06-10, a 164 cm TL male, was captured by rod and reel by aquarium staff on August 31, 2006 and exhibited for 137 d. It was PAT-tagged on January 16, 2007 and released into Monterey Bay. Geolocations indicate that it traveled initially to the southwest (Figure 28.3c), then toward the end of February entered a warmer subtropical (>22 °C) water mass, and turned back toward the coast. The shark's PAT tag transmitted on April 15, 2007 near Cabo San Lucas and was recovered.

Shark 07-04, a 143 cm TL male, was captured in a commercial gillnet, brought to the aquarium on August 28, 2007 and exhibited for 152 d. It was double-tagged (PAT and SLRT) on February 5, 2008 and released into Monterey Bay. Argos positions from the SLRT tag revealed that as it moved south it remained near the coast, within or proximal to the 1,000-m-depth contour (Figure 28.3d). It reached waters near Cabo San Lucas in mid-March and then continued into the Gulf of California. The SLRT tag showed it moving two-thirds of the way into the Gulf. On June 11, 2008, west of Isla Tortuga in the middle Gulf of California, shark 07-04's PAT tag descended at constant speed from near the surface to a depth greater than the sensor limit (>1,000 m) and remained there for 4 d (99 h), before it returned to the surface at a constant speed and began transmitting. The tag sank at 4.1 m/min. for 4 h 9 min. and ascended at 24 m/min. for 48 min. Initial transmission showed the tag was on the western slope of the Guymas Basin, a deep trough in the Gulf of California with a maximum depth of 2,000 m. The PAT tag began transmitting on June 16, 2008 and was recovered after drifting for several days. The SLRT tag continued transmitting for a further 17 d, in positions near the coast of Baja California Sur. The single position obtained on land was an Argos LC-B position and therefore had a high error.

Shark 08-11, a 137 cm TL female, was caught by a purse-seine net in Santa Monica Bay on August 16, 2008 and exhibited for 11 d. The shark was removed from display, transported to Goleta, California (near its initial capture location), and released on September 7, 2008. PAT-tag-derived geolocations suggested that the animal remained within the Southern California Bight during the track (Figure 28.3e). Four days after release, it was accidentally recaptured in a commercial

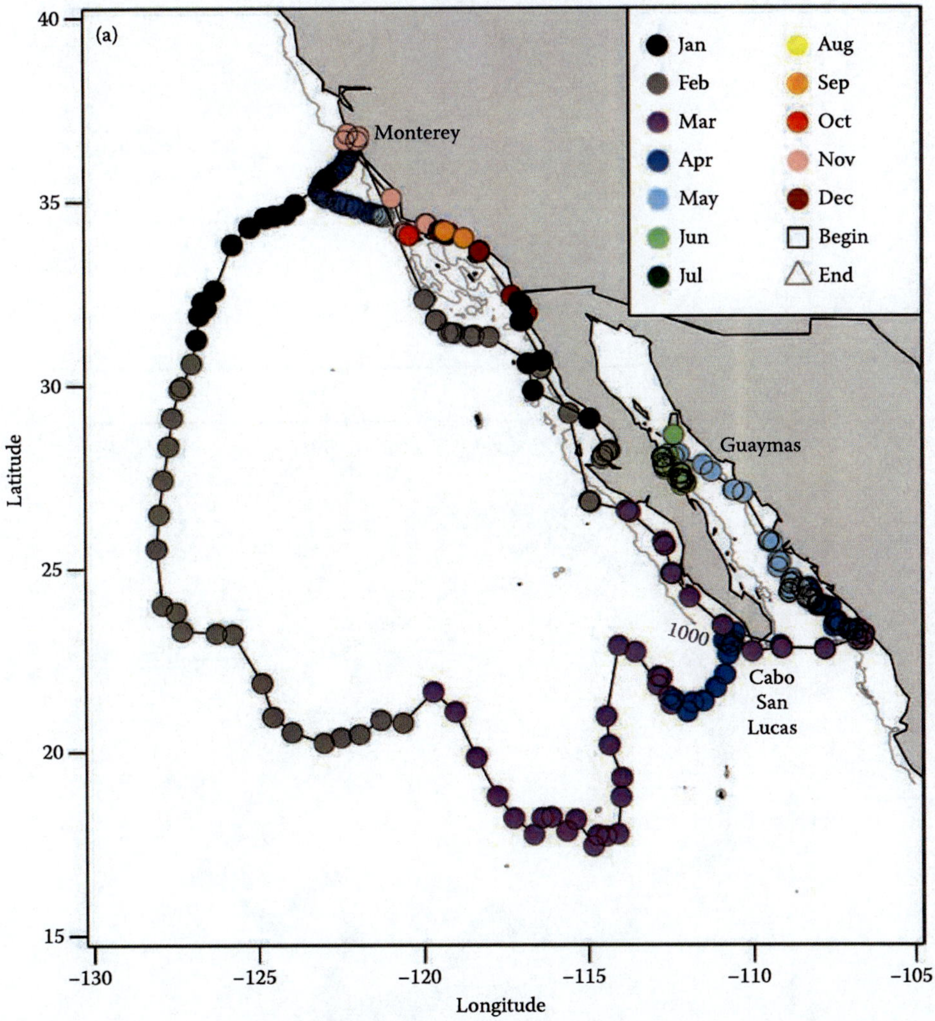

Figure 28.3 Movements of juvenile White Sharks released from the Monterey Bay Aquarium. (a) All animals. (b) Shark 04-01. (c) Shark 06-10. (d) Shark 07-04. (e) Inset showing shark 07-04 inside the Gulf of California. (f) Shark 08-11. (g) Shark 09-11. Colors denote month [legend in (a)], and numbers denote days of month. Tracks for sharks 04-01 and 06-10 were derived by light-SST geolocation from PAT tag records using uKFSST; track for shark 08-11 is not presented because of the small area occupied by the animal in comparison to light-SST geolocations errors; and tracks of sharks 07-04 and 09-11 are based on SLRT-tag Argos positions. The gray line shows the 1000-m depth contour.

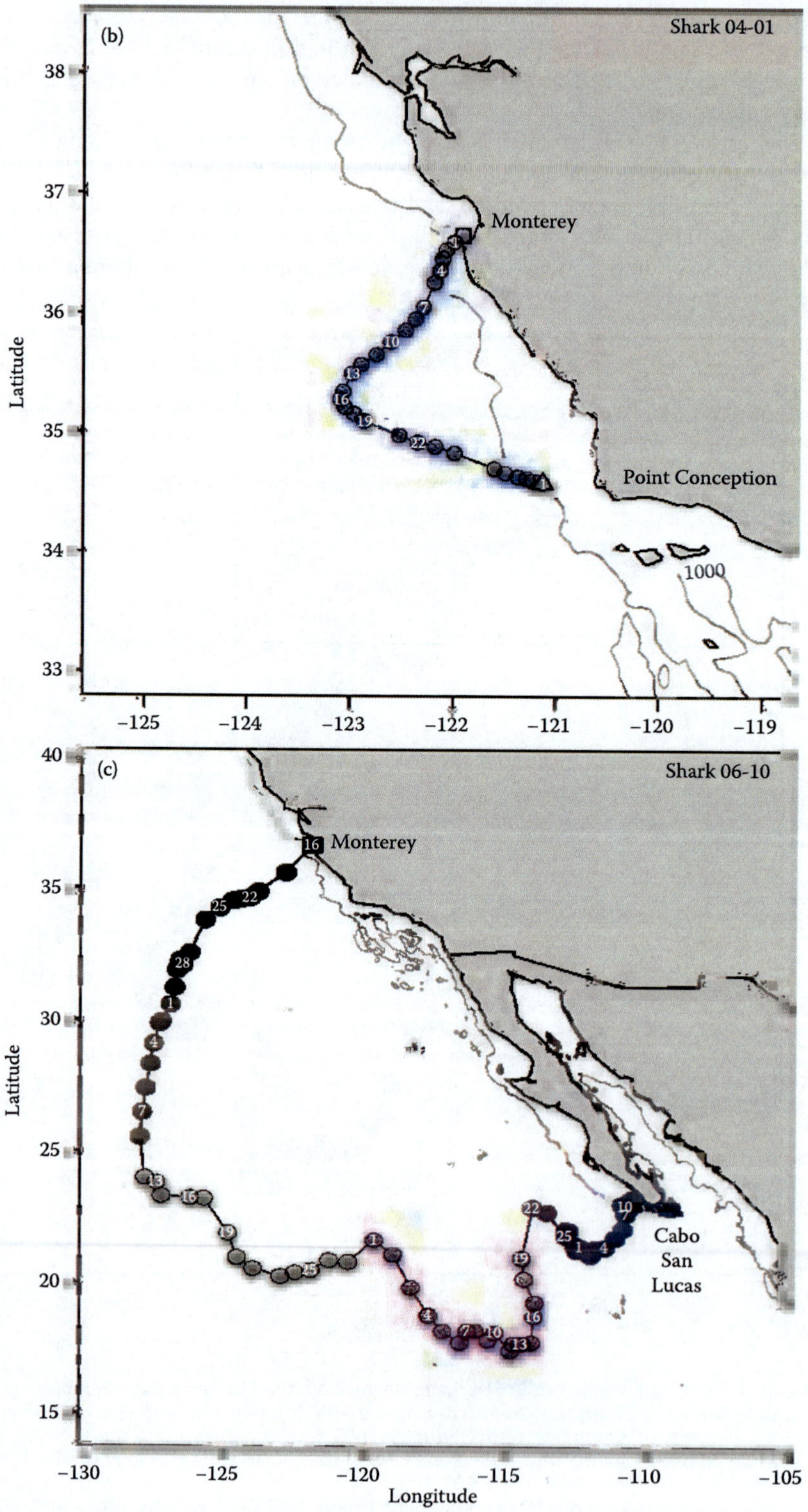

Figure 28.3 (Continued)

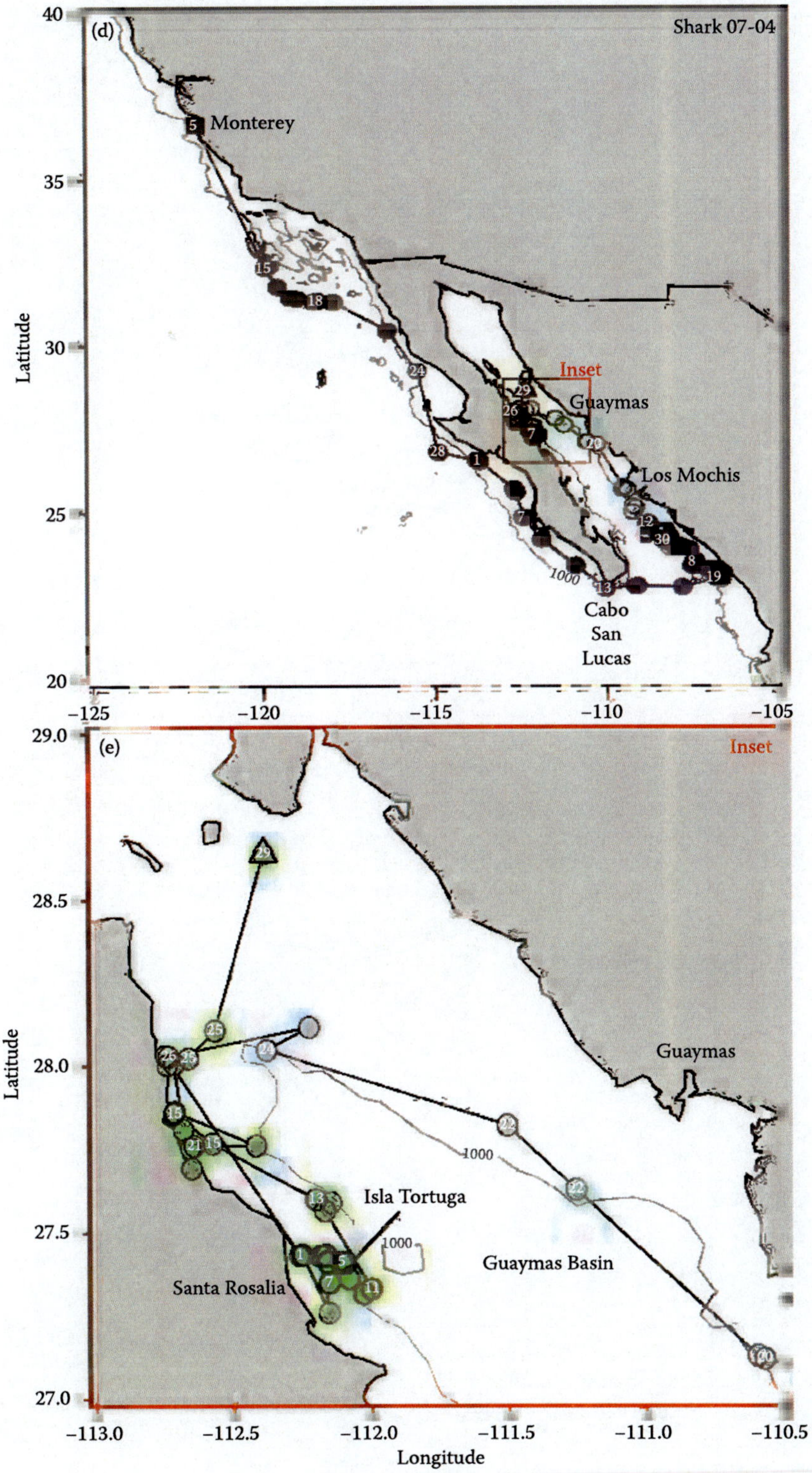

Figure 28.3 (Continued)

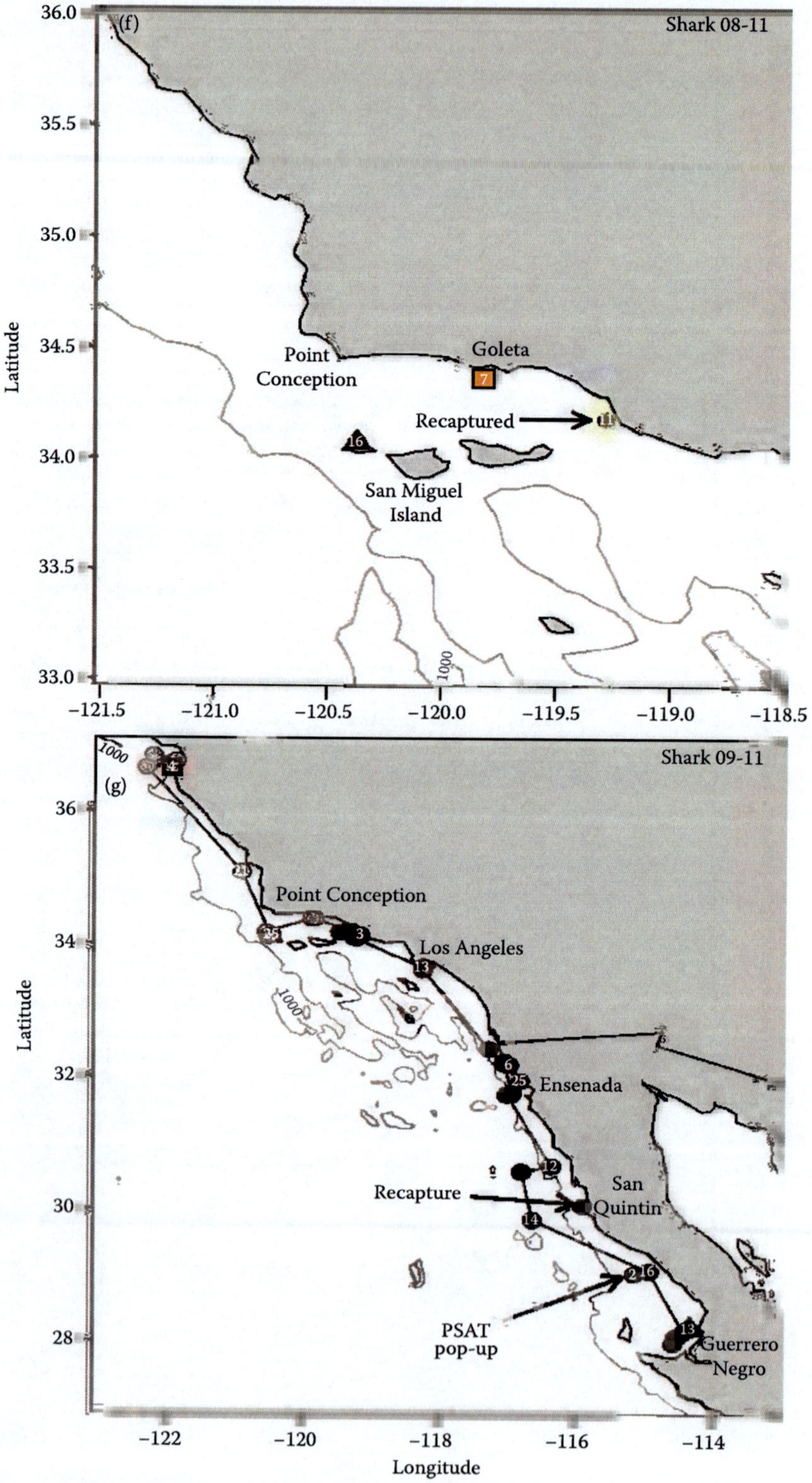

Figure 28.3 (Continued)

gillnet 35 km south of Goleta. The fisherman released the shark from the net, with the PAT tag still attached, and reported that the shark was very energetic and appeared full, presumably from feeding. The tag popped up 35 d later near San Miguel Island, California, and was recovered.

Shark 09-11, a 157 cm TL female, was captured by purse-seine net in Santa Monica Bay on August 13, 2009 and exhibited for 70 d. It was double-tagged (PAT and SLRT) on November 4, 2009 and released into Monterey Bay. The animal moved south in coastal waters reaching Bahia Sebastian Vizcaino (near Guerrero Negro), Baja California, Mexico, and then returned north (Figure 28.3f). The PAT tag released from the animal prematurely, drifted at the surface for 96 h, and then began transmitting. The SLRT tag began transmitting from a terrestrial location and was subsequently recovered from a fisherman near San Quintin, Baja California, Mexico. Events in the track of shark 09-11 are summarized in Table 28.3 (Figures 28.3 and 28.4).

Distance to Coast

Across all postcaptive sharks, 29% of all locations (SLRT tag and PAT tag combined, standardized to one location per animal per day) were within 37 km (~20 nautical miles) of the mainland coast (United States and Mexico), and 77% were within 370 km (~200 nautical miles) of the mainland coast (United States and Mexico), including a maximum offshore distance of 1,350 km (~730 nautical miles) during February 2007 (Figure 28.5).

Depth and Temperature

Postcaptive sharks showed similar overall habitat preferences to wild sharks, with similar median depths and temperatures during each diel period (Table 28.4). For all diel periods, for both depth and temperature, wild and postcaptive animals were not significantly different (rank-sum tests; Table 28.4). Given the small sample sizes (four captive and three wild sharks), the statistical power is low, and visual inspection of the distributions does indicate differences (Figure 28.6). Both wild and postcaptive animals showed a preference for near-surface waters and engaged in diving behavior to depths up to 300 m (Figures 28.6 and 28.7).

There were differences in upper temperature range and diel behavior between the two groups (Figure 28.6). The upper limit of the ambient temperature range was higher for postcaptive sharks (26°C) than wild sharks (22.5°C) because of their greater penetration into subtropical waters. The lower limit of temperature was similar for wild (9.5°C) and postcaptive (8.5°C) sharks. Diel behavior (Figure 28.6) was similar for wild and postcaptive sharks during day, dawn, and dusk, while postcaptive sharks utilized a broader depth range during night, largely because of nocturnal diving behavior.

The archival depth record for shark 04-01 revealed diving during both day and night. The track did not contain any evidence of demersal swimming (Carey and Robison, 1981; Carey, 1990) or close association with the benthos (Figure 28.7), providing further indication that the shark was offshore (Figure 28.3). The offshore movement of shark 04-01 resulted in it occupying warmer waters than if it had stayed near to shore (Figure 28.8), because the upwelling effect of the California Current depresses coastal water temperatures (Bograd et al., 2000).

Shark 06-10 displayed a pattern of diving at night for much of its track and showed no indication of swimming along the bottom, consistent with a position in deep offshore waters (Figure 28.7). During the first phase of the track, when the animal was in Monterey Bay and cool upwelled waters, it exhibited little diving, and upon entering warmer offshore waters, it began diving at night (Figures 28.7 and 28.8). Visual inspection of the detailed archival record showed no evidence of swimming along the bottom.

Shark 07-04 exhibited nocturnal diving during much of its track (Figure 28.7). Beginning in mid-April, there were numerous instances where the archival depth record suggested that the

Table 28.3 Events in the Track of Shark 09-11

Date	Event	Location
8/13/2009	Caught by purse seine	Santa Monica
9/1/2009	On display	Monterey Bay Aquarium
11/5/2009	Released to wild, swam to south (Figure 28.3)	Monterey Bay
1/23/2010 to 2/5/2010	Time-at-depth remains primarily at 75 m for 10 d, long before gillnet recapture, vertical excursions to surface occur rarely (Figure 28.4)	
2/6/2010 to 2/25/2010	PAT depth sensor shows depths from 30 to 75 m (8)	
2/13/2010	Last spot transmission from water and furthest south point in the track	Guerrero Negro
2/26/2010	PAT tag floats to surface	Gillnet site, El Rosario de Arriba
3/2/2010	PAT tag begins transmitting 132 km south of gillnet site (after 4 d drifting at surface with north winds). Pin was broken prior to initiation of release. PAT tag washed up at Isla de Arena in Guerrero Negro region, not recovered	North of Guerrero Negro
3/7/2010	SLRT tag begins transmitting from land	San Quintin
4/13/2010	Team member recovers SLRT tag at fisherman's house. Fisherman reported that gillnet was left out 1 week because of bad weather. Upon haul, shark was found partially eaten and decomposing with SLRT tag on dorsal fin.	San Quintin

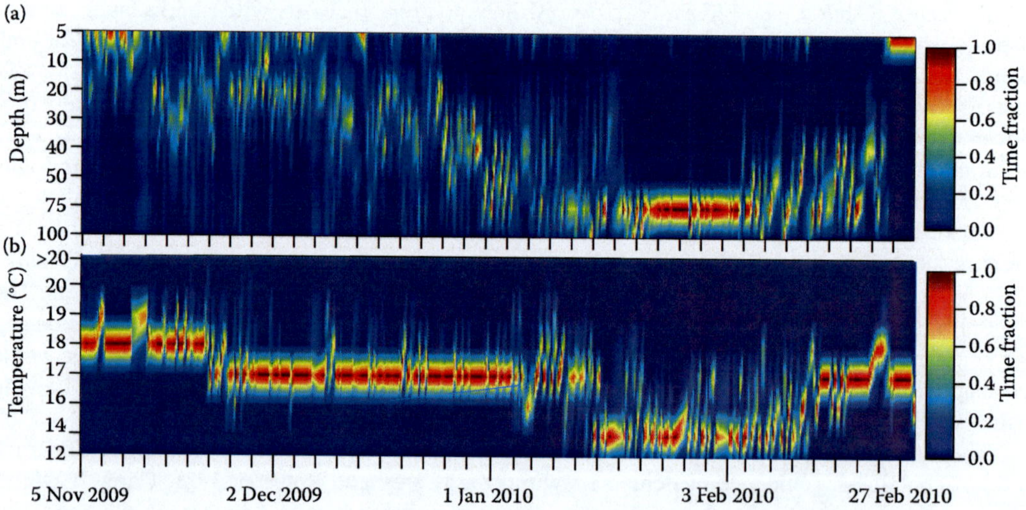

Figure 28.4 Proportion of time spent at depth (a) and temperature (b) during track of shark 09-11. The *x* axis shows time, and the *y* axis shows depth and temperature bins, whereas the colors represent times spent in given ranges. A period at nearly constant depth (75 m) occurs during January. In the days prior to the surfacing of the tag, there is no record of constant depth, as would occur if the tag were on a shark in a gillnet. Therefore, it is likely that the tag was pulled from the shark during the initial struggle.

animal was swimming along the bottom (Figure 28.7). During this period, the shark was near the mainland coast of Mexico between Mazatlan and Los Mochis, and in May and June it was near the eastern coast of the Baja Peninsula in the area of Mulege and Santa Rosalia (Figure 28.3d).

Shark 08-11 showed diving behavior consistent with swimming along the bottom during all periods of the track (Figure 28.7). It inhabited waters typical of the Southern California Bight and showed the greatest amount of diving during day and crepuscular periods (Figure 28.8). Archival data were used to perform geolocations, which indicated that the shark remained within the Southern California Bight region during the track.

Shark 09-11's tag was not recovered, but the transmitted record provided summary information (Figure 28.4). For its first 10 d at liberty, this shark spent most of its time in the top 10 m of the water column where water temperatures were 18–19°C. During the following 5 weeks, its depths oscillated between the surface and 40 m, with water temperatures remaining consistently around 17°C. From late December 2009, it began spending increasing amounts of time between 40- and 100-m depth, with water temperatures at these depths dropping to 14–16°C. After nearly 2 weeks spent predominantly at 75–100 m, the animal began oscillating again between 40 and 100 m with water temperatures increasing to 17–18°C.

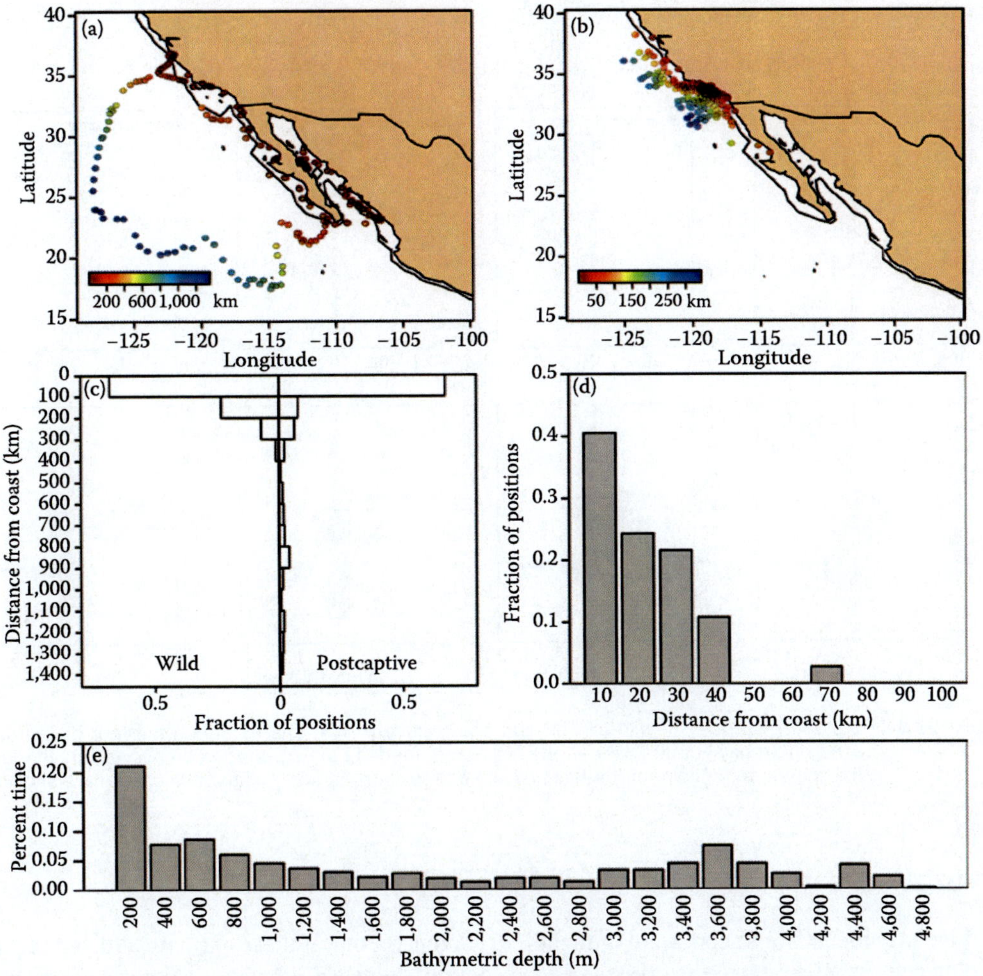

Figure 28.5 Distance from mainland coast for the tracks of juvenile White Sharks. The black line is the 1,000-m-depth contour. (a) Distance from mainland for postcaptive sharks (n = 5). (b) Distance from mainland for animals tagged in the wild (n = 6; Weng et al., 2007b). (c) Distance from mainland for wild versus postcaptive sharks. (d) Distance from mainland for SLRT tags, which provide accurate geopositions (n = 2 postcaptive sharks). (e) Bathymetric depths for the locations estimated for postcaptive sharks. Panels c through e are standardized to one location per animal per day to remove bias of multiple SLRT-tag locations available on many days.

Table 28.4 Median (Interquartile Range) Depths and Temperatures for Postcaptive and Wild Sharks during the Four Periods of the Diel Cycle

Period	Postcaptive	Wild	Rank-Sum *p*
Dawn	13 (27–10) m	13 (20–4) m	0.86
Day	5 (7–5) m	11 (13–1) m	0.51
Dusk	24 (37–15) m	14 (12–4) m	0.23
Night	18 (37–11) m	5 (8–2) m	0.06
Dawn	16 (15–17) ºC	16 (15–17) ºC	1
Day	18 (15–19) ºC	16 (16–18) ºC	0.63
Dusk	16 (15–17) ºC	16 (16–17) ºC	0.29
Night	17 (15–18) ºC	18 (17–19) ºC	0.4

The differences between the postcaptive and wild sharks were tested with the rank-sum test.

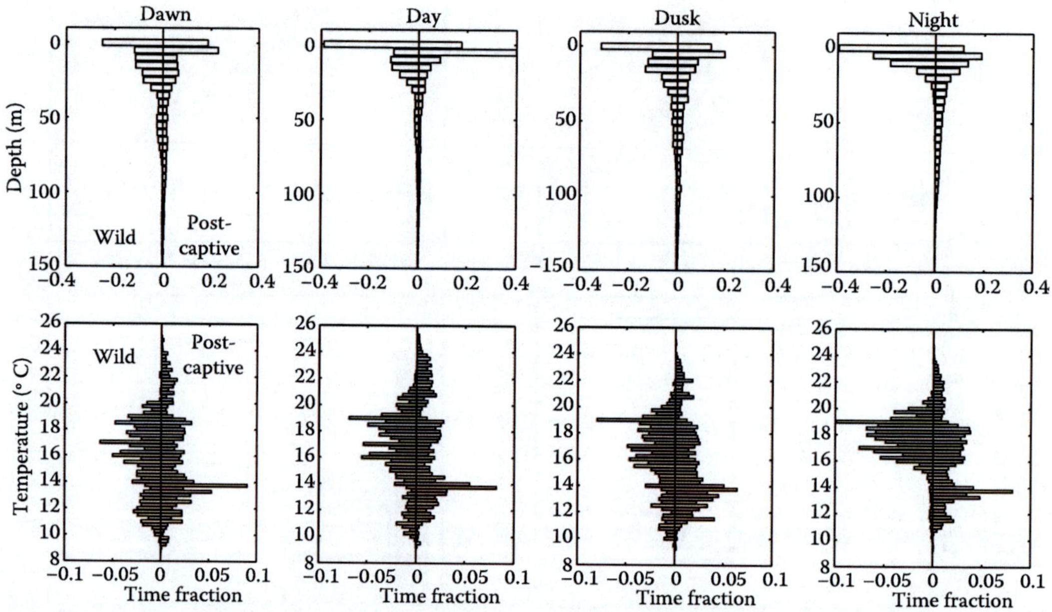

Figure 28.6 Habitat preferences of wild (n = 3) and postcaptive (n = 4) sharks during the four diel periods based on archival data. Diel periods are determined via the light records collected by the tags. Wild sharks are shown on the left side of each histogram, and postcaptive sharks are shown to the right.

DISCUSSION

Five juvenile White Sharks were brought into captivity. Four of these animals initiated regular feeding and assimilated into the diverse pelagic community displayed within the aquarium's Outer Bay Wing. A fifth animal did not initiate regular feeding and was released after 11 d. The exhibit of these animals contributed to large increases in the number of people reached through education and outreach, thereby having a positive impact on public awareness of White Sharks and some of the conservation issues facing them. These sharks were later released back to the wild. In order to evaluate whether postcaptive sharks showed natural behaviors, we compared their geographic

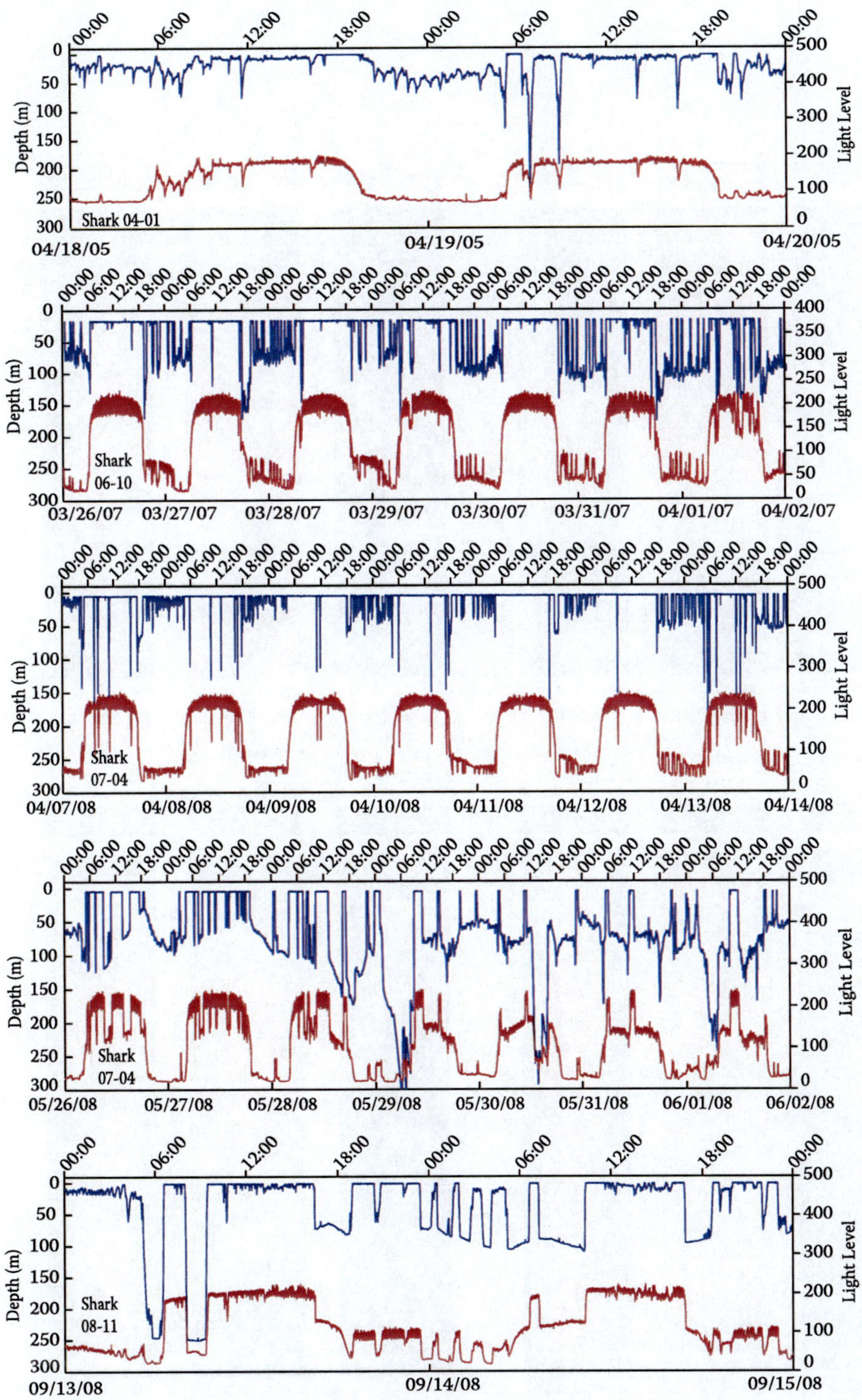

Figure 28.7 Vertical-movement pattern of juvenile White Sharks based on archival data. Lower *x* axis is date; upper *x* axis is time of day; left *y* axis is water column depth, shown by the blue line; right *y* axis is ambient light level (arbitrary units) shown by the magenta line. No data are presented for shark 09-11 because this tag was not recovered.

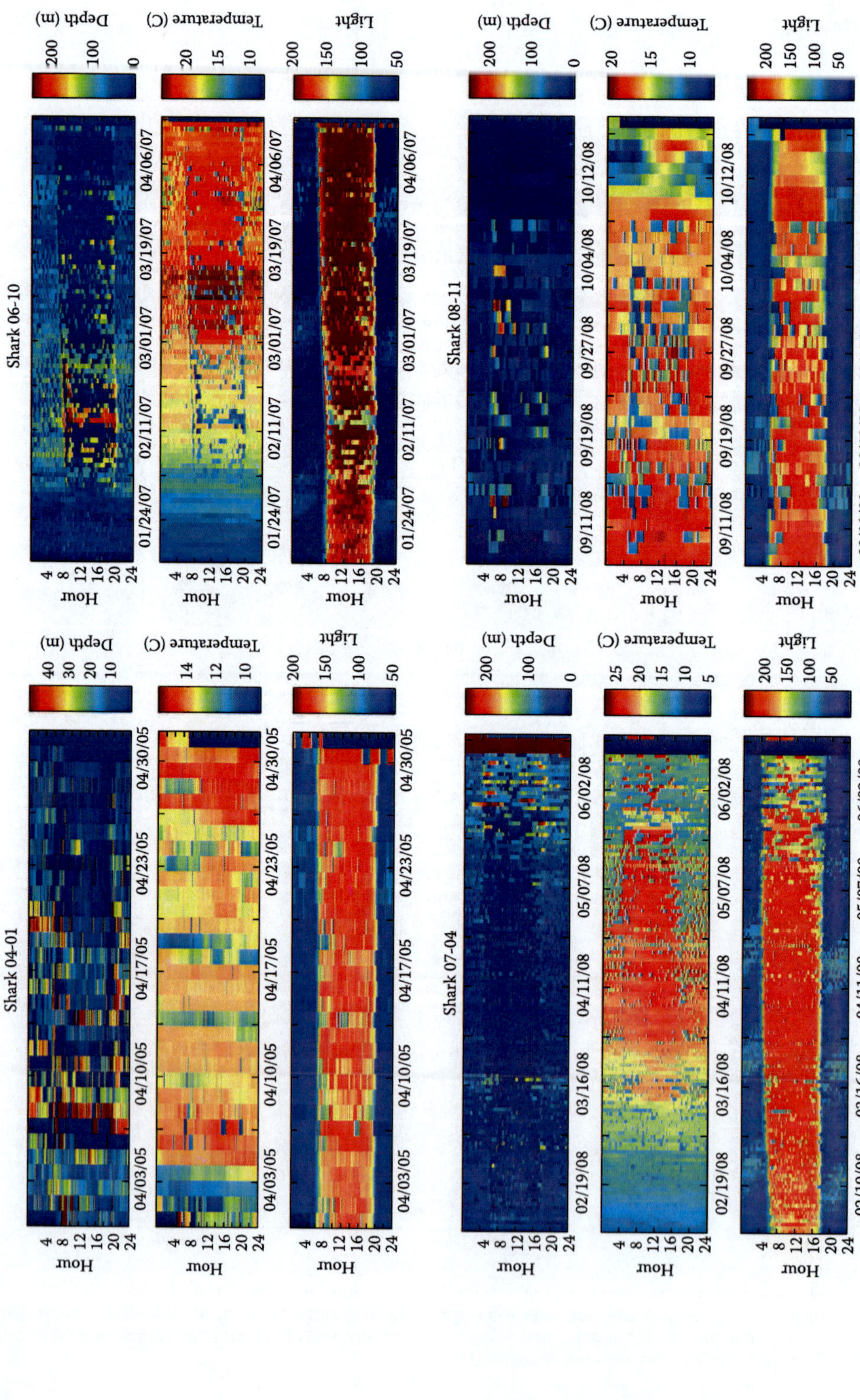

Figure 28.8 Hovmoller-style time series plots of depth, temperature, and light for juvenile White Sharks released from the Monterey Bay Aquarium. The *x* axes show date of track observations, the *y* axes show the 24-h diel cycle, and colors represent depth within the water column, ambient temperature, or light. The central band of each plot corresponds to daytime, and the upper and lower bands correspond to nighttime. High values for depth in the central band mean diurnal diving and vice versa.

movements, as well as depth and temperature preferences, with wild sharks (Bruce et al., 2006; Dewar et al., 2004; Klimley et al., 2002; Weng et al., 2007b). Once released back to the wild, the animals undertook southward movements similar to juvenile White Sharks tagged in the wild (Weng et al., 2007b). Both wild and postcaptive sharks preferred near-surface waters, undertook dives to 300 m, and at times showed close association with the benthos. Postcaptive sharks had a warmer upper temperature range, because of the longer tracks and greater penetration into the subtropics. Diel patterns differed between the two groups, potentially because of occupancy of different regions.

Geolocation

Short-duration PAT tags were attached to the first two captive-released animals to investigate survivorship for 1–2 months following release. Subsequent PAT tags were programmed to remain on the sharks for longer durations to investigate larger-scale movements. SLRT tag attachment techniques developed for Salmon Sharks (*Lamna ditropis*) (Weng et al., 2005) were adapted for these juvenile White Sharks, and double-tagging provided both higher spatial-resolution movement data, and a benchmark for comparison to PAT-tag-derived geolocations in the California Current.

Errors for light-SST based geolocation in this study were comparable to those in the study of Weng et al. (2007b). Longitude errors were 70.4 ± 61 km compared to 54 ± 36 km (mean ± SD), and latitude errors were 228 ± 39 km compared to 231 ± 159 km.

Coastal SST fields differ from pelagic fields in that they tend to be less meridionally structured, which affects their utility in geolocation models. Furthermore, as shown here, juvenile White Sharks spend much of their time proximal to the shoreline and the ocean's surface where solar heating increases temperatures recorded by the tags. These biased water temperatures are included in summary data transmitted by PAT tags. Warm bias in SST data causes equatorward bias in latitude estimates, which is southward in the northern hemisphere. The removal of uKFSST position estimates on land and those removed by the speed filter resulted in positions with a southerly bias, and this southerly bias has been noted previously (Weng et al., 2007b).

The recovery of four PAT tag archival data sets contributed high-resolution data for geolocation estimation and assessment of animal distribution through the water column. The summarized depth data transmitted by PAT tags was comparable to the archival depth data, indicating that the transmitted summary data were generally representative of patterns in the archival data.

The uncertainties of geolocations estimated from environmental data time series collected by archival tags are well known (Musyl et al., 2001). However, advances in the application of time-series methods using light and sea-surface temperature data have made strides toward better light-based geolocations (Lam et al., 2010). Galuardi et al. (2010) developed a useful method of post hoc correction of geolocations with bathymetry data. Efforts to incorporate more environmental data like depth and subsurface temperature directly into state-space methods should improve likelihoods (Patterson et al., 2010), particularly in oceanographically complex regions like the California Current.

Survival following Release

Geographic movement information along with diving behavior recorded by PAT tags confirmed that sharks 04-01, 06-10, 07-04, and 08-11 survived for the duration of tracking (30, 89, 132, and 39 d, respectively). Shark 09-11 survived following release and moved south to Baja California but was captured and died in a set gillnet 113 d after release.

Postcaptive behaviors likely to be benthic foraging were similar to those noted in previous studies (Dewar et al., 2004; Weng et al., 2007b). These data indicated that the postcaptive sharks were likely to be feeding and thus had been successfully reintroduced to the wild. Evidence of feeding was also obtained from the accidental recapture of shark 08-11 4 d after release. This animal was reported to be in energetic condition and to have a full stomach by the fisherman. Given the short duration since this shark's release, it appears it required very little time to readapt to the wild.

The PAT tag attached to shark 07-04 sank to the bottom, but available the data indicate that the animal itself did not sink and survived beyond this event. In order to reach this conclusion, we evaluated possible scenarios as follows:

1. The animal died and sank.
2. The animal was attacked by a predator, and the tag sank with a portion of the body or alone because of bio-fouling.
3. The animal was captured by fishermen, the tag was discarded, and it sank because of bio-fouling.
4. The dart came out or the tether broke, and the tag sank because of bio-fouling.
5. The tag was bitten by a predator and pulled from the shark and then sank due to bio-fouling.

The pin on the tag was intact at the time the release command was sent, meaning the tag had probably not been forcefully detached from the leader, and the leader was likely still attached. The tag was programmed to sever linkage to its tether after 96 h at constant depth, thus removing the weight of the leader and dart, potentially allowing the tag to return to the surface, explaining the 4 d at >1,000 m.

The continuation of SLRT-tag transmissions confirmed the animal's survival after PAT-tag detachment, eliminating scenarios 1 and 2. The SLRT-tag positions were not characteristic of a Baja fishing operation, such as straight-line tracks or high-quality points on land near a settlement or encampment, eliminating scenario 3. These data leave us with scenarios 4 and 5.

In order to determine whether the sinking rate was plausible for a tag with barnacles growing on it, an identical PAT tag with identical leader was used for a tank experiment. A small *Balanus* spp. barnacle of 5.5 g (weight in air), similar to barnacles that have been observed on other PAT tags, produced a sinking rate of 5.1 m/min. Therefore, a slightly smaller barnacle on the tag could have produced the sinking rate of 4.1 m/min. We are unable to distinguish between options 4 and 5 with the available data. As such, the most parsimonious interpretation of the data is that the tag pulled from the shark with leader and dart attached; the weight of leader, dart, and biofouling resulted in negative buoyancy causing it to sink to the ocean floor; and the leader and dart were disconnected when the tag's program severed the pin, allowing it to resurface and transmit.

Shark 09-11 also presented an unusual data set, although in this case we are certain that the shark was captured and killed in a commercial set gillnet. Based on the records from the PAT tag, the SLRT tag, and an interview with the fisherman who caught the shark, we believe the following is the most likely sequence of events. The animal reached a southernmost point in Guerrero Negro waters where transmissions were received from the SLRT tag, then returned north, and was captured in a set gillnet near El Rosaria de Arriba. The PAT tag came off the shark in the initial struggle in the net, because there is no period at a constant depth in the record (the fisherman reported the bottom-set net was in waters 64 m deep), and the pin was broken before transmission, suggesting it was ripped off the animal forcefully, leaving dart and leader behind. The PAT tag drifted to the south because of strong north winds (which were reported during this period) and transmitted after 96 h (as programmed to detect constant depth). The SLRT tag began transmitting from land on March 7, 2010, likely a short time after the gillnet was pulled. Therefore the shark was most likely in the gillnet from February 26 (when the PAT tag floated to the surface) to March 7, 2010 (when the SLRT tag resumed transmissions).

Connectivity of Eastern Pacific Nursery Area, Phenology, and Neritic/Pelagic Habitats

Postcaptive sharks moved further into subtropical habitats than wild individuals, reaching waters of Baja California Sur and the Gulf of California. Detection of these greater southerly movements was possible because of longer tracking durations and higher quality SLRT-tag data for postcaptive fish than was available for wild sharks. Wild sharks may also show movements to southern Baja and the Gulf of California if longer records are obtained. The existence of juvenile White Sharks in the Gulf of California is known from fishery records and other observations (Galvan-Magana et al., 2010; Chapter 15, this book). While YOY sharks have not been recorded in the Gulf of California, the movements of an adult female lead the authors of Chapter 16, this book to hypothesize that the area was a pupping ground. The connectivity of the Southern California Bight region with the Gulf of California that is demonstrated by these movements indicates that these areas may comprise a single nursery for White Sharks in the Eastern Pacific.

In the Southern California Bight, young-of-the-year White Sharks appear to undertake southern movements during summer and autumn (Weng et al., 2007b). Sharks 04-01, 06-10, and 07-04 were released during winter and into Central California waters colder than would normally be experienced by sharks of this age-class south of Point Conception (Klimley, 1985; Weng et al., 2007b). Sharks released into Monterey Bay may have been strongly motivated to move southwards toward warmer waters. Sharks 04-01 and 06-10 undertook southward movements but did not stay near the coast. Shark 08-11 was released in the Southern California Bight (Goleta) in the autumn, and southerly movements were not detected during its 39-d track. Because it was released into the typical habitat and temperature for an animal of this age class, motivations for large-scale movements may have been different than those released into cooler waters. It was recorded to be actively feeding during an accidental recapture and thus appears to have quickly readapted to the wild. Shark 09-11 was released into Monterey Bay in the autumn and undertook a southward movement to Pacific waters of Baja California, remaining near the coast.

Juvenile White Sharks appear to primarily occupy neritic habitats in the eastern Pacific (Klimley, 1985; Weng et al., 2007b; Chapters 14 and 16, this book), although in other regions they have been shown to occupy pelagic regions (Bruce et al., 2006). Sharks 04-01 and 06-10 showed greater offshore movements than wild sharks. Geolocations from uKFSST (this study), the method of (Teo et al., 2004), as well as manual matching of daily SST images to temperature measurements from the tag, all concurred that the animals made offshore movements. These offshore movements may have been a response to the winter release season: southwest movements from the California coast bring animals into warmer offshore waters, once they cross the relatively narrow zone of upwelling. Alternately, these offshore movements may have resulted from disorientation following release. Shark 07-04 did not provide data until south of Point Conception, but the remaining data from its track indicated that it remained near the coast, as did sharks 08-11 and 09-11.

Habitat Preferences and Behavior

Periods of nocturnal diving, with shallower distributions during the day, were observed for sharks 06-10 and 07-04. Previous studies have shown that diving by pelagic fishes could provide directional cues (Klimley et al., 2002; Westerberg, 1984). Diving exposes fish to horizontal current shear layers that contain different olfactory information and also brings animals closer to the highly variable electromagnetic field of the earth's crust (and reduces the signal of the main dipole). Returning to the surface may provide access to celestial cues or polarized light directionality (Klimley et al., 2002). Hence, vertical movements may enhance navigation, and if visual cues to navigation were available during the day, then vertical movements would be unnecessary.

This reversed diel behavior, in which more dives occur during night than during day, did not occur for shark 08-11, which was transported south and released in the Southern California Bight. Because it did not undertake a directed movement from the region, it may not have been searching for navigational cues. Nocturnal diving was observed during traveling phases for adult White Sharks moving through the California Current between California and the central Pacific, typically in the winter (Weng et al., 2007a), and may serve to provide orientation cues at this later life stage also.

The association of young-of-the-year White Sharks with the benthos has been observed in prior studies (Bruce et al., 2006; Dewar et al., 2004; Klimley et al., 2002; Weng et al., 2007b) and was first observed in Swordfish (Carey and Robison, 1981; Carey, 1990). This behavior is believed to be indicative of foraging, because diet analysis suggests that this life stage targets demersal resources more than pelagic (Casey and Pratt, 1985; Dewar et al., 2004; Klimley, 1985; Tricas and McCosker, 1984). Potential prey species accessible to young-of-the-year White Sharks are reviewed by Dewar et al. (2004) and Weng et al. (2007b).

Klimley at al. (2002) collected bathymetric data concurrent with active acoustic tracking, demonstrating a close association with the benthos. Detailed archival records are available from four animals from this project, and of these, two displayed behaviors indicative of demersal swimming. These two individuals, sharks 07-04 and 08-11, exhibited the behavior while in waters south of Point Conception. Although detailed dive data are not available, summary data suggest that shark 09-11 also spent nearly 2 weeks in an apparent demersal swimming mode (Figure 28.4).

The absence of demersal swimming north of Point Conception suggests that movements of juvenile White Sharks in the region are primarily driven by the search for warmer southerly waters. Sharks 04-01 and 06-10 both fit this pattern. For both sharks, the records end as soon as the animal nears the coast, so we are unable to determine whether demersal swimming ensued once the animal was in a nearshore habitat.

Vulnerability to Fishing Gear and Management of White Shark Nurseries

Studies of White Sharks in the eastern Pacific indicate that juvenile White Sharks are the most vulnerable to commercial fishing, and thus mortality is higher for this age class than for adults. Studies of adults and subadults have not revealed the capture of tagged individuals in fisheries (Jorgensen et al., 2009; Klimley, 1985; Weng et al., 2007a), whereas capture of young-of-the-year and small juveniles in the eastern Pacific is routine (Weng et al., 2007b; Chapters 14, 15, and 16, this book). Our research program has identified gillnets as the primary risk to young-of-the-year and juvenile White Sharks in the eastern Pacific (Chapter 14, this book).

In addition to understanding their geographic range, documenting the vertical distribution of White Sharks in the water column contributes to a better understanding of their vulnerability to different types of fishing gear. Vertical-habitat data showing a preference for the surface mixed layer and upper thermocline indicate that juvenile White Sharks may be most susceptible to fishing gear deployed at these depths. Dewar et al. (2004) suggested that because most diving occurs during the day, nocturnal gillnet effort should have lower bycatch of juvenile White Sharks than diurnal gillnet effort. Our results suggest that nocturnal diving may occur, so that this strategy may have limited benefit. In regions where mixed layer temperatures are cooler than 16°C, juvenile White Sharks rarely make vertical excursions and thus should be less vulnerable to deeper gear. Northward movements of larger juveniles may reduce the risk of encountering bottom-set gillnets, because this gear is presently banned north of Point Conception, California.

Our results show that juvenile White Sharks utilize near-coast waters as important nursery areas. In Southern California waters, the set-gillnet fishery was closed in waters inside 3 nautical miles of the coast in 1994 (California State Proposition 132). Because the set-gillnet fishery is excluded from waters within 3 nautical miles of the California coast, the remaining fishery occurs in federal waters. Under federal and state agreement, the State of California regulates the set-gillnet fishery, and among the mandates is the release of captured White Sharks. Our data reveal that the 3-nautical-mile gillnet ban is likely to have had a strongly positive impact on juvenile White Sharks. Anecdotal evidence based on interviews with long-time fisherman is consistent with this, because catches of juvenile White Sharks were routine when gillnets operated within 3 nautical miles of the coast (Chapter 14, this book).

Survival in Gillnets

Of the 38 White Sharks historically taken for display by other aquariums (Table 28.2), fewer than five were targeted by aquarium staff in the field searching for this species (I. Gordon, personal communication; Hewitt, 1984). Sharks obtained opportunistically from fishermen, which were held on tethered lines or in an inadequate tank until aquarium staff could retrieve them, probably suffered from high capture stress (Hewitt, 1984; Stevens, 1994). Of these, there is only one record of a White Shark observed feeding in an aquarium (Malcolm et al., 2001). The prevailing wisdom is that capture, transport, and captivity stresses, inability to induce feeding, and inadequate tank design have contributed to past failures (Stevens, 1994). Capture stress appears to result in acidosis (respiratory and metabolic), hypoglycemia, and possibly even cellular damage (Manire et al., 2001), with entanglement duration and degree of struggle being among the important factors in the likelihood of survivorship.

Two of the five animals that were held in captivity were originally captured in commercial gillnets. The long duration of stay in the Monterey Bay Aquarium, as well as successful return to the wild that these two animals displayed, indicates that juvenile White Sharks can survive capture in gillnets. White Sharks, at least at this stage, are not obligate ram ventilators, and our team has observed individuals buccal pumping to ventilate their gills while in livewells. The most important factor in survivorship in gillnets appears to be the soak time. In instances where animals were recovered dead in gillnets, fishermen reported that weather or mechanical problems delayed their hauling time, causing a soak time greater than 24 h. In instances where soak time is less than 24 h, survivorship of juvenile White Sharks appears favorable.

Conclusions

From 2004 to 2008, the Monterey Bay Aquarium brought five juvenile White Sharks into captivity, allowing the public to view this rare species. All five individuals were released back to the wild and readapted to wild conditions, assuming behaviors similar to wild sharks. One of these individuals was later captured and killed in a set gillnet. The successful feeding of these sharks and their integration into an established captive community are very likely important factors in their captive well-being. Results from this and other studies show that the Southern California Bight, waters off Pacific Baja California, and the Gulf of California comprise a key nursery area for White Sharks in the eastern Pacific Ocean. The primary management concern for the species in this region is the bycatch of juveniles in set-gillnet fisheries; however, tagging data suggest relatively high postrelease survivorship. Management should be coordinated between the United States and Mexico.

ACKNOWLEDGMENTS

Funding was provided by the Monterey Bay Aquarium. This research would not have been possible without the close cooperation and assistance of California gillnet fishermen. We thank Nick Guglielmo, Chris Williams, Ben Henke, Terry Wilmarth, Richard Deluna, Mike Ettinger, Kip Whited, Steve Pantos, Bob Ball, Tommy Ptak, Curtis Hebert, Andy Rasmussen, and other gillnet fisherman who have helped. Assistance with fieldwork and logistics was provided by Ken Bates, Carolina Espinoza, Scott Reid, Joe Welsh, Chuck Farwell, Annette Lenneman, and Dick Winkler. We thank the students at the California State University Long Beach Shark Lab who assisted with field work for this program: Erica Jarvis, Kim Anthony, Lyall Bellquist, Tom Mason, Chris Martin, Carlos Mirelese, Kate Jirik, Misty Paig, Heather Gliniak, Hayley Zemel, Chris Mull, Bonnie Rogers, Traci De La Torre, Lu Acosta, Derek Castleman, Lyall Bellquist, Eugene Chae, Lori Hale,

Jeremy Vaudo, Darin Topping, Barbara Zeigler, JoAnn May, George Sims, Kady Lyons, Carrie Espasandin, Megan McKinzie, Thomas Farrugia, Erika Fox, Mario Espinosa, Barrett Wolf, and Kim Johnson. Tag recoveries were made possible by Shane Anderson, Eddy Kisfaludy, and Oscar Sosa-Nishizaki. Chris Perle assisted with tag programming and logistics, and Johnoel Ancheta and Jeff Muir assisted with figures. Information on aquarium attendance and programs was provided by Randy Hamilton and Jim Covel, and information on previous captivity attempts was provided by Henry Mollet. Jeannette Post, Tiffany Davies, and Dodie Lau provided administrative support. J.O. managed the program; C.L., C.W., M.B., and K.L. managed the rapid response team; J.O., C.L., C.W., K.W., M.M., J.E., and S.J. conducted fieldwork; K.W. and T.S. analyzed data; and K.W. wrote the chapter.

REFERENCES

Argos. 2003. *Argos User Manual*. Largo, MD: Service Argos, Inc.

Bestley, S., Patterson, T.A., Hindell, M.A., and Gunn, J.S. 2008. Feeding ecology of wild migratory tunas revealed by archival tag records of visceral warming. *Journal of Animal Ecology* 77: 1223–1233.

Bograd, S.J., Digiacomo, P.M., Durazo, R., Hayward, T.L., Hyrenbach, K.D., Lynn, R.J., Mantyla, A.W., Schwing, F.B., Sydeman, W.J., Baumgartner, T., Lavaniegos, B., and Moore, C.S. 2000. The state of the California Current, 1999–2000: Forward to a new regime? *California Cooperative Oceanic Fisheries Investigations Reports* 41: 26–52.

Brennan, N.P., Darcy, M.C., and Leber, K.M. 2006. Predator-free enclosures improve post-release survival of stocked common snook. *Journal of Experimental Marine Biology and Ecology* 335: 302–311.

Bruce, B.D., Stevens, J.D., and Malcolm, H. 2006. Movements and swimming behaviour of white sharks (*Carcharodon carcharias*) in Australian waters. *Marine Biology* DOI 10.1007/s00227-006-0325-1.

Campana, S.E., Joyce, W., and Manning, M.J. 2009. Bycatch and discard mortality in commercially caught blue sharks *Prionace glauca* assessed using archival satellite pop-up tags. *Marine Ecology-Progress Series* 387: 241–253.

Carey, F.G. 1990. Further acoustic telemetry observations of swordfish. In *Planning the Future of Billfishes*, edited by R. Stroud, Second International Billfish Symposium, Part 2. August 1–5, 1988: Kailua-Kona, Hawaii, National Coalition for Marine Conservation.

Carey, F., and Robison, B. 1981. Daily patterns in the activities of swordfish, *Xiphias gladius*, observed by acoustic telemetry. *Fishery Bulletin* 79: 277–292.

Casey, J.G., and Pratt, H.L., Jr. 1985. Distribution of the white shark, *Carcharodon carcharias*, in the western North Atlantic. *Memoirs Southern California Academy of Sciences* 9: 2–14.

Dewar, H., Domeier, M., and Nasdy-Lucas, N. 2004. Insights into young of the year white sharks (*Carcharodon carcharias*) behavior in the Southern California Bight. *Environmental Biology of Fishes* 70: 133–143.

Domeier, M.L., Dewar, H., and Nasby-Lucas, N. 2003. Mortality rate of striped marlin (*Tetrapturus audax*) caught with recreational tackle. *Marine and Freshwater Research* 54: 435–446.

Ellis, R., McCosker, J.E., and Giddings, A. 1995. *Great White Shark*. Stanford, CA: Stanford University Press.

Falk, J.H., Reinhard, E.M., Vernon, C.L., Bronnenkant, K., Heimlich, J.E., and Deans, N.L. 2007. *Why Zoos & Aquariums Matter: Assessing the Impact of a Visit to a Zoo or Aquarium*, p. 24. Silver Spring, MD: Association of Zoos and Aquariums.

Farwell, C. 2001. *Tunas in Captivity. Volume 19: Fish Physiology*, pp. 391–412. Orlando, FL: Academic Press.

Fergusson, I., Compagno, L., and Marks, M. 2000. Carcharodon carcharias. In *2007 IUCN Red List of Threatened Species*. http://www.iucnredlist.org.

Frick, L.H., Reina, R.D., and Walker, T.I. 2010. Stress related physiological changes and post-release survival of Port Jackson sharks (*Heterodontus portusjacksoni*) and gummy sharks (*Mustelus antarcticus*) following gill-net and longline capture in captivity. *Journal of Experimental Marine Biology and Ecology* 385: 29–37.

Galuardi, B., Royer, F., Golet, W., Logan, J., Neilson, J., and Lutcavage, M. 2010. Complex migration routes of Atlantic bluefin tuna (*Thunnus thynnus*) question current population structure paradigm. *Canadian Journal of Fisheries and Aquatic Sciences* 67: 966–976.

Galván-Magaña, F., Hoyos-Padilla, E.M., Navarro-Serment, C.J., and Márquez-Farías, F. 2010. Records of white shark, *Carcharodon carcharias*, in the Gulf of California, Mexico. *Marine Biodiversity Records* 3: e111.

Graves, J.E., Luckhurst, B.E., and Prince, E.D. 2002. An evaluation of pop-up satellite tags for estimating postrelease survival of blue marlin (*Makaira nigricans*) from a recreational fishery. *Fishery Bulletin* 100: 134–142.

Hampton, J. 1986. Effect of tagging on the condition of southern bluefin tuna, *Thunnus maccoyii. Australian Journal of Marine and Freshwater Research* 38: 699-705.

Hanan, D.A., Holts, D.B., and Coan Jr., A.L. 1993. *California Department of Fish and Game Fish Bulletin 175: The California Drift Gill Net Fishery for Sharks and Swordfish 1981–82 through 1990–91*. California Department of Fish and Game.

Hays, G.C., Akesson, S., Godley, B.J., Luschi, P., and Santidrian: 2001. The implications of location accuracy for the interpretation of satellite-tracking data. *Animal Behaviour* 61: 1035–1040.

Hewitt, J.C. 1984. The great white shark in captivity: A history and prognosis. *AAZPA Annual Conference Proceedings* pp. 317–323.

Hill, R.D., and Braun, M.J. 2001. Geolocation by light level. In *Electronic Tagging and Tracking in Marine Fisheries: Proceedings of the Symposium on Tagging and Tracking Marine Fish with Electronic Devices*, p. 315.

Hovmoller, E. 1949. The trough and ridge diagram. *Tellus* 1: 62–66.

Inskipp, T., and Gillett, H.J. 2005. *Checklist of CITES Species and Annotated CITES Appendices and Reservations*, p. 339. Geneva, Switzerland: CITES Secretariat & Cambridge, UK: UNEP-WCMC.

Jorgensen, S.J., Reeb, C.A., Chapple, T.K., Anderson, S., Perle, C., Van Sommeran, S.R., Fritz-Cope, C., Brown, A.C., Klimley, A.P., and Block, B.A. 2009. Philopatry and migration of Pacific white sharks. *Proceedings of the Royal Society B Biological Sciences* 277: 679–688.

Kerstetter, D.W., Luckhurst, B.E., Prince, E.D., and Graves, J.E. 2003. Use of pop-up satellite archival tags to demonstrate survival of blue marlin (*Makaira nigricans*) released from pelagic longline gear. *Fishery Bulletin* 101: 939–948.

Klimley, A.P. 1985. The areal distribution and autecology of the white shark, *Carcharodon carcharias*, off the West Coast of North America. *Memoirs Southern California Academy of Sciences* 9: 15–40.

Klimley, A., Beavers, S., Curtis, T., and Jorgensen, S. 2002. Movements and swimming behavior of three species of sharks in La Jolla Canyon, California. *Environmental Biology of Fishes* 63: 117–135.

Lam, C.H., Nielsen, A., and Sibert, J.R. 2010. Incorporating sea-surface temperature to the light-based geolocation model, TrackIt. *Marine Ecology-Progress Series* 419: 71–84.

Malcolm, H., Bruce, B.D., and Stevens, J.D. 2001. *A Review of the Biology and Status of White Sharks in Australian Waters*, p. 114. Hobart, Australia: CSIRO Marine Research.

Manire, C., Hueter, R., Hull, E., and Spieler, R. 2001. Serological changes associated with gill-net capture and restraint in three species of sharks. *Transactions of the American Fisheries Society* 130: 1038–1048.

Mollet, H.F., and Cailliet, G.M. 2002. Comparative population demography of elasmobranchs using life history tables, Leslie matrices and stage-based matix models. *Marine and Freshwater Research* 53: 503–516.

Musyl, M.K., Brill, R.W., Curran, D.S., Gunn, J.S., Hartog, J.R., Hill, R.D., Welch, D.W., Eveson, J.P., Boggs, C.H., and Brainard, R.E. 2001. Ability of archival tags to provide estimates of geographical position based on light intensity. In *Electronic Tagging and Tracking in Marine Fishes*, edited by J.R. Sibert and J.L. Nielsen, pp. 89–108. The Netherlands: Kluwer Academic Publishers.

Musyl, M.K., Moyes, C.D., Brill, R.W., and Fragoso, N.M. 2009. Factors influencing mortality estimates in post-release survival studies. *Marine Ecology-Progress Series* 396: 157–159.

O'Brien, J.W., and Sunada, J.S. 1994. A review of the southern California experimental drift longline fishery for sharks 1988–1991. *California Cooperative Oceanic Fisheries Investigations Reports* 35: 222–229.

Patterson, T.A., McConnell, B.J., Fedak, M.A., Bravington, M.V., and Hindell, M.A. 2010. Using GPS data to evaluate the accuracy of state-space methods for correction of Argos satellite telemetry error. *Ecology* 91: 273–285.

Powell, D.C. 2001. *A Fascination for Fish. Adventures of an Underwater Pioneer*. University of California Press: Berkley/Los Angeles, CA.

Reidarson, T., and McBain, J. 1994. Hyperglycemia in two great white sharks. *IAAAM Newsletter* 25: 4.

R. 2008. *R: A language and environment for statistical computing*. Vienna, Austria.

Sherrill-Mix, S.A., and James, M.C. 2008. Evaluating potential tagging effects on leatherback sea turtles. *Endangered Species Research* 4: 187–193.

Skomal, G.B. 2007. Evaluating the physiological and physical consequences of capture on post-release survivorship in large pelagic fishes. *Fisheries Management and Ecology* 14: 81–89.

Smith, W., and Sandwell, D. 1997. Global sea floor topography from satellite altimetry and ship depth soundings. *Science* 277: 1956–1962.

Stevens, J.D., Bonfil, R., Dulvy, N.K., and Walker, P.A. 2000. The effects of fishing on sharks, rays, and chimeras (chondrichthyans), and the implications for marine ecosytems. *ICES Journal of Marine Science* 57: 476–494.

Stevens, J.E. 1994. The delicate constitution of sharks. *Bioscience* 44: 661–664.

Teo, S.L.H., Boustany, A., Blackwell, S., Walli, A., Weng, K.C., and Block, B.A. 2004. Validation of geolocation estimates based on light level and sea surface temperature from electronic tags. *Marine Ecology Progress Series* 283: 81–98.

Tricas, T.C., and McCosker, J.E. 1984. Predatory behavior of the white shark (*Carcharodon carcharias*), with notes on its biology. *Proceedings of the California Academy of Sciences*. 43: 221–238.

Weng, K.C., Castilho, P.C., Morrissette, J.M., Landiera-Fernandez, A., Holts, D.B., Schallert, R.J., Goldman, K.J., and Block, B.A. 2005. Satellite tagging and cardiac physiology reveal niche expansion in salmon sharks. *Science* 310: 104–106.

Weng, K., Boustany, A., Pyle, P., Anderson, S., Brown, A., and Block, B. 2007a. Migration and habitat of white sharks (*Carcharodon carcharias*) in the eastern Pacific Ocean. *Marine Biology* 152: 877–894.

Weng, K.C., O'Sullivan, J.B., Lowe, C.G., Winkler, C.E., Dewar, H., and Block, B.A. 2007b. Movements, behavior and habitat preferences of juvenile white sharks *Carcharodon carcharias* in the eastern Pacific. *Marine Ecology Progress Series* 338: 211–224.

Westerberg, H. 1984. The orientation of fish and the vertical stratification at fine- and micro-structure scales, In *Mechanisms of Migration in Fishes*, edited by J. McCleave, G. Arnold, J. Doddon, and W. Neill, pp. 179–203. New York, NY: Plenum Press.

White, N.A., and Sjoberg, M. 2002. Accuracy of satellite positions from free-ranging grey seals using Argos. *Polar Biology* 25: 629–631.

Wood, C.M., Turner, J.D., and Graham, M.S. 1983. Why do fish die after severe exercise? *Journal of Fish Biology* 22: 189–201.

Yalowitz, S.S. 2004. *White Shark Survey 2004*. Monterey, CA: Monterey Bay Aquarium.

Yuen, H.S.H., Dizon, A.E., and Uchiyama, J.H. 1975. *Notes on the Tracking of the Pacific Blue Marlin, Makaira nigricans, International Billfish Symposium 1972*, pp. 265–268. NOAA Technical Report, NMFS, SSRF-675.

CHAPTER 29

Shark Spotters

A Pioneering Shark Safety Program in Cape Town, South Africa

Alison Kock*
University of Cape Town, Soetwater Education Centre, and Save Our Seas Foundation Shark Centre

Sarah Titley
Soetwater Education Centre

Wally Petersen
Soetwater Education Centre

Monwabisi Sikweyiya
Soetwater Education Centre

Sakhile Tsotsobe
City of Cape Town, South Africa

Darryl Colenbrander
City of Cape Town, South Africa

Howard Gold
City of Cape Town, South Africa

Gregg Oelofse
City of Cape Town, South Africa

ABSTRACT

In Cape Town, South Africa, White Sharks (*Carcharodon carcharias*) are responsible for the majority of bites on recreational water users. Although relatively rare, shark bites that result in human injury or death threaten existing shark-protective measures through negatively influencing public perception regarding sharks and shark conservation, the possible introduction of culling programs by authorities and illegal hunting. Furthermore, shark bites can negatively impact local business and tourism. A unique shark safety program was initiated in Cape Town in 2004 as a partnership between the City of Cape Town and local community organizations with the aim of finding a balance between White Shark conservation and recreational water-user safety. This initiative, called Shark Spotters, employs 14 to 28 spotters at 4 to 10 of Cape Town's popular beaches (numbers are season-dependent). The Shark Spotters scan coastal waters for sharks from elevated positions during daylight hours, 7 days a week, 365 days a year. A protocol using four informational flags and a shark siren warn water users of the nearby presence of sharks. Upon hearing the warning, water users are encouraged to exit the ocean and wait until the Shark Spotters give the "all clear" before returning to the water. This proactive measure aims to reduce the risk of a shark bite and the associated negative impacts for both humans and sharks. Data are recorded on the number of sharks detected, shark behavior, sea conditions, and the number of water users. During the period November 2004 to December 2009, 619 shark sightings were recorded. Data are presented on the seasonal occurrence, duration, and distance from shore of shark sightings for each beach, in addition to the frequency of sounding the shark alarm and resultant beach closures. Limitations of the program include the ability to detect sharks in poor environmental conditions (water visibility, wind, and glare), human error, and operating hours. Despite these limitations, the program has proven to be an effective shark safety system at appropriate beaches. Additional advantages of the program are job creation and skills development for historically disadvantaged South Africans, environmental education and awareness, and the collection of information that contributes to a better understanding of White Shark coastal occurrence and behavior.

INTRODUCTION

Globally, human-wildlife conflict is a growing obstacle to achieving conservation goals (Gore et al., 2008; Madden, 2008; Nyhus et al., 2003). Although relatively rare, shark bites that result in human injury or death threaten existing shark-protective measures through negatively influencing public perception regarding sharks and shark conservation and through the possible introduction of culling programs by authorities and illegal hunting. Furthermore, shark bites can negatively impact business and tourism, particularly at the very local level of affected coastal towns and beaches, while also impacting negatively on the broad perceived amenity value of the coast by the public. These economic impacts and negative public perceptions may compel authorities to take action with the aim of alleviating the perceived conflict.

Traditionally, solutions to wildlife-human conflict are the extermination of so-called "problem animals" (Sillero-Zubiri and Stwizer, 2004). Worldwide methods of reducing shark attack include the implementation of barriers, culling programs, chemical and electronic deterrents, acoustic telemetry, or a combination of these (Dudley et al., 2006; Chapter 31, this book). In South Africa, along the KwaZulu-Natal coast, an extractive program has been employed since 1952 to reduce the number of sharks in the region, which has resulted in a significant reduction in the number of shark attacks along that stretch of coast (Cliff and Dudley, 1992; Davies, 1963; Dudley et al., 2006). This extraction program uses a combination of large-mesh gillnets (shark nets) and drumlines (Cliff and Dudley, 1992; Davies, 1963; Dudley et al., 1998, 2006). However, shark nets and drumlines carry high environmental costs (Dudley and Cliff, 1993), and globally there is a move away from extraction programs like these, with emphasis on finding more environmentally responsible solutions of mitigating human-wildlife conflict (Dudley et al., 2006; Macdonald and Sillero-Zubiri, 2002).

White Sharks are a threatened species and fully protected in South Africa (Compagno, 1991; Compagno et al., 1997). Seal Island in False Bay is a known White Shark aggregation area (Ferreira and Ferreira, 1996; Kock and Johnson, 2006; Laroche et al., 2008; Martin et al., 2005). Cape Town, with a thriving tourism industry and a local population of 3.5 million people, with special reference to False Bay, is a large recreational area for thousands of swimmers, surfers, divers, kayakers, spearfishermen, kiteboarders, and windsurfers (Dotchim, 2006). This combination of relatively high numbers of White Sharks present in a bay used annually by thousands of recreational water users results in a capricious situation with increased opportunities for conflict between sharks and humans.

In the last decade, Cape Town has experienced a significant increase in White Shark–human interactions, including a total of 13 recorded shark bites between January 2000 and December 2009, of which three were fatal attacks between September 2003 and December 2009 (Cliff, 2006; G. Oelofse and Y. Kamp, unpublished data). This directly led to increased pressure on authorities to address the issue and find ways to reduce the risk of shark encounters. In 2006, after a spate of attacks, the City of Cape Town convened a special workshop aimed at addressing a range of issues around shark bites in an attempt to manage the situation (Nel and Peschak, 2006). After this extensive review, including evaluating the implementation of shark barriers, shark nets and drumlines, and electronic repellents (Dudley et al., 2006; Nel and Peschak, 2006), it was recommended that a novel community initiative, called the Shark Spotters program, be endorsed as a method to mitigate shark-human encounters in Cape Town (Oelofse and Kamp, 2006).

MATERIALS AND METHODS

In Cape Town, Shark Spotting started on an *ad hoc* basis by local community groups in Muizenberg and Fish Hoek in 2004 (Oelofse and Kamp, 2006). Through a formal partnership with the City of Cape Town and the establishment of a structured nonprofit organization, named Shark Spotters, the program was extended to an additional 10 beaches over a 3-yr. period on the Cape Peninsula and False Bay (Figure 29.1). The program is managed and administered by the Kommetjie Environmental Awareness Group and employs one project manager, one field manager, one research manager, and 14 permanent Shark Spotters. An additional six Shark Spotters are hired over the summer period (September to April). The annual operational budget for the program is R1.5 million. The program has been dependent on funding by the City of Cape Town since September 2005 with official funding of R1 million guaranteed per annum to June 2012. A further 5 yrs. of funding may be available pending a review of the program's performance for the 2007–2012 period. In 2010, the Save Our Seas Foundation became a primary funder, allocating R300,000 per annum for a minimum of 2 yrs. Local businesses and individuals provide support, equipment, or supplementary funding to make up the shortfall.

Shark Spotters are permanently based (7 days a week, 365 days a year) at four of Cape Town's popular beaches, namely Muizenberg, St. James, Fish Hoek, and Noordhoek, and at an additional two beaches, Clovelly and Glencairn, over the summer period to coincide with the influx of seasonal tourists and increase in beach usage over the summer months (Table 29.1). Between 2006 and 2009, a further four spotting initiatives were operational at Strandfontein, Bluewaters, Mnandi, and Monwabisi during the summer months (September to April), in association with South African Surf Lifesaving (Table 29.1). However, because of funding limitations and inadequate vantage points close to these beaches, combined with low numbers of shark sightings, these four programs were discontinued in the summer of 2009. Spotting programs were also attempted at Kommetjie and the Strand, but after less than 6 months of operation, it was determined that the logistics of operation

Figure 29.1 Map of the Cape Peninsula and False Bay in the Western Cape, South Africa showing the location of Shark Spotters programs at popular beaches. Muizenberg, Fish Hoek, St. James, and Noordhoek are operational throughout the year (P); Bluewaters, Clovelly, Strandfontein, Mnandi, Monwabisi, and Glencairn are operational only during the summer months (S); and the Strand and Kommetjie were temporary programs that are no longer operational (T).

were not workable because of an inadequate vantage point at Kommetjie, and the Strand was too far away from the base of operations. Thus, these two programs were also discontinued.

At beaches where Shark Spotters programs are operational, two Shark Spotters are assigned to each beach. One spotter is positioned at an elevated point above the beach (Figure 29.2), while the second spotter is positioned on the beach. All spotters on duty are in radio contact with one another. The spotter spotting for sharks is equipped with polarized sunglasses to reduce glare, binoculars to scan waters below, a two-way radio, and a remote to set off a beach siren, which functions as a shark alarm. Once the elevated spotter detects a shark, the siren is sounded, warning flags are changed, and water users are actively encouraged to leave the water. The spotter on the beach is responsible for changing the flags, assisting water users out of the water, and providing information to the public. To reduce spotting fatigue and human error, workdays are broken down into two shifts per day of no longer than six hours per shift. In summer months (September to April) shifts are from 7 a.m. to 1 p.m. and 1 p.m. to 7 p.m., whereas winter shifts (May to August) are from 8 a.m. to 1 p.m. and 1 p.m. to 6 p.m.

Each beach has individual characteristics that affect spotting conditions, e.g., elevation of lookout, position in relation to the sun, gradient and topography of the beach, and size of the surf zone. Some beaches, e.g., Muizenberg (Figure 29.3a), have large surf zones (>300 m wide), whereas other beaches, e.g., Fish Hoek (Figure 29.3b), have smaller surf zones (<100 m wide). This has an impact on the type of water activities that are practiced in the area. At Muizenberg, surfing is one of the most popular activities, whereas very little surfing takes place in Fish Hoek, but it is very popular for swimming and kayaking.

Table 29.1 Description of the Status, Duration, Beach Type, and Elevation of Shark Spotters Programs at 10 of Cape Town's Beaches

Beach	Date Started	Status	Operational Period	Number of Spotting Days	Primary Beach Type	Elevation Provided by	Distance above Seal Level (m)	Total Number of Shark Sightings
Bluewaters	12/15/2005	Ended 2009	September to April	523	Sand	Lifesaving tower/ building	10–15	9
Clovelly	10/10/2007	Current	September to April	479	Sand/ reef	Mountainside	40–50	3
Fish Hoek	9/12/2004	Current	All year	1935	Sand/ reef	Mountainside	110–120	138
Glencairn	12/8/2006	Current	September to April	622	Sand/ reef	Mountainside	70–80	10
Mnandi	12/15/2005	Ended 2009	September to April	523	Sand	Lifesaving tower/ building	10–15	5
Monwabisi	12/15/2005	Ended 2009	September to April	523	Sand	Lifesaving tower/ building	10–15	4
Muizenberg	11/14/2004	Current	All year	1872	Sand	Mountainside	90–100	341
Noordhoek	1/1/2007	Current	All year	1095	Sand	Mountainside	40–50	33
St. James	8/19/2006	Current	All year	1229	Sand/ reef	Mountainside	70–80	75
Strandfontein	12/15/2005	Ended 2009	September to April	523	Sand/ reef	Lifesaving tower/ building	10–15	1

New Shark Spotters are trained by experienced Shark Spotters and fishermen on how to detect sharks and fish and what to do in the event of a sighting in the field. Further training is provided on medic first aid and how to provide immediate medical care for shark bite victims. Workshops and seminars on sharks and associated issues are held yearly. Field experience is provided through White Shark research trips and outings on White Shark cage diving trips to increase the learning experience and knowledge of sharks, in particular White Sharks. The experience of interacting very closely with the sharks assists the spotters in species identification and an in depth knowledge of the species, which they in turn can convey to the public. Seminars to improve shark spotter

Figure 29.2 Photograph of a Shark Spotter at a lookout above Fish Hoek beach.

Figure 29.3 Aerial view of two of Cape Town's most popular beaches where Shark Spotters programs are in place, Muizenberg (a) and Fish Hoek (b).

communication skills are also provided to increase the spotter's ability to communicate effectively with the public.

Effective communication between the Shark Spotters and recreational beach users at the relevant beaches are crucial to the success of the program. If a shark is sighted near the beach, it is imperative that there are adequate mechanisms in place to ensure that water users are made aware of the presence of the shark(s) and how to respond accordingly. The program uses a combination of four flags, a siren, and signage located at various points on the beaches to communicate shark activity and spotting conditions to the public (Figure 29.4). Color-coded flags are used to communicate the conditions of spotting ability as well as shark sightings to the public:

1. A green flag means that the spotting conditions are good, and no sharks have been seen.
2. A black flag means that the spotting conditions are poor, but no sharks have been seen.

Figure 29.4 Example of the multilingual signage used at beaches in Cape Town to communicate to the public the existence of the Shark Spotters program, the operational protocol, and the daily status of shark activity and visibility conditions at a particular beach.

3. A white flag with a black shark diagram means that a shark is currently near the beach.
4. A red flag serves as a warning that a shark was recently sighted in the area and/or that there are conditions existing that may mean an increase in shark activity, such as large schools of fish in the area or whale strandings, which may attract increased numbers of sharks.

In addition to providing warnings to the public to mitigate shark encounters, a daily log of activities and sightings information is kept. In addition to documenting shark sightings, Shark Spotters' record times of duty, person on duty, weather and sea conditions, flags flown, and the number of water users present (Figure 29.5a). Sightings of dolphins, whales, schools of fish, and fishing activities are also recorded. In the event of a shark sighting, detailed information is collected and filled in on a separate sheet (Figure 29.5b). The date, time of day, estimated (rough) shark size (small, medium, or large), duration of sighting (start and end), behavior of shark (e.g., present in or behind the surf zone, direction of movement, exhibiting patrolling behavior, or simply swimming past the beach) while visible, and a map of the route the shark swam are recorded. Shark sightings are recorded as one sighting until the spotter loses visual contact with the shark. If a shark is sighted a few minutes later again, but the spotter is not completely confident that it is the same shark seen earlier, it is then recorded as a separate sighting.

To aid in identification of sharks, Shark Spotters take note of the animal's size, coloration, shape, position of dorsal, pelvic, and caudal fins and swimming style. The dark dorsal color of the shark is illuminated against the light background of the sandy bottom at many of the beaches (Figure 29.6). It was important to verify that the species of shark being sighted by Shark Spotters were indeed White Sharks (*Carcharodon carcharias*) as was believed to be, thus independent verification using expert identification, video, and photographs were made on an opportunistic basis. All shark sightings and related information are archived in a Microsoft Access database. We standardized the total number of shark sightings, when comparing sighting number between beaches, with the number of spotting days to account for differences in effort.

RESULTS

Shark Spotting

Sharks were frequently sighted at Shark Spotter beaches between November 2004 and December 2009 with a total of 619 sightings. Sighting duration ranged from just a few seconds to 240 min, with an average of 19.98 min for Fish Hoek, 19.93 min for Muizenberg, 24.40 min for St. James, and 18.13 min for Noordhoek. On 23 separate occasions (3.7% of total number of sightings), we were able to verify the species being sighted by spotters through direct observation (by boat or from the land) or through video and photographs (Figure 29.7). On all these occasions, the sightings were confirmed to be *Carcharodon carcharias* based on the relatively large size of the sharks compared with the other possible species, their distinct silhouette (broad head and spindle shaped body), large dorsal fin, and their swimming behavior (swimming at the surface for extended periods of time in a relatively nonsinusoidal swimming pattern).

Certain beaches had significantly higher sightings than others (Figure 29.8). The beach with the highest number of sightings was Muizenberg (341 sightings), whereas the lowest number of sightings was recorded at Strandfontein (one sighting) just 8 km away. The four permanent Shark Spotter beaches, Muizenberg, St. James, Fish Hoek, and Noordhoek, accounted for the most (94.8%) sightings, whereas the six seasonal beaches, Bluewaters, Clovelly, Glencairn, Mnandi, Monwabisi, and Strandfontein, accounted for the rest (5.2%). The beaches with the lowest number of sightings were also the beaches with the lowest elevation points (<15 m), namely Bluewaters, Mnandi, Monwabisi, and Strandfontein, and/or those with extensive reef systems in the spotting area, namely Clovelly and Glencairn. Interesting to note are the very

(a)

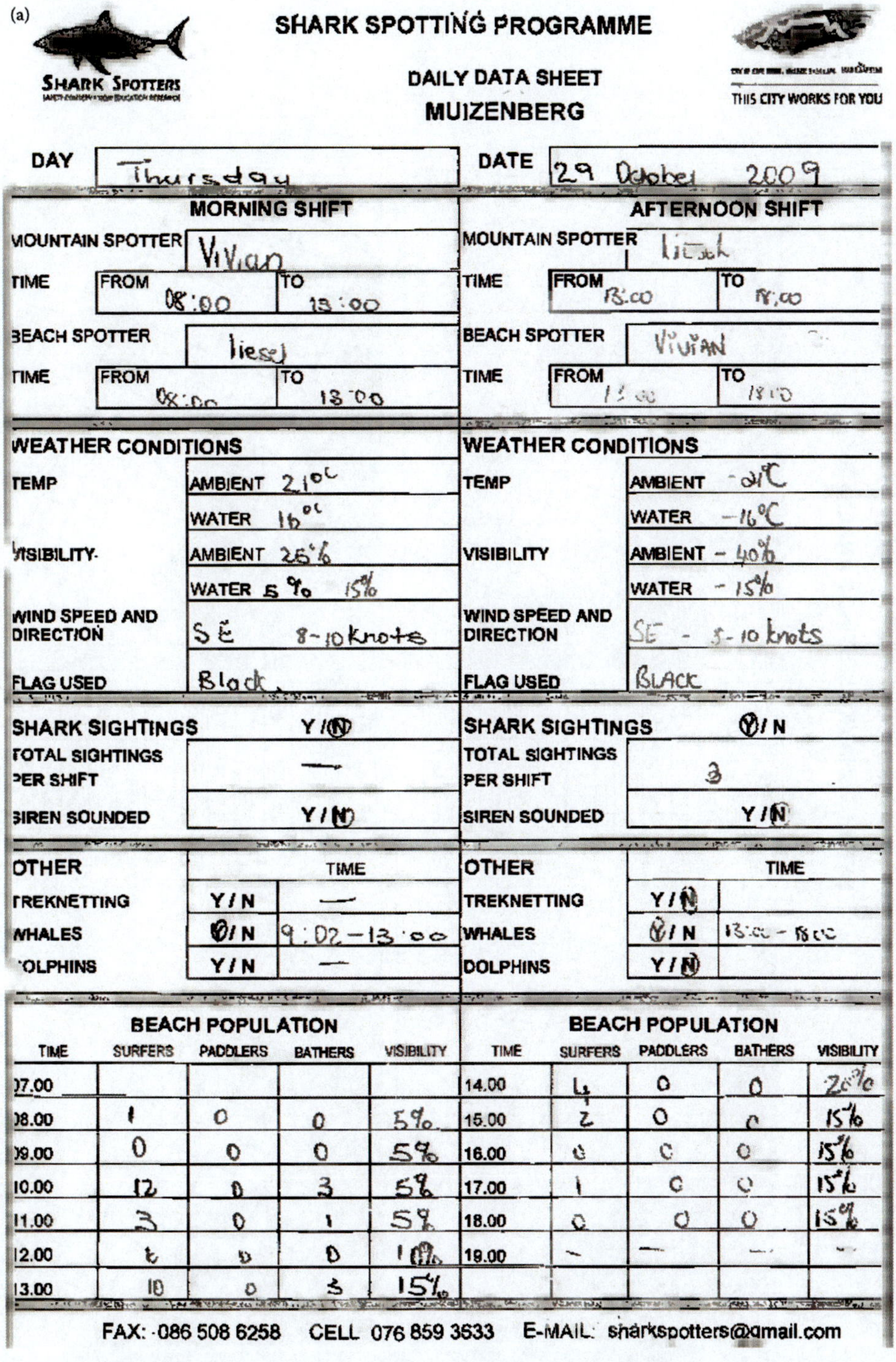

SHARK SPOTTERS

SHARK SPOTTING PROGRAMME

DAILY DATA SHEET

MUIZENBERG

THIS CITY WORKS FOR YOU

DAY	Thursday	DATE	29 October 2009

MORNING SHIFT			AFTERNOON SHIFT		
MOUNTAIN SPOTTER	Vivian		MOUNTAIN SPOTTER	liesel	
TIME	FROM 08:00	TO 13:00	TIME	FROM 13:00	TO 18:00
BEACH SPOTTER	liesel		BEACH SPOTTER	VIVIAN	
TIME	FROM 08:00	TO 13:00	TIME	FROM 13:00	TO 18:00

WEATHER CONDITIONS			WEATHER CONDITIONS		
TEMP	AMBIENT	21°C	TEMP	AMBIENT	- 21°C
	WATER	16°C		WATER	- 16°C
VISIBILITY	AMBIENT	25%	VISIBILITY	AMBIENT	- 40%
	WATER	5% 15%		WATER	- 15%
WIND SPEED AND DIRECTION	SE 8-10 knots		WIND SPEED AND DIRECTION	SE - 5-10 knots	
FLAG USED	Black		FLAG USED	BLACK	

SHARK SIGHTINGS	Y/N (N circled)	SHARK SIGHTINGS	Y/N (Y circled)
TOTAL SIGHTINGS PER SHIFT	—	TOTAL SIGHTINGS PER SHIFT	3
SIREN SOUNDED	Y/N (N circled)	SIREN SOUNDED	Y/N (N circled)

OTHER		TIME	OTHER		TIME
TREKNETTING	Y/N	—	TREKNETTING	Y/N (N circled)	
WHALES	Y/N (Y circled)	9:02–13:00	WHALES	Y/N (Y circled)	13:00 – 18:00
DOLPHINS	Y/N	—	DOLPHINS	Y/N (N circled)	

BEACH POPULATION					BEACH POPULATION				
TIME	SURFERS	PADDLERS	BATHERS	VISIBILITY	TIME	SURFERS	PADDLERS	BATHERS	VISIBILITY
07.00					14.00	4	0	0	20%
08.00	1	0	0	5%	15.00	2	0	0	15%
09.00	0	0	0	5%	16.00	0	0	0	15%
10.00	12	0	3	5%	17.00	1	0	0	15%
11.00	3	0	1	5%	18.00	0	0	0	15%
12.00	6	0	0	10%	19.00	—	—	—	—
13.00	10	0	3	15%					

FAX: 086 508 6258 CELL 076 859 3633 E-MAIL: sharkspotters@gmail.com

Figure 29.5 Examples of the datasheets filled in daily by Shark Spotters. Detailed information on water users, environmental conditions, shark sightings, and other marine animal sightings is collected each day by Shark Spotters (a). In the event of a shark sighting, the time of the sighting, duration, size of the animal, and whether the shark alarm was sounded are recorded, and where applicable the path traveled by the shark is filled in on the appropriate maps (b).

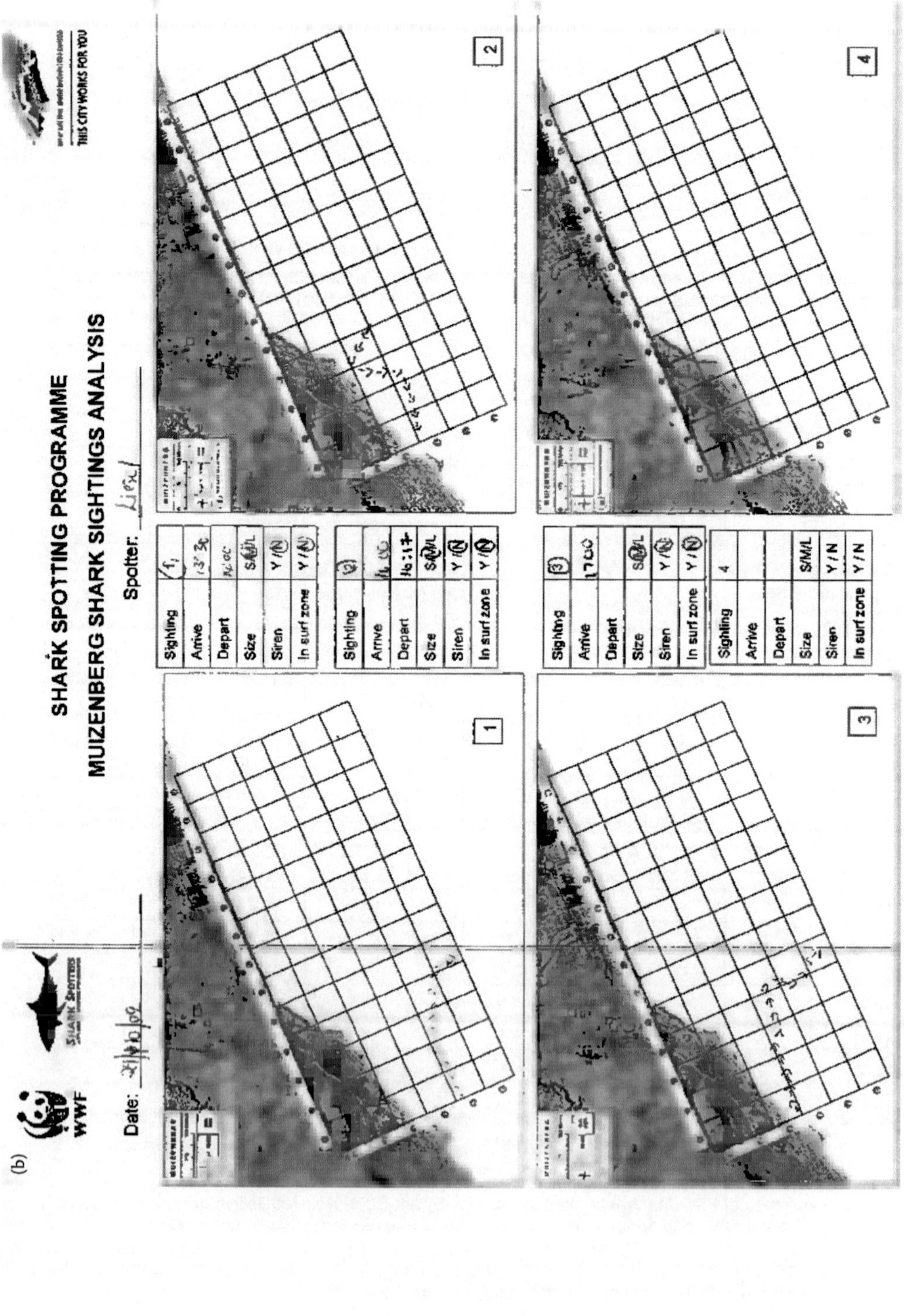

Figure 29.5 (Continued)

Figure 29.6 Photograph of a White Shark near Surfer's Corner, Muizenberg, taken from the Shark Spotters lookout (A). The public and untrained observers regularly mistake floating kelp (B) or other objects for sharks. (Courtesy of Myfanway Rowlands, Our World-Underwater Scholarship Society.)

low sightings for Clovelly and Glencairn, 3 and 10 sightings, respectively; both of these sites have good elevation points and are adjacent to Fish Hoek and St. James, beaches with high numbers of sightings.

Shark sightings showed a strong seasonal trend at all four permanent Shark Spotter beaches, which were operational throughout the year (Table 29.2). Most, 96.7% (599 sightings) of shark sightings were made over the summer period (September to April), whereas only 20 (3.3%) sightings were recorded between May and August. The months with the highest total number of sightings were October (140), November (100), December (136), and January (92).

Frequency of Sounding the Shark Alarm

When a shark is sighted within a few hundred meters of water users, the shark alarm is sounded to alert the public to the presence of the shark(s), and water users are encouraged to leave the

Figure 29.7 Photograph of a ~5-m female White Shark taken on September 28, 2006 during a research trip to verify the species of shark(s) being detected by Shark Spotters. The shark was swimming in less than 2 m of water at Fish Hoek beach. (Courtesy of Enrico Gennari and Alison Kock, Save Our Seas Shark Centre.)

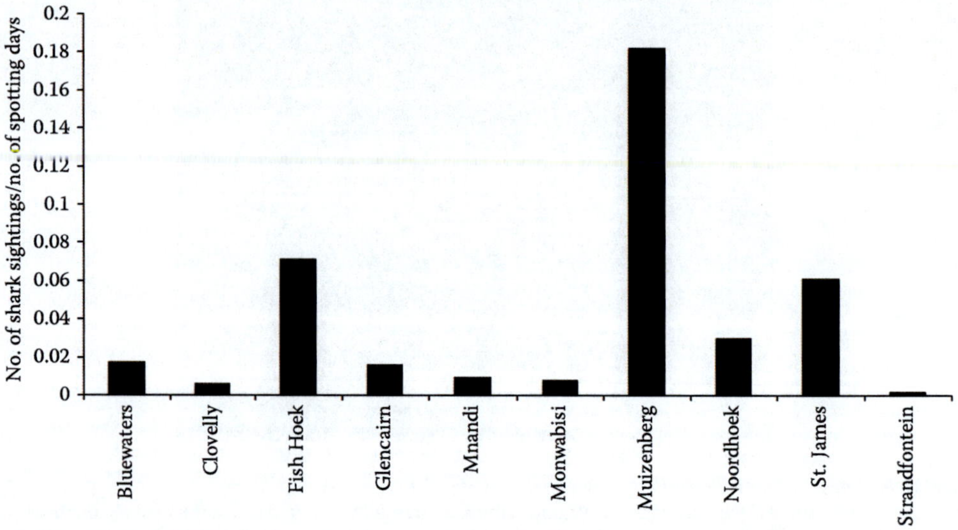

Figure 29.8 The total number of shark sightings, standardized by effort, recorded at each beach with a Shark Spotter program.

water immediately. Between November 2004 and December 31, 2009, shark alarms were sounded 276 times at Shark Spotter beaches (Table 29.3). Water users were present most (66.7–100%) of the times when sharks were sighted. Fish Hoek sounded the alarm the most number of times (84.4%), and Fish Hoek, St. James, and Noordhoek beaches were closed almost every time a shark was sighted in the area. Muizenberg, although it had the highest number of shark sightings, only sounded the alarm 90 (27%) times.

Distance from Shore Sharks Spotted

At all permanent Shark Spotter beaches, namely Muizenberg, Fish Hoek, St. James, and Noordhoek, sharks were sighted more frequently swimming behind the surf zone (73.8% of shark sightings) in deeper and calmer water than inside the surf zone.

Table 29.2 Shark Sightings by Month

Beach	Jan	Feb	Mar	Apr	May	Jun	Jul	Aug	Sep	Oct	Nov	Dec	Total
Bluewaters	7	0	0	0	—	—	—	—	0	0	0	2	9
Clovelly	0	0	1	0	—	—	—	—	0	0	0	2	3
Fish Hoek	13	7	8	0	0	1	0	2	9	55	12	31	138
Glencairn	5	0	0	0	—	—	—	—	2	0	0	3	10
Mnandi	5	0	0	0	—	—	—	—	0	0	0	0	5
Monwasbisi	4	0	0	0	—	—	—	—	0	0	0	0	4
Muizenberg	38	17	22	8	4	2	0	1	26	71	75	77	341
Noordhoek	12	5	4	2	0	0	2	1	0	3	2	2	33
St. James	7	5	7	1	2	2	2	1	7	11	11	19	75
Strandfontein	1	0	0	0	—	—	—	—	0	0	0	0	1
Total	92	34	42	11	6	5	4	5	44	140	100	136	619

Shark sightings showed a strong seasonal trend, with most sightings made over the summer months (September to April). The months with the highest total numbers of sightings were October, November, December, and January. Dashed lines indicate that no shark spotting took place.

Table 29.3 **The Number (and Percentage) of Times Water Users Were Present and the Shark Siren Was Sounded When a Shark Was Sighted at a Shark Spotter Beach**

Beach	Total Number of Shark Sightings	Number of Times Shark Siren Sounded	Percentage of Times Water Users Were Present	Percentage of Times Beach Closed
Bluewaters	9	8	100	88.8
Clovelly	3	2	66.7	66.7
Fish Hoek	138	115	88.6	84.4
Glencairn	10	8	80	80
Mnandi	5	3	100	60
Monwabisi	4	0	75	0
Muizenberg	341	90	94.8	27
Noordhoek	33	18	73.3	73.3
St. James	75	32	86.9	46.3
Strandfontein	1	0	0	0

Communication and Awareness

Signage at appropriate beaches has been instrumental in communicating shark information and the operational protocol of the Shark Spotters program to the public. In addition to the signage, brochures and posters explaining the operational protocol were also provided to the public. Regular press releases on shark activity and the promotion of positive shark-related media articles and news reports (not only based on shark bites) are actively promoted and encouraged. A dedicated website (http://www.sharkspotters.org.za) is used to convey the relevant information on the program, in addition to reporting the most recent shark sightings to the public. In addition to Shark Spotter signage, it was recognized that additional signage was necessary to warn water users of the presence of White Sharks and to provide more detailed information on sharks. Thus, additional signage has been erected at most of Cape Town beaches where White Sharks are likely to occur (Chapter 31, this book) (Figure 29.9).

DISCUSSION

Shark-human conflict poses a threat to both White Shark conservation and coastal tourism and economies. Effective management and mitigation of shark-human conflict is therefore important for both wildlife-conservation purposes and socioeconomic imperatives. Removal (extraction) programs, such as the shark nets installed along the KwaZulu-Natal coastline or culling, were not recommended for Cape Town for a combination of reasons, including their possible negative impact on the White Shark population and potential high levels of bycatch of whales, dolphins, and seals in the case of shark nets (Nel and Peschak, 2006). Furthermore, removal programs are being subjected to increased scrutiny by environmentalists because of the high nonshark bycatch and global shark population declines (Dudley et al., 2006; Chapter 31, this book). Shark hunts as a solution to reducing shark bites have been shown to be ineffective (Chapter 31, this book). Although exclusion barriers and nets were a possibility at certain beaches in Cape Town, concerns were raised about the high cost and maintenance, poor aesthetics, and possible negative effects on tourism and historical fisheries; electronic deterrents were deemed a possibility in the future after more research and development (Nel and Peschak, 2006). In light of these issues and an expert review, the City of Cape Town endorsed a pioneering Shark Spotters program, which functions as an early warning system, to mitigate shark bites in its waters.

Figure 29.9 Example of warning signage erected at beaches in Cape Town where the relative risk of encountering sharks is high.

Shark Spotting

At least four species of shark, namely Bronze Whaler (*Carcharhinus brachyrus*), Ragged Tooth Shark (*Carcharias taurus*), and, to a lesser extent, Dusky Shark (*Carcharhinus obscurus*) and Spotted Gully Shark (*Triakis megalopterus*), can possibly be mistaken for juvenile (<3 m) White Sharks in Cape waters by inexperienced observers, particularly False Bay (Lamberth, 2006). Basking Sharks (*Cetorhinus maximus*) are occasionally sighted on the Atlantic seaboard side of the Cape Peninsula and may also possibly be mistaken for White Sharks at Noordhoek beach. However, because of their large size, stocky and spindle-shaped body, and especially their tendency to swim near the surface for extended periods of time as documented by the spotters, the authors are confident that White Sharks are the species being sighted most frequently by Shark Spotters. This information is consistent with acoustic tagging studies concurrently being conducted on White Sharks in the area, which indicate their common presence in the near shore waters of False Bay (A. Kock, unpublished data).

White Sharks were frequently sighted at Cape Town beaches where spotters are present. The ability to detect sharks is dependent on two key factors, namely an elevated spotting position (greater than 50 m) and a predominately gentle beach gradient with a sandy bottom. Beaches that had the highest number of sightings were beaches with the highest elevation points and a predominantly sandy beach substrate. The beaches with the highest number of sightings are two of Cape Town's most popular recreational swimming and surfing beaches, i.e., Muizenberg and Fish Hoek (Figure 29.8). The lowest sightings were recorded at beaches that had less than a 15-m elevation platform for sighting sharks, i.e., Strandfontein, Bluewaters, Mnandi, and Monwabisi beaches. However, White Sharks are known to be relatively common in these areas (Ferreira and Ferreira, 1996; A. Kock, unpublished data) over summer periods; therefore it is likely that the low sightings recorded at these beaches are not indicative that few sharks are in the area but rather because the spotting platform (<15 m) is simply not adequate to detect sharks reliably in these areas. The beaches are characteristically sandy, and therefore if higher platforms could be built, it may be possible in the future to reintroduce spotting in these areas.

Although differences in the number of sightings at the different beaches are most likely a reflection of the ability of spotters to detect sharks at certain beaches, there is a possibility that the differences in sightings could be as a result of sharks showing an affinity for certain areas. Evidence for this comes from the data collected at Clovelly and Glencairn. The low sightings at Clovelly are interesting given that it is situated in between two beaches (Fish Hoek and St. James) that have high shark sightings. The spotters' ability to detect sharks at Clovelly may possibly be negatively effected by the presence of an extensive reef system adjacent to the beach, which could obscure a shark swimming over the dark bottom, in addition to the presence of a river mouth that drains into the ocean and results in murky water in the area. However, one would expect higher sightings in this area even with these conditions. Glencairn, on the other hand, which also had very low sightings and is adjacent to Fish Hoek, has a gently sloping sandy beach, and overall shark-spotting ability is good; thus, the low sightings recorded there are of particular interest and could provide evidence that White Sharks prefer certain beaches over others, perhaps because of the presence of prey or certain environmental conditions.

Shark sightings at all spotting beaches showed a strong seasonal trend, with significantly more sightings over the summer period from September to April, with little or no sightings at beaches over the winter months. However, spotting conditions are better over the winter period (May to August) as the predominant northwesterly (offshore) wind results in cleaner water at these coastal areas (Atkins, 1970). The winter seasonal (May to September) presence of White Sharks at Seal Island, False Bay, is well documented (Laroche et al., 2008; Martin et al., 2005; Hammerschlag et al., 2006). The inshore presence of White Sharks is less well known but was first documented by Wallet (1978) and subsequently by Ferreira and Ferreira (1996) for one of the beaches in False Bay, namely Macassar. Therefore, the peak in White Shark sightings close to shore as documented by the Shark Spotters is probably representative of a seasonal increase in White Sharks and not a reflection of the ability to detect sharks.

Shark Sightings and Shark Bites

The high numbers of shark sightings recorded by the Shark Spotters program is indicative that it is successful at detecting sharks at appropriate beaches and can provide a measure of safety to water users by reducing the number of encounters with sharks. However, the program can never guarantee 100% safety. This was especially evident on January 12, 2010 when a White Shark killed a swimmer in Fish Hoek beach (Oelofse et al., 2010). It was concluded that spotting conditions were not ideal at the time of the attack with a strong southerly wind and intermittent cloud, which negatively impacted the spotter's ability to detect the shark. Oelofse et al. (2010) concluded that in this incident, the spotter could not have prevented the bite, because the shark was not seen by any of the witnesses on the beach beforehand, implying that it was swimming at a depth out of view of the spotter. It is therefore imperative that it be conveyed to the public that the program can never guarantee 100% safety and has a number of limitations, which are discussed in more detail below. Entering the water will always be at the individual's own risk, but people are encouraged to use the information provided by the program on higher-risk months and areas for encountering sharks to make informed decisions when using the water.

Between November 2004 and December 2009, no shark bites were recorded at Shark Spotter beaches even though there were 619 shark sightings. The increase in sightings recorded by Shark Spotters coincides with a peak in the number of water users over the same period (Dotchim, 2006; Van Herwerden et al., 1989). This means that the potential for interactions between White Sharks and people is increased over this time. However, Cliff (2006) analyzed the 25 shark bites in Cape

Town between 1960 and 2005 and found no significant seasonal trend in the frequency of shark bites, despite the fact that September was the month with the most bites (16% of the total). This peak in the number of shark bites, although not statistically significant, coincides with the transition from winter to summer, when more recreational water users start to use the ocean. It also coincides with the increase in sightings from winter months to summer months as recorded by Shark Spotters (Table 29.2) but does not coincide with the peak in the number of water users or the peak in the number of shark sightings. The beaches with the highest number of shark bites were Muizenberg and Fish Hoek (Cliff, 2006), which are also the beaches with the highest number of swimmers and surfers (Van Herwerden, 1989) and also the beaches with the highest number of shark sightings (Figure 29.8). One expects the highest number of shark bites to coincide with the highest number of shark sightings and water users. However, this does not seem to always be the case in Cape Town. More research on the relationship between the presence of sharks and their behavioral state when near water users is warranted to better understand why shark bites occasionally occur, when there are so many occasions that sharks and water users are present simultaneously with no negative interactions.

Frequency of Sounding the Shark Alarm

Shark alarms are used to warn water users of the nearby presence of sharks, and people are encouraged to exit the water until the shark has swum away. These alarms are regularly sounded over the summer months when water use is at its highest and sharks are regularly sighted. Sharks are more frequently (73.8% of the time) sighted swimming behind the surf zone in deeper, calmer water at the four full-time Shark Spotter beaches. This has two consequences, firstly that at beaches like Fish Hoek and St. James, where the surf zone is about a third the size of Muizenberg, the shark alarm is sounded almost every time a shark is sighted because it is closer to swimmers and surfers. Whereas the situation at Muizenberg is very different because the surf zone is so wide (>300 m) that the sharks are often swimming far enough away from the water users so that it is obvious to the spotters that they do not pose an immediate threat. Initially beaches were closed every time a shark was sighted, but this caused much irritation and frustration amongst regular water users and impacted local businesses like surf schools which rely on the beach being open to conduct business (Kock et al., personal communication). Thus, closing the beach unnecessarily is not desired in Cape Town and creates unnecessary hype when it is obvious the shark poses no threat and is simply swimming past.

An interesting observation has been that when the program first started in 2004, people would leave the water after the shark alarm was sounded and would not return to the water that day (Shark Spotters, unpublished data). However, over recent years, a change in the reaction to sightings has been documented by the Shark Spotters programs, and now many people leave the water when the shark siren sounds, but instead of going home, they wait for the Shark Spotters to give the "all clear" before getting back into the water again. In fact, some locals are even reluctant to get out the water when the siren goes off, and regular press releases are issued to remind the locals that they should leave the water when the siren is sounded. This shift in the public's reaction to shark sightings may be indicative of changing attitudes toward sharks in Cape Town and an increased understanding that sharks are not mindless man-eaters out to kill people, as they are still so often portrayed.

Communication and Awareness

Shark awareness and education are fundamental components of the Shark Spotters program and the overall shark safety strategy for the City of Cape Town. The success of the Shark Spotters program relies on an effective communication strategy. As a result of the Shark Spotters program and the daily monitoring of in-shore shark activity, the city and its partners

(National Sea Rescue Institute and Save Our Seas Foundation Shark Centre) are able to provide public warnings via a communication network between various stakeholder organizations and the media when high shark presence is encountered. This communication network also provides immediate and organized responses from rescue, government, scientists, and media to a shark bite.

Shark sightings near beaches around the world still trigger sensational media reports of "killer," "rogue," or "monster" sharks, "lurking" close to shore, giving the public the impression that sharks are uncommon in coastal areas, when they are not. The Shark Spotters program provides a fresh perspective on the presence of sharks at popular beaches and the reactions of the public to these shark sightings. It illustrates that when sharks and their presence are handled in an informative and objective way, one can reduce the hype and hysteria generated by what is natural shark behavior. The information collected by Shark Spotters, such as the seasonal presence of sharks (e.g., summer months), the beaches that have higher sightings than others, and the fact that sightings are higher beyond the surf zone in deeper water, can contribute to current and applicable shark safety information related to reducing the risk of encounters with sharks. In Cape Town, when more than five shark sightings are recorded on a particular day at a single beach, an alert of "high shark activity" is communicated to the public. Text messages of each sighting are also sent to a network of affected agencies, and the public can subscribe to receive shark sightings by mobile phone. However, finding the balance between providing the public with responsible and current White Shark information, such as issuing warnings when sightings are high or closing beaches when cetaceans strand while not contributing to the shark "hype," is not easy, and the information is still sometimes used by media or public to sensationalize the situation.

Limitations of Shark Spotting

Although proven to be effective, the Shark Spotters program has limitations that include the following:

1. Environmental conditions affecting spotting ability (water quality, glare wind)
2. Human error or fatigue
3. Operating hours beyond which no safety is provided to water users (i.e., early morning and late evening)
4. Only works at beaches with associated high elevation points and is less effective at beaches with in-shore reefs and deep in-shore troughs

These limitations can be minimized but are never removed. As such, the Shark Spotters program can never be considered 100% effective for guaranteeing a safe swimming zone. Means to reduce the limitations are as follows:

1. To have spotting shifts of no longer than 6 h
2. The use of polarized glasses
3. Regular training of spotters

Ongoing communication to the public informing them of the limitations of the program is essential to ensure that they remain aware of the relative risk of encountering sharks and still consider all safety tips and advice when deciding to enter the sea.

Stable funding is a limiting factor in the growth of the Shark Spotter program. Oelofse and Kamp (2006) recommended the project be funded through the three tiers of government, but to date only municipal government has allocated funding. Increased funding will allow for more training opportunities for Shark Spotters, better equipment, and rollout at more beaches where spotting could work.

Additional Benefits of Shark Spotting

Not only has the spotting program been shown to be an effective early warning system to mitigate shark encounters, the program also provides employment and skill development opportunities for previously disadvantaged South Africans. As a result of their medic first aid training, they are able to assist with minor injuries sustained by the public while on the beach and assist with small cuts, jellyfish stings, and similar incidents. Shark Spotters are available to talk to the public about sharks and related issues on a daily basis and thereby provide continuous awareness and education. As part of a network with the National Sea Rescue, Cape Town's Disaster Risk Management, Department of Environmental Affairs, and nonprofit organizations, Shark Spotters are often the ones called to respond to a stranding or other marine emergency. The collection of information on shark sightings, other marine animal sightings, and the number of water users provides reliable and continuous observations of Cape Town's coastal environment. Over time, these shark sightings can provide an independent monitoring system of nearshore White Shark coastal behavior. The data can be used to issue high shark alert warnings and advise people against any increased risk in shark encounters. Shark Spotters are able to assist scientists in the field attempting to tag White Sharks close to shore and add to the knowledge base of White Shark behavior.

Conclusion

A sustainable future requires that we look for solutions to human-wildlife conflicts that provide a balance between human needs and those of our natural assets. The Shark Spotters program fulfills this directive because it attempts to balance the needs of both recreational water users and those of White Shark conservation and preservation. It incorporates safety, research, response, and awareness. The Shark Spotters program has proven to be an effective program for proactively reducing the risk of shark encounters by alerting water users to their presence and getting them out of the water as a precautionary action. In addition to being a warning system and fostering awareness on shark related issues, the information collected by Shark Spotters can provide valuable information on White Shark presence and behavior close to shore, in an environment that is not well studied.

ACKNOWLEDGMENTS

Special thanks go to all the dedicated Shark Spotters, past and present, namely Patrick Davids, Agnes Murema, Alex Msithweni, Eric Myeko, Ethel Tshandu, Faeze Gordon, Ghaliep Mohammed, Innocent Leoto, Jerry Matiso, Liesel Lott, Matthews Sikweyiya, Seferino Gelderbloem, Shameef Cornielson, Steven Bonkolo, Valentino Simmerie, Vivian Kwebana, Ashley Sullivan, Thandile Xhalabile, Bongani Mlonyeni, Vusi Dhlamini, and Patrick Hayes; to Greg Bertish of Tru Blu Travel and Dave and Fiona Chudleigh from Surf Shark for initiating the Shark Spotter concept; our funders, the City of Cape Town, Save Our Seas Foundation, Reef, William Simpson, Pick 'n Pay, Wavescapes, EMT, African Shark Eco-Charters, Ballistic, Sun Valley Pharmacy, Deepblue, Kuhuna Surfing Academy, Puma SA, Monzamedia, and WWF. Special thanks to all our supporters National Sea Rescue Institute, with special mention of Ian Klopper and Craig Lambinon, SA Surf Lifesaving, Cape Town White Shark Working Group members, Fish Hoek treknet fishermen, Table Mountain National Parks, and Cape Peninsula University of Technology students. Thanks to Lauren Best, Adrian Hewitt, and Brocq Maxey for data management. Additional support was provided by Morne Hardenberg, Sherri Bell, Clive Wakeford, Amy Davidson, Cheryl-Samantha Owen, and Yvonne Kamp. Thank you to Myfanwy Rowlands from the Our World-Underwater Scholarship Society for use of her photographs. We are grateful to Charlie Huveneers and Tobey Curtis for helpful comments on the manuscript.

REFERENCES

Atkins, G. R. 1970. Winds and current patterns in False Bay. *Trans Royal Society South Africa* 39: 139–148.

Cliff, G. 2006. A review of shark attacks in False Bay and the Cape Peninsula between 1960 and 2005. In *Finding a balance: White shark conservation and recreational safety in the inshore waters of Cape Town, South Africa*, edited by D.C. Nel and T.P. Peschak, pp. 20–31. Proceedings of a specialist workshop. WWF South Africa Report Series – 2006/Marine/001 Annexure 1.

Cliff, G., and S.F.J. Dudley. 1992. Protection against shark attack in South Africa, 1952 to 1990. Sharks: Biology and fisheries. *Aust J Mar Freshw Res* 43: 263–272.

Compagno, L.J.V. 1991. Government protection for the great white shark (*Carcharodon carcharias*) in South Africa. *S Afr J Sci* 87: 284–285.

Compagno, L.J.V., M.A. Marks, and I.K. Fergusson. 1997. Threatened fishes of the world: *Carcharodon carcharias* (Linnaeus, 1758) (Lamnidae). *Environ Biol Fishes* 50: 61–62.

Davies, D.H. 1963. Shark attack and its relationship to temperature, beach patronage and the seasonal abundance of dangerous sharks. *Investig Rep Oceanogr Res Inst, Durban* 6: 1–43.

Dotchim, M. 2006. Recreational trends and safety in waters of the Cape as it relates to risk of interactions with sharks. In *Finding a balance: White shark conservation and recreational safety in the inshore waters of Cape Town, South Africa*, edited by D.C. Nel and T.P. Peschak, pp. 130–137. Proceedings of a specialist workshop. WWF South Africa Report Series – 2006/Marine/001 Annexure 1.

Dudley, S.F.J., and G. Cliff. 1993. Some effects of shark nets in the Natal nearshore environment. *Environ Biol Fishes* 36: 243–255.

Dudley, S.F.J., R.C. Haestier, K.R. Cox, et al. 1998. Shark control: Experimental fishing with baited drumlines. *Mar Freshw Res* 49: 653–661.

Dudley, S.F.J., G. Cliff, M.D. Anderson-Reade, et al. 2006. Shark deterrent options for Cape Town. In *Finding a balance: White shark conservation and recreational safety in the inshore waters of Cape Town, South Africa*, edited by D.C. Nel and T.P. Peschak, pp. 109–120. Proceedings of a specialist workshop. WWF South Africa Report Series – 2006/Marine/001 Annexure 1.

Ferreira, C.A., and T.P. Ferreira. 1996. Population dynamics of white sharks in South Africa. In *Great white sharks: The biology of Carcharodon carcharias*, edited by A.P. Klimley and D.G. Ainley, pp. 381–391. San Diego, CA: Academic Press.

Gore, M.L., B.A. Knuth, C.W. Scherer, and P. D. Curtis. 2008. Evaluating a conservation investment designed to reduce human–wildlife conflict. *Conserv Lett* 1: 136–145.

Hammerschlag, N., R.A. Martin, and C. Fallows. 2006. Effects of environmental conditions on predator-prey interactions between white sharks (*Carcharodon carcharias*) and Cape fur seals (*Arctocephalus pusillus pusillus*) at Seal Island, South Africa. *Environ Biol Fishes* 76: 341–350.

Kock, A., and R.L. Johnson. 2006. White shark abundance: Not a causative factor in numbers of shark bite incidents. In *Finding a balance: White shark conservation and recreational safety in the inshore waters of Cape Town, South Africa*, edited by D.C. Nel and T.P. Peschak, pp. 1–19. Proceedings of a specialist workshop. WWF South Africa Report Series – 2006/Marine/001 Annexure 1.

Lamberth, S.J. 2006. White shark and other chondrichthyan interactions with the beach-seine (treknet) fishery in False Bay, South Africa. *Afr J Mar Sci* 28: 723–727.

Laroche, R.K., A.A. Kock, L.M. Dill, et al. 2008. Running the gauntlet: A predator-prey game between sharks and two age classes of seals. *Anim Behav* 76: 1901–1917.

Macdonald, D.W., and C. Sillero-Zubiri. 2002. Large carnivores and conflict: Lion conservation in context. In *Lion conservation research. Workshop 2: Modelling conflict*, edited by A.J. Loveridge, T. Lynam, and D.W. Macdonald, pp. 1–8. Wildlife Conservation Research Unit, Oxford University. http://www.peopleandwildlife.org.uk/crmanuals/CarnivoreConflictP&WManual.

Madden, F. M. 2008. The growing conflict between humans and wildlife: Law and policy as contributing and mitigating factors. *J Int Wild Law Policy* 1548–1476, 11: 189–206.

Martin, R.A., N. Hammerschlag, R.S. Collier, and C. Fallows. 2005. Predatory behavior of white sharks (*Carcharodon carcharias*) at Seal Island, South Africa. *J Mar Biol Assoc UK* 85: 1121–1135.

Nel, D.C., and T.P. Peschak (eds). 2006. *Finding a balance: White shark conservation and recreational safety in the inshore waters of Cape Town, South Africa*. Proceedings of a specialist workshop. WWF South Africa Report Series – 2006/Marine/001.

Nyhus, P. J., H. Fisher, F.M. Madden, et al. 2003. Taking the bite out of wildlife damage: The challenges of wildlife compensation schemes. *Conserv Pract* 4: 39–41.

Oelofse, G. 2010. *Findings from the City's review of the fatal shark attack at Fish Hoek.* http://www.capetown.gov.za/en/Documents/sharkattackattachment.pdf.

Oelofse, G., and Y. Kamp. 2006. Shark spotting as a water safety program in Cape Town. In *Finding a balance: White shark conservation and recreational safety in the inshore waters of Cape Town, South Africa*, edited by D.C. Nel and T.P. Peschak, pp. 121–129. Proceedings of a specialist workshop. WWF South Africa Report Series – 2006/Marine/001 Annexure 1.

Sillero-Zubiri, C., and D. Stwizer. 2004. Management of wild canids in human-dominated landscapes. People and Wildlife Initiative. Wildlife Conservation Research Unit, Oxford University. http://www.peopleandwildlife.org.uk.

Van Herwerden, L., C.L. Griffiths, R. Bally, et al. 1989. Patterns of shore utilization in a metropolitan area: The Cape Peninsula, South Africa. *Ocean Shore Mgt* 12: 331–346.

Wallett, T. 1978. *Shark attack and treatment of victims in southern African waters*. Cape Town, South Africa: Purnell.

CHAPTER 30

The Use of a Nonlethal Technique to Assess the Reproductive Biology of the White Shark, *Carcharodon carcharias*

James A. Sulikowski*
University of New England

Laura J. Williams
University of New England

Michael L. Domeier
Marine Conservation Science Institute

ABSTRACT

The need to answer basic life history questions for White Sharks, *Carcharodon carcharias*, is well established. However, sampling has been problematic because of the size, mobility, and stock status of this species. In recent years, circulating concentrations of plasma steroid hormones, such as 17-β-estradiol (E_2), progesterone (P_4), and testosterone (T), have been used to evaluate the events associated with reproductive cycles and sexual maturity in several elasmobranch species. To begin understanding the reproductive cycle of White Sharks, blood samples were collected from four female and five male adult sharks. Circulating levels of plasma E_2 ranged from 248 to 1,826 pg/ml and levels of P_4 ranged from 210 to 593 pg/ml in females, and levels of T ranged from 2,519 to 7,484 pg/ml in male White Sharks. Additionally, hormone values were assessed in juvenile White Sharks. Circulating levels of T in four juvenile males ranged between 172 and 338 pg/ml, whereas E_2 levels in seven juvenile females were all below detectible levels (5 pg/ml). Although the limited sample size precludes any direct interpretation of these results, collectively, the data suggest that these hormones can be quantified and used as noninvasive reproductive end-points for this species.

INTRODUCTION

The White Shark, *Carcharodon carcharias*, is a large apex predator with a worldwide distribution primarily inhabiting temperate and subtropical oceans (Compagno, 1984). Notable declines in its global biomass prompted the listing of the White Shark as vulnerable by the International Union for Conservation of Nature (2010) and the Convention on International Trade in Endangered Species Appendix II (2002). However, effective management of this species has been limited because of its highly migratory nature and general lack of knowledge surrounding its behavior, ecology, and biology (Adams et al. 1994; Wintner and Cliff, 1999; Bruce et al., 2006; Weng et al., 2007; Nasby-Lucas et al., 2009). White Sharks have been reported to undertake seasonal migrations of varying scales, fine scale to transoceanic, to access coastal and pelagic regions. Migrations to coastal regions often correlate with shark specific site fidelity and prey distributions, and historically it was hypothesized that these regions are mainly used as foraging sites (Ainley et al., 1985; Bruce, 1992; Goldman and Anderson, 1999; Bonfil et al., 2005; Domeier and Nasby-Lucas, 2007); a new hypothesis, however, has emerged suggesting coastal White Shark aggregations may also serve as mating sites (Chapters 11 and 16, this book). The function of the pelagic regions is unclear, with hypotheses encompassing mating, parturition, foraging, or any combination of the above (Anderson and Pyle, 2003; Carrier et al., 2004; Weng et al., 2007; Domeier and Nasby-Lucas, 2008, Nasby-Lucas et al., 2009; Chapters 11, 12, and 16, this book). Although much progress has recently been made, many aspects of White Shark biology, including habitat utilization and life history, remain largely unknown (Boustany et al., 2002; Dewar et al., 2004; Bonfil et al., 2005; Bruce et al., 2006; Weng et al., 2007; Domeier and Nasby-Lucas, 2008; Chapter 16, this book).

In order to ensure proper management of White Shark populations, it is imperative to understand the life history of this species. In previous studies, postmortem examinations were conducted, and vertebral ring counts were used to assess age, growth, and size at maturity from a limited number of individuals. Collectively, these studies suggest that White Sharks exhibit common elasmobranch life history characteristics, including late maturity, slow growth, low fecundity, and long life spans (Cailliet et al., 1985; Adams et al., 1994; Compagno et al., 1997; Wintner and Cliff, 1999), which sufficiently reduce its ability to withstand exploitation (Dulvy et al., 2008). Moreover, information regarding the reproductive biology of this species, an important aspect of the life history necessary for management, is lacking. Because of the conservation status of White Sharks, the use of conventional techniques, i.e., sacrificing specimens, is prohibited.

Fortunately, crucial components of the reproductive biology can still be obtained through the use of steroid hormones. Previous research suggests that physiological and morphological processes central to elasmobranch reproductive biology (as in other vertebrates) are regulated by endocrine factors (Koob and Callard, 1999). Steroid-hormone analysis represents a nonlethal technique that can be used to elucidate reproductive status and cyclicity (Sulikowski et al., 2007a; Awruch et al., 2008) and has been verified in a number of skate (Rasmussen et al., 1999; Sulikowski et al., 2004, 2005, 2007a; Kneebone et al., 2007), ray (Snelson et al., 1997; Tricas et al., 2000), and shark species (Manire et al., 1995; Heupel et al., 1999; Awruch et al., 2008; Hoffmayer et al., 2010). The use of this technique on White Sharks could potentially offer critical insight into the relative timing and locations of mating and parturition, key parameters that would allow the solidification of our understanding of this species' overall life history. The current study represents the first use of plasma steroid hormone concentrations as biochemical markers to assess the reproductive biology of White Sharks by nonlethal means.

MATERIALS AND METHODS

Sampling Techniques

Adult White Sharks were captured via hook-and-line (Chapters 10, 11, and 16, this book) off the coast of Guadalupe Island, Mexico and Southeast Farallon Island, Central California coast, United

States. After each shark was removed from the water, a 10-ml heparinized syringe fitted with a 14-gauge needle was used to collect 10 ml of blood from the caudal vein. Blood samples were then centrifuged at 1,300 rpm for 5 min. The separated plasma was stored in a standard freezer until it was shipped to the University of New England where radioimmunoassay (RIA) was performed. The presence of seminal fluid and spermatophores in males was determined by examining the claspers and viewing the seminal fluid under a compound microscope (Olympus BX51) (Chapter 11, this book). Juvenile White Shark samples were obtained from the Monterey Bay Aquarium and from targeted fishing off the coast of Southern California.

Hormone Preparation and Analysis

The full procedural details for plasma steroid extraction and RIA are found in Tsang and Callard (1987) and Sulikowski et al. (2004). Briefly, each White Shark plasma sample was extracted twice with 10 volumes of diethyl ether (anhydrous), which was evaporated under a stream of nitrogen before the dried extracts were reconstituted in phosphate-buffered saline with 0.1% gelatin and stored at –20°C until assay. Correction for procedural losses during the extraction process was accomplished by adding 1,000 counts per minute of the appropriate tritiated steroid into each plasma sample. The samples were run in duplicate, and radioactivity was determined in a Perkin Elmer Tri-Carb 2900 PR liquid scintillation counter.

Because of the short period of time between sampling dates, the average hormone concentrations of adult females captured on November 18–20, 2009 were combined to give an average value. Similarly, the hormone concentrations of adult males captured on October 29 and November 2 and the three males captured November 19–20 were pooled in order to provide an average value for each of those grouped dates. In addition, hormone concentrations for juvenile sharks were grouped by sex, and the average value was recorded.

RESULTS

Plasma samples were obtained from adult females collected on December 9, 2008 (n = 1); November 18, 2009 (n = 1); November 19, 2009 (n = 1); and November 20, 2009 (n = 1) (Table 30.1). The values for E_2 and P_4 ranged from 248 to 1,827 pg/ml and from 210 to 593 pg/ml, respectively (Table 30.1). Although the low sample size precludes the use of any statistical analyses and limits the ability to conclusively report mean values for each hormone, the results do indicate that this methodology (RIA) can measure the plasma E_2 and P_4 concentrations in adult female White Sharks (Table 30.2). Despite the short time period between sampling dates, a large disparity in adult E_2 concentrations, a nearly five-fold increase of this hormone, was observed when comparing the female captured in December 2008 with those captured in November 2009. In contrast, P_4 values were similar in all female specimens.

Blood samples from juvenile females were collected in July (n = 1), August (n = 3), September (n = 2), and October (n = 1). Despite several attempts at concentrating juvenile-female E_2 levels, values obtained for this hormone were all less than 5 pg/ml in each female [170 ± 40 cm total length (TL); Table 30.2]. Progesterone was not measured in juvenile females because of low aliquots of available plasma.

Plasma steroid concentrations of T were collected for eight mature male White Sharks captured in December 2008 (n = 3), October 2009 (n = 1), and November 2009 (n = 4) (Table 30.1). The values for T ranged from 2,519 to 7,484 pg/ml in these sharks. As was the case with females, the low sample size prevented the use of any statistical analyses for monthly data interpretation in males. However, like females, the results that indicate RIA can measure the plasma T concentrations in adult male White Sharks (Table 30.1). Interestingly, the T concentrations in mature males showed

Table 30.1 Plasma Estradiol, Progesterone, and Testosterone Concentrations for Individual White Sharks

Date	Location	Total Length (cm)	Reproductive Status	E_2 (pg/ml)	P_4 (pg/ml)	T (pg/ml)	Spermatophores Present
11/18/2009	29 08.989°N, 118.16.997°W	485	Mature	248.33	435.25	NS	NA
11/19/2009	29 08.989°N, 118.16.997°W	462	Mature	484.26	593.17	NS	NA
11/20/2009	29 08.989°N, 118.16.997°W	462	Mature	780.61	209.76	NS	NA
12/9/2008	29 08.989°N, 118.16.997°W	495	Mature	1,826.99	227.67	NS	NA
10/29/2009	37 41.625°N, 122 59.882°W	427	Mature	NS	NS	6,665.38	Yes
11/2/2009	37 41.644°N, 122 59.878°W	452	Mature	NS	NS	5,661.70	No
11/19/2009	29 08.989°N, 118.16.997°W	439	Mature	NS	NS	7,483.62	No
11/20/2009	29 08.989°N, 118.16.997°W	472	Mature	NS	NS	2,519.12	Yes
11/20/2009	29 08.989°N, 118.16.997°W	541	Mature	NS	NS	5,019.27	Yes
12/4/2008	29 08.989°N, 118.16.997°W	457	Mature	NS	NS	3842.70	No
12/4/2008	29 08.989°N, 118.16.997°W	442	Mature	NS	NS	4,780.96	No
12/6/2008	29 08.989°N, 118.16.997°W	447	Mature	NS	NS	3,828	No
7/16/2009	34 14.560°N, 119 17.250°W	168	Immature	NS	NS	307.75	NS
8/1/2009	34 23.180°N, 119 39.788°W	154	Immature	NS	NS	171.92	NS
8/23/2009	34 08.870°N, 119 14.415°W	134	Immature	NS	NS	243.27	NS
9/23/2009	34 01.837°N, 118 33.633°W	213	Immature	NS	NS	337.79	NS

Estradiol (E_2) and progesterone (P_4) concentrations for individual mature female White Sharks and testosterone (T) concentrations for immature and mature male White Sharks captured from July to December. Entries designated with NS indicate that a measurement was not sampled for that particular parameter, and entries with NA indicate that the particular parameter is not applicable.

Table 30.2 Plasma Estradiol and Progesterone Concentrations for Individual Female White Sharks

Sample Date	*n*	E_2 (pg/ml)	P_4 (pg/ml)	Total Length (cm)
July to October 2009	7	< 5	NS	170 ± 40
November 2009	3	471 ± 249	413 ± 210	471 ± 12
December 2008	1	1,827	228	495

Plasma estradiol (E_2) and progesterone (P_4) concentrations for individual female White Sharks captured in November and December. The values are given as means ± SD. Entries designated with NS indicate that a measurement was not sampled for that particular parameter.

a general decline between October and December. In addition, the aforementioned ends of the range were collected in November, the low from a 472-cm male and the high from a 439-cm male. Spermatophores were present in one of the two males sampled in late October/early November and in two of the three males sampled in late November 2009 (Table 30.1).

Blood samples from juvenile males were collected in July ($n = 1$), August ($n = 2$), and September ($n = 1$). Testosterone levels from juvenile males ranged from 172 pg/ml (154 cm TL) to 338 pg/ml (213 cm TL) and were statically lower (analysis of variance; $p < 0.001$) than the T levels in the adult male White Sharks (Table 30.1).

DISCUSSION

The need to answer basic life questions for White Sharks is well established (Weng et al., 2007; Nasby-Lucas et al., 2009; Chapter 16, this book) yet has been problematic because of the size, mobility, and stock status of this species (Convention on International Trade in Endangered Species, 2004). Although limited by sample sizes, the information provided in this manuscript represents the first use of steroid hormones as a noninvasive biomarker to investigate reproductive biology of this vulnerable elasmobranch species (International Union for Conservation of Nature, 2010).

Estradiol concentrations in the four female White Sharks analyzed were comparable with those of other elasmobranch species where hormones have been used to assess reproductive biology (Table 30.3). Interestingly, the highest E_2 value that was observed in White Sharks was similar to preovulatory values in other elasmobranchs (Table 30.3), where a strong association between follicle size and estradiol concentrations has been documented (Koob and Callard, 1999; Sulikowski et al., 2007a). Although E_2 values for only four females were available for use in this study, the relatively lower values in November ($n = 3$) compared with the relatively higher value obtained from the December specimen suggest the possibility that these changes are indicative of a biologically significant event. However, defining whether or not this marked increase in E_2 is representative of progression toward the preovulatory follicle stage and ovulation in White Sharks will require the further collection of blood samples.

The dynamics of elevated progesterone levels is complex and appears to differ among the elasmobranch reproductive modes (Koob and Callard et al., 1999; Table 30.3). The consistently low P_4 values along with the concurrent E_2 concentrations could potentially indicate several functions of P_4 in White Sharks (Table 30.2). Possible functions include acting as a substrate for E_2 synthesis and preparing the reproductive tract for ovulation. However, the comparison of these low P_4 values with those of other gravid elasmobranch species suggests that none of the sampled adult female White Sharks were gravid (Table 30.3). However, future blood samples are needed in order to understand the importance of this hormone in female White Shark reproduction.

In male elasmobranchs, in vitro studies have demonstrated that T, produced by Sertoli cells, is associated with spermatocyst development (Cuevas and Callard, 1992) and with distinct stages of spermatogenesis (Sulikowski et al., 2004). Thus, T levels can be used as a proxy to assess male reproductive cyclicity in the absence of morphological (i.e., testes) data. In the current study, the plasma T concentrations of the eight adult male White Sharks are considerably lower than T levels of other male elasmobranchs, particularly during the mating season (Table 30.4). In addition, peaks in T are generally correlated with spermatocyst formation in the testes and/or the presence of mature sperm when the claspers are articulated (Table 30.4). The relatively low T levels in male White Sharks (Table 30.1) in conjunction with both the decline in testosterone values from October to December and the fact that spermatophores were observed in only three out of the eight sampled male White Sharks, suggests that the time frame of sampling for the current study may have missed the peak in the male reproductive cycle, especially considering that this cycle usually occurs over a very limited time frame in most sharks (Driggers et al., 2004; Sulikowski et al., 2007b). Our results

Table 30.3 Estradiol and Progesterone Values Compared with Reproductive Events in Female Elasmobranch Species

				Reproductive Event			
Species	**Common Name**	**Reproductive Mode**	**Sample Size (*n*)**	**Preovulatory (pg/ml)**	**Gestation/Egg Encapsulation (pg/ml)**	**Parturition/ Oviposition (pg/ml)**	**References**
Cephaloscyllium laticeps	Draughtboard Shark	Oviparous	23 6[pec]; 14[lpo]	E_2 3,500[a] P_4 3,000[a]	E_2 2,000[m, peca] P_4 8,000[peca]	E_2 3,800[m, cecoa] P_4 4,000[m, cecoa]	Awruch et al. (2008)
Amblyraja radiata	Thorny Skate	Oviparous	E_2: 47 P_4: 5[pec], 11[cec]	nr	P_4 4,183 ± 1,594[m, pec] P_4 1,853 ± 404[m, cec]	nr	Kneebone et al. (2007)
Leucoraja erinacea	Little Skate	Oviparous	7	E_2 2,000[a] P_4 17,000[ha]	E_2 1,000[a] P_4 3,600[a]	E_2 200[a] P_4 1,000[a]	Koob et al. (1986)
Leucoraja ocellata	Winter Skate	Oviparous	66	E_2 5,143 ± 521[m] P_4 109 ± 28[m]	E_2 2,628 ± 283[m, eeo] P_4 638 ± 140[m, eeo]	E_2 2,628 ± 283[m, eeo] P_4 638 ± 140[m, eeo]	Sulikowski et al. (2004)
Malacoraja senta	Smooth Skate	Oviparous	P_4: 2[pec], 5[cec]	nr	P_4 262[m, pec] P_4 41 ± 17[m, cec]	nr	Kneebone et al. (2007)
Squalus acanthias	Spiny Dogfish	Ovoviviparous	E_2: 5 P_4: 7	nr	E_2 1,000[la] P_4 4,500[ea]	nm	Tsang and Callard (1987)
Carcharias taurus	Sand Tiger Shark	Adelphophagy	1	E_2 3,800[a] P_4 2,750[a]	nr	E_2 ≤ 100[a] P_4 ≤ 100[a]	Henningsen et al. (2008)
Dasyatis sabina	Atlantic Stingray	Histotrophy	10	E_2 643; 5,013[ov] P_4 220; 630[ov]	E_2 202 P_4 140	E_2 9,666 P_4 290	Tricas et al. (2000)
Carcharhinus acronotus	Blacknose Shark	Placental Viviparous	4	E_2 1,364 P_4 119	nr	nr	Rasmussen and Gruber (1990)
Carcharhinus limbatus	Blacktip Shark	Placental Viviparous	3	E_2 370 P_4 4,880	E_2 4 P_4 113	nr	Rasmussen and Gruber (1990)
Negaprion brevirostris	Lemon Shark	Placental viviparous	4	nr	nr	E_2 892 P_4 216	Rasmussen and Gruber (1990)
Rhizoprionodon porosus	Caribbean Sharpnose	Placental viviparous	5		E_2 4,499 P_4 180	nr	Rasmussen and Gruber (1990)
Sphyrna tiburo	Bonnethead Shark	Placental viviparous	44	E_2 37,018 P_4 8,850 ± 5,230[m]	E_2 200 ± 80[m, e]; 920 ± 130[m, l] P_4 2,500 ± 740[m, e]; 390 ± 30[m, l]	E_2 1,970 ± 200[m] P_4 920 ± 180[m]	Manire et al. (1995)

Estradiol (E_2) and progesterone (P_4) values (pg/ml) compared with reproductive events in female elasmobranch species. All of the values indicate reported peaks in steroid hormone concentrations unless otherwise noted. cec, complete egg case; ceco, complete egg case and oviposition; e, early pregnancy/embryos; eeo, egg encapsulation and oviposition; h, 24–48 h prior to ovulation; l, late pregnancy; m, mean value; nr, not reported; pec, partial egg case; ov, ovulation.

[a] Values estimated from a graph.

Table 30.4 Testosterone Values Compared with Spermatogenesis in Male Elasmobranch Species

Species	Common Name	Reproductive Mode	Spermatogenesis	Sample size (*n*)	Testosterone (pg/ml)	References
Hemiscyllium ocellatum	Epaulette Shark	Oviparous	50% in stage V or VI	5	68,000[a]	Heupel et al. (1999)
Cephaloscyllium laticeps	Draughtboard Shark	Oviparous	13% stage VI 11% stage VII	113	5,700[ma]	Awruch et al. (2008)
Amblyraja radiata	Thorny Skate	Oviparous	Average 42% stage VI[a]	3	72,000[a]	Kneebone et al. (2007)
Leucoraja ocellata	Winter Skate	Oviparous	stage III–VI	64	60,670 ± 6,642[m]	Sulikowski et al. (2004)
Malacoraja senta	Smooth Skate	Oviparous	Average 27% stage VI[a]	5	23,146	Kneebone et al. (2007)
Raja eglanteria	Clearnose Skate	Oviparous	nr	20	42,706 ± 16,787[m]	Rasmussen et al. (1999)
Carcharias taurus	Sand Tiger Shark	Adelphophagy	nr	3	40,450	Henningsen et al. (2008)
Dasyatis sabina	Atlantic Stingray	Histotrophy	nr (stages IV–VI)	5 248	9,526 ± 1,047.1[m] 17,270	Snelson et al. (1997) Tricas et al. (2000)
Carcharhinus acronotus	Blacknose Shark	Placental viviparous	nr	2	78,455	Rasmussen and Gruber (1990)
Carcharhinus leucas	Bull Shark	Placental viviparous	nr	2	357,540	Rasmussen and Gruber (1990)
Carcharhinus perezi	Caribbean Reef Shark	Placental viviparous	nr	5	3,086	Rasmussen and Gruber (1990)
Negaprion brevirostris	Lemon Shark	Placental viviparous	nr	5	200,000	Rasmussen et al. (1992)
Rhizoprionodon terraenovae	Atlantic Sharpnose	Placental viviparous	nr	76	128,800 ± 14,400[m]	Hoffmayer et al. (2010)
Sphyrna tiburo	Bonnethead Shark	Placental viviparous	Spermatozoa present	4	132,000 ± 9,200[m]	Gelsleichter et al. (2002)

Testosterone (T) values (pg/ml) compared with spermatogenesis in male elasmobranch species. All values indicate reported peaks in steroid hormone concentration unless otherwise noted. m, mean value; nr, not reported.

[a] Values estimated from a graph.

contrast with the observation of presumed mating wounds on female White Sharks during the same period the blood was sampled (Chapter 16, this book), and we have not established peak T levels for this species; thus, additional blood and spermatophore samples must be collected to place our current results into a comprehensive reproductive cycle context.

ACKNOWLEDGMENTS

We thank Chris Fischer and the crew of the *M/V Ocean* for capture of adult White Sharks. We also thank Dr. Chuck Winkler, Monterey Bay Aquarium, and Dr. Chris Lowe, California State University Long Beach, for providing plasma samples from immature White Sharks. This chapter represents University of New England's Marine Science Centers contribution number 42.

REFERENCES

Adams D.H., M.E. Mitchell, and G.R. Parsons. 1994. Seasonal occurrence of the white shark, *Carcharodon carcharias*, in waters off the Florida west coast, with notes on its life history. *Mar Fish Rev* 56: 24–28.

Ainley D.G., R.P. Henderson, H.R. Huber, R.J. Boekelheide, S.G. Allen, and T.L. McElroy. 1985. Dynamics of white shark/pinniped interactions in the Gulf of the Farallones. *Mem South Calif Acad Sci* 9: 109–122.

Anderson S.D., and P. Pyle. 2003. A temporal, sex-specific occurrence pattern among white sharks at the South Farallon Islands, California. *Calif Fish Game* 89: 96–101.

Awruch C.A., S.D. Frusher, N.W. Pankurst, and J.D. Stevens. 2008. Non-lethal assessment of reproductive characteristics for management and conservation of sharks. *Mar Ecol Prog Ser* 355: 277–285.

Bonfil R., M. Meÿer, M.C. Scholl, R. Johnson, S. O'Brien, H. Oosthuizen, S. Swanson, D. Kotze, and M. Paterson. 2005. Transoceanic migration, spatial dynamics, and population linkages of white sharks. *Science* 310: 100–103.

Boustany A.M., S.F. Davis, P. Pyle, S.D. Anderson, B.J. Le Boeuf, and B.A. Block. 2002. Satellite tagging: Expanded niche for white sharks. *Nature* 415: 35–36.

Bruce B.D. 1992. Preliminary observations on the biology of the white shark *Carcharodon carcharias* in south Australian waters. *Aust J Mar Freshw Res* 43: 1–11.

Bruce B.D., J.D. Stevens, and H. Malcolm. 2006. Movements and swimming behaviour of white sharks (*Carcharodon carcharias*) in Australian waters. *Mar Biol* 150: 161–172.

Cailliet G.M., L.J. Natanson, B.A. Weldon, and D.A. Ebert. 1985. Preliminary studies on the age and growth of the white shark *Carcharodon carcharias*, using vertebral bands. *South Calif Acad Sci Mem* 9: 49–60.

Carrier J.C., H.L. Pratt, and J.I. Castro. 2004. Reproductive biology of elasmobranchs. In *Biology of Sharks and Their Relatives*, edited by J.C. Carrier, J.A. Musick, and M.R. Heithaus, pp. 269–286. Boca Raton, FL: CRC Press.

Convention on International Trade in Endangered Species. 2002. Notification to the Parties No. 2002/042 of 24 June 2002, transmitting *The Role of CITES in the Conservation and Management of Sharks* (a revised and updated version of document AC18 Doc. 19.2 – *Report on Implementation of the International Plan of Action for Sharks*, prepared by the IUCN Species Survival Commission's Shark Specialist Group and TRAFFIC).

Convention on International Trade in Endangered Species. 2004. *White Shark Carcharodon carcharias: Status and Management Challenges*. Document AC20 Inf. 1. Conclusions of the workshop on great white shark conservation research. New York, NY: Wildlife Conservation Society, Central Park Zoo.

Compagno L.J.V. 1984. FAO species catalogue: Vol. 4. Sharks of the world. An annotated and illustrated catalogue of shark species known to date. *FAO Fisheries Synopsis No. 125*. 4: 1–250.

Compagno L.J.V., M.A. Marks, and I.K. Fergusson. 1997. Threatened fishes of the world: *Carcharodon carcharias* (Linnaeus, 1758) (Lamnidae). *Environ Biol Fish* 50: 61–62.

Cuevas M.E., and G. Callard. 1992. In vitro steroid secretion by staged spermatocysts (sertoli/germ cell units) of dogfish (*Squalus acanthias*) testis. *Gen Comp Endocrinol* 88: 151–165.

Dewar H., M. Domeier, and N. Nasby-Lucas. 2004. Insights into young of the year white shark, *Carcharodon carcharias*, behavior in the Southern California Bight. *Environ Biol Fish* 70: 133–143.

Domeier M.L., and N. Nasby-Lucas. 2007. Annual re-sightings of photographically identified white sharks (*Carcharodon carcharias*) at an eastern Pacific aggregation site (Guadalupe Island, Mexico). *Mar Biol* 150: 977–984.

Domeier M.L., and N. Nasby-Lucas. 2008. Migration patterns of white sharks (*Carcharodon carcharias*) tagged at Guadalupe Island, Mexico, and identification of an eastern Pacific shared offshore foraging area. *Mar Ecol Prog Ser* 370: 221–237.

Driggers W.B., D.A. Oakley, G. Ulrich, J.K. Carlson, B.J. Cullum, and J.M. Dean. 2004. Reproductive biology of *Carcharhinus acronotus* in the coastal waters of South Carolina. *J Fish Biol* 64: 1540–1551.

Dulvy N.K., J.K. Baum, S. Clarke, L.J.V. Compagno, E. Cortés, A. Domingo, S. Fordham, S. Fowler, M.P. Francis, C. Gibson, J. Martínez, J.A. Musick, A. Soldo, J.D. Stevens, and S. Valenti. 2008. You can swim but you can't hide: The global status and conservation of oceanic pelagic sharks and rays. *Aquatic Conserv: Mar Freshw Ecosyst* 18: 459–482.

Gelsleichter J., L.E.L. Rasmussen, C.A. Manire, J. Tyminski, B. Chang, and L. Lombardi-Carlson. 2002. Serum steroid concentrations and development of reproductive organs during puberty in male bonnethead sharks, *Sphyrna tiburo*. *Fish Physiol Biochem* 26: 389–401.

Goldman K.J., and S.D. Anderson. 1999. Space utilization and swimming depth of white sharks, *Carcharodon carcharias*, at the South Farallon Islands, central California. *Environ Biol Fish* 56: 351–364.

Henningsen A.D., F.L. Murru, L.E.L. Rasmussen, B.R. Whitaker, and G.C. Violetta. 2008. Serum levels of reproductive steroid hormones in captive sand tiger sharks, *Carcharias taurus* (rafinesque), and comments on their relation to sexual conflicts. *Fish Physiol Biochem* 34: 437–446.

Heupel M.R., J.M. Whittier, and M.B. Bennett. 1999. Plasma steroid hormone profiles and reproductive biology of the epaulette shark, *Hemiscyllium ocellatum*. *J Exp Zool* 284: 586–594.

Hoffmayer E.R., J.A. Sulikowski, J.M. Hendon, and G.R. Parsons. 2010. Plasma steroid concentrations of adult male Atlantic sharpnose sharks, *Rhizoprionodon terraenovae*, in the northern Gulf of Mexico, with notes on potential long term shifts in reproductive timing. *Environ Biol Fish* 88: 1–7

International Union for Conservation of Nature. 2010. IUCN Red List of Threatened Species. Version 2010.3. http://www.iucnredlist.org.

Kneebone J., D.E. Ferguson, J.A. Sulikowski, and P.C.W. Tsang. 2007. Endocrinological investigation into the reproductive cycles of two sympatric skate species, *Malacoraja senta* and *Amblyraja radiata*, in the western Gulf of Maine. *Environ Biol Fish* 80: 257–265.

Koob T.J., and I.P. Callard. 1999. Reproductive endocrinology of female elasmobranchs: Lessons from the little skate (*Raja erinacea*) and spiny dogfish (*Squalus acanthias*). *J Exp Zool* 284: 557–574.

Koob T.J., P. Tsang, and I.P Callard. 1986. Plasma estradiol, testosterone and progesterone levels during the ovulatory cycle of the skate (*Raja erinacea*). *Biol Reprod* 35: 267–275.

Manire C.A., L.E.L. Rasmussen, D.L. Hess, and R.E. Hueter. 1995. Serum steroid hormones and the reproductive cycle of the female bonnethead shark, *Sphyrna tiburo*. *Gen Comp Endocrinol* 97: 366–376.

Nasby-Lucas N., H. Dewar, C.H. Lam, K.J. Goldman, and M.L. Domeier. 2009. White shark offshore habitat: A behavioral and environmental characterization of the Eastern Pacific Shared Offshore Foraging Area. *PLoS One* 4: 1–14.

Rasmussen, L.E.L., and S.H. Gruber. 1990. Serum levels of circulating steroid hormones in free ranging carcharhinid sharks. *NOAA Tech Rpt NMFS* 90: 143–155.

Rasmussen L.E.L., D.L. Hess, and S.H. Gruber. 1992. Serum steroid hormones during reproduction in elasmobranchs. In *Reproductive Biology of South American Vertebrates*, edited by W.C. Hamlett, pp. 19–42. New York, NY: Springer-Verlag.

Rasmussen L.E.L., D.L. Hess, and C.A. Luer. 1999. Alterations in serum steroid concentrations in the clearnose skate, *Raja eglanteria*: Correlations with season and reproductive status. *J Exp Zool* 284: 575–585.

Snelson F.F., Jr., L.E.L. Rasmussen, M.R. Johnson, and D.L. Hess. 1997. Serum concentrations of steroid hormones during reproduction in the Atlantic stingray, *Dasyatis sabina*. *Gen Comp Endocrinol* 108: 67–79.

Sulikowski J.A., P.C.W. Tsang, and W.H. Howell. 2004. An annual cycle of steroid hormone concentrations and gonad development in the winter skate, *Leucoraja ocellata*, from the western Gulf of Maine. *Mar Biol* 144: 845–853.

Sulikowski J.A., P.C.W. Tsang, and W.H. Howell. 2005. Age and size at sexual maturity for the winter skate, *Leucoraja ocellata*, in the western Gulf of Maine based on morphological, histological and steroid hormone analyses. *Environ Biol Fish* 72: 429–441.

Sulikowski J.A., W.B. Driggers III, G.W. Ingram J., Jr., Kneebone, D.E. Ferguson, and P.C.W. Tsang. 2007a. Profiling plasma steroid hormones: A non-lethal approach for the study of skate reproductive biology and its potential use in conservation management. *Environ Biol Fish* 80: 285–292.

Sulikowski J.A., W.B. Driggers, T.S. Ford, R.K. Boonstra, and J.K. Carlson. 2007b. Reproductive cycle of the blacknose shark *Carcharhinus acronotus* in the Gulf of Mexico. *J Fish Biol* 70: 428–440.

Tricas T.C., K.P. Maruska, and L.E.L. Rasmussen. 2000. Annual cycles of steroid hormone production, gonadal development, and reproductive behavior in the Atlantic stingray. *Gen Comp Endocrinol* 118: 209–225.

Tsang P.C.W., and I.P. Callard. 1987. Morphological and endocrine correlates of the reproductive cycle of the aplacental viviparous dogfish, *Squalus acanthias*. *Gen Comp Endocrinol* 66: 182–189.

Weng K.C., A.M. Boustany, P. Pyle, S.D. Anderson, A. Brown, and B.A. Block. 2007. Migration and habitat of white sharks (*Carcharodon carcharias*) in the eastern Pacific Ocean. *Mar Biol* 152: 877–894.

Wintner S.P., and G. Cliff. 1999. Age and growth determination of the white shark, *Carcharodon carcharias*, from the east coast of South Africa. *Fish Bull* 97: 153–169.

CHAPTER 31

Responding to the Risk of White Shark Attack: Updated Statistics, Prevention, Control Methods, and Recommendations

Tobey H. Curtis*
National Oceanic and Atmospheric Administration

Barry D. Bruce
CSIRO Marine and Atmospheric Research

Geremy Cliff
KwaZulu-Natal Sharks Board and University of KwaZulu-Natal

Sheldon F. J. Dudley
KwaZulu-Natal Sharks Board and University of KwaZulu-Natal

A. Peter Klimley
University of California, Davis

Alison Kock
University of Cape Town, Soetwater Education Centre, and Save Our Seas Foundation Shark Centre

Robert N. Lea
California Academy of Sciences

Christopher G. Lowe
California State University, Long Beach

John E. McCosker
California Academy of Sciences

Gregory B. Skomal
Massachusetts Division of Marine Fisheries

Jonathan M. Werry
Griffith University and Ocean and Coast Research

John G. West
Taronga Conservation Society Australia

ABSTRACT

Unprovoked attacks by sharks on humans are traumatic events that may sometimes have tragic consequences. One of the species most commonly associated with shark attacks is the White Shark (*Carcharodon carcharias*), the world's largest predatory fish. The White Shark's seasonal distribution along some coastlines and natural predatory tactics provide ample potential for interactions with humans utilizing the ocean, and its large size gives it the potential to inflict serious injuries if an attack is initiated. Unprovoked attacks by White Sharks on humans, however infrequent, tend to draw significant and often exaggerated attention from the media and general public, and pressure is often placed on local governmental organizations (public safety, law enforcement, fisheries agencies, etc.) to respond in some manner to prevent further attacks. The purpose of this chapter is to review records of White Shark attacks and provide recommendations to help organizations make informed decisions when responding to White Shark attacks. Included are facts on White Shark biology, attack statistics, and a review of response plans that have been implemented in various sites around the world. The recommendations are intended to help reduce sensationalistic, irrational, or ineffective responses by decision makers and benefit the beach-going public as well as vulnerable White Shark populations.

INTRODUCTION

The marine environment is a vast wilderness filled with many thousands of species, most of which are benign to humans or indifferent to their occasional intrusions. As human utilization of the ocean has increased over the last century, encounters with more dangerous species have become more frequent. Few marine species are more feared than sharks. Unprovoked attacks by sharks on humans are infrequent, but they can be extremely traumatic events. In general, the risk of shark attack is exceptionally low when compared with other dangers potentially encountered by beachgoers (e.g., drowning, rip currents, surfboard accidents, stingrays, jellyfish, etc.) (Klimley and Curtis, 2006; Burgess et al., 2010). However, similar to other animal attacks, they draw a disproportionate amount of public and media attention because of their dramatic circumstances. Millions of people engage in swimming, surfing, boating, snorkeling, or scuba diving in the ocean each year, providing billions of dollars in revenues to coastal communities worldwide. Repeated shark attacks within a certain area that result in injuries or deaths are not only extremely traumatic to those involved but can also lead to adverse economic impacts on coastal communities in close proximity to attack locations (Hazin et al., 2008). This may result in considerable public pressure to take action to reduce the risk of shark attacks in such areas. There are over 500 shark species in the world's oceans, yet only about 30 species have been documented to attack humans (International Shark Attack File, unpublished data: http://www.flmnh.ufl.edu/fish/sharks/statistics/species2.htm).

Worldwide, one of the species most often associated with attacks is the White Shark (*Carcharodon carcharias*) (Baldridge, 1996; Burgess and Callahan, 1996). The White Shark is the world's largest predatory fish, growing to lengths exceeding 6 m. It is sparsely distributed in tropical to temperate seas around the world (Compagno, 2001) but is known to seasonally aggregate in certain coastal locations to feed on preferred prey species, including marine mammals (seals, sea lions, whale carcasses, etc.) and a variety of pelagic and demersal fishes (Tricas and McCosker, 1984; Casey and Pratt, 1985; Klimley, 1985; Ellis and McCosker, 1991). Known aggregations of White Sharks occur off the coasts of South Africa, Australia, New Zealand, and the west coast of North America (Ellis and McCosker, 1991; Compagno, 2001). Therefore, it is not unexpected that these regions experience the greatest frequency of unprovoked White Shark attacks on humans (Burgess and Callahan, 1996).

Despite the White Shark's reputation as a powerful predator and "man-eater," this species is actually quite sensitive to exploitation by fisheries. As apex predators in the marine environment, they are naturally low in abundance (Chapple et al., 2011). Additionally, because of their slow growth, late maturity, and low fecundity (Francis, 1996), the ability of White Shark populations to rebound from depletion is limited. White Shark fins, jaws, teeth, and meat are all considered highly valued products, and there has been and continues to be considerable trade in these products (Compagno, 2001). Bycatch of White Sharks in commercial fisheries has occurred throughout its range, whereas in some regions, trophy hunting was also common practice. However, because of evidence of declining populations around the world, the White Shark is afforded some of the highest protections of any marine fish. Trade in White Shark products is now internationally regulated through their listings on Appendix II of the Convention on International Trade in Endangered Species of Wild Fauna and Flora, in the Convention of Migratory Species, and in the United Nations Convention on Law of the Sea. The International Union for Conservation of Nature Red List of Threatened Species identifies White Sharks as globally "vulnerable." Furthermore, they are protected from harvest in territorial waters of numerous countries, including South Africa, the United States, Canada, Australia, New Zealand, Namibia, and some Mediterranean nations.

Responses to White Shark attacks and attack-prevention efforts have varied over time and by region. A variety of fishing (shark removal) and nonfishing (no shark removal) methods have been employed with varying degrees of costs and benefits (Dudley and Cliff, 1993, 2010), but standard-response guidelines are lacking. In addition to the potential trauma inflicted on a shark bite victim, the hours following an attack by a White Shark can be extremely chaotic for those involved in the response (emergency medical personnel, law enforcement, news media, and local officials and scientists). Repeated attacks, uncontrolled shark hunts, or other unfortunate situations may arise in the absence of a formal response from those in authority. There are clear incentives to reduce the risk of shark attack where White Sharks and humans potentially interact, but organized response protocols are also necessary to protect the public.

The purpose of this chapter is to inform the beach-going public, managers, and authorities (e.g., beach management, law enforcement, fisheries agencies) about White Shark attacks, including attack statistics, and to review the methods of attack prevention and response. We identify the pros and cons of these methods at reducing the risk of shark attack and provide managers with the facts necessary to make informed decisions when responding to White Shark attacks. By reducing irrational or ineffective responses, the resulting recommendations are intended to benefit the safety of the beach-going public as well as sustain vulnerable White Shark populations.

WHITE SHARK ATTACK STATISTICS

Definition of Shark Attack

A shark attack for the purpose of this chapter follows the definitions of Cliff (1991, 2006), West (1996), and Burgess and Callahan (1996) as any unprovoked, physical contact between shark and victim or the victim's diving equipment, if worn on the body, or the victim's personal craft, even if the rider was uninjured. Such craft include a surfboard, body-board, sailboard, kiteboard, or surf-ski (resembles a long, narrow kayak) but exclude motorized or larger sailing craft. We have included attacks on spearfishers as unprovoked, although some researchers argue that the act of spearing fish may elicit an attack and thus should be regarded as provocation to do so.

Identifying a White Shark Attack

One of the primary objectives in a shark attack investigation is to determine the species responsible. Two important sources of information are a description of the attacking shark from the victim or eyewitnesses and a forensic examination of the bite. This information should be combined with knowledge of the species, size, diet, and behavior of sharks occurring at the attack site when attempting to ascertain its identity.

White sharks have several anatomical features that set them apart from most other sharks, especially those members of the family Carcharhinidae, such as the Tiger (*Galeocerdo cuvier*), Bull (*Carcharhinus leucas*), Blacktip (*C. limbatus*), and Gray Reef Shark (*C. amblyrhynchos*), which are also implicated in attacks. These include the White Shark's lunate tail, with a lateral keel at the base; large, strap-like gill slits; a relatively large dark eye and a prominent conical snout. The White Shark is also one of very few large sharks with sharp demarcation between the dorsal (gray) and ventral pigmentation (white).

The most conclusive evidence of a White Shark attack is the discovery of a tooth or part thereof, often only a small fragment, with its distinctive serrations (Figure 31.1). Such fragments may be embedded in the victim's equipment, such as a board or paddle craft, or lodged in larger bones, such as the femur. Because shark teeth are radio-opaque, it may be beneficial to x-ray the bitten area for the presence of such fragments.

White Sharks have serrated, cutting teeth in both their upper and lower jaws (Figure 31.1). Their bites are usually relatively clean-cut, leaving a crescent-shaped perimeter. This is most evident when the shark bites a firm, inanimate object such as rider's board. Here the result is often a neat arc, in which the position of the individual teeth can be determined, and consequently intertooth distances and bite circumference can be measured. When a White Shark bites into soft tissue, the victim invariably pulls away, thereby increasing the ripping or tearing effect, making measuring far more difficult. Lowry et al. (2009) looked at interdental distances in 10 species of sharks, including the White, Tiger, and Bull Sharks, and related them to shark length. They found that of the sharks with serrated, cutting teeth, White Sharks have the highest interdental distances, up to 48 mm in the case of extremely large individuals (5.6 m total length), although there is some overlap with large Tiger Sharks. Depth of penetration is another important indicator, because White Shark teeth are very long in comparison with other species such as the Tiger and Bull Shark. West (2011) provides the criteria used to establish the identity of sharks responsible for attacks in Australian waters.

Global Trends

Data on White Shark attacks worldwide are collected and cataloged by the International Shark Attack File (ISAF) based at the Florida Museum of Natural History, University of Florida (http://www.flmnh.ufl.edu/fish/sharks/isaf/isaf.htm). The ISAF is a compilation of all known shark attacks

Figure 31.1 White Shark tooth fragment with characteristic serrations recovered from the wetsuit of an Australian attack victim. (Courtesy of Terry Walker.)

and currently includes over 4,000 individual investigations dating as far back as the mid-1500s. A comprehensive analysis of worldwide patterns in unprovoked White Sharks was presented by Burgess and Callahan (1996). This section provides a brief summary and update from this work and also incorporates recent records compiled by ISAF collaborators in California, Australia, and South Africa.

White sharks have been implicated in a total of 346 unprovoked attacks on humans worldwide since 1839, including 102 fatalities. Although the majority of shark attacks around the world can be attributed to more common carcharhinid sharks (e.g., *Carcharhinus leucas*, *C. limbatus*, etc.) (Burgess et al., 2010), the White Shark is most frequently cited as responsible where the identity of the attacking species is ascertained. This is due to the White Shark's readily identifiable physical features and the generally diagnostic nature of bite patterns and tooth impressions described above. Attacks by White Sharks, therefore, have a higher likelihood of being correctly identified compared with attacks by other species, even though the numbers of attacks by other sharks are more frequent.

Over the last century, the frequency of White Shark attacks each decade has been gradually increasing (Burgess and Callahan, 1996). Between 1900 and 1909, there were less than five attacks reported, but between 1950 and 1959 there were over 20 attacks, and from 1990 to 1999 there were at least 60 attacks. This increase is largely due to increasing human populations and growing use of the ocean for recreation, as well as increased communications in reporting of attacks (Klimley and Curtis, 2006; Burgess et al., 2010). However, over the same period, the percentage of White Shark attacks resulting in a fatality has significantly declined. In the early 1900s, nearly 100% of White

Table 31.1 Number of Unprovoked White Shark Attacks and Fatalities by World Region (1876–2008)

Region	Total Attacks	Fatal Attacks
Australia	126	50
West United States	112	12
South Africa	56	19
Mediterranean Sea	23	11
New Zealand	10	3
East United States	8	4
South America	3	0
Japan	2	2
Hawaii	1	0
Mexico	1	1
Caribbean Islands	1	0
South Korea	1	0
Other	2	0
Total	346	102

Shark attacks resulted in fatality, but by the end of the twentieth century, with improved medical response, the fatality rate has declined to less than 20%. The White Shark is still considered to be the most potentially dangerous shark in the world, given that it is responsible for more human fatalities than any other species.

The number of attacks by country/region is given in Table 31.1 (N.B., these numbers may differ slightly from those given in the subsequent sections because of regional variations in recordkeeping). Approximately 85% (294 attacks) of White Shark attacks have occurred in three regions: Australia (126 attacks), the Pacific coast of the United States (112 attacks), and South Africa (56 attacks). White Shark attack trends in these regions are described in more detail below. A number of attacks have also been reported in the Mediterranean Sea (mostly Italy and Greece), New Zealand, Chile, and the Atlantic coast of the United States (from South Carolina to Massachusetts) (Table 31.1). The higher frequency of White Shark attacks in these areas can be attributed to a combination of three primary factors:

1. High human use of coastal waters (swimming, surfing, diving, etc.)
2. Water temperatures consistent with the White Shark's natural preferences (~15–22°C; Casey and Pratt, 1985; Compagno, 2001)
3. High density of natural prey, resulting in high local White Shark abundance

Regional Trends

Northeast Pacific

The White Shark is essentially the only shark species responsible for unprovoked attacks on humans in the northeastern (NE) Pacific. Since 1950, at least 100 attacks have occurred, primarily in California, and considering the dramatic increase in human recreational ocean use during the last four decades, it is remarkable that shark attacks are so uncommon. Reviews and extensive treatments of eastern Pacific White Shark attacks include Miller and Collier (1981), Engaña and McCosker (1984), Lea and Miller (1985), McCosker and Lea (1996, 2006), and Collier (2003).

The trends, location, months, and human activity during eastern Pacific attacks are listed in Tables 31.2 and 31.3. In summary, humans have been attacked during each month (Table 31.2); however, the majority of attacks have occurred during August (16%), September (20%), and October

Table 31.2 Monthly Record of Confirmed Attacks by White Sharks in the Eastern North Pacific (1950–2009)

Month	California and Baja	Oregon and Washington	Total
January	5	1	6
February	3	1	4
March	2	1	3
April	4	2	6
May	7	0	7
June	2	0	2
July	10	0	10
August	16	2	18
September	18	4	22
October	14	1	15
November	8	2	10
December	8	1	9
Total	97	15	112

Source: Adapted from McCosker, J. E. and Lea, R. N., *Proceedings of the California Academy of Sciences*, 57, 479–501, 2006.

Table 31.3 Confirmed Unprovoked Attacks by White Sharks upon Humans in the Eastern North Pacific (1950–2009)

	Swimmer	Surface Diver	Surfer	Hookah	Scuba	Kayaker	WS	BB
Washington			1		1			
Oregon			14					
Central and Northern California	6	33	32	6	5	5	1	2
Southern California	1					2		
Guadalupe Island		2						
Total attacks	7	35	47	6	6	7	1	2
Total fatalities	5	4	1	1		1		

Source: Adapted from McCosker, J. E. and Lea, R. N., *Proceedings of the California Academy of Sciences*, 57, 479–501, 2006.
"Surface divers" includes scuba divers, breathhold divers, and a hookah diver who were attacked while at the surface. WS, windsurfer; BB, body boarder.

(13%). Geographically (Table 31.3), the vast majority of attacks have occurred north of Point Conception, California (34°30' N latitude), and only a single attack has occurred off Washington State. Attacks south of Point Conception are rare because of the rarity of nearshore pinniped colonies, and the abundance of attacks further north is related to the abundance of pinnipeds and the high level of water use by humans in that area. Within California, 80% of attacks have occurred from Humboldt County south to Monterey County, and 62% of total attacks have taken place along an approximately 160-km stretch of coast between Marin County and Monterey County and out to the Farallon Islands (43 km offshore), an area called the "red triangle" by the mainstream media. Attacks have occurred between 7 a.m. and 7 p.m., peaking at 1 p.m. (McCosker and Lea, 2006). Victims are mostly men; however, we lack information about water-use activity by sex and therefore apply no significance. Nearly all attacked California water users were wearing neoprene (usually black) wet suits. However, this is no doubt a function of black being the most common color used in commercially produced wetsuits. We are unable to correlate attack behavior with the color of the wet suits, surfboards, or the kayaks used by the victims.

The activities of humans when attacked (Table 31.3) has shifted each decade, as new and different recreational activities have become more popular. The majority of attacks in California between 1950 and 1980 were upon swimmers and surface divers (often spearfishers or abalone divers). Attacks on hookah divers and scuba divers are rare. The increasing popularity of surfing, particularly using shorter and more maneuverable surfboards, has made surfers the primary target since the 1980s. Kayakers, windsurfers, and body-boarders (also called "boogie-boarders" or "belly-boarders") are becoming more common.

The trend in attack frequency during 5-yr. intervals has decreased since 1991–1995 (19 attacks). During this century, there are typically one to three attacks per year. We are unable to correlate an increase in attack occurrence with extreme El Niño (Southern Oscillation) events. Only 12 fatalities have been recorded since 1950. The rarity of consumption of human victims in the NE Pacific remains unexplained. Consumption has been recorded in South African and Australian waters (Martin, 2003; West, 1996) but not to our knowledge in California, Oregon, or Washington. However, three bodies that we have reported in previous papers have not been recovered. It is certainly possible that humans reported as "missing at sea" or "drowned but not recovered" in the NE Pacific might have been consumed by White Sharks. It is difficult to fully explain the low fatality rate in the NE Pacific (~10% of victims), as compared with much higher rates in Chile (80%), Australia (34%), and South Africa (24%), but it is likely linked to differences in available medical response (McCosker and Lea, 2006) or variation in human behavior (e.g., use of the "buddy system") or possibly shark behavior between regions.

South Africa

White sharks occur along the entire South African coast (Compagno, 2001) and have been implicated in attacks in three of the four coastal provinces, namely KwaZulu-Natal (KZN, formerly Natal), Eastern Cape, and Western Cape (Wallett, 1983; Cliff, 1991; Levine, 1996). Information on this species in southern Africa is summarized in Chapter 32, this book.

In the last two decades (1990–2009), there were 120 shark attacks in the region. This equates to, on average, 6.0 unprovoked attacks per annum on the South African coast, of which 15% were fatal (Table 31.4). White Sharks were responsible for nearly half (47%) of these attacks at 2.9 per annum; 21% of White Shark attacks were fatal (n = 56). Most (59%) of the White Shark attacks and the majority (67%) of those that were fatal took place in the Western Cape. This province is the center of the species' distribution in South Africa (Bass et al., 1975). It has all but three of South Africa's 13 breeding and nonbreeding colonies of Cape Fur Seal *Arctocephalus pusillus pusillus* (Kirkman et al., 2006), many of which are on small islands within sight of the coast. The waters around these

Table 31.4 Annual Number of Shark Attacks in Each of the Coastal Provinces of South Africa (1990–2009) and for the Entire South African Coast (1960–1990)

	All Shark Attacks per Annum	All Fatal Shark Attacks per Annum	White Shark Attacks per Annum	Fatal White Shark Attacks per Annum
KwaZulu-Natal	1	0.1	0.2	0
Eastern Cape	3.1	0.4	1	0.2
Western Cape	2	0.5	1.7	0.5
Northern Cape	0	0	0	0
Total (1990–2009)	6	0.9	2.9	0.6
Total (1960–1990)	3.3	0.4	1	0.2

colonies are important feeding grounds for larger (>3 m total length) White Sharks (Ferreira and Ferreira, 1996; Martin et al., 2005; Hammerschlag et al., 2006; Kock and Johnson, 2006; Laroche et al., 2008). Given that larger White Sharks are more likely to regard humans as prey, it is not surprising that the incidence of both White Shark attacks and White Shark-inflicted mortalities is highest in the Western Cape, where these sharks are most common (Kock and Johnson, 2006), and decreases through the Eastern Cape to KZN.

In the warm temperate water of the Western Cape, White Shark attacks took place throughout the year with some evidence of a seasonal pattern; 51% took place between April and July. Of the victims, 39% were snorkel divers, 30% were surfers or other wave riders, and 15% were surf-ski riders paddling well beyond the surf zone. Only 9% were scuba divers. Of the snorkel divers, 69% were engaged in spearfishing. Spearfishers appear to be at greater risk than other recreational users of the sea, in that they often venture into very deep water, well beyond the surf zone, and handle struggling, bleeding fish that are highly attractive to sharks.

White Shark attacks in the Eastern Cape averaged 1.0 per annum (Table 31.4), of which 16% were fatal (n = 19). All but one of the victims were board riders, most of whom were on a surfboard. The attacks took place throughout the year with a peak in July (32%). The Eastern Cape, which lies between the subtropical waters of KZN and the warm temperate Western Cape, has a single seal colony and is also thought to include the pupping grounds for this species (Cliff et al., 1996a; Dicken, 2008), although no pregnant female has yet been examined (Chapter 32, this book).

In KZN there were, on average, only 0.2 White Shark attacks per annum, none of which were fatal (n = 4; Table 31.4). White sharks entering nearshore KZN waters are largely immature individuals and appear to be highly nomadic (Cliff et al., 1989, 1996a, 1996b). In the first half of the twentieth century, most South African shark attacks took place in KZN, with the bull shark probably responsible for many of them (Wallett, 1983; Cliff, 1991). The longstanding presence of shark nets, first introduced in Durban in 1952 (Cliff and Dudley, 1992), some recently replaced with drumlines and currently deployed at 38 locations, has reduced the rate of shark attack at protected beaches by over 90% (Dudley, 1997). Only two serious but nonfatal attacks have taken place at protected beaches since 1980.

For South Africa as a whole, but not in KZN, there has been an increase in the total number of shark attacks with time. Between 1960 and 1990, South Africa experienced 3.3 total attacks per annum, 11% of which were fatal (n = 103; Cliff, 1991; Table 31.4). In this period, there were 1.0 White Shark incidents per annum, with a 16% fatality rate (n = 31). The increase from 1.0 to 2.9 White Shark attacks (3.3 to 6.0 total shark attacks per annum) is a feature not confined to South Africa and reflects a global trend in which more and more people are using the sea for recreation (Burgess and Callahan, 1996). By using wetsuits, board riders and divers can spend long periods in relatively cold water (Compagno, 1999). Both factors increase the exposure to sharks. The frequency of shark attack is not, however, simply related to the number of people in the water. The premier school holidays in South Africa are in midsummer—December and January—when beaches along the entire coast are at their busiest. Despite this, the number of White Shark attacks in these 2 months was only 14% of the total.

Australia

The first attack attributed to the White Shark in Australian waters was a fatality recorded in Victoria in 1839. Since then, White Sharks have been identified or implicated as the species involved in 147 attacks (126 unprovoked to April 2010), and they represent the species responsible for the highest number of unprovoked attacks for any shark at 19.5% of all shark attacks in Australian waters. Of the attacks recorded, 50 (34.0%) have resulted in fatality, 56 victims (38.1%) sustained nonfatal injuries, and 41 (27.9%) received no injuries. Attacks have been primarily confined to the southern regions of the continent, ranging from the vicinity of Brisbane in southeast Queensland

(27.5° S latitude), around the south coast to the Abrolhos Islands in Western Australia (28.5° S latitude). This is more restricted than the Australian range of the species, which extends from at least Mackay in eastern Queensland to the Montebello Islands in northwest Western Australia (Bruce et al., 2006; Last and Stevens, 2009; Chapter 19, this book; Figure 31.2). The highest numbers of White Shark attacks have been recorded in the state of New South Wales (NSW) followed by South Australia, Victoria, and Western Australia, respectively. However, the highest rate of fatality occurs in South Australia, with over 50% of attacks being fatal (Table 31.5).

White sharks are widely but not evenly distributed across their range in Australia, with South Australian waters generally having both a higher abundance relative to other states (based on commercial bycatch rates; Malcolm et al., 2001) and a relatively high percentage of medium to large sharks (>3.0 m) (Bruce, 1992; Malcolm et al., 2001). This probably contributes to the observed

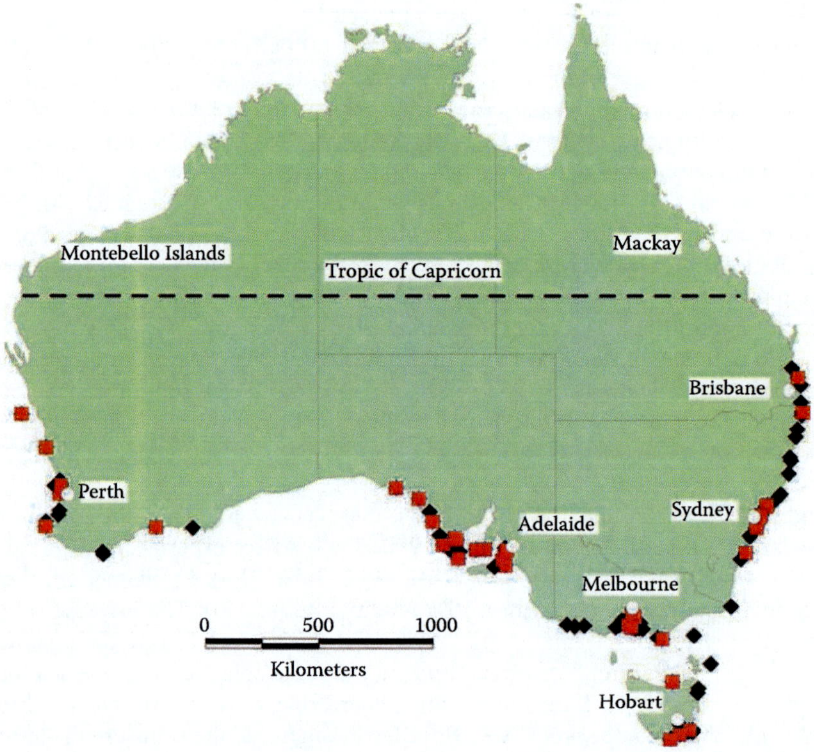

Figure 31.2 The locations of recorded White Shark attacks in Australian waters (1836–2010). Red squares indicate fatalities; black diamonds indicate that the victim was not injured or sustained injuries and survived.

Table 31.5 White Shark Attacks in Australian Waters by State (1839–2010)

State	Fatal (%)	Injured (%)	Uninjured (%)	Total
Queensland	1 (33.3)	2 (66.7)	0	3
New South Wales	13 (26.5)	19 (38.8)	17 (34.7)	49
Victoria	9 (32.1)	12 (42.9)	7 (25.0)	28
Tasmania	4 (30.8)	3 (23.1)	6 (46.2)	13
South Australia	17 (51.5)	10 (30.3)	6 (18.2)	33
Western Australia	6 (28.6)	10 (47.6)	5 (23.8)	21
Total	50 (34.0)	56 (38.1)	41 (27.9)	147

differences in fatality rate for that state. The high number of attacks by White Sharks in NSW is in part because of its populous nature, with the state holding approximately 33% of Australia's population, of which a large proportion is coastally located. Hence the number of people using the ocean for recreation is large relative to other states.

The veracity of lengths for White Sharks involved in attacks is difficult to confirm and notoriously overestimated, particularly when reported in the media. Length estimates of sharks involved in attacks range from <2.0 m to >6.0 m (Figure 31.3). However, unlike areas such as California (McCosker and Lea, 1996), a significant number (24%) of White Shark attacks have been attributed to juveniles (<3.0 m) and in most cases along open ocean, sandy beaches. White sharks more commonly appear at pinniped colonies at sizes >3.0 m in Australian waters, although occasional observations of sharks <3.0 m are made. The smallest White Shark examined with seal remains in its gut contents was a 2.7-m specimen examined by Bruce (1992). Thus, attacks in surf zones by juvenile White Sharks are unlikely to represent an attack scenario supportive of the mistaken identity theory proposed by Tricas and McCosker (1984) and McCosker (1985). The highest percentage of attacks attributed to juvenile White Shark in Australian waters, some as small as 1.8–2.0 m, occurs in surf zone localities in NSW (52%); the lowest occurs in South Australia (19%), again supporting encounters with generally larger sharks in the latter.

The occurrence of juvenile White Sharks in surf zones is well documented (Klimley, 1985; Chapters 14 and 17, this book). In some areas of eastern Australia (e.g., the Port Stephens region of NSW and the Corner Inlet region of Victoria), juvenile White Sharks aggregate seasonally along restricted areas of the coast (Chapters 18 and 19, this book). However, despite the abundance of juvenile White Sharks within size ranges capable of attack, very few attacks have been recorded at these aggregation sites despite the presence of swimmers and surfers and frequent encounters between sharks and people. On the other hand, higher numbers of attacks are recorded in areas where juvenile White Sharks are not known to be seasonally or temporarily resident. Clearly the abundance of White Sharks alone is not always a good predictor of shark attack risk. This may be due to a difference in prey focus of the sharks in different localities and is likely to vary with shark size as well as location. In any event, White Sharks over 2 m in length should be considered potentially dangerous.

As in most other jurisdictions, the number of total shark attacks has risen per decade since records began (Burgess and Callahan, 1996; West, 2011). The number of attacks attributed to White Sharks in Australia has also risen significantly in the last decade from what was a reasonably stable figure of approximately 1–1.5 per annum in the 1920s through the 1990s ($n = 99$) to an average of 4.7 per annum in the 2000s ($n = 48$). West (2011) estimated that the number of beach users in Australia

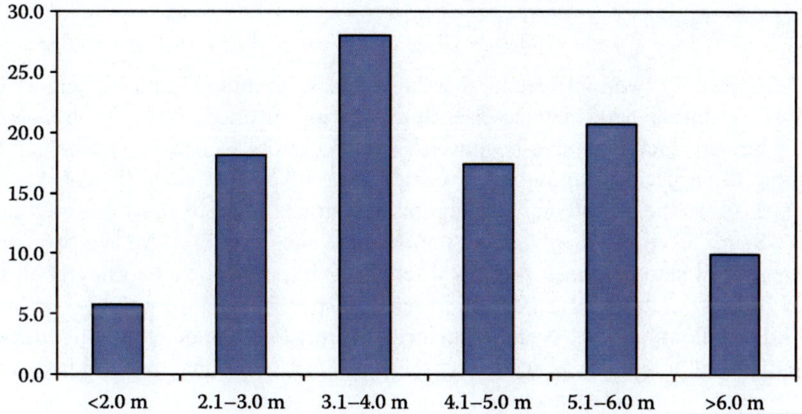

Figure 31.3 Frequency of size categories of White Sharks involved in attacks in Australian waters.

Table 31.6 Activity of Australian White Shark Attack Victims at Time of Attack

Activity	Total Number of Attacks	Number of Fatal Attacks	Percentage of Fatal Attacks
Diving	40	18	45
Swimming	37	20	54.1
Surfing	52	6	11.5
Water craft	14	2	14.3

rose by approximately 20% between 1999–2000 and 2008–2009 based on data on beach visitations and surf rescues from Surf Life Saving Australia.

The significant rise in attacks attributed to White Sharks in the 2000s cannot be accounted for by increasing population and sea-based recreational lifestyle changes alone. The rise equates to over a doubling of the rate of attacks per million persons since the 1990s. Increasing reporting and recording rates combined with the advent of the internet, mobile phones, digital cameras, more reliable identification of White Sharks as being the attacking species, as well as a growing awareness of the Australian Shark Attack File (affiliated with the global ISAF) may all have played a role in increasing the reports over the last decade. There were four main activities of victims involved in White Shark attacks: diving (including scuba, snorkeling, spear-fishing, and on hookah); swimming, surfing (surfboard or body-board), and watercraft use (including kayaks, surf skis, sailboards, and boats; Table 31.6). Interestingly, although surfing as an activity sees the highest number of attacks, it records the lowest frequency of fatalities, whereas swimming records the highest frequency of fatalities. This may in part be due to a surfboard providing some protection to the victim such that injuries are either lessened or are not sustained, whereas a swimmer without a board is more unprotected. However, there is a temporal component to these overall statistics, in that attacks on swimmers and swimming fatalities were primarily confined to decades prior to the 1950–1960s. This was also prior to the more recent popularity of surfing and to the development of current emergency medical response practices. The increasing efficacy of medical response teams, the rise of surf life-saving organizations—with patrolled beaches, as well as the propensity to swim or surf in groups—whether intentional or driven by larger crowds at swim/surf localities, has no doubt decreased response times to victims after an attack and has lead to an increase in survival rate. The high level of fatalities in attacks on divers corresponds with the large size of sharks generally involved with those attacks and, to some extent, the remoteness of this activity to emergency services compared with most swimming and surfing activities.

Motivation for Attacks

There are essentially two motivations that cause sharks to attack humans: defense and hunger. When animals feel threatened, they experience conflicting instincts: one is to escape and another is to fight. If they are unable to flee because their opportunity to escape is blocked, they do not always fight but often perform an agonistic or defensive display. If such displays are not heeded, it can trigger, in some species of shark, an aggressive attack. Such displays are well described for the Gray Reef Shark, *C. amblyrhynchos*, by Johnson and Nelson (1973). An agonistic display in the form of an aerial tail slap has been described for the White Shark by Klimley et al. (1996b) and was recorded by these authors directed both toward conspecifics and toward a vessel in the vicinity of pinniped kills or floating baits. Such behavior has rarely been associated with attacks by White Sharks on humans, although it is possible that the significance of some displays may not always be recorded by eyewitness or victim accounts.

The second, and more common, motivation to attack is hunger. Humans have rarely been consumed by White Sharks. However, the behavior of White Sharks during attacks on humans frequently reflects their known predatory behaviors. Human divers are often seized by White Sharks but usually released without being consumed, as explained by Tricas and McCosker (1984). Those authors proposed the "bite and spit hypothesis," whereby a feeding shark swims in mid-water, looking upward and mistakes a surface diver or surfer on a short surfboard for a pinniped (i.e., seals, sea lions). It then swims rapidly upward to seize and immobilize its prey and then will often retreat allowing the victim to lapse into shock or bleed to death, thereby reducing the potential injury to the shark caused by the teeth and nails of its presumed pinniped victim. Once a pinniped victim has been exsanguinated, the shark will return to feed on the body at leisure.

Although White Sharks may have a search image for pinnipeds, the variety of colors and shapes of objects struck at the surface in the turbid waters of Central California, such as flotsam, a crab-pot buoy, an inflatable raft (Collier et al., 1996), or a marine kayak (McCosker and Lea, 1996), would indicate that they are not extremely selective. Although a particular object may not exactly resemble normal prey, the stimulus may be sufficient to trigger a predatory reaction (Burgess et al., 2010). Their ability to discriminate has been demonstrated during an experiment carried out in the clear waters of the Gulf of Spencer, South Australia, in which White Sharks were presented a square and seal-shaped object at the surface with a preference being shown toward investigating the latter (Strong, 1996). Available evidence indicates that when surfboards or kayaks of different lengths are in the water together, White Sharks show a higher rate of attack at the shortest of the group (McCosker and Lea, 2006).

In Central California, the water conditions are turbid, with limited visibility unlike sites such as Guadalupe Island, Mexico, and the Neptune Islands, South Australia. Hence, it would be more difficult for a shark to see objects clearly at the surface in the former region, and this would likely support a feeding tactic of directing indiscriminate predatory strikes at any object at the surface. However, in situations where a shark can discriminate a novel object at the surface, it may attempt to gauge its palatability with low-impact investigatory bites. If the texture, taste, or odor of the object deems it edible, the shark may commence more typical predatory feeding behavior. White Shark bites on humans are often consistent with "investigative strikes," but because of the White Shark's large size and dentition, such bites can still cause severe damage (Burgess and Callahan, 1996).

The observations of natural predatory events on seals have been used extensively in the literature to interpret attack patterns on humans. Much of this work has been carried out at the Farallon Islands off Central California (Klimley et al., 1992, 1996a). Researchers observing feeding White Sharks often noticed first the appearance of blood-stained water and only later could see the seal float to the surface. What was significant was that the carcass did not exude blood when it rose to the surface. This lead to the hypothesis that the sharks carry their prey under water until it exhausts its blood supply and dies due to asphyxiation, a predatory strategy termed exsanguination (Klimley, 1994; Klimley et al., 1996a). The remaining carcass of the prey, which at this time was immobile, was consumed without hesitation. This behavior contrasts with the seeming hesitation in consuming a human. Humans have been partially consumed by White Sharks (Engaña and McCosker, 1984; Martin, 2003), and there are attacks where the victim's body has not been retrieved and presumed to have been completely consumed (West, 2011). However, an entire human has yet to be found in the stomach of a White Shark (Burgess and Callahan, 1996). For all practical purposes, however, the capture of a White Shark after an attack has occurred is rare, particularly in cases where the victim is not recovered. Over 50% of the divers seized either are released without taking a bite or with a single bite inflicted to the body. This is consistent with the sharks not having a preference for humans.

An explanation proposed was that White Sharks prefer to feed on seals and sea lions because of their energy-rich, external fatty layer. This layer comprises roughly half of the body weight of returning yearlings and has twice the caloric value of muscle tissue (Klimley et al., 1996b). Evidence cited to support this hypothesis is drawn from observations that White Sharks tend not to feed upon lean prey at certain locations. A White Shark was observed to strike a pelican at South Farallon Island, but it refrained from feeding further on it (Klimley et al., 1996a). The carcasses of sea otters often wash up on the beaches of Santa Cruz, California, with White Shark tooth fragments embedded in their skin, yet they do not have any part of their body missing from a bite (Ames and Morejohn, 1980). Although White Sharks commonly scavenge whale carcasses in many oceans (Curtis et al., 2006), White Sharks prefer not to feed on rotting prey at the Farallon Islands (Klimley et al., 1996a). White sharks at the Farallones fed on fatty tissue, in which transmitters were embedded, removed from a seal (Goldman et al., 1996). However, a shark did not feed on the muscular tissue of a seal, from which the fat was removed (Klimley et al., 1996b). The propensity for White Sharks to target seal prey around the site of seal colonies is not unexpected, and in such locations, their preference to do so over other prey may be driven by feeding on seals being the reason for sharks visiting such areas in the first place. More experiments are needed to support the above mentioned hypotheses in order to establish their veracity under different situations and locations.

White sharks, even large individuals, are not obligate pinniped predators, and recent analyses of chemical signatures of vertebrae suggest that some sharks may rarely feed on pinnipeds (Estrada et al., 2006; Kerr et al., 2006). Large sharks (>3.0 m) are known to target pinniped prey during certain seasons, but they also feed on a variety of finfish, elasmobranchs, and invertebrates that lack fatty tissues in other locations and at other times of the year (Bruce, 1992; Casey and Pratt, 1985; Klimley, 1985). White sharks make regular long-distance migrations away from seal colonies into areas where pinnipeds are rare (although cetaceans may be common), moving through a wide variety of marine ecosystems from coastal to open ocean, cool temperate to tropical waters (Boustany et al., 2002; Bonfil et al., 2005; Bruce et al., 2006; Weng et al., 2007; Domeier and Nasby-Lucas, 2008; Jorgensen et al., 2009; Nasby-Lucas et al., 2009; Chapters 11, 12, 16, 21, and 22, this book). Predatory strategies and prey preferences consequently vary depending on where sharks occur. Therefore, the behavior of White Sharks during attacks on humans may not always reflect pinniped-predation strategies. Elucidating the reasons why White Sharks attack humans across the range of shark habitats and their behavior during and after attack sequences is a fertile area for new research that may provide some clues as to how to further minimize the risk of White Shark attack.

METHODS OF RESPONSE TO THE RISK OF SHARK ATTACK OR IN THE AFTERMATH OF SUCH ATTACKS

Removal of Sharks

Shark "Hunts"

Historically, the most common response to high-profile or multiple shark attacks in an area was to initiate a public shark hunt. A number of fishing vessels would voluntarily go out to sea, deploy a variety of gear (rod and reel, longline, gillnet, harpoon, etc.), and attempt to harvest as many sharks as possible in the hope of capturing the "culprit" shark. This type of response was portrayed in the fictional Peter Benchley novel and Steven Spielberg film *JAWS* and has been documented to occur in the United States (Fernicola, 2001), Australia, and South Africa in the past. These tactics typically lacked organization or any coordination of fishing effort. Larger-scale hunting efforts would typically subside after days or weeks, providing only temporary results. Sharks of several species would sometimes be landed, often including small individuals or species not likely to attack humans.

The capture of the shark responsible for an attack is exceptionally rare and is difficult to confirm, given the low likelihood of finding human remains even if the appropriate shark was caught. In general, the motivation to undertake such hunts are due to a perceived risk that the shark may remain active in the area and may therefore strike again or that having attacked a human in the area the individual shark may become predisposed to doing so again. White Sharks are highly mobile animals that can travel 70–100 km over the course of a day (Boustany et al., 2002; Bruce et al., 2006) and may quickly leave the general vicinity of an attack. Multiple White Sharks may also be present in the area of an attack or may visit the area under their normal movement patterns over time. The chances of catching the shark responsible after an attack by fishing in the immediate vicinity can therefore be low, and the risk of killing the wrong shark may be high under some situations. There is no evidence to support the notion that White Sharks having once attacked a human are any more or less likely to do so again. Therefore, the effectiveness of shark hunts for preventing subsequent attacks is likely to be low and in most cases is difficult to ascertain (Wetherbee et al., 1994).

In more recent years, White Sharks have been protected in many parts of the world under either fisheries or conservation-based legislation, and their indiscriminate killing is either restricted or illegal. Some jurisdictions where White Sharks are fully protected maintain the option to kill individual sharks should they represent a clear public danger (e.g., Western Australia and South Australia), but such actions are only considered a last resort and generally require case by case Government approval under strict protocols (see section below on shark response plans). Given the lack of evidence for the effectiveness of shark hunts, this approach is not currently considered an efficient means to consistently reduce the risk of shark attack, and support for such actions has been short-lived in instances where it has been trialed (Holland et al., 1999).

Shark Culling

A shark "culling," or control, program is defined here as a government-controlled shark-fishing effort, designed to reduce the number of potentially dangerous sharks in an area of human use and thus reduce the risk of human-shark encounters and attack frequency. These programs are typically carried out by government departments or contracted fishing crews and vessels. They tend have more standardized effort than shark hunts. They may be intensive programs run over relatively short-term periods in the form of commercial-style longlining or the deployment of nets or baited lines (e.g., drum-lines) in specific localities on an ongoing basis.

Short-Term Shark Control Programs

One of the most well-documented short-term shark-culling programs was the series of fishing activities conducted in the Hawaiian Islands (Wetherbee et al., 1994). The primary targets and the drivers for initiating this program were attacks by species such as Tiger Sharks (*Galeocerdo cuvier*). However, White Sharks occur in the Hawaiian Islands (Taylor, 1985; Wetherbee et al., 1994; Boustany et al., 2002); they have been implicated in attacks on humans there (Taylor, 1993; Wetherbee et al., 1994), and thus they were featured as one of the species of concern. Human remains were found in a 3.5-m White Sharks caught off Oahu in 1926, and a surfer was confirmed bitten by a White Shark off the island of Oahu in 1969 (Taylor, 1993). As a result of several notable shark attacks in the late 1950s and early 1970s, the State of Hawaii funded two major shark-culling programs around the main Hawaiian Islands (MHI) between 1959 and 1976. The objectives of these programs were to determine what species of sharks were found in Hawaiian waters and to reduce their population numbers based on the assumption that reducing the numbers of sharks in near-shore waters would reduce the probability of attacks on humans (Wetherbee et al., 1994). Extensive longline fishing for sharks was conducted around all the MHI, resulting in the culling of over 2,849 sharks, comprising mostly Sandbar (*Carcharhinus plumbeus*) and Tiger Sharks. Out of a total of 36,122 hooks set throughout both control programs around all MHI, only two White Sharks were captured during the 1959–1960 program (Wetherbee et al., 1994).

Despite large numbers of sharks caught and killed, Wetherbee et al. (1994) found no evidence of a subsequent reduction in shark attacks. The 1990s saw a renewed call for shark culling in Hawaii after a series of widely publicized attacks. However, because of the expense of mounting a repeat operation, the recognition of the potential ecological ramifications of removing large numbers of sharks, and cultural opposition to shark culling, the state opted to fund studies of shark behavior as an alternative, as well as smaller-scale "targeted" hunts, as discussed above, the latter in attempts to capture sharks potentially responsible. Public education about sharks, their ecology, and their behavior obtained from targeted research significantly reduced public outcry following attacks and eliminated the state's support of targeted hunts, particularly because movement data indicated that Tiger Sharks could travel up to 30 km/day and easily move between islands (Holland et al., 1999, also see discussion under "Electronic Tags" below).

Ongoing Shark Control Programs

Permanent or semipermanent deployment of shark-fishing gear off high-use beaches as a means of reducing the risk of shark attack has been utilized in several countries. Two types of gear are generally used: large-meshed, anchored gillnets (Figure 31.4) and drumlines, which consist of a large baited hook suspended from an anchored float (Queensland Department of Primary Industries and Fisheries, 2006; Dudley, 1997; Green et al., 2009). Neither of these gear types create barriers between the open sea and the beach front but are used as fishing apparatus to catch and remove a percentage of the sharks that occur in the area. Nets are typically up to 200 m in length, may be set singularly or in multiple configurations, are either surface or bottom set, and do not extend throughout the entire water column. Nets were first deployed off certain beaches in New South Wales, Australia, in 1937 (Collins, 1972; Coppleson, 1962; Reid and Krogh, 1992) and off KZN, South Africa, in 1952 (Davies, 1961, 1964; Dudley, 1997 and references therein). A mix of nets and drumlines was introduced off the beaches of Queensland, Australia, in 1962 (Dudley, 1997; Paterson, 1979), and nets were first deployed off Dunedin, New Zealand, in 1969 (Francis, 1998). In 2004, a combination of drumlines and two longlines with 100 hooks each was deployed off Recife, Brazil (F.H.V. Hazin, personal communication in Dudley, 2006). In Queensland, such gear is deployed permanently, but deployment is seasonal in NSW (September to April) and Dunedin (December to February), and in KZN, the gear is removed at affected beaches during the annual winter "run" of sardines (Armstrong et al., 1991) in order to avoid catches of predators accompanying sardine shoals (Dudley and Simpfendorfer, 2006).

The rationale behind these programs is that long-term fishing reduces local numbers of potentially dangerous sharks, thereby reducing the likelihood of an encounter between sharks and humans, hence achieving a reduction in the risk of shark attack (New South Wales Shark Menace Committee, 1929; Cliff and Dudley, 1992; Davies, 1961, 1964; Dudley, 1997, 2006; Last and Stevens, 1994; Paterson, 1979; Springer and Gilbert, 1963). The two Australian operations and that in South Africa are the largest, each operating at a number of discrete beach locations distributed over large stretches

Figure 31.4 Staff of the KwaZulu-Natal Sharks Board removes a White Shark from the protective beach nets off Durban, South Africa. (Courtesy of KZNSB.)

of coastline (Paterson, 1990; Queensland Department of Primary Industries and Fisheries, 2006; Dudley, 1997; Green et al., 2009). Numbers of shark attacks prior to the introduction of protective measures were not high in absolute terms (Collins, 1972; Wallett, 1983; Paterson, 1986; Cliff, 1991; Queensland Department of Primary Industries, 1992), but each program was introduced because sufficient attacks had occurred to elicit concerns for public safety and because of the potential negative economic effect on tourism, which is important to each region (Dudley, 2006 and references therein).

All of the programs have been very successful in reducing the incidence of shark attack at the protected beaches (Paterson, 1990; Queensland Department of Primary Industries and Fisheries, 2006; Dudley, 1997, 2006; Green et al., 2009). In KZN, for example, 21 attacks occurred at the beaches of the largest city, Durban, in the decade prior to the introduction of nets in 1952 (Wallett, 1983), but there were no incidents involving large sharks in the subsequent 58 yrs. (KZN Sharks Board, unpublished data). Elsewhere in KZN, nets reduced by over 90% the number of incidents occurring per annum. Similar success has been reported for the other programs, but given that neither nets nor drumlines physically prevent sharks from entering a protected bathing area, shark-fishing gear does not eliminate attacks completely.

The Bull Shark was probably responsible for the bulk of attacks in the years prior to the introduction of shark control programs at each of the above locations other than Dunedin, with a lesser number of incidents probably attributable to the White Shark and the Tiger Shark (Cliff, 1991). Through long-term fishing, the nets and drumlines are thought to have removed bull sharks that were locally resident in the vicinity of each protected beach (Wallett, 1983; Cliff and Dudley, 1991). The gear does also have the potential to catch itinerant White and Tiger Sharks moving through an area and, indeed, given the near-cessation of all shark attacks at protected beaches subsequent to the introduction of fishing gear, incidents attributable to these species have also declined. At Dunedin, all attacks prior to the introduction of nets are believed to have been by White Sharks, and these attacks ceased with the introduction of nets, although no captures of White Sharks were reported by the program after accurate records were established from 1986 to 1991 (Francis, 1998). White Sharks were, however, reportedly caught by the program during the 1970s. It is probably not coincidental, however, that the two attacks by large sharks that occurred at protected beaches in KwaZulu-Natal in the 30-yr. period 1980–2009 were attributable to itinerant White Sharks rather than resident Bull Sharks (KwaZulu-Natal Sharks Board, unpublished data), suggesting that the effectiveness of the gear is lower in the case of nonresident animals.

None of the programs operate at known White Shark aggregating sites, and catch rates of White Sharks in the KZN program, for example, showed no trend over the period 1978–2003 (Dudley and Simpfendorfer, 2006), suggesting that catches in that program are sustainable or do not provide an effective index of population size. The deployment of shark-fishing gear, whether nets or drumlines, at White Shark aggregating sites would almost certainly reduce the number of White Sharks and hence overall risk of attack, but it is likely also that long-term fishing for White Sharks at such sites would have a significant effect on the population and may be unsustainable.

The benefit of the long-term deployment of shark nets and drumlines, particularly in locations where locally resident Bull Sharks are responsible for the majority of shark attacks, is that the rate of occurrence of attacks is reduced significantly, with positive consequences for public safety and tourism. There are costs, however, that include the following:

1. The financial cost of maintenance of the gear, which requires regular servicing and, in the case of drumlines, rebaiting of the hooks
2. The fact that the mechanism for providing protection is shark fishing (i.e., it is an extractive process)
3. The bycatch of animals that pose no potential threat to people in the water

(Cliff and Dudley, 1992; Collins, 1972; Dudley and Cliff, 1993; Dudley and Simpfendorfer, 2006; Keith et al., 2002; Krogh, 1996; Natoli et al., 2008; Paterson, 1979, 1986, 1990; Reid and Krogh,

1992; Simpfendorfer, 1993). In the case of nets, this bycatch includes dolphins, rays, turtles, and, occasionally, large teleosts. Drumlines are more selective than nets and take a considerably reduced bycatch (Dudley et al., 1998), but an installation consisting of drumlines only may not provide the same level of protection as an installation consisting of mixed gear (Gribble et al., 1998).

In KZN, various measures have been implemented to reduce catches of marine animals, including:

1. An overall reduction of the quantity of gear deployed per beach
2. The replacement of some nets with drumlines
3. The removal of fishing gear during the annual sardine run
4. The release of all live animals, with most released sharks being tagged for research purposes
5. The attachment of acoustic devices to the nets in an attempt to deter marine mammals

Also, sharks that do not survive capture are retained for research purposes, thereby ensuring that the biological material is not wasted. Some of these measures have also been implemented in one or more of the other programs (Green et al., 2009).

Nonfishing Methods for Minimizing Risk of Shark Encounter and Attack

Beach Closures

Beach closures are a precautionary action aimed at reducing the risk of shark attack when the risk of attack is higher than normal. In most cases, beaches are closed in the following instances:

1. Following a shark attack in the vicinity
2. When shark sightings are higher than usual for areas that have regular sightings
3. When a shark is sighted in an area that is not well known for shark sightings

Beaches have also been closed when there has been an abundance of potential prey in an area (i.e., schools of fish, seal haulouts, or when a large cetacean has stranded because these are well known for attracting sharks). Beach closures can last from a few minutes to days, depending on the situation and managing authorities. Beach closures are relatively easy to implement provided adequate staff are available to alert swimmers to get out of the water, and temporary signs should be erected to warn water users that the beach is closed. Beach closures are a low-cost way to reduce the risk of shark attack. However, beach closures are only a temporary solution and may cause locals and visitors to leave the area. Guidelines for beach closures should be drawn up for each area to ensure that beaches are closed only when appropriate, because frequent beach closures can frustrate water users and impact local business (Chapter 29, this book). However, the effect of a shark attack on local business would be far greater, and so beach closures are justified. Examples of such guidelines can be found at http://www.slsa.com.au/site/_content/resource/00000298-docsource.pdf.

Shark Barriers

Mechanical barriers, or shark fences, have been erected in various parts of the world as a means of physically excluding sharks from bathing areas. Barriers have been constructed of various materials, including small mesh nets (too small to capture animals and thus different from mesh nets used to catch sharks), concrete, wooden or steel pylons, rocks, etc. These have tended to be erected at locations in sheltered waters given the difficulty of maintaining such structures in surf conditions. Probably the most well-known examples, consisting of small mesh or vertical steel bars, are at beaches found in Sydney Harbor (Green et al., 2009). Barriers also exist in Queensland's Gold Coast canal system and at other Queensland locations (Queensland Department of Primary

Industries and Fisheries, 2006; Dudley and Gribble, 1999). Shark barriers have been constructed in the past at a number of other sheltered locations, including Florida, Croatia [I.K. Fergusson, personal communication (in Dudley and Gribble 1999)], Panama (Springer and Gilbert, 1963), and Maputo, Mozambique (da Graça, 2005).

Attempts have been made to construct shark barriers at surf beaches. In 1907, a semicircular enclosure with a diameter of 183 m (600 ft) was constructed of steel piles and vertical steel grids in the surf at Durban, KZN (Davies, 1964). It was demolished in 1928 because of damage caused by wave action and corrosion, and because of high maintenance costs (Davies, 1964). Barriers of various designs were constructed at several beaches on the KZN south coast after a number of attacks occurred in the late 1950s, and again maintenance in surf conditions was difficult (Davies, 1964), with the barriers eventually proving impractical (Wallett, 1983). A fence was built at the surf beach at Coogee, New South Wales, in 1929 (New South Wales Legislative Assembly, 1935) but again proved impractical and costly to maintain and ultimately was abandoned (Coppleson and Goadby, 1988). There is one location, Hong Kong, where shark barriers currently are maintained in surf conditions, although swell size is typically small at most of the protected beaches (Dudley, 2006). Maintenance is difficult, however, and costs are high (Dudley, 2006; Green et al., 2009).

Shark Spotters and Aerial Surveys

Another method used to reduce the risk of attacks is to use visual surveys in the form of land-based "spotters" or aerial surveys from planes or helicopters at beaches where the propensity for encounter with sharks is high. In areas where water visibility is sufficient to see a large shark approaching a bathing area, spotters on the ground or in the air can relay the sighting to lifeguards and warn swimmers of the potential danger. Historically, temporary spotting has been conducted by lifeguards and through aerial surveys at particular beaches, but within the last 6 yrs., a permanent (the first of its kind) land-based spotter program has been implemented in Cape Town, South Africa (Chapter 29, this book). A summary of the program is described below.

"Shark Spotters" is a shark safety and outreach program used in Cape Town, South Africa. The objective is to reduce the risk of shark attack by warning water users when sharks are in close proximity and then closing the beach. Shark Spotters is part of an overall safety strategy that includes research on White Shark spatial and temporal patterns and public education. The initiative was driven at the community level after a spate of fatal and major shark attacks on recreational water users (Oelofse and Kamp, 2006). Shark Spotters are positioned at strategic points along the Cape Peninsula, primarily along the False Bay coastline. The program currently employs 14 to 28 spotters at 5 to 10 of Cape Town's popular beaches (numbers are seasonally dependent) throughout the year (Chapter 29, this book). The shark spotters scan coastal waters for sharks from an elevated platform above the beach during daylight hours, 7 d a week. A standardized protocol using four informational flags and a shark siren warn water users of the nearby presence of sharks. Upon hearing the warning, the beach is closed, and water users exit the ocean and wait until the shark spotters give the "all clear" sign before returning to the water. Two spotters are positioned at each beach. One spotter is placed on the mountainside equipped with polarized sunglasses to remove glare and improve spotting conditions and binoculars for close-up scans. This spotter is in radio contact with a second spotter located on the beach. The spotter on the beach is in charge of changing the flags to communicate that a shark has been seen, assisting water users in getting out of the water, and providing on-site explanation of the situation. Both spotters have the ability to set off the warning siren when a shark is in close proximity to water users. Daily data are recorded on the number of sharks detected, shark behavior, sea conditions, and the number of water users (Chapter 29, this book).

The Shark Spotters program has proven to be an effective warning system with a total of 619 shark sightings recorded since the start of the program in November 2004 until December 31, 2009. Consistent data collected on shark sightings provide seasonal and regional information on shark sightings, which can be used to predict higher risk areas and times of the presence of sharks and therefore the likelihood of shark attack. Constant monitoring allows for the identification of high shark activity and can be used to issue shark alerts and press releases or close beaches. Information collected on shark behavior (e.g., distance sharks swim from shore) can be used to keep safety tips current for specific areas. The program provides job opportunities and skill development for previously disadvantaged nationals. Shark spotters are accessible to the public on a daily basis, which plays a valuable role in education and outreach to both locals and visitors. Scientists can benefit from the program by being notified of shark sightings close to shore in those cases where tagging of animals is necessary. The annual budget to run a shark-spotting program is relatively low compared with other methods (i.e., exclusion nets or antishark gillnets) (Oelofse and Kamp, 2006).

One of the limitations of the program is that not all sharks may be spotted, which means there is still a risk of shark attack. However, this is arguably the case in all other methods implemented to reduce risk with the exception of shark barriers. White Sharks in the False Bay area are large, may spend lengthy periods near the surface, and are often readily visible over open sandy areas in the vicinity of surf beaches which makes detection easier for spotters (Figure 31.5). Spotting smaller sharks or species that do not spend a lot of time at the surface could be more difficult (e.g., Bull Sharks). Poor weather and sea conditions and glare influence the ability of shark spotters to see sharks and can reduce the effectiveness of the spotters. The program is restricted to beaches with an elevated platform nearby and ideally a sandy bottom to allow for the dark silhouette of a shark to be visible from above. The program is open to human error, because constant observation and concentration are difficult. The spotters often have to operate in poor conditions (i.e., rain, cold, heat, or wind). To address the problem of spotter fatigue, daily shifts should be as short as possible. The program is most beneficial for swimmers and surfers, whereas there is limited benefit for divers and kayakers using waters further away. Managing public expectation of the program is crucial to address concerns of liability and to make sure that water users do not feel a false sense of security. It needs to be communicated adequately that shark spotters are not responsible for water-user safety, and entering the ocean will be at the individual's own risk.

Aerial surveys of high-risk beach areas using small planes or helicopters can also be used to inform beachgoers of the presence of a potentially dangerous shark (Figure 31.5). Such surveys have been attempted sporadically in the United States, South Africa, and Australia, but they have similar

Figure 31.5 Aerial photograph of a White Shark (see circle) patrolling a Massachusetts beach. (Courtesy of Dan McKiernan, Massachusetts Division of Marine Fisheries.)

limitations to land-based shark spotters and can be more costly, and because of the nature of aerial observation, these programs are limited in the amount of time that continuous observation is possible.

Research

Increased knowledge can clearly be an effective tool when considering a variety of management strategies. The field of research on White Sharks has dramatically expanded over the last 50 yrs., providing us with many of the insights that we are able to share throughout this book. Much of shark research, particularly in the early years, was carried out with shark attack prevention as its primary goal. Continued research in regions with relatively high shark attack frequencies can therefore be very beneficial for developing strategies to reduce the risk of attack. In many cases, shark attack prevention efforts themselves have helped improve the body of knowledge on the attacking species (Cliff et al., 1989, 1996a; Wetherbee et al., 1994).

Data collection, of course, is the key to successful research efforts. Data from shark attacks or from any shark monitoring or control efforts should be collected in a standardized fashion. Any available information from a shark attack incident should be compiled and shared with the ISAF, or regional shark scientists that collaborate with the ISAF. As the number of cataloged attack events increases over time, more patterns, correlations, or activities can be identified that are associated with the risk of shark attack (Baldridge, 1996; Burgess and Callahan, 1996; Burgess et al., 2010). The same is true for data collected from any of the shark control methods described here. Data collection protocols should be an integral part of any program upon its implementation.

Focusing research in areas that have a high risk of shark attack can help identify shark behaviors or environmental conditions that may contribute to the presence of sharks in high-use areas and hence the risk of interactions and attacks. Such research can be purely observational, such as visually monitoring natural predatory activity (Klimley et al., 1992, 1996a), photographing or tagging individual sharks for mark-recapture analyses (Klimley and Anderson, 1996; Domeier and Nasby-Lucas, 2007; Chapter 25, this book), or more technologically sophisticated research using electronic tags to track the movements and behaviors of individual sharks (Goldman et al., 1996; Klimley et al., 2001; Jorgensen et al., 2010; Chapters 11, 12, 13, 16, 17, 18, 21, and 22, this book). Using data derived from well-designed research efforts, scientists and managers will have the tools to implement meaningful measures to

1. Control anthropogenic activities that may impact the sharks.
2. Establish protocols and procedures for the prevention of shark attack.
3. Educate the public to allay the fear of shark attack.

Electronic Tags

The rapid development of fish-tagging technologies in recent years has provided the scientific community with exciting new insights into fish movements. There are currently a variety of acoustic, archival, and satellite-linked telemetry devices that can be attached to sharks to study migration, site fidelity, physiology, behavior, and ecology (Goldman et al., 1996; Lowe and Goldman, 2001; Klimley et al., 2001; Boustany et al., 2002; Bonfil et al., 2005; Weng et al., 2005; Bruce et al., 2006; Domeier and Nasby-Lucas, 2008; Sims, 2010). Results from such studies have not only vastly improved our understanding of the biology of sharks but have also proven useful in fisheries management and conservation.

As noted above, electronic tags also clearly have the potential to aid in shark attack response decision making. For example, Holland et al. (1999) tracked the movements of Tiger Sharks off Hawaii to determine their level of site fidelity. Managers wanted to know whether fishing for sharks near the site of a recent attack would be likely to capture the shark responsible for the attack. If Tiger

Sharks demonstrated a degree of site attachment, then fishing near the attack location would make capturing the "culprit" shark more likely. However, the tracking study demonstrated that Tiger Sharks ranged widely and made long-distance offshore movements within 24 h of tagging near the coast. Given the Tiger Shark's expansive home range, Holland et al. (1999) concluded that shark-culling programs would not be effective at reducing the frequency of attacks from this species, and fishing near an attack location would be unlikely to result in capture of the culprit shark. Hawaii has since suspended its shark control programs.

Satellite tagging of large White Sharks has revealed that they also make long-distance movements into offshore environments (Boustany et al., 2002; Bonfil et al., 2005; Bruce et al., 2006; Weng et al., 2007; Domeier and Nasby-Lucas, 2008; Nasby-Lucas et al., 2009; Jorgensen et al., 2009; Chapters 11, 12, 13, 16, and 21, this book), but during certain times of the year they demonstrate site fidelity to feeding grounds near pinniped colonies (Goldman et al., 1996; Klimley et al., 2001). Beach areas in close proximity to natural White Shark feeding grounds, therefore, may have a higher than average risk of encounter and attack. Tagging White Sharks with transmitters can provide long-term data on patterns of seasonal occurrence, as well as daily movements in and out of these beach areas. It can also enable areas of temporal residency versus transit beach areas to be identified and thus inform focused efforts for monitoring beachgoer safety (Chapters 17 and 19, this book). Correlations with physical and biological parameters can also be identified to provide a potential predictive ability for the movement of White Sharks into beach areas (Chapter 19, this book).

Tagging technology is currently available to provide beachgoers with an "early warning system," utilizing receivers that detect the presence of tagged sharks and immediately alert beach safety authorities onshore. Although there are numerous potential benefits in the use of this technology, both for beachgoer safety as well as scientific advancement, the costs are quite high. For such an early warning system to be effective, large numbers of White Sharks would have to be tagged, but monitoring every individual shark in the population is not possible. There is always the potential for an untagged shark to move about undetected. The life expectancy of these tags is also finite (up to 10 yrs. with current batteries), so sharks have to regularly be re-tagged throughout their lifespan, which can be a costly and time-consuming process. However, such programs offer opportunities to identify conditions that may predispose higher risk of shark encounters and hence direct surveillance resources during such periods. As electronic tag technology continues to improve, such efforts may become more cost-effective. Research both past and present on shark movement patterns, temporary residency, and behavior provide considerable information for decision makers, and continued research should be encouraged.

Education, Outreach, and Preparedness

Providing current information on sharks, shark attack, and safety tips are a crucial part of any shark safety strategy. By increasing the level of understanding in the public, one can reduce the hype and fear that goes along with a shark attack. Education can also help beachgoers be aware of high-risk behaviors. Timely shark alerts can reduce the risk of shark encounter and attack. It is important to provide information on sharks to the public throughout the year and not only in the event of a shark attack. Education and outreach is relatively low cost to implement. However, outreach programs can be time consuming, and the information content of such programs needs to be kept current. Shark researchers should be encouraged to share their research findings with the public in the interest of outreach.

Various media are available to communicate information to the public. These include the following:

1. Warning and educational signage at beaches (Figure 31.6)
2. Brochure or pamphlets easily available in local stores or in parking areas
3. Regular newsletters and/or Internet blogs

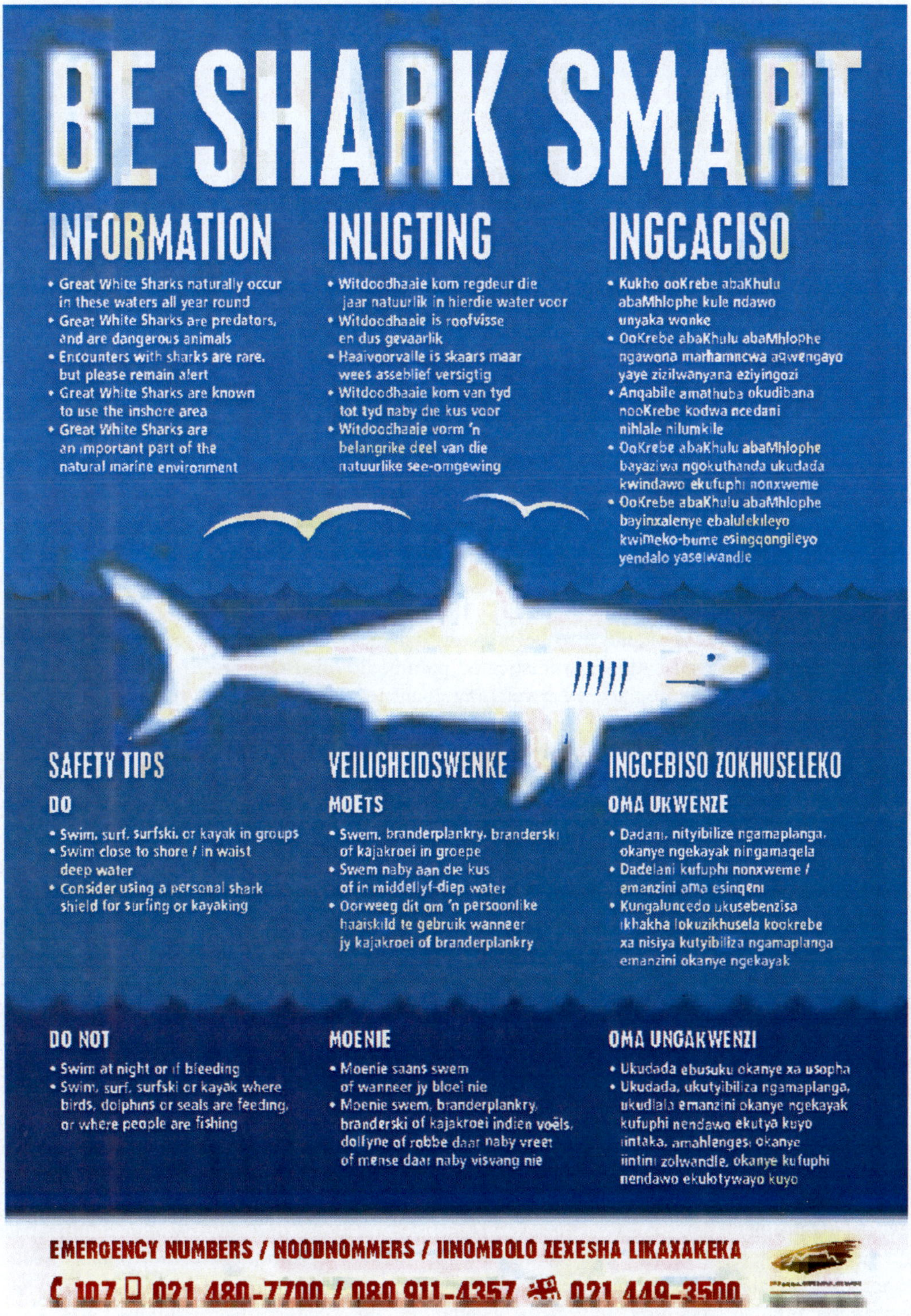

Figure 31.6 Multilingual signage used to educate beachgoers about shark attack risk in South Africa. (Courtesy of City of Cape Town.)

4. Press releases
5. Newspaper articles
6. Popular articles in magazines
7. Websites with up-to-date information
8. Presentations to various clubs and organizations
9. Radio and/or TV reports
10. Informational stands at beach and surfing festivals
11. Social-networking websites (e.g., Twitter or Facebook)

In many cases, shark behavior cannot be predicted or controlled. However, beachgoers can control their own behavior and should be encouraged to avoid high-risk activities that may increase the probability of encounter with White Sharks. There are numerous documents on government and nongovernment agency websites that provide guidance for beachgoers, and the common themes include simple and arguably effective guidelines for minimizing the risk of encounter and shark attack as follows:

1. Swim at a patrolled beach and between the safety flags.
2. Leave the water immediately if a shark is sighted.
3. Do not swim alone or too far from shore.
4. Do not swim at dawn, at dusk, or at night.
5. Do not swim near seal haulouts or near schools of fish.
6. Do not swim near, or interfere with, shark control equipment.

High-risk beach areas should also be prepared for emergency medical response in the event of an attack. In industrial nations, sophisticated paramedic infrastructure to deal with traumatic injuries is already largely in place. In developing countries, however, timely responses to medical emergencies like shark attack may be lacking. Locations with a high risk of shark attack should consider putting together a medical trauma kit, or "shark attack pack," that would include materials to control blood loss (e.g., sterile gauze dressings, elastic bandages, tape, forceps), reduce the effects of shock (e.g., blankets, painkillers, IV fluids), and provide sufficient first aid until a victim can be transported to a hospital. These packs should be kept in close proximity to beaches, such as in a lifeguard facility or nearby hotel or shop. Coordinated, timely responses with appropriate medical equipment and skills can significantly improve the chances of saving life or limb.

In addition to receiving information on beach safety and shark attack prevention, the public also needs to be educated about the importance of White Sharks to marine ecosystems. White sharks are apex predators that are essential to maintaining ocean ecosystem health. They are well-developed hunters, not the mindless killers often portrayed in the popular media. Similar to numerous terrestrial predators (tigers, lions, wolves, etc.), their conservation and continued existence should be valued. Although interactions between White Sharks and humans can never be fully eliminated, educating the public is key to limiting emotional or fearful responses subsequent to a shark incident.

In addition to proactive outreach and emergency planning, an official shark incident protocol may be a beneficial component to shark attack preparedness. These protocols are essentially emergency response plans implemented by governmental officials that clearly outline the roles and responsibilities of agencies and personnel in the event of a shark attack. Shark incident protocols have been implemented in some form in several locations, including parts of Australia, South Africa, and the United States. Official staff that are typically included in a protocol include police, medical responders, lifeguards, beach managers, coast guard/marine patrol, local fisheries agencies, local government officials, scientists, and media contacts. The protocol is often maintained and coordinated by a fisheries agency that employs scientists familiar with shark research and management.

It is important for all parties involved in a shark incident protocol to cooperate in its development and implementation. A common policy statement should be agreed upon regarding if and when a shark can be killed following an attack. A comprehensive roster of official contacts in each organization should be established and maintained. When a shark attack is reported, the first responders should be familiar with the contacts necessary to start the official response according to the protocol.

Existing shark incident protocols include specific instructions and contacts for the following response steps:

1. Immediate response: activating a rescue and emergency medical response
2. Controlling the scene: making the incident site safe, establishing access for emergency vehicles and personnel, providing first aid to the victim, and interviewing witnesses
3. Informing the public and media: official statements released regarding incident details, and the formal response
4. Short-term response: closing beaches, shark removal (if deemed appropriate), data collection, and continued media communications
5. Long-term response: shark control or attack prevention efforts, outreach, and refinement of procedures

In regions with higher shark attack risk, effective implementation of a clearly defined shark incident protocol can save lives, improve public safety, and reduce the chaos that often follows a shark attack. It can demonstrate a unified and coordinated response by local government, allowing control of the response, and help avoid miscommunication and negative impacts on the public, local economy, and the public's attitude to White Shark conservation. Examples of government agency response plans are available at http://www.fish.wa.gov.au/sec/env/sharks and http://www.pir.sa.gov.au/__data/assets/pdf_file/0017/13094/gw_shark1.pdf.

Dealing with the Media

Shark attacks receive a disproportionate amount of media coverage given the rarity of such events compared with other traumatic events. An effective communication strategy that is kept current is crucial to the dissemination of factual and objective information to the public. Traditionally, the headlines and associated stories are hyped-up accounts of the attacks, fraught with inaccuracies on sharks and perpetuating the reputation of the shark as a man-eater. Common problems in news articles and news reports are as follows:

1. Inaccuracies in information on sharks and the attack
2. Misquotes
3. Information taken out of context
4. Sensationalism
5. Interviews with self-proclaimed shark experts

These kinds of reports create fear and confusion and do not constitute responsible reporting. Reporters and media agencies, like most of the public, know very little about sharks and need to be educated as much as the public does.

A clear communication strategy should be in place within the delegated authorities (Preen and Richards, 2006; Peschak, 2006). Stakeholder agencies should take a proactive approach to liaising with the media and not just rely on being reactive in the event of a shark attack (Preen and Richards, 2006). Proactive steps can include the following:

1. Fostering relationships between stakeholders and local reporters
2. Providing ongoing information on sharks and behavior to the public through seminars and written reports to media outlets
3. Issuing regular press releases making the public aware of the times of year when shark activity is higher and therefore there is an increased risk

4. Providing the press with a list of people whom they can contact to get information from
5. Development of a press kit that includes the above information, as well as appropriate copyright-free shark photographs (Peschak, 2006)

Providing both information and images to the media appropriate to the event (in case of an attack) can minimize the use of incorrect images (e.g., the wrong species of shark) and thus make for more reliable reporting that has stronger educational outcomes.

Within each stakeholder organization, the responsibility of dealing with the media should be delegated to an experienced communications person. A support structure is essential in dealing with shark attack, and it will take a coordinated response to deal with the situation effectively. A strong recommendation based on a recent fatal attack in Cape Town, South Africa, on January 12, 2010 (http://www.fishhoek.com/sharks.html), demonstrated the effectiveness of a timely press conference at the scene of the attack. Media representatives from around the country, including international correspondents flocked to the scene and attempted to interview eyewitnesses and gather information from various bystanders. The National Sea Rescue Institute coordinated a press conference with representation from emergency response, a local shark scientist, and the lifesaving club. The information provided was factual and timely and was broadcast around the world. In the absence of such press conferences or media releases, the reporters will use whatever information they can gather, even if it includes statements from unqualified people, because they are under pressure to report on the story and are often under impending deadlines to get the story in the earliest edition.

Another recommendation is to as quickly as possible gather the facts about the shark attack and formally report on the findings of the incident. People are encouraged not to make assumptions about the event or offer up information that is not directly linked to the incident. In many cases, information is taken out of context to make the story even more newsworthy. It is important to remember that shark attacks are traumatic events where somebody has either been hurt or has died. Responsible communications personnel need to be sympathetic to the victim and their families and friends when disseminating information on sharks and shark attack.

Trying to find the balance between acknowledging the tragedy and putting shark attack in perspective is not an easy task. Only through a planned communications strategy can this be achieved. Being caught off guard and not prepared in the event of a shark attack is a recipe for disaster. Continued communication between government, local law enforcement, lifesaving clubs, sea rescue organizations, shark scientists, nongovernment organizations, and the media will have a positive influence on the way shark attack is reported in the media.

CONCLUSIONS

The risk of shark attack is low; however, it is an unavoidable risk that must be considered when entering the ocean for recreation. As human utilization of the ocean increases, together with possible increases in White Shark numbers where protection of populations is enforced, so will the frequency of shark encounters and thus the likely risk of attacks on humans. Certain locations around the world have higher frequencies of attacks by White Sharks than others, including the Pacific coast of the United States, Australia, and South Africa. These regions have both significant populations of White Sharks and increasing human populations using coastal waters for recreation. We have identified a number of methods that have been used to help reduce the risk of White Shark attacks in these regions and may be applied in other regions should the risk of shark attack render such actions prudent. Each approach has its own costs and benefits that need to be considered, and no single method is applicable in all scenarios.

Some of the methods presented here use fishing methods to remove potentially dangerous sharks from the waters near beach areas. Of these, fixed or semipermanent beach meshing and drumlines have proven effective at reducing the risk of shark attack at the beaches they protect, whereas shark hunts and state-run shark control programs have been less effective. The successful programs have provided the benefits of protection to beachgoers and minimized the potential economic harm to coastal communities affected by shark attacks. The consistent maintenance of the fishing gear, however, can be very costly and have long-term ecological impacts that must be considered. The effectiveness of such extractive measures may be higher for shark species that develop residency patterns. In general, White Sharks are not resident to areas commonly used by beachgoers and are transient visitors. Shark control programs are likely to be less effective in these cases because they depend entirely on the chance encounter of transient sharks and fishing gear rather than providing an effective barrier between sharks and people. There are some areas, however, where White Sharks are more common and will develop periods of temporary residency (e.g., around seal colonies and in some surf zone locations; see examples above). Deployment of shark control methods in such areas is likely to have significant impacts on both local and regional White Shark populations and is not recommended.

Nonfishing methods attempt to use more of a preventative approach and have more positive conservation results for White Sharks. Temporary beach closures in areas or during seasons of high shark attack risk are the simplest methods to protect beachgoers without impacting shark populations. Such closures, however, can result in adverse economic impacts on local communities. In these cases, other methods should be considered, including shark spotters, aerial surveys, or—if coastal conditions allow—shark barriers. Research efforts, particularly acoustic and/or satellite tagging studies, can help inform decision making as well as provide direct indications of White Shark presence in specific areas (e.g., "early warning systems"). Depending on the approach, research can be costly, but some level of scientific investigation near potential attack sites can provide numerous benefits and should be encouraged. Finally, formal education and outreach programs about White Sharks should be implemented to help reduce ignorance and hysteria in the general public. With a well-informed public, responses to a potential White Shark attack will be less emotional or irrational, and shark attack prevention efforts will be more effective.

RECOMMENDATIONS

Given all of the information described above, we offer the following recommendations to managers who may be considering implementation of a program to reduce the risk of White Shark attacks in their jurisdiction:

1. **Study the area and the sharks within it.** By examining the available data on the occurrence of shark attack and initiating a research program to investigate local shark species, their seasonal patterns of occurrence, environmental factors, and the typical behavior of sharks in the area, managers will have the information they need to select the most appropriate shark attack minimization strategy.
2. **Develop a shark incident protocol.** This protocol would be an official document that describes in detail the agencies, personnel, procedures, and responsibilities involved in the response to a shark attack or shark sighting. Medical response procedures should be prioritized. Such documentation will help avoid confusion and increase the efficiency and coordination of governmental responses.
3. **Educate the public.** Beachgoers need to be made aware of the potential for shark attack around high-risk areas. High-risk human behaviors also need to be identified. Proactive outreach efforts can also help allay public fears by dispelling many of the popular myths about the nature of White Sharks.
4. **Avoid public shark hunts.** There is little evidence that supports indiscriminate hunting of sharks as a means to reduce shark attack risk. Standardized, state-controlled fishing efforts or a variety of nonfishing methods are considered more effective. In regions where White Sharks are legally protected, only government officials have the authority to kill White Sharks. Although this is a stated

option in, for example, Australian state shark response plans, these plans have a series of prescriptive actions to avoid reaching this endpoint, and such an action would only ever be enforced as an unlikely last resort.

5. **Conserve White Shark populations to the extent practicable.** White Sharks play important ecological roles in the marine environment, but their populations are vulnerable to overexploitation. Permanent beach meshing or other shark control gears should not be used in natural White Shark aggregation sites, because they can quickly deplete regional populations. Finding a balance between public safety and shark conservation is essential.

By following these recommendations and applying the information provided in this chapter, governmental organizations will be able to develop appropriate methods to deal with shark attack risk within their affected communities. Effective shark attack prevention programs will serve to benefit the safety of the public, while sustaining ecologically important White Shark populations.

ACKNOWLEDGMENTS

We appreciate the thoughtful discussions and input on this topic from participants of the International White Shark Symposium. Thanks to George H. Burgess and the American Elasmobranch Society for sharing data from the International Shark Attack File. We are also grateful to the City of Cape Town and Shark Spotters for sharing information on shark sightings and shark attacks in Cape Town.

REFERENCES

Ames, J.A., and G.W. Morejohn. 1980. Evidence of white shark, *Carcharodon carcharias*, attacks on sea otters, *Enhydra lutris*. *Calif Fish Game* 66: 196–209.

Armstrong, M.J., P. Chapman, S.F.J. Dudley, I. Hampton, and P.E. Malan. 1991. Occurrence and population structure of pilchard *Sardinops ocellatus*, round herring *Etrumeus whiteheadi* and anchovy *Engraulis capensis* off the east coast of South Africa. *S Afr J Mar Sci* 11: 227–249.

Baldridge, H. D., Jr. (1996). Comments on means for avoidance or deterrence of white-shark attacks on humans. In *Great white sharks: The biology of Carcharodon carcharias*, edited by A.P. Klimley and D.G. Ainley, pp. 477–479. San Diego, CA: Academic Press.

Bass, A.J., J.D. D'Aubrey, and N. Kistnasamy. 1975. Sharks of the east coast of southern Africa: IV. The families Odontaspididae, Scapanorhynchidae, Isuridae, Cetorhinidae, Alopiidae, Orectolobidae and Rhiniodontidae. *Investig Rep Oceanogr Res Inst Durban* 39: 1–102.

Bonfil, R., M. Meyer, M. C. Scholl, R. Johnson, S. O'Brien, H. Oosthuizen, S. Swanson, D. Kotze, and M. Paterson. 2005. Transoceanic migration, spatial dynamics, and population linkages of white sharks. *Science* 310: 100–103.

Boustany, A.M., S.F. Davis, P. Pyle, S.D. Anderson, B.J. LeBoeuf, and B.A. Block. 2002. Satellite tagging: Expanding niche for white sharks. *Nature* 415: 35–36.

Bruce, B.D. 1992. Preliminary observations on the biology of the white shark, *Carcharodon carcharias*, in South Australian waters. In *Sharks: Biology and Fisheries*, edited by J. G. Pepperell, *Aust J Mar and Freshw Res* 43: 1–11.

Bruce, B.D., J.D. Stevens, and H. Malcolm. 2006. Movements and swimming behaviour of white sharks (*Carcharodon carcharias*) in Australian waters. *Mar Biol* 150: 161–172.

Burgess, G.H., and M. Callahan. 1996. Worldwide patterns of white shark attacks on humans. In *Great white sharks: The biology of Carcharodon carcharias*, edited by A.P. Klimley and D.G. Ainley, pp. 457–469. San Diego, CA: Academic Press.

Burgess, G.H., R.H. Buch, F. Carvalho, B.A. Garner, and C.J. Walker. 2010. Factors contributing to shark attacks on humans: A Volusia County, Florida, case study. In *Sharks and their relatives: II. Biodiversity, adaptive physiology, and conservation*, edited by J.C. Carrier, J.A. Musick, and M.R. Heithaus, pp. 541–565. Boca Raton, FL: CRC Press.

Casey, J.G., and H.L. Pratt, Jr. 1985. Distribution of the white shark, *Carcharodon carcharias*, in the western North Atlantic. *Mem South Calif Acad Sci* 9: 2–14.

Chapple, T.K., S.J. Jorgensen, S.D. Anderson, P.E. Kanive, A.P. Klimley, L.W. Botsford, and B.A. Block. 2011. A first estimate of white shark, *Carcharodon carcharias*, abundance off central California. *Biol Lett* doi: 10.1098/rsbl.2011.0124.

Cliff, G. 1991. Shark attacks on the South African coast between 1960 and 1990. *S Afr J Sci* 87: 513–518.

Cliff, G. 2006. A review of shark attacks in False Bay and the Cape Peninsula between 1960 and 2005. In *Finding a balance: White shark conservation and recreational safety in the inshore waters of Cape Town, South Africa*, edited by D.C. Nel and T.P. Peschak, pp. 20–31. Proceedings of a specialist workshop. WWF South Africa Report Series – 2006/Marine/001 Annexure 1.

Cliff, G., and S.F.J. Dudley. 1991. Sharks caught in the protective nets off Natal, South Africa: 4. The bull shark *Carcharhinus leucas* Valenciennes. *S Afr J Mar Sci* 10: 253–270.

Cliff, G., and S.F.J. Dudley. 1992. Protection against shark attack in South Africa, 1952 to 1990. In *Sharks: Biology and fisheries*, edited by J.G. Pepperell. *Aust J Mar Freshw Res* 43: 263–272.

Cliff, G., S.F.J. Dudley, and B. Davis. 1989. Sharks caught in the protective gill nets off Natal, South Africa. 2. The great white shark *Carcharodon carcharias* (Linnaeus). *S Afr J Mar Sci* 8: 131–144.

Cliff, G., S.F.J. Dudley, and M.R. Jury. 1996a. Catches of white sharks in KwaZulu-Natal, South Africa and environmental influences. In *Great white sharks: The biology of Carcharodon carcharias*, edited by A.P. Klimley and D.G. Ainley, pp. 351–362. San Diego, CA: Academic Press.

Cliff, G., R.P. Van der Elst, A. Govender, T.K. Witthuhn, and E.M. Bullen. 1996b. First estimates of mortality and population size of white sharks on the South African coast. In *Great white sharks: The biology of Carcharodon carcharias*, edited by A.P. Klimley and D.G. Ainley, pp. 393–400. San Diego, CA: Academic Press.

Collier, R. 2003. *Shark attacks of the twentieth century from the Pacific coast of North America*. Chatsworth, CA: Scientia Publishing.

Collier, R.S., M. Marks, and R.W. Warner. 1996. White shark attacks on inanimate objects along the Pacific coast of North America. In *Great white sharks: The biology of Carcharodon carcharias*, edited by A.P. Klimley and D.G. Ainley, pp. 217–228. San Diego, CA: Academic Press.

Collins, A.J. 1972. Shark meshing off New South Wales surfing beaches. *Fisherman* Spring: 11–19.

Compagno, L.J.V. 1999. Encounters with sharks in South Africa. In *Sharks* (second edition), edited by J.D. Stevens, pp. 134–147. New York, NY: Checkmark Books.

Compagno, L.J.V. 2001. Sharks of the world: An annotated and illustrated catalogue of shark species known to date. Bullhead, mackerel and carpet sharks (Heterodontiformes, Lamniformes and Orectolobiformes). *FAO Species Cat Fish Purposes* 2: 1–269.

Coppleson, V.M. 1962. *Shark attack* (second edition). Sydney, Australia: Angus and Robertson.

Coppleson, V.M., and P. Goadby. 1988. *Shark attack: How, why, when, and where sharks attack humans*. Sydney, Australia: Angus and Robertson.

Curtis, T.H., J.T. Kelly, K.L. Menard, R.K. Laroche, R.E. Jones, and A.P. Klimley. 2006. Observations on the behavior of white sharks scavenging from a whale carcass at Point Reyes, California. *Calif Fish Game* 92: 113–124.

da Graça, M. 2005. Polana Beach. *Indico* (in-flight magazine of Linhas Aéras de Moçambique) Série II 30: 36–40.

Davies, D.H. 1961. The shark problem. *Bull S Afr Assoc Mar Biolog Res* 2: 23–27.

Davies, D.H. 1964. *About sharks and shark attack*. Pietermaritzburg, South Africa: Shuter and Shooter.

Dicken, M.L. 2008. First observations of young of the year and juvenile great white sharks (*Carcharodon carcharias*) scavenging from a whale carcass. *Mar Freshw Res* 59: 596–602.

Domeier, M.L., and N. Nasby-Lucas. 2006. Annual re-sightings of photographically identified white sharks (*Carcharodon carcharias*) at an eastern Pacific aggregation site (Guadalupe Island, Mexico). *Mar Biol* 150: 977–984.

Domeier, M.L., and N. Nasby-Lucas. 2008. Migration patterns of white sharks *Carcharodon carcharias* tagged at Guadalupe Island, Mexico, and identification of an eastern Pacific shared offshore foraging area. *Mar Ecol Prog Ser* 370: 221–237.

Dudley, S.F.J. 1997. A comparison of the shark control programs of New South Wales and Queensland (Australia) and KwaZulu-Natal (South Africa). *Ocean Coast Manag* 34: 1–27.

Dudley, S.F.J. 2006. International review of responses to shark attack. In *Finding a balance: White shark conservation and recreational safety in the inshore waters of Cape Town, South Africa*, edited by D.C. Nel and T.P. Peschak, pp. 95–108. Proceedings of a specialist workshop. WWF South Africa Report Series – 2006/Marine/001 Annexure 1.

Dudley, S.F.J., and G. Cliff. 1993. Some effects of shark nets in the Natal nearshore environment. *Environ Biol Fishes* 36: 243–255.

Dudley, S.F.J., and G. Cliff. 2010. Shark control: Methods, efficacy, and ecological impact. In *Sharks and their relatives: II. Biodiversity, adaptive physiology, and conservation*, edited by J.C. Carrier, J.A. Musick, and M.R. Heithaus, pp. 567–591. Boca Raton, FL: CRC Press.

Dudley, S.F.J., and N.A. Gribble. 1999. Management of shark control programs. In *Case studies of the management of elasmobranch fisheries*, edited by R. Shotton. *FAO Fish Tech Pap* 378: 819–859.

Dudley, S.F.J., and C.A. Simpfendorfer. 2006. Population status of 14 shark species caught in the protective gillnets off KwaZulu-Natal beaches, South Africa, 1978–2003. *Mar Freshw Res* 57: 225–240.

Dudley, S.F.J., R.C. Haestier, K.R. Cox, and M. Murray. 1998. Shark control: Experimental fishing with baited drumlines. *Mar Freshw Res* 49: 653–661.

Engaña, A.C., and J.E. McCosker. 1984. Attacks on divers by white sharks in Chile. *Calif Fish Game* 70: 173–179.

Ellis, R.E., and J.E. McCosker. 1991. *Great white shark*. New York, NY: HarperCollins Publishers.

Estrada, J.A., A.N. Rice, L.J. Natanson, and G.B. Skomal. 2006. Use of isotopic analysis of vertebrae in reconstructing ontogenetic feeding ecology in white sharks. *Ecol* 87: 829–834.

Fernicola, R.G. 2001. *Twelve days of terror: A definitive investigation of the 1916 New Jersey shark attacks*. Guilford, CT: The Lyons Press.

Ferreira, C.A., and T.P. Ferreira. 1996. Population dynamics of white sharks in South Africa. In *Great white sharks: The biology of Carcharodon carcharias*, edited by A.P. Klimley and D.G. Ainley, pp. 381–391. San Diego, CA: Academic Press.

Francis, M.P. 1996. Observations on a pregnant white shark with a review of reproductive biology. In *Great white sharks: The biology of Carcharodon carcharias*, edited by A.P. Klimley and D.G. Ainley, pp. 157–172. San Diego, CA: Academic Press.

Francis, M.P. 1998. New Zealand shark fisheries: Development, size and management. *Mar Freshw Res* 49: 579–591.

Goldman, K.J., J.E. McCosker, S.D. Anderson, and A.P. Klimley. 1996. Temperature, swimming depth, and movements of a white shark at the South Farallon Islands, California. In *Great white sharks: The biology of Carcharodon carcharias*, edited by A.P. Klimley and D.G. Ainley, pp. 111–120. San Diego, CA: Academic Press.

Green, M., C. Ganassin, and D.D. Reid. 2009. *Report into the NSW Shark Meshing (Bather Protection) Program*. New South Wales Department of Primary Industries.

Gribble, N.A., G. McPherson, and B. Lane. 1998. Shark control: A comparison of meshing with set drumlines. In *Shark and management conservation*, edited by N.A. Gribble, G. McPherson, and B. Lane, pp. 98–124. Proceedings from the Sharks and Man Workshop of the Second World Fisheries Congress. Queensland Department of Primary Industries.

Hammerschlag, N., R.A. Martin, and C. Fallows. 2006. Effects of environmental conditions on predator-prey interactions between white sharks (*Carcharodon carcharias*) and Cape fur seals (*Arctocephalus pusillus pusillus*) at Seal Island, South Africa. *Environ Biol Fishes* 76: 341–350.

Hazin, F.H.V., G.H. Burgess, and F.C. Carvalho. 2008. A shark attack outbreak off Recife, Pernambuco, Brazil: 1992–2006. *Bull Mar Sci* 82: 199–212.

Holland, K.N., B.W. Wetherbee, C.G. Lowe, C.G. Meyer. 1999. Movements of tiger sharks (*Galeocerdo cuvier*) in coastal Hawaiian waters. *Mar Biol* 134: 665–673.

Johnson, R.H., and D.R. Nelson. 1973. Agonistic display in the gray reef shark, *Carcharhinus menisorrah*, and its relationship to attacks on man. *Copeia* 1973: 76–84.

Jorgensen, S.J., C.A. Reeb, T.K. Chapple, S. Anderson, C. Perle, S.R. Van Sommeran, C. Fritz-Cope, A.C. Brown, A.P. Klimley, and B.A. Block. 2010. Philopatry and migration of Pacific white sharks. *Proc Royal Soc B* 277: 679–688.

Keith, M., V.M. Peddemors, M.N. Bester, and J.W.H. Ferguson. 2002. Population characteristics of Indo-Pacific humpback dolphins at Richards Bay, South Africa: Implications for incidental capture in shark nets. *S Afr J Wildl Res* 32: 153–162.

Kerr, L.A., A.H. Andrews, G.M. Cailliet, T.A. Brown, and K.H. Coale. 2006. Investigations of $\Delta^{14}C$, $\delta^{13}C$, and $\delta^{15}N$ in vertebrae of white shark (*Carcharodon carcharias*) from the eastern North Pacific Ocean. *Environ Biol Fish* 77: 337–353.

Kirkman, S.P., W.H. Oosthuizen, and M.A. Meÿer. 2006. The seal population of Seal Island, False Bay. In *Finding a balance: White shark conservation and recreational safety in the inshore waters of Cape Town, South Africa*, edited by D.C. Nel and T.P. Peschak, pp. 83–94. Proceedings of a specialist workshop. WWF South Africa Report Series – 2006/Marine/001 Annexure 1.

Klimley, A. P. 1985. The areal distribution and autecology of the white shark, *Carcharodon carcharias*, off the west coast of North America. *Mem South Calif Acad Sci* 9: 15–40.

Klimley, A.P. 1994. The predatory behavior of the white shark. *Am Sci* 82: 122–133.

Klimley, A.P., and S.D. Anderson. 1996. Residency patterns of white sharks at the South Farallon Islands, California. In *Great white sharks: The biology of Carcharodon carcharias*, edited by A.P. Klimley and D.G. Ainley, pp. 365–373. San Diego, CA: Academic Press.

Klimley, A.P., and T.H. Curtis. 2006. Shark attack versus ecotourism: Negative and positive interactions. In *Proceedings of the 22nd Vertebrate Pest Conference*, edited by R.M. Timm and J.M. O'Brien, pp. 33–44. University of California, Davis.

Klimley, A.P., S.D. Anderson, P. Pyle, and R.P. Henderson. 1992. Spatio-temporal patterns of white shark (*Carcharodon carcharias*) predation at the South Farallon Islands, California. *Copeia* 1992: 680–690.

Klimley, A.P., P. Pyle, and S.D. Anderson. 1996a. The behavior of white shark and prey during predatory attacks. In *Great white sharks: The biology of Carcharodon carcharias*, edited by A.P. Klimley and D.G. Ainley, pp. 175–191. San Diego, CA: Academic Press.

Klimley, A.P., B.J. Le Boeuf, K.M. Cantara, J.E. Richert, S.F. Davis, S. Van Sommeran, and J.T. Kelly. 2001. The hunting strategy of white sharks at a pinniped colony. *Mar Biol* 13: 617–636.

Klimley, A.P., P. Pyle, and S.D. Anderson. 1996b. Tail slap and breach: Agonistic displays among white sharks. In *Great white sharks: The biology of Carcharodon carcharias*, edited by A.P. Klimley and D.G. Ainley, pp. 241–255. San Diego, CA: Academic Press.

Kock, A., and R.L. Johnson. 2006. White shark abundance: Not a causative factor in numbers of shark bite incidents. In *Finding a balance: White shark conservation and recreational safety in the inshore waters of Cape Town, South Africa*, edited by D.C. Nel and T.P. Peschak, pp. 1–19. Proceedings of a specialist workshop. WWF South Africa Report Series – 2006/Marine/001 Annexure 1.

Krogh, M. 1996. Bycatch in the protective shark meshing program off south-eastern New South Wales, Australia. *Biol Conserv* 77: 219–226.

Laroche, R.K., A.A. Kock, L.M. Dill, and W.H. Oosthuizen. 2008. Running the gauntlet: A predator-prey game between sharks and two age classes of seals. *Anim Behav* 76: 1901–1917.

Last, P.R., and J.D. Stevens. 1994. *Sharks and rays of Australia*. Hobart, Australia: CSIRO Division of Fisheries.

Last, P.R., and J.D. Stevens. 2009. *Sharks and rays of Australia* (second edition). Hobart, Australia: CSIRO Division of Fisheries.

Lea, R.N., and D.J. Miller. 1985. Shark attacks off the California and Oregon coasts: An update, 1980–84. *Mem South Calif Acad Sci* 9: 136–150.

Levine, M. 1996. Unprovoked attacks by white sharks off the South African coast. In *Great white sharks: The biology of Carcharodon carcharias*, edited by A.P. Klimley and D.G. Ainley, pp. 435–448. San Diego, CA: Academic Press.

Lowe, C.G., and K.J. Goldman. 2001. Thermal and bioenergetics of elasmobranchs: Bridging the gap. *Environ Biol Fish* 60: 251–266.

Lowry, D., A.L.F. de Castro, K. Mara, L.B. Whitenack, B. Delius, G.H. Burgess, and P. Motta. 2009. Determining shark size from forensic analysis of bite damage. *Mar Biol* 156: 2483–2492.

Malcolm, H., B.D. Bruce, and J.D. Stevens 2001. *A review of the biology and status of white sharks in Australian waters*. Final report to Environment Australia, Marine Species Protection Program. Hobart, Australia: CSIRO.

Martin, R.A. 2003. *Field guide to the great white shark*. ReefQuest Centre for Shark Research, spec. pub. 1.

Martin, R.A., N. Hammerschlag, R.S. Collier, and C. Fallows. 2005. Predatory behaviour of white sharks (*Carcharodon carcharias*) at Seal Island, South Africa. *J Mar Biolog Assoc UK* 85: 1121–1135.

McCosker, J.E. 1985. White shark attack behavior: Observations of and speculations about predator and prey strategies. *Mem South Calif Acad Sci* 9: 123–135.

McCosker, J.E., and R.N. Lea. 1996. White shark attacks in the eastern Pacific Ocean: An update and analysis. In *Great white sharks: The biology of Carcharodon carcharias*, edited by A.P. Klimley and D.G. Ainley, pp. 419–434. San Diego, CA: Academic Press.

McCosker, J.E., and R.N. Lea. 2006. White shark attacks upon humans in California and Oregon, 1993–2003. *Proc Calif Acad Sci* 57: 479–501.

Miller, D.J., and R.S. Collier. 1981. Shark attacks in California and Oregon, 1926–1979. *Calif Fish Game* 67: 76–104.

Nasby-Lucas, N., H. Dewar, C.H. Lam, K.J. Goldman, and M.L. Domeier. 2009. White shark offshore habitat: A behavioral and environmental characterization of the eastern Pacific shared offshore foraging area. *PLoS One* 4: e8163.

Natoli, A., V.M. Peddemors, and A.R. Hoelzel. 2008. Population structure of bottlenose dolphins (*Tursiops aduncus*) impacted by bycatch along the east coast of South Africa. *Conserv Genet* 9: 627–636.

New South Wales Legislative Assembly. 1935. *Report of the Shark Menace Advisory Committee on suggested methods of protecting beaches from shark attack.* Sydney, Australia: Government Printer.

New South Wales Shark Menace Committee. 1929. *General Report of Sub-Committee.* Sydney, Australia: Government Printer.

Oelofse, G., and Y. Kamp. 2006. Shark spotting as a water safety program in Cape Town. In *Finding a balance: White shark conservation and recreational safety in the inshore waters of Cape Town, South Africa*, edited by D.C. Nel and T.P. Peschak, pp. 121–129. Proceedings of a specialist workshop. WWF South Africa Report Series – 2006/Marine/001 Annexure 1.

Paterson, R. 1979. Shark meshing takes a heavy toll of harmless marine animals. *Aust Fish* 38: 17–23.

Paterson, R. 1986. Shark prevention measures working well. *Aust Fish* 45: 12–18.

Paterson, R.A. 1990. Effects of long-term anti-shark measures on target and non-target species in Queensland, Australia. *Biol Conserv* 52: 147–159.

Peschak, T. P. 2006. Sharks and shark bite in the media. In *Finding a balance: White shark conservation and recreational safety in the inshore waters of Cape Town, South Africa*, edited by D.C. Nel and T.P. Peschak, Proceedings of a specialist workshop. WWF South Africa Report Series – 2006/Marine/001 Annexure 1.

Preen, M., and Richards, R. 2006. The role of government in demystifying sharks and shark attacks. In *Finding a balance: White shark conservation and recreational safety in the inshore waters of Cape Town, South Africa*, edited by D.C. Nel and T.P. Peschak. Proceedings of a specialist workshop. WWF South Africa Report Series – 2006/Marine/001 Annexure 1.

Queensland Department of Primary Industries. 1992. *Review of the Operation and Maintenance of Shark Meshing Equipment in Queensland Waters*. Report of the Committee of Enquiry.

Queensland Department of Primary Industries and Fisheries. 2006. *A Report on the Queensland Shark Safety Program.*

Reid, D.D., and M. Krogh. 1992. Assessment of catches from protective shark meshing off New South Wales beaches between 1950 and 1990. In *Sharks: Biology and fisheries*, edited by J.G. Pepperell. *Aust J Mar Freshw Res* 43: 283–296.

Simpfendorfer, C. 1993. The Queensland Shark Meshing Program: Analysis of the results from Townsville, North Queensland. In *Shark conservation*, edited by J.G. Pepperell, J. West, and P. Woon, pp. 71–85. Proceedings of an International Workshop on the Conservation of Elasmobranchs. Mosman, Australia: Zoological Parks Board of New South Wales.

Sims, D.W. 2010. Tracking and analysis techniques for understanding free-ranging shark movements and behavior. In *Sharks and their relatives: II. Biodiversity, adaptive physiology, and conservation*, edited by J.C. Carrier, J.A. Musick, and M.R. Heithaus, pp. 351–392. Boca Raton, FL: CRC Press.

Springer, S., and P.W. Gilbert. 1963. Anti-shark measures. In *Sharks and survival*, edited by P.W. Gilbert, pp. 465–476. Boston, MA: D.C. Heath and Co.

Strong, W.R. 1996. Shape discrimination and visual predatory tactics in white sharks. In *Great white sharks: The biology of Carcharodon carcharias*, edited by A.P. Klimley and D.G. Ainley, pp. 229–240. San Diego, CA: Academic Press.

Taylor, L.R., Jr. 1985. White sharks in Hawaii: Historical and contemporary records. *Mem South Calif Acad Sci* 9: 41–48.

Taylor, L.R., Jr. 1993. *Sharks of Hawaii: Their biology and cultural significance*. Honolulu, HI: University of Hawaii Press.

Tricas, T.C., and J.E. McCosker. 1984. Predatory behavior of the white shark (*Carcharodon carcharias*), with notes on its biology. *Proc Calif Acad Sci* 43: 221–238.

Wallett, T. 1983. *Shark attack in Southern African waters and treatment of victims*. Cape Town, South Africa: C. Struik.

Weng, K.C., P.C. Castilho, J.M. Morrissette, A.M. Landeira-Fernandez, D.B. Holts, R.J. Schallert, K.J. Goldman, and B.A. Block. 2005. Satellite tagging and cardiac physiology reveal niche expansion in salmon sharks. *Science* 310: 104–106.

Weng, K.C., A.M. Boustany, P. Pyle, S.D. Anderson, A. Brown, and B.A. Block. 2007. Migration and habitat of white sharks (*Carcharodon carcharias*) in the eastern Pacific Ocean. *Mar Biol* 152: 877–894.
West, J.G. 1996. White shark attacks in Australian waters, In *Great white sharks: The biology of Carcharodon carcharias*, edited by A.P. Klimley and D.G. Ainley, pp. 449–455. San Diego, CA: Academic Press.
West, J.G. (2011). Changing patterns of shark attacks in Australian waters. *J Mar Freshw Res* 62: 744–754.
Wetherbee, B.M., C.G. Lowe, G.L. Crow. 1994. A review of sharks control in Hawaii with recommendations for future research. *Pac Sci* 48: 95–115.

CHAPTER 32

A Review of Research on the White Shark, *Carcharodon carcharias*, in Southern Africa

Sheldon F. J. Dudley*
KwaZulu-Natal Sharks Board and University of KwaZulu-Natal

ABSTRACT

Research on the White Shark, *Carcharodon carcharias* (Linnaeus), in southern African waters has spanned more than 150 yrs. Peer-reviewed publications, supplemented by published workshop proceedings, postgraduate theses, and selected books or book chapters, total more than 70. The bulk of the output (58 publications) has been produced in the past 2 decades, with 40 publications since the turn of the 21st century. Approximately one-third of the output has included data or samples from animals caught in nets set to reduce the risk of shark bites at the tourist beaches of KwaZulu-Natal, South Africa, and another third has centered on animals associated with island-based colonies of Cape Fur Seals, *Arctocephalus pusillus pusillus*, in the Western Cape. Research topics facilitated by access to specimens from the nets have included age and growth, stomach content analysis, morphology, genetics, and parasitology. No adult females and very few mature males have been caught, however, so there has been little research on reproductive biology. A long-term time series of catch and effort data has enabled an assessment of population status, and a first attempt at a population estimate has been made using tag-recapture data. In the Western Cape, predator-prey interactions between White Sharks and seals have been examined by means of visual observation and electronic tagging. Tracking technology and photographic identification have been used to describe fine-scale and long-range movements, as well as site fidelity. The development of ecotourism based on cage diving, with associated provisioning, has led to research into the potential for sharks to be conditioned and for there to be consequences either for local ecosystems or for human safety. Future research priorities include the following:

1. The continued use of genetic techniques and tracking technology to further investigate movements and global population structure
2. Investigation of the possibility that a nursery area may exist in Eastern Cape waters
3. Use of photo-identification databases to obtain population estimates of sharks visiting the islands

INTRODUCTION

Research on the White Shark, *Carcharodon carcharias* (Linnaeus), in southern Africa began in the mid-nineteenth century. In 1849, Andrew Smith published morphological and behavioral descriptions of what he named *Carcharodon capensis*, the holotype of which is in the Natural History Museum in London (Compagno, 1999a). Later, Thompson (1914) listed *Carcharodon rondeleti* Müller and Henle, the Man-eater, the Great White or Great Blue Shark, and Barnard (1925) listed *Carcharodon carcharias* (Linnaeus), describing the species as formidable. J.L.B. Smith (1950) included a colorful account of the species' fearsome reputation, emphasizing the large serrated teeth and adding White Pointer and Blue Pointer to the list of common names.

Peer-reviewed publications, supplemented by published workshop proceedings, postgraduate theses, and selected books or book chapters now total more than 70. The bulk of the output (58 publications) has been produced in the past two decades, with 40 publications since the turn of the 21st century (Figure 32.1). Approximately one-third of the output has included data or samples from animals caught in large-mesh gillnets, or shark nets, that are in place off popular swimming beaches on the KwaZulu-Natal (KZN) coast of South Africa to reduce the risk of shark bite incidents, the first of which were installed in 1952 [see Cliff and Dudley (1992) and references therein]. The quality of catch data improved with time and was regarded as reliable from 1978 (Cliff and Dudley, 1992). From that time, all sharks caught in the nets were brought ashore and dissected, other than those that were released alive or were badly decomposed (Dudley and Simpfendorfer, 2006). The information collected has facilitated research on a number of topics (Figure 32.2). Another third of the output has centered on the behavior of animals associated with island-based colonies of Cape Fur Seals, *Arctocephalus pusillus pusillus*, primarily at Mossel Bay, Gansbaai, and False Bay in the Western Cape (WC) (Figures 32.1 and 32.2). Total length (TL) is used throughout this review. The equations of Mollet and Cailliet (1996) were used to convert precaudal length to TL and to estimate TL from total mass.

AGE AND GROWTH

Wintner and Cliff (1999) counted vertebral growth rings to determine the age and growth of White Sharks caught in the KZN nets between 1984 and 1995 (Figure 32.3). The size range of females was 160–360 cm ($n = 61$) and 42–442 kg ($n = 60$), and that of males was 177–449 cm ($n = 53$) and 46–882 kg ($n = 53$). Growth ring (GR) counts for females ranged from zero to eight, and that for males ranged from 1 to 13. Von Bertalanffy and Gompertz growth curves were generated. Of 16 White Sharks injected with oxytetracycline, one individual that measured 175 cm was recaptured after 942 d at liberty (Figure 32.4). During that period, GR deposition appeared to be annual. In a separate study, a female of 567 cm from Gansbaai in the WC was reported to have 23 GR (Mollet et al., 1996).

Size at birth for White Sharks was estimated from the age-length curve to be 126–169 cm (Wintner and Cliff, 1999), and size at 50% maturity for males was estimated to be 375 cm (Dudley and Simpfendorfer, 2006), although Bass et al. (1975) reported mature males of 304, 310, and 339 cm, respectively. The only "apparently" mature female reported from South African waters measured 445 cm (Bass et al., 1975), but Francis (1996) suggested that it was probably longer because the method used underestimates TL. Large immature females measured 381 and 391 cm (Bass et al., 1975), 420 cm (Cliff et al., 1989), and 437 and 469 cm (KwaZulu-Natal Sharks Board, unpublished data; Figure 32.5). Size at 50% maturity of females may be approximately 475 cm [deduced from information in this review and from Compagno (2001) and Francis (1996)]. Applying the growth curve of Wintner and Cliff (1999) and assuming annual GR deposition, age at 50% maturity would be between 8 and 9 yrs. for males and about 16 yrs. for females.

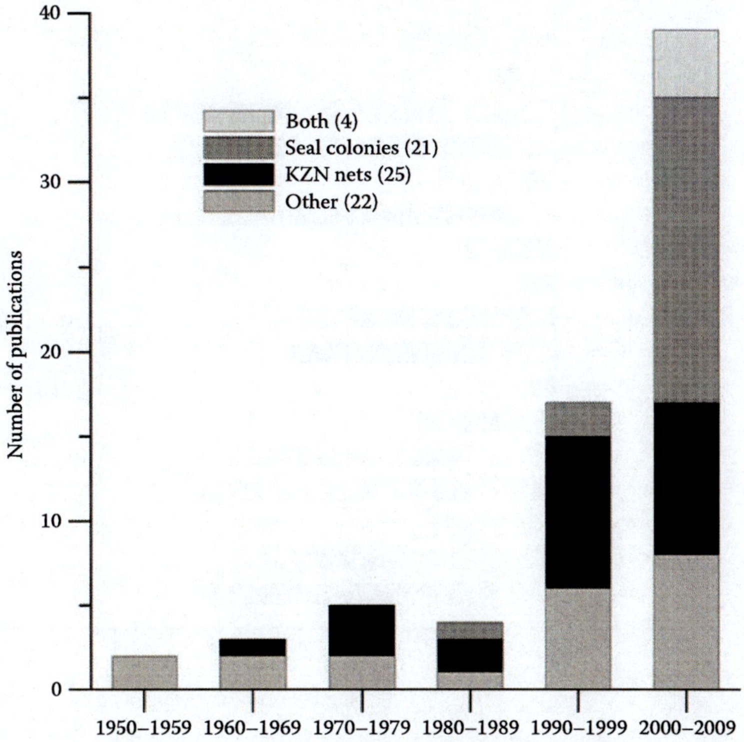

Figure 32.1 Research output on southern African White Sharks (1950–2009) by research location (n = 72 publications) (excludes three publications pre-1950 and includes one publication dated 2010 that was in press in 2009).

DISTRIBUTION BY SIZE AND LIFE HISTORY STAGE

The regional center of abundance of the White Shark encompasses the east coast of South Africa, from False Bay in the WC to KZN (Compagno, 2001), but the species has been recorded from the west coast of South Africa as well as Namibia and Angola; from Mozambique, Tanzania (Zanzibar), Kenya, Madagascar, the Seychelles, Réunion, and Mauritius; and even from the cold waters of Gough Island in the South Atlantic (Bass et al., 1975; Cliff et al., 2000; Compagno, 1999a, 2001; Zuffa et al., 2002).

It has been suggested that Eastern Cape (EC) waters, and specifically Algoa Bay, may be a nursery ground for the species (Cliff et al., 1996a; Dicken, 2008). Smith (1951) reported the capture of a small male White Shark of 140 cm (19.95 kg) in Algoa Bay, and Ferreira and Ferreira (1996), citing L.J.V. Compagno and the Port Elizabeth Museum (personal communication), reported that the mean size of 18 specimens caught in the same region was 180 cm. Also, young of the year, the smallest being 150 cm, and juvenile White Sharks were observed feeding on a whale carcass in the bay, and up to 15 small (<250 cm) White Sharks were hooked during a shore-angling competition, also in the bay (Dicken, 2008). Randall (1973) received information, apparently from angling writer C.L. Biden, about two specimens of 152 cm from Durban, KZN. In the KZN shark nets, a small number of White Sharks of <187 cm TL (estimated to be ≤1 yr. old) were caught (Cliff et al., 1996a). The smallest White Sharks caught in the nets since accurate record-keeping began in 1978 measured 153 and 157 cm, respectively, and were caught in 2009 and 2008 (KwaZulu-Natal Sharks Board, unpublished data).

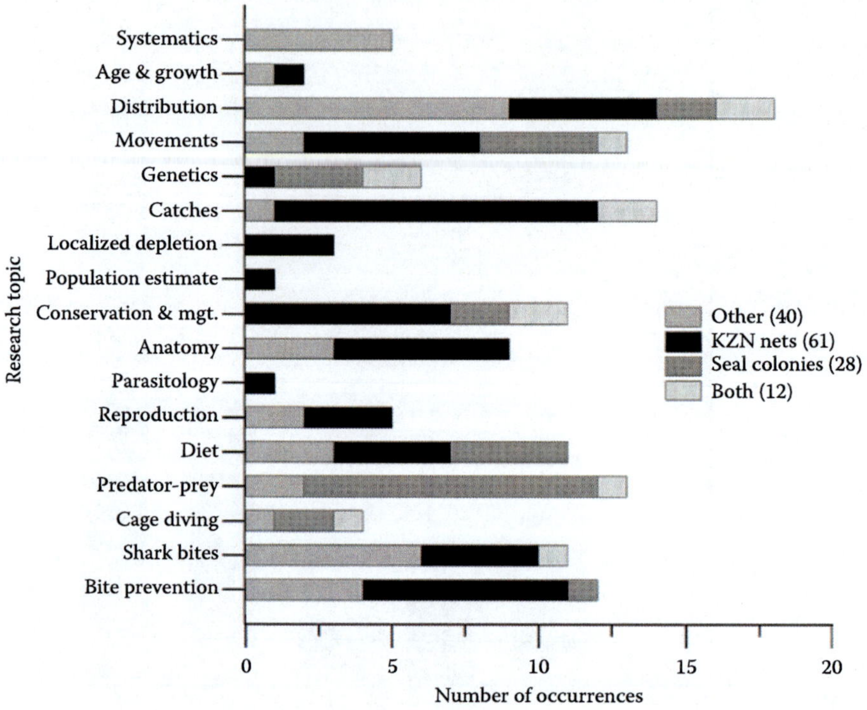

Figure 32.2 Output by research topic and location (1849–2009) ($n = 141$ topic occurrences in 75 publications; includes one publication dated 2010 that was in press in 2009).

Figure 32.3 X-radiograph of White Shark vertebral centrum showing growth rings consisting of alternating opaque and translucent band pairs. (Courtesy of Sabine Wintner, KwaZulu-Natal Sharks Board.)

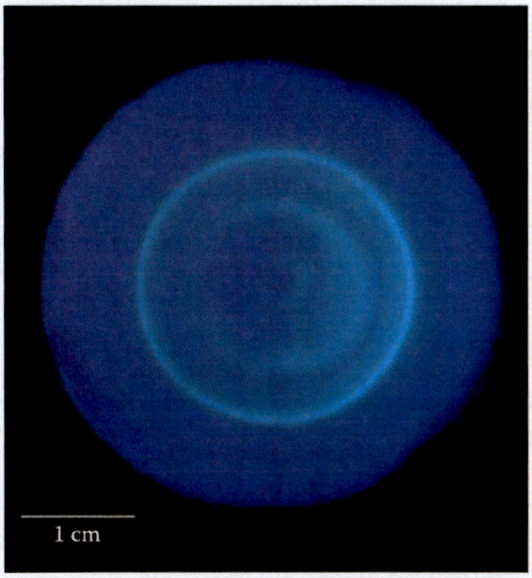

Figure 32.4 Vertebral centrum from a White Shark that had been injected with oxytetracycline and released 942 d prior to recapture. The growth band containing the oxytetracycline fluoresces under ultraviolet light, providing a marker of known date. (Courtesy of Sabine Wintner, KwaZulu-Natal Sharks Board.)

Figure 32.5 Immature female White Shark of 469 cm (1,160 kg) caught on an experimental drumline at Richards Bay, KwaZulu-Natal, in February 2002. (Courtesy of KwaZulu-Natal Sharks Board.)

Based on size, neither "juvenile" (presumably referring to young of the year) nor large adult White Sharks were caught in the KZN shark nets during 1966–1972, and all of the specimens examined were immature (Wallett, 1973). Of 132 males and 197 females caught in the nets and examined between 1978 and 1988, none was mature, with the modal size class of male White Sharks being 265–270 cm and that of females 253–258 cm (Cliff et al., 1989). Subsequently, three mature males were captured between 1989 and 1993, and measured 350, 384 and 452 cm, respectively (Cliff et al., 1996a). Wallett (1978) reported a female of 480 cm from the nets but did not comment on its maturity.

Also in KZN, and during the era of commercial whaling, recreational anglers fishing from the piers at the entrance to Durban Harbor targeted the large sharks associated with whale carcasses (Davies, 1964; Mara, 1985; Ellis and McCosker, 1991). Fourteen sharks caught between 1948 and 1964 weighed 1,000 pounds (454 kg) or more, and the largest weighed 1,660 pounds (753 kg) (Mara,

1985). From the morphometric equations of Mollet and Cailliet (1996), the calculated lengths would be 370 and 436 cm TL, respectively, although it is acknowledged that variations in girth have a substantial effect on such estimates. At least some of these sharks would probably have been mature if male, but even the largest was probably immature if female. More recently (1993), three males estimated to be 387, 387, and 446 cm, respectively, and hence probably mature, were observed scavenging on a whale carcass near Durban (Cliff et al., 1996a). In 1998, a female of an estimated 400 cm was documented scavenging on another whale carcass, and three sharks of unknown sex and ranging between 350 and 500 cm were seen in the vicinity (Dudley et al., 2000).

It has been suggested that the small number of adult specimens from KZN may be a function of gear selectivity (Bass et al., 1975; Cliff et al., 1989), and there is some evidence, as detailed above, that sharks larger than those typically caught in the protective nets do occur in the region. It also appears, however, that animals exceeding 500 cm, which would probably include most mature females, have not been observed in the shelf waters of KZN.

In the WC, shark size is usually estimated against, for example, boat length (Gubili et al., 2009; Martin et al., 2005, 2009), although stereophotography has also been used (Gubili, 2004). The modal size class of White Sharks (sexes combined) reported from Mossel Bay and Gansbaai is 276–325 cm, and that in False Bay is 326–375 cm (Kock and Johnson, 2006). Hence these animals tend to be larger than those caught in the KZN nets, but most animals of this size are probably still immature. In all three WC locations, approximately 20% of sightings are of animals >375 cm, the approximate size at maturity of males (Dudley and Simpfendorfer, 2006), but no animals >475 cm, which may be the approximate size at maturity of females, have been seen at Mossel Bay, and animals of this size constitute only 1 and 2%, respectively, of sightings at Gansbaai and in False Bay (Kock and Johnson, 2006). Similarly, Ferreira and Ferreira (1996) reported that only 2–3% of sharks sighted at locations between False Bay and Mossel Bay were >450 cm. They suggested that this may be a result of targeting by fishermen prior to the species receiving protection (Compagno, 1991) but, given the difficulty of capturing sharks of that size, it seems more likely that they seldom occur in South African coastal waters.

A small number of particularly large individuals has been reported from South Africa. A specimen weighing approximately 1,361 kg (calculated length, 527 cm) was harpooned from a whaler on the whaling grounds about 160 km off Durban (Davies, 1963, 1964). Davies (1964) reported a capture on rod and line of an animal of 1,208 kg (calculated length, 507 cm) but did not provide its locality. The largest on record is a female of 567 cm captured off Gansbaai [L.J.V. Compagno, personal communication in Cliff et al. (1989); Mollet et al. (1996)].

Specimens recorded from the tropical western Indian Ocean tend to be large: Mauritius, male, ~500 cm; Zanzibar, unknown sex, >400 cm; Madagascar, female, 380 cm; unknown sex, ≥500 cm; two "large" sharks; pregnant female, unknown length; and Kenya, pregnant female, estimated length, 640 cm (Cliff et al., 2000). In addition, Zuffa et al. (2002) documented animals from Madagascar (17 records), Zanzibar (two records), Kenya (one record), and Réunion (two records), most of which were large (450–600 cm), indicating that it is the larger animals that occasionally enter tropical waters, as suggested by Bass et al. (1975).

MOVEMENTS AND SITE FIDELITY

Barnard (1925) described the South African locality of the White Shark as "Cape seas" (p. 33), and although most published research on the species over the next 70 yrs. emanated from KZN, it was recognized that the southern African distribution is centered in Cape waters (Bass et al., 1975).

Davies (1963) described the species off KZN as being oceanic rather than coastal, and Davies (1964) suggested that, based on catches taken in the shark nets off Durban 1952–1961, it was not very common inshore, other than those individuals that followed whale carcasses towed to Durban

Harbor from the offshore whaling grounds. Bass et al. (1975), however, stated that, rather than being a tropical oceanic species as "commonly reputed" (p. 25), it is a regular visitor from Cape waters that is normally found on the continental shelf and often comes into shallow inshore waters. Both Wallett (1973) and Bass et al. (1975) investigated the occurrence by season and latitude on the KZN coast of immature White Sharks within two broad size categories and attempted to relate size-based differences in occurrence to prevailing sea surface temperature. Cliff et al. (1989), however, with access to a larger data set of animals caught in the shark nets, concluded that both small (<275 cm) and large (≥275 cm) immature White Sharks were more common in winter throughout the netted region. They suggested that the latitudinal temperature gradient within the netted region (approximately 2°C in summer and 1°C in winter) was insufficiently large to demonstrate any difference in temperature tolerance between the small and large immature sharks. The chosen size classes may have been too broad, however, because Cliff et al. (1996a) subsequently reported that captures of particularly small White Sharks <187 cm tend to take place in August and September and at the southern net installations, which, from information in Cliff et al. (1989), would be characterized by relatively cool sea surface temperatures (<20°C).

Ferreira and Ferreira (1996) provided the first description of White Shark centers of abundance, or hot spots, in the WC and EC, including, from west to east, False Bay (specifically Seal Island and Macassar Beach), Gansbaai (Dyer Island and Geyser Rock), Struis Bay (several sites), Mossel Bay (Seal Island and Hartenbos Beach), all of which are in the WC, and Algoa Bay (Bird Island group) in the EC. Struis Bay is the only one of the five centers that has no island-based colony of Cape Fur Seals, *A. p. pusillus*. Most observations were conducted at Dyer Island and Struis Bay, where sharks were most common in winter. Ferreira and Ferreira (1996) suggested that shark movements tend to be affected more by prey movements than by water temperature. Subsequent work by Martin et al. (2005) at Seal Island, False Bay, confirmed that most predatory activity by White Sharks feeding on seals occurred from late May to late August and that the sharks subsequently seem to move away from the island and closer to shore, where they prey on various fish species common in summer. This period of maximum activity around the island corresponds with the period from when seal pups of the year become active to the time of weaning (Kirkman et al., 2006). The subsequent departure from the island would explain the observation by Wallett (1978) that, in summer, White Sharks come inshore at various locations in the Cape and that Macassar Beach, for example, becomes "heavily populated" with them.

Sharks tagged with conventional streamer tags but also larger plaque tags identifiable from a distance moved between Dyer Island and Struis Bay in the WC (Ferreira and Ferreira, 1996). Also, longer-distance movements by White Sharks between the waters of the Cape and KZN were confirmed by means of conventional tagging (Cliff et al., 1996b). Six sharks, ranging in size from 187 to 364 cm, were recaptured, with time at liberty 1–527 d and distances traveled 0–1409 km, including 774 km from KZN to the EC and 1409 km from the WC to KZN. Four of the six recaptures of tagged White Sharks were made at or close to the tagging locality (Cliff et al., 1996b). Excluding an animal caught 1 day after tagging, the other three, two in the WC and one in KZN, were caught after about 1 yr., suggesting site fidelity or repeated use of migratory routes.

Although movements by White Sharks along the South African coast had been demonstrated through the use of conventional tags, as described above, Compagno (2001) suggested that, given the size and swimming abilities of the White Shark, together with scattered records from locations in the tropics and off oceanic islands, intercontinental movement is likely. He suggested further that it would be likely to be revealed through a combination of improved tracking technologies and the development of bodyprint libraries. Several years later the work of Bonfil et al. (2005), summarized below, fulfilled these predictions, using a combination of acoustic tags (Figure 32.6), satellite tags, and photographic identification to reveal White Shark movements. Bonfil et al. (2005) identified four behavioral patterns: "rapid transoceanic return migrations, frequent long-distance coastal return migrations, smaller-scale patrolling, and site fidelity" (p. 100). On November 7, 2003

Figure 32.6 White Shark with an acoustic tag. The pattern of notches on the trailing edge of the dorsal fin can be used to identify individual animals. (Courtesy of Alison Kock.)

a pop-up archival satellite tag was attached to a female of ~380 cm off Gansbaai, and the shark then traveled ~11,100 km in 99 d to a location near the Exmouth Gulf in Western Australia. En route, she dived to depths of 980 m and occupied water of 3.4°C, demonstrating that the species can tolerate very low temperatures. Photographic records of dorsal fin profiles (Figure 32.6) indicated that she had been a repeat winter visitor to Gansbaai from 1999, and she returned to Gansbaai on August 20, 2004, demonstrating site fidelity. Long-distance coastal return migrations of immature White Sharks to the productive Tugela Banks of northern KZN and into southern Mozambique suggest that these migrations might be feeding-related. Other satellite-tagged sharks showed smaller-scale movement between locations on the WC coast, and some stayed in their tagging locations for 29–75 d. Photographic and acoustic-tagging evidence showed extended residence of White Sharks in Gansbaai (up to 68 d) and Mossel Bay (up to 211 d). Sharks that depart these locations tend to return after several months of absence, again demonstrating site fidelity.

GENETICS

Whereas tracking coupled with photo identification provides direct observation of movements (Gubili, 2008), genetic techniques provide an indirect method of assessing population connectivity within and between ocean basins (Gubili, 2004, 2008: Pardini et al., 2000, 2001). Pardini et al. (2001) found, through analysis of the maternally inherited mitochondrial genome, that female-mediated gene flow between White Sharks from South Africa and Australia/New Zealand is rare. They contrasted this with the outcome of a study of nuclear DNA by Pardini et al. (2000), who found that male-mediated gene flow occurs at a level such that there is low differentiation between populations from those locations. Hence, Pardini et al. (2001) concluded that female sharks are probably philopatric and that males may undertake transoceanic excursions.

Bonfil et al. (2005) argued, based on their demonstration of transoceanic movement by a female White Shark, that, as an alternative explanation, both sexes may migrate but:

1. Only the males reproduce in the recipient population.
2. Females might mate with males in the recipient population but then return to their original location to give birth.

It seems unlikely, however, that the movement of this particular shark was linked to mating because, at a length of 380 cm, she was probably immature. Gubili (2004), analyzing both microsatellite loci and mitochondrial DNA from White Shark muscle and skin samples collected in Cape waters, differed with Pardini et al. (2001) in that she found significant differences in both types of DNA within the South African population, although she cautioned that sample sizes were small. She suggested that transoceanic movements are limited. Subsequently, using a larger sample, Gubili (2008) found one matrilineal clade in the Atlantic-western Indian Ocean and another in the Pacific Ocean, with further substructuring within oceans in terms of both mitochondrial and nuclear DNA. She concluded that this supports the hypothesis that females are philopatric and that gene flow is mediated by both sexes.

Gubili et al. (2009) used microsatellite genotyping of individual White Sharks at Mossel Bay to assess the accuracy of photographic dorsal-fin profiling and found a match in about 85% of cases over a 5-yr. period. They caution that natural marks may change over a longer time period.

In a different application of genetic techniques, Chapman et al. (2003) used White Shark tissues from various parts of the world, including South Africa, to develop a molecular diagnostics assay to identify White Shark body parts, including dried fins. The assay enables improved trade monitoring and enforcement of moratoria.

CATCHES AND POPULATION STATUS

Known sources of fishing mortality in South African waters have included catches in the KZN protective nets, incidental catches in commercial fishing gear, and catches by recreational and commercial fishermen targeting White Sharks for jaws, teeth, and fins (Compagno, 1991). In addition, some recreational spearfishermen carry an explosive "powerhead" for self-defense purposes (Cliff et al., 1996b). Of these sources of mortality, only catches in the KZN nets have been documented extensively. From 1991, no White Sharks could be caught or killed except on the authority of a permit (Compagno, 1991; see "Conservation and Management").

In the WC, catches of White Sharks by boat- or shore-based anglers included "many specimens" (p. 38) in Mossel Bay, Hermanus, False Bay, and Saldanha Bay, including eighteen sharks by a single angler in False Bay between 1974 and 1978 and 55 sharks by another angler in Mossel Bay (Wallett, 1978). It has been estimated that, prior to the protection of the species in 1991, catches taken in WC waters numbered at least 50 sharks per annum (T. Ferreira personal communication in Cliff et al., 1996a). Lamberth (2006) reported that 20 White Sharks were taken as bycatch in the beach-seine fishery in False Bay between 1974 and 2006.

In 1952, large-mesh gillnets, or shark nets, were installed off the beaches of Durban, KZN (Davies, 1963), and at other KZN beaches from the early 1960s (Holden, 1977; Wallett, 1978). Total effort increased from less than 2 km of netting in 1952 to a maximum of 44.7 km in 1989, with a period of rapid increase between 1964 and 1970, but then declined to 28.2 km by 2003 (Dudley and Simpfendorfer, 2006; see also "Conservation and Management"). From 1952 to 1961, the mean annual catch of White Sharks at Durban was 11.0 sharks ± 1.6 (SE) [range, 6–22; data from Davies (1963)]. Catch data are incomplete for the period 1962–1965, and although data are available from 1966 onwards, they are regarded as reliable in terms of both completeness and species identification only from 1978 onwards (Cliff and Dudley, 1992). From published time series, the mean annual catch from 1966 to 1972 was 41.0 sharks ± 6.1 (SE) [range, 24–74; data from Holden (1977) and Cliff et al. (1996a)], and that from 1978 to 2003 was 35.8 sharks ± 2.6 (SE) (Dudley and Simpfendorfer, 2006).

Holden (1977) observed that the KZN shark nets provided a longer time series of catch and effort data than would have been the case if the bather-protection program were a commercial fishery. Dudley and Simpfendorfer (2006) concurred, noting that, although such programs affect shark populations directly, they also represent a form of fishery-independent monitoring tool. Time series of White Shark catch rates in the KZN nets have been reported by various authors (Cliff and Dudley, 1992; Cliff et al., 1989, 1996a; Dudley 1997; Dudley and Cliff 1993; Dudley and Simpfendorfer, 2006; Holden 1977). From data presented by Holden (1977), there was some indication of an initial decline in catch rate, but it was not significant over the period 1952–1961 ($p = 0.12$). With the rapid increase in the number of net installations and total length of netting in the 1960s, catch rate clearly declined between 1966 and 1972 (Cliff et al., 1996a). Dudley and Simpfendorfer (2006) found that, over the period 1978–2003 when data quality was regarded as good, there was no trend in catch rate, in the mean or median size of males, or in the median size of females, but the mean size of females showed a decline, possibly explained by the sex-biased philopatry described by Pardini et al. (2001). The KZN nets were assessed to have a medium potential effect on the White Shark population, based on a low intrinsic rate of population increase (r), an assumed small population size, and a moderate annual catch (Dudley and Simpfendorfer, 2006). They noted, however, that an assessment of South African abundance indices by the Food and Agriculture Organization of the United Nations (2004) concluded that catches in recent decades appear to be sustainable.

Compagno (2001) reported that there had been claims that White Shark numbers were increasing in certain areas, including South Africa, but that there was no "hard evidence" (p. 102). An increase in encounters between White Sharks and humans in waters near Cape Town between 2000 and 2005 was documented by Cliff (2006), but Kock and Johnson (2006) argued that there was no evidence to suggest that this could be explained by a change in South African White Shark abundance.

LOCALIZED STOCK DEPLETION

Holden (1977) described independent trends in catch rate at Durban and Brighton Beach, just 10 km apart, in several shark species or species groups. He tentatively attributed this to territoriality, but it probably represents localized stock depletion resulting from site fidelity. In the White Shark, the phenomenon was less marked than in other species but was nevertheless apparent (Holden, 1977) and subsequently appeared to be supported by longer time series of catch rate data from those two locations and from Umhlanga Rocks, some 10 km north of Durban (Dudley, 1997). If localized depletion of White Sharks did occur in KZN, it may explain why the marked decline in catch rate between 1966 and 1972 coincided with the period of rapid expansion of the KZN program and then stopped (Dudley and Simpfendorfer, 2006). In other words, the decline evident in the spatially aggregated data may represent "serial depletions of localized stocks" (p. 237), and the southern African population as a whole may have been affected less than the decline would suggest (Dudley and Simpfendorfer, 2006).

POPULATION ESTIMATE

Cliff et al. (1996b) used tag-recapture data and a modified Petersen method to make a first attempt at quantifying the White Shark population on South Africa's east coast. Most tagged sharks were caught either in the KZN nets (22 sharks) or, in the WC, on rod and line (46 sharks). Of a total of 73 sharks tagged during 1989–1993, six (8.2%) were recaptured, but one was caught just 1 day after tagging and was excluded from the analysis. The estimated population size was 1,279 sharks (CV 24%). The study also yielded an estimate of fishing mortality (≤ 0.055 yr.$^{-1}$) which, in the authors' opinion, did not represent overfishing of the stock.

CONSERVATION AND MANAGEMENT

In 1991, protective legislation for the White Shark in South Africa was announced as a preemptive measure, given the species' status as a "large, unabundant, poorly known apical superpredator" (Compagno, 1991, p. 284). It was announced at the time that the protective measures would be reviewed as results of research on population biology became available (Compagno, 1991), but it appears likely that they will remain on precautionary grounds. Although there had been no trend in catch rate in the KZN nets in recent decades (Dudley and Simpfendorfer, 2006) and fishing mortality did not appear to be high (Cliff et al., 1996b), Wintner and Cliff (1999) noted that growth is slow, the population size is probably small (citing Cliff et al., 1996b), and little was known about reproduction in South African waters. In addition, the level of mixing of White Sharks from southern Africa with populations elsewhere is probably low (Cliff et al., 1996b; Gubili, 2004), although, from the work of Pardini et al. (2001) and Gubili (2004), there seems to be some uncertainty about its extent. There is growing consensus from genetic evidence that females are philopatric (Gubili, 2004, 2008; Pardini et al., 2001) which, coupled with low immigration, could lead to the rapid decline of an exploited stock (Pardini et al., 2001). The fact that White Sharks do undertake long-distance—even transoceanic—movements has been demonstrated directly, however, and this emphasizes the need for global protective measures such as listing on CITES Appendix 2 (Bonfil et al., 2005).

In South Africa, exceptionally high market values have been associated with White Shark parts, including jaws, teeth, and fins, which are likely to encourage illegal trade (Compagno, 2001). Other potential threats to the local White Shark population include plastics and organochlorine pesticides. Cliff et al. (2002) documented the ingestion of, as well as entanglement in, plastic debris by sharks, including White Sharks, caught in the KZN shark nets between 1978 and 2000 (Figure 32.7). Schlenk et al. (2005) found elevated levels of various organochlorines, including DDT, in the tissues of White Sharks sampled off KZN, but population effects are unknown.

Figure 32.7 White Shark entangled in a polypropylene band, a form of plastic pollution. (Courtesy of KwaZulu-Natal Sharks Board.)

From the late 1980s, all White Sharks found alive in the KZN nets were released and tagged where possible (Cliff and Dudley, 1992). In addition, from the mid-1990s, total effort in the KZN beach protection program was reduced, with the primary objective being to reduce catches of marine animals (Dudley and Cliff, 2010; Dudley and Simpfendorfer, 2006). By December 2005, the total length of netting on the KZN coast was 27.3 km (Dudley and Cliff, 2010), a reduction of nearly 39% from the peak of 44.7 km in 1989 reported by Dudley and Simpfendorfer (2006).

ANATOMY/MORPHOLOGY/ALBINISM

Lingham-Soliar (2003) compared collagen fibers from cetacean blubber with those in the skin of the White Shark and the Tiger Shark (*Galeocerdo cuvier*). In a subsequent study he concluded that the nature and arrangement of collagen fibers in the skin of the first dorsal fin of the White Shark are crucial to the role of the fin as a dynamic stabilizer, with the tautness of the fibers varying according to whether the shark is swimming quickly or slowly (Lingham-Soliar, 2005a). Lingham-Soliar (2005b) described the anatomy and histology of the caudal fin and caudal peduncle of the White Shark and concluded that they jointly formed a dynamic locomotory structure. Tensile-load tests indicated high passive resistance to bending by the skin imparted by a complex fiber architecture, with implications for energy conservation.

Various morphometric relationships were presented by Bass et al. (1975), Cliff et al. (1989, 1996a), Mollet et al. (1996), and Smale and Heemstra (1997). Lingham-Soliar (2005c) found negative allometry in the caudal fin span versus fork length in White Sharks but positive allometry in liver weight versus body weight. He suggested that, in large White Sharks, the relatively larger liver provides greater buoyancy, allowing energy-efficient slow swimming. In smaller sharks, the relatively larger caudal fin span provides greater hydrodynamic lift and enables prolonged, intermediate swimming speeds, facilitating feeding on fast-moving fish. An albino White Shark of 167 cm (34.5 kg), with white skin and red eyes and lacking melanin in the dermis, iris, and retina, was caught by an angler in EC waters and described by Smale and Heemstra (1997) (Figure 32.8).

PARASITOLOGY

Dippenaar et al. (2008) investigated the spatial distribution of the copepod, *Nemesis lamna*, on the gills of 11 White Sharks "in an attempt to elucidate the ecological factors" (p. 143) that may

Figure 32.8 Albino female White Shark of 167 cm caught off Boknes in the Eastern Cape in 1996. (Courtesy of Malcolm Smale, Port Elizabeth Museum at Bayworld.)

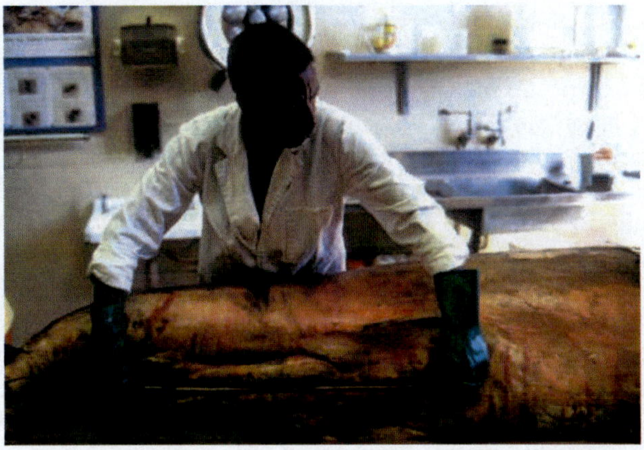

Figure 32.9 Measuring the inner clasper length of a relatively small mature male White Shark of 350 cm caught in a shark net at Mzamba, Eastern Cape, in December 1989. (Courtesy of KwaZulu-Natal Sharks Board.)

determine their distribution on the gills. They found no relationship between shark size and number of *N. lamna*. Preferred areas of attachment and orientation on the gills appeared to be related to water flow over the gills, with regions of strongest flow being avoided.

REPRODUCTION

Given the very small number of mature animals examined in southern African waters (Bass et al., 1975; Cliff et al., 1996a), little has been published pertaining to reproduction in the species, other than an estimate of male size at 50% maturity (Dudley and Simpfendorfer, 2006; Figure 32.9). Elsewhere in the western Indian Ocean, in July 1996 a pregnant female reported to be an estimated 640 cm was caught in an artisanal net fishery near Malindi, Kenya, at a depth of about 35 m and 8–10 km offshore (Cliff et al., 2000). She was reported to have had 7 embryos of 110 cm and 10–20 kg and may have aborted another 10 embryos on capture. A second pregnant shark of unknown length was caught in a bottom-set gillnet off northeastern Madagascar in about 1998, with an unknown number of embryos each weighing about 17 kg (Cliff et al., 2000). Based on a mean mass of near-term embryos of 26.3 kg reported by Uchida et al. (1996), both of these litters were probably preterm. The possible existence of a nursery ground in EC waters is discussed in "Distribution by Size and Life History Stage."

DIET

In KZN, Bass et al. (1975) documented the stomach contents of 20 White Sharks. Cliff et al. (1989) then examined the contents of 299 noneverted stomachs sampled from the shark nets 1978–1988 and found that 41% contained food. An ontogenetic change in diet, expressed in terms of frequency of occurrence, was apparent between sharks <300 cm (elasmobranchs, 44%; teleosts, 35%; marine mammals, 26%) and sharks >300 cm (elasmobranchs, 28%; teleosts, 32%; marine mammals, 40%). This increase in occurrence of marine mammals with size of predator was not observed, however, in a supplementary analysis of stomach contents collected 1989–1993, in which 9 of 10 sharks that had ingested marine mammals were <300 cm (Cliff et al., 1996a).

Important prey items within the total sample of Cliff et al. (1989) included small sharks (<126 cm), which were found in 27% of nonempty stomachs, and cetaceans, found in 22%. Cockcroft et al. (1989) concluded that the occurrence of dolphin flukes and vertebrae in shark stomachs probably represented active predation rather than scavenging. Pinnipeds were found in only 7% of nonempty stomachs, but the nearest seal colony to KZN is some 600 km away in the EC. During the annual "sardine run," the winter influx of Sardines (*Sardinops sagax*) into coastal waters of southern KZN, Sardines are seasonally important in White Shark diet (Cliff et al., 1989, 1996a), being found in 40% of 50 nonempty stomachs sampled during the run between 1978 and 2005 (Dudley and Cliff, 2010). In 10 stomachs for which a count of prey items was obtained, the median number per stomach was 70 Sardines (Dudley and Cliff, 2010), with one stomach containing 477 individuals (Cliff et al., 1996a).

White Shark dietary items are also documented through observation of feeding. During the time of commercial whaling, White Sharks would feed on carcasses being towed to whaling stations (Davies, 1964). In the postwhaling era, White Sharks were observed scavenging on floating carcasses of Humpback Whales (*Megaptera novaeangliae*) in KZN and the EC (Dicken, 2008; Dudley et al., 2000) and on the carcass of a Bryde's Whale (*Balaenoptera brydei*) in KZN (Dudley et al., 2000). Numerous observations, many of them detailed, have been made of White Sharks feeding on Cape Fur seals in Cape waters (Hammerschlag et al., 2006; Johnson et al., 2006; Laroche et al., 2008; Martin et al., 2005; Stewardson, 1999). Based on an analysis of injuries to African Penguins (*Spheniscus demersus*) in the EC, Randall et al. (1988) suggested that shark-inflicted mortalities, inflicted primarily by White Sharks, are probably an important cause of natural mortality at the islands in Algoa Bay.

PREDATOR-PREY BEHAVIOR

It has been suggested (Compagno et al., 1997) that the popular image of the White Shark as an "unknown, solitary, anthropophagous marine monster" (p. 62) has masked the realities of the species' behavior. Given this background, it is encouraging that readily observed predator-prey interactions between White Sharks and Cape Fur Seals at island-based seal colonies in Cape waters (Figure 32.10) have given rise to the most prolific field of White Shark research in southern African waters in the past decade. Compagno (2001) noted that observations at Dyer Island had revealed that the larger animals are "anything but inept, clumsy predators" (p. 104), a statement supported by the work of Johnson (2003). At Seal Island in False Bay, observations of nearly 2,100 natural predations

Figure 32.10 Predation by a White Shark on a Cape Fur Seal near Seal Island, False Bay. (Courtesy of Alison Kock, Save Our Seas Foundation.)

Figure 32.11 White Shark near Seal Island, False Bay, exhibiting a predatory behavior known as a Polaris Breach, in which the shark leaps from the water in a vertical or nearly vertical orientation. (Courtesy of Dr. Neil Hammerschlag, http://www.neilhammer.com.)

and over 120 strikes on towed decoys between 1997 and 2003 were used to formally document the predatory behavior of White Sharks feeding on seals (Martin et al., 2005). An ethogram comprising 20 behavioral units was constructed, and the Polaris Breach was found to be the most commonly employed initial strike at this particular location (Figure 32.11). The sharks appeared to select lone, incoming young-of-the-year seals at or near the surface. Hammerschlag et al. (2006), with an additional year's observations, described environmental factors that affect both the ability of White Sharks to detect and approach seals and the ability of seals to avoid predation. They concluded, *inter alia*, and in agreement with Martin et al. (2005), that attack frequency and success rate increased during low light levels, when a seal silhouetted at the surface is visible to White Sharks below but the shark's dark dorsal surface is difficult for the seal to detect. At the same location, Laroche et al. (2008) used a combination of empirical research and game modeling to investigate predator-prey interactions. They found that adult seals, when moving to and from the island, tended to select behavioral options, in terms of factors such as time of day, swimming depth, group size, and others, to minimize predation risk. Pup seals, by contrast, adopted more risky behavior, and White Sharks tended to adopt behavioral tactics to take advantage of this. Laroche et al. (2008) referred to an intergenerational behaviorally mediated indirect interaction, in which pup-seal behavior influences White Shark behavior, which in turn influences adult seal behavior. Martin et al. (2009), using a technique known as geographic profiling, found that White Sharks at Seal Island used an anchor point, or a point at which search behavior began. This point was in water about 24 m deep and some 100 m from the Launch Pad, the seal entry-exit point at the island. The chances of encountering and intercepting a seal at the anchor point were less than at the Launch Pad, but the location of the anchor point was where White Sharks optimized their predatory success rate, which was affected by a combination of the probability of encountering prey, capture rate, and intraspecific competition.

At Mossel Bay, Johnson et al. (2009) used a combination of active and passive acoustic telemetry to investigate fine-scale movements of White Sharks at and between three persistent aggregation sites. They proposed that the high rate of movement at Seal Island, the aggregation site associated with pinniped presence, was related to maximizing search area when hunting Cape Fur Seals.

Seal aggression toward White Sharks was recorded in two locations. At the Rondeklippe seal colony of Plettenberg Bay, in the WC, Stewardson and Brett (2000) documented the threatening of a 350–400-cm White Shark by an adult male seal. They acknowledged that this behavior may occur infrequently but suggested that it is adaptive for reducing risk of predation. By contrast, Johnson et al. (2008) documented frequent mobbing of White Sharks by rafting seals in the channel between

Dyer Island and Geyser Rock. They agreed that mobbing reduces predation risk and suggested that the risk associated with such behavior is lower in the channel than, for example, in the vicinity of Seal Island, False Bay, because the water is shallow and clear and hence precludes the ambush approach from depth.

On two occasions when White Sharks were observed scavenging on whale carcasses, a number of sharks were seen in the vicinity of the carcass, but simultaneous feeding by conspecifics was not observed (Dicken, 2008; Dudley et al., 2000), suggesting intraspecific competitive exclusion. Simultaneous feeding by a White Shark and several Tiger Sharks was seen, however (Dudley et al., 2000). Displacement by larger sharks of sharks hunting seals was similarly documented by Martin et al. (2005) at Seal Island, False Bay. At Dyer Island, predation attempts by White Sharks on kelp gulls *Larus dominicanus* included a breach toward two hovering gulls that did not result in contact, and the killing of a floating gull that was not consumed (Johnson et al., 2006).

CAGE DIVING

Commercial White Shark cage diving began in South Africa in the early 1990s (Compagno, 2001) and experienced a "meteoric rise" (Compagno, 1999b, p. 139) (Figure 32.12). As described by Laroche et al. (2007), this is a form of ecotourism, or nature-based tourism, that is based on the use of provisioning (chum and bait) to attract White Sharks to the surface and to keep them at the boat for as long as possible [see also Compagno (1999b)]. The South African government limited access to White Shark sites through the introduction of licensing and placed restrictions on methods of attraction (Compagno, 2001), although Johnson and Kock (2006) expressed concern about poor compliance. Cage-diving sites include Seal Island (False Bay), Dyer Island (Gansbaai), and Seal Island (Mossel Bay), all in the WC, with two additional sites proposed, Quoin Rock (WC) and Algoa Bay (EC) (A.J. Boyd, personal communication).

A spike in the number of shark bite incidents in Cape waters in the late 1990s (http://www.shark.co.za/attack.htm) led to a media-fueled assumption that there may be a link with cage diving (Compagno, 2001). Johnson and Kock (2006) argued, however, that, provided that a prohibition on the intentional feeding of sharks was adhered to, habituation of White Sharks to the baits provided by cage-dive operators would be unlikely to occur. They further argued that if conditioning were to occur, the association would be with cage-diving vessels and would be unlikely to result in increased risk to other water users.

Figure 32.12 A White Shark investigates a shark cage near Dyer Island in the Western Cape. (Courtesy of Dan Beecham, Save Our Seas Foundation.)

Laroche et al. (2007) noted that the *modus operandi* of cage diving, using provisioning to attract sharks to the surface as described above, conflicts with the typical predatory behavior described by Martin et al. (2005) at Seal Island, False Bay, which is to launch from depth attacks upon seals silhouetted at the surface (Figure 32.11). Although any change in predator-prey interactions could potentially have broader ecosystem effects, experimental investigation at Seal Island demonstrated that individual sharks tended to show a decrease in response with time to the rewards offered (Laroche et al., 2007). Hence the authors concluded that moderate levels of ecotourism probably have a minor impact on shark behavior.

SHARK BITES ON HUMANS AND BOATS

Interactions between White Sharks and humans are one of the most frequently documented aspects of White Shark behavior in the region. Initially, bites on humans that occurred in KZN were considered unlikely to be attributable to the White Shark because it was thought to be locally uncommon (Davies, 1963), but it was recognized subsequently that the species may have been responsible for a small percentage of incidents (Bass et al., 1975; Davies, 1964; Wallett, 1973). Bass et al. (1975) noted that, in general, the number of incidents attributed to White Sharks may be biased upwards by easy recognition. Both Ellis and McCosker (1991) and Levine (1996) suggested that, during the mid-twentieth century, there was a link between White Shark bites and commercial whaling in KwaZulu-Natal, but Davies (1964) did not hold this view, pointing out that whaling tended to occur in autumn and winter, and shark bites tended to occur in summer. Cliff (1991) reviewed statistics on shark bites collected on the South African coast 1960–1990 and noted that most KZN incidents were attributable to the Bull Shark *Carcharhinus leucas*, whereas White Sharks were responsible for many of the incidents in the cooler Cape waters. He documented an increasing trend in the number of Cape incidents, which he suggested was attributable to increased participation in surfing and spearfishing, facilitated by increased use of wetsuits. Levine (1996) reviewed 63 unprovoked incidents that occurred on the South African coast between 1922 and 1994 and that she believed were reliably attributable to the White Shark: 31 in the WC, 17 in the EC, and 15 in KZN. Consistent with the findings of Cliff (1991), Levine documented an increase in the number of board riders and divers since the 1970s. White Shark incidents tended to occur throughout the day (Cliff, 1991; Levine, 1996). The survival of bite victims improved from 45.4% pre-1958 to 82.7% during 1958–1994, apparently because of the introduction of the Feinberg trauma kit and appropriate training of lifeguards (Levine, 1996).

Between 1936 and 1977, various accounts were documented of boats being bitten by White Sharks, in some cases while in pursuit of a fish hooked by an occupant of the boat but in other cases with no apparent provocation (Davies, 1961; Wallett, 1978). All of these incidents occurred in WC waters, and most were in False Bay. Despite the introduction of White Shark cage diving in WC waters after 1992, which entails daily chumming and use of baits to lure sharks to boats (Compagno, 1999b), there have been no further reports of serious damage to boats (Compagno, 1999b, 2001).

MEASURES TO REDUCE THE RISK OF SHARK BITE INCIDENTS

Large-mesh gillnets, or shark nets, the first of which were installed in 1952, were set to reduce the risk of shark bites at the tourist beaches of KZN [Cliff and Dudley (1992) and references therein]. They have been effective in reducing the occurrence of shark bite incidents at those beaches (Dudley, 1997), with the principal mechanism believed to be localized depletion of the number of sharks, primarily Bull Sharks, in the vicinity of the beaches [Dudley and Simpfendorfer (2006) and references therein]. The nets and their effects on the White Shark population are discussed further

Figure 32.13 Field testing of an electrical shark repellent near Dyer Island. (Courtesy of Vic Peddemors, KwaZulu-Natal Sharks Board.)

in "Catches and Population Status" and "Conservation and Management." In early 2007, some nets were replaced with drumlines (KwaZulu-Natal Sharks Board, unpublished data), a more selective shark-fishing device that takes a reduced nonshark bycatch but continues to catch a number of shark species, including the White Shark, Bull Shark, and Tiger Shark (Dudley et al., 1998).

Smit and Peddemors (2003) conducted field trials of an electrical shark repellent at Dyer Island in the WC (Figure 32.13). They concluded that the electrical field reduced the probability of a White Shark being able to access a static bait "in at most 5 minutes … from about 0.70 in power-off mode to about 0.08 in power-on mode and in a period of at most 10 minutes from 0.90 to 0.16" (p. 59). They pointed out that the test conditions represent a "worst-case" scenario in terms of the potential for a shark bite on a human, in that the sharks were probably at the test location (a pinniped colony) in order to feed and hence were probably in a heightened predatory mode. Also, the sharks were stimulated by the chum slick used to attract them to the baits.

An increase in shark bite incidents between 2000 and 2005 in waters adjacent to the Cape Peninsula in the WC (Cliff, 2006) led to an increase in public concern and the convening of a workshop to "guide the management of the issue" (Nel and Peschak, 2006, p. 1). Options available for reducing the risk of shark-human encounters were reviewed (Dudley et al., 2006). Amongst a number of recommendations arising from the workshop were the following:

1. Shark-fishing devices should not be introduced.
2. The Shark Spotter program, a surveillance and early-warning system that was introduced in 2004 (Oelofse and Kamp, 2006; Chapter 29, this book), should be endorsed (Nel and Peschak, 2006).

FUTURE RESEARCH QUESTIONS

Compagno (2001) noted that the White Shark is threatened by a lack of knowledge of important aspects of its biology, and he lamented the fact that there was limited interest in addressing this. Previously, Compagno et al. (1997) had called for increased research on aspects such as behavioral ecology and population biology. Research output in the past decade has gone some way to addressing these shortcomings, but Bonfil et al. (2005) noted that there remains, for example, a lack of evidence of sex- or size-related patterns of space utilization. They suggested that multidisciplinary studies are needed that integrate genetic analyses and electronic tagging and that utilize improved monitoring instruments. It is evident from this review that there is little knowledge about the location and movements of adult female sharks, in particular, in southern African waters. In the northeast Pacific, aggregation sites with a high proportion of adult sharks of both sexes have been described off the coast of Central California (Ainley et al., 1985; Weng et al., 2007a) and at Guadalupe Island, Mexico (Domeier and Nasby-Lucas, 2007). Although aggregation sites have been described in the waters of South Africa's Western Cape, only 20% of the animals sighted are estimated to be larger than the male size at maturity, and only about 2% are larger than the female size at maturity. A question that arises is whether the long-distance, offshore patterns of movement that have been described by various authors in both the northeast and southwest Pacific (Bonfil et al., 2010; Boustany et al., 2002; Domeier and Nasby-Lucas, 2008; Jorgensen et al., 2009; Weng et al., 2007a) occur regularly off the South African coast, or whether longshore movement is more typical, as appears to be the case off the Australian coast (Bruce et al., 2006). The answer may vary according to life history stage, as in the northeast Pacific. Weng et al. (2007b) described the movements of young-of-the-year and juvenile White Sharks in the coastal waters of the northeast Pacific and listed presumed nursery grounds that have been reported in both the northern and southern hemispheres. Although it has been suggested that Algoa Bay in the Eastern Cape is a nursery area for White Sharks in South African waters, very little sampling has occurred there (Dicken, 2008), and this should be addressed.

Gubili (2004) suggested the investigation of parasites as biological tags and the increased use of geographic information systems to identify correlations between shark behavior and biology and the physical environment. Laroche et al. (2008) asked to what extent the intergenerational indirect interactions described at Seal Island, False Bay, may drive the evolution of behavioral traits, and what the long-term effects may be on population and community dynamics.

Finally, between October 1997 and February 2004, 4,024 h distributed over 900 d were spent in the field observing and photographing sharks (Bonfil et al., 2005). The database has been used to demonstrate long-term site fidelity (Bonfil et al., 2005), but the suitability of this and other databases of photo-identified White Sharks collected in Western Cape waters should be assessed for purposes of generating population estimates or indices of abundance in a similar manner to the studies of Chapple et al. (2011) and Chapter 26, this book in the northeast Pacific.

REFERENCES

Ainley, D.G., R.P. Henderson, H.R. Huber, R.J. Boekelheide, S.G. Allen, and T.L. McElroy. 1985. Dynamics of white shark/pinniped interactions in the Gulf of the Farallones. *Mem South Calif Acad Sci* 9: 109–122.

Barnard, K.H. 1925. A monograph of the marine fishes of South Africa: Part I. *Amphioxus*, Cyclostomata, Elasmobranchii, and Teleostei: Isospondyli to Heterosomata. *Ann S Afr Mus* 21: 1–418.

Bass, A.J., J.D. D'Aubrey, and N. Kistnasamy. 1975. Sharks of the east coast of southern Africa: IV. The families Odontaspididae, Scapanorhynchidae, Isuridae, Cetorhinidae, Alopiidae, Orectolobidae, and Rhiniodontidae. *Investig Rep Oceanogr Res Inst, Durban* 39: 1–102.

Bonfil, R., M. Meÿer, M.C. Scholl, R. Johnson, S. O'Brien, H. Oosthuizen, S. Swanson, D. Kotze, and M. Paterson. 2005. Transoceanic migration, spatial dynamics, and population linkages of white sharks. *Science* 310: 100–103.

Bonfil, R., M.P. Francis, C. Duffy, M.J. Manning, and S. O'Brien. 2010. Large-scale tropical movements and diving behavior of white sharks *Carcharodon carcharias* tagged off New Zealand. *Aquat Biol* 8: 115–123.

Boustany, A.M., S.F. Davis, P. Pyle, S.D. Anderson, B.J. Le Boeuf, and B.A. Block. 2002. Expanded niche for white sharks. *Nature* 415: 35–36.

Bruce, B.D., J.D. Stevens, and H. Malcolm. 2006. Movements and swimming behaviour of white sharks (*Carcharodon carcharias*) in Australian waters. *Mar Biol* 150: 161–172.

Chapman, D.D., D.L. Abercrombie, C.J. Douady, E.K. Pikitch, M.J. Stanhope, and M.S. Shivji. 2003. A streamlined, bi-organelle, multiplex PCR approach to species identification: Application to global conservation and trade monitoring of the great white shark, *Carcharodon carcharias*. *Conserv Genet* 4: 415–425.

Chapple, T.K., S.J. Jorgensen, S.D. Anderson, P.E. Kanive, A.P. Klimley, L.W. Botsford, and B.A. Block. 2011. A first estimate of white shark, *Carcharodon carcharias*, abundance off Central California. *Biol Lett* doi:10.1098/rsbl.2011.0124.

Cliff, G. 1991. Shark attacks on the South African coast between 1960 and 1990. *S Afr J Sci* 87: 513–518.

Cliff, G. 2006. A review of shark attacks in False Bay and the Cape Peninsula between 1960 and 2005. In *Finding a balance: White shark conservation and recreational safety in the inshore waters of Cape Town, South Africa*, edited by D.C. Nel and T.P. Peschak, pp. 20–31. Proceedings of a specialist workshop. WWF South Africa Report Series – 2006/Marine/001 Annexure 1.

Cliff, G., and S.F.J. Dudley. 1992. Protection against shark attack in South Africa, 1952 to 1990. In *Sharks: Biology and fisheries*, edited by J.G. Pepperell, pp. 263–272. *Aust J Mar Freshw Res* 43.

Cliff, G., S.F.J. Dudley, and B. Davis. 1989. Sharks caught in the protective gill nets off Natal, South Africa: 2. The great white shark *Carcharodon carcharias* (Linnaeus). *S Afr J Mar Sci* 8: 131–144.

Cliff, G., S.F.J. Dudley, and M.R. Jury. 1996a. Catches of white sharks in KwaZulu-Natal, South Africa and environmental influences. In *Great white sharks: The biology of Carcharodon carcharias*, edited by A.P. Klimley and D.G. Ainley, pp. 351–362. San Diego, CA: Academic Press.

Cliff, G., R.P. Van der Elst, A. Govender, T.K. Witthuhn, and E.M. Bullen. 1996b. First estimates of mortality and population size of white sharks on the South African coast. In *Great white sharks: The biology of Carcharodon carcharias*, edited by A.P. Klimley and D.G. Ainley, pp. 393–400. San Diego, CA: Academic Press.

Cliff, G., L.J.V. Compagno, M.J. Smale, R.P. van der Elst, and S.P. Wintner. 2000. First records of white sharks, *Carcharodon carcharias*, from Mauritius, Zanzibar, Madagascar and Kenya. *S Afr J Sci* 96: 365–367.

Cliff, G., S.F.J. Dudley, P.G. Ryan, and N. Singleton. 2002. Large sharks and plastic debris in KwaZulu-Natal, South Africa. *Mar Freshw Res* 53: 575–581.

Cockcroft, V.G., G. Cliff, and G.J.B. Ross. 1989. Shark predation on Indian Ocean bottlenose dolphins *Tursiops truncatus* off Natal, South Africa. *S Afr J Zool* 24: 305–310.

Compagno, L.J.V. 1991. Government protection for the great white shark (*Carcharodon carcharias*) in South Africa. *S Afr J Sci* 87: 284–285.

Compagno, L.J.V. 1999a. An overview of chondrichthyan systematics and biodiversity in Southern Africa. *Trans R Soc S Afr* 54: 75–120.

Compagno, L.J.V. 1999b. Encounters with sharks in South Africa. In *Sharks* (second edition), edited by J.D. Stevens, pp. 134–147. New York, NY: Checkmark Books.

Compagno, L.J.V. 2001. Sharks of the world: An annotated and illustrated catalogue of shark species known to date. Bullhead, mackerel and carpet sharks (Heterodontiformes, Lamniformes and Orectolobiformes). *FAO Species Cat Fish Purposes* No. 1 2: 1–269.

Compagno, L.J.V., M.A. Marks, and I.K. Fergusson. 1997. Threatened fishes of the world: *Carcharodon carcharias* (Linnaeus, 1758) (Lamnidae). *Environ Biol Fishes* 50: 61–62.

Davies, D.H. 1961. Shark attack on fishing boat in South Africa. *Investig Rep Oceanogr Res Inst Durban* 1: 1–2.

Davies, D.H. 1963. Shark attack and its relationship to temperature, beach patronage and the seasonal abundance of dangerous sharks. *Investig Rep Oceanogr Res Inst Durban* 6: 1–43.

Davies, D.H. 1964. *About sharks and shark attack*. Pietermaritzburg, South Africa: Shuter and Shooter.

Dicken, M.L. 2008. First observations of young of the year and juvenile great white sharks (*Carcharodon carcharias*) scavenging from a whale carcass. *Mar Freshw Res* 59: 596–602.

Dippenaar, S., R. van Tonder, S. Wintner, and P. Zungu. 2008. Spatial distribution of *Nemesis lamna* Risso 1826 (Copepoda: Siphonostomatoida: Eudactylinidae) on the gills of white sharks *Carcharodon carcharias* off KwaZulu-Natal, South Africa. *Afr J Mar Sci* 30: 143–148.

Domeier, M.L., and N. Nasby-Lucas. 2007. Annual re-sightings of photographically identified white sharks (*Carcharodon carcharias*) at an eastern Pacific aggregation site (Guadalupe Island, Mexico). *Mar Biol* 150: 977–984.

Domeier, M.L. and N. Nasby-Lucas. 2008. Migration patterns of white sharks *Carcharodon carcharias* tagged at Guadalupe Island, Mexico, and identification of an eastern Pacific shared offshore foraging area. *Mar Ecol Prog Ser* 370: 221–237.

Dudley, S.F.J. 1997. A comparison of the shark control programs of New South Wales and Queensland (Australia) and KwaZulu-Natal (South Africa). *Ocean Coast Manag* 34: 1–27.

Dudley, S.F.J., and G. Cliff. 1993. Some effects of shark nets in the Natal nearshore environment. *Environ Biol Fishes* 36: 243–255.

Dudley, S.F.J., and G. Cliff. 2010. The influence of the annual sardine run on catches of large sharks in the protective gill nets off KwaZulu-Natal, South Africa, and the occurrence of *Sardinops sagax* in shark diet. *Afr J Mar Sci* 32: 383–397.

Dudley, S.F.J., and C.A. Simpfendorfer. 2006. Population status of 14 shark species caught in the protective gillnets off KwaZulu-Natal beaches, South Africa, 1978–2003. *Mar Freshw Res* 57: 225–240.

Dudley, S.F.J., R.C. Haestier, K.R. Cox, and M. Murray. 1998. Shark control: Experimental fishing with baited drumlines. *Mar Freshw Res* 49: 653–661.

Dudley, S.F.J., M.D. Anderson-Reade, G.S. Thompson, and P.B. McMullen. 2000. Concurrent scavenging off a whale carcass by great white sharks, *Carcharodon carcharias*, and tiger sharks, *Galeocerdo cuvier*. *Fish Bull* 98: 646–649.

Dudley, S.F.J., G. Cliff, M.D. Anderson-Reade, G.E. Charter, and P.W. von Blerk. 2006. Shark deterrent options for Cape Town. In *Finding a balance: White shark conservation and recreational safety in the inshore waters of Cape Town, South Africa*, edited by D.C. Nel and T.P. Peschak, pp. 109–120. Proceedings of a specialist workshop. WWF South Africa Report Series – 2006/Marine/001 Annexure 1.

Ellis, R., and J.E. McCosker. 1991. *Great white shark*. Stanford, CA: Stanford University Press.

Ferreira, C.A., and T.P. Ferreira. 1996. Population dynamics of white sharks in South Africa. In *Great white sharks: The biology of Carcharodon carcharias*, edited by A.P. Klimley and D.G. Ainley, pp. 381–391. San Diego, CA: Academic Press.

Food and Agriculture Organization of the United Nations. 2004. Report of the FAO ad hoc expert advisory panel for the assessment of proposals to amend Appendices I and II of CITES concerning commercially-exploited aquatic species. *FAO Fish Rep* 748: 1–51. ftp://ftp.fao.org/docrep/fao/007/y5523e/y5523e00.pdf.

Francis, M.P. 1996. Observations on a pregnant white shark with a review of reproductive biology. In *Great white sharks: The biology of Carcharodon carcharias*, edited by A.P. Klimley and D.G. Ainley, pp. 157–172. San Diego, CA: Academic Press.

Gubili, C. 2004. *Application of molecular genetic techniques to the great white shark (Carcharodon carcharias L.): A revision of population genetic assessments*. MSc dissertation, University of Aberdeen, Aberdeen, UK.

Gubili, C. 2008. *Application of molecular genetics for conservation of the great white shark, Carcharodon carcharias, L. 1758*. PhD dissertation, University of Aberdeen, Aberdeen, UK.

Gubili, C., R. Johnson, E. Gennari, W.H. Oosthuizen, D. Kotze, M. Meÿer, D.W. Sims, C.S. Jones, and L.R. Noble. 2009. Concordance of genetic and fin photo identification in the great white shark, *Carcharodon carcharias*, of Mossel Bay, South Africa. *Mar Biol* 156: 2199–2207.

Hammerschlag, N., R.A. Martin, and C. Fallows. 2006. Effects of environmental conditions on predator-prey interactions between white sharks (*Carcharodon carcharias*) and Cape fur seals (*Arctocephalus pusillus pusillus*) at Seal Island, South Africa. *Environ Biol Fishes* 76: 341–350.

Holden, M.J. 1977. Elasmobranchs. In *Fish population dynamics*, edited by J.A. Gulland, 187–215. New York, NY: John Wiley and Sons.

Johnson, R.L. 2003. *Behavioural ecology of the white shark (Carcharodon carcharias) at Dyer Island, South Africa*. MSc dissertation, University of Pretoria, Pretoria, South Africa.

Johnson, R.L., and A. Kock. 2006. South Africa's white shark cage-diving industry: Is there cause for concern? In *Finding a balance: White shark conservation and recreational safety in the inshore waters of Cape Town, South Africa*, edited by D.C. Nel and T.P. Peschak, pp. 40–59. Proceedings of a specialist workshop. WWF South Africa Report Series – 2006/Marine/001 Annexure 1.

Johnson, R.L., A. Venter, M.N. Bester, and W.H. Oosthuizen. 2006. Seabird predation by white shark, *Carcharodon carcharias*, and Cape fur seal, *Arctocephalus pusillus pusillus*, at Dyer Island. *S Afr J Wildl Res* 36: 23–32.

Johnson, R., T. Keswick, M.N. Bester, and W.H. Oosthuizen. 2008. Encounters between white sharks and Cape fur seals in a shallow channel. *J Mar Biolog Assoc UK - Biodivers Rec*. http://www.mba.ac.uk/jmba/biodiversityrecords.php?jmbaref=6182.

Johnson, R., M.N. Bester, S.F.J. Dudley, W.H. Oosthuizen, M. Meÿer, L. Hancke, and E. Gennari. 2009. Coastal swimming patterns of white sharks (*Carcharodon carcharias*) at Mossel Bay, South Africa. *Environ Biol Fishes* 85: 189–200.

Jorgensen, S.J., C.A. Reeb, T.K. Chapple, S. Anderson, C. Perle, S.R. Van Sommeran, C. Fritz-Cope, A.C. Brown, A.P. Klimley, and B.A. Block. 2009. Philopatry and migration of Pacific white sharks. *Proc R Soc Lond B Biol Sci*. http://rspb.royalsocietypublishing.org/content/277/1682/679.full.pdf+html?sid=7a744019-8c44-4a5c-9bf9-285a023bec6b.

Kirkman, S.P., W.H. Oosthuizen, and M.A. Meÿer. 2006. The seal population of Seal Island, False Bay. In *Finding a balance: White shark conservation and recreational safety in the inshore waters of Cape Town, South Africa*, edited by D.C. Nel and T.P. Peschak, pp. 83–94. Proceedings of a specialist workshop. WWF South Africa Report Series – 2006/Marine/001 Annexure 1.

Kock, A., and R.L. Johnson. 2006. White shark abundance: Not a causative factor in numbers of shark bite incidents. In *Finding a balance: White shark conservation and recreational safety in the inshore waters of Cape Town, South Africa*, edited by D.C. Nel and T.P. Peschak, pp. 1–19. Proceedings of a specialist workshop. WWF South Africa Report Series – 2006/Marine/001 Annexure 1.

Lamberth, S.J. 2006. White shark and other chondrichthyan interactions with the beach-seine (treknet) fishery in False Bay, South Africa. *Afr J Mar Sci* 28: 723–727.

Laroche, R.K., A.A. Kock, L.M. Dill, and W.H. Oosthuizen. 2007. Effects of provisioning ecotourism activity on the behaviour of white sharks *Carcharodon carcharias*. *Mar Ecol Prog Ser* 338: 199–209.

Laroche, R.K., A.A. Kock, L.M. Dill, and W.H. Oosthuizen. 2008. Running the gauntlet: A predator-prey game between sharks and two age classes of seals. *Anim Behav* 76: 1901–1917.

Levine, M. 1996. Unprovoked attacks by white sharks off the South African coast. In *Great white sharks: The biology of Carcharodon carcharias*, edited by A.P. Klimley and D.G. Ainley, pp. 435–448. San Diego, CA: Academic Press.

Lingham-Soliar, T. 2003. The dinosaurian origin of feathers: Perspectives from dolphin (Cetacea) collagen fibers. *Naturwissenschaften* 90: 563–567.

Lingham-Soliar, T. 2005a. Dorsal fin in the white shark, *Carcharodon carcharias:* A dynamic stabilizer for fast swimming. *J Morphol* 263: 1–11.

Lingham-Soliar, T. 2005b. Caudal fin in the white shark, *Carcharodon carcharias* (Lamnidae): A dynamic propeller for fast, efficient swimming. *J Morphol* 264: 233–252.

Lingham-Soliar, T. 2005c. Caudal fin allometry in the white shark *Carcharodon carcharias:* Implications for locomotory performance and ecology. *Naturwissenschaften* 92: 231–236.

Mara, J. 1985. The shark fishermen. In *A fisherman's tale: Fifty years of angling along the Natal coast*, edited by D. Rorke, pp. 47–55. Durban, South Africa: Angler Publications and Promotions.

Martin, R.A., N. Hammerschlag, R.S. Collier, and C. Fallows. 2005. Predatory behaviour of white sharks (*Carcharodon carcharias*) at Seal Island, South Africa. *J Mar Biolog Assoc UK* 85: 1121–1135.

Martin, R.A., D.K. Rossmo, and N. Hammerschlag. 2009. Hunting patterns and geographic profiling of white shark predation. *J Zool* 279: 111–118.

Mollet, H.F., and G.M. Cailliet. 1996. Using allometry to predict body mass from linear measurements of the white shark. In *Great white sharks: The biology of Carcharodon carcharias*, edited by A.P. Klimley and D.G. Ainley, 81–89. San Diego, CA: Academic Press.

Mollet, H.F., G.M. Cailliet, A.P. Klimley, D.A. Ebert, A.D. Testi, and L.J.V. Compagno. 1996. A review of length validation methods and protocols to measure large white sharks. In *Great white sharks: The biology of Carcharodon carcharias*, edited by A.P. Klimley and D.G. Ainley, 91–108. San Diego, CA: Academic Press.

Nel, D.C., and T.P. Peschak (eds). 2006. *Finding a balance: White shark conservation and recreational safety in the inshore waters of Cape Town, South Africa*. Proceedings of a specialist workshop. WWF South Africa Report Series – 2006/Marine/001.

Oelofse, G., and Y. Kamp. 2006. Shark spotting as a water safety programme in Cape Town. In *Finding a balance: White shark conservation and recreational safety in the inshore waters of Cape Town, South Africa*, edited by D.C. Nel and T.P. Peschak, pp. 121–129. Proceedings of a specialist workshop. WWF South Africa Report Series – 2006/Marine/001 Annexure 1.

Pardini, A.T., C.S. Jones, M.C. Scholl, and L.R. Noble. 2000. Isolation and characterization of dinucleotide microsatellite loci in the great white shark, *Carcharodon carcharias*. *Mol Ecol* 9: 1171–1193.

Pardini, A.T., C.S. Jones, L.R. Noble, B. Kreiser, H. Malcolm, B.D. Bruce, J.D. Stevens, G. Cliff, M.C. Scholl, M. Francis, C.A.J. Duffy, and A.P. Martin. 2001. Sex-biased dispersal of great white sharks. *Nature* 412: 139–140.

Randall, J.E. 1973. Size of the great white shark (*Carcharodon*). *Science* 181: 169–170.

Randall, B.M., R.M. Randall, and L.J.V. Compagno. 1988. Injuries to jackass penguins (*Spheniscus demersus*): Evidence for shark involvement. *J Zool* 214: 589–599.

Schlenk, D., Y. Sapozhnikova, and G. Cliff. 2005. Incidence of organochlorine pesticides in muscle and liver tissues of South African great white sharks *Carcharodon carcharias*. *Mar Pollut Bull* 50: 208–211.

Smale, M.J., and P.C. Heemstra. 1997. First record of albinism in the great white shark, *Carcharodon carcharias* (Linnaeus, 1758). *S Afr J Sci* 93: 243–245.

Smit, C.F., and V. Peddemors. 2003. Estimating the probability of a shark attack when using an electric repellent. *S Afr Stat J* 37: 59–78.

Smith, A. 1849. *Pisces: Illustrations of the zoology of South Africa. 4.* London, UK: Smith, Elder, and Co.

Smith, J.L.B. 1950. *The sea fishes of southern Africa* (first edition, second impression). South Africa: Central News Agency.

Smith, J.L.B. 1951. A juvenile of the man-eater *Carcharodon carcharias*. *Ann Mag Nat Hist* 12: 729–736.

Stewardson, C.L. 1999. Preliminary investigations of shark predation on Cape fur seals *Arctocephalus pusillus pusillus* from the eastern Cape coast of South Africa. *Trans R Soc S Afr* 54: 191–203.

Stewardson, C.L., and M. Brett. 2000. Aggressive behaviour of an adult male Cape fur seal (*Arctocephalus pusillus pusillus*) towards a great white shark (*Carcharodon carcharias*). *Afr Zool* 35: 147–150.

Thompson, W.W. 1914. Catalogue of fishes of the Cape Province. *Union of South Africa, Province of the Cape of Good Hope, Marine Biological Report, Cape Town* 2: 132–167.

Uchida, S., M. Toda, K. Teshima, and K. Yano. 1996. Pregnant white sharks and full-term embryos from Japan. In *Great white sharks: The biology of Carcharodon carcharias*, edited by A.P. Klimley and D.G. Ainley, pp. 139–155. San Diego, CA: Academic Press.

Wallett, T.S. 1973. *Analysis of shark meshing returns off the Natal coast*. MSc dissertation, University of Natal, Durban, South Africa.

Wallett, T. 1978. *Shark attack and treatment of victims in southern African waters*. Cape Town, South Africa: Purnell.

Weng, K.C., A.M. Boustany, P. Pyle, S.D. Anderson, A. Brown, and B.A. Block. 2007a. Migration and habitat of white sharks (*Carcharodon carcharias*) in the eastern Pacific Ocean. *Mar Biol* 152: 877–894.

Weng, K.C., J.B. O'Sullivan, C.G. Lowe, C.E. Winkler, H. Dewar, and B.A. Block. 2007b. Movements, behavior and habitat preferences of juvenile white sharks *Carcharodon carcharias* in the eastern Pacific. *Mar Ecol Prog Ser* 338: 211–224.

Wintner, S.P., and G. Cliff. 1999. Age and growth determination of the white shark, *Carcharodon carcharias*, from the east coast of South Africa. *Fish Bull* 97: 153–169.

Zuffa, M., G. Van Grevelynghe, A. De Maddalena, and T. Storai. 2002. Records of the white shark, *Carcharodon carcharias* (Linnaeus, 1758), from the western Indian Ocean. *S Afr J Sci* 98: 347–349.